ADVANCES IN PHYTOPLANKTON ECOLOGY

ADVANCES IN PHYTOPLANKTON ECOLOGY

Applications of Emerging Technologies

Edited by

LESLEY A. CLEMENTSON
CSIRO Oceans and Atmosphere
Hobart, TAS, Australia

RUTH S. ERIKSEN
CSIRO Oceans and Atmosphere
Hobart, TAS, Australia

ANUSUYA WILLIS
Australian National Algae Culture Collection
CSIRO National Collections and Marine Infrastructure
Hobart, TAS, Australia

ELSEVIER

Cover Image description (left to right):
1. Scanning Electron Microscopy image of girdle view of centric diatom *Thalassiosira gravida*, from Antarctica.
2. Light microscopy image of the centric diatom *Chaetoceros dichaeta*, from Antarctica.
3. Scanning Electron Microscopy image of the coccolithophorid *Emiliania huxleyi* from the sub-Antarctic.
Photographer - Ruth S. Eriksen; the authors thank the Australian Antarctic Division for the use of their Electron Microscope to obtain images 1 and 3.

Elsevier
Radarweg 29, PO Box 211, 1000 AE Amsterdam, Netherlands
The Boulevard, Langford Lane, Kidlington, Oxford OX5 1GB, United Kingdom
50 Hampshire Street, 5th Floor, Cambridge, MA 02139, United States

Library of Congress Cataloging-in-Publication Data
A catalog record for this book is available from the Library of Congress

British Library Cataloguing-in-Publication Data
A catalogue record for this book is available from the British Library

ISBN: 978-0-12-822861-6

For information on all Elsevier publications
visit our website at https://www.elsevier.com/books-and-journals

Publisher: Candice Janco
Acquisitions Editor: Louisa Munro
Editorial Project Manager: Andrae Akeh
Production Project Manager: R. Vijay Bharath
Designer: Christian J. Bilbow

Typeset by VTeX

Dedication

We dedicate this book to the individuals who have inspired each of us to pursue a career in freshwater and marine research.

- *Barry Hart*, who provided a university course that initiated my journey through the ever-amazing world of environmental science;
- *Darius Stramski*, whose help and support was invaluable when I ventured into marine bio-optics;
- All members of the Hobart-based CSIRO Microalgal Team, past and present, who have all individually added to my knowledge of phytoplankton.

Lesley

Three women have been influential and inspirational in my journey with phytoplankton:

- The late *Gerry Nash* for introducing me to the captivating world of electron microscopy (she always said "think small");
- *Leanne Armand* for teaching me to deep dive into the world of Antarctic diatoms;
- *Fiona Scott* for sharing her time, expertise, methods, and joy of working with phytoplankton.

Ruth

- The late *Jeremy Pickett-Heaps* for inspiring my fascination with algae cell biology. Jeremy's lectures were a revelation on the intricate details within cells and the research behind these discoveries;
- *Rick Wetherbee*, whose guidance facilitated an unusual freedom in the direction of my research, and for his on-going support;
- *Michele Burford* for introducing me to the field of cyanobacteria and whose support and encouragement enable me to achieve my research and leadership goals.

Anusuya

Contents

List of contributors

Manal Al-Kandari

Kuwait Institute for Science Research, Jamal Abdul Nasser-Shuwaikh, Kuwait City, Safat, Kuwait

Andrew E. Allen

Scripps Institution of Oceanography, University of California, San Diego, CA, United States

Microbial and Environmental Genomics Group, J. Craig Venter Institute, La Jolla, CA, United States

Catharina Alves-de-Souza

Algal Resources Collection, MARBIONC, Center for Marine Science, University of North Carolina Wilmington, Wilmington, NC, United States

Mark Baird

CSIRO Oceans and Atmosphere, Hobart, TAS, Australia

John A. Berges

Department of Biological Sciences and School of Freshwater Sciences, University of Wisconsin-Milwaukee, Milwaukee, WI, United States

Levente Bodrossy

CSIRO Oceans and Atmosphere, Hobart, TAS, Australia

Christopher J.S. Bolch

Institute for Marine and Antarctic Studies, University of Tasmania, Launceston, TAS, Australia

Chris Bowler

Fédération de Recherche Global Ocean Systems Ecology and Evolution, FR2022/Tara Oceans GOSEE, Paris, France

Institut de Biologie de l'ENS (IBENS), Département de biologie, École normale supérieure, CNRS, INSERM, Université PSL, Paris, France

Astrid Bracher

Physical Oceanography of Polar Seas, Climate Sciences, Alfred Wegener Institute, Helmholtz Centre for Polar and Marine Research, Bremerhaven, Germany

Institute of Environmental Physics, Faculty of Physics and Electrical Engineering, University Bremen, Bremen, Germany

Robert J.W. Brewin

Centre for Geography and Environmental Science, College of Life and Environmental Sciences, University of Exeter, Cornwall, United Kingdom

Michele A. Burford

Australian Rivers Institute, Griffith University, Brisbane, QLD, Australia

Lisa Campbell

Department of Oceanography, Texas A&M University, College Station, TX, United States

Aurea M. Ciotti

Center for Marine Biology, University of São Paulo, São Sebastião, SP, Brazil

Alejandro Clément

Plancton Andino Spa., Puerto Varas Laboratory, Puerto Varas, Chile

Lesley A. Clementson

CSIRO Oceans and Atmosphere, Hobart, TAS, Australia

Nicole Correa

CAICAI Foundation, Puerto Varas, Los Lagos, Chile

Plancton Andino Spa., Puerto Varas Laboratory, Puerto Varas, Chile

Peter D. Countway

Bigelow Laboratory for Ocean Sciences, East Boothbay, ME, United States

Kathryn J. Coyne

University of Delaware, Lewes, DE, United States

Pascal Craw

CSIRO Oceans and Atmosphere, Hobart, TAS, Australia

Stephanie Dutkiewicz

Department of Earth, Atmospheric and Planetary Sciences, Massachusetts Institute of Technology, Cambridge, MA, United States

Chetan C. Gaonkar
Department of Oceanography, Texas A&M University, College Station, TX, United States

Laurence Garczarek
UMR 7144, ECOMAP, CNRS, Sorbonne Université, CNRS, Station Biologique de Roscoff, Roscoff, France

Catherine Gérikas Ribeiro
GEMA Center for Genomics, Ecology & Environment, Faculty of Sciences, Universidad Mayor, Santiago, Chile

Sydney M. Greenlee
University of Maine, Orono, ME, United States

Andres Gutiérrez-Rodríguez
National Institute of Water and Atmospheric Research (NIWA), Wellington, New Zealand

Matthew J. Harke
Gloucester Marine Genomics Institute, Gloucester, MA, United States

Darren W. Henrichs
Department of Oceanography, Texas A&M University, College Station, TX, United States

Anna Hickman
University of Southampton, National Oceanography Centre Southampton, Southampton, United Kingdom

Andrea Highfield
The Marine Biological Association of the UK, Plymouth, United Kingdom

Takafumi Hirata
Arctic Research Center, Hokkaido University, Sapporo, Japan
Graduate School of Environmental Science, Hokkaido University, Sapporo, Japan

Sharon E. Hook
CSIRO Oceans and Atmosphere, Hobart, TAS, Australia

Federico M. Ibarbalz
Universidad de Buenos Aires, CONICET, Centro de Investigaciones del Mar y la Atmósfera (CIMA), Buenos Aires, Argentina

Bethany C. Kolody
Innovative Genomics Institute, University of California Berkeley, Berkeley, CA, United States

Tihomir S. Kostadinov
Department of Liberal Studies, California State University, San Marcos, CA, United States

Bernd Krock
Alfred Wegener Institut-Helmholtz Zentrum für Polar- und Meeresforschung, Bremerhaven, Germany

Adriana Lopes dos Santos
Asian School of the Environment, Nanyang Technological University, Singapore

Lara Marcus
CAICAI Foundation, Puerto Varas, Los Lagos, Chile
Department of Zoology, Faculty of Natural and Oceanographic Sciences, University of Concepción, Concepción, Chile

Jorge I. Mardones
Center for Harmful Algae Studies (CREAN), Fisheries Development Institute (IFOP), Puerto Montt, Los Lagos, Chile
FONDAP Research Center Dynamics of High Latitude Marine Ecosystem (IDEAL), Valdivia, Chile
CAICAI Foundation, Puerto Varas, Los Lagos, Chile

Glenn B. McGregor
Queensland Department of Environment and Science, Brisbane, QLD, Australia

Mathieu Mongin
CSIRO Oceans and Atmosphere, Hobart, TAS, Australia

Colleen B. Mouw
University of Rhode Island, Graduate School of Oceanography, Narragansett, RI, United States

Satoshi Nagai
Japan Fisheries Research and Education Agency, Fisheries Stock Assessment Center, Bioinformatics and Biosciences Division, Genome Structure Analysis Group, Yokohama, Kanagawa, Japan

Brett A. Neilan

School of Environmental and Life Sciences, The University of Newcastle, Newcastle, NSW, Australia

Scott D. Nodder

National Institute of Water and Atmospheric Research (NIWA), Wellington, New Zealand

Denise Ong

Asian School of the Environment, Nanyang Technological University, Singapore

Emanuele Organelli

National Research Council (CNR), Institute of Marine Sciences (ISMAR), Rome, Italy

Javier Paredes-Mella

Center for Harmful Algae Studies (CREAN), Fisheries Development Institute (IFOP), Puerto Montt, Los Lagos, Chile

CAICAI Foundation, Puerto Varas, Los Lagos, Chile

Anthony Richardson

CSIRO Oceans and Atmosphere, Queensland Biosciences Precinct, St. Lucia, QLD, Australia

School of Mathematics and Physics, The University of Queensland, St. Lucia, QLD, Australia

Declan Schroeder

The Marine Biological Association of the UK, Plymouth, United Kingdom

Veterinary Population Medicine, The University of Minnesota, St Paul, MN, United States

School of Biological Sciences, University of Reading, Reading, United Kingdom

Barbara C. Sendall

Queensland Department of Health, Brisbane, QLD, Australia

Xiao Li Shi

State Key Laboratory of Lake Science and Environment, Nanjing Institute of Geography and Limnology, Chinese Academy of Sciences, Nanjing, China

Sebastian Silva

Department of Ecology, The Pontifical Catholic University of Chile, Santiago, Chile

Jennifer Skerratt
CSIRO Oceans and Atmosphere, Hobart, TAS, Australia

Matthew C. Smith
Sandy Bay, TAS, Australia

Monika Soja-Wozniak
CSIRO Oceans and Atmosphere, Hobart, TAS, Australia

Rowena Stern
The Marine Biological Association of the UK, Plymouth, United Kingdom

Andrew D.L. Steven
CSIRO Oceans and Atmosphere, Queensland Biosciences Precinct, University of Queensland, Brisbane, QLD, Australia

Alison R. Taylor
Department of Biology and Marine Biology, The University of North Carolina Wilmington, Wilmington, NC, United States

Kimberlee Thamatrakoln
Department of Marine and Coastal Sciences, Rutgers University, New Brunswick, NJ, United States

Daniel Vaulot
UMR 7144, ECOMAP, CNRS, Sorbonne Université, CNRS, Station Biologique de Roscoff, Roscoff, France
Asian School of the Environment, Nanyang Technological University, Singapore

Luigi Vezzulli
Department of Earth, Environmental and Life Sciences (DISTAV), University of Genoa, Genoa, Italy

Flora Vincent
Department of Plant and Environmental Sciences, Weizmann Institute of Science, Rehovot, Israel

Peter Von Dassow
Department of Ecology, The Pontifical Catholic University of Chile, Santiago, Chile
Millenium Institute of Oceanography, Universidad de Concepción, Concepción, Chile

Yanfei Wang

University of Delaware, Lewes, DE, United States

Karen Wild-Allen

CSIRO Oceans and Atmosphere, Hobart, TAS, Australia

Anusuya Willis

Australian National Algae Culture Collection, CSIRO National Collections and Marine Infrastructure, Hobart, TAS, Australia

Susanna A. Wood

Cawthron Institute, Nelson, New Zealand

Jason N. Woodhouse

Department of Experimental Limnology, Leibniz-Institute of Freshwater Ecology and Inland Fisheries, Berlin, Germany

Kyoko Yarimizu

Office of Industry-Academia-Government and Community Collaboration, Hiroshima University, Higashi-Hiroshima, Hiroshima, Japan

Erica B. Young

Department of Biological Sciences and School of Freshwater Sciences, University of Wisconsin-Milwaukee, Milwaukee, WI, United States

Acknowledgment

The editors would like to sincerely thank CSIRO and in particular Dr. Andy Steven for providing funds to bring this book to reality.

Introduction

Phytoplankton ecology is the study of processes influencing the abundance and distribution of phytoplankton, their interactions with the environment, and their role in the flow and transformation of energy in an ecosystem. Phytoplankton are microscopic algae which form the base of aquatic food chains; they are fundamentally important in providing substantial carbon fixation and influencing global carbon and nutrient cycles. Productivity associated with phytoplankton supports higher trophic levels through diverse food webs and underpins environmentally and economically important resources, in both marine and freshwater systems.

Traditionally morphological and physiological characteristics were the only way to identify and classify these microscopic organisms, which limited the extent and breadth of possible research. Advances in microscopic observation, such as transmission and scanning electron microscopy in the 1960's and 1970's, allowed a select few to observe, with exquisite detail, the fine-scale of the phytoplankton cell, advancing our classification systems as more and more species and genera were discovered. Although these instruments are now more widely available, the cost of purchasing and the specialist expertise required for maintenance has meant that universities often provide vital hubs for access to students and researchers.

The development of remote sensing and molecular techniques dramatically changed the approaches to investigating phytoplankton ecology, in terms of both the scale and the detail of observations. The introduction of routine satellite remote sensing in the late 20th century allowed the observation of chlorophyll concentration, and by proxy phytoplankton biomass, on global spatial and temporal scales as had never been seen previously. Although limited to surface waters, satellite remote sensing images provided a visual explanation of how currents, upwellings, and eddies could transport phytoplankton from one region to another and respond to changing environmental drivers, such as nutrients and sea surface temperature. Today, researchers have more than 20 years of satellite-based observations to determine changes in annual and decadal signals of phytoplankton phenology in any region of the globe.

During a similar time-period, the development of moored and towed instruments and sensors advanced rapidly, allowing monitoring at multiple timescales, multiple locations, difficult-to-access locations, and during periods of severe weather events. Initially, instruments logged and stored data, which was downloaded at a later date, but as telecommunications improved, these instruments were able to provide near real-time observations. Today it is common for studies to have several instruments in situ that monitor sites 24/7, with data available in near real-time, enabling quick response to events of interest.

The 1970s saw the invention of Sanger sequencing, which led to the first publication of molecular sequences for phytoplankton; this dramatically altered the landscape for researchers, who were now able to move beyond testing theory to biodiversity discovery. In the early 21st century, the development of next generation sequencing, or massive parallel sequencing, created an exponential growth in omics techniques and their applications. The significant increase in speed at which sequencing could be performed enabled the time frame for analyses to move from years to overnight. Sequencing was now available to a wide range of studies and the development of omics techniques, such as metabarcoding, comparative genomics, and transcriptomics, has followed. These omics techniques have changed the approaches used in phytoplankton ecology for detailed species detection and functional analysis of populations.

Each of these major lines of enquiry (taxonomy, remote sensing and monitoring systems, and omics) have expanded our understanding of the role of phytoplankton ecology in the natural world. Insights into phytoplankton ecology can be used to harness the unique properties of phytoplankton, such that it will play an important role in our future. Current research and some commercializations include waste water remediation, mitigation of hydrocarbon oil spills in the aquatic environment, the production of biofuels, renewable energy, added nutrition to foods for human and animal consumption, and sustainable food sources.

Increasingly, citizen science is now being used to expand the spatial and temporal coverage of observations. Partnerships between interested and motivated members of the public and phytoplankton experts are on the increase, enabled by almost global coverage afforded by mobile phone cameras, apps and low-cost sampling equipment. Projects such as "Citclops" (European Union) and "EyeOnWater Australia" (CSIRO Australia), are just two examples of where everyday citizens, without a formal science background, can contribute to building timeseries and databanks of information, which can prove extremely useful in monitoring changes in water quality, and hence phytoplankton ecology. Such projects will contribute greatly to the observation of effects due to climate change on both local and global scales.

We believe this book provides a timely link between traditional and new and emerging techniques, illustrating how combined approaches provide new insights and broader ecological knowledge. The book is written for students as well as more experienced practitioners to gain an understanding of how a combination of techniques or unfamiliar technologies can be applied. Case studies in each chapter provide examples of how a range of techniques can be applied and how multidisciplinary approaches to understanding phytoplankton ecology are needed. This has expanded collaborative work, facilitating the creation of holistic studies, an exciting space to be in. In this book, each of the three sections has a focus on a particular approach: (1) taxonomy, (2) remote monitoring/modeling, and (3) omics techniques; however, there are overlapping techniques within the chapters, demonstrating the multidisciplinary nature of phytoplankton ecological studies. The introduction to each section provides insights into their chapter content.

We hope you enjoy the considerable breadth and depth of experience and expertise presented here and can apply learnings to create even greater advances in phytoplankton ecology.

Lesley A. Clementson, Ruth S. Eriksen, and Anusuya Willis

PART 1

Phytoplankton taxonomy

Ruth S. Eriksen

Of all the tools available to the modern student or scholar of taxonomy, no one single technique is sufficient to capture and enumerate the true diversity of the phytoplankton. Ranging in size from a few microns to more than a millimeter, phytoplankton are uniquely adapted to a range of aquatic habitats, across large gradients in sunlight, depth, nutrient availability, and chemical composition. Taxa display a beguiling array of forms (solitary, chain-forming, colonial, cysts) and features (flagella, setae, processes, areola, spines, organic threads, lists, plates, scales, and more) much to the delight of microscopists over the centuries, with minute details now readily revealed by powerful electron microscopy techniques.

It is difficult to state the true number of phytoplankton taxa: estimates for the diatoms alone range from tens of thousands, to over a hundred thousand species. Despite a prevalence of richly informative morphologies, typically more than one approach is necessary for conclusive identification. The "democratization" of molecular tools has been a significant step forward in understanding diversity. The advances made by combining old and new approaches mean that weaknesses can be acknowledged, and strengths combined to lead us towards the most robust and stable taxonomy. This in turn supports our understanding of evolutionary relationships, biodiversity, and ecological functioning in aquatic environments.

The assignation of a name to describe and distinguish a single entity is not a simple process. In addition to strict and detailed guidance on the correct use of names through international protocols, there must be a suite of evidence published that will allow advances in the field of taxonomy to be accepted within the broader scientific community. A name is more than just an identity for an organism, and each contributing chapter in this part demonstrates how a name infers ecological and habitat preferences, along with groupings of similar organisms, and their evolutionary relationships. The intent of this part is to help readers, new to these emerging techniques for taxonomic resolution, see how they can be used across a range of applications.

Species that form blooms have been the focus of much ecological research, yet only a small percentage of the total known phytoplankton species dominate blooms. Harmful algal blooms (HABs, both marine and freshwater) commonly feature in the case studies in this part, because of the implications of making an incorrect diagnosis. Correct identification and enumeration of causative species, and the detection of toxins, or other biomarkers are critical for protection of human and ecosystem health, but also provide high quality data for predictive modeling of environmental triggers for HAB events.

All chapters provide a summary of the future challenges in the discovery and description of new taxa and the application of that information to routine monitoring, ecological research, and management of our natural resources. These challenges are not insignificant, and often revolve around the need for standardized approaches to the collection, sharing, and processing of genetic information and the storing of reference material in DNA repositories. Visions of automated approaches to "all-of-community" identification and enumeration are realistic, once these challenges have been addressed.

McGregor and Sendall in Chapter 1 introduce cyanobacteria taxonomy and systematics, describing the development and acceptance of rigorous polyphasic approaches to both the validation of existing names, and proposals for new taxa. The importance of correct nomenclature, and the application of names in a clear, consistent, and unambiguous manner is fundamental to describing and understanding this diverse group. A name confers a set of ecological and functional attributes, information that is often acquired slowly and incrementally, by combinations of detailed field and laboratory observations. In the case of cyanobacteria, naming taxa has been complicated by debate about which International Code for Nomenclature (prokaryotes ICNP or algae, fungi and plants ICN) has priority, resulting in a problematic number of invalid proposals for new cyanobacteria taxa. The authors provide examples of the collection of morphological and molecular evidence for novel taxa and outline the steps for formal assignation according to the current nomenclatural system. Part of this process is rigorous review of the rich historical literature to ensure publication advances, rather than confuses, cyanobacteria systematics.

Two Australian case studies are presented on the application of both microscopy and molecular methods in (i) the routine monitoring of picocyanobacteria for water quality assessments; (ii) polyphasic characterization of a filamentous cyanobacteria, subsequently identified as a novel species, and thus influencing the management response to bloom events.

Stern et al. in Chapter 2 show how a global-scale plankton monitoring program, traditionally based on morphological analysis of phytoplankton, can be enhanced by the application of molecular techniques. Taxa that are difficult to discriminate by light microscopy (due to preservation artifacts, mechanical damage, small size or/and lack of distinguishing features) may be resolved on the CPR silks with molecular tools. The development of specific techniques for working with formalin-preserved samples has unlocked a new level of capability for the world's longest and spatially extensive marine plankton survey. "Hidden taxa" from the previously unseen planktonic world are revealed, and analysis of preserved samples from an archive extending back decades can now be probed for understanding long-term changes in phytoplankton and microbes in the open ocean. As one of the few methodologically consistent, and continuous surveys monitoring the open oceans, the archive contains approximately 500,000 samples from 1958 onwards, opening up a world of opportunities for studying past events, decadal-scale changes in the world's oceans, and, importantly, providing predictive power for models of future oceans. Microscopic analysis of the CPR silks has resulted in records of many hundreds of phytoplankton taxa; this will only increase with the continued development and application of molecular taxonomy approaches.

The CPR case studies in this chapter focus on (i) genotypic and phenotypic diversity of the coccolithophore *Emiliania huxleyi* within a bloom event, and over a multiyear timescale; (ii) examining species diversity within the morphologically challenging genus *Pseudo-nitzschia*, including samples from different stages of the Pacific decadal oscillation off North America; (iii) applications of the CPR sample archive to investigate the causative fungal species sampled in the weeks following a massive dust storm originating in Central Australia that transported gigatons of iron-rich dust into the oceanic waters East of Australia.

Bolch in Chapter 3 continues the theme of the influence of molecular techniques on taxonomy and systematics. Dinoflagellate taxonomy has historically been underpinned by the rich morphodiversity of the group, particularly for thecate species, and a parallel line of enquiry of the extensive fossil record for those taxa that form resting stages or cysts. The chapter outlines the timeline of advances in taxonomy and systematics, and evolutionary theory, and the eventual integration of the taxonomy of fossil cysts and extant living dinoflagellate cysts through polyphasic approaches. As with the cyanobacteria, the refinement and now widespread availability of molecular techniques has fueled developments in dinoflagellate taxonomy and systematics. It is now standard practice that new descriptions and revisions are supported by molecular data and phylogenies, and these are most valuable when combined with morphology, ultrastructure, and chemotaxonomic data. This polyphasic approach has often (but not always) resulted in the "quick" acceptance of revisions to traditional taxonomy and an overall strengthening of dinoflagellate systematics.

Cryptic speciation of microreticulate cyst-forming species of the genus *Gymnodinium* is explored in the case study, further strengthening the application of molecular, morphological, biochemical and ultrastructural data that are needed to resolve the presence of cryptic species. Distinction between the cysts of toxic and nontoxic species is critical for the management of harmful algal bloom events, and microscopy-based monitoring alone is insufficient to accurately and reliably distinguish between the cysts of harmful and harmless taxa. This chapter also offers the reader insight into the fast-evolving world of ancient DNA and paleomics. Exciting advances in the sequencing of sediment ancient DNA (sedaDNA) from marine cores mean it is now possible to look back 10,000 years or more at the structure of past plankton communities.

Mardones et al. in Chapter 4 discuss taxonomic research in Chile, where large-scale finfish and shellfish aquaculture operations across several regions along the Chilean coast are affected by frequent and large HAB events. Management of HABs requires a vast depth of knowledge about the environmental niche, behavior, nutritional strategies, toxicity, genetics, life cycle, physiology, and biogeography of the causative organisms, all of which hinge on assigning the correct name, consistently. In the event of a developing bloom, time is of the essence, and thus reliable, flexible, and affordable analytical methods are crucial to protect the natural resource, producers and consumers of seafood.

Microscopy-based monitoring is widespread, yet poorly known species, rare, poorly sampled or preserved species, and those with little taxonomic characterization can be difficult to detect. Other lines of evidence are required for rapid risk assessment and appropriate management responses, especially where human health and export markets are at risk. *Mardones et al.* outline several new approaches to advance traditional microscopy-based monitoring in Chile by the inclusion of (i) molecular methods, (ii) cytological assays, (iii) flow cytometry, (iv) phycotoxin analysis, and (v) citizen science into aquaculture monitoring programs as case studies. They provide detailed examples of each of these approaches in the Chilean context, and present the case for a powerful suite of taxonomic tools, which can be used together to detect the presence of causative species, and monitor the development and passing of bloom events. The combination of these approaches within the HAB monitoring programs has revealed higher levels of diversity than previously known.

CHAPTER 1

Cyanobacterial diversity and taxonomic uncertainty: polyphasic pathways to improved resolution

Glenn B. McGregor[a] and Barbara C. Sendall[b]
[a]Queensland Department of Environment and Science, Brisbane, QLD, Australia
[b]Queensland Department of Health, Brisbane, QLD, Australia

Introduction

The cyanobacteria are a morphologically diverse monophyletic group of oxygenic photosynthetic prokaryotes within the domain Bacteria encompassing over 400 genera. As a phylum they range from microscopic unicells and colonial groups to macroscopic mats and unstructured aggregations, occupying a diverse range of habitats, including extreme environments. They make a major contribution to the biosphere in terms of carbon and nitrogen production (Garcia-Pichel et al., 2003), support aquatic food webs, and many species produce biologically active compounds, which are of both great interest and potential concern to human, agricultural, and ecosystems values (Bláha et al., 2009; Svirčev et al., 2019). In the aquatic domain, cyanobacteria occupy a wide variety of niches, and in subtropical and tropical regions they are often the dominant component of phytoplankton communities in lakes and reservoirs, accounting for most of the picoplanktonic and nanoplanktonic species (McGregor, 2013). In Australia and elsewhere, some species commonly form blooms, produce harmful cyanotoxins, and taste and odour compounds (O'Neil et al., 2012; Dittmann et al., 2013; Huisman et al., 2018). Impacts of blooms on commercial fishing, aquaculture, water treatment for potable use, tourism, and recreational activities have imposed significant economic impacts (Hamilton et al., 2014; He et al., 2016). Equally, they pose a risk to aquatic ecosystems through the creation of anoxic conditions, alteration of the light regime, production of toxins, and via habitat alteration (Zanchett and Oliveira-Filho, 2013; Rastogi et al., 2015; Metcalf and Codd, 2020). Conditions supporting bloom development have been exacerbated by anthropogenic eutrophication, hydrological modification of river systems, and climate change (Burford et al., 2020; Visser et al., 2016; Wells et al., 2015), such that blooms have become a recurrent seasonal feature in many parts of the world. Although the common bloom-forming and toxin-producing species are generally well characterized, their number is likely to be an underestimate and is expected to increase as a result of enhanced population screening, particularly from the benthos and novel environments, and the discovery of new toxins (Humpage et al., 2012; Minowa et al., 2019; Conklin et al., 2020). In this context, names are important; they confer a set of ecological, physiological, and toxicological characteristics of an organism which are used in the management of water blooms to assess risk and implement strategies for their control. Therefore the correct identification and designation of cyanobacterial taxa is a critical component of this process. Equally, they confer a set of ecological and functional attributes discerned through autecological or physiological observations, not necessarily transferable to other taxa. In this chapter, we describe the contemporary approaches for recognizing and describing cyanobacterial diversity, and through two case studies, highlight some of the ongoing challenges to their practical application.

Background

Taxonomy provides a framework to understand cyanobacterial diversity and facilitate their study and identification in nature. Historically, the taxonomy and systematics of cyanobacteria (as blue-green algae) were considered along with the eukaryotic microalgae in the botanical tradition with a strong focus on morphological, and to a lesser extent, ecological attributes. This tradition dates back over a century to seminal floras by Thuret (1875), Bornet and Flahault (1887, 1888a,b), and Gomont (1892), which are recognized as the starting points for the taxonomy of this group. These works provide a valuable record comprising careful observation of natural material using light microscopy, along with detailed descriptions and illustrations of critical morphological features, such as the type of branching, filament polarity, sheath formation, structure, and location of specialized cells, such as heterocytes and akinetes, hormogonia, hairs, and vegetative cell shape, dimensions, and planes of cell division. Subsequent monographs by Geitler (1925, 1932, 1942), Elenkin (1936 & 1949), Desikachary (1959), Starmach (1966), Kondrateva (1968), and Bourrelly (1970), focussing largely on European and North and Central American regions, provided the essential basis for cyanobacterial identification until the end of the twentieth century. Subsequent recognition of the cyanobacteria as prokaryotes in the early 1960s (Stanier and Van Niel, 1962) ultimately lead to proposals for them to be subject to the taxonomic criteria of bacteria. Initially Rippka et al. (1979) recommended cyanobacteria to be divided into five sections, which became the primary basis for the nomenclatural classification in Bergey's Manual of Systematic Bacteriology (Castenholz et al., 2001). The most substantial taxonomic revisions to incorporate ultrastructural and early molecular data spanning both the botanical and bacterial domains was undertaken from late 1980s through to the early 2000s (Anagnostidis and Komárek, 1985; Komárek and Anagnostidis, 1999, 2005; Komárek, 2013); these works have continued to provide the basis for subsequent revisions over the past two decades.

Nomenclature

As cyanobacteria were traditionally recognized as "algae," and identified as such, their classification developed under the botanical system, and their nomenclature covered by the provisions of the International Code of Nomenclature for algae, fungi, and plants (ICN) (McNeill et al., 2012) (formerly the International Code of Botanical Nomenclature). Since the proposal by Stanier et al. (1978) for them to be covered by the provisions of the International Code for Nomenclature of Prokaryotes (ICNP) (Parker et al., 2019), debate has continued on which nomenclatural system has priority, with many authors highlighting the incongruences between the two codes; importantly, difference relating to typification, effective publication, and valid publication (for a summary of these issues see reviews by Oren, 2004; Hoffmann et al., 2005; Oren and Tindall, 2005; Oren and Ventura, 2017; Oren, 2020). Cyanobacteria were formally included in the ICNP in 1999 by modification of General consideration 5, which relates to the definition of prokaryotes (De Vos and Trüper, 2000). The requirements and differences between these codes continue to confuse authors, with a recent literature survey revealing 29% of 170 new cyanobacterial taxa published in the past 20 years were invalid, due to violation of one or more nomenclatural rule (Johansen et al., 2019). Consequently, to date, very few taxa have been validly published under the provisions of the ICNP, with most authors continuing to publish under the ICN.

Species concepts

The species remains the fundamental unit of our taxonomic system, the most resolved entity of the classification hierarchy. Despite a renewed effort in the description of cyanobacterial diversity over the past two decades, the species concept remains a long-standing key issue, which continues to be debated (Wheeler and Meier, 2000). It is obvious that species concepts, as they relate to sexual organisms, do not apply to cyanobacteria. Furthermore, many argue that prokaryote species do not exist (Goldenfield and Woese, 2007; Lakeman et al., 2009); whereas some maintain that this issue is a "non-problem," contending that lineages can be studied effectively without any definition of "species" or any requirement that all prokaryotic lineages match such a definition (Doolittle, 2019). In practical terms, defining a prokaryotic species remains key to resolving uncertainty in estimates of cyanobacteria biodiversity (Ribeiro et al., 2018). In many ways, this is an historical issue resulting from a discordance in taxonomic concepts between the two nomenclatural systems, with competing species concepts hotly debated since the nineteenth century. Dvořák et al. (2015) emphasized the important difference between species concepts and species definitions, which is worth reinforcing here. A species concept is a theoretical demarcation of a species boundary broadly applicable across organisms (ontological notion). Whereas, a species definition is a practical demarcation of a species boundary applied within the context of a specific group (epistemological criteria); for example, a genomic threshold used to distinguish between taxa (i.e., DNA–DNA hybridization, average nucleotide identity used by bacteriologists, % similarity in 16S rRNA sequence, etc.).

Many species concepts have been proposed. The classic phenic concept, based on subjective morphological and ecological traits, is insufficient to adequately characterize cyanobacterial biodiversity and fails to recognize crypsis and phenotypic variability characteristics of this group (Castenholz, 1992; Rosselló-Mora and Amann, 2001). Concepts such as the biological species concept (Mayr and Ashlock, 1991), which relates to interbreeding individuals, is clearly not applicable to obligately asexual cyanobacteria. Concepts of species as lineages considered by the evolutionary species concept, or the phylogenetic species concept, aim for a classification approach, which groups species based on their evolutionary history. Johansen and Casamatta (2005) reviewed these concepts as they relate to cyanobacteria and advocated the phylogenetic species concept *sensu* Mishler and Theriot (2000) as the most appropriate. In summary, this concept recognizes species as the least inclusive taxon in a phylogenetic classification, grouped into species due to their monophyly. Species are the smallest monophyletic groups deemed worthy of formal recognition, because of the amount of support for their monophyly, and/or because of their importance in biological processes operating on the lineage in question (Mishler and Theriot, 2000). They emphasize that although autapomorphies are the clearest way to reveal monophyly, and designate a species, unambiguous characters are often scarce in the character-poor cyanobacteria. Consequently, they proposed a broad set of potential autapomorphies, including cladistic or maximum likelihood analysis of molecular sequence data, or a combination of molecular, biochemical, and morphological data under the moniker of the monophyletic species concept for their cyanobacterial-specific modified construct (Johansen and Casamatta, 2005). This concept continues to have wide support within the cyanobacterial literature as the basis for the description of new taxa (e.g., Dvořák et al., 2015; Machado-De-Lima et al., 2017; Shalygin et al., 2018; González-Resendiz et al., 2019).

Pathways from morphology to molecules

Although the seminal floras of the late nineteenth and early twentieth centuries, which established a sound morphological baseline for the taxonomy of cyanobacteria, were generally well considered, richly illustrated, and, for the most part, are highly descriptive; these early taxonomists were nonetheless subjective in their choice of diagnostic features. Subsequently, both in nature and controlled culture conditions, many of these "diagnostic features" have been shown to be plastic, altering in response to prevailing environmental conditions or life cycle stages over time. Furthermore, factors such as the limited range of morphological features of some species or convergent evolutionary events, may obscure genetic diversity such that taxa cannot be reliably distinguished by morphology alone. The accelerated application of molecular techniques has greatly improved our understanding of cyanobacterial diversity, using objective sequence-based approaches to resolve monophyletic lineages. This has led to an invigorated taxonomic practice and a significant increase in revisions, amendments, and new taxa descriptions (e.g., Komárek et al., 2014, 2020) informed by this revolution in molecular phylogenetics.

Polyphasic approach

Most contemporary taxonomic revisions and descriptions of new taxa use a combination of morphological, ultrastructural, ecological, and genetic data, widely referred to as the *polyphasic* or *total evidence approach* (*sensu* Komárek, 2016, 2020). Through the application of this approach, there has been a paradigm shift from continuing a historical legacy of creating polyphyletic morphospecies/genera, where phylogenetic relationships were obscured, to the establishment of monophyletic taxa characterized by a set of objective autapomorphies, which aid in the delimitation and recognition of cyanobacterial diversity. In this section, we describe the polyphasic approach as summarized by González-Resendiz et al. (2019), who framed its use in taxonomic practice as a series of recommendations relating to the description of new cyanobacterial taxa.

Morphology and ecology

There is a vast historical literature against which newly proposed taxa should be carefully compared. As previously discussed, morphological traits were often subjectively chosen as critical markers based on the experience of the observer or chance presentation at the time of collection, with many early authors not considering aspects of temporal or spatial variability. Key morphological features include filament polarity, presence and type of branching, colony formation, sheath characteristics, vegetative cell dimensions, presence, position and dimensions of akinetes and heterocytes, presence and location of aerotopes, etc. Cell ultrastructure, particularly the arrangement of thylakoid membranes, has also long been used as a diagnostic trait in cyanobacterial taxonomy (Komárek and Cáslavská, 1991). Cyanobacteria are plastic in terms of morphological expression of diacritical features both in the wild and in culture. These changes may occur as a species moves through defined life cycle stages (Komárková and Šimek, 2003), in response to seasonal changes in the biophysical environment, grazing pressure (Yang et al., 2008; Rangel et al., 2020), or following nutrient perturbations/stress, leading to the expression of aberrant features rarely observed in nature (McGregor and Sendall, 2021). Additionally, and contrary to previous studies, the taxonomic value of thylakoid morphology has recently been shown to be limited to only a few cases where thylakoid absence or their parallel arrangement could be used as an informative apomorphy (Mareš et al., 2019). In practice, ideally there should be a clear hiatus in the morphological traits of the proposed species and existing species, however, this is not always possible for some of the reasons already highlighted. Molecular data continues to reveal cryptic species/genera within traditional polyphyletic taxa across the algal domain, including the cyanobacteria (Casamatta et al., 2003; Hofmann et al., 2010; Bohunická et al., 2015; Kim et al., 2016; Nakayama et al., 2019). Cryptic taxa are indistinguishable based on morphological traits alone, due to convergent evolutionary events, where similar features evolve in independent lineages

(Dvořák et al., 2014; Komárek et al., 2014). Within the cyanobacteria this is common; simple trichal forms from the Pseudoanabaenales provide many such examples. Osorio-Santos et al. (2014) described seven new species from the genus *Oculatella*. The strains were clearly separated based upon a concatenated alignment of the 16S rRNA and 16S–23S ITS, and differences in the secondary structure of the conserved domains of the ITS region. However, despite some strains possessing granules at the cross walls, there was considerable overlap in size ranges for all measured features such that a clear morphological separation was possible for only two of the seven species. They argue that these cryptic species are worthy of recognition as publication facilitates further exploration of genetic, morphological, and ecological differentiation in this group, which is geographically widespread and occupies a range of ecological niches (Osorio-Santos et al., 2014). Examples of genus-level crypsis include *Lyngbya*-like types, which span nine genera (Engene et al., 2012, 2013, 2015, 2018; Komárek et al., 2013; McGregor and Sendall, 2015, 2019; Caires et al., 2018a,b), and small *Leptolyngbya*-like types, which span ca. 30 genera across several families (Abed et al., 2002; Taton et al., 2011; Perkerson et al., 2011; Dadheech et al., 2012a,b; Song et al., 2015; Vaz et al., 2015; Sciuto and Moro, 2016; Zimba et al., 2020; Dvořák et al., 2017; Jahodářová et al., 2017, 2018; Sciuto et al., 2017; Zammit, 2018; Becerra-Absalon et al., 2018; Cellamare et al., 2018; Mai et al., 2018; Zhou et al., 2018; Chakraborty et al., 2019), most erected over the past decade. All these genera were based on compelling phylogenetic data; however, many of these new entities are not readily distinguishable from one another in nature based on morphology alone.

Ecological information accompanying taxonomic treatments of new taxa varies in quality, but generally there is a paucity of detailed evaluations, and globally data is uneven in terms of habitat coverage and geography. However, this information is critical as it provides an important foundation to test assumptions about the dispersal of cyanobacteria and geographic boundaries, niche preferences, and evolutionary forces that provide cohesion to shape species (Vinuesa et al., 2005). Niche preference and the ecological divergence are regarded major drivers of prokaryote speciation (Coutinho et al., 2015; Tromas et al., 2018), particularly when these niches are sufficiently isolated from one another in space, through geographic separation or habitat heterogeneity (Pietrasiak et al., 2019), or time, as reduced encounter rates within a single lake due to nutrient preferences (Willis et al., 2018) tend to act as barriers to gene exchange (Schmutzer and Barraclough, 2019). While there is a growing consensus amongst taxonomists that other information sources that may be useful as species diagnostic features, such as morphology and ecology, are subordinate to DNA sequence data (Leliaert and De Clerck, 2017), the inclusion of detailed ecological information in the polyphasic approach is vital for testing congruence with phylogenetic data and providing support for the recognition of new species in nature.

Molecular evidence of monophyly

Molecular markers have been used since the 1980s to identify cyanobacteria and infer their phylogeny. Of the 145 cyanobacterial genera validly described between January 2000 and July 2020, 96% were supported by at least one gene sequence, and 62% include more than one sequence; only 6 were described without any molecular support (Fig. 1.1).

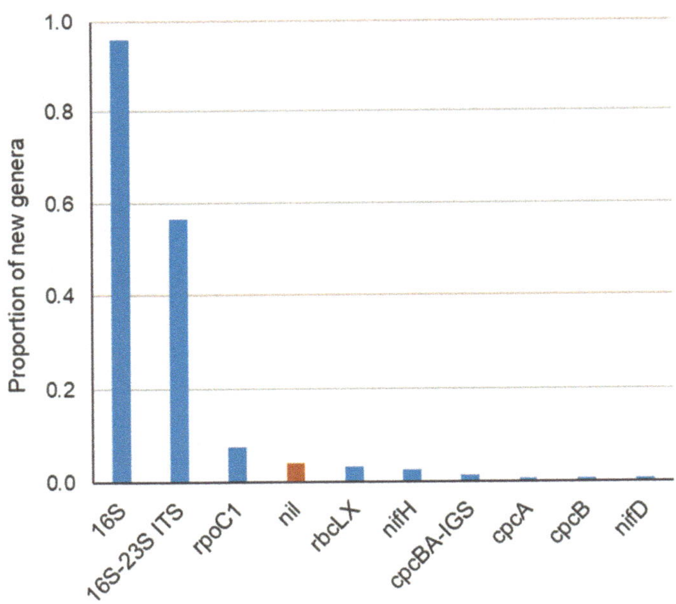

Figure 1.1 Proportion of new cyanobacteria genera published between January 2000 and July 2020 that used various genes as supporting evidence.

Most phylogenetic evidence was based on either the 16S rRNA alone, or in combination with the 16S–23S rRNA internal transcribed spacer region (ITS). The 16S rRNA gene has been the most frequently used gene in cyanobacteria phylogenetic reconstructions since its utility was revealed for resolving phylogeny of the Archaea and Bacteria (Woese, 1987; Woese et al., 1990). It is a component of the small subunit of the prokaryote ribosome, which acts as a scaffold defining the positions of the ribosomal proteins (Stackebrandt and Goebel, 1994; Iteman et al., 2000; Boyer et al., 2001). Due to its functional role, the 16S gene is highly conserved, with infrequent variable regions (Johansen et al., 2017a). Furthermore, as compared to other loci, the rate of horizontal gene transfer (HGT) is relatively low (Tian et al., 2015), making it a reliable candidate for developing phylogenies assuming genes are vertically inherited (Wilmotte et al., 2017). The total number of cyanobacterial sequences for the 16S rRNA gene in GenBank as of February 2021 was approximately 30,000. Yarza et al. (2014) recommended thresholds of 16S gene sequence identity be used for discriminating prokaryote taxa, and these values are commonly applied to cyanobacterial taxonomy; 98.7% for species discrimination, 94.5% for genera, and 86.5% for families.

Despite the useful features and widespread application of the 16S rRNA gene in taxonomic practice, it is not always sensitive in resolving closely related taxa (Komárek, 2006; Mareš, 2018). For example, the 16S-rRNA sequence similarity within the Aphanizomenonaceae (Nostocales) is high, and within the *Dolichospermum*/*Aphanizomenon* complex, which contains many widespread and well characterized planktonic species, similarity values often exceed the generally accepted species-level threshold. In these cases, less conserved housekeeping (e.g., *rnpAB*, *rpoB*) and protein-coding (e.g., *rpoC1*, *rbcL*, *cpcBA*) loci have been used for resolving taxa (e.g., Singh et al., 2015; Dvořák et al., 2014). Used together in multilocus sequence analysis (MLSA), concatenated gene sequences from several loci can improve phylogenetic resolution, particularly for infrageneric diversity (Maiden et al., 1998; Stackebrandt et al., 2002; Tanabe et al., 2007; Mazard et al., 2012; Sciuto et al., 2012; da Silva Malone et al., 2015). For example, Melendrez et al. (2011) was able to resolve up to 14 more ecotypes of *Synechococcus* from a thermal spring using a three locus MLSA compared with 16S rRNA and 16S–23S ITS analysis alone. However, there is not always additional value in this approach. Komárek et al. (2020) found that *Thermosynechococcus* strains were more resolved as a separate clade using a phylogeny derived from MLSA, using 260 ubiquitous concatenated housekeeping loci than using 16S rRNA alone, supporting the creation of a new family, whereas Mareš (2018) showed broadly similar phylogenies when comparing 16S and a 23 multilocus analysis, showing the strength of the 16S at distinguishing cyanobacterial genera (Yarza et al., 2014; Komárek et al., 2014). This highlights the need to carefully consider the best configuration of genes and approaches for each phylogenetic application based on the profile of the taxa being investigated.

For resolving species-level associations, the use of the ITS region between the 16S and 23S rRNA genes has increased rapidly over the past decade. The ITS is more variable than the 16S in sequence, length, and secondary structure (Iteman et al., 2000; Boyer et al., 2001). The simplest ITS regions have two transfer RNA genes ($tRNA^{Ala}$ and $tRNA^{Ile}$), one or other of the tRNA genes, or neither (Boyer et al., 2001). Many cyanobacterial strains have multiple ribosomal operons, that is, one or more operons that vary in sequence, length, or composition (Iteman et al., 2000; Gugger et al., 2005; Sendall and McGregor, 2018). In more complex situations, fragments of different 16S–23S rRNA operons within the same strain may be highly divergent (Boyer et al., 2001; Iteman et al., 2000), potentially resulting from horizontal transfer events (Stüken et al., 2009; Johansen et al., 2017b), rather than the product of simple vertical transmission expected in asexual organisms. Because of the potential presence of multiple rRNA operons in a single strain and the possibility of a paralogous origin of one or other rRNA operons, or differential amplification of operons by PCR (Boyer et al., 2001), it is recommended that cloning and sequencing of 16S-ITS-23S PCR products be performed in preference to direct sequencing. Therefore ITS sequences can be used to construct meaningful phylogenies only when orthology of multiple operons (if present) is established.

The percent dissimilarity of aligned ITS regions in orthologous ribosomal operons has increasingly been used to discriminate cyanobacteria at the species level (Erwin and Thacker, 2008; Osorio-Santos et al., 2014; Pietrasiak et al., 2014, 2019; Bohunická et al., 2015; González-Resendiz et al., 2019; Shalygin et al., 2019; Jung et al., 2020). Within species dissimilarity values are generally $<3\%$, whereas different species have dissimilarity values $>7\%$; values between are considered ambiguous (Erwin and Thacker, 2008; Osorio-Santos et al., 2014). Iteman et al. (2000, 2002) established the foundation for the use of the secondary structure of the ITS to distinguish closely related cyanobacteria. They showed that ITS contained regions where the secondary structure was both conserved (D1, D1′, D2, D3, D4, box A) and variable (V2, V3, box B). Johansen and Casamatta (2005) subsequently proposed the taxonomic utility of the secondary structure of the semi-conserved regions of the ITS. They showed clear congruences between 16S rRNA gene-based phylogenies and autapomophies in the secondary structure of various ITS helices. Many authors have subsequently applied this approach to the recognition of intra- and intergeneric limits, particularly to closely related taxa (e.g., Casamatta et al., 2006; Řeháková et al., 2007; Johansen et al., 2011; Bohunická et al., 2015; Alvarenga et al., 2016; Casamatta et al., 2020; Jung et al., 2020). Interestingly, despite its widespread use, no formal schema has been developed to standardize the way in which secondary structure motifs are presented, hence it often varies between authors.

As we discussed above, collectively the application of molecular methods has led to a rapid increase in the number of new or emended descriptions in the past two decades. However, in many cases, this has fostered a unidirectional or forward-looking approach to species designations with few examples of rigorous evaluation against "historical taxa." In many cases, novel placement within a taxonomic tree has quickly become the preeminent criterion for designating a strain as a new species (De Clerck et al., 2013). González-Resendiz et al. (2019) noted an example of this in *Desertifilum tharense*, which they consider an exact match to *Phormidium californicum*, but was not mentioned by the authors as such. For most taxa described prior to 2000, there exists no reference strains or sequences with reliable environmental provenance that represent the holotype. For most of these taxa, the holotype is either an illustration or a preserved herbarium specimen. Sequencing of dried herbarium specimens has been attempted with some success (Dvořák et al., 2020; Mareš et al., 2013; Palinska et al., 2006), but many are too degraded for sequencing, or have been lost or destroyed. Without sequenced types, these taxa are essentially orphaned in time, with little prospect of future evaluation (Komárek, 2020). Establishment of new phylogenetic reference points for old taxa provides the best way forward, however there are few cases of this work being undertaken. Some recent examples include the establishment of an epitype and reference strain for *Oscillatoria princeps*, the type species for the genus *Oscillatoria* (Mühlsteinová et al., 2018), and the epitypification of *Geitlerinema splendidum* (Johansen et al., 2017b), *Cyanothece aeruginosa* (Mareš et al., 2019), *Crinalium magnum* (Mikhailyuk et al., 2019), and *Pleurocapsa minor* (Shalygin et al., 2019). Proposals now exist to more broadly remedy this situation, which involves finding strains or environmental populations synonymous to holotypes in terms of their location, morphology, and ecology, sequencing them and creating epitypes or neotypes (Johansen et al., 2019; Soo et al., 2019). These strains can then be deposited into culture collections for other researchers to access. In the interim, authors are encouraged to carefully consider novel material against the historical literature to avoid descriptions of new taxa, which are synonymous with existing taxa.

Formal designation of taxa according to the nomenclatural system

If authors are satisfied that their taxon is indeed novel based on one or more of the evidences discussed above, the next step in the pathway is formal designation and publication. We have briefly outlined some of the differences in the two nomenclatural systems covering cyanobacteria, importantly those relating to typification, effective publication, and valid publication. Given the ongoing developments in harmonizing the two systems, most new or emended cyanobacterial taxa published over the past decade have been in accordance with the rules of the ICN (Turland et al., 2018), which we recommend authors become familiar with. Incorrect typification is the most common problem leading to invalid publication (Johansen et al., 2019). A holotype must be one specimen or illustration (Art. 9.1); for specimens they must be dead and preserved (dried, a permanent slide, preserved, or a gathering), or a living culture in an immobilized state (cryopreserved), and it must be specifically stated that it is inactive (Art. 8.4). Additionally, the holotype must be from a single locality, collected on a single date by a single person or team (Art. 8.2), and must be deposited with a single accession number in a single herbarium or institutional collection (Art. 8.1). The protologue should explicitly nominate the type by designating "typus", "holotypus," or the modern equivalent "holotype." The protologue must also include a detailed unambiguous morphological description of the new taxon (Art. 38); ideally this will include observations of both natural material and strains where applicable, such that it distinguishes it from other taxa and facilitates its identification, noting that critical features may change over time or not be present in culture. Unlike the ICNP, effective publication can be in any printed matter to the public or at least to scientific institutions with generally accessible libraries (Art. 29.1), including electronically and online as a PDF. Published taxa that do not conform to this and the other requirements of the ICN risk being invalid and joining the ranks of many others without any formal recognition or status in taxonomy.

In the following case studies, we further explore two of the most pressing challenges to the recognition of cyanobacterial diversity in nature: cryptic diversity and phenotypic plasticity with examples.

Case Study 1.1: Unexplored and cryptic diversity in picocyanobacteria

Conventional light microscopy continues to be the principal method for the estimation of phytoplankton diversity and abundance in lakes and reservoirs across the globe. Recent data from Australia has shown that most routine cyanobacterial bloom monitoring continues to rely on light microscopy as the principal diagnostic tool with approximately 25,000 samples analyzed by commercial laboratories annually (McGregor unpublished data). The approach relies on considerable skills in discerning key morphological characteristics, such as cell size and shape, trichome polarity and shape of the terminal cell, shape, size, and position of specialized cells such as akinetes and heterocytes, and the presence of aerotopes. The variability of these features over time, and within and between populations leads to uncertainty in identifications. Samples containing vegetative material that lack akinetes/heterocytes or are dominated by small coccoid types are particularly problematic. Additionally, cryptic diversity within the character poor cyanobacteria, is a challenge to the recognition of taxa based solely on morphology (Casamatta et al., 2003; Dvořák et al., 2014). This problem is compounded by the absence of comprehensive regional floras based on morphological features known to have congruence with cyanobacterial phylogeny. The widespread adoption of gene-based technologies to support routine phytoplankton surveillance programs and laboratory analyses has been relatively slow and is hampered by (i) the lack of a comprehensive and reliable baseline (both regional and global), against which derived sequences can be compared; (ii) many cyanobacterial genera have not been established using molecular criteria, therefore derived sequences cannot be associated with these "morphological taxa," and (iii) the absence of rapid, reliable, and cost-effective molecular approaches amenable to routine application.

Picocyanobacteria are amongst the most difficult group to confidently identify or separate from closely related species using routine microscopy, due to their small size and ambiguous and variable morphology. This may be compounded where the life cycle comprises both single cells and colonies, making them almost impossible to distinguish (Huber et al., 2017). Komárek et al. (2001) described the life cycle in *Aphanothece stratus* from a lake in Brazil, which comprised a colonial benthic phase and a single-celled or microcolony planktonic phase, each with overlapping cell morphology. Others have shown that colony formation or disaggregation into single cells may occur due to changing environmental conditions (Koblížek et al., 2000; Callieri et al., 2011) or in response to predator interactions (Jezberová and Komárková, 2007; Huber et al., 2017). This phenotypic plasticity has significant implications for the resolution of mixed species within a population using morphological attributes alone.

Collectively, picocyanobacteria have a global distribution, which includes most of the world's oceans and freshwater lakes (Callieri et al., 2012; Dvořák et al., 2014). In the freshwater domain, picocyanobacteria commonly inhabit most eutrophic lakes (Komárková, 2002; Ernst et al., 2003; Ivanikova et al., 2007; Callieri et al., 2012), and they have been known to permanently dominate the phytoplankton of several hypertrophic shallow lakes, often forming dense blooms (Iachetti and Llames, 2015; Izaguirre et al., 2015; Cabello-Yeves et al., 2017; Huber et al., 2017; Laux et al., 2018). In tropical and subtropical regions of Australia, they are ubiquitous, and often dominate the phytoplankton of lakes and reservoirs during the warmer months October–May (McGregor, 2013). Some species have been implicated in the production of cyanotoxins. Recently *Synechococcus* populations from shallow lakes in Singapore were shown to produce both cylindrospermopsin and anatoxin-a (Yew-Hoong et al., 2021). Colonial picocyanobacteria range in size from 5 to >500 cells arranged in simple groups (i.e., *Aphanocapsa*, *Anathece*, *Cyanocatena*, *Cyanodictyon*, *Cyanogranis*), or morphologically diverse shapes with cells arranged in uniform shapes or at the end of, or around colorless structured mucilage (i.e., *Cyanonephron*, *Merismopedia*, *Snowella*, *Woronichinia*). Single-celled taxa vary from 0.3–3 μm in diameter and in freshwater; *Synechococcus* and *Cyanobium* are the dominant genera.

To demonstrate the need for improved practical approaches to quantify community diversity, we assessed the contribution of taxonomic uncertainty in estimates of planktonic picocyanobacteria communities. Data derived from identifications and enumeration using conventional light microscopy from ten mesotrophic reservoirs along a tropical–subtropical gradient in north-eastern Australia over the period 2010–2020 was analyzed. The frequency and relative contribution of two groupings of picocyanobacteria were analyzed (Fig. 1.2): (1) Synechococcales/Chroococcales–unicellular or colonial picocyanobacteria (ca. <2 μm in diameter), for which there are no or few gene sequences available from the following genera: *Aphanocapsa*, *Anathece*, *Cyanocatena*, *Cyanonephron*, *Cyanogranis*, *Gloeocapsa*, *Gloeothece*, *Merismopedia*, *Myxobaktron*, *Rhabdoderma*, *Synechococcus*. (2) Unicellular *Synechococcus*-like picocyanobacteria–planktonic, solitary, elongated rod-shaped cells (L:W ratio 12) without apparent mucilaginous envelopes, dividing in a single plane perpendicular to the longitudinal

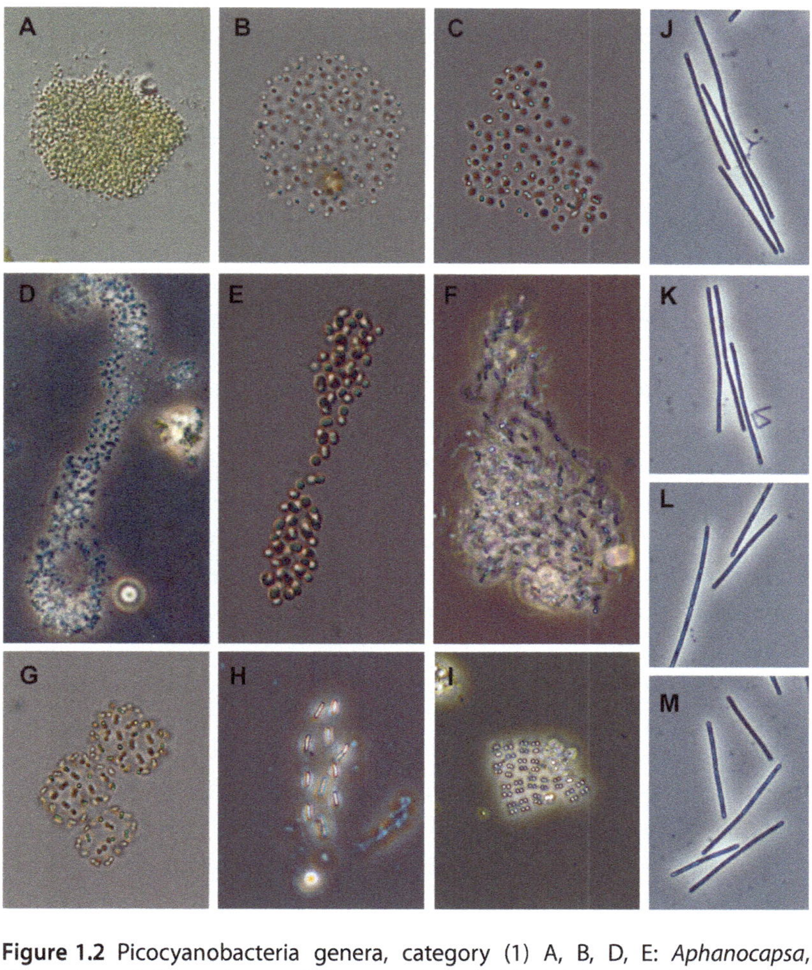

Figure 1.2 Picocyanobacteria genera, category (1) A, B, D, E: *Aphanocapsa*, C: *Anathece*, F: *Cyanodictyon*, G: *Cyanonephron*, H: *Rhabdoderma*, I: *Merismopedia*, category (2) J–M: *Synechococcus*-like. Images: G. McGregor.

axis; cells 8.8–21.7 × 0.9–1.3 μm. Picocyanobacteria group 1, comprising mostly colonial coccoid genera, contributed between 10 and 45% of the total cyanobacterial cell biovolume and 40 to 70% of the total species richness across all reservoirs independent of their longitude; at times in nine of the ten reservoirs, they constituted the entire cyanobacterial population (Fig. 1.3). *Synechococcus*-like material was most dominant in the five subtropical reservoirs, where it often contributed between 25 and 55% of the total cell biovolume. Taxonomic uncertainty for picocyanobacteria of this scale for these 10 lakes hampers robust estimates of phytoplankton community assemblages more generally and contributes to reduced reliability of results for application in both autecological studies and public health surveillance monitoring. This may represent a significant risk when toxigenic species are present. Most of these genera have no type strains or sequences, or where sequences are available in public repositories, they are often poorly or incorrectly designated, therefore it is difficult to reliably evaluate the morphologically based identifications.

Compared to their marine relatives, freshwater picocyanobacteria remain poorly studied at the genomic level. Analysis of the 16S rDNA of freshwater *Synechococcus* shows it is a polyphyletic genus (Urbach et al., 1998; Robertson et al., 2001; Crosbie et al., 2003), and in many cases of its now general usage, far removed from the original freshwater benthic concept (Nägeli, 1849). Komárek et al. (2020) recently reviewed the taxonomy of *Synechococcus*-like cyanobacteria using a polyphasic approach and found that 44 genotypes used in the phylogenetic analysis should be considered as seven distinct genera spanning four orders across marine, freshwater, and thermal environments. To accommodate the diversity of these *Synechococcus*-like types, they established one new order and three new families, and solved several taxonomic issues associated with invalid taxa. Importantly, they established a neotype for *Synechococcus elongatus* (Nägeli, 1849), the type species for the genus *Synechococcus*, based on strain PCC 6301. This material provides a definitive anchor for the genus and along with types for the other freshwater single-celled picocyanobacteria, including species from *Cyanobium* and *Picosynechococcus*, will provide greater certainty to future phylogenetic analyses of novel freshwater lineages.

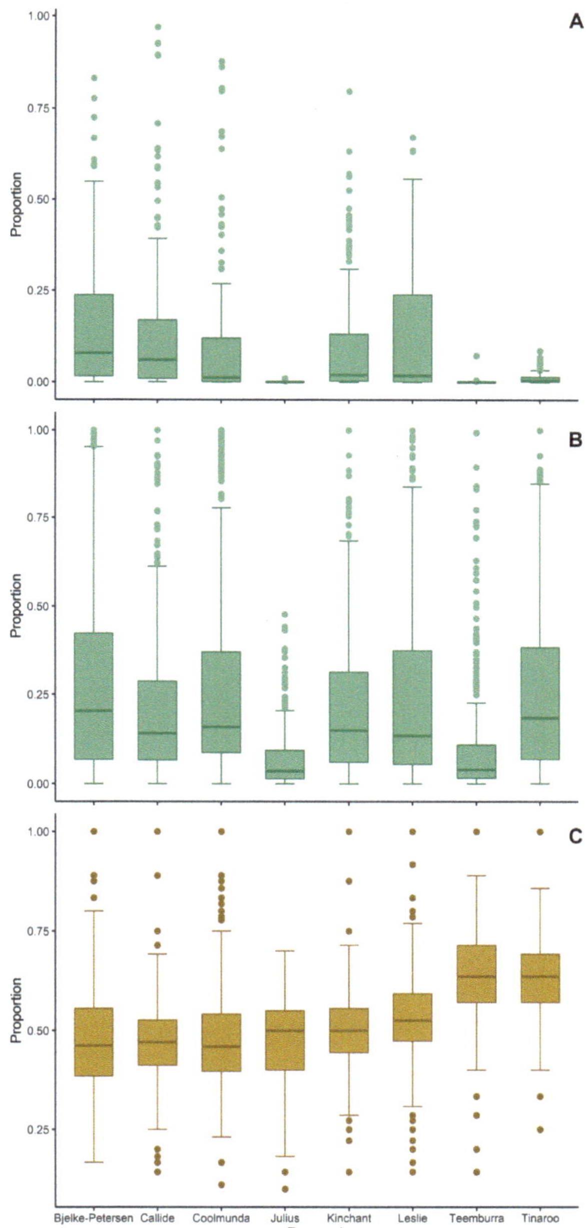

Figure 1.3 Abundance of picocyanobacteria across eight reservoirs (A) proportional abundance of *Synechococcus*, (B) proportional abundance of all picocyanobacteria genera, (C) proportion of the total cyanobacteria taxa comprised by picocyanobacteria genera.

Case Study 1.2: Phenotypic plasticity in *Dolichospermum brachiatum*

The polyphasic approach typically combines observations of wild material from its natural habitat with phenotypic and genotypic evaluation of cultivated material grown under controlled laboratory conditions. It has long been observed that cyanobacteria maintained in culture may undergo changes in morphology from the wild type over time; this often includes modification or loss of critical taxonomic features, such as cell size, heterocytes and akinete size and location, presence of aerotopes, filament coiling, presence and type of branching, and colony formation (de Souza Santos et al., 2011; Zapomělová et al., 2008; Otsuka et al., 2000). Cultivated strains may also exhibit aberrant morphology when grown under certain light and nutrient regimes such that they no longer remotely resemble the wild type. Many species within natural populations, and indeed between populations, are known to exhibit highly variable phenotypes, particularly in relation to characters such as trichome morphology (Willis et al., 2016, 2018), facultative cyanotoxin production (Sendall and McGregor, 2018; Wood et al., 2012), and toxin quota (Willis et al., 2016, 2018; Heath et al., 2011; Wood et al., 2012). For example, nine strains of *Raphidiopsis raciborskii* (Woloszynska) Aguilera, Berrendero Gómez, Kastovsky, Echenique & Salerno isolated from a single population showed significant diversity in morphology, toxin cell quota, growth rate, and genome variability (Willis et al., 2018); collectively traits that allow different genotypes to adapt and respond to changes in their natural environment. Intraspecific diversity in genes controlling P uptake has been identified in *Microcystis*, suggesting that individual genotypes may exhibit phenotypic plasticity in their adaptation to low phosphorus (Jackrel et al., 2019; Willis and Woodhouse, 2020). Phenotypic plasticity is underpinned by genes that encode functions that selectively enable a genotype to survive under a wide range of prevailing and changing environmental conditions. Such phenotypic plasticity is an important aspect of the successful colonization of cyanobacteria across diverse aerophytic, terrestrial, and aquatic habitats.

Understanding the full range of intraspecific variability across a species range is important for its delimitation and recognition in nature. Therefore careful observation of morphological and physiological phenotypes under a range of conditions likely for a given species' biophysical niche is critical to progress this understanding. We examined a population of filamentous cyanobacteria that dominated the plankton of a temperate lake in Western Victoria, Australia; based on features observed using light microscopy, it was putatively identified as *Chrysosporum ovalisporum* (Forti) E. Zapomelová, O. Skácelová, P. Pumann, R. Kopp & E. Janecek. This species is widely distributed in Australia and is a facultative producer of the cyanotoxin cylindrospermopsin (Shaw et al., 1999), hence its accurate identification is critical to support management strategies to minimize the risk of human exposure to the toxin. Interestingly, the population showed facultative T-type true branching, atypical of *C. ovalisporum* as well as highly variable vegetative cell, apical cell, and akinete morphology (Fig. 1.4). A polyphasic approach to characterization of the wild material led to the isolation of four strains, which were subjected to *in vitro* growth studies and genotypic characterization of the primary sequence and secondary structure of cloned 16S-23S rDNA amplicons. Strains were cultivated in three types of media (CT after Suda et al. (2002): phosphorus-free CT, nitrogen-free BG-11 after Rippka et al. (1979)), and revealed morphology and T-type branching consistent with wild-type material only in phosphorus-free CT media, suggesting that

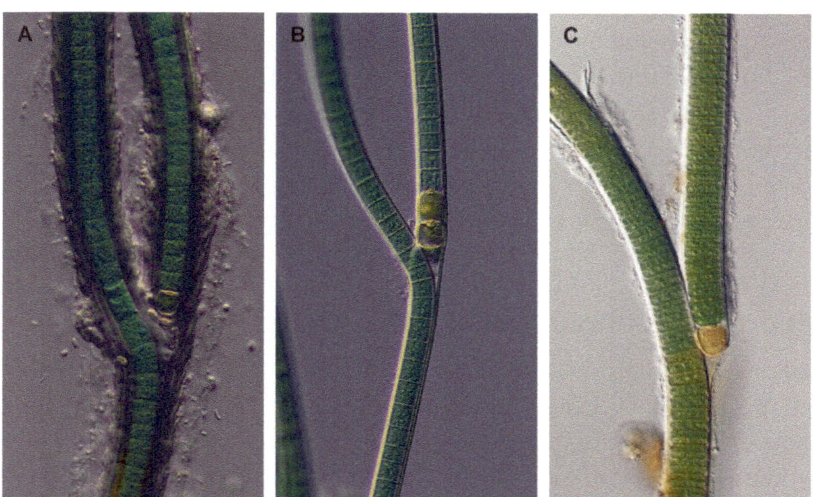

Figure 1.4 Example of cryptic diversity, false-branching in three genera: (A) *Ewamiania*, (B) *Tolypothrix*, (C) *Coelodesmium*. Images: G. McGregor.

the morphological features were a phenotypic response of the strains to P-limitation. A low concentration of soluble reactive phosphorus was observed in the lake when the wild material was collected, providing support for our hypothesis. Phylogenetic analyses of the 16Sr-RNA gene showed that all four strains were putatively *Dolichospermum planctonicum* (Brunnthaler) Wacklin, L. Hoffmann & Komárek 2009. However, p-distance similarity in the ITS of the 16S-23S rRNA gene, polymorphisms in the secondary structure of the ITS, combined with unique morphological characteristics indicated the strains represented a novel species, which we designated as *Dolichospermum brachiatum* (McGregor and Sendall, 2021). Our observation of true branching in *D. brachiatum* is the first record for *Dolichospermum* and the second observation from the family Aphanizomenonaceae, in which this feature has previously been observed in *Umezakia natans* (M. Watanabe, 1987). More importantly our study has shown that both wild material and laboratory strains of *D. brachiatum* are capable of significant morphological plasticity under varying nutrient regimes. Unrecognized intraspecific diversity is as much of a hindrance to the practice of morphology-based identification of cyanobacterial taxa as is cryptic diversity. In this case study, the management responses to the bloom from which this material originated was based on an understanding developed for another toxigenic species from a different genus, leading to unnecessary toxin analysis and monitoring expenditure; it's likely that this has been repeated elsewhere in Australia over the past decade (Crawford et al., 2017). This study has demonstrated the value of polyphasic characterization in expanding our knowledge of the spectrum of intraspecific diversity across a species' range to inform accurate recognition of that species in different natural environments.

Opportunities through whole genome sequencing

Cyanobacteria reproduce asexually and undergo homologous recombination and horizontal gene transfer (HGT). Consequently, evolutionary relationships within the group are complicated and often do not relate to the limited morphological features upon which they are identified and speciated, hence the value in establishing robust phylogenetic relationships based on established genetic markers. Although phylogenetic analysis of the 16S-23S rRNA operon, as a vital component of the polyphasic approach, has become the gold standard in identifying and defining monophyletic taxa, there still exists several problems in applying this operon singly, or in phylogenomic/multilocus analyses with other genes, to resolve taxa of cyanobacteria. These include: (i) poor resolution of highly polyphyletic taxa; (ii) difficulty in establishing cultures of species that are not amenable to culture (e.g. picocyanobacteria); (iii) lack of reference sequences for many taxa; (iv) incorrectly or incompletely identified strains in culture collections, which are not suitable as reference strains; (v) unreliable provenance of many sequences in GenBank and other nucleotide sequence databases due to incorrectly/incompletely assigned taxa; and (vi) nucleotide sequence databases contain a myriad of "uncultured cyanobacterium" and "uncultured bacterium" clones that are unsuitable for establishing robust and reliable phylogenetic/phylogenomic relationships as they lack a taxonomic anchor. Increasingly, taxonomists are analyzing whole genomes to answer questions related to differentiation of taxa at both a high level (Shih et al., 2013) and fine scale (Willis and Woodhouse, 2020). Whole genome sequencing (WGS) has increased in popularity as technology to generate and analyze complete genome sequences of cyanobacteria has become more cost- and time-effective. The technology is a potentially powerful tool to resolve many issues in cyanobacterial taxonomy, namely taxonomic uncertainty, polyphyly, and cryptic diversity. It also has the potential to broaden our understanding of functional aspects of cyanobacterial ecology such as toxin production and bloom dynamics (Li et al., 2016) and adaptation to their environment (Chrismas et al., 2015).

Many studies have demonstrated a high level of congruence between 16S rRNA gene phylogenies and those generated from core genes identified from WGS (Shih et al., 2013). In practice, robust phylogenomic trees are generated by first identifying from WGS a core set of single copy genes common to all taxa of interest as these orthologous genes in the core genome encode essential metabolic processes and are more stable than genes in the "flexible genome," which are believed to undergo substantial evolutionary changes, including HGT (Daubin et al., 2002; Shi and Falkowski, 2008; Hess, 2011). Comparison of whole genome sequences allows an average nucleotide identity (ANI) to be calculated and applied to differentiate taxa at the genus and species-level. Measures of ANI are applied in the same way measures of DNA-DNA hybridization (Stackebrandt and Goebel, 1994) and 16S rRNA gene sequence identities (Yarza et al., 2014) have been used historically to define cut-off thresholds for species delineation in prokaryotes. Richter and Rosselló-Móra (2009) defined the species boundary in prokaryotes (based on sequences mainly from proteobacteria and archaea) as approximately 95–96% ANI, suggesting that ANI of genes in the core genome is a comparable measure to DNA-DNA hybridization, and therefore can be used as a surrogate. Kim et al. (2014) compared the ANI of 6787 prokaryotic genomes, concluding that there was an overall ANI between intra- and interspecies relationships at around 95–96%. They further determined 98.65% sequence similarity in the16S rRNA gene corresponded to the 95–96% ANI value for separation of species. Figueras et al. (2014) suggested using a combination of ANI and multilocus phylogenetic analysis to correctly identify species, particularly when they are closely related.

WGS applications

Phylogenomic analysis has been proven as a useful tool in differentiating closely related species and genera that are difficult to distinguish by morphology and analysis of the 16S rRNA gene. For example, Willis and Woodhouse (2020) used phylogenomic analysis of core genes identified by comparative WGS to resolve 22 genomes of *Raphidiopsis* into three species at an ANI boundary of 96%. Similarly, phylogenomic analysis of single copy genes in 158 genomes of *Microcystis* identified at least 15 species at 96% ANI. At the intergeneric level, direct comparisons of ANI were used to resolve the highly monophyletic genera *Dolichospermum* and *Aphanizomenon*, in which the between-genus similarity in the 16S rRNA gene is as high (ca. 99.4%) as it is for the within-genus similarity (99.4–99.5%; McGregor and Sendall, 2021). Analysis of both partial and near-complete genomes of *Dolichospermum* and *Aphanizomenon* revealed low ANIs (85.3–87.6%) between the genera, resolving the high level of monophyly otherwise detected by single-gene analysis. Phylogenomic analysis has also been applied for elucidation of another issue raised in this review, that of cryptic diversity (Li et al., 2016). Komárek et al. (2020) resolved the genus *Synechococcus* using a combination of phylogenomic, morphological, and ecological attributes into at least seven genera.

The polyphasic approach is typically applied to non-axenic unispecific strains, obtained by culture-dependent approaches; this process is therefore biased towards "weedy" or readily cultivated types. There are few published works that have overcome the need to use cultured strains, amplifying 16S rRNA genes from single colonies or filaments for phylogenetic analyses (Mareš et al., 2015; Kurmayer et al., 2018), the phycocyanin intergenic spacer for genotyping (Hayes and Barker, 1997; Barker et al., 2000), or toxin genes for ecological analyses (Grindberg et al., 2011; Lara et al.,

2013). A technique called isothermal multiple strand displacement (MDA; Woyke et al., 2011; Kvist et al., 2007) is an amplification technology that allows complete or near complete amplification of a whole genome from a single cell or filament (Grindberg et al., 2011; Jones et al., 2011). It has been used primarily to identify functional genes, such as those encoding the apratoxin pathway in *Lyngbya boullioni* Hoffmann & Demoulin (Grindberg et al., 2011), but this technology might prove useful for differentiating taxa that cannot be cultured or those that are difficult to maintain in culture. Other culture-independent methods, such as flow cytometry cell sorting coupled with WGS or 16S rRNA ± ITS sequencing provide useful workflows for studying picocyanobacterial community dynamics and unraveling the phylogeny of their population constituents (Crosbie et al., 2003; Huber et al., 2017).

Many cyanobacteria genomes available on public databases, such as Genbank have been generated from metagenome sequencing of natural populations; these genomes are useful when working with unculturable species and undertaking ecological and functional studies. Metagenome sequences can be assembled into individual single genome sequences and correlated with specific taxa if appropriate reference sequences are available. Alneberg et al. (2018) compared data from single-amplified genomes of bacteria and metagenome sequences. They found there was strong agreement between the single-amplified and metagenome-assembled genomes, so there is promise that uncultured cyanobacterial taxa can be identified from natural habitats by amplification of single cells, colonies, or filaments.

Metagenomics can also be used to assess cyanobacterial diversity in natural populations by identification of OTUs derived from amplification of specific genes, such as the 16S rRNA gene (Casero et al., 2019; Liu et al., 2019) and toxin genes (Lezcano et al., 2017), potentially revealing the presence of cryptic toxin-producers in low-abundance and detecting picocyanobacteria. Metagenome data can therefore reveal more taxon diversity than detected by morphological identification (Casero et al., 2019; Liu et al., 2019), informing the need for further characterization of cryptic taxa in natural populations.

WGS has the potential to resolve many issues in cyanobacterial taxonomy, but its utility is contingent on the establishment of a reference genome of a single strain of each previously and contemporarily described genus/species (described using a polyphasic approach with a designated type/holotype/epitype). Ideally sequencing the genomes of well-supported type species (holotypes/epitypes) would create reference genomes against which strains of the same or closely related species could be compared. This would make differentiation of new species of an established taxon or identification of novel taxa simpler and more reliable as well as facilitating further studies of inter- and intraspecific diversity at the population level (Willis and Woodhouse, 2020). However, only approximately 10% of type species have available genome sequences (R. Soo, pers. comm.) but improved access and further uptake of WGS in the future will provide a continuous expansion in the number of available reference genomes.

Conclusions and future challenges

The description of new cyanobacterial taxa continues at a rapid pace invigorated by increased access to molecular technologies, expanding databases of gene sequences with reliable provenance, and the coalescence of a consensus approach to the recognition of monophyletic species. However, there remains several challenges towards developing a coherent classification for the cyanobacteria and the application of a routine taxonomic practice for their identification in nature.

More than two thirds of cyanobacterial genera do not have 16S rRNA gene sequence data for their generitype; this includes almost all genera established prior to 2000. Priority must be given to establishing epitypes for these genera from environmental samples from the type localities corresponding closely to the original description. These epitypes will provide the necessary molecular baseline against which newly recognized taxa can be compared, thus avoiding "redescribing" of traditional taxa as new entities. These epitypes will also offer the possibility of revising older taxa established prior to the inclusion of sequence data as standard approach, which are otherwise "lost in time."

There has been increased reliance on sequences lodged in public databases as the polyphasic approach has been rapidly accepted as a standard taxonomic practice. Consequently, these databases are now immense, containing hundreds of thousands of sequences; however many sequences are either incorrectly or incompletely described, resulting in anomalous outcomes in phylogenetic reconstructions. There is a need for improved confidence in these sequences, which can be achieved through better documentation, including the inclusion of ecological and habitat related information. Additionally, authors are encouraged to derive putative identifications associated with strains/sequences by taking into consideration the full range of morphological expressions observed in nature compared against the diacritical features of types, with less reliance on the observations of strains, which may present with aberrant features resulting from responses to growth conditions in culture.

Microscopy remains the principal tool for the recognition of cyanobacterial diversity in nature across much of the world, particularly in supporting routine lake and marine algal bloom monitoring programs. Both cryptic diversity and phenotypic plasticity diminish the reliability of identifications derived through observations of morphology alone; however, for many of the common and well characterized taxa, this approach is still useful, particularly in developing countries, where access to molecular methods is not feasible. Through the increased application of the polyphasic approach, we now have many studies with parallel cytomorphological and molecular evaluations of cyanobacterial taxa. This provides us with an opportunity to critically evaluate the utility of features traditionally used to discriminate them in nature. At a minimum, estimates of uncertainty for many groups can be established, such that their identification may not be reliably established using microscopy alone. This is likely to be the case for cryptic genera and simple filamentous and coccoid forms with few distinguishable features. Many of these types will only be reliably identifiable using gene sequencing approaches.

Culture-dependent methods to establish strains for sequencing is the most common method used in the polyphasic approach. This continues to limit the range of taxa that can be studied, as many are sensitive to environmental variability, have host-specific relationships with other taxa, or are from extreme environments, which cannot be effectively mimicked in the laboratory. The ability to amplify complete/near-complete genomes from single cells, colonies or filaments has the potential to identify unculturable taxa. Culture-independent approaches offered through whole genome sequencing directly from the environment also provide great promise in identifying cryptic or unculturable taxa. Metagenomic technologies have advanced to facilitate assembly of single whole genome sequences, which can then be compared against accessions of WGS in databases, such as Genbank and ascribed to characterized taxa. At a population level, the entire consortium of cyanobacteria could potentially be identified, extending our knowledge of the distribution of taxa across geographically diverse habitats. The major constraint to the broad-scale application of culture-independent approaches is that many taxa are not represented in genomic databases. As of July 2020, there were 222 WGS sequences of cyanobacteria available on Genbank. Almost 50% of these belonged to the genus *Prochlorococcus*, approximately 10% were from *Synechococcus* and 6.6% belonged to unclassified cyanobacteria, the latter mainly representing data collected from metagenomic studies. Therefore many genera are represented by a single species and a single WGS. An additional concern is that WGS from holotypes/epitypes of cyanobacterial taxa are not widely available as reference genomes. Apart from *Prochlorococcus* and *Synechococcus*, the majority of WGS belong to genera/species, which produce cyanotoxins and collectively encompass only a narrow range of described taxa.

We have shown that descriptions and revisions at the level of genus and species is progressing at a rapid pace, however higher-level taxonomic revisions have not received as much attention. 16S rRNA sequence thresholds have been suggested for delimiting families (Yarza et al., 2014), but alone, this gene has not been sufficiently sensitive to resolve families within some orders. It is likely that more information from other regions of the genome will be necessary. Though well-defined family-level clades may be identified as clusters of highly related genera in phylogenetic analyses, identifying morphological synapomorphies across taxa within these groups has not always been possible, due to divergent characteristics across some families or relatively simple features across others. This contrasts with the traditional approach, where families have largely been characterized by a set of common and discrete cytomorphological features, which facilitate their recognition in nature. Recent attempts have therefore moved towards distinguishing family-level nucleotide synapomorphies with various levels of success (e.g. Řeháková et al., 2014). Mai et al. (2018) presented two options for taxonomists relating to family-level recognition: (i) either combine several established families into fewer families to create monophyletic taxa or (ii) recognize new phylogenetically distinct lineages as separate families. They advocated for the latter option as it provides a potentially more finely divided hierarchy recognizing the many distinct lineages of taxa. There will undoubtedly be more work on this issue in the future focussing on reconciling the many new genera into higher taxonomic groupings to maintain monophyly across families and orders.

Finally, authors are encouraged to familiarize themselves with the current nomenclatural systems covering the cyanobacteria and their requirements for naming new taxa prior to their publication. The prevalence of invalidly described taxa continues to plague cyanobacterial taxonomy and add unnecessary uncertainty. This situation has been further complicated by several recent papers, which have completely ignored nomenclatural rules, opting for their own system (e.g., Walter et al., 2017). This has resulted in over 120 new genera and species, all invalid with no standing in either nomenclatural code. As we highlighted at the beginning of this chapter, names are important; they confer a unique set of ecological, physiological, and toxicological characteristics of an organism. Their name and taxonomic rank places them in a long-established framework, which we use to give order to and make sense of earth's biodiversity. The polyphasic approach continues to provide the best way forward to discover and describe cyanobacterial biodiversity, incorporating over a century of traditional botanical studies with the very best of modern molecular gene-based technologies.

References

Abed, R.M., Safi, N.M., Köster, J., De Beer, D., El-Nahhal, Y., Rullkötter, J., Garcia-Pichel, F., 2002. Microbial diversity of a heavily polluted microbial mat and its community changes following degradation of petroleum compounds. Applied and Environmental Microbiology 68 (4), 1674–1683.

Alneberg, J., Karlsson, C.M., Divne, A.M., Bergin, C., Homa, F., Lindh, M.V., Hugerth, L.W., Ettema, T.J., Bertilsson, S., Andersson, A.F., Pinhassi, J., 2018. Genomes from uncultivated prokaryotes: a comparison of metagenome-assembled and single-amplified genomes. Microbiome 6 (1), 1–14.

Alvarenga, D.O., Rigonato, J., Branco, L.H.Z., Melo, I.S., Fiore, M.F., 2016. *Phyllonema aviceniicola* gen. nov., sp. nov. and *Foliisarcina bertiogensis* gen. nov., sp. nov., epiphyllic cyanobacteria associated with *Avicennia schaueriana* leaves. International Journal of Systematic and Evolutionary Microbiology 66 (2), 689–700.

Anagnostidis, K., Komárek, J., 1985. Modern approach to the classification system of cyanophytes. 1-Introduction. Algological Studies/Archiv für Hydrobiologie, Supplement Volumes, 291–302.

Barker, G.L., Konopka, A., Handley, B.A., Hayes, P.K., 2000. Genetic variation in *Aphanizomenon* (cyanobacteria) colonies from the Baltic Sea and North America. Journal of Phycology 36 (5), 947–950.

Becerra-Absalon, I., Johansen, J.R., Munoz-Martin, M.A., Montejano, G., 2018. *Chroakolemma* gen. nov. (Leptolyngbyaceae, Cyanobacteria) from soil biocrusts in the semi-desert Central Region of Mexico. Phytotaxa 367 (3), 201–218.

Bláha, L., Babica, P., Maršálek, B., 2009. Toxins produced in cyanobacterial water blooms - toxicity and risks. Interdisciplinary Toxicology 2 (2), 36–41.

Bohunická, M., Pietrasiak, N., Johansen, J.R., Gómez, E.B., Hauer, T., Gaysina, L.A., Lukešová, A., 2015. *Roholtiella*, gen. nov. (Nostocales, Cyanobacteria) - a tapering and branching cyanobacteria of the family Nostocaceae. Phytotaxa 197 (2), 84–103.

Bornet, É., Flahault, C., 1887. Revision des Nostocacées hétérocystées contenues dans les principaux herbiers de France (Troisième fragment). Annales des Sciences Naturelles, Botanique, Septième Série 5, 51–129.

Bornet, E., Flahault, C., 1888a. Note sur deux nouveaux genres d'algues perforantes. J. Mersch.

Bornet, E., Flahault, C., 1888b. Revision des Nostocacées hétérocystées contenues dans les principaux herbiers de France.

Bourrelly, P., 1970. Les Algues D'eau Douce. Initiation à la Systématique. Tome III: Les Algues bleues et rouges, Les Eugléniens, Peridiniens et Cryptomonadines. Boubée & Cie., Paris VI. 512 pp.

Boyer, S.L., Flechtner, V.R., Johansen, J.R., 2001. Is the 16S–23S rRNA internal transcribed spacer region a good tool for use in molecular systematics and population genetics? A case study in cyanobacteria. Molecular Biology and Evolution 18 (6), 1057–1069.
Burford, M.A., Carey, C.C., Hamilton, D.P., Huisman, J., Paerl, H.W., Wood, S.A., Wulff, A., 2020. Perspective: advancing the research agenda for improving understanding of cyanobacteria in a future of global change. Harmful Algae 91, 101601.
Cabello-Yeves, P.J., Haro-Moreno, J.M., Martin-Cuadrado, A.B., Ghai, R., Picazo, A., Camacho, A., Rodriguez-Valera, F., 2017. Novel *Synechococcus* genomes reconstructed from freshwater reservoirs. Frontiers in Microbiology 8, 1151.
Caires, T.A., de Mattos Lyra, G., Hentschke, G.S., de Gusmão Pedrini, A., Sant'Anna, C.L., de Castro Nunes, J.M.D.C., 2018a. *Neolyngbya* gen. nov. (Cyanobacteria, Oscillatoriaceae): a new filamentous benthic marine taxon widely distributed along the Brazilian coast. Molecular Phylogenetics and Evolution 120, 196–211.
Caires, T.A., Lyra, G.D.M., Hentschke, G.S., da Silva, A.M.S., de Araujo, V.L., Sant'Anna, C.L., Nunes, J.M.D.C., 2018b. Polyphasic delimitation of a filamentous marine genus, *Capillus* gen. nov. (Cyanobacteria, Oscillatoriaceae) with the description of two Brazilian species. Algae 33 (4), 291–304.
Callieri, C., Lami, A., Bertoni, R., 2011. Microcolony formation by single-cell *Synechococcus* strains as a fast response to UV radiation. Applied and Environmental Microbiology 77 (21), 7533–7540.
Callieri, C., Cronberg, G., Stockner, J.G., 2012. Freshwater picocyanobacteria: single cells, microcolonies and colonial forms. In: Whitton, B.A. (Ed.), Ecology of Cyanobacteria II. Springer, Dordrecht, pp. 229–269.
Casamatta, D.A., Vis, M.L., Sheath, R.G., 2003. Cryptic species in cyanobacterial systematics: a case study of *Phormidium retzii* (Oscillatoriales) using RAPD molecular markers and 16S rDNA sequence data. Aquatic Botany 77 (4), 295–309.
Casamatta, D.A., Gomez, S.R., Johansen, J.R., 2006. *Rexia erecta* gen. et sp. nov. and *Capsosira lowei* sp. nov., two newly described cyanobacterial taxa from the Great Smoky Mountains National Park (USA). Hydrobiologia 561 (1), 13–26.
Casamatta, D.A., Villanueva, C.D., Garvey, A.D., Stocks, H.S., Vaccarino, M., Dvořák, P., Hašler, P., Johansen, J.R., 2020. *Reptodigitus chapmanii* (Nostocales, Hapalosiphonaceae) gen. nov.: a unique Nostocalean (Cyanobacteria) genus based on a polyphasic approach. Journal of Phycology 56 (2), 425–436.
Casero, M.C., Velázquez, D., Medina-Cobo, M., Quesada, A., Cirés, S., 2019. Unmasking the identity of toxigenic cyanobacteria driving a multi-toxin bloom by high-throughput sequencing of cyanotoxins genes and 16S rRNA metabarcoding. Science of the Total Environment 665, 367–378.
Castenholz, R.W., 1992. Species usage, concept, and evolution in the cyanobacteria (blue-green algae). Journal of Phycology 28 (6), 737–745.
Castenholz, R.W., Wilmotte, A., Herdman, M., Rippka, R., Waterbury, J.B., Iteman, I., Hoffmann, L., 2001. Phylum BX. cyanobacteria. In: Bergey's Manual of Systematic Bacteriology. Springer, New York, NY, pp. 473–599.
Cellamare, M., Duval, C., Drelin, Y., Djediat, C., Touibi, N., Agogué, H., Leboulanger, C., Ader, M., Bernard, C., 2018. Characterization of phototrophic microorganisms and description of new cyanobacteria isolated from the saline-alkaline crater-lake Dziani Dzaha (Mayotte, Indian Ocean). FEMS Microbiology, Ecology 94 (8), 1–24.
Chakraborty, S., Verma, E., Singh, S.S., 2019. Cyanobacterial siderophores: ecological and biotechnological significance. In: Cyanobacteria. Academic Press, pp. 383–397.
Chrismas, N.A.M., Anesio, A.M.B., Sanchez-Baracaldo, P., 2015. Multiple adaptations to polar and Alpine environments within cyanobacteria: a phylogenomic and Bayesian approach. Frontiers in Microbiology 6, 1070.
Conklin, K.Y., Stancheva, R., Otten, T.G., Fadness, R., Boyer, G.L., Read, B., Zhang, X., Sheath, R.G., 2020. Molecular and morphological characterization of a novel

dihydroanatoxin-a producing *Microcoleus* species (cyanobacteria) from the Russian River, California, USA. Harmful Algae 93, 101767.
Coutinho, F.H., Meirelles, P.M., Moreira, A.P.B., Paranhos, R.P., Dutilh, B.E., Thompson, F.L., 2015. Niche distribution and influence of environmental parameters in marine microbial communities: a systematic review. PeerJ 3, e1008.
Crawford, A., Holliday, J., Merrick, C., Brayan, J., van Asten, M., Bowling, L., 2017. Use of three monitoring approaches to manage a major *Chrysosporum ovalisporum* bloom in the Murray River, Australia, 2016. Environmental Monitoring and Assessment 189 (4), 202.
Crosbie, N.D., Pöckl, M., Weisse, T., 2003. Dispersal and phylogenetic diversity of nonmarine picocyanobacteria, inferred from 16S rRNA gene and *cpcBA*-intergenic spacer sequence analyses. Applied and Environmental Microbiology 69 (9), 5716–5721.
da Silva Malone, C.F., Rigonato, J., Laughinghouse IV, H.D., Schmidt, E.C., Bouzon, Z.L., Wilmotte, A., Fiore, M.F., Sant'Anna, C.L., 2015. *Cephalothrix* gen. nov. (Cyanobacteria): towards an intraspecific phylogenetic evaluation by multilocus analyses. International Journal of Systematic and Evolutionary Microbiology 65 (9), 2993–3007.
Dadheech, P.K., Abed, R.M., Mahmoud, H., Mohan, M.K., Krienitz, L., 2012a. Polyphasic characterization of cyanobacteria isolated from desert crusts, and the description of *Desertifilum tharense* gen. et sp. nov. (Oscillatoriales). Phycologia 51 (3), 260–270.
Dadheech, P.K., Mahmoud, H., Kotut, K., Krienitz, L., 2012b. *Haloleptolyngbya alcalis* gen. et sp. nov., a new filamentous cyanobacterium from the soda lake Nakuru, Kenya. Hydrobiologia 691 (1), 269–283.
Daubin, V., Gouy, M., Perriere, G., 2002. A phylogenomic approach to bacterial phylogeny: evidence of a core of genes sharing a common history. Genome Research 12 (7), 1080–1090.
De Clerck, O., Guiry, M.D., Leliaert, F., Samyn, Y., Verbruggen, H., 2013. Algal taxonomy: a road to nowhere? Journal of Phycology 49 (2), 215–225.
de Souza Santos, K.R., Jacinavicius, F.R., Leite, C., 2011. Effects of the pH on growth and morphology of *Anabaenopsis elenkinii* Miller (Cyanobacteria) isolated from the alkaline shallow lake of the Brazilian Pantanal. Fottea 11 (1), 119–126.
De Vos, P., Trüper, H.G., 2000. In: Judicial Commission of the International Committee on Systematic Bacteriology; IXth International (IUMS) Congress of Bacteriology and Applied Microbiology. Minutes of the meetings, 14, 15 and 18 August 1999, Sydney, Australia. International Journal of Systematic and Evolutionary Microbiology 50, 2239–2244.
Desikachary, T.V., 1959. Cyanophyta (Vol. 2). Indian Council of Agricultural Research, New Delhi.
Dittmann, E., Fewer, D.P., Neilan, B.A., 2013. Cyanobacterial toxins: biosynthetic routes and evolutionary roots. FEMS Microbiology Reviews 37 (1), 23–43.
Doolittle, W.F., 2019. Speciation without species: a final word. Philosophy, Theory, and Practice in Biology 11 (014).
Dvořák, P., Hindák, F., Hašler, P., Hindakova, A., Poulíčková, A., 2014. Morphological and molecular studies of *Neosynechococcus sphagnicola*, gen. et sp. nov. (Cyanobacteria, Synechococcales). Phytotaxa 170 (1), 24–34.
Dvořák, P., Poulíčková, A., Hašler, P., Belli, M., Casamatta, D.A., Papini, A., 2015. Species concepts and speciation factors in cyanobacteria, with connection to the problems of diversity and classification. Biodiversity and Conservation 24 (4), 739–757.
Dvořák, P., Casamatta, D.A., Hašler, P., Jahodářová, E., Norwich, A.R., Poulíčková, A., 2017. Diversity of the cyanobacteria. In: Hallenbeck, P. (Ed.), Modern Topics in the Phototrophic Prokaryotes. Springer, Cham.
Dvořák, P., Hašler, P., Poulíčková, A., 2020. New insights into the genomic evolution of cyanobacteria using herbarium exsiccatae. European Journal of Phycology 55 (1), 30–38.

Elenkin, A.A., 1936 & 1949. Sinezelenye vodorosli SSSR Monographia algarum cyanophycearum aquidulcium et terrestrium in finibus URSS inventarum. Pars spec 1, 2. Akademia Nauk SSSR, Moscow-Leningrad.

Engene, N., Rottacker, E.C., Kaštovský, J., Byrum, T., Choi, H., Ellisman, M.H., Komárek, J., Gerwick, W.H., 2012. *Moorea producens* gen. nov., sp. nov. and *Moorea bouillonii* comb. nov., tropical marine cyanobacteria rich in bioactive secondary metabolites. International Journal of Systematic and Evolutionary Microbiology 62 (5), 1171–1178.

Engene, N., Paul, V.J., Byrum, T., Gerwick, W.H., Thor, A., Ellisman, M.H., 2013. Five chemically rich species of tropical marine cyanobacteria of the genus *Okeania* gen. nov. (Oscillatoriales, Cyanoprokaryota). Journal of Phycology 49 (6), 1095–1106.

Engene, N., Tronholm, A., Salvador-Reyes, L.A., Luesch, H., Paul, V.J., 2015. *Caldora penicillata* gen. nov., comb. nov. (Cyanobacteria), a pantropical marine species with biomedical relevance. Journal of Phycology 51 (4), 670–681.

Engene, N., Tronholm, A., Paul, V.J., 2018. Uncovering cryptic diversity of *Lyngbya*: the new tropical marine cyanobacterial genus *Dapis* (Oscillatoriales). Journal of Phycology 54 (4), 435–446.

Ernst, A., Becker, S., Wollenzien, U.I., Postius, C., 2003. Ecosystem-dependent adaptive radiations of picocyanobacteria inferred from 16S rRNA and ITS-1 sequence analysis. Microbiology 149 (1), 217–228.

Erwin, P.M., Thacker, R.W., 2008. Phototrophic nutrition and symbiont diversity of two Caribbean sponge–cyanobacteria symbioses. Marine Ecology. Progress Series 362, 139–147.

Figueras, M.J., Beaz-Hidalgo, R., Hossain, M.J., Liles, M.R., 2014. Taxonomic affiliation of new genomes should be verified using average nucleotide identity and multilocus phylogenetic analysis. Genome Announcements 2 (6).

Garcia-Pichel, F., Belnap, J., Neuer, S., Schanz, F., 2003. Estimates of global cyanobacterial biomass and its distribution. Algological Studies 109 (1), 213–227.

Geitler, L., 1925. Cyanophyceae. SuBwasser-Flora Mitteleuropas 12, 1–450.

Geitler, L., 1932. Cyanophyceae. In: Kolkwitz, R. (Ed.), Rabenhorst's Kryptogamen Flora von Deutschland, Österreich und der Schweiz, Vol. 14. Akademische Verlagsgesellschaft, Leipzig, pp. 1–1196.

Geitler, L., 1942. Schizophyta. In: Die natuichen Pflanzenfamilien, p. 21.

Goldenfield, N., Woese, C.R., 2007. Connections: biology's next revolution. Nature 445, 369.

Gomont, M., 1892. Monographie des Oscillatoriées (Nostocacées homocystées). Ann Sci. Nat. Bot. Ser. 7 (15), 263–368.

González-Resendiz, L., Johansen, J.R., León-Tejera, H., Sánchez, L., Segal-Kischinevzky, C., Escobar-Sánchez, V., Morales, M., 2019. A bridge too far in naming species: a total evidence approach does not support recognition of four species in *Desertifilum* (Cyanobacteria). Journal of Phycology 55 (4), 898–911.

Grindberg, R.V., Ishoey, T., Brinza, D., Esquenazi, E., Coates, R.C., Liu, W.T., Gerwick, L., Dorrestein, P.C., Pevzner, P., Lasken, R., Gerwick, W.H., 2011. Single cell genome amplification accelerates identification of the apratoxin biosynthetic pathway from a complex microbial assemblage. PLoS ONE 6 (4), 18565.

Gugger, M., Molica, R., Le Berre, B., Dufour, P., Bernard, C., Humbert, J.F., 2005. Genetic diversity of *Cylindrospermopsis* strains (Cyanobacteria) isolated from four continents. Applied and Environmental Microbiology 71 (2), 1097–1100.

Hamilton, D.P., Wood, S.A., Dietrich, D.R., Puddick, J., 2014. Costs of harmful blooms of freshwater cyanobacteria. In: Cyanobacteria: an Economic Perspective. John Wiley & Sons, pp. 247–256.

Hayes, P.K., Barker, G.L., 1997. Genetic diversity within Baltic sea populations of *Nodularia* (cyanobacteria) 1. Journal of Phycology 33 (6), 919–923.

He, X., Liu, Y.L., Conklin, A., Westrick, J., Weavers, L.K., Dionysiou, D.D., Lenhart, J.J., Mouser, P.J., Szlag, D., Walker, H.W., 2016. Toxic cyanobacteria and drinking water: impacts, detection, and treatment. Harmful Algae 54, 174–193.

Heath, M.W., Wood, S.A., Ryan, K.G., 2011. Spatial and temporal variability in *Phormidium* mats and associated anatoxin-a and homoanatoxin-a in two New Zealand rivers. Aquatic Microbial Ecology 64 (1), 69–79.

Hess, W.R., 2011. Cyanobacterial genomics for ecology and biotechnology. Current Opinion in Microbiology 14 (5), 608–614.

Hoffmann, L., Komárek, J., Kaštovský, J., 2005. System of cyanoprokaryotes (cyanobacteria) state in 2004. Algological Studies/Archiv für Hydrobiologie, Supplement Volumes, 95–115.

Hofmann, L.C., Nettleton, J.C., Neefus, C.D., Mathieson, A.C., 2010. Cryptic diversity of *Ulva* (Ulvales, Chlorophyta) in the Great Bay Estuarine System (Atlantic USA): introduced and indigenous distromatic species. European Journal of Phycology 45 (3), 230–239.

Huber, P., Diovisalvi, N., Ferraro, M., Metz, S., Lagomarsino, L., Llames, M.E., Royo-Llonch, M., Bustingorry, J., Escaray, R., Acinas, S.G., Gasol, J.M., 2017. Phenotypic plasticity in freshwater picocyanobacteria. Environmental Microbiology 19 (3), 1120–1133.

Huisman, J., Codd, G.A., Paerl, H.W., Ibelings, B.W., Verspagen, J.M., Visser, P.M., 2018. Cyanobacterial blooms. Nature Reviews. Microbiology 16 (8), 471–483.

Humpage, A., Falconer, I., Bernard, C., Froscio, S., Fabbro, L., 2012. Toxicity of the cyanobacterium *Limnothrix* AC0243 to male Balb/c mice. Water Research 46 (5), 1576–1583.

Iachetti, C.M., Llames, M.E., 2015. Light limitation helps stabilize the phytoplankton assemblage steady-state in a temperate and highly turbid, hypertrophic shallow lake (Laguna Chascomús, Argentina). Hydrobiologia 752 (1), 33–46.

Iteman, I., Rippka, R., de Marsac, N.T., Herdman, M., 2000. Comparison of conserved structural and regulatory domains within divergent 16S rRNA–23S rRNA spacer sequences of cyanobacteria. Microbiology 146 (6), 1275–1286.

Iteman, I., Rippka, R., de Marsac, N.T., Herdman, M., 2002. rDNA analyses of planktonic heterocystous cyanobacteria, including members of the genera *Anabaenopsis* and *Cyanospira*. The GenBank accession numbers of the 16S rDNA gene sequences reported in this paper are AY038032–AY038037. Microbiology 148 (2), 481–496.

Ivanikova, N.V., McKay, R.M.L., Bullerjahn, G.S., Sterner, R.W., 2007. Nitrate utilization by phytoplankton in Lake Superior is impaired by low nutrient (P, Fe) availability and seasonal light limitation — a cyanobacterial bioreporter study. Journal of Phycology 43, 475–484.

Izaguirre, I., Sánchez, M.L., Schiaffino, M.R., O'Farrell, I., Huber, P., Ferrer, N., Zunino, J., Lagomarsino, L., Mancini, M., 2015. Which environmental factors trigger the dominance of phytoplankton species across a moisture gradient of shallow lakes? Hydrobiologia 752 (1), 47–64.

Jackrel, S.L., White, J.D., Evans, J.T., Buffin, K., Hayden, K., Sarnelle, O., Denef, V.J., 2019. Genome evolution and host-microbiome shifts correspond with intraspecific niche divergence within harmful algal bloom-forming *Microcystis aeruginosa*. Molecular Ecology 28 (17), 3994–4011.

Jahodářová, E., Dvořák, P., Hašler, P., Poulíčková, A., 2017. Revealing hidden diversity among tropical cyanobacteria: the new genus *Onodrimia* (Synechococcales, Cyanobacteria) described using the polyphasic approach. Phytotaxa 326 (1), 28–40.

Jahodářová, E., Dvořák, P., Hašler, P., Holušová, K., Poulíčková, A., 2018. *Elainella* gen. nov.: a new tropical cyanobacterium characterized using a complex genomic approach. European Journal of Phycology 53 (1), 39–51.

Jezberová, J., Komárková, J., 2007. Morphometry and growth of three *Synechococcus*-like picoplanktic cyanobacteria at different culture conditions. Hydrobiologia 578 (1), 17–27.

Johansen, J.R., Casamatta, D.A., 2005. Recognizing cyanobacterial diversity through adoption of a new species paradigm. Algological Studies 117 (1), 71–93.

Johansen, J.R., Kovacik, L., Casamatta, D.A., Iková, K.F., Kastovský, J., 2011. Utility of 16S-23S ITS sequence and secondary structure for recognition of intrageneric and intergeneric limits within cyanobacterial taxa: *Leptolyngbya corticola* sp. nov. (Pseudanabaenaceae, Cyanobacteria). Nova Hedwigia 92 (3), 283–302.

Johansen, J.R., Mareš, J., Pietrasiak, N., Bohunická, M., Zima Jr, J., Štenclová, L., Hauer, T., 2017a. Highly divergent 16S rRNA sequences in ribosomal operons of *Scytonema hyalinum* (Cyanobacteria). PLoS ONE 12 (10), 0186393.

Johansen, J.R., Strunecký, O., Bohunická, M., Čapková, K.Č., Raabová, L., Dvořák, P., Komárek, J., 2017b. A revision of the genus *Geitlerinema* and a description of the genus *Anagnostidinema* gen. nov. (Oscillatoriophycidae, Cyanobacteria). Fottea 17 (1), 114–1126.

Johansen, J.R., Hauer, T., Kaštovský, J., 2019. The explosion of nomenclaturally invalid cyanobacterial genera: navigating two nomenclatural codes. In: Proceedings of the 21st Symposium of the International Association of Cyanophyte/Cyanobacteria Research. Moreton Bay Research Station, Minjerribah, Australia, 12–16 August 2019.

Jones, A.C., Monroe, E.A., Podell, S., Hess, W.R., Klages, S., Esquenazi, E., Niessen, S., Hoover, H., Rothmann, M., Lasken, R.S., Yates, J.R., Reinhardt, R., Kube, M., Burkart, M.D., Allen, E.A., Dorrestein, P.C., Gerwick, W.H., Gerwick, L., 2011. Genomic insights into the physiology and ecology of the marine filamentous cyanobacterium *Lyngbya majuscule*. Proceedings of the National Academy of Sciences 108 (21), 8815–8820.

Jung, P., Mikhailyuk, T., Emrich, D., Baumann, K., Dultz, S., Büdel, B., 2020. Shifting boundaries: ecological and geographical range extension based on three new species in the cyanobacterial genera *Cyanocohniella*, *Oculatella*, and *Aliterella*. Journal of Phycology 56 (5), 1216–1231.

Kim, J.I., Linton, E.W., Shin, W., 2016. Morphological and genetic diversity of *Euglena deses* group (Euglenophyceae) with emphasis on cryptic species. Algae 31 (3), 219–230.

Kim, M., Oh, H.S., Park, S.C., Chun, J., 2014. Towards a taxonomic coherence between average nucleotide identity and 16S rRNA gene sequence similarity for species demarcation of prokaryotes. International Journal of Systematic and Evolutionary Microbiology 64 (2), 346–351.

Koblížek, M., Komenda, J., Masojídek, J., Pechar, L., 2000. Cell aggregation of the cyanobacterium *Synechococcus elongatus*: role of the electron transport chain. Journal of Phycology 36 (4), 662–668.

Komárek, J., 2006. Cyanobacterial taxonomy: current problems and prospects for the integration of traditional and molecular approach. Algae 21, 349–375.

Komárek, J., 2013. Susswasserflora von Mitteleuropa: Bd. 19/3. Cyanoprokaryota 3. Teil. Heterocystous Genera. Springer Spektrum, Berlin.

Komárek, J., 2016. A polyphasic approach for the taxonomy of cyanobacteria: principles and applications. European Journal of Phycology 51 (3), 346–353.

Komárek, J., 2020. Quo vadis, taxonomy of cyanobacteria. Fottea 20 (1), 104–110.

Komárek, J., Anagnostidis, K., 1999. Cyanoprokaryota. 1.Teil: Chroococcales. Gustav Fischer Verlag, Jena. 548 pp.

Komárek, J., Anagnostidis, K., 2005. Bd. 19/2: Cyanoprokaryota: teil 2: Oscillatoriales. Elsevier, München.

Komárek, J., Cáslavská, J., 1991. Thylakoidal patterns in oscillatorialean genera. Archiv für Hydrobiologie 64, 267–270.

Komárek, J., Azevedo, S.M., Domingos, P., Komárková, J., Tichý, M., 2001. Background of the Caruaru tragedy; a case taxonomic study of toxic cyanobacteria. Algological Studies/Archiv für Hydrobiologie, Supplement Volumes, 9–29.

Komárek, J., Zapomělová, E., Šmarda, J., Kopecký, J., Rejmánková, E., Woodhouse, J., Neilan, B.A., Komarkova, J., 2013. Polyphasic evaluation of *Limnoraphis robusta*, a water-bloom forming cyanobacterium from Lake Atitlán, Guatemala, with a description of *Limnoraphis* gen. nov. Fottea 13 (1), 39–52.

Komárek, J., Kaštovský, J., Mareš, J., Johansen, J.R., 2014. Taxonomic classification of Cyanoprokaryotes (cyanobacterial genera) 2014, using a polyphasic approach. Preslia 86 (4), 295–335.

Komárek, J., Johansen, J.R., Šmarda, J., Strunecký, O., 2020. Phylogeny and taxonomy of *Synechococcus*-like cyanobacteria. Fottea 20 (2), 171–191.

Komárková, J., 2002. Cyanobacterial picoplankton and its colonial formations in two eutrophic canyon reservoirs (Czech Republic). Archiv für Hydrobiologie 154 (4), 605–623.

Komárková, J., Šimek, K., 2003. Unicellular and colonial formations of picoplanktonic cyanobacteria under variable environmental conditions and predation pressure. Algological Studies/Archiv für Hydrobiologie, Supplement Volumes, 327–340.

Kondrateva, N.V., 1968. Sin'o-zeleni vodorosti-Cyanophyta (Blue-green algae-Cyanophyta). Vizn. Prisnov. Vodorost. Ukr. RSR 1 (2), 524.

Kurmayer, R., Christiansen, G., Holzinger, A., Rott, E., 2018. Single colony genetic analysis of epilithic stream algae of the genus *Chamaesiphon* spp. Hydrobiologia 811 (1), 61–75.

Kvist, T., Ahring, B.K., Lasken, R.S., Westermann, P., 2007. Specific single-cell isolation and genomic amplification of uncultured microorganisms. Applied Microbiology and Biotechnology 74, 926–935.

Lakeman, M.B., Von Dassow, P., Cattolico, R.A., 2009. The strain concept in phytoplankton ecology. Harmful Algae 8 (5), 746–758.

Lara, Y., Lambion, A., Menzel, D., Codd, G.A., Wilmotte, A., 2013. A cultivation-independent approach for the genetic and cyanotoxin characterization of colonial cyanobacteria. Aquatic Microbial Ecology 69 (2), 135–143.

Laux, M., Werner, V.R., Vialle, R.A., Ortega, J.M., Giani, A., 2018. Planktonic cyanobacteria from a tropical reservoir of southeastern Brazil: a picocyanobacteria rich community and new approaches for its characterization. Nova Hedwigia 107 (1/2), 229–256.

Leliaert, F., De Clerck, O., 2017. Refining species boundaries in algae. Journal of Phycology 53 (1), 12–16.

Lezcano, M.A., Velázquez, D., Quesada, A., El-Shehawy, R., 2017. Diversity and temporal shifts of the bacterial community associated with a toxic cyanobacterial bloom: an interplay between microcystin producers and degraders. Water Research 125, 52–61.

Li, X., Dreher, T.W., Li, R., 2016. An overview of diversity, occurrence, genetics and toxin production of bloom-forming *Dolichospermum* (*Anabaena*) species. Harmful Algae 54, 54–68.

Liu, Z., Li, J., Wei, B., Huang, T., Xiao, Y., Peng, Z., Xie, M., Xiong, T., 2019. Bacterial community and composition in Jiang-shui and Suan-cai revealed by high-throughput sequencing of 16S rRNA. International Journal of Food Microbiology 306, 108271.

Machado-De-Lima, N.M., Martins, M.D., Branco, L.H., 2017. Description of a tropical new species of *Wilmottia* (Oscillatoriales, Cyanobacteria) and considerations about the monophyly of *W. murrayi*. Phytotaxa 307 (1), 43–54.

Mai, T., Johansen, J., Pietrasiak, N., Bohunická, M., Martin, M.P., 2018. Revision of the *Synechococcales* (Cyanobacteria) through recognition of four families including Oculatellaceae *fam. nov.* and Trichocoleaceae *fam. nov.* and six new genera containing 14 species. Phytotaxa 365 (1), 1–59.

Maiden, M.C., Bygraves, J.A., Feil, E., Morelli, G., Russell, J.E., Urwin, R., Zhang, Q., Zhou, J., Zurth, K., Caugant, D.A., Feavers, I.M., 1998. Multilocus sequence typing: a portable approach to the identification of clones within populations of pathogenic microorganisms. Proceedings of the National Academy of Sciences 95 (6), 3140–3145.

Mareš, J., 2018. Multilocus and SSU rRNA gene phylogenetic analyses of available cyanobacterial genomes, and their relation to the current taxonomic system. Hydrobiologia 811, 19–34.
Mareš, J., Hrouzek, P., Kaňa, R., Ventura, S., Strunecký, O., Komárek, J., 2013. The primitive thylakoid-less cyanobacterium *Gloeobacter* is a common rock-dwelling organism. PLoS ONE 8 (6), e66323.
Mareš, J., Lara, Y., Dadáková, I., Hauer, T., Uher, B., Wilmotte, A., Kaštovský, J., 2015. Phylogenetic analysis of cultivation-resistant terrestrial cyanobacteria with massive sheaths (*Stigonema* spp. and *Petalonema alatum*, Nostocales, Cyanobacteria) using single-cell and filament sequencing of environmental samples. Journal of Phycology 51 (2), 288–297.
Mareš, J., Strunecký, O., Bučinská, L., Wiedermannová, J., 2019. Evolutionary patterns of thylakoid architecture in cyanobacteria. Frontiers in Microbiology 10, 1–22.
Mayr, E., Ashlock, P.D., 1991. Principles of Zoological Systematics, 2nd edition. McGraw-Hill, New York.
Mazard, S., Ostrowski, M., Partensky, F., Scanlan, D.J., 2012. Multi-locus sequence analysis, taxonomic resolution and biogeography of marine *Synechococcus*. Environmental Microbiology 14 (2), 372–386.
McGregor, G.B., 2013. Freshwater cyanobacteria of north-eastern Australia: 2. Chroococcales. Phytotaxa 133 (1), 1–130.
McGregor, G.B., Sendall, B.C., 2015. Phylogeny and toxicology of *Lyngbya wollei* (Cyanobacteria, Oscillatoriales) from north-eastern Australia, with a description of *Microseira* gen. nov. Journal of Phycology 51 (1), 109–119.
McGregor, G.B., Sendall, B.C., 2019. *Potamosiphon australiensis* gen. nov., sp nov. (Oscillatoriales), a new filamentous cyanobacterium from subtropical north-eastern Australia. Phytotaxa 387 (2), 77–93.
McGregor, G.B., Sendall, B.C., 2021. True branching and phenotypic plasticity in the planktonic cyanobacterium *Dolichospermum brachiatum* sp. nov. (Nostocales, Aphanizomenonaceae), from south-eastern Australia. Phytotaxa 491 (2), 93–114.
McNeill, J., Barrie, F.R., Buck, W.R., Demoulin, V., Greuter, W., Hawksworth, D.L., Herendeen, P.S., Knapp, S., Marhold, K., Prado, J., Prud'homme van Reine, W.F., Smith, G.F., Wiersema, J.H., Turland, N.J., 2012. International Code of Nomenclature for Algae, Fungi, and Plants (Melbourne Code): Adopted by the Eighteenth International Botanical Congress, Melbourne, Australia, July 2011. [Regnum Vegetabile. V. 154: 1–208]. Koeltz Scientific Books.
Melendrez, M.C., Lange, R.K., Cohan, F.M., Ward, D.M., 2011. Influence of molecular resolution on sequence-based discovery of ecological diversity among *Synechococcus* populations in an alkaline siliceous hot spring microbial mat. Applied and Environmental Microbiology 77 (4), 1359–1367.
Metcalf, J.S., Codd, G.A., 2020. Co-occurrence of cyanobacteria and cyanotoxins with other environmental health hazards: impacts and implications. Toxins 12 (10), 629.
Mikhailyuk, T., Vinogradova, O., Holzinger, A., Glaser, K., Samolov, E., Karsten, U., 2019. New record of the rare genus *Crinalium* Crow (Oscillatoriales, Cyanobacteria) from sand dunes of the Baltic Sea, Germany: epitypification and emendation of *Crinalium magnum* Fritsch et John based on an integrative approach. Phytotaxa 400 (3), 165–181.
Minowa, T., Cho, Y., Oshima, Y., Konoki, K., Yotsu-Yamashita, M., 2019. Identification of a novel saxitoxin analogue, 12β-deoxygonyautoxin 3, in the cyanobacterium, *Anabaena circinalis* (TA04). Toxins 11 (9), 539.
Mishler, B.D., Theriot, E.C., 2000. The phylogenetic species concept (sensu Mishler and Theriot): monophyly, apomorphy and phylogenetic species concepts. In: Wheeler, Q.D., Meier, R. (Eds.), Species Concepts and Phylogenetic Theory, a Debate. Columbia University Press, New York, pp. 44–54.
Mühlsteinová, R., Hauer, T., De Ley, P., Pietrasiak, N., 2018. Seeking the true *Oscillatoria*: a quest for a reliable phylogenetic and taxonomic reference point. Preslia 90 (2), 151–169.

Nägeli, C., 1849. Gattungen einzelliger Algen: physiologisch und systematisch bearbeitet. Friedrich Schulthess.

Nakayama, T., Nomura, M., Takano, Y., Tanifuji, G., Shiba, K., Inaba, K., Inagaki, Y., Kawata, M., 2019. Single-cell genomics unveiled a cryptic cyanobacterial lineage with a worldwide distribution hidden by a dinoflagellate host. Proceedings of the National Academy of Sciences 116 (32), 15973–15978.

O'Neil, J.M., Davis, T.W., Burford, M.A., Gobler, C.J., 2012. The rise of harmful cyanobacteria blooms: the potential roles of eutrophication and climate change. Harmful Algae 14, 313–334.

Oren, A., 2004. A proposal for further integration of the cyanobacteria under the Bacteriological Code. International Journal of Systematic and Evolutionary Microbiology 54 (5), 1895–1902.

Oren, A., 2020. Three alternative proposals to emend the Rules of the International Code of Nomenclature of Prokaryotes to resolve the status of the Cyanobacteria in the prokaryotic nomenclature. International Journal of Systematic and Evolutionary Microbiology 70 (7), 4406–4408.

Oren, A., Tindall, B.J., 2005. Nomenclature of the Cyanophyta/Cyanobacteria/Cyanoprokaryotes under the International Code of Nomenclature of Prokaryotes. Algological Studies/Archiv für Hydrobiologie, Supplement Volumes, 39–52.

Oren, A., Ventura, S., 2017. The current status of cyanobacterial nomenclature under the "Prokaryotic" and the "Botanical" code. Antonie Van Leeuwenhoek 110 (10), 1257–1269.

Osorio-Santos, K., Pietrasiak, N., Bohunická, M., Miscoe, L.H., Kováčik, L., Martin, M.P., Johansen, J.R., 2014. Seven new species of *Oculatella* (Pseudanabaenales, Cyanobacteria): taxonomically recognizing cryptic diversification. European Journal of Phycology 49 (4), 450–470.

Otsuka, S., Suda, S., Li, R., Matsumoto, S., Watanabe, M.M., 2000. Morphological variability of colonies of *Microcystis* morphospecies in culture. The Journal of General and Applied Microbiology 46 (1), 39–50.

Palinska, K.A., Thomasius, C.F., Marquardt, J., Golubic, S., 2006. Phylogenetic evaluation of cyanobacteria preserved as historic herbarium exsiccata. International Journal of Systematic and Evolutionary Microbiology 56 (10), 2253–2263.

Parker, C.T., Tindall, B.J., Garrity, G.M. (Eds.), 2019. International Code of Nomenclature of Prokaryotes: Prokaryotic Code (2008 revision). International Journal of Systematic and Evolutionary Microbiology 69 (1A), S1–S111.

Perkerson III, R.B., Johansen, J.R., Kovacik, L., Brand, J., Kaštovský, J., Casamatta, D.A., 2011. A unique Pseudanabaenalean (Cyanobacteria) genus *Nodosilinea* gen. nov. based on morphological and molecular data. Journal of Phycology 47 (6), 1397–1412.

Pietrasiak, N., Mühlsteinová, R., Siegesmund, M.A., Johansen, J.R., 2014. Phylogenetic placement of *Symplocastrum* (Phormidiaceae, Cyanophyceae) with a new combination *S. californicum* and two new species: *S. flechtnerae* and *S. torsivum*. Phycologia 53 (6), 529–541.

Pietrasiak, N., Osorio-Santos, K., Shalygin, S., Martin, M.P., Johansen, J.R., 2019. When is a lineage a species? A case study in *Myxacorys* gen. nov. (Synechococcales: Cyanobacteria) with the description of two new species from the Americas. Journal of Phycology 55 (5), 976–996.

Rangel, L.M., Silva, L.H., Faassen, E.J., Lürling, M., Ger, K.A., 2020. Copepod prey selection and grazing efficiency mediated by chemical and morphological defensive traits of cyanobacteria. Toxins 12 (7), 465.

Rastogi, R.P., Madamwar, D., Incharoensakdi, A., 2015. Bloom dynamics of cyanobacteria and their toxins: environmental health impacts and mitigation strategies. Frontiers in Microbiology 6, 1254.

Řeháková, K., Johansen, J.R., Casamatta, D.A., Xuesong, L., Vincent, J., 2007. Morphological and molecular characterization of selected desert soil cyanobacteria: three species new to science including *Mojavia pulchra* gen. et sp. nov. Phycologia 46 (5), 481–502.
Řeháková, K., Johansen, J.R., Bowen, M.B., Martin, M.P., Sheil, C.A., 2014. Variation in secondary structure of the 16S rRNA molecule in cyanobacteria with implications for phylogenetic analysis. Fottea 14 (2), 161–178.
Ribeiro, K.F., Duarte, L., Crossetti, L.O., 2018. Everything is not everywhere: a tale on the biogeography of cyanobacteria. Hydrobiologia 820 (1), 23–48.
Richter, M., Rosselló-Móra, R., 2009. Shifting the genomic gold standard for the prokaryotic species definition. Proceedings of the National Academy of Sciences 106 (45), 19126–19131.
Rippka, R., Deruelles, J., Waterbury, J.B., Herdman, M., Stanier, R.Y., 1979. Generic assignments, strain histories and properties of pure cultures of cyanobacteria. Microbiology 111 (1), 1–61.
Robertson, B.R., Tezuka, N., Watanabe, M.M., 2001. Phylogenetic analyses of *Synechococcus* strains (cyanobacteria) using sequences of 16S rDNA and part of the phycocyanin operon reveal multiple evolutionary lines and reflect phycobilin content. International Journal of Systematic and Evolutionary Microbiology 51 (3), 861–871.
Rosselló-Mora, R., Amann, R., 2001. The species concept for prokaryotes. FEMS Microbiology Reviews 25 (1), 39–67.
Schmutzer, M., Barraclough, T.G., 2019. The role of recombination, niche-specific gene pools and flexible genomes in the ecological speciation of bacteria. Ecology and Evolution 9 (8), 4544–4556.
Sciuto, K., Moro, I., 2016. Detection of the new cosmopolitan genus *Thermoleptolyngbya* (Cyanobacteria, Leptolyngbyaceae) using the 16S rRNA gene and 16S–23S ITS region. Molecular Phylogenetics and Evolution 105, 15–35.
Sciuto, K., Andreoli, C., Rascio, N., La Rocca, N., Moro, I., 2012. Polyphasic approach and typification of selected *Phormidium* strains (Cyanobacteria). Cladistics 28 (4), 357–374.
Sciuto, K., Moschin, E., Moro, I., 2017. Cryptic cyanobacterial diversity in the Giant Cave (Trieste, Italy): the new genus *Timaviella* (Leptolyngbyaceae). Cryptogamie, Algologie 38 (4), 285–323.
Sendall, B.C., McGregor, G.B., 2018. Cryptic diversity within the *Scytonema* complex: characterization of the paralytic shellfish toxin producer *Heteroscytonema crispum*, and the establishment of the family *Heteroscytonemataceae* (Cyanobacteria/Nostocales). Harmful Algae 80, 158–170.
Shalygin, S., Pietrasiak, N., Gomez, F., Mlewski, C., Gerard, E., Johansen, J.R., 2018. *Rivularia halophila* sp. nov. (Nostocales, Cyanobacteria): the first species of *Rivularia* described with the modern polyphasic approach. European Journal of Phycology 53 (4), 537–548.
Shalygin, S., Huang, I.S., Allen, E.H., Burkholder, J.M., Zimba, P.V., 2019. *Odorella benthonica* gen. & sp. nov. (Pleurocapsales, Cyanobacteria): an odor and prolific toxin producer isolated from a California aqueduct. Journal of Phycology 55 (3), 509–520.
Shaw, G.R., Sukenik, A., Livne, A., Chiswell, R.K., Smith, M.J., Seawright, A.A., Norris, R.L., Eaglesham, G.K., Moore, M.R., 1999. Blooms of the cylindrospermopsin containing cyanobacterium, *Aphanizomenon ovalisporum* (Forti), in newly constructed lakes, Queensland, Australia. Environmental Toxicology 14 (1), 167–177.
Shi, T., Falkowski, P.G., 2008. Genome evolution in cyanobacteria: the stable core and the variable shell. Proceedings of the National Academy of Sciences 105 (7), 2510–2515.
Shih, P.M., Wu, D., Latifi, A., Axen, S.D., Fewer, D.P., Talla, E., Calteau, A., Cai, F., Tandeau de Marsac, N., Rippka, R., Herdman, M., Sivonen, K., Coursin, T., Laurent, T., Goodwin, L., Nolan, M., Davenport, K.W., Han, C.S., Rubin, E.M., Eisen, J.A., Woyke, T., Gugger, M., Kerfeld, C.A., 2013. Improving the coverage of the cyanobacterial phylum using diversity-driven genome sequencing. Proceedings of the National Academy of Sciences 110 (3), 1053–1058.

Singh, P., Fatma, A., Mishra, A.K., 2015. Molecular phylogeny and evogenomics of heterocystous cyanobacteria using *rbcl* gene sequence data. Annals of Microbiology 65 (2), 799–807.
Song, G., Jiang, Y., Li, R., 2015. *Scytolyngbya timoleontis*, gen. et sp. nov. (Leptolyngbyaceae, Cyanobacteria): a novel false branching Cyanobacteria from China. Phytotaxa 224 (1), 72–84.
Soo, R.M., James Hemp, J., Hugenholtz, P., 2019. Evolution of photosynthesis and aerobic respiration in the cyanobacteria. Free Radical Biology & Medicine 140, 200–205.
Stackebrandt, E., Goebel, B.M., 1994. Taxonomic note: a place for DNA-DNA reassociation and 16S rRNA sequence analysis in the present species definition in bacteriology. International Journal of Systematic and Evolutionary Microbiology 44 (4), 846–849.
Stackebrandt, E., Frederiksen, W., Garrity, G.M., Grimont, P.A., Kämpfer, P., Maiden, M.C., Nesme, X., Rosselló-Mora, R., Swings, J., Trüper, H.G., Vauterin, L., 2002. Report of the ad hoc committee for the re-evaluation of the species definition in bacteriology. International Journal of Systematic and Evolutionary Microbiology 52 (3), 1043–1047.
Stanier, R., Sistrom, W., Hansen, T., Whitton, B., Castenholz, R., Pfennig, N., Gorlenko, V., Kondratieva, E., Eimhjellen, K., Whittenbury, R., Gherna, R., Truper, H., 1978. Proposal to place the nomenclature of the Cyanobacteria (Blue-Green Algae) under the rules of the International Code of Nomenclature of Bacteria. International Journal of Systematic Bacteriology 28 (2), 335–336.
Stanier, R.Y., Van Niel, C.B., 1962. The concept of a bacterium. Archiv für Mikrobiologie 42 (1), 17–35.
Starmach, K., 1966. Cyanophyta, Glaucophyta. Panstwowe Wydaconictwo Naukowe.
Stüken, A., Campbell, R.J., Quesada, A., Sukenik, A., Dadheech, P.K., Wiedner, C., 2009. Genetic and morphologic characterization of four putative cylindrospermopsin producing species of the cyanobacterial genera *Anabaena* and *Aphanizomenon*. Journal of Plankton Research 31 (5), 465–480.
Suda, S., Watanabe, M.M., Otsuka, S., Mahakahant, A., Yongmanitchai, W., Nopartnaraporn, N., Liu, Y., Day, J.G., 2002. Taxonomic revision of water-bloom-forming species of oscillatorioid cyanobacteria. International Journal of Systematic and Evolutionary Microbiology 52 (5), 1577–1595.
Svirčev, Z., Dulić, T., Obreht, I., Codd, G.A., Lehmkuhl, F., Marković, S.B., Hambach, U., Meriluoto, J., 2019. Cyanobacteria and loess—an underestimated interaction. Plant and Soil 439 (1), 293–308.
Tanabe, Y., Kasai, F., Watanabe, M.M., 2007. Multilocus sequence typing (MLST) reveals high genetic diversity and clonal population structure of the toxic cyanobacterium *Microcystis aeruginosa*. Microbiology 153 (11), 3695–3703.
Taton, A., Wilmotte, A., Šmarda, J., Elster, J., Komárek, J., 2011. *Plectolyngbya hodgsonii*: a novel filamentous cyanobacterium from Antarctic lakes. Polar Biology 34 (2), 181–191.
Thuret, G., 1875. Essai de classification des Nostochinées. Annales des Sciences Naturelles, Botanique, Série 6 1, 372–382.
Tian, R., Lin, C., Zhang, W., Cao, H., Qian, P., 2015. Rare events of intragenus and intraspecies horizontal transfer of the 16S rRNA gene. Genome Biology and Evolution 7 (8), 2310–2320.
Tromas, N., Taranu, Z.E., Martin, B.D., Willis, A., Fortin, N., Greer, C.W., Shapiro, B.J., 2018. Niche separation increases with genetic distance among bloom-forming cyanobacteria. Frontiers in Microbiology 9, 438.
Turland, N.J., Wiersema, J.H., Barrie, F.R., Greuter, W., Hawksworth, D.L., Herendeen, P.S., Knapp, S., Kusber, W.-H., Li, D.-Z., Marhold, K., May, T.W., McNeill, J., Monro, A.M., Prado, J., Price, M.J., Smith, G.F. (Eds.), 2018. International Code of Nomenclature for Algae, Fungi, and Plants (Shenzhen Code) Adopted by the Nineteenth Inter-

national Botanical Congress Shenzhen, China, July 2017. Regnum Vegetabile, vol. 159. Koeltz Botanical Books, Glashütten.

Urbach, E., Scanlan, D.J., Distel, D.L., Waterbury, J.B., Chisholm, S.W., 1998. Rapid diversification of marine picophytoplankton with dissimilar light-harvesting structures inferred from sequences of *Prochlorococcus* and *Synechococcus* (Cyanobacteria). Journal of Molecular Evolution 46 (2), 188–201.

Vaz, M.G.M.V., Genuario, D.B., Andreote, A.P.D., Malone, C.F.S., Sant'Anna, C.L., Barbiero, L., Fiore, M.F., 2015. *Pantanalinema* gen. nov. and *Alkalinema* gen. nov.: novel Pseudanabaenacean genera (Cyanobacteria) isolated from saline–alkaline lakes. International Journal of Systematic and Evolutionary Microbiology 65 (1), 298–308.

Vinuesa, P., Silva, C., Werner, D., Martínez-Romero, E., 2005. Population genetics and phylogenetic inference in bacterial molecular systematics: the roles of migration and recombination in *Bradyrhizobium* species cohesion and delineation. Molecular Phylogenetics and Evolution 34 (1), 29–54.

Visser, P.M., Verspagen, J.M., Sandrini, G., Stal, L.J., Matthijs, H.C., Davis, T.W., Paerl, H.W., Huisman, J., 2016. How rising CO_2 and global warming may stimulate harmful cyanobacterial blooms. Harmful Algae 54, 145–159.

Walter, J.M., Coutinho, F.H., Dutilh, B.E., Swings, J., Thompson, F.L., Thompson, C.C., 2017. Ecogenomics and taxonomy of Cyanobacteria phylum. Frontiers in Microbiology 8, 2132.

Watanabe, M., 1987. Studies on the planktonic blue-green algae. 2. *Umezakia natans* gen. et sp. nov. (Stigonemataceae) from the Mikata lakes, Fukui prefecture. Bulletin of the National Science Museum. Series B: Botany 13 (3), 81–88.

Wells, M.L., Trainer, V.L., Smayda, T.J., Karlson, B.S., Trick, C.G., Kudela, R.M., Ishikawa, A., Bernard, S., Wulff, A., Anderson, D.M., Cochlan, W.P., 2015. Harmful algal blooms and climate change: learning from the past and present to forecast the future. Harmful Algae 49, 68–93.

Wheeler, Q.D., Meier, R.D. (Eds.), 2000. Species Concepts and Phylogenetic Theory: a Debate. Columbia Univ. Press, New York.

Willis, A., Woodhouse, J.N., 2020. Defining cyanobacterial species: diversity and description through genomics. Critical Reviews in Plant Sciences 39 (2), 101–124.

Willis, A., Chuang, A.W., Woodhouse, J.N., Neilan, B.A., Burford, M.A., 2016. Intraspecific variation in growth, morphology and toxin quotas for the cyanobacterium, *Cylindrospermopsis raciborskii*. Toxicon 119, 307–310.

Willis, A., Woodhouse, J.N., Ongley, S.E., Jex, A.R., Burford, M.A., Neilan, B.A., 2018. Genome variation in nine co-occurring toxic *Cylindrospermopsis raciborskii* strains. Harmful Algae 73, 157–166.

Wilmotte, A., Laughinghouse, H.D.I., Capelli, C., Rippka, R., Salmaso, N., 2017. Taxonomic identification of cyanobacteria by a polyphasic approach. In: Molecular Tools for the Detection and Quantification of Toxigenic Cyanobacteria. Wiley, Hoboken, NJ, USA, pp. 79–119.

Woese, C.R., 1987. Bacterial evolution. Microbiological Reviews 51 (2), 221.

Woese, C.R., Kandler, O., Wheelis, M.L., 1990. Towards a natural system of organisms: proposal for the domains Archaea, Bacteria, and Eucarya. Proceedings of the National Academy of Sciences of the United States of America 87, 4576–4579.

Wood, S.A., Smith, F.M., Heath, M.W., Palfroy, T., Gaw, S., Young, R.G., Ryan, K.G., 2012. Within-mat variability in anatoxin-a and homoanatoxin-a production among benthic *Phormidium* (Cyanobacteria) strains. Toxins 4 (10), 900–912.

Woyke, T., Sczyrba, A., Lee, J., Rinke, C., Tighe, D., Clingenpeel, S., Rex Malmstrom, R., Stepanauskas, R., Cheng, J., 2011. Decontamination of MDA reagents for single cell whole genome amplification. PLoS ONE 6 (10), e26161.

Yang, Z., Kong, F., Shi, X., Zhang, M., Xing, P., Cao, H., 2008. Changes in the morphology and polysaccharide content of *Microcystis aeruginosa* (cyanobacteria) during flagellate grazing 1. Journal of Phycology 44 (3), 716–720.

Yarza, P., Yilmaz, P., Pruesse, E., Glöckner, F.O., Ludwig, W., Schleifer, K.H., Whitman, W.B., Euzéby, J., Amann, R., Rosselló-Móra, R., 2014. Uniting the classification of cultured and uncultured bacteria and archaea using 16S rRNA gene sequences. Nature Reviews. Microbiology 12 (9), 635–645.

Yew-Hoong, K., Yang, Z., Chien, K., Kit, J.W., Harn, S., Han, N., Li, W., He, Y., 2021. Novel Cyanotoxin-Producing *Synechococcus* in Tropical Lakes. Water Research, p. 116828.

Zammit, G., 2018. Systematics and biogeography of sciophilous cyanobacteria; an ecological and molecular description of *Albertania skiophila* (Leptolyngbyaceae) gen. & sp. nov. Phycologia 57 (5), 481–491.

Zanchett, G., Oliveira-Filho, E.C., 2013. Cyanobacteria and cyanotoxins: from impacts on aquatic ecosystems and human health to anticarcinogenic effects. Toxins 5 (10), 1896–1917.

Zapomělová, E., Hrouzek, P., Řeháková, K., Šabacká, M., Stibal, M., Caisová, L., Komárková, J., Lukešová, A., 2008. Morphological variability in selected heterocystous cyanobacterial strains as a response to varied temperature, light intensity and medium composition. Folia Microbiologica 53 (4), 333.

Zhou, W., Ding, D., Yang, Q.S., Ahmad, M., Zhang, Y.Z., Lin, X.C., Zhang, Y.Y., Ling, J., Dong, J., 2018. *Marileptolyngbya sina* gen. nov., sp. nov. and *Salileptolyngbya diazotrophicum* gen. nov., sp. nov. (Synechococcales, Cyanobacteria), species of cyanobacteria isolated from a marine ecosystem. Phytotaxa 383 (1), 75–92.

Zimba, P.V., Shalygin, S., Huang, I.S., Momčilović, M., Abdulla, H., 2020. A new boring toxin producer - *Perforafilum tunnelli* gen. & sp. nov. (Oscillatoriales, Cyanobacteria) isolated from Laguna Madre, Texas, USA. Phycologia, 1–15.

CHAPTER 2

Uses of molecular taxonomy in identifying phytoplankton communities from the Continuous Plankton Recorder Survey

Rowena Stern[a], Declan Schroeder[a,b,c], Andrea Highfield[a], Manal Al-Kandari[d], Luigi Vezzulli[e], and Anthony Richardson[f,g]

[a]The Marine Biological Association of the UK, Plymouth, United Kingdom
[b]Veterinary Population Medicine, The University of Minnesota, St Paul, MN, United States
[c]School of Biological Sciences, University of Reading, Reading, United Kingdom
[d]Kuwait Institute for Science Research, Jamal Abdul Nasser-Shuwaikh, Kuwait City, Safat, Kuwait
[e]Department of Earth, Environmental and Life Sciences (DISTAV), University of Genoa, Genoa, Italy
[f]CSIRO Oceans and Atmosphere, Queensland Biosciences Precinct, St. Lucia, QLD, Australia
[g]School of Mathematics and Physics, The University of Queensland, St. Lucia, QLD, Australia

2.1 The Continuous Plankton Recorder Survey across large time- and space scales

The Continuous Plankton Recorder (CPR) is a mechanical plankton collection device designed by Sir Alister Hardy and first deployed in 1925–1926 in Antarctica (reviewed in Reid et al., 2003). To date, this is the longest and most spatially extensive marine plankton survey in the world, surveying for nearly 90 years (Edwards et al., 2010) and now records nearly 700 taxonomic entities. The CPR is deployed off "ships of opportunity" (commercial or research vessels), where it collects plankton onto a silk mesh that is continuously moving at a constant speed, powered by an impellor that spins as the CPR is towed behind the vessel. The original aim of the survey was to map plankton distribution over large time- and space scales, and record and interpret changes as a tool for fisheries and to correlate plankton distribution with hydrographic changes. Though the end users may have changed, the reasons for its continuation remains the same to this day. CPR surveys are deployed world-wide, surveying offshore and oceanic regions of the North Atlantic Ocean, Pacific Ocean, Indian Ocean, and Southern Ocean (Batten et al., 2019). In the North Atlantic, the CPR has surveyed 7 million nautical miles of ocean, collecting 272,700 plankton samples and over 3,400,000 abundance entries, of which more than 300 taxa are phytoplankton (MBA, 2021). The scale of data has provided accurate long-term changes for individual and aggregated plankton taxa. Over 1000 peer-reviewed publications have arisen from its data (reviewed by Batten et al., 2019). This ranges from climate change impacts on planktonic zooplankton distribution over multiple decades (Beaugrand et al., 2002, 2009), non-native diatom introduction as a result of Arctic ice melt (Reid et al., 2007), climate change environmental influences of harmful algae taxa (Hinder et al., 2012), and food web links of plankton to shellfish and seabirds (Wulf et al., 2018; MacDonald et al., 2019). The CPR survey also records phytoplankton color index (PCI), a visual estimation of green color against a standard color chart from 0 (no color) to 3 (green). This represents a semilogarithmic scale of increasing color intensity (reviewed by Batten et al., 2003a) that a significant positive relationship with remotely sensed chlorophyll a and with in situ fluorometrically measured chlorophyll and cell numbers (Raitsos et al., 2005; Batten et al., 2003b). The PCI dataset was instrumental in determining long-term changes in picocyanobacteria in the Atlantic shelf waters with negative implications for copepod nutrition (Schmidt et al., 2020). The CPR dataset has been instrumental in the development of plankton biodiversity indicators for policy applications (McQuatters-Gollop et al., 2019; Bedford et al., 2020) and is used to support EU's Marine strategy framework directive (MSFD).

Each sample filters 1.5–3 m^3 of plankton onto a silk with mesh size of 270 μm at 10 m depth. Plankton taxa are counted in a semiquantitative manner and show consistency with other methods over season scales (Lynam et al., 2013; Batten et al., 2003a), but less so over decadal scales. This is thought to be influenced by increasing ship speeds that slightly reduces the volume of water filtered by the CPR (reviewed by Richardson et al., 2006). The relatively large size of the silk mesh undersamples phytoplankton, particularly for smaller species (Richardson et al., 2006), although the CPR regularly captures phytoplankton down to 10 μm in size or smaller. Larger phytoplankton can reduce the effective mesh size or phytoplankton are caught on the leno weave (a single strand in one direction and a double twisted strand in the other) of the relatively thick silk strands of the mesh, which constitute 30–40% of the mesh area (Batten et al., 2003a). The silk is wound onto a spool at the rear of the CPR body, which contains formalin to immediately fix the plankton trapped between the two layers of silk. Each CPR deployment can survey up to 450 nautical miles, making it ideal for large-scale surveys. Costs of running CPR surveys are relatively cheap compared with research cruises, because of the use of commercial vessels, which has sustained the survey over multiple decades (Reid et al., 2003; Batten et al., 2019). The great strength of the CPR is its methodological consistency across time and the multidecadal record based on morphological identification. The strengths and limitations of the CPR, reviewed by Richardson et al. (2006), are a trade-off between sample analysis time and the provision of the highest taxonomic identification and is responsive to new research, recording unusual sightings in its notes. Many phytoplankton and microbes are unidentifiable by light microscopy due to their size or fragility, or may be broken up on the mesh upon collection. The greatest diversity and abundance of microbes are below 10 μm (De Vargas et al., 2015), beyond the capabilities of most light microscopes. Thus molecular approaches are critical to fully understanding diversity of these unrecorded taxa, although there is a need to integrate historical data using the original methods into contemporary and similarly data-rich techniques. The challenge/opportunity is to take advantage of the samples archive that has used a preservation method that is not optimal for molecular analyses of phytoplankton.

2.2 The challenge of using formalin-preserved CPR phytoplankton samples in molecular work

Formalin is commonly used for preservation of medical and biological samples, including those from the CPR. Formalin is an effective preservative for zooplankton, but is not ideal for phytoplankton as it can lead to the disintegration of some smaller naked species, although it does preserve the larger armored species well. Few other preservatives at the low concentration used in the CPR have the ability to preserve plankton for decades. However, for DNA-based studies, formalin can cause structural degradation of DNA molecules. Unbuffered formalin is acidic and can cause DNA strand breakage, leaving shorter pieces of DNA, generally less than 300 bp in length. It also induces chemical modifications, including formaldehyde-DNA adducts, DNA-protein cross-links, and DNA-DNA cross-links. One particular problem is the deamination of cytosine bases to deoxyuracil, leading to a C-T conversion in sequencing reactions and the loss of bases, leaving abasic sites (Do and Dobrovic, 2015). Until recently, this long list of structural changes to DNA has led to many researchers to ignore samples fixed in formalin. CPR samples are preserved in formaldehyde (4% final concentration by volume) and buffered with sodium tetraborate, resulting in a pH of 7.4 to 7.5, thus aiding the preservation of pH sensitive taxa, such as calcifying coccolithophorids and foraminifera (Hallegraeff et al., 2014). The inclusion of borax buffer in the formalin mixture to keep the pH neutral has meant DNA integrity in CPR-preserved samples has been relatively well maintained, in addition to being stored in cool, dark conditions in humid boxes. The generally poor condition of DNA together with poorer performing PCR amplification enzymes have previously made PCR amplification too difficult. However, with the development of improved PCR enzymes and high throughput sequencing that generate thousands to millions of short read sequences, samples fixed and stored in formalin are now accessible to genetic manipulation. This method is increasingly used to re-analyze clinical formalin-fixed paraffin-embedded (FFPE) tissue to reassess disease-causing mutations, such as single nucleotide polymorphisms (SNPs), where a single nucleotide change causes a disease state. In most cases, 80% (70–99.8% range) of single nucleotide polymorphisms found in freshly frozen tissue can be reliably recovered from FFPE tissue (Zhang et al., 2017, review). False SNPs were found, but these were randomly distributed, and increasing coverage of sequencing using high-throughput sequencing (HTS) significantly reduces the detection of false nucleotide calls (reviewed by Zhang et al., 2017). This has spurred improvements to computer-based methods based on short sequences that are compatible with sequence outputs from formalin-fixed DNA (Caporaso et al., 2010) and methods to detect and purge formalin-induced mutations from HTS datasets (Kato et al., 2018).

Various methods have been developed to improve the quality of DNA from formalin-fixed CPR samples in other contexts, allowing partial mitochondrial genomic fragments from formalin-fixed animal specimen tissues to be recovered and partially mapped to their corresponding species (Hykin et al., 2015). One method (Vezzulli et al., 2012, 2016) has been developed for bacterial analysis and specifically for the marine bacterial pathogen, *Vibrio* spp. However, this method can be adapted to any plankton (Fig. 2.1). The filtering silk is sandwiched by silk above it to keep the plankton in place. Random parts of the silk are cut and used as replicates. The plankton is then washed off the replicate silk samples to dilute and remove any residual formalin in the sample, and best results are obtained with extended washing for 12 hours as the silk is sticky. The plankton removed from the silk is then collected by centrifugation and washed again in new buffer. To obtain intracellular contents, the lysozyme enzyme stage is performed for between 1 to 2 hours to digest bacterial cells and Proteinase K stages

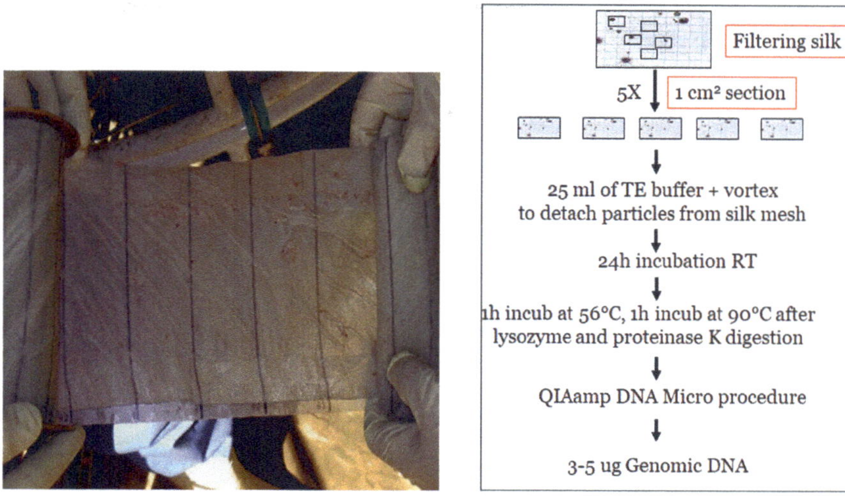

Figure 2.1 Uncut CPR sample taken off its roll (A). These are cut into approximately 10 cm-long samples and placed in plastic covers and stored in cool, dark conditions. Example protocol for extracting planktonic bacteria from CPR samples (B), adapted from Vezzulli et al. (2012).

for 4–24 hours to digest eukaryotic planktonic walls that are tougher to degrade. Formalin often makes cell walls tougher, so an extended period increases the efficiency of extraction of intracellular contents. After this stage, a high temperature (90 °C) incubation is used to remove protein-DNA cross-linking. Cross-linking is a major inhibitor of PCR amplification. At this stage, DNA extraction can be carried out through chemical means, such as phenol-chloroform (Ripley et al., 2008) and/or a DNA extraction kit, such as formalin-fixed paraffin embedded (FFPE). Although these kits often loose much DNA, it is of high quality and amenable to downstream processes. Kits designed for formalin-preserved samples often contain chemicals to remove cross linkages, such as uracil-DNA glycosylase (UNG) is often employed to remove artificially-induced uracils from the DNA obtained from the FFPE sample. Using this methodology and others (see Case Studies 2.3.1 and 2.3.2), DNA has been extracted from approximately 600 CPR samples, mostly from the North Atlantic ocean (Fig. 2.2). The integrity of CPR DNA analyzed on a bioanalyzer (Fig. 2.3) shows that larger fragment sizes mostly >700 bp are still retained even after decades. Older samples (1961, 1976), contained smaller fragments ranging from 200–800 base pairs, although the rate of DNA fragmentation was minimal from 1961 to 1976 compared to the period from 1976 to 1998.

The figure shows that the DNA appears to undergo degradation until a time-point where it remains relatively stable in storage over a 43 year period.

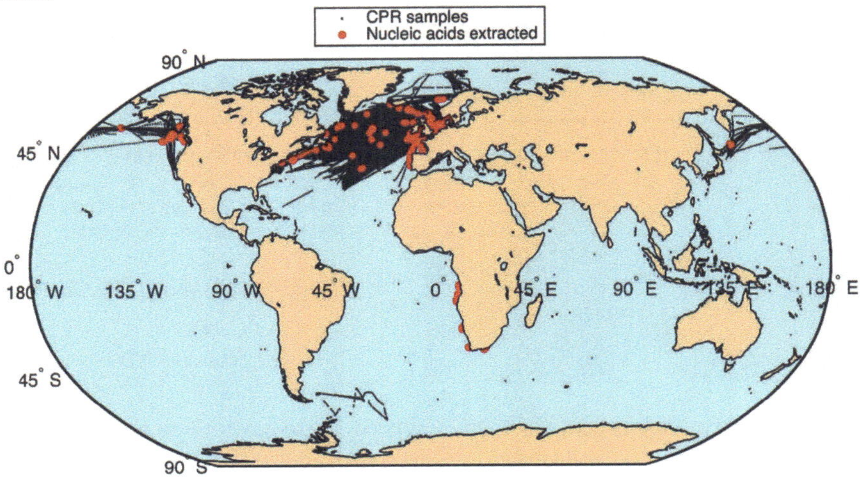

Figure 2.2 Global map of nearly 600 CPR samples that have had nucleic acids (DNA or RNA) extracted from the North Atlantic and sister surveys in the Pacific and Africa. Credit. C. Ostle, MBA.

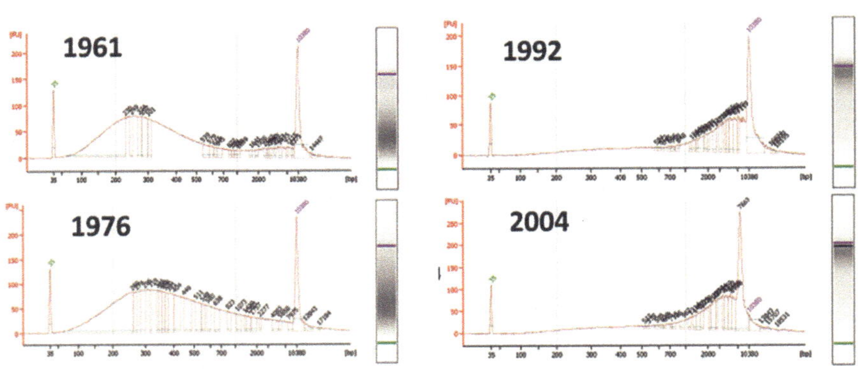

Figure 2.3 Electropherograms of DNA extracted from selected CPR samples collected in front of the Rhine Estuary from 1961 to 2004, analyzed by an Agilent Bioanalyser using the high sensitivity DNA kit (Agilent Technologies). A dominant DNA smears in the 200–800 bp size range was found in genomic DNA recovered from ~50-years old CPR samples. The older the sample. Size markers are included for reference and summary of fragment sizes are shown in boxes next to the plots. Adapted from Vezzulli et al. (2011). ISME J.: 1–10.

DNA regions commonly used to identify taxa are known as phylogenetic markers. By far, the most common DNA marker used to identify eukaryotic taxa is the small ribosomal subunit, SSU rRNA gene or the 18S or 16S rRNA for eukaryotes and prokaryotes, respectively, which assists in peptide synthesis for all cellular proteins in all living organisms (reviewed by De la Cruz et al., 2015; Woese and Fox, 1977). The large subunit ribosomal RNA (LSU rRNA gene (28S or 23S rRNA for eukaryotes and prokaryotes, respectively)) is also part of the peptide synthesis machinery and is commonly used. Both the SSU and LSU genes are highly conserved, so are ideal to assess evolutionary relationships among distant taxa. The SSU and LSU are adjacent to each other, separated by two internal transcribed spacer (ITS1 and ITS2) units and a smaller ribosomal encoding region, known as the 5.8S region. The ITS1-5.8S-ITS2 region are also used as a marker for many eukaryotic phytoplankton. However, bacteria and archaeal species have a different ITS structure (Rogers, 2011). It is often used to resolve species or populations within species, as the noncoding parts are not as conserved and contain more variation. These differences can be used to separate species that are too similar to be distinguished by the SSU or LSU alone. The accurate taxonomic identification of many phytoplankton is supported formally by full-length sequences of the SSU (~1800 bp) rRNA genes in combination with morphological data and often other molecular data (Woese and Fox, 1977, and elegantly reviewed by Sogin and Silberman, 1998). However, the DNA barcoding project (Hebert et al., 2003) proposed a genetic identification tool for species identification based on a short, standardized gene fragment of ~750 bp based on the cytochrome oxidase I (COI) gene located in the mitochondria of eukaryotes. A new specimen can be crosschecked against a known profile of reference voucher sequences corresponding to that species (Hebert et al., 2003) to confirm its identity. The size of the sequence conveniently fitted into one sanger sequencing run if performed in both directions, improving efficiency of the process and allowing larger sequence libraries to be created. This approach has been successful for many animals highlighting cryptic, synonymous, or extinct species and matching juveniles to adults, but is not a method suitable for resolving higher taxa (Waugh, 2007). Several phytoplankton groups, including brown algae, red algae, diatoms, dinoflagellates, and haptophytes have successfully been identified using COI or an alternative DNA marker, such as 18S or ITS1-ITS2 regions (Pawlowski et al., 2012), However, there are specific challenges in applying DNA barcoding to microbes due to their genetic diversity, physiology (some anaerobes lack the COI gene), and the requirement for associated physical specimens in microbes that cannot be cultured. Thus direct barcoding directly from environmental samples can capture much more diversity than from cultures.

The small fragment sizes of DNA from formalin-preserved CPR samples make genetic identification challenging from full length SSU or LSU, although 500–700 bp PCR products can still be obtained from DNA derived from CPR samples (Kirby and Lindley, 2005). However, the development of HTS or next-generation sequencing (reviewed by Goodwin et al., 2016) have made it possible to sequence and assemble from short fragments of DNA, thereby enabling genetic investigation from older samples. High-throughput sequencing technology can generate thousands of reads from a shorter DNA fragment of 300 bp (Illumina, 2020). These deep-level sequences allow even the rarest of species to be detected. Although ribosomal genes are highly conserved, there are relatively more variable regions interspersed amongst them that can be informative in identifying microbial taxa (Fuks et al., 2018; Chakravorty et al., 2007). PCR fragments of the 16S-V6 hypervariable region marker from prokaryotes (98 bp, Sogin et al., 2006) were sequenced and quantified (Vezzulli et al., 2012). Metabarcoding is the application of DNA barcoding to mixed environmental samples, allowing multiple taxa to be barcoded simultaneously. This is usually performed using high-throughput sequencing and successfully applied to CPR samples from bacteria (Vezzulli et al., 2012, 2016), phytoplankton (see Case Study 2.3.2) and zooplankton using both CO1 and 18S metabarcoding of Australian CPR samples in the Southern Ocean (Suter et al., 2020). It can also be applied to detect the dispersal of nonmarine functional genes. For example, the long-term presence of sulphonamide antibiotic resistance genes, was confirmed in North Sea CPR samples from 1970–2011 using droplet digital PCR (Di Cesare et al., 2018). Such information is not possible using microscopy alone, and invaluable in tracing the long-term effects of anthropogenic pollution of antibiotic resistance genes carried by marine bacteria.

Recent advances have made it possible to reconstruct genomes from metagenomic data derived from environmental aquatic samples using a combined approach of whole genome enrichment with HTS (Fig. 2.4). Vezzulli et al. (2017) enriched *Vibrio cholerae* bacterial DNA using whole genome enrichment approaches, using an array of baits derived from known genomes to bind to target DNA. Here baits are ~100mer long synthetic RNA probes labeled with a biotin molecule and designed to target genomic regions of interest. The probes hybridize to their corresponding target genomic DNA. The genomic DNA can be captured on

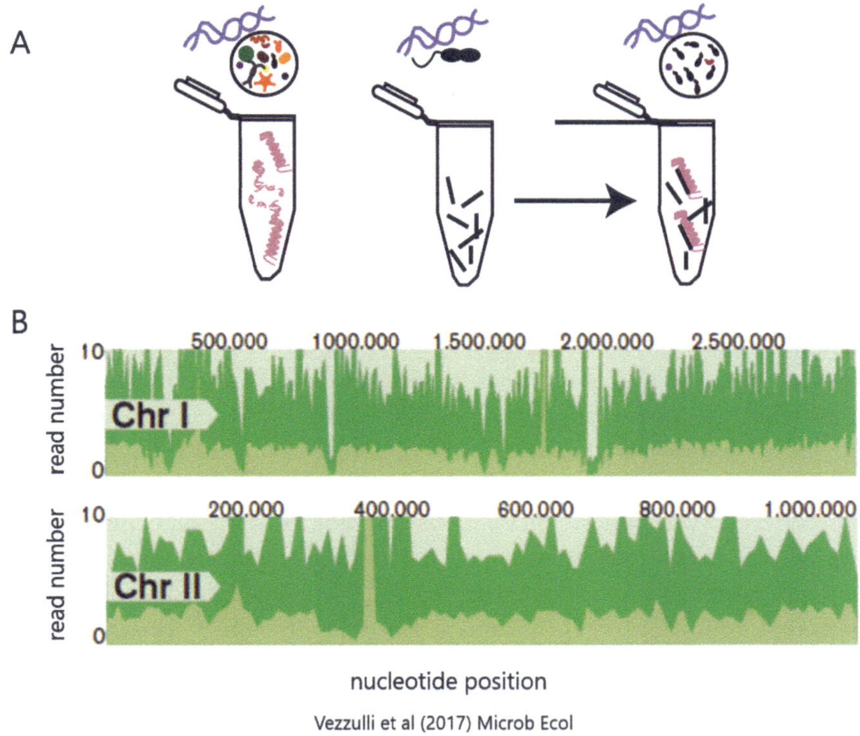

Figure 2.4 *Vibrio cholerae* draft genome sequence reconstructed from WGE enrichment and sequencing of eDNA from an African river showing four times genome coverage depth achieved using this method. Panel B adapted from Fig. 2.3, Arboscience (2020). Targeted pathogen genomics via NGS hybridisation capture. Application note, itself redrawn from Vezzulli et al. (2017). Microb. Ecol. 73: 732–738.

streptavidin-coated magnetic beads. Streptavidin binds to the biotin on the probes. Nonspecific DNA can be washed away leaving a targeted enriched genomic library that can be amplified and ready for HTS (Fig. 2.4). It was found to be 2500 times more successful in detecting *Vibrio* signals of DNA from an African river than shotgun genomic sequencing alone (Vezzulli et al., 2017). This approach has been also successful in reconstructing genomes of *Vibrio* from DNA derived from formalin-preserved CPR samples in Lake Tanganyika (Vezzulli et al., 2021). As formalin-induced damage increases with time (see Fig. 2.3), these samples were processed immediately, thus limiting the damage formalin might do. Such approaches are ideal for bacteria, archaeal, virus, or eukaryotic mitochondrial or plastid genomes, which have relatively small genomes, but may be more challenging for the larger nuclear genomes of some eukaryotic phytoplankton.

This unique repository can be used to understand long-term changes in phytoplankton and microbes (unicellular plankton down to viruses) in the open oceans. Of future significance is its use in developing baselines of genetic information of organisms under past climates to compare with current genetic studies for predictions on future movements of plankton. Such efforts could track changes in space and time of known or previously unrecorded plankton species that could mark adaptation at genetic scales (e.g., of populations), their physiological range and long-term alterations in their phenology. Combined with other physical datasets, long-term genetic data from the CPR survey could provide the basis for predictive modeling of the previously unseen planktonic world. As the CPR sampling programme is one of the few surveys that routinely monitors the open oceans, this is a crucial resource to understand open-ocean plankton dynamics. The unique manner of sample collection, which differs from traditional plankton tows or water collection devices, means the CPR work will need to be aligned with other genetic studies, but this is a problem for many long-term marine stations. With a response by the research community to increase of comparative studies between the CPR and typical survey sampling methods over several years, the development of standardized methodology could be developed that can meaningfully be compared with other genetic data.

2.3 Case studies

Case Study 2.3.1: Measuring *Emiliania huxleyi* population diversity in CPR samples

Coccolithophores are a group of phytoplankton characterized by calcified organic body scales called coccoliths; they encase the cells in a coccosphere (Young et al., 2003). *Emiliania huxleyi* is a common coccolithophore and is ubiquitous in the worlds' oceans, frequently forming large blooms containing millions of cells, extending thousands of kilometers (Holligan et al., 1993). Coccolithophores contribute significantly to the total global carbon fixation in the oceans (Balch, 2004), yet their role in the marine carbon cycle is more complex, because their calcification results in the net release of carbon (Rost and Riebesell, 2004). They are therefore a key component in regulating the carbon cycle and are also critical in maintaining the exchange of CO_2 between the ocean and atmosphere. For these reasons, *E. huxleyi* is a model species for studies unravelling coccolithophore responses to global environmental change.

Through scanning electron microscopic imagery, seven different *E. huxleyi* morphotypes, which vary in their levels of calcification, have been characterized (Young et al., 2003; Hagino et al., 2011), and within these morphotype groups, there is considerable intraspecific diversity (Schroeder et al., 2005) that can be resolved through molecular analyses. Distribution of different morphotypes have been linked to different oceanic conditions. For example, in a study of the waters of the Patagonian shelf, Poulton et al. (2011) reported *E. huxleyi* type B/C to be dominant; however, morphotype A abundance was greater in warmer waters characterized by low nitrate and phosphate concentrations.

There is considerable interest in what the response of *E. huxleyi* to rising atmospheric CO_2 will be and whether this will be reflected in changes in the morphotype composition of *E. huxleyi* communities. Laboratory studies have investigated the effects increased CO_2 on *E. huxleyi* with strain-specific responses often reported (e.g., Langer et al., 2009), making it difficult to predict the response of this species to elevated CO_2.

For these reasons, long-term time series, such as the CPR archive provide a potentially unique insight into how *E. huxleyi* populations, including the distribution of different genotypes/morphotypes, have changed over time. This may ultimately help us to predict how this species will respond to future environmental change. Indeed, a study by Rickaby et al. (2016) recently predicted that the *E. huxleyi* morphotype A was more likely to outcompete the B/C and R morphotypes in response to higher levels of CO_2 in the worlds' oceans. Data from the CPR archive could aid in these predictions.

With the CPR mesh size of 270 μm (Batten et al., 2003a), relatively few *E. huxleyi* cells (5–7 μm diameter) are retained, especially in low density areas. However, the silk mesh tends to retain a relatively constant fraction of small cells of different species (Richardson et al., 2006), and when plankton density is high, clogging of the silk mesh can result in a reduction of the mesh size meaning smaller cells are retained (Hays et al., 1995). In particular, where there is an *E. huxleyi* bloom, cell density can reach 10^5 mL^{-1} (e.g., Castberg et al., 2001; Schroeder et al., 2003) resulting in clogging of the mesh and reducing the effective mesh size retaining cells that can be counted by the CPR analysts and now used for molecular analyses.

E. huxleyi diversity has been studied using various molecular tools, including random amplification of polymorphic DNA (RAPD) and microsatellite analysis (Medlin et al., 1996; Krueger-Hadfield et al., 2014). Schroeder et al. (2005) identified the *gpa* gene as a useful tool for resolving different *E. huxleyi* morphotypes, as well as providing more resolution on different genotypes within these groups. This molecular marker has been used successfully in studies investigating *E. huxleyi* bloom dynamics in situ and in mesocosm experiments (Martinez-Martinez et al., 2007), and has subsequently been employed for use on formalin-preserved CPR samples by Ripley et al. (2008). Ripley et al. (2008) successfully extracted DNA and amplified the *gpa* gene from a range of samples from the CPR archive that were identified by CPR analysts as having "high" coccolithophore numbers, the oldest sample analyzed dating back to 1972. Using a PCR-sequencing approach, the authors revealed a genetically diverse *E. huxleyi* bloom population within the sample collected in 1997, with three prevalent genotypes corresponding to a single morphotype.

As stated in the introductory section, due to the CPR silks being stored in buffered formalin, this has raised concerns about the integrity of genetic data, i.e., not only can DNA be amplified, but does formalin fixation introduce nucleotide modifications to the DNA during subsequent PCR? (reviewed by Schander and Halanych, 2003). To address this, Highfield et al. (2014) conducted a validation experiment of molecular data acquired from CPR silks. An *E. huxleyi* bloom in the English Channel was sampled using a traditional CTD approach, where biomass was collected on filters immediately following sampling and the DNA extracted. The bloom was also sampled using a CPR tow, corresponding to four samples that were preserved in formalin according to Batten et al. (2003a). CPR DNA was extracted according to Ripley et al. (2008) and the *E. huxleyi* bloom diversity was analyzed utilizing the *gpa* gene, using PCR and sequence analysis for both the CPR and CTD samples. The population was dominated by one genotype, observed using both the CTD and CPR sampling methods. This supported the use of CPR samples for generating an accurate picture of the *E. huxleyi* population diversity as it was in agreement with freshly processed samples. This study also compared the *E. huxleyi* diversity from two different CPR samples: one located at the beginning of the tow and classified as the "edge" of the bloom (defined by lower *E. huxleyi* cell numbers, determined by flow cytometry and being on the periphery of the area of reflectance based on MODIS satellite imagery). The other sample was located centrally within the bloom (characterized by high *E. huxleyi* cell concentrations and within an area of high reflectance). Both spatially separated samples provided a similar picture of *E. huxleyi* diversity by this genetic approach, illustrating that the bloom was spatially dominated by the same genotype. This supported the use of a single CPR sample (corresponding to 10 nautical miles of tow) to provide an accurate picture on the *E. huxleyi* diversity within the bloom.

A new fluorescent-based probe assay' developed by Balestreri (2015) potentially offers a new way to analyze *E. huxleyi* populations within the CPR archive; it is more efficient compared to traditional PCR/cloning and sequencing methodology. Using this technique, Balestreri (2015) looked at the distribution of different *E. huxleyi* morphotypes over a six-year period from the English Channel collected by the Western channel observatory time series (https://www.westernchannelobservatory.org.uk/). Genotypes corresponding to the A morphotype dominated the community here, with the B morphotype being rare, despite this morphotype previously blooming here (Young and Westbroek, 1991). By applying this technique to the CPR archive, the potential to track changes in *E. huxleyi* populations over long time-scales across the globe is possible. This will help researchers predict how global environmental change may impact *E. huxleyi* and its implications for the global carbon cycle and food webs.

In 2014, Krueger-Hadfield et al. (2014) used microsatellites (repetitive genetic markers used to follow genetic inheritance of genes in species (Richard et al., 2008) in conjunction with the *gpa* gene, to genotype 93 clonal isolates collected from across the world. Of these, 52 came from a single bloom event in the North Sea collected on the D366 United Kingdom ocean acidification cruise in June–July 2011 (Poulton et al., 2014). One key observation from this study was that those isolates in the D366 data set could be divided into three main *gpa* genotypes or coccolith morphology motif (CMM) groups, all belonging to morphotype A. No CMM IIs, were detected in the D366 dataset, a genotype only associated with B /C and C morphotypes now restricted to the Southern Hemisphere (Balestreri, 2015). Given the observations made by Rickaby et al. (2016), the dominance of morphotypes A's over B or B/C or overcalcifying A's (morphotype R) (Krueger-Hadfield et al., 2014) could be an indicator of a system adapting to higher CO_2 scenarios. This is currently happening more in the Northern Hemisphere with an intraspecies regime shift that has never been observed before. The true combined power of these new genetic tools combined with the broad CPR geographical range has yet to realized.

Case Study 2.3.2: Identifying harmful algae from CPR samples using metabarcoding approaches

Harmful algae are a diverse range of taxa that are known to cause problems for humans or the aquatic environment, via a wide variety of mechanisms and impacts natural or anthropogenic. Some grow excessively in huge blooms that when they decompose can lead to deoxygenation of the water, suffocating marine animals and outcompeting other algae (Anderson et al., 2012). Others produce toxins that bioaccumulate in shellfish capable of killing or severely damaging humans and marine animals when consumed (Anderson et al., 2012). Toxin production in harmful algae varies across taxa, and even within a species (Anderson et al., 2012; Bates et al., 2018), and it remains a challenge to link harmful algae taxa to toxin production, which is required for management action in affected areas. To protect against human poisoning, most countries have a policy of closures of wild fisheries or aquaculture zones when toxic algal species are above threshold levels set by management authorities. A major obstacle is the difficulty in speciating them by light microscopy, the most convenient monitoring tool. Molecular approaches allow for fine-scale identification of toxin-forming species, or indeed the expression of known toxin-encoding genes, which can lead to improved decision making in wild and cultured fisheries. Harmful algal blooms (HABs) occur naturally, but also in response to anthropogenic factors, such as aquaculture, nutrient input, and dredging activities, which are increasing in some coastal areas, such as those in the USA (Hallegraeff, 2004; Anderson et al., 2021; Hallegraeff et al., 2021).

Pseudo-nitzschia produces the neurotoxin domoic acid that causes gastro-intestinal illness, amnesia, and neurological effects in marine animals and humans (Costa et al., 2014). It is a long, thin, chain-forming diatom first described by Peragallo and Peragallo (1897–1908). *Pseudo-nitzschia* cannot routinely be delineated to species by light microscopy (Bresnan et al., 2015), and these do not map well to species as revealed by ultrastructural features using electron microscopy and DNA nucleotide variation through genetics. There are 53 *Pseudo-nitzschia* species described (Guiry and Guiry, 2020), of which 26 species are known to be toxigenic (Bates et al., 2018) and their taxonomy, toxicity, distribution, and ecology is well reviewed by Trainer et al. (2012). As in all diatoms, *Pseudo-nitzschia* spp. is capable of sexual reproduction, and population substructure is observable in several species, such as *P. multistriata* (Tesson et al., 2014).

A systematic study of two similar but distinct *Pseudo-nitzschia* species: *P. delicatissima and P. pseudodelicatissima*, found ITS genetic marker corresponded to ultrastructural features in some species (Amato et al., 2007). The LSU performed reasonably well at separating many, but not all, *Pseudo-nitzschia* species (Amato et al., 2007). Because of their distinctive morphological characteristic of forming long chains, the genus is well sampled by the CPR to two morphospecies discernable by light microscopy (Richardson et al., 2006). A study by Stern et al. (2018a) evaluated *Pseudo-nitzschia* diversity using partial ITS and LSU marker regions, which would be compatible with HTS approaches in CPR samples. The LSU region was much more successful than ITS and continued to generate sequences with novel outcomes. The study focused on coastal and open waters off the Northwest coast of North America over three years (2002, 2005, 2008) from the same sampling transect. These samples were collected at different temperature regimes in response to the Pacific decadal oscillation, which tends to result in warmer water in positive phases (2005) and cooler waters in negative phases (2002 and 2008, see Fig. 2.5). The treatment of genetic data is discussed below, but the overall findings were unexpected. No real pattern was found between the Pacific decadal oscillation and the types of *Pseudo-nitzschia* found in warmer years. Few *Pseudo-nitzschia* species were found, but the genetic diversity within species was enormous, indicating populations adapting to different environments and providing each with an ecological advantage. *Pseudo-nitzschia multiseries* was found in open waters, in contrast to previous descriptions in coastal environments only, although there is a paucity of open-water surveys of this genus in the Pacific Ocean.

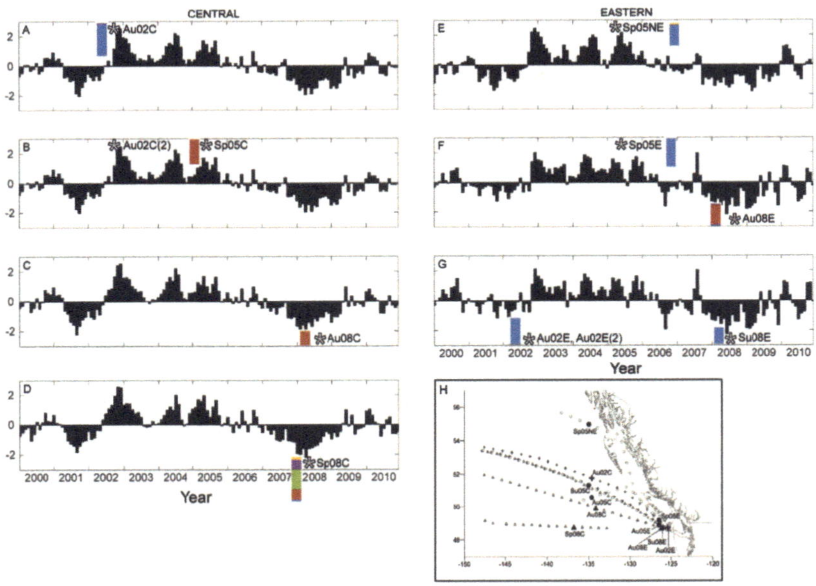

Figure 2.5 Panels A–G represent Pacific decadal oscillation (PDO) anomalies from 2000 to 2010 in warm, neutral, and cool sea surface temperature phases in Central Pacific and eastern Pacific regions. Asterisks indicate the months when DNA sequencing data was extracted from CPR samples. Superimposed are separate bars (not relating to anomaly plot axes) showing the number of *Pseudo-nitzschia* OTUs in the sample. Blue represent *Pseudo-nitzschia fraudulenta*, Red: *Pseudo-nitzschia multiseries*, Green: *Pseudo-nitzschia hasleana*, purple and yellow represent distinct novel environmental types. Panel H shows geographic locations of each sample in the Pacific Ocean, taken from the coast of North America and further offshore. Triangles, crosses, circles show the CPR routes. Figure adapted from Stern, R., Moore, S. et al. (2018). Mar. Ecol. Prog. Ser. 606:7–28.

A key technical challenge had to be overcome in this study. PCR amplifications worked better with diluted DNA samples, indicating that inhibitors were preventing PCR enzymes, common in DNA from mixed samples. Direct PCR with a 362 bp marker did not yield products, so a two-step approach was adopted: firstly, amplifying a larger

600 bp LSU product (D1R-D2C, Scholin et al., 1994). This yielded a product that was used as a template for a second PCR step, amplifying a 362 bp marker region within that 600 bp region using primers from McDonald et al. (2007). Only half of the original set of samples were amplified successfully, which may reflect formalin-degradation status of DNA inhibiting PCR reaction, but is still surprisingly successful for samples collected 6–12 years before DNA extraction. Thus there are clear advantages in applying molecular techniques to CPR samples to better delineate community composition and identify whether harmful species are producing toxins.

In a study by Stern et al. (2018a), PCR products were sequenced using older 454 GS Titanium FLX pyrosequencing technology (Roche). The same PCR assay was carried out using clone library and sequencing approaches. Results revealed similar species identified by clone-library and 454 pyrosequencing methods, although greater rarer diversity was found with the latter, whereas more resolution was afforded with the former with longer sequencing reads. Interestingly, there was a lack of correspondence between electron-microscopic species recorded on those same samples compared with the genetically identified species. Only *P. fraudulenta* was common to both taxonomic datasets. This issue is not uncommon in many morphological and genetic comparisons and can have many causes, including biases in methodologies, and points to the need for and benefits of combined monitoring approaches. Other sampling programmes along the Pacific Northwest coast of North America also captured species not observed by morphological methods (Fernandes et al., 2014; Hubbard et al., 2008). Contamination was considered as the potential sources of the discrepancy in the study by Stern et al. (2018a), as CPR samples are collected on a roll of silk and are close to each other in the device. However, this is unlikely because each sample is covered by an additional silk layer, minimizing mixing among samples and between the layers of the rolls, and the sample locations studied were hundreds of miles apart. Thus their corresponding silks would not be in direct contact. Another intriguing explanation is that the DNA was recovered from free eDNA, where cells would not be visible. The presence of *P. multiseries*-free eDNA from the coast is unlikely to survive the weeks of time it would take to transport it to the open ocean, but they might arise out of local open-water populations fertilized by silicon from upwelling events. An independent study that tracked *Clausocalanus* zooplankton species using cytochrome oxidase I (COI) barcodes and by microscopy (Suter et al., 2020) indicates little eDNA retention at least in this situation. *Clausocalanus* species undergo diel migration, feeding at the surface at night and resting at depths in the day. However, two species revealed the same night-time detection in CPR samples when they were at the surface in the same deployment, indicating little cross contamination or eDNA retention. However, the survival of free eDNA or contamination within CPR or other marine samples from other origins is a complex issue and could always be a possibility.

The methodology used in the Stern et al. (2018a) study is typical of many such studies using a common bioinformatic pipeline (Caporaso et al., 2010), and then followed with a custom phylogenetic identification process using a custom diatom LSU sequence alignment and phylogeny. Within this process, many choices can vary the outcome. Of broader significance, although many studies follow similar methodologies, there are few explicit standards in genetic methodologies or reporting (Stern et al., 2018b), which could create false differences with other studies or make it difficult to carry out metaanalyses of sequencing studies. Of interest is the treatment of operational taxonomic units (OTUs), of which there is much debate. The OTU is a way to classify a taxonomic group, usually by defining a genetic threshold that captures a group of unique sequences, analogous to many DNA barcoding methods. Such methods were used to overcome diversity inflation because of sequencing errors. However, this creates inaccuracies in species richness if false groupings are created that contain too many unique taxa together or too few taxa. As sequencing methodologies mature and errors are reduced, exact sequence variants provide a consistent genetic definition that would aid in their reuse in future studies (Callahan et al., 2017). In this case, OTU identification was performed at 90 and 99%, although the latter was much more informative, but may have removed some rarer diversity. Both sequencing methods did not find any altered nucleotide composition, especially for thymine (one type of formalin-induced alteration) from sequences of *P. fraudulenta* or *P. multiseries* (the most common species) derived from CPR sample DNA, compared to their counterparts in public databases. This indicates that nucleotide mutations from formalin within the sequencing reads of this relatively small dataset were minimal, at least for these forms of base modifications. It is now possible to map abasic sites (Liu et al., 2019), arising as artifacts when nucleotides undergo hydrolysis from formalin damage (Do and Dobrovic, 2015) to assess damage and develop optimal strategies to overcome or minimize damage.

The Stern et al. (2018a) study revealed how combining genetic, microscopy, and oceanographic data could enhance our understanding of harmful algae dynamics. It was noteworthy that this study identified the presence of multiple populations that supports previous hypotheses of ecotypes (Thessen, 2007) for *Pseudo-nitzschia* that could assist in delineating physiological preferences, and phenological and geographic boundaries. The genetics revealed new diversity, whereas microscopic abundance data showed a potential link to temperature that could be explored further. However, this and similar efforts would benefit from more improved comparative morpho-genetic databases, especially focusing on systematic bias and quantifying detection limits moving towards standardizing genetic monitoring (reviewed by Stern et al., 2018b). This could infer additional characteristics to taxonomic databases, improve confidence in genetic data, and potentially extend morphological or genetic long-term datasets in time and space. Such comparisons have been done with SeaWiFS remotely sensed chlorophyll-a and the CPR PCI, (Raitsos et al., 2005) to extend datasets of phytoplankton biomass back to 1960s.

Case Study 2.3.3: Using molecular approaches to identify the fungus *Aspergillus sydowii* from CPR samples

A major strength of molecular approaches is that they often enable identification of species that cannot be discriminated by conventional light microscopy. Every year millions of tons of dust blow off deserts into the sea, carrying up to 10,000 microbes per gram of soil with them (Kellogg and Griffin, 2003). Air samples taken during dust storms show 3–10 times as many microbes than during clear conditions in US Virgin Islands (Kellogg and Griffin, 2003). This could have implications for marine health. Dust-contaminated air samples from the Caribbean Sea contain *Aspergillus syndowii*, which is known to cause Aspergillosis diseases in Caribbean Sea fan gorgonian corals (reviewed by Kim and Rypien, 2015). The dust provides micronutrients to the surface layer and can fertilize phytoplankton blooms directly through nutrients as demonstrated by increased remotely sensed chlorophyll-a (Bali et al., 2019). Many phytoplankton are iron-limited, and iron delivered by these dust storms were associated with a 100-fold increase in the harmful algal bloom forming cyanobacteria *Trichodesmium* (Lenes et al., 2001). Dust was also claimed to cause localized cooling, deepening the mixed surface layer and extending nutrient availability (Bali et al., 2019). In September 2009, following a severe 10-year drought, there was a massive dust storm that originated in central Australia. The dust storm transported 16 Tg of dust throughout eastern Australia, traversing the Tasman Sea to reach New Zealand (Hallegraeff et al., 2014). Serendipitously, the Australian Continuous Plankton Recorder survey (AusCPR) was towing along the east coast within 3 weeks of the dust storm. Once returned to the laboratory, CPR samples were black and covered in unknown 5 μm spheres. Maximum concentrations were up to 150,000 m^3 and the "bloom" was estimated to cover an area 25 times the size of England (Hallegraeff et al., 2014). On closer inspection, the spheres were identified as fungal spores, but the species could not be ascertained using light microscopy. This is where molecular approaches were enlisted, but not on the formalin-preserved samples.

Spores of some fungal species are exceptionally stress-resistant. Some spores survive pasteurization and other treatments, and temperatures up to 85 °C for several minutes (Dijksterhuis, 2019). Fungal spores from the CPR samples were thus rinsed of formaldehyde and grown on agar, where they miraculously germinated. Based on molecular sequencing of three different genes (the large subunit rRNA gene [LSU], internal transcribed spacer [ITS], and the beta-tubulin gene), the fungal spores were identified as *Aspergillus sydowii* (99–100% sequence match). The beta-tubulin gene sequences had a 99% match (391/392 bp), similar to the variation in the number of single nucleotide polymorphisms for this gene in other *A. sydowii* strains in GenBank. A phylogram (50% majority-rule consensus tree) showed that in all trees sampled, the beta-tubulin gene sequence grouped with *A. sydowii* (clade support value, 100%) (Fig. 2.6). No other species, out of all sequences available in GenBank, fell into this group. Relationships among the four *A. sydowii* strains and the spores from CPR samples were poorly resolved (clade support values of 53 and 72). The more conserved ITS gene, however, showed a 100% match to *Aspergillus versicolor*, *Aspergillus carneus*, and *A. sydowii*.

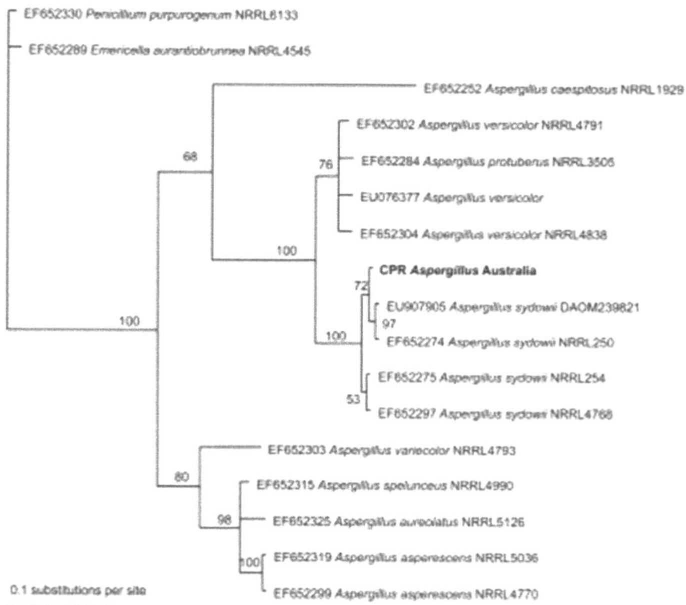

Figure 2.6 Bayesian phylogenetic analysis of the (A) beta-tubulin gene from publically available fungal sequences. The Aspergillus fungi sequence (bold type) recovered from Australian CPR survey samples show they are closely related to *Aspergillus syndowii* sequences. Image from Hallegraeff, G. et al. (2014). Appl Environ Microbiol, 80(11), 3315–3320.

The potential ecosystem implications of the massive *A. sydowii* bloom were investigated. *A. sydowii* has been circumstantially associated with gorgonian coral fan disease in the Caribbean Sea and shown to be transmissible between gorgonians (Kim and Rypien, 2015) and with other climate-change driven stressors may have a sizeable impact on the spread of the disease. Fungal cultures were nontoxic to fish gill cells, but reduced the motility of the HAB species *Alexandrium catenella*, and dramatically reduced the motility and the photophysiological performance of the dinoflagellate *Symbiodinium*, a common coral algal symbiont (Hayashi et al., 2016). Molecular approaches were the only way to conclusively identify the fungus to a species. There is likely to be other eukaryotes or prokaryotes that could withstand weak buffered formaldehyde, and could then be resurrected in cultures for molecular analysis. This study highlights the opportunities to capture microbial responses to these rare, large-scale atmospheric events.

2.4 The future of CPR sample archive repository as a tool for global ecological and taxonomic research

The UN Decade of Ocean Science 2021–2030 highlights that marine observations are key to mapping and predicting the ocean environment in a changing world. Many people depend on the ocean—global fish production was 179 million tonnes in 2018 (worth $US 401 billion), of which approximately 45% is from aquaculture, highlighting increasing dependency on farmed fish and the role of phytoplankton and microbes in these altered environments (FAO, 2020).. Many of these activities will alter microbiological communities. We still lack understanding of phytoplankton distribution and their functional capacity in a changing marine environment. The value of long-term time series for describing baseline observations and detecting changes in the distribution of planktonic species has been well proven (Beaugrand and Ibanez, 2004; McLeod et al., 2012). However, many long-term time series relied on light microscopy for species identification, which can only tell a part of the story. There remains key data gaps to manage the oceans effectively. For example, the Marine strategy framework directive is European legislation to maintain healthy and biologically diverse seas. Indicators have been developed for fish, zooplankton and larger phytoplankton (McQuatters-Gollop et al., 2019), however there is a paucity of indicators for pico- to nano-sized plankton (Moffat et al., 2011; Gowen et al., 2011) that are captured effectively using genetic approaches and are difficult or impossible to measure by light microscopy due to their size. These cells are the most diverse and abundant in the ocean (De Vargas et al., 2015) and can inform many ecological processes, such as feeding relationships (Boyce et al., 2015), sensitive responses to abiotic environmental factors (Vadrucci et al., 2013), and biogeochemical cycles through sinking rates (Falkowski et al., 2003). Phytoplankton prey size relationships with their predators are important signifiers of food chain length and predator control mechanisms (Boyce et al., 2015). In legal assessments of water quality for MSFD, chlorophyll-a and water transparency are the main biological indicators for eutrophication MSFD descriptor 5 (HELCOM, 2014) that are reported on a pan-European scale, although many regional studies incorporate a suite of other variables, such as indicator species and levels of nuisance or toxic algae (OSPAR, 2013), albeit less frequently (Ferreira et al., 2011).

A recognition of the importance of biological diversity can be seen through efforts to integrate biodiversity as an essential ocean variable or EOV (Pearlman et al., 2017) in a similar way that sea surface temperature or salinity are core EOVs. Essential biodiversity variables (EBD) proposed include phytoplankton biomass and diversity and would have explicit taxonomic records associated with them and could be measured alongside other EOVs. Microbial biomass and diversity are likely to become an EBD soon. The success of eDNA will be based on its ease of use, confident interpretation, and global accessibility. However, to be useful and meaningful, DNA and RNA data need to be intercomparable and operable with other global datasets to connect genetic data with temperature, nutrients, salinity, or even accurate location data. (Miloslavich et al., 2018).

Thus the CPR archive can complement other datasets with over 50 years of historical plankton samples over an unparalleled scale, covering a large range of biomes from polar to tropical waters. Its use as a record of biodiversity information can be exploited to extract long-term patterns of formerly unrecorded microbes. Examples in this chapter show that the CPR survey samples have the capacity to reveal past diversity patterns in ecologically important phytoplankton and microbes.

Phytoplankton microscopy monitoring programmes are well established in Europe, Australia, New Zealand, and North America, but are challenging in regions with less monitoring funding, because they are time-consuming, requiring specialist training, and there is a declining pool of expertise available (Wood et al., 2013; Culverhouse, 2015). However, genetic-based phytoplankton analysis can complement microscopy data, but requires significant infrastructure and improved tools for broad uptake by nonspecialists. For example, phytoplankton sequences are not sufficiently represented in public DNA sequence databases (Abad et al., 2016) and lack sufficient quality control or associated metadata. Several repositories exist, such as Genbank (Benson et al., 2018), holding approximately 7 million genetic taxa records. The Barcode of life database, BoLD, (International Barcode of Life Consortium, 2021) offers additional curation tools and incorporates image and metadata and that data mines Genbank records, but is much smaller (~296,000 records) and, crucially, is not readily linked to commonly used bioinformatics pipelines that process HTS data. The Ocean Biodiversity Information System (https://obis.org) offers in situ surface ocean records, but there are update delays of a year, and it lacks interoperability with common genetics databases, such as Genbank. Most users are researchers that use specialist computer-based identification tools and databases that are suitable for their original purpose, but problematic when comparing genetically-identified taxa from multiple literature sources. There is tremendous reuse value in genetic data in new contexts, particularly as only a few genes markers are used to identify taxa. Various tool kits are being developed, such as UniEuk (Berney et al., 2017) and MgNify at EBI (Mitchell et al., 2019; Korobeynikov et al., 2019), that offer common workflows to process or reprocess metabarcoding and metagenomic datasets respectively. As many species are identified using both genetic and morphological characters; UniEuk also offers a common taxonomic framework agreed by the expert research community to avoid confusion over which identification system has been used. Other global projects such as the Earth microbiome project (Thompson et al., 2017) have developed a sampling system and analysis pipeline that are being adopted and may end up becoming the standard method. Such tools, which are now fragmented could be adopted in long-term monitoring organizations to become standardized. This would reduce barriers to its use and improve the uptake in genetic monitoring tools in long-term monitoring organizations.

To develop genetic approaches routinely into the CPR survey and many other long-term marine monitoring programmes will require substantial efforts to standardize and compare genetic-based taxa with morphological counterparts over a long-period. However, the efforts would be well worth it. Many marine monitoring programs in the USA use eDNA to track marine organisms of monitoring interest (U.S Fish & Wildlife Service, 2019) such as invasive fish species in lakes. The development of genetic-based microbial diversity monitoring requires careful thought to allow routine uptake and unambiguous interpretation. For example, the Marine Microbes programme in Australia has integrated DNA collection and analysis alongside other oceanographic measurements that reflect improved understanding of *Phaeocystis* population dynamics over large latitudinal ranges (Sow et al., 2020), where its genomic approaches noted that a more complex distribution pattern was strongly connected to polar and subantarctic fronts in the Southern Ocean. Using a multimarker metabarcoding approach, zooplankton were consistently detected and regular seasonal assemblages could be distinguished from an anomalous biotic shift response during a marine heatwave (Berry et al., 2019). Such signs could act as early warning indicators for marine managers. As samples are collected, they become archive material for use in future research investigating past spatio-temporal distributions of phytoplankton. The information gathered through eDNA will be increasingly valuable as long-term microscopic datasets. Thus a key goal is to develop standardized methods for long-term biobanking of DNA and RNA, so they will be of use to later scientific enquiry. This is not a trivial task and is often overlooked. Organisms respond uniquely to different preservation methods, and even well agreed methods, such as freezing, are disputed in that they also fragment genomic DNA (Klingström et al., 2018), which is often ignored when using sequencing methods requiring shorter (<35kb) fragments, but making it inaccessible to direct sequencing methods that can directly sequence stretches of DNA thousands of nucleotide long. Thus the value of long-term biobanked samples needs to be highlighted through compeling case studies and supported through long-term stakeholder engagement to ensure funding from grant and government agencies. Biobanking is also important to ensure continuity of phytoplankton data. With the rapid changes in genetic technology, parallel testing with old methods will be required. The value of unambiguous standards that link phenotype with genotype is essential to establish taxonomic equivalencies and detection limits using voucher or stable isolates. There are multiple projects that are contributing to this effort, but the challenge will be to unify them.

Automated platforms that currently measure sea surface temperature and salinity over large space and fine time-scales are currently being developed to sample seawater for molecular analyses. Currently, they are only deployed from static sites and are large, energy-intensive, and require assistance from research ships (Boss et al., 2018). Whilst CPR surveys continue, there is space for eDNA sampling and genetic detection alongside historic plankton sampling on a silk that has continued since the 1920s. The CPR platform has successfully utilized the Water and Microplankton Sampler (WaMS) to detect smaller microbes (Stern et al., 2015) as a proof of principle although the water volume collected was small due to the battery size. Miniaturization is essential to reduce the size of samplers for the CPR device or underwater autonomous vehicles. Using microfluidic engineering systems, such as the lab on a chip would allow for probe-based species detection in the space on a microscope slide (e.g., Jha et al., 2012; Foudeh et al., 2012). New sequencing technology, such as MinION (Oxford Nanopore Ltd), are little bigger than hand-held devices and could be deployed within a larger autonomous sampling device. Further, wireless technology devices can instantly transmit data to shore. Such sampling could offer near real-time observation of species succession and community dynamics, or processes involved in primary production and be more readily comparable to other eDNA surveys.

Connecting old and new datasets and methodologies will be the most challenging aspect to improve biodiversity monitoring, but will enhance our understanding of microbial ecology and identify data gaps. Out of all efforts, the development of common databases and ease of access to genetic data will be key to this transition.

References

Abad, D., Albaina, A., Aguirre, M., Laza-Martinez, A., Uriarte, I., et al., 2016. Is metabarcoding suitable for estuarine plankton monitoring? A comparative study with microscopy. Mar. Biol. 163. https://doi.org/10.1007/s00227-016-2920-0.

Amato, A., Kooistra, W.H.C.F., Levialdi Ghiron, J.H., Mann, D.G., Pröschold, T., et al., 2007. Reproductive isolation among sympatric cryptic species in marine diatoms. Protist 158, 193–207. https://doi.org/10.1016/j.protis.2006.10.001.

Anderson, D.M., Cembella, A.D., Hallegraeff, G.M., 2012. Progress in understanding harmful algal blooms: paradigm shifts and new technologies for research, monitoring, and management. Annu. Rev. Mar. Sci. 4, 143–176. https://doi.org/10.1146/annurev-marine-120308-081121.

Anderson, D.M., Fensin, E., Gobler, C.J., Hoeglund, A.E., Hubbard, K.A., et al., 2021. Marine harmful algal blooms (HABs) in the United States: history, current status and future trends. Harmful Algae 102, 101975. https://doi.org/10.1016/j.hal.2021.101975.

Balch, W.M., 2004. Re-evaluation of the physiological ecology of the coccolithophores. In: Thierstein, H.R., Young, J.R. (Eds.), Coccolithophores. From Molecular Processes to Global Impact, 1 ed. Springer.

Balestreri, C., 2015. Genetic variability of the microalga *Emiliania huxleyi* (Haptophyta): a temporal and geographical study. University of Oxford thesis.

Bali, K., Mishra, A.K., Singh, S., Chandra, S., Lehahn, Y., 2019. Impact of dust storm on phytoplankton bloom over the Arabian Sea: a case study during March 2012. Environ. Sci. Pollut. Res. 26, 11940–11950. https://doi.org/10.1007/s11356-019-04602-7.

Bates, S., Hubbard, K., Lundholm, N., Montresor, M., Leaw, C.P., 2018. Pseudo-nitzschia, Nitzschia, and domoic acid: new research since 2011. Harmful Algae 79. https://doi.org/10.1016/j.hal.2018.06.001.

Batten, S.D., Abu-Alhaija, R., Chiba, S., Edwards, M., Graham, G., et al., 2019. A global plankton diversity monitoring program. Front. Mar. Sci. 6, 321. https://doi.org/10.3389/fmars.2019.00321.

Batten, S.D., Clark, R., Flinkman, J., Hays, G., John, E., et al., 2003a. CPR sampling: the technical background, materials and methods, consistency and comparability. Prog. Oceanogr. 58 (2–4), 193–215. https://doi.org/10.1016/j.pocean.2003.08.004.

Batten, S.D., Walne, A.W., Edwards, M., Groom, S.B., 2003b. Phytoplankton biomass from continuous plankton recorder data: an assessment of the phytoplankton colour index. J. Plankton Res. 25 (7), 697–702. https://doi.org/10.1093/plankt/25.7.697.

Beaugrand, G., Ibanez, F., 2004. Monitoring marine plankton ecosystems. II: long-term changes in North Sea calanoid copepods in relation to hydro-climatic variability. Mar. Ecol. Prog. Ser. 284, 35–47.

Beaugrand, G., Luczac, C., Edwards, M., 2009. Rapid biogeographical plankton shifts in the North Atlantic Ocean. Glob. Change Biol. 15, 1790–1803. https://doi.org/10.1111/j.1365-2486.2009.01848.x.

Beaugrand, G., Reid, P.C., Ibañez, F., Lindley, J.A., Edwards, M., 2002. Reorganization of North Atlantic marine copepod biodiversity and climate. Science 296, 1692–1694. https://doi.org/10.1126/science.1071329.

Bedford, J., Ostle, C., Johns, D.G., Atkinson, A., Best, M., et al., 2020. Lifeform indicators reveal large-scale shifts in plankton across the North-West European shelf. Glob. Change Biol. 26, 3482–3497. https://doi.org/10.1111/gcb.15066.

Benson, D.A., Cavanaugh, M., Clark, K., Karsch-Mizrachi, I., Ostell, J., et al., 2018. GenBank. Nucleic Acids Res. 46 (D1), D41–D47. https://doi.org/10.1093/nar/gkx1094.

Berney, C., Ciuprina, A., Bender, S., Brodie, J., Edgcomb, V., et al., 2017. *UniEuk*: time to speak a common language in protistology! J. Eukaryot. Microbiol. 64, 407–411. https://doi.org/10.1111/jeu.12414.

Berry, T.E., Saunders, B.J., Coghlan, M.L., Stat, M., Jarman, S., et al., 2019. Marine environmental DNA biomonitoring reveals seasonal patterns in biodiversity and identifies ecosystem responses to anomalous climatic events. PLoS Genet. 15 (2), e1007943. https://doi.org/10.1371/journal.pgen.1007943.

Boss, E., Waite, A., Muller-Karger, F., Yamazaki, H., Wanninkhof, R., et al., 2018. Beyond chlorophyll fluorescence: the time is right to expand biological measurements in ocean observing programs. Limnol. Oceanogr. Bull. 27 (3), 89–90.

Boyce, Frank, Leggett, 2015. From mice to elephants: overturning the "one size fits all" paradigm in marine plankton food chains. Ecol. Lett. 18, 504–515.

Bresnan, E., Kraberg, A., Fraser, S., Brown, L., Hughes, S., et al., 2015. Diversity and seasonality of *Pseudo-nitzschia* (Peragallo) at two North Sea time-series monitoring sites. Helgol. Mar. Res. 69, 193–204.

Callahan, B.J., McMurdie, P.J., Holmes, S.P., 2017. Exact sequence variants should replace operational taxonomic units in marker-gene data analysis. ISME J. 11 (12), 2639–2643. https://doi.org/10.1038/ismej.2017.119.

Caporaso, J.G., Kuczynski, J., Stombaugh, J., Bittinger, K., Bushman, F.D., 2010. QIIME allows analysis of high-throughput community sequencing data. Nat. Methods 7 (5), 335–336. https://doi.org/10.1038/nmeth.f.303.

Castberg, T., Larsen, A., Sandaa, R.A., Brussaard, C.P.D., Egge, J.K., et al., 2001. Microbial population dynamics and diversity during a bloom of the marine coccolithophorid *Emiliania huxleyi* (Haptophyta). Mar. Ecol. Prog. Ser. 221, 39–46.

Chakravorty, S., Helb, D., Burday, M., Connell, N., Alland, D., 2007. A detailed analysis of 16S ribosomal RNA gene segments for the diagnosis of pathogenic bacteria. J. Microbiol. Methods 69 (2), 330–339. https://doi.org/10.1016/j.mimet.2007.02.005.

Costa, L.G., Giordano, G., Aschner, M., 2014. Domoic acid. In: Aminoff, M.J., Daroff, R.B. (Eds.), Encyclopedia of the Neurological Sciences, 2nd ed. Academic Press, pp. 1016–1017.

Culverhouse, P.F., 2015. Biological Oceanography needs new tools to automate sample analysis. J. Mar. Biol. Aquacult. 1, 1–2. https://doi.org/10.15436/2381-0750.15.e002.

De la Cruz, J., Karbstein, K., Woolford Jr, J.L., 2015. Functions of ribosomal proteins in assembly of eukaryotic ribosomes in vivo. Annu. Rev. Biochem. 84, 93–129. https://doi.org/10.1146/annurev-biochem-060614-033917.

De Vargas, C., Audic, S., Henry, N., Decelle, J., Mahé, F., et al., 2015. Eukaryotic plankton diversity in the sunlight ocean. Science 348 (6237), 1261605. https://doi.org/10.1126/science.1261605.

Di Cesare, A., Petrin, S., Fontaneto, D., Losasso, C., Eckert, E.M., et al., 2018. ddPCR applied on archived Continuous Plankton Recorder samples reveals long-term occurrence of class 1 integrons and a sulphonamide resistance gene in marine plankton communities. Environ. Microbiol. Rep. 10 (4), 458–464.

Dijksterhuis, J., 2019. Fungal spores: highly variable and stress-resistant vehicles for distribution and spoilage. Food Microbiol. 81, 2–11.

Do, H., Dobrovic, A., 2015. Sequence artifacts in DNA from formalin-fixed tissues: causes and strategies for minimization. Clin. Chem. 61 (1), 64–71. https://doi.org/10.1373/clinchem.2014.223040.

Edwards, M., Beaugrand, G., Hays, G.C., Koslow, J.A., Richardson, A.J., 2010. Multidecadal oceanic ecological datasets and their application in marine policy and management. Trends Ecol. Evol. 25, 602–610. https://doi.org/10.1016/j.tree.2010.07.007.

Falkowski, P.G., Laws, E.A., Barber, R.T., Murray, J.W., 2003. Phytoplankton and their role in primary, new, and export production. In: Fasham, M.J.R. (Ed.), Ocean Biogeochemistry: the Role of the Ocean Carbon Cycle in Global Change. Springer-Verlag Berlin Heidelberg New York, pp. 99–121.

FAO, 2020. The State of World Fisheries and Aquaculture 2020. Sustainability in action. Rome.

Fernandes, L.F., Hubbard, K.A., Richlen, M.l., Smith, J., Bates, S.S., et al., 2014. Diversity and toxicity of the diatom Pseudo-nitzschia Peragallo in the Gulf of Maine, Northwestern Atlantic Ocean. Deep-Sea Res., Part 2, Top. Stud. Oceanogr. 103, 139–162.

Ferreira, J.G., Andersen, J.H., Borja, A., Bricker, S.B., Camp, J., et al., 2011. Overview of eutrophication indicators to assess environmental status within the European Marine Strategy Framework Directive. Estuar. Coast. Shelf Sci. 93, 117–131.

Foudeh, A.M., Fatanat Didar, T., Veres, T., Tabrizian, M., 2012. Microfluidic designs and techniques using lab-on-a-chip devices for pathogen detection for point-of-care diagnostics. Lab Chip 12, 3249–3266. https://doi.org/10.1039/c2lc40630f.

Fuks, G., Elgart, M., Amir, A., et al., 2018. Combining 16S rRNA gene variable regions enables high-resolution microbial community profiling. Microbiome 6, 17. https://doi.org/10.1186/s40168-017-0396-x.

Goodwin, S., McPherson, J., McCombie, W., 2016. Coming of age: ten years of next-generation sequencing technologies. Nat. Rev. Genet. 17, 333–351. https://doi.org/10.1038/nrg.2016.49.

Gowen, R.J., et al., 2011. The Development of UK Pelagic (Plankton) Indicators and Targets for the MSFD. Report to the Department for Environment, Food and Rural Affairs: Belfast. 41 pp.

Guiry, M.D., Guiry, G.M., 2020. *AlgaeBase*. World-Wide Electronic Publication. National University of Ireland, Galway. http://www.algaebase.org. (Accessed 6 June 2021).

Hagino, K., Bendif, E.M., Young, J.R., Kogame, K., Probert, I., et al., 2011. New evidence for morphological and genetic variation in the cosmopolitan coccolithophore *Emiliania huxleyi* (Prymnesiophyceae) from the COX1b-ATP4 genes. J. Phycol. 47 (5), 1164–1176.

Hallegraeff, G., Coman, F., Davies, C., Hayashi, A., McLeod, D., et al., 2014. Australian dust storm associated with extensive *Aspergillus sydowii* fungal "bloom" in coastal waters. Appl. Environ. Microbiol. 80 (11), 3315–3320.

Hallegraeff, G.M., 2004. Manual on Harmful Marine Microalgae. UNESCO, Paris.

Hallegraeff, G.M., Anderson, D.M., Belin, B., Bottein, M-Y., Bresnan, E., et al., 2021. Are harmful marine microalgal blooms and their societal impacts expanding and intensifying? A 35-year global data analysis. Nat. Commun. In press.

Hayashi, A., Crombie, A., Lacy, E., Richardson, A.J., Vuong, D., et al., 2016. Aspergillus sydowii marine fungal bloom in Australian coastal waters, its metabolites and potential impact on *Symbiodinium* dinoflagellates. Marine Drugs 14, 59.

Hays, G.C., Warner, A.J., John, A.W.G., Harbour, D.S., Holligan, P.M., 1995. Coccolithophores and the Continuous Plankton Recorder Survey. J. Mar. Biol. Assoc. UK 75 (2), 503–506.

Hebert, P.D., Cywinska, A., Ball, S.L., deWaard, J.R., 2003. Biological identifications through DNA barcodes. Proc. R. Soc. Lond. B, Biol. Sci. 270 (1512), 313–321. https://doi.org/10.1098/rspb.2002.2218.

HELCOM, 2014. Eutrophication status of the Baltic Sea 2007–2011 - a concise thematic assessment. In: Baltic Sea Environment Proceedings No. 143.

Highfield, A.C., Evans, C., Walne, A.W., Miller, P.I., Schroeder, D.C., 2014. How many Coccolithovirus genotypes does it take to terminate an *Emiliania huxleyi* bloom? Virology 466–467, 138–145.

Hinder, S., Hays, G., Edwards, M., et al., 2012. Changes in marine dinoflagellate and diatom abundance under climate change. Nat. Clim. Change 2, 271–275. https://doi.org/10.1038/nclimate1388.

Holligan, P.M., Fernàndez, E., Aiken, J., Balch, W., 1993. A biogeochemical study of the coccolithophore, Emiliania huxleyi, in the North Atlantic. Glob. Biogeochem. Cycles 7 (4), 879–900.

Hubbard, K.A., Rocap, G., Armbrust, E.V., 2008. Inter- and intraspecific community structure within the diatom genus *Pseudo-nitzschia* (Bacillariophyceae). J. Phycol. 44 (3). https://doi.org/10.1111/j.1529-8817.2008.00518.x.

Hykin, S.M., Bi, K., McGuire, J.A., 2015. Fixing formalin: a method to recover genomic-scale DNA sequence data from formalin-fixed museum specimens using high-throughput sequencing. PLoS ONE 10 (10), e0141579. https://doi.org/10.1371/journal.pone.0141579.

Illumina Inc, 2020. Maximum read length for Illumina sequencing platforms. https://emea.support.illumina.com/bulletins/2020/04/maximum-read-length-for-illumina-sequencing-platforms.html. (Accessed 2 June 2021).

International Barcode of Life Consortium, 2021. BIOSCAN programme overview. https://ibol.org/site/wp-content/uploads/2019/07/BIOSCAN-overviewJuly9.pdf. (Accessed 24 May 2021).

Jha, S.K., Chand, R., Han, D., Jang, Y.C., Ra, G.S., et al., 2012. An integrated PCR microfluidic chip incorporating aseptic electrochemical cell lysis and capillary electrophoresis amperometric DNA detection for rapid and quantitative genetic analysis. Lab Chip 12, 4455–4464. https://doi.org/10.1039/c2lc40727b.

Kato, M., Nakamura, H., Nagai, M., Kubo, T., Elzawahry, A., et al., 2018. A computational tool to detect DNA alterations tailored to formalin-fixed paraffin-embedded samples in cancer clinical sequencing. Gen. Med. 10 (44). https://doi.org/10.1186/s13073-018-0547-0.

Kellogg, C.A., Griffin, D.W., 2003. African dust carries microbes across the ocean: Are they affecting human and ecosystem health? USGS Open-File Report 03-028. pp. 1–4. https://pubs.usgs.gov/of/2003/0028/report.pdf.

Kim, K., Rypien, K., 2015. Aspergillosis of Caribbean Sea Fan Corals, *Gorgonia* spp. In: Woodley, C.M., Downs, C.A., Bruckner, A.W., Porter, J.W., Galloway, S.B. (Eds.), Diseases of Coral. John Wiley and Sons Inc. Chapter 16.

Kirby, R.R., Lindley, A., 2005. Molecular analysis of Continuous Plankton Recorder samples, an examination of echinoderm larvae in the North Sea. J. Mar. Biol. Assoc. UK 2005, 451–459.

Klingström, T., Bongcam-Rudloff, E., Pettersson, O.V., 2018. A comprehensive model of DNA fragmentation for the preservation of High Molecular Weight DNA. BioRXiv preprint. https://doi.org/10.1101/254276.

Korobeynikov, A., Shlemov, A., Kunyavskaya, O., Lapidus, A., et al., 2019. MGnify: the microbiome analysis resource in 2020. Nucleic Acids Res. https://doi.org/10.1093/nar/gkz1035.

Krueger-Hadfield, S.A., Balestreri, C., Schroeder, J., Highfield, A., Helaouet, P., et al., 2014. Genotyping an *Emiliania huxleyi* (Prymnesiophyceae) bloom event in the North Sea reveals evidence of asexual reproduction. Biogeosciences 11, 5215–5234. https://doi.org/10.5194/bg-11-3919-2014.

Langer, G., Nehrke, G., Probert, I., Ly, J., Ziveri, P., 2009. Strain-specific responses of *Emiliania huxleyi* to changing seawater carbonate chemistry. Biogeosciences 6, 2637–2646.

Lenes, J.M., Darrow, B.P., Cattrall, C., Heil, C.A., Callahan, M., et al., 2001. Iron fertilization and the *Trichodesmium* response on the West Florida shelf. Limnol. Oceanogr. 6. https://doi.org/10.4319/lo.2001.46.6.1261.

Liu, Z.J., Martínez Cuesta, S., van Delft, P., Balasubramanian, S., 2019. Sequencing abasic sites in DNA at single-nucleotide resolution. Nat. Chem. 11, 629–637. https://doi.org/10.1038/s41557-019-0279-9.

Lynam, C.P., Halliday, N.C., Höffle, H., Wright, P.J., van Damme, C., et al., 2013. Spatial patterns and trends in abundance of larval sandeels in the North Sea: 1950–2005. ICES J. Mar. Sci. 70, 540–553.

MacDonald, A., Heath, M.R., Greenstreet, S.P.R., Speirs, D.C., 2019. Timing of sandeel spawning and hatching off the East coast of Scotland. Front. Mar. Sci. 6. https://doi.org/10.3389/fmars.2019.00070. 70 pp. https://www.frontiersin.org/article/10.3389/fmars.2019.00070.

Martinez-Martinez, J., Schroeder, D.C., Larsen, A., Bratbak, G., Wilson, W.H., 2007. Molecular dynamics of *Emiliania huxleyi* and co-occurring viruses during two separate mesocosm studies. Appl. Environ. Microbiol. 73 (2), 554–562.

McDonald, S.M., Sarno, D., Zingone, A., 2007. Identifying Pseudo-nitzschia species in natural samples using genus-specific PCR primers and clone libraries. Harmful Algae 6 (6), 849–860.

McLeod, D.J., Hallegraeff, G.M., Hosie, G.W., Richardson, A.J., 2012. Climate-driven range expansion of the red-tide dinoflagellate Noctiluca scintillans into the Southern Ocean. J. Plankton Res. 34, 332–337. https://doi.org/10.1093/plankt/fbr112.

McQuatters-Gollop, A., Atkinson, A., Aubert, A., Bedford, J., Best, M., et al., 2019. Plankton lifeforms as a biodiversity indicator for regional-scale assessment of pelagic habitats for policy. Ecol. Indic. 101, 913–925.

Medlin, L.K., Barker, G.L.A., Campbell, L., Green, J.C., Hayes, P.K., et al., 1996. Genetic characterisation of *Emiliania huxleyi* (Haptophyta). J. Mar. Syst. 9 (1–2), 13–31.

Miloslavich, P., Bax, N.J., Simmons, S.E., Klein, E., Appeltans, W., et al., 2018. Essential ocean variables for global sustained observations of biodiversity and ecosystem changes. Glob. Change Biol. 24, 2416–2433. https://doi.org/10.1111/gcb.14108.

Mitchell, A.L., Almeida, A., Beracochea, M., Boland, M., Burgin, J., et al., 2019. MGnify: the microbiome analysis resource in 2020. Nucleic Acids Res. https://doi.org/10.1093/nar/gkz1035.

Moffat, C., Aish, A., Hawkridge, J.M., Miles, H., Mitchell, P.I., et al., 2011. Advice on United Kingdom Biodiversity Indicators and Targets for the Marine Strategy Framework Directive. Department for Environment Food and Rural Affairs.

OSPAR Commission, 2013. Common procedure for the identification of the Eutrophication status of the OSPAR Maritime area. OSPAR Commission document Agreement 2013-8. https://mcc.jrc.ec.europa.eu/main/document.py?Num=0&mot=Common%20Procedure%20for%20the%20Identification%20of%20the%20Eutrophication%20Status%20of%20the%20OSPAR%20Maritime%20Area&classement=&code=201606235331.

Pawlowski, J., Audic, S., Adl, S., Bass, D., Belbahr, i L., et al., 2012. CBOL protist working group: barcoding eukaryotic richness beyond the animal, plant, and fungal kingdoms. PLoS Biol. 10 (11), e1001419.

Pearlman, J., Buttigieg, P.L., Simpson, P., Muñoz, C., Hesop, E., et al., 2017. Accessing existing and emerging best practices for ocean observation, a new approach for end-to-end management of best practices. In: Proceedings of the Oceans'17. Anchorage: Conference, MTS/IEEE. IEEE, Anchorage, AK.

Poulton, A.J., Stinchcombe, M.C., Achterberg, E.P., Bakker, D.C.E., Dumousseaud, C., Lawson, H.E., Lee, G.A., Richier, S., Suggett, D.J., Young, J.R., 2014. Coccolithophores on the north-west European shelf: calcification rates and environmental controls. Biogeosciences 11, 3919–3940. https://doi.org/10.5194/bg-11-3919-2014.

Poulton, A.J., Young, J.R., Bates, N.R., Balch, W.M., 2011. Biometry of detached *Emiliania huxleyi* coccoliths along the Patagonian Shelf. Mar. Ecol. Prog. Ser. 443, 1–17. https://doi.org/10.3354/meps09445.

Raitsos, D.E., Reid, P.C., Lavender, S.J., Edwards, M., Richardson, A.J., 2005. Extending the SeaWiFS chlorophyll dataset back 50 years in the North-East Atlantic. Geophys. Res. Lett. 32, L06603.

Reid, P.C., Colebrook, J.M., Matthews, J.B.L., Aiken, J., 2003. The Continuous Plankton Recorder: concepts and history, from Plankton Indicator to undulating recorders. Prog. Oceanogr. 58, 117–173. https://doi.org/10.1016/j.pocean.2003.08.002.

Reid, P.C., Johns, D.G., Edwards, M., Starr, M., Poulin, M., et al., 2007. A biological consequence of reducing Arctic ice cover: arrival of the Pacific diatom *Neodenticula seminae* in the North Atlantic for the first time in 800000 years. Glob. Change Biol. 13, 1910–1921. https://doi.org/10.1111/j.1365-2486.2007.01413.x.

Richard, G.F., Kerrest, A., Dujon, B., 2008. Comparative genomics and molecular dynamics of DNA repeats in eukaryotes. Microbiol. Mol. Biol. Rev. 72 (4), 686–727. https://doi.org/10.1128/MMBR.00011-08.

Richardson, A.J., Walne, A.W., John, A.W.G., Jonas, T.D., Lindley, J.A., et al., 2006. Using continuous plankton recorder data. Prog. Oceanogr. 68, 27–74. https://doi.org/10.1016/j.pocean.2005.09.011.

Rickaby, R.E.M., Hermoso, M., Lee, R.B.Y., Rae, B.D., Heureux, A.M.C., et al., 2016. Environmental carbonate chemistry selects for phenotype of recently isolated strains of *Emiliania huxleyi*. Deep-Sea Res., Part 2, Top. Stud. Oceanogr. 127, 28–40.

Ripley, S.J., Baker, A.C., Miller, P.I., Walne, A.W., Schroeder, D.C., 2008. Development and validation of a molecular technique for the analysis of archived formalin-preserved phytoplankton samples permits retrospective assessment of *Emiliania huxleyi* communities. J. Microbiol. Methods 73, 118–124.

Rogers, S.O., 2011. The Central Dogma and Beyond in Integrated Molecular Evolution. CRC Press. ISBN 978-1-4398-1995-1, pp. 65–66. Chapter 4. https://books.google.co.uk/books?id=xywnbQMKzXwC&pg=PA66&redir_esc=y#v=onepage&q&f=false. (Accessed 2 June 2021).

Rost, B., Riebesell, U., 2004. Coccolithophores and the biological pump: responses to environmental changes. In: Coccolithophores. Springer, Berlin, Heidelberg, pp. 99–125.

Schander, C., Halanych, K.M., 2003. DNA, PCR and formalinised animal tissue- a short review and protocols. Org. Divers. Evol. 3, 195–205.

Schmidt, K., Birchill, A.J., Atkinson, A., Brewin, R.J.W., Clark, J.R., et al., 2020. Increasing picocyanobacteria success in shelf waters contributes to long-term food web degradation. Glob. Change Biol. 26, 5574–5587. https://doi.org/10.1111/gcb.15161.

Scholin, C.A., Villac, M.C., Buck, K.R., Krupp, J.M., Powers, D.A., Fryxell, G.A., Chavez, F.P., 1994. Ribosomal DMA sequences discriminate among toxic and non-toxic *Pseudonitzschia* species. Nat. Toxins 2, 152–165. https://doi.org/10.1002/nt.2620020403.

Schroeder, D.C., Biggi, G.F., Hall, M., Davy, J., Martínez, J.M., et al., 2005. A genetic marker to separate *Emiliania huxleyi* (prymnesiophyceae) morphotypes1. J. Phycol. 41, 874–879. https://doi.org/10.1111/j.1529--8817.2005.04188.x.

Schroeder, D.C., Oke, J., Hall, M., Malin, G., Wilson, W.H., 2003. Virus succession observed during an *Emiliania huxleyi* bloom. Appl. Environ. Microbiol. 69 (5), 2484–2490.

Sogin, M.L., Morrison, H.G., Huber, J.A., Welch, M.D., Huse, S.M., et al., 2006. Microbial diversity in the deep sea and the underexplored 'rare biosphere'. Proc. Natl. Acad. Sci. USA 103, 12115–12120.

Sogin, M.L., Silberman, J.D., 1998. Evolution of the protest and protistan parasites from the perspective of molecular systematics. Intl. J. Parasitol. 28, 11–20.

Sow, S.L., Trull, T., Bodrossy, L., 2020. Oceanographic fronts shape *Phaeocystis* assemblages: a high-resolution 18S rRNA gene survey from the ice-edge to the equator of the South Pacific. Front. Microbiol. 11, 1847. https://doi.org/10.3389/fmicb.2020.01847.

Stern, R.F., Kraberg, A., Bresnan, E., Kooistra, W.H.C.F., Lovejoy, C., et al., 2018b. Molecular analyses of protists in long-term observation programmes—current status and future perspectives. J. Plankton Res. 40 (5), 519–536. https://doi.org/10.1093/plankt/fby035.

Stern, R.F., Moore, S., Trainer, V., Bill, B., Fischer, A., Batten, S., 2018a. Spatial and temporal patterns of *Pseudo-nitzschia* genetic diversity in the North Pacific Ocean from the Continuous Plankton Recorder survey. Mar. Ecol. Prog. Ser. 606, 7–28.

Stern, R.F., Picard, K., Hamilton, K.M., Walne, A., McQuatters-Gollop, A., Mills, D., Edwards, M., 2015. An automated water sampler from Ships of Opportunity detects new boundaries of marine microbial biodiversity. Prog. Oceanogr. 137, 409–420.

Suter, L., Polanowski, A.M., Clarke, L.J., Kitchener, J.A., Deagle, B.E., 2020. Capturing open ocean biodiversity: comparing environmental DNA metabarcoding to the continuous plankton recorder. Mol. Ecol. 00, 1–18. https://doi.org/10.1111/mec.15587.

Tesson, S.V.M., Montresor, M., Procaccini, G., Kooistra, W.H.C.F., 2014. Temporal changes in population structure of a marine planktonic diatom. PLoS ONE 9 (12), e114984. https://doi.org/10.1371/journal.pone.0114984.

Thessen, A., 2007. Taxonomy and ecophysiology of *Pseudo-nitzschia* in the Chesapeake *Bay*. PhD thesis. University of Maryland. Available at http://hdl.handle.net/1903/7707. (Accessed 28 August 2020).

Thompson, L.R., Sanders, J.G., McDonald, D., Amir, A., Ladau, J., et al., 2017. The Earth Microbiome Project Consortium. A communal catalogue reveals Earth's multiscale microbial diversity. Nature 551, 457–463. https://doi.org/10.1038/nature24621.

Trainer, V., Bates, S.S., Lundholm, N., 2012. *Pseudo-nitzschia* physiological ecology, phylogeny, toxicity, monitoring and impacts on ecosystem health. Harmful Algae 14, 271–300.

U.S Fish & Wildlife Service, 2019. Fisheries program eDNA. Published by Department of Interior, USA government https://www.fws.gov/midwest/fisheries/edna.html. (Accessed 2 June 2021).

Vadrucci, M.R., Stanca, E., Mazziotti, C., Fonda Umani, S., Assimakopoulou, G., et al., 2013. Ability of phytoplankton trait sensitivity to highlight anthropogenic pressures in Mediterranean lagoons: a size spectra sensitivity index (ISS- phyto). Ecol. Indic. 34, 113–125.

Vezzulli, L., et al., 2012. Long-term effects of ocean warming on the prokaryotic community: evidence from the vibrios. ISME J. 6, 21–30.

Vezzulli, L., Grande, C., Reid, P.C., Hélaouët, P., Edwards, M., et al., 2016. Climate influence on Vibrio and associated human diseases during the past half-century in the coastal North Atlantic. Proc. Natl. Acad. Sci. USA 113 (34), E5062–E5071.

Vezzulli, L., Grande, C., Tassistro, G., Brettar, I., Höfle, M.G., et al., 2017. Whole-genome enrichment provides deep insights into Vibrio cholerae metagenome from an African river. Microb. Ecol. 73, 734–738.

Vezzulli, L., Oliveri, C., Borello, A., Gregory, L., Kimirei, I., et al., 2021. Aquatic reservoir of Vibrio cholerae in an African Great Lake assessed by large scale plankton sampling and ultrasensitive molecular methods. ISME J. In press.

Waugh, J., 2007. DNA barcoding in animal species: progress, potential and pitfalls. BioEssays 29 (2), 188–197. https://doi.org/10.1002/bies.20529. PMID: 17226815.

Woese, C.R., Fox, G.E., 1977. Phylogenetic structure of the prokaryotic domain: the primary kingdoms. Proc. Natl. Acad. Sci. 74 (11), 5088–5090. https://doi.org/10.1073/pnas.74.11.508.

Wood, S.A., Smith, K.F., Banks, J.C., Tremblay, L.A., Rhodes, L., et al., 2013. Molecular genetic tools for environmental monitoring of New Zealand's aquatic habitats, past, present and the future. N.Z. J. Mar. Freshw. Res. 47 (1), 90–119. https://doi.org/10.1080/00288330.2012.745885.

Wulf, F.G., Andersson, B.C., Trofimova, T., Hátún, H., 2018. Links between phytoplankton dynamics and shell growth of *Arctica islandica* on the Faroe Shelf. J. Mar. Syst. 179, 72–87. https://doi.org/10.1016/j.jmarsys.2017.11.005.

Young, J., Westbroek, P., 1991. Genotypic variation in the coccolithophorid species Emiliania huxleyi. Mar. Micropaleontol. 18 (1–2), 5–23.

Young, J.R., Geisen, M., Cros, L., Kleijne, A., Spreingel, C., et al., 2003. A guide to extant coccolithophore taxonomy. J. Nannoplankton Res. (Special Issue 1), 1–125.

Zhang, P., Lehmann, B.D., Shyr, Y., Guo, Y., 2017. The utilization of formalin fixed-paraffin-embedded specimens in high throughput genomic studies. Int. J. Genomics 2017, 1926304. https://doi.org/10.1155/2017/1926304.

CHAPTER 3

Impact of molecular approaches on dinoflagellate taxonomy and systematics

Christopher J.S. Bolch
Institute for Marine and Antarctic Studies, University of Tasmania, Launceston, TAS, Australia

3.1 Molecular systematics and diversity of dinoflagellates: a historical perspective

3.1.1 Dinoflagellates: rich morphological and functional diversity

Dinoflagellates are eukaryote cells possessing two flagella, one conventional, and the other a complex ribbon-like flagellum composed of a series of waves giving a distinctive helical appearance (Fensome et al., 1993). Almost all have a distinctive and visible nucleus (dinokaryon) with permanently condensed chromosomes.

Like many microeukaryotes (protists), dinoflagellate cells exhibit an array of morphological and functional diversity, occupy diverse, and sometimes extreme environments, and pursue a wide range of complex life-histories, ranging from free-living to symbiotic and parasitic (Taylor et al., 2008). They are nutritionally diverse; approximately half photosynthetic (autotrophic), but many species are capable of mixotrophy (Taylor et al., 2008). The remainder are non-photosynthetic heterotrophs displaying a remarkable variety of complex intracellular organelles/structures depending on their nutritional strategy: osmotrophy, phagotrophic feeding via engulfment, via tubular feeding appendages (peduncles), or a feeding veil (pallium) (Jacobson, 1999). The full range of these strategies are distributed across most current dinoflagellate orders. Most but not all dinoflagellates are haplontic: a dominant haploid phase that reproduces via vegetative cell division, with sexual reproduction producing a diploid zygote that is short-lived or, in cyst forming species, the zygote transforms to a long-lived, highly-resistant benthic resting cyst (dinocyst) that sink to bottom and remain dormant to seed future populations. Cysts may remain dormant for many decades (Lewis et al., 1999) and, perhaps, more than a century under ideal conditions (Lundholm et al., 2011). Some species produce highly resistant cysts, whose walls can fossilize, providing a rich but selective geological record of dinoflagellate evolution that extends back to the mid-Triassic, over 240 million years ago (Fensome et al., 1996, 1999).

The structural and functional diversity of dinoflagellates provides an extraordinarily rich set of features for taxonomy and classification. Current classification of both living dinoflagellate and fossil resting cysts is based primarily on features of the cell wall (cortex or amphiesma), a complex multilayered structure composed of numerous vesicles that contain cellulose-like plates that form a shell (theca) in thecate species, or no plates at all in naked (athecate) species. The number and arrangement of plates/vesicles distinguishing higher level taxa (e.g., Order, Genus), and the size/shape of specific diagnostic plates are important for distinguishing species. For athecate species, the amphiesma has fewer distinguishing features except the acrobase, a distinctive groove across or around the cell apex. The shape and form of this groove, combined with presence/absence and shape/position of organelles, cell size and shape, all form the primary characters for distinguishing and classifying athecate dinoflagellates. Amphisemal wall patterns are also reflected to a varying extent in resistant walls of resting cysts of both thecate and athecate species, usually evident as surface ridges and relief, and the position of spines and the shape and relative position of the excystment pore after germination.

The uncertain taxonomy and affinities of the dinoflagellates extends back 250 years to the first investigations of living protists. Pascher (1914) correctly considered them to be one of many algal lineages and placed dinoflagellates in a new phylum (the Pyrrophyta) and those with a girdle in the class Dinophyceae. Studies of fossilized dinocysts developed over a similar period, with the first descriptions of dinocysts as "Xanthidia" Ehrenberg (1838) from Cretaceous flint, before Mantell (1850) erected the genus *Spiniferites* to encompass what we now know to be cysts of gonyaulacoid dinoflagellates (*Gonyaulax* and related genera). Later work erected a new genus *Hystricosphaera* (Wetzel, 1933), and "hystricopheres" became the widely used term for organic-walled spiny fossil cysts. Since these early studies, more than 2500 fossil dinoflagellate cysts species have been described (Taylor et al., 2008) using palynological taxonomy and classification systems that developed largely separate from those for living dinoflagellates. Compounding an already confusing situation, the taxonomy and classification of dinoflagellates developed within phycology under the botanical nomenclatural code, alongside the classification of predominantly non-photosynthetic species in protistology under the zoological nomenclatural code. Building from studies of plankton biologists Braarud, Erdtman, and Nordli, the link between fossil cysts and extant living dinoflagellate cysts was finally recognized and confirmed (Evitt, 1961; Evitt and Davidson, 1964; see Dale, 2021), thus paving the way for integration of both evolutionary theory and more gradually, dinoflagellate systematics. Though still remaining partially separate, an integrated view has greatly assisted the formation and testing of competing evolutionary hypotheses and a more holistic polyphasic approach to taxonomic revision.

Many groups of single-celled organisms show little morphological diversity, and even fewer leave extensive fossil records. In contrast, the wealth of graded forms of apparently homologous structures displayed by both living and fossilized dinoflagellates provide a uniquely rich data to understand the evolution of morphological and functional diversity over geological timescales. As a result, sophisticated but essentially opposing living-derived and fossil-calibrated evolutionary theories were already well developed by the time the first molecular biological studies of the group began in the early- and mid-1980s. Since that time, the scale and diversity of application to dinoflagellate systematics has grown logarithmically, driven by technological innovations that massively increased availability and scale and molecular data collection at increasingly low cost for both equipment and analyses.

This review tracks the application and impact of molecular biology on dinoflagellate systematics and taxonomy from the mid-80s to the present day. Development and growth over the past 35 years are characterized by four molecular technology-driven phases. The emerging problems and potential solutions required to enable high-throughput molecular approaches to systematics and taxonomy are also considered.

3.1.2 Growth of dinoflagellate molecular systematics

The first dinoflagellate molecular sequences, from *Crypthecodinium cohnii* and *Prorocentrum micans*, were published through the 1980s (Hinnebusch et al., 1981; Maroteaux et al., 1985; Lenaers et al., 1989) and totaled 5 sequences by 1990. The invention of polymerase chain reaction (PCR) and widespread availability of PCR-based DNA sequencing chemistry in the early 1990s dramatically increased the number researchers able to generate molecular data, a logarithmic growth in published molecular systematics papers and molecular sequences over the following decade (Fig. 3.1).

Decreased equipment costs, higher throughput capillary-based sequencing, and reduced sequencing costs fueled continued acceleration in data collection from 2002–2005. Growth has been maintained by continued invention, development, widespread availability, and massively decreasing cost, of high-throughput and next-generation sequencing (NGS) (Fig. 3.1; Panel A). Step-wise jumps from 2012–14 and 2019–20 represent bulk accession of several large genomic/transcriptomic studies to public databases.

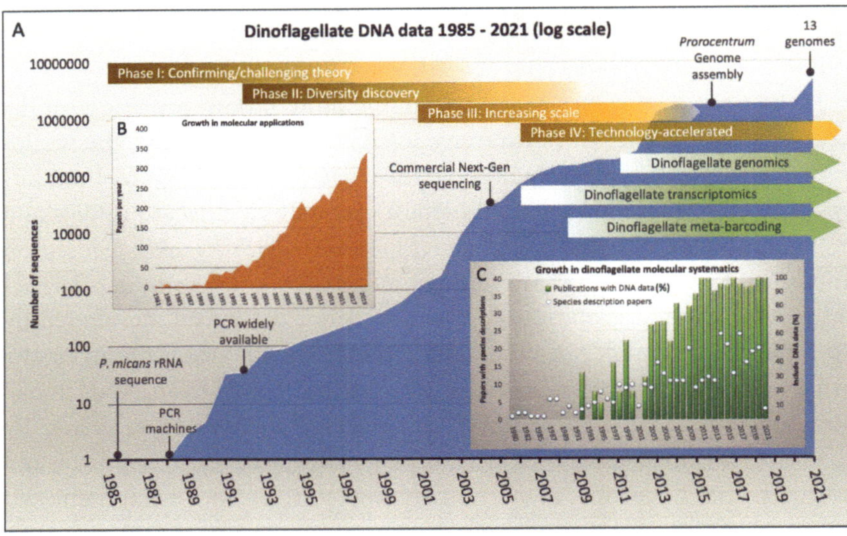

Figure 3.1 Growth and impact of molecular data on dinoflagellate systematics from 1985 to March 2021. A: The cumulative increase in published dinoflagellate DNA sequences on Genbank (nr) database on a logarithmic scale, and relative to major technical/data throughput landmarks. The four main phases of application and impact are indicated by orange arrows. Uptake of key advances in data collection by the dinoflagellate taxonomists community are indicated by green arrows; B: Growth in the number of dinoflagellate research papers per year (ISI web of science) utilizing molecular approaches; C: Growth in dinoflagellate molecular systematics from 1980 showing publications per year (ISI web of science) that include new dinoflagellate species descriptions, and the proportion (%) that include DNA sequence as supporting data.

The impact of changing molecular technology is also evident from some simple statistics: At the time of writing in 2021, almost 5 million dinoflagellate DNA sequences existed on major public molecular databases. More than 60% have been added since 2020, 96% in the last 10 years, and 99% since the advent of commercial NGS technology. The vast bulk are nuclear DNA sequences resulting from dinoflagellate transcriptome sequencing, and a small number of genome sequencing projects on a few selected taxa; predominantly *Symbiodinium* (zoothanthellae) and related taxa, plus selected toxic and harmful species, such as the dinoflagellates of the genus *Alexandrium*. Despite the strong bias in species coverage, sequences from all dinoflagellate orders are represented on public databases (Fig. 3.2).

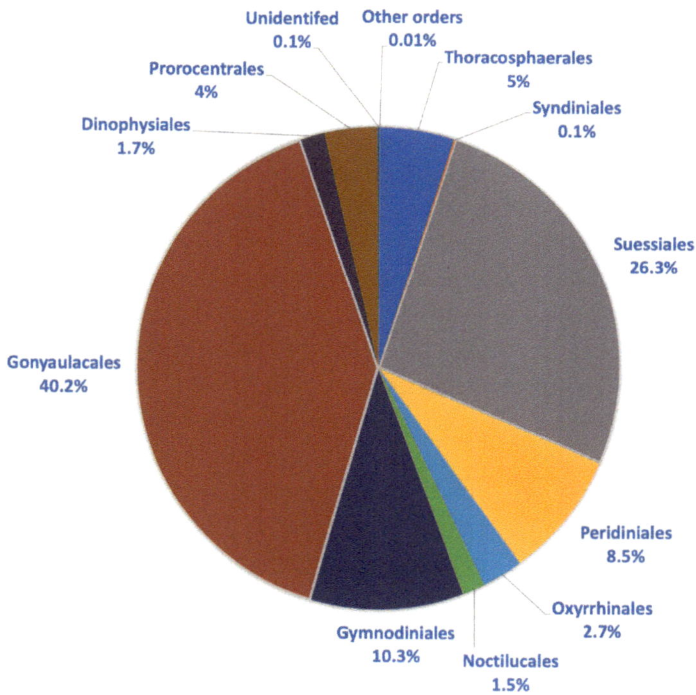

Figure 3.2 Taxonomic distribution (%) of total sequence data, representing different dinoflagellate orders, available on NCBI nucleotide (nr) database as of March 2021.

Approximately 3500 dinoflagellate mitochondrial sequences exist, with the majority being sequences of cytochrome genes (e.g., CO1, CytB) used for DNA barcoding studies. Approximately 4000 plastid/chloroplast gene DNA sequences are available, primarily the psbA gene (D1 protein, photosystem II) and a smaller number of other genes. Between 90,000 and 140,000 sequences are from uncultured dinoflagellates (stringent vs. relaxed search criteria), predominantly from DNA transcriptome sequencing of coral zooxanthellae (*Symbiodinium*), the endoparasitic species *Amoebophyra*, and a small range of harmful/toxic species. Approximately 2000 of these uncultured/unknown sequences are derived from DNA metabarcoding, and hence the morphology and functional characteristics cannot be easily determined.

Due to their comparatively broad taxon coverage and the presence of both highly conserved as highly variable regions, ribosomal RNA (rRNA) genes remain the most common target gene to support systematic studies. Between 19,000 and 24,000 rRNA sequences from a known species source are listed on public databases (excluding uncultured and environmental samples). Approximately 11,000 of these rRNA gene sequences were collected using Sanger dye-terminator DNA sequencing; the remaining are generated from genome and transcriptome sequencing with NGS-technology. In less than a decade, transcriptomic/genomic-derived rRNA sequences almost equal the sequencing effort (in number if not diversity) gathered by hundreds of researchers using Sanger sequencing over the prior 30 years.

If we consider a representative sample of publications, including new dinoflagellate genus and species descriptions (Fig. 3.1; panel C), the rate of description papers has steadily increased since the 1990s. Though countering a recognized trend of decreasing recruitment and funding for taxonomic research (see Wheeler, 2014), the trend in microeukaryote taxonomy is more indicative of increasing contribution of molecular data to resolving cryptic speciation and morphological convergence among phylogenetically diverse taxa. More importantly, molecular approaches have substantially modified the practice of dinoflagellate systematics and taxonomy. It is now standard practice to include molecular data and phylogenies to support new descriptions and revisions (Fig. 3.1; panel C). Almost all dinoflagellate systematics papers since the mid-2000s adhere to this practice, the only exception being fossil resting cyst descriptions, where molecular data cannot be obtained.

3.2 Phase I: challenging existing evolutionary theories

3.2.1 Establishing the phylogenetic position of the dinoflagellates

The first ultrastructural studies of dinoflagellates during the 1950s and 60s uncovered a range of unique nuclear and cell-division features. As a result, Dodge (1965) proposed the Mesocaryota and suggested they represent an ancient lineage with features intermediate between prokaryotes and eukaryotes, and no close relatives (Dodge, 1983). However, other researchers noted the cell wall (cortical) similarities and shared features with ciliates and Apicomplexa, proposing an evolutionary affinity between these three groups (e.g. Corliss, 1975; Taylor, 1976).

The earliest molecular investigations of dinoflagellates in the 1980s formed part of "tree-of-life" research using ribosomal RNA gene sequences, as a by-product of work to determine deep-branching relationships among prokaryotes and eukaryotes. Support for competing dinoflagellate evolutionary hypotheses yielded conflicting outcomes. Sequences from *Crypthecodinium cohnii* (5S rRNA, Hinnebusch et al., 1981) initially cast doubt on the ancient origin hypothesis, whereas small-subunit (SSU-rRNA) sequences from *Prorocentrum micans* supported a divergent and ancient origin among eukaryotes (Herzog and Maroteaux, 1986; Lenaers et al., 1989, 1991). In retrospect, the true position and affinities of dinoflagellates among eukaryotes could not be clearly established by these early analyses, but more detailed secondary structure-driven analysis of 25 eukaryote lineages showed dinoflagellates diverged near or within the eukaryotic "crown radiation" (Sogin, 1989), dispelling the ancient mesokaryote hypothesis. Wolters' (1991) analysis of >40 eukaryote taxa subsequently confirmed the hypothesized affinity of dinoflagellates with ciliates and apicomplexans (Dinozoa, *sensu* Cavalier-Smith, 1981), and their close relationship to the Chromista (golden brown algae). While new lineages and sub-groupings have emerged, along with various controversies, the majority of work affirms dinoflagellates evolutionary position among the Alveolata. More fine-scale rRNA and nuclear gene studies within the alveolates now group the apicomplexans and dinoflagellates together as the Myzozoa (see Cavalier-Smith and Chao, 2004), allied with the parasitic Perkinsid apicomplexans (*Perkinsus* spp.), and the heterotrophic bi-flagellate *Oxyrrhis marina* as the most basal early-branching dinoflagellate (Saldarriaga et al., 2003); more recent multigene and protein phylogenies continue to affirm these relationships (Bachvaroff et al., 2014). Research has largely moved to questions of deeper rooted relationships among the chromophyte algae, such as Heterokonts alveolates and Rhizaria (so-called SAR supergroup, or Harosa sensu Cavalier-Smith), the evolution and relationships among of dinoflagellate families and orders (e.g., Zardoya et al., 1995; Grzebyk et al., 1998; Fensome et al., 1999; Janouskovec et al., 2017), evolution of their astonishing plastid diversity (e.g., Saldarriaga et al., 2001; reviewed in Dorrell and Howe, 2015), and integration with how various cortical (thecal plate) patterns evolved (e.g., Orr et al., 2012; Price and Bhattacharya, 2017; Janouskovec et al., 2017).

3.2.2 Evolution and phylogeny of the dinoflagellates

Well before molecular biology, the high level of morphodiversity of established families and genera created considerable debate that existing morphotaxonomy did not reflect true dinoflagellate evolutionary relationships. Several early evolutionary schemes were devised, with the most sophisticated well-supported models derived from careful analysis of cortical structure: the amphiesmal vesicles and thecal plates that were the basis of both extant and fossil dinoflagellate taxonomy. By the 1970s, two competing but diametrically opposed models had been devised (Fig. 3.3): the plate reduction model derived from and supported by fossil cyst evidence, placed Gymnodinales (naked species, lots of vesicles) as ancestral and Prorocentrales (two main plates) as most recently derived; and two plate increase models derived from observations of living dinoflagellates that placed Prorocentrales as ancestral, and Gymnodinales as most recently derived (Bujak and Williams, 1981).

A series of rRNA gene studies during the 1990s added molecular data from a range of dinoflagellate lineages (e.g., Zardoya et al., 1995), but the use of short sequences (<400 bp) and limited/selective taxon sampling made interpretation difficult and raised more questions than answers. The first serious test of dinoflagellate evolutionary theory with molecular approaches was Saunders et al. (1997), who included 30 to 44 small- and large-subunit RNA sequences from multiple species representing most family lineages known at the time. Their trees exhibited features common of many subsequent and, even recent, dinoflagellate phylogenies based on rRNA and other single nuclear genes: (1) early branching of *Noctiluca*; (2) a basal position of *Amphidinium* as ancestral among the "core dinoflagellates," both consistent with more current evidence, and (3) early branching of Gonyaulacales and Suessiales (represented by *Symbiodinium* spp.), more likely to be artifacts of highly divergent evolutionary rates among lineages; (4) the crown of the dinoflagellate tree populated by interleaving clusters of gymnodinoids, peridinoids and prorocentroids—the "GPP complex"—with weakly supported higher-order relationships (Fig. 3.3C). The position of both ancestral and recently derived lineages in the crown of the tree conflicted with all existing evolutionary models, and another model was proposed, where the majority of thecate and athecate lineages in the GPP complex remained unresolved (Fig. 3.3C). These early molecular findings were subsequently integrated with fossil dinoflagellate evidence by Fensome et al. (1999), proposing an evolutionary scenario (Fig. 3.3D) that remained largely intact until recent transcriptome-derived multigene phylogenies began to resolve additional complexities (Fig. 3.3E).

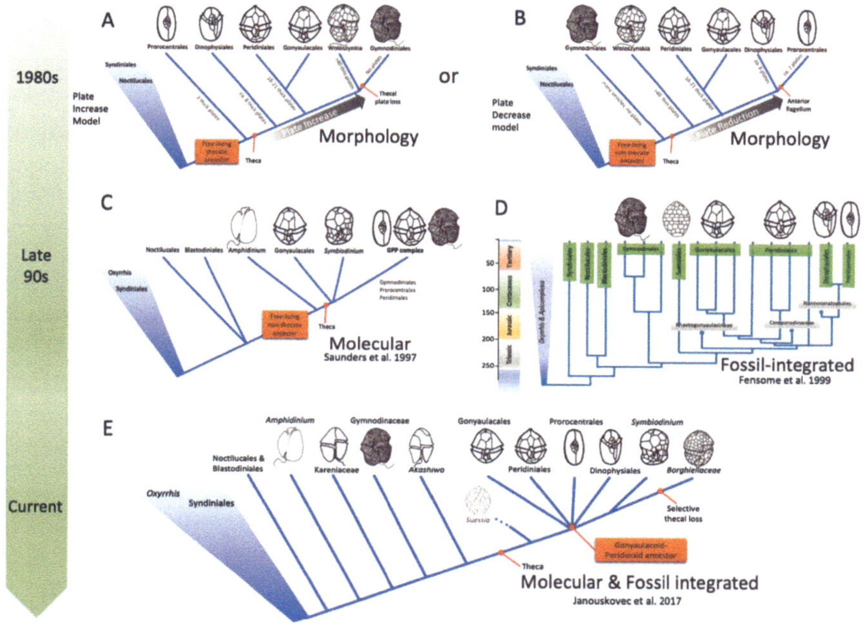

Figure 3.3 Impact of molecular approaches on dinoflagellate evolutionary theory. A–B: Main competing evolutionary hypotheses of the 1980s were the plate-increase and the plate-decrease models, both based on morphological features and the structure of the dinoflagellate cortex; C–D: Competing/conflicting molecular-informed hypotheses of the late-90s based on molecular data. C: Phylogeny of dinoflagellate orders derived from Saunders et al. (1997); D: Evolutionary hypothesis derived from Fensome et al. (1999), integrating molecular evidence with fossil cyst evidence; E: Multi-gene dinoflagellate phylogenetic hypothesis, integrating both molecular and fossil evidence adapted from Janouskovec et al. (2017), based on alignments of 101 protein sequences and 29,400 sites.

Much of the following two decades has steadily addressed the evolutionary conundrum posed by the GPP-complex. Fossil evidence supports the case that poor resolution among major lineages is due to an early (Jurassic) and very rapid radiation over a short period of time. However, early branching of gonyaulacoids and the interleaving of several apparently quite distinct lineages in the GPP complex also indicate artifacts and biases inherent in phylogenetic analysis, such as short sequences with too few phylogenetically informative sites, analytical models not well-suited to taxa with very different evolutionary rates, effects of evolutionary noise and mutational saturation, and the combined effect of limited taxon sampling and long-branch clustering artifacts. Improvements to desktop computing power have since allowed widespread application of increasingly sophisticated evolutionary models and analysis of larger taxon datasets and increasingly long sequences, yet many contemporary dinoflagellate phylogenies yield similarly confused and poorly resolved order-level clustering.

Recent progress and improvement has come from a combination of increased taxon coverage/selection with multiple nuclear gene phylogenies. Initial efforts with rRNA and nuclear genes more clearly resolved the basal position of athecate gymnodinoid taxa (Bachvaroff et al., 2011; Orr et al., 2012), suggesting a single origin of the theca, an inference confirmed using analyses based on >70 ribosomal protein sequences (Bachvaroff et al., 2014). Similar evolutionary patterns and lineage affiliations emerge from phylogenies using large numbers (>200) nuclear genes mined from dinoflagellate transcriptomes (Price and Bhattacharya, 2017; Janouskovec et al., 2017). These studies reinforce a single evolutionary origin of thecate species, clarify the early evolutionary development of several athecate lineages, but still only partially resolve branch order in the main thecate lineages that now form the crown of the dinoflagellate tree (see Fig. 3.3). They also indicate additional evolutionary complexities, such as: (1) extant suessioids (e.g., *Symbiodinium*, *Polarella*) arising from within the gonyauloacoid/peridinoid lineage independent of the Triassic fossil Suessiales (e.g., *Suessia*), and (2) multiple independent evolution of a complex gymnodinoid-like tabulation in early branching *Amphidinium* and *Gymnodinium*, as well as more derived lineages, such as *Borghiella* (Janouskovec et al., 2017) that were formerly part of *Woloszynskia*, a genus shown to be polyphyletic. Despite the impressive scale of informative sequences, recent transcriptome-derived phylogenies still suffer from selective biases caused by low-taxon sampling. Of the 40–50 species included, only 12–20 thecate genera are represented, with several lineages only by a single species. However multi-gene phylogenomics will soon be the standard approach, steadily solving taxon-bias issues as more species/genera are added to public databases and revealing further surprising complexities in the evolution of dinoflagellates.

3.3 Phase II: discovering widespread cryptic diversity

Molecular approaches provide data with a more direct link to the evolutionary processes driving diversification speciation of organisms less affected by evolution of convergent morphology, and less reliant on individual interpretation of complex features into an evolutionary order. Though useful on its own, it is most valuable and convincing when combined with traditional morphology, ultrastructure, and chemotaxonomic data; a polyphasic approach already widely adopted in plankton taxonomy. For these reasons, molecular data and phylogenetics was quickly embraced by phytoplankton taxonomists.

A major consequence of molecular data integration has been increasing recognition of the polyphyletic nature of many dinoflagellate lineages, in particular the athecate order Gymnodiniales, and species formerly allocated to *Woloszynskia* that possess many thin thecal plates. Both exhibit complex cortical patterns difficult to interpret in an evolutionary context, and both have gradually been sub-divided into new families and genera over the past 20 years (e.g. Lindberg et al., 2005; Moestrup et al., 2006, 2008, 2009a,b, 2014). For example, Daugbjerg et al. (2000) split the genus *Gymnodinium* into four genera and created *Karenia*, *Karlodinium*, and *Akashiwo* based on features of the apical groove. The revisions were supported by a convincing LSU-rRNA phylogeny, and relatively quickly accepted and integrated into dinoflagellate taxonomy. Over the past two decades a multitude of studies have erected new orders and families (e.g., Torodiniales, Boutrup et al., 2016), and more than 70 new genera, all with support of detailed rRNA gene phylogenies. Though some revisions have been controversial and not all universally accepted (e.g., revisions to *Alexandrium* by Gómez and Artigas, 2019; opposed by Mertens et al., 2020), the vast majority of systematic revisions are strengthened, rather than weakened, by inclusion of molecular phylogenetics.

3.3.1 The rise of reverse taxonomy: sequence first, identify later

Rapid uptake and growth of molecular methods has also facilitated the speed with which sequence data can be collected (now a few hours to a few days) compared to other forms of data requiring hours of detailed light microscopy, or electron microscopy with specialized fixatives, preparation, and imaging equipment. Increased scale/coverage of molecular databases has also increased ease in placing a new species within the existing systematic and phylogenetic framework. Even for dinoflagellates that have relatively rich and informative morphology, it can be difficult to quickly establish species-level identity from microscopy and in some genera (e.g.,

Gambierdiscus); conclusive identification of some species can only be established from comparative analysis of rRNA or other gene sequence data.

For most unknown species it is now cheaper and more efficient to practice *reverse taxonomy*: sequence first and identify later, an approach that underpins on-going discovery of cryptic diversity and regional genetic variation in dinoflagellates. One can extract and sequence tens to hundreds of cultures or single cells within a few days, then use the data to identify and place sequences into the existing taxonomy using database searches (BLAST) verified by rapid phylogenies. This approach allows researchers to quickly select species/samples of particular interest for further examination, and also target novel material for new taxonomic descriptions. Initially unexpected or puzzling DNA sequence variation often initiates a closer examination of morphology for evidence of cryptic speciation; searching for morphological/cytological features that support the phylogenetic groupings uncovered using DNA sequences. Example of genera where sequence-guided approaches have been widely applied include the following: *Amphidinium*, a genus displaying limited morpho-diversity and cryptic speciation, now shown using molecular data to contain at least 18 species (e.g. Murray et al., 2004, 2012; Karafas et al., 2017); *Gambierdiscus*, initially represented by a single species (*G. toxicus*), but now containing 20 species largely distinguished using sequence data (e.g., Litaker et al., 2009; Kretzschmar et al., 2017, 2019); *Karenia* (e.g., De Salas et al., 2004a,b; Haywood et al., 2004) established with three species in 2000, but now containing at least 12 species; *Karlodinium*, a genus considered to contain potentially 3 species when established, but now containing 16 species (e.g., De Salas et al., 2008); *Takayama*, established with four species using sequence-guided principles (De Salas et al., 2003), now containing seven species; distinguishing cryptic species among the smaller members of *Phalacroma* (Daugbjerg et al., 2019); and the microreticulate cyst group of *Gymnodinium* species (Bolch et al., 1999; Gu et al., 2013) highlighted in further detail in the case study. A more extreme example of sequence-guided taxonomy is the "tamarensis-group" in the genus *Alexandrium*, where five species were established based solely on genotypic criteria (John et al., 2014).

Case Study 3.1: Cryptic speciation among microreticulate cyst forming *Gymnodinium* species

The chain-forming *Gymnodinium catenatum* is of notable interest as the only Gymnodinoid to produce paralytic shellfish toxins (PST). First described from Mexico by Graham (1943), it was rarely reported until identified as the causative species of major paralytic shellfish poisoning (PSP) in Mexico and NW Spain in the late 70s and mid-80s (see Hallegraeff et al., 2012 for detail). The ability to form long chains of 2 to >100 cells means it is easily recognized (Fig. 3.4) but, ironically, this feature has actually fueled considerable confusion; essentially any chain-forming photosynthetic gymnodinoid was often been erroneously reported as *G. catenatum* (Fig. 3.4; examples of similar species). In the mid-90s, description of the superficially similar *Gymnodinium impudicum* (Fraga et al., 1995) prompted more careful examination of chain-forming species, and eventual re-examination of global historical reports of *Gymnodinium catenatum* (e.g., Bolch et al., 1999; Bolch and Reynolds, 2002). Similarly, the uniquely microreticulated red-brown resting cyst (Fig. 3.4) also appeared to provide little scope for confusion. Yet the existence of multi-modal and overlapping size distributions at some sites (Bolch and Hallegraeff, 1990; Bravo and Ramilo, 1999; Amorim et al., 2001) and markedly different size fossil cysts in the Baltic Sea (Thorsen et al., 1995) suggested the presence of potentially unresolved cryptic species. Molecular approaches rapidly resolved the question, Bolch et al. (1999), Ellegaard and Oshima (1998), and Ellegaard and Moestrup (1999) combining molecular, morphology, biochemical, and ultrastructural data to resolve two new non-toxic species (*G. nolleri* and *G. microreticulatum*) responsible for the smaller classes of resting cyst. Two additional non-toxic species, *G. trapeziforme* (Attaran-Fariman et al., 2007) and *G. inusitatum* (Gu et al., 2013) have since been resolved with the aid of molecular data. Though the cyst forms are quite distinctive (Fig. 3.4), the plankton cell size ranges and morphology overlap with *G. nolleri* and *G. catenatum*, respectively, and mis-classification is likely if identified using morphology alone. Indeed, comparative analysis of all available microreticulate-group LSU-rRNA sequences (Fig. 3.4) indicate that Japanese source sequences currently attributed to *G. nolleri* are in fact *G. trapeziforme*. Additionally, the scale of sequence divergence among *G. microreticulatum*-related sequences suggests perhaps as many as three additional cryptic species among the smaller representatives in the group.

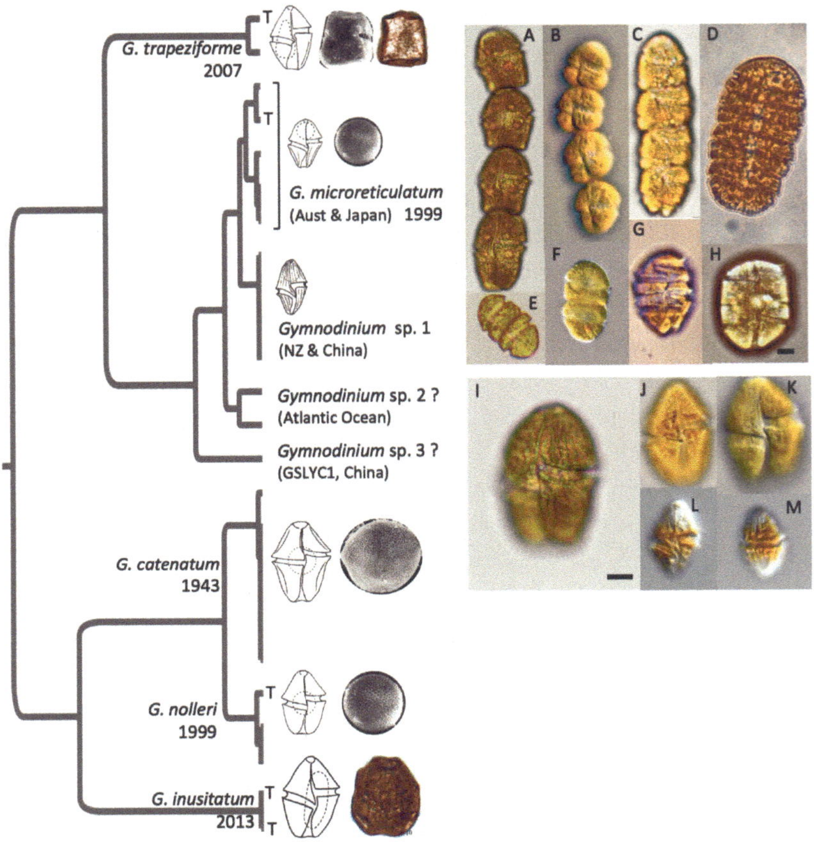

Figure 3.4 Case study: Cryptic speciation in the microreticulate cyst group of *Gymnodinium*. Phylogenetic relationships (D1–D3, LSU rRNA) of *Gymnodinium catenatum* and related species forming microreticulate resting cysts. T = from type material or location. Cells and cysts are drawn at same relative scale. Figures A–H. Examples of chain-forming gymnodinoid species often confused or misidentified. A: *Gymnodinium catenatum*; B: *Gymnodinium impudicum*; C: *Cochlodinium polykrikoides*; D: *Pheopolykrikos beauchampii*; E: *Gymnodinium litoralis;* F: *Gymnodinium* sp. (undescribed); G: *Polykrikos tanit*; H: *Phaeopolykrikos hartmanni*. Figures I–M. Comparative single-cell morphology of four microreticulate cyst-forming species. I: *G. catenatum*; J: *G. trapeziforme*; K: *G. nolleri*; L–M: *Gymnodinium* sp. 1 (CAWD191). Scale bars = 10 μm. (Images: 4C and cyst of *G. inusitatum* from Gu et al., 2013; Image 4D by M. DeSalas, unpublished; *G. trapeziforme* cyst SEM by Gilan Attaran-Fariman, unpublished; *G. catenatum* cyst SEM by Gustaaf Hallegraeff; all other images by author.)

3.4 Phase III: increasing taxon and gene sampling

3.4.1 Solving the culture bias problem

Though increasing penetration of molecular systematics into protistan taxonomy led to considerable progress from 2000 onward, the vast majority of dinoflagellate studies still relied mainly on uni-algal cultures leaving increasingly gaps in taxon coverage of phylogenetic studies. Our inability to culture but a small fraction of the evident diversity, and the problem of culture selection/bias, were both well known prior to the invention of PCR, therefore it is not surprising that the technique was applied to environmental samples and uncultured material quite rapidly after it became widely available. Single-cell PCR has, in particular, contributed substantially to broadening taxon coverage of many uncultured and heterotrophic species/groups of dinoflagellates. Genetic identification of single cells retrieved from mixed plankton samples is possible using micromanipulation, followed by PCR directly from un-processed single cells of many planktonic organisms, including dinoflagellates (e.g., Marín et al., 2001; Bolch, 2001). Further adaptations and improvements demonstrated that single-cell PCR was also possible from material subjected to common plankton fixatives (Godhe et al., 2002; Auinger et al., 2008) and from living as well as apparently non-viable resting cysts with little or no pre-extraction of DNA (Bolch, 2001). However, single-cell PCR is painstakingly slow, requires considerable practice, and is inevitably destructive sampling. Even in the hands of the highly skilled, moving single cells through the steps of isolation, washing, imaging, and successful PCR often has a low success rate. Considerable enhancement of throughput is possible through increasing integration of high resolution imaging with flow cytometry and other micro-fluidic technologies (see Section 3.6; Bolch and Percy, 2013).

3.4.2 Increasing gene targets and multigene phylogenies

Although the focus of most systematic studies of phytoplankton remains on rRNA genes, they have the same limitations suffered by most single gene phylogenies, particularly in poor resolution of more deeply rooted evolutionary relationships. Over the period 2000–2010, several studies sought to expand the range of nuclear or other gene targets for phylogenetic studies, including actin, beta-tubulin, and heat-shock proteins (hsp70, hsp90) (e.g., Fast et al., 2002; Saldarriaga et al., 2003; Hoppenrath and Leander, 2010), or mitochondrial genes cytochrome oxidase (*cox*1), and cytochrome B (*cob*) (e.g., Zhang et al., 2007; Stern et al., 2010; Murray et al., 2012). These studies consistently indicated early branching of the gymnodinoids, tended to recover stronger support for monophyly of the various orders, but otherwise suffered some of the same issues of low support and interleaving of lineages characteristic of single-gene rRNA phylogenies. In contrast, combining multiple sequences into concatenated, multi-gene phylogenies improved resolution and support of relationships at higher taxonomic levels (e.g., Zhang et al., 2007; Orr et al., 2012), and paved the way for larger multi-gene phylogenies as dinoflagellate transcriptomics and genomics became more feasible and affordable.

The first forays of environmental metagenomics using rRNA gene cloning and DNA barcoding methods indicated orders of magnitude higher eukaryote planktonic diversity existed in nature (e.g., Moon-Van Der Staay et al., 2001). Many sequences displayed only distant relationships to existing taxa, further highlighting the highly selective nature of cultures and the consequence of restricted and patchy taxon coverage in sequence databases. Initial work to develop and apply DNA bar-coding approaches to dinoflagellates explored a range of other gene targets, such as cytochrome oxidase 1 (CO1) and cytochrome B (CytB) (e.g., Lin et al., 2006, 2009; Stern et al., 2010) or rRNA-ITS regions (Stern et al., 2012) as taxonomic tools for rapid classification. However, when applied to environmental samples, high diversity exposed even more limited comparative databases for many barcode sequences, and produced an emerging problem of *orphaned sequences*: sequences or entire clusters without a home, and no link to morphotaxonomic or functional identity in the existing systematic structure. Though usefully documenting levels of genetic diversity, such sequences cannot contribute to a coherent systematic picture until neighboring taxa are identified either from cultured sources, or linked with morphological/functional data, such as cell images, or genome/metagenomic data.

3.5 Phase IV: technology-accelerated diversity discovery

As in all fields of biology, technology continues to massively accelerate the collection of molecular data and application to dinoflagellate diversity and systematics. Over the last decade to 2020, commercial sequencing services have developed cheap end-to-end sequencing services that take material from crude/preserved samples through to sequence data collection and bioinformatic processing and analysis services. Though cost and technical expertise are no longer a barrier, the bottleneck is now taxonomic and bioinformatic expertise essential to integration into the evolutionary and systematic framework. Of the wide range of molecular tools now available, three have particular potential to revolutionize the way we document diversity and how we approach taxonomic and phylogenetic studies: (1) DNA barcoding; (2) transcriptomics; and (3) ancient DNA (aDNA).

3.5.1 DNA barcoding and meta-barcoding in taxonomy

First demonstrated for animal taxa using partial mitochondrial CO1 sequences (Hebert et al., 2003), DNA barcoding is conceptually simple: essentially, sequences of a standard region of DNA can provide an unambiguous and diagnostic identification of species, and thus serve as a rapid global bioidentification system to supplement the limited capacity provided by traditional taxonomy. The idea was quickly taken up by dinoflagellate taxonomy in the mid-late 2000s, and several mitochondrial gene targets were explored and shown to have potential, but difficult, to universally amplify, and have insufficient sequence variation to fully resolve dinoflagellate species in some genera (e.g. Stern et al., 2010). Other targets include chloroplast genes for particular taxa (e.g., *psbA* for *Symbiodinium*; LaJeunesse and Thornhill, 2011), and the internal transcribed spacer region of rRNA (rRNA-ITS; Stern et al., 2012). However, universal application is hampered by existence of differing sequence classes and paralogous rRNA genes in some genera, and a high number of indels that create difficulty with unambiguous DNA alignment. The SSU-rRNA and LSU-rRNA remain the default target for systematic studies, but with more variable regions (D1–D2 region of LSU-rRNA, rRNA-ITS) used in a similar manner to that conceived for DNA barcoding. However, even these regions show limited resolution of existing morphospecies in some genera (e.g., *Dinophysis* spp. Hart et al., 2007).

The true power of the DNA barcoding comes from combining it with *meta-barcoding*: the PCR amplification and high-throughput sequencing of bar-coding targets from natural environmental samples. The ultimate promise is rapid and simultaneous identification of potentially all taxa in a sample, but significant hurdles remain.

A recent example is Le Bescot et al. (2016), who used SSU-rRNA gene (V9 region) as a barcode to examine global dinoflagellate diversity across differing size classes. Though effective at genus-level, the V9 has limited species-level resolution. Of particular note was the increasing level of dinoflagellate unassigned metabarcodes in the nano- and picoplankton fractions (35 and 49%, respectively), particularly lineages allied with Gymnodiniales. A similarly high proportion of barcodes unassigned to genus level or below (64%) are also evident in studies using the V4–V5 region (e.g., Sze et al., 2018). These studies again emphasize the massive undescribed diversity that, without additional source-linked data, remains difficult to integrate into the existing taxonomic framework. Combining multiple meta-barcodes (e.g., SSU-rRNA and D1–D2 of LSU-rRNA) offers both improved species resolution (e.g., Esenkulova et al., 2020) and improved taxonomic coverage, and, if limited to highly typical target species, can offer detection equal to or superior to the morphology for most phytoplankton groups.

Targeted metabarcoding offers similar promise for detection of specific groups, such as toxic or harmful species (e.g., *Gambierdiscus*; Smith et al., 2017a). Though proving to have high sensitivity of detection in a presence/absence sense, quantitative or relative abundance assessment remains problematic, due to combined problems of PCR-bias and/or differing gene copy number between species (Smith et al., 2017b). Although still offering considerable promise, the significant problems of species resolution, undocumented diversity, and database coverage must be addressed before metabarcoding can be integrated into reliable automated detection and identification systems.

3.5.2 Transcriptomics in dinoflagellate taxonomy

A transcriptome is a collection of RNA (or DNA) molecules, which represents the full range of messenger RNA (mRNA) expressed by an organism; essentially a catalogue of genes expressed by the cell (under the conditions) and the relative proportions of each gene. Initially conceived as tool to examine changes in gene expression, transcriptome data also provide a rich source of sequence data for phylogenetics. This is particularly the case for dinoflagellates, where genome sequencing approaches are difficult, due to their massive genomes (up to 245 Gb) containing high proportions of noncoding and repetitive DNA sequence (Hou and Lin, 2009).

Over the past decade a series of funding initiatives (e.g., The Moore Foundation Marine Microbiology Initiative (MMI); https://www.moore.org/initiative-strategy-detail?initiativeId=marine-microbiology-initiative) sought to "fill in the gaps" in the eukaryotic evolutionary tree through the generation of high-quality transcriptomes from hundreds of microeukaryotes taxa, including many species representing a wide range of dinoflagellates lineages. Mining and analysis of available public data has already generated several surprising and novel interpretations of dinoflagellate evolution (e.g., Price and Bhattacharya, 2017; Janouskovec et al., 2017;

Cooney et al., 2020) that are now informing revisions to dinoflagellate taxonomy, especially relationships at ordinal level and above. As sequencing costs continue to decline, transcriptome data is steadily becoming the preferred approach for systematics research. Compared to traditional sequencing, the additional sample preparation time required is outweighed by capacity to simultaneously generate sequence data of hundreds to thousands of genes, from which to assess relationships and reconstruct well-supported phylogenies.

3.5.3 Ancient DNA and palaeomics

Modern molecular approaches based on PCR are reliant on availability of living or carefully archived material, and thus restricted to analysis of the present (extant species) and the very recent past. Despite many dinoflagellates having a long-lived fossilizable cysts, viability in most in coastal sediments decreases to only a few percent within a few decades (Lewis et al., 1999), but under continuous undisturbed anoxia (Lundholm et al., 2011) or plant hormone stimulation, live recovery is possible from cysts thought to be 100–150 years old (Delebecq et al., 2020). Intact DNA can also be recovered from fixed material (Marín et al., 2001), and also sequenced from apparently non-viable resting stages (Bolch, 2001; Bolch and Percy, 2013); but DNA degradation means that existing PCR-based methods are only successful over no more than a few hundred years.

Although these limitations are not necessarily a problem for taxonomy, it presents a major hurdle to addressing species biogeography (the study of current distributions and species dispersal) and their response to environmental change over longer timescales. However, under suitable conditions (e.g., dark, non-acidic, dry or anoxic), DNA can persist over thousands of years (Briggs et al., 2007). Degradation still occurs steadily via fragmentation into shorter lengths caused by hydrolysis and depurination. Chemical damage (lesions) also accumulates (e.g., cytosine deamination), but it does so in a characteristic pattern that provides a tool to distinguish ancient fragments from modern contaminant DNA (Dabney et al., 2013). This degradation typically limits application of PCR to material less than a few hundred years old (e.g., Klouch et al., 2016). Though PCR can be applied over much longer timescales (e.g., Coolen et al., 2013), it is subject to considerable taphonomic biases, due to selective and uneven DNA degradation down-core, and an inability to distinguish whether PCR-amplified DNA has arisen from contamination with modern DNA.

Recent adaption of existing terrestrial ancient (aDNA) approaches to extract and sequence sedimentary ancient DNA (*seda*DNA) from marine cores now offers a detailed window into plankton communities over the past 10,000 years and beyond (Armbrecht et al., 2020; Armbrecht, 2020; Armbrecht et al., 2021). The approach is outlined in Fig. 3.5.

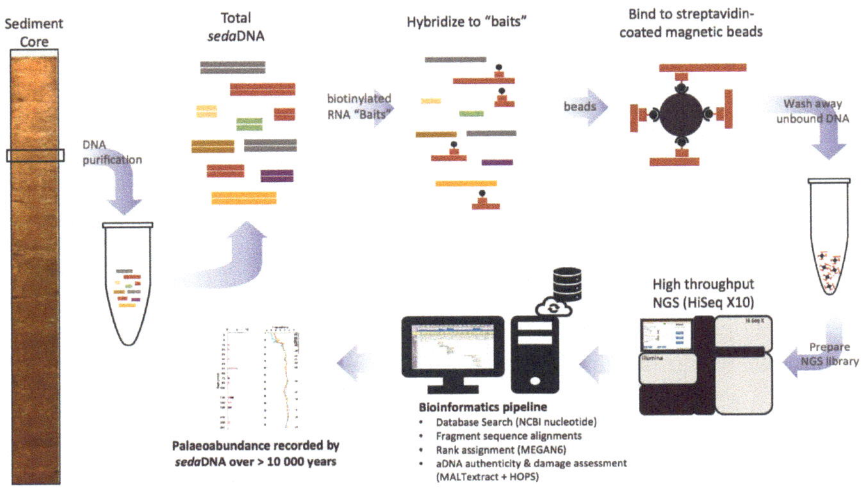

Figure 3.5 Hybridization-capture enrichment and analysis of plankton and harmful dinoflagellate sequences from ancient sedimentary DNA (*seda*DNA) in marine sediment cores (adapted and redrawn from Armbrecht et al., 2020). Targeted *seda*DNA sequences and complementary bait-sequences are shown in red.

Briefly, fragmented total *seda*DNA is extracted from core sections, and a series of pre-designed oligonucleotides, termed "*baits*," are added to enable a targeted, semi-specific enrichment called *hybridization capture* (HC). The HC process selectively binds and retains bait-complementary sequences from the extracted total *seda*DNA. The enriched fragments are then prepared and sequenced at high sequencing depth using NGS-based approaches (Armbrecht et al., 2021). The HC process is also adaptable and flexible; different groups of organisms can be targeted by design of additional baits-sets to capture sedaDNA of any group of interest. Importantly, the process does not rely on PCR amplification, and therefore is not reliant on preservation of long/intact DNA or affected by amplification biases. Therefore it can provide a semi-quantitative proportional representation of the target sequences (species) groups included in the bait-set. Resulting fragment data can also be subjected to verification of ancient origin using

DNA damage analysis, and instances of modern contamination detected and removed from the data (Armbrecht, 2020; Armbrecht et al., 2021).

Though a major step forward, working with aDNA requires considerable care to reduce modern contamination, beginning from the point of core collection. It also requires facilities with considerable DNA contamination control, and a series of laboratory plasticware, extraction, and other controls to detect systematic and casual contamination with modern DNA fragments that may enter the sequencing and analytical pipeline (Armbrecht et al., 2019). Other barriers to effective sedaDNA analysis in marine systems include the following: (1) potential biases and efficiency of HC-enrichment process; (2) sufficient sequencing depth (mostly cost-related) to encounter enough fragments of the target species/groups, and (3) biases and limitations of fragment assignment to genus/species level, due to patchy or limited coverage of sequence databases, a problem for any molecular approach relying on database sequence search/matching for identification.

3.6 Future perspectives: where to from here?

A long-standing primary goal of taxonomists is that taxonomy should reflect the true evolutionary relationship among the organisms (Mayr, 1981); a particular shared conviction of microbial taxonomists faced with huge genetic diversity, but few morpho- and phenotypic characters (Hugenholtz et al., 2021). Secondarily, the taxonomy must be a sufficiently complete catalogue of the true diversity to form the basis of a robust and reliable classification and identification system. Over the past 30 years, integration of a molecular sequence and phylogeny-driven approach has brought us within reach of the first goal, but also continues to increasingly indicate that we are far away from the goal of completeness. The seemingly unlimited potential of molecular biology has more recently created a third goal: creation of automated systems for accurate identification and abundance assessment of entire communities directly from the environment. Though there are considerable technical and engineering challenges in developing such systems for long-term marine deployment, the following major hurdles need to be addressed before all-of-community molecular identification becomes a realistic goal.

3.6.1 Data quality control

Reliable molecular identification is reliant on accurate high-quality reference data against which any unknown sequence can be assessed. These systems/algorithms already exist (e.g., GenBank/BLAST) in manual and automated form integrated into bioinformatic analytical pipelines; however, large public databases exercise little control over quality and source/identity of data submitted. Minor base-call errors/differences between submissions from the same source material are common and inevitable, given differing/changing sequencing technology, different stringency of base-correction, and quality criteria. In most cases, the variation is sufficiently small relative to species/level sequence variation, and thus does not impact unduly on automated or manual species assignment, especially if species assignments are manually verified by phylogenetic clustering and inspection (e.g., Grzebyk et al., 2017).

A more significant problem is misattribution of species and/or source information associated with sequence data. An all-inclusive GenBank phylogeny of almost any dinoflagellate genus quickly reveals instances of species-level clusters liberally interspersed with "rogue sequences"; seemingly out-of-place, due to (1) misidentification due to insufficient expertise or effort, (2) valid species assignment, but misleading, due to subsequent species/genus revisions or cryptic species resolution; (3) lab sample or sequence labeling errors not recognized and corrected before submission. Another barrier to a practical molecular identification is the often substantial lag-time (years in some cases) between widely accepted taxonomic revisions, and revision to taxonomic meta-data associated with sequences. Taxonomic assignments can be both confused and out-of-date, unless the end-user is intimately aware of the relevant taxonomic literature and history of revisions.

The issue of data quality is now well-recognized and addressed through broad annotation and quality initiatives (e.g., RefSeq, NCBI) and through the construction of independent curated reference databases: databases exercising strict control of both data quality and the veracity of sequence source/identity information associated with submissions. Examples include the following: (1) the prokaryotic diversity database EZBioCloud (https://www.ezbiocloud.net/) that contains >84,000 species, >65,000 16SrRNA sequences, and >200,000 genomes; (2) the SILVA ribosomal rRNA gene database, containing >11 million rRNA sequences, and (3) the PR^2 database of eukaryote 18S rRNA sequences (https://pr2-database.org/; Guillou et al., 2013), the most complete and recognized curated database for microeukaryotes and phytoplankton taxa.

For dinoflagellates, the recent DINOREF SSU-rRNA database, based on the V4 region (Mordret et al., 2018), provides an excellent base from which to develop a more comprehensive curated dinoflagellate database. Importantly, the detailed quality assurance and annotation criteria used in its construction can now guide additional submissions that increase taxon coverage, and also database scope through addition of other barcoding loci with better capacity to resolve at species-level (e.g., LSU-rRNA; Grzebyk et al., 2017).

3.6.2 Taxon coverage and species assignment

A major limitation of all identification methods reliant on sequence comparative alignment is the completeness of comparative databases. If we confine the context to only dinoflagellate "dark-matter," recent global scale meta-barcoding studies indicate at least 50% of the marine dinoflagellate community remains undocumented beyond sequence information (Le Bescot et al., 2016), even using a barcode (SSU-rRNA, V9) more suited to genera-level assignment and a generous OTU assignment cut-off of 90% similarity. Unsurprisingly, the highest proportions are in the nano- and pico-plankton community, which is difficult to sample and examine using standard morphological approaches, and to a large extent ignored in most morphotaxonomic studies. Species-level divergences in this region of the SSU are an order of magnitude lower in most genera. By comparison, clustering the more informative SSU-rRNA V4 sequences, even at 98% similarity (commonly used in metabarcoding), still lumps multiple species into the same OTU (Mordret et al., 2018), considerably underestimating species diversity. Thus the global oceanic proportion of unknown species is likely to be 95% or more. Even the most complete curated dinoflagellate database, DINOREF, (circa 1670 sequences, 422 species from 149 genera; Mordret et al., 2018) is modest in relative scale, being estimated to cover approximately 18% of recognized species. It does, however, cover 60% of current genera, significantly improving the capacity to assign a majority of reads at least to known genera.

The effect of low database completeness manifests in meta-barcoding as increasing proportions of taxa of uncertain identity at species level, not only the most biologically informative unit, but most often the level of identity required to determine ecological or human consequences. For example, marine biotoxin production is in most cases associated with particular species, not entire genera. Unfortunately many barcoding studies fail to report proportions of assigned taxa making it difficult to determine how much diversity has been excluded from subsequent analyses.

Though complete taxa coverage is long way from our current grasp, much better coverage exists for larger, more conspicuous or dominant coastal marine species, especially harmful/toxic HAB taxa. Several recent studies demonstrate its value as a broad-sweep HAB species detection tool (e.g., Grzebyk et al., 2017; Esenkulova et al., 2020; Manning et al., 2021), by mining meta-barcode data for particular sequences of species/genera of interest. This more targeted approach reduces the task to a scale where it becomes feasible to undertake additional validation of assignment/identity by phylogenetic clustering (e.g., Grzebyk et al., 2017), an important step in ensuring species assignments can be interpreted in the appropriate context, not affected by taxon coverage bias, or not limited resolution of the barcode used.

The same meta-barcode studies also highlight current limitations in terms of deriving relative abundance or quantitative assessment, especially when attempting to move beyond broad assessment of dominant taxa. For example, the majority of taxa in most meta-barcoding datasets are present at very low abundance (far less than 1%). Though this mirrors well-documented species abundance of microbial communities, it also raises questions of how rare taxa (low abundance reads) should be interpreted. The emerging consensus appears to treat the rare (or sometimes all) meta-barcoded taxa as presence-absence data, and most studies include notable "new records" among their study highlights or conclusions. Although this seems like a logical outcome, most barcoding datasets also include puzzling anomalies, such as reads assigned to well-documented tropical species (e.g., *Alexandrium cohorticula, A. tamiyavanichi*) from cool-temperate sampling locations, cool-temperate/polar species (e.g., *A. catenella*) from tropical coral reefs (e.g., Manning et al., 2021), and often at higher relative abundance than samples from areas the species are known and abundant. Though DNA meta-records may be indicative of cryptic presence (of DNA if not living cells), other explanations are equally or more likely, such as (1) artifacts of database coverage and species assignment, especially where several species have identical barcode sequences (Jin et al., 2020); (2) lab/sample cross contamination, or (3) barcode "cross-talk," due to intraspecific/clonal variation among loci (e.g., Scholin and Anderson, 1994) from past/on-going hybridization/introgression between species. Hence, all new distribution records derived from rare meta-barcode taxa should be interpreted with considerable caution until assignment artifacts are excluded, or their presence can be independently verified using other methods.

3.6.3 Filling in the blanks: the taxon coverage problem

To enable automated sequence-based identity, significant throughput improvement is essential for improving both taxon coverage and coverage of additional barcoding regions for taxa, where existing SSU-rRNA barcodes have insufficient resolution. Of critical importance is the generation of sequence data that can be reliably linked to phenotypic and functional data representing a single biological entity. Single-cell PCR has advantage of producing data with a direct link to identity data, but is simply too slow to tackle the problem. However, the recent integration of advanced microfluidics with single-cell genome and transcriptome amplification enables high-throughput single-cell analysis.

First demonstrated from marine bacteria by Stepanauskas and Sieracki (2007), single-cell genomics combines fluorescent activated cells sorting (FACS) with multiple displacement amplification (MDA) to generate plate-libraries of single-amplified genomes (SAGs) composed of multiple copied genome fragments. Each SAG can subsequently be processed in a number of ways, such as NGS, to enable whole genome sequencing and assembly and functional prediction (e.g., Rinke et al., 2013; Stepanauskas et al., 2017), or investigated multiple times for individual targets. The same approach has also been successfully applied to protistan cells (Heywood et al., 2011; Sieracki et al., 2019). Conceptually similar combinations of reverse transcriptase PCR and NGS approaches (RNA-Seq), with FACS, can also generate single-cell transcriptome profiles (scRNA-seq; Tang et al., 2009), as recently demonstrated in microeukaryotes to examine transcriptional responses to viral infection in the haptophyte *Emiliana huxleyi* (Ku et al., 2020). Single-cell 'omics can also be achieved using droplet-based fluidics to detect and sort cells (Caen et al., 2017), with the advantages of increased scalability and capacity for integration in micro-scale lab-on-chip systems (see Girault et al., 2019).

Improved spectral resolution of sorting flow cytometers by incorporation of multiple lasers, offering >30 spectral channels, now offer unprecedented capacity to distinguish and sort different cell types in mixed marine plankton samples. Targeted sorting of individual plankton species is feasible (see Fig. 3.6), either as a bulk sort into one container, or sorting selected cell types into individual wells of multi-well plates.

From a taxonomic perspective, to link molecular data to morpho/phenotype by hi-resolution imaging is also desirable, but the majority of current imaging flow cytometers (e.g., FlowCam, FlowCytobot, Amnis FlowSight) are analytical (non-sorting) instruments. Though cells may be

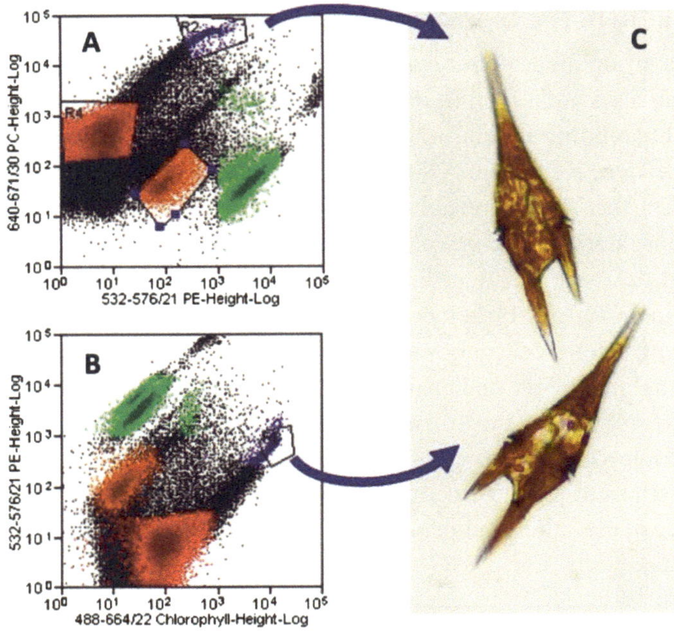

Figure 3.6 Example of targeted flow-cytometric sorting of dinoflagellate cells from mixed phytoplankton communities using a Beckman-Coulter Astrios Mo-Flo FACS equipped with four lasers. A, B: Sample spectral plots showing different gated-regions corresponding to different phytoplankton groups/species, identified in by test-sorting. C: Live *Tripos furca* cells sorted from gated region R2 (Blue). (Images and data: Bolch and Pinfold, unpublished.)

imaged after sorting using high-resolution plate imaging, instruments incorporating high-resolution, in-flow imaging, and sorting capacity are an optimal technical solution for the creation of image-linked single-cell libraries, and enable high-throughput morpho-identity linked single-cell ‘omics (Fig. 3.7).

3.7 Conclusions

Without doubt, molecular biology has transformed dinoflagellate taxonomy and systematics, not only the type and scale of data we collect, but also fundamentally altered taxonomic practice: sequence-led discovery is now the norm rather than the exception. The majority of new species/genus discovery, and major revisions to dinoflagellate taxonomy in the past 25 years have been supported, or initiated, by molecular data. However, the rich morpho- and life-cycle diversity of dinoflagellates, combined with

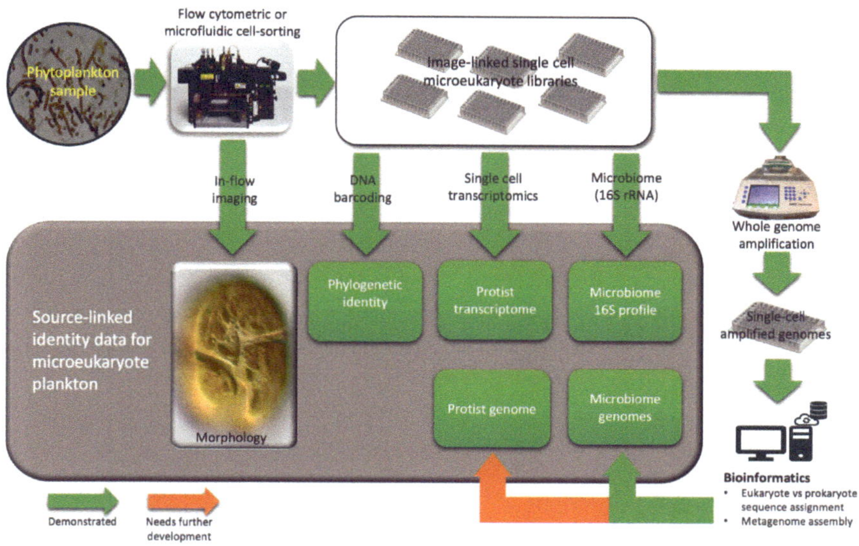

Figure 3.7 Conceptual plan for morpho-identity linked phylogenomics for high-throughput plankton systematics. Green arrows indicate steps where published microeukaryote protocols are well developed; orange arrows indicate steps published for other cell types (e.g., human), but not yet fully demonstrated for single-celled microeukaryotes. (Images: *Karenia* cell by M. DeSalas.)

the evident scale of undocumented diversity, means most practicing taxonomists are understandably reluctant to embrace a taxonomic and identification systems based entirely on sequence data and phylogenetic principles.

Regardless of the form of data, taxonomists are inevitably faced with the same question: how different is different enough for new species? Applying a genospecies concept by defining OTUs based purely on relative sequence divergence continues to be problematic, so, where do we stop? Some authors argue that we are already sub-dividing protist diversity beyond "species" level to identify population-level variation, risking making taxonomy potentially unworkable in an ecological context (Caron and Hu, 2019). Ultimately, taxonomy has to offer a workable means of identifying and classifying organisms. The same authors argue that taxonomy should be grounded by an integrated species concept, incorporating morphological, biological, genotypic, and ecological data. Essentially, this endorses polyphasic approaches that integrate multiple data types, an approach common among dinoflagellate taxonomists, but not easily adapted to either automation or high-throughput. The quest needs to shift toward enabling integrated polyphasic data collection in high-throughput format.

Developments in high-throughput sequencing approaches have created new taxonomic applications, such as meta-barcoding, offering almost infinite promise, and, with future integration of microfluidic and molecular technologies, can offer solutions to both throughput and the value of data collection through enabling identity-linked single-cell 'omics. However, the same technological advances also highlight how much we do not know about dinoflagellate diversity in the world's oceans, and raises questions of how best to catalogue and interpret all-of-community molecular data. This review proposes that increased linkage of multiple data-types to morpho- and functional identity will greatly improve our capacity to interpret the ever-increasing wealth of data.

Acknowledgments

I thank Dr. Miguel De Salas (Univ. of Tasmania) for LM images of *Pheopolykrikos* and *Karenia* species in Figs. 3.4 and 3.7. Professor Gustaaf Hallegraeff (University of Tasmania) and Associate Professor Gilan Attaran-Fariman (Chabahar Maritime University, Iran) provided original SEM images of *G. trapeziforme* and *G. catenatum* cysts, respectively, shown in Fig. 3.4. Dr. Linda Armbrecht (University of Adelaide) provided data and images included in Fig. 3.5, and kindly commented on drafts of sections describing ancient DNA approaches. Dr. Terry Pinfold (Menzies Institute, University of Tasmania) assisted with design and execution of flow cytometry and FACS experiments, and provided the fluorescent flow-plots, shown in Fig. 3.6. Finally, I thank numerous research students who have shared a common passion for the remarkable and ever-surprising dinoflagellates.

References

Amorim, A., Dale, B., Godinho, R., Brotas, V., 2001. *Gymnodinium catenatum*-like cysts (Dinophyceae) in recent sediments from the coast of Portugal. Phycologia 40 (6), 572–582. https://doi.org/10.2216/i0031-8884-40-6-572.1.

Armbrecht, L., Hallegraeff, G., Bolch, C.J.S., Woodward, C., Cooper, A., 2021. Hybridisation capture allows DNA damage analysis of ancient marine eukaryotes. Scientific Reports 11 (1), 3220. https://doi.org/10.1038/s41598-021-82578-6.

Armbrecht, L., Herrando-Pérez, S., Eisenhofer, R., Hallegraeff, G.M., Bolch, C.J.S., Cooper, A., 2020. An optimized method for the extraction of ancient eukaryote DNA from marine sediments. Molecular Ecology Resources 20 (4), 906–919. https://doi.org/10.1111/1755-0998.13162.

Armbrecht, L.H., 2020. The potential of sedimentary ancient DNA to reconstruct past ocean ecosystems. Oceanography 33 (2), 116–123. https://doi.org/10.5670/oceanog.2020.211.

Armbrecht, L.H., Coolen, M.J.L., Lejzerowicz, F., George, S.C., Negandhi, K., Suzuki, Y., Young, J., Foster, N.R., Armand, L.K., Cooper, A., Ostrowski, M., Focardi, A., Stat, M., Moreau, J.W., Weyrich, L.S., 2019. Ancient DNA from marine sediments: precautions and considerations for seafloor coring, sample handling and data generation. Earth-Science Reviews 196. https://doi.org/10.1016/j.earscirev.2019.102887.

Attaran-Fariman, G., De Salas, M.F., Negri, A.P., Bolch, C.J.S., 2007. Morphology and phylogeny of *Gymnodinium trapeziforme* sp. nov. (Dinophyceae): a new dinoflagellate from the southeast coast of Iran that forms microreticulate resting cysts. Phycologia 46 (6), 644–656. https://doi.org/10.2216/07-05.1.

Auinger, B.M., Pfandl, K., Boenigk, J., 2008. Improved methodology for identification of protists and microalgae from plankton samples preserved in Lugol's iodine solution: combining microscopic analysis with single-cell PCR. Applied and Environmental Microbiology 74 (8), 2505–2510. https://doi.org/10.1128/AEM.01803-07.

Bachvaroff, T.R., Gornik, S.G., Concepcion, G.T., Waller, R.F., Mendez, G.S., Lippmeier, J.C., Delwiche, C.F., 2014. Dinoflagellate phylogeny revisited: using ribosomal proteins to resolve deep branching dinoflagellate clades. Molecular Phylogenetics and Evolution 70 (1), 314–322. https://doi.org/10.1016/j.ympev.2013.10.007.

Bachvaroff, T.R., Handy, S.M., Place, A.R., Delwiche, C.F., 2011. Alveolate phylogeny inferred using concatenated ribosomal proteins. Journal of Eukaryotic Microbiology 58 (3), 223–233. https://doi.org/10.1111/j.1550-7408.2011.00555.x.

Bolch, C., Percy, L., 2013. DINO 9 Workshop Report: linking dinoflagellate life-cycle stages using single-cell isolation and molecular typing. In: Lewis, J., et al. (Eds.), Biological and Geological Perspectives of Dinoflagellates, Vol. 5. Geological Society of London.

Bolch, C.J., Hallegraeff, G.M., 1990. Dinoflagellate cysts in recent marine sediments from Tasmania, Australia. Botanica Marina 33, 173–192.

Bolch, C.J.S., 2001. PCR protocols for genetic identification of dinoflagellates directly from single cysts and plankton cells. Phycologia 40 (2), 162–167. https://doi.org/10.2216/i0031-8884-40-2-162.1.

Bolch, C.J.S., Negri, A.P., Hallegraeff, G.M., 1999. *Gymnodinium microreticulatum* sp. nov. (Dinophyceae): a naked, microreticulate cyst-producing dinoflagellate, distinct from *Gymnodinium catenatum* and *Gymnodinium nolleri*. Phycologia 38 (4), 301–313. https://doi.org/10.2216/i0031-8884-38-4-301.1.

Bolch, C.J.S., Reynolds, M.J., 2002. Species resolution and global distribution of microreticulate dinoflagellate cysts. Journal of Plankton Research 24 (6), 565–578. https://doi.org/10.1093/plankt/24.6.565.

Boutrup, P.V., Moestrup, O., Tillmann, U., Daugbjerg, N., 2016. *Katodinium glaucum* (Dinophyceae) revisited: proposal of new genus, family and order based on ultrastructure and phylogeny. Phycologia 55 (2), 147–164. https://doi.org/10.2216/15-138.1.

Bravo, I., Ramilo, I., 1999. Distribution of microreticulate dinoflagellate cysts from the Galician and Portuguese coast. Scientia Marina 63 (1), 45–50. https://doi.org/10.3989/scimar.1999.63n145.

Briggs, A.W., Stenzel, U., Johnson, P.L.F., Green, R.E., Kelso, J., Prüfer, K., Meyer, M., Krause, J., Ronan, M.T., Lachmann, M., Pääbo, S., 2007. Patterns of damage in genomic DNA sequences from a Neandertal. Proceedings of the National Academy of Sciences of the United States of America 104 (37), 14616–14621. https://doi.org/10.1073/pnas.0704665104.

Bujak, J.P., Williams, G.L., 1981. The evolution of dinoflagellates. Canadian Journal of Botany 59, 2077–2087.

Caen, O., Lu, H., Nizard, P., Taly, V., 2017. Microfluidics as a strategic player to decipher single-cell omics? Trends in Biotechnology 35 (8), 713–727. https://doi.org/10.1016/j.tibtech.2017.05.004.

Caron, D.A., Hu, S.K., 2019. Are we overestimating protistan diversity in nature? Trends in Microbiology 27 (3), 197–205. https://doi.org/10.1016/j.tim.2018.10.009.

Cavalier-Smith, T., 1981. Eukaryote kingdoms: seven or nine? Biosystems 14 (3), 461–481. https://doi.org/10.1016/0303-2647(81)90050-2.

Cavalier-Smith, T., Chao, E.E., 2004. Protalveolate phylogeny and systematics and the origins of Sporozoa and dinoflagellates (phylum *Myzozoa* nom. nov.). European Journal of Protistology 40 (3), 185–212. https://doi.org/10.1016/j.ejop.2004.01.002.

Coolen, M.J.L., Orsi, W.D., Balkema, C., Quince, C., Harris, K., Sylva, S.P., Filipova-Marinova, M., Giosan, L., 2013. Evolution of the plankton paleome in the Black Sea from the Deglacial to Anthropocene. Proceedings of the National Academy of Sciences of the United States of America 110 (21), 8609–8614. https://doi.org/10.1073/pnas.1219283110.

Cooney, E.C., Okamoto, N., Cho, A., Hehenberger, E., Richards, T.A., Santoro, A.E., Worden, A.Z., Leander, B.S., Keeling, P.J., 2020. Single-cell transcriptomics of *Abedinium* reveals a new early-branching dinoflagellate lineage. Genome Biology and Evolution 12 (12), 2417–2428. https://doi.org/10.1093/GBE/EVAA196.

Corliss, J.O., 1975. Nuclear characteristics and phylogeny in the protistan phylum Ciliophora. Biosystems 7 (3–4), 338–349. https://doi.org/10.1016/0303-2647(75)90012-x.

Dabney, J., Meyer, M., Pääbo, S., 2013. Ancient DNA damage. Cold Spring Harbor Perspectives in Biology 5 (7). https://doi.org/10.1101/cshperspect.a012567.

Dale, B., 2021. From hystrichospheres to dinoflagellate cysts: scandinavian contributions to Evitt's pivotal recognition of fossil dinoflagellate cysts. Palynology 45 (1), 165–170. https://doi.org/10.1080/01916122.2020.1771623.

Daugbjerg, N., Hansen, G., Larsen, J., Moestrup, Ø., 2000. Phylogeny of some of the major genera of dinoflagellates based on ultrastructure and partial LSU rDNA sequence data, including the erection of three new genera of unarmoured dinoflagellates. Phycologia 39 (4), 302–317. https://doi.org/10.2216/i0031-8884-39-4-302.1.

Daugbjerg, N., Hansen, S.A., Richardson, K., 2019. Cryptic diversity of small-sized species of Phalacroma (Dinophysales, Dinophyceae) from Denmark Strait (Eastern Arctic Greenland). Phycological Research 67 (3), 244–249. https://doi.org/10.1111/pre.12366.

De Salas, M.F., Bolch, C.J.S., Botes, L., Nash, G., Wright, S.W., Hallegraeff, G.M., 2003. *Takayama* gen. nov. (Gymnodiniales, Dinophyceae), a new genus of unarmored dinoflagellates with sigmoid apical grooves, including the description of two new species. Journal of Phycology 39 (6), 1233–1246. https://doi.org/10.1111/j.0022-3646.2003.03-019.x.

De Salas, M.F., Bolch, C.J.S., Hallegraeff, G.M., 2004a. *Karenia asterichroma* sp. nov. (Gymnodiniales, Dinophyceae), a new dinoflagellate species associated with finfish aquaculture mortalities in Tasmania, Australia. Phycologia 43 (5), 624–631. https://doi.org/10.2216/i0031-8884-43-5-624.1.

De Salas, M.F., Bolch, C.J.S., Hallegraeff, G.M., 2004b. *Karenia umbella* sp. nov. (Gymnodiniales, Dinophyceae), a new potentially ichthyotoxic dinoflagellate species from Tasmania, Australia. Phycologia 43 (2), 166–175. https://doi.org/10.2216/i0031-8884-43-2-166.1.

De Salas, M.F., Laza-Martínez, A., Hallegraeff, G.M., 2008. Novel unarmored dinoflagellates from the toxigenic family Kareniaceae (Gymnodiniales): five new species of *Karlodinium* and one new *Takayama* from the Australian sector of the Southern Ocean. Journal of Phycology 44 (1), 241–257. https://doi.org/10.1111/j.1529-8817.2007.00458.x.

Delebecq, G., Schmidt, S., Ehrhold, A., Latimier, M., Siano, R., 2020. Revival of ancient marine dinoflagellates using molecular biostimulation. Journal of Phycology 56 (4), 1077–1089. https://doi.org/10.1111/jpy.13010.

Dodge, J.D., 1965. Chromosome structure in the dinoflagellates and the problem of the mesokaryotic cell. Excerpta Medica International Congress Series 91, 339.

Dodge, J.D., 1983. The functional and phylogenetic significance of dinoflagellate eyespots. Biosystems 16 (3), 259–267. https://doi.org/10.1016/0303-2647(83)90009-6.

Dorrell, R.G., Howe, C.J., 2015. Integration of plastids with their hosts: lessons learned from dinoflagellates. Proceedings of the National Academy of Sciences of the United States of America 112 (33), 10247–10254. https://doi.org/10.1073/pnas.1421380112.

Ehrenberg, C.G., 1838. Die Infusionsthierchen als volkommene Organismen. Ein Blick in das Tiefere organische Leben der Natur. Leopold Voss, Leipzig. xvii, 547 pp.

Ellegaard, M., Moestrup, Ø., 1999. Fine structure of the flagellar apparatus and morphological details of *Gymnodinium nolleri* sp. nov. (Dinophyceae), an unarmored dinoflagellate producing a microreticulate cyst. Phycologia 38 (4), 289–300. https://doi.org/10.2216/i0031-8884-38-4-289.1.
Ellegaard, M., Oshima, Y., 1998. *Gymnodinium nolleri* Ellegaard *et* Moestrup sp. ined. (Dinophyceae) from Danish waters, a new species producing *Gymnodinium catenatum*-like cysts: molecular and toxicological comparisons with Australian and Spanish strains of *Gymnodinium catenatum*. Phycologia 37 (5), 369–378. https://doi.org/10.2216/i0031-8884-37-5-369.1.
Esenkulova, S., Sutherland, B.J.G., Tabata, A., Haigh, N., Pearce, C.M., Miller, K.M., Smol, J.P., 2020. Comparing metabarcoding and morphological approaches to identify phytoplankton taxa associated with harmful algal blooms. Facets 5 (1), 784–811. https://doi.org/10.1139/facets-2020-0025.
Evitt, W.R., 1961. Observations on the morphology of fossil dinoflagellates. Micropaleontology 7 (4), 385–420.
Evitt, W.R., Davidson, S.E., 1964. Dinoflagellate studies I. Dinoflagellate cysts and thecae. Stanford University Publications, Geological Sciences 10 (1), 1–12.
Fast, N.M., Xue, L., Bingham, S., Keeling, P.J., 2002. Re-examining alveolate evolution using multiple protein molecular phylogenies. Journal of Eukaryotic Microbiology 49 (1), 30–37. https://doi.org/10.1111/j.1550-7408.2002.tb00336.x.
Fensome, R.A., MacRae, R.A., Moldowan, J.M., Taylor, F.J.R., Williams, G.L., 1996. The early Mesozoic radiation of dinoflagellates. Paleobiology 22 (3), 329–338. https://doi.org/10.1017/S0094837300016316.
Fensome, R.A., Saldarriaga, J.F., Taylor, M.F.J.R., 1999. Dinoflagellate phylogeny revisited: reconciling morphological and molecular based phylogenies. Grana 38 (2–3), 66–80. https://doi.org/10.1080/00173139908559216.
Fensome, R.A., Taylor, F.J.R., Norris, G., Sarjeant, W.A.S., Wharton, D.L., Williams, G.L., 1993. A Classification of Living and Fossil Dinoflagellates. Micropaleontology, Special Publication. ISSN 0160-2071, vol. 7. Hanover Pennsylvania, USA. 351 pp.
Fraga, S., Bravo, I., Delgado, M., Franco, J.M., Zapata, M., 1995. *Gyrodinium impudicum* sp. nov. (Dinophyceae), a non toxic, chain-forming, red tide dinoflagellate. Phycologia 34 (6), 514–521. https://doi.org/10.2216/i0031-8884-34-6-514.1.
Girault, M., Beneyton, T., del Amo, Y., Baret, J.-C., 2019. Microfluidic technology for plankton research. Current Opinion in Biotechnology 55, 134–150. https://doi.org/10.1016/j.copbio.2018.09.010.
Godhe, A., Anderson, D.M., Rehnstam-Holm, A.-S., 2002. PCR amplification of microalgal DNA for sequencing and species identification: studies on fixatives and algal growth stages. Harmful Algae 1 (4), 375–382. https://doi.org/10.1016/S1568-9883(02)00049-5.
Gómez, F., Artigas, L.F., 2019. Redefinition of the dinoflagellate genus *Alexandrium* based on *Centrodinium*: reinstatement of *Gessnerium* and *Protogonyaulax*, and *Episemicolon* gen. nov. (Gonyaulacales, Dinophyceae). Journal of Marine Sciences 1284104. https://doi.org/10.1155/2019/1284104.
Graham, H.W., 1943. *Gymnodinium catenatum*, a new dinoflagellate from the Gulf of California. Transactions of the American Microscopical Society 62 (3), 259–261.
Grzebyk, D., Audic, S., Lasserre, B., Abadie, E., de Vargas, C., Bec, B., 2017. Insights into the harmful algal flora in northwestern Mediterranean coastal lagoons revealed by pyrosequencing metabarcodes of the 28S rRNA gene. Harmful Algae 68, 1–16. https://doi.org/10.1016/j.hal.2017.06.003.
Grzebyk, D., Sako, Y., Berland, B., 1998. Phylogenetic analysis of nine species of *Prorocentrum* (Dinophyceae) inferred from 18S ribosomal DNA sequences, morphological comparisons, and description of *Prorocentrum panamensis*, sp. nov. Journal of Phycology 34 (6), 1055–1068. https://doi.org/10.1046/j.1529-8817.1998.341055.x.

Gu, H., Liu, T., Vale, P., Luo, Z., 2013. Morphology, phylogeny and toxin profiles of *Gymnodinium inusitatum* sp. nov., *Gymnodinium catenatum* and *Gymnodinium microreticulatum* (Dinophyceae) from the Yellow Sea, China. Harmful Algae 28, 97–107. https://doi.org/10.1016/j.hal.2013.06.001.

Guillou, L., Bachar, D., Audic, S., Bass, D., Berney, C., Bittner, L., Boutte, C., Burgaud, G., De Vargas, C., Decelle, J., Del Campo, J., Dolan, J.R., Dunthorn, M., Edvardsen, B., Holzmann, M., Kooistra, W.H.C.F., Lara, E., Le Bescot, N., Logares, R., Mahé, F., Massana, R., Montresor, M., Morard, R., Not, F., Pawlowski, J., Probert, I., Sauvadet, A.L., Siano, R., Stoeck, T., Vaulot, D., Zimmermann, P., Christen, R., 2013. The Protist Ribosomal Reference Database (PR2): a catalog of unicellular eukaryote Small Sub-Unit rRNA sequences with curated taxonomy. Nucleic Acids Research 41 (D1), D597–D604. https://doi.org/10.1093/nar/gks1160.

Hallegraeff, G.M., Blackburn, S.I., Doblin, M.A., Bolch, C.J.S., 2012. Global toxicology, ecophysiology and population relationships of the chainforming PST dinoflagellate *Gymnodinium catenatum*. Harmful Algae 14, 130–143. https://doi.org/10.1016/j.hal.2011.10.018.

Hart, M.C., Green, D.H., Bresnan, E., Bolch, C.J., 2007. Large subunit ribosomal RNA gene variation and sequence heterogeneity of *Dinophysis* (Dinophyceae) species from Scottish coastal waters. Harmful Algae 6 (2), 271–287. https://doi.org/10.1016/j.hal.2006.10.001.

Haywood, A.J., Steidinger, K.A., Truby, E.W., Bergquist, P.R., Bergquist, P.L., Adamson, J., Mackenzie, L., 2004. Comparative morphology and molecular phylogenetic analysis of three new species of the genus *Karenia* (Dinophyceae) from New Zealand. Journal of Phycology 40 (1), 165–179. https://doi.org/10.1111/j.0022-3646.2004.02-149.x.

Hebert, P.D.N., Cywinska, A., Ball, S.L., deWaard, J.R., 2003. Biological identifications through DNA barcodes. Proceedings of the Royal Society of London. Series B 270, 313–321. https://doi.org/10.1098/rspb.2002.2218.

Herzog, M., Maroteaux, L., 1986. Dinoflagellate 17S rRNA sequence inferred from the gene sequence: evolutionary implications. Proceedings of the National Academy of Sciences 83 (22), 8644. https://doi.org/10.1073/pnas.83.22.8644.

Heywood, J.L., Sieracki, M.E., Bellows, W., Poulton, N.J., Stepanauskas, R., 2011. Capturing diversity of marine heterotrophic protists: one cell at a time. The ISME Journal 5 (4), 674–684. https://doi.org/10.1038/ismej.2010.155.

Hinnebusch, A.G., Klotz, L.C., Blanken, R.L., Loeblich III, A.R., 1981. An evaluation of the phylogenetic position of the dinoflagellate *Crypthecodinium cohnii* based on 5S rRNA characterization. Journal of Molecular Evolution 17 (6), 334–347. https://doi.org/10.1007/BF01734355.

Hoppenrath, M., Leander, B.S., 2010. Dinoflagellate phylogeny as inferred from heat shock protein 90 and ribosomal gene sequences. PLoS ONE 5 (10). https://doi.org/10.1371/journal.pone.0013220.

Hou, Y., Lin, S., 2009. Distinct gene number-genome size relationships for eukaryotes and non-eukaryotes: gene content estimation for dinoflagellate genomes. PLoS ONE 4 (9), e6978. https://doi.org/10.1371/journal.pone.0006978.

Hugenholtz, P., Chuvochina, M., Oren, A., Parks, D.H., Soo, R.M., 2021. Prokaryotic taxonomy and nomenclature in the age of big sequence data. The ISME Journal 15, 1879–1892. https://doi.org/10.1038/s41396-021-00941-x.

Jacobson, D.M., 1999. A brief history of dinoflagellate feeding research. Journal of Eukaryotic Microbiology 46 (4), 376–381. https://doi.org/10.1111/j.1550-7408.1999.tb04616.x.

Janouskovec, J., Gavelis, G.S., Burki, F., Dinh, D., Bachvaroff, T.R., Gornik, S.G., Bright, K.J., Imanian, B., Strom, S.L., Delwiche, C.F., Waller, R.F., Fensome, R.A., Leander, B.S., Rohwer, F.L., Saldarriaga, J.F., 2017. Major transitions in dinoflagellate evolution unveiled by phylotranscriptomics. Proceedings of the National Academy of Sciences of the United States of America 114 (2), E171–E180. https://doi.org/10.1073/pnas.1614842114.

Jin, S., Kim, K.Y., Kim, M.S., Park, C., 2020. An assessment of the taxonomic reliability of dna barcode sequences in publicly available databases. Algae 35 (3), 293–301. https://doi.org/10.4490/algae.2020.35.9.4.

John, U., Litaker, R.W., Montresor, M., Murray, S., Brosnahan, M.L., Anderson, D.M., 2014. Formal revision of the *Alexandrium tamarense* species complex (Dinophyceae) taxonomy: the introduction of five species with emphasis on molecular-based (rDNA) classification. Protist 165 (6), 779–804. https://doi.org/10.1016/j.protis.2014.10.001.

Karafas, S., Teng, S.T., Leaw, C.P., Alves-de-Souza, C., 2017. An evaluation of the genus *Amphidinium* (Dinophyceae) combining evidence from morphology, phylogenetics, and toxin production, with the introduction of six novel species. Harmful Algae 68, 128–151. https://doi.org/10.1016/j.hal.2017.08.001.

Klouch, K.Z., Schmidt, S., Andrieux-Loyer, F., Le Gac, M., Hervio-Heath, D., Qui-Minet, Z.N., Quéré, J., Bigeard, E., Guillou, L., Siano, R., 2016. Historical records from dated sediment cores reveal the multidecadal dynamic of the toxic dinoflagellate *Alexandrium minutum* in the Bay of Brest (France). FEMS Microbiology, Ecology 92 (7), 1–16. https://doi.org/10.1093/femsec/fiw101.

Kretzschmar, A.L., Larsson, M.E., Hoppenrath, M., Doblin, M.A., Murray, S.A., 2019. Characterisation of Two Toxic *Gambierdiscus* spp. (Gonyaulacales, Dinophyceae) from the Great Barrier Reef (Australia): *G. lewisii* sp. nov. and *G. holmesii* sp. nov. Protist 170 (6), 125699. https://doi.org/10.1016/j.protis.2019.125699.

Kretzschmar, A.L., Verma, A., Harwood, T., Hoppenrath, M., Murray, S., 2017. Characterization of *Gambierdiscus lapillus* sp. nov. (Gonyaulacales, Dinophyceae): a new toxic dinoflagellate from the Great Barrier Reef (Australia). Journal of Phycology 53 (2), 283–297. https://doi.org/10.1111/jpy.12496.

Ku, C., Sheyn, U., Sebé-Pedrós, A., Ben-Dor, S., Schatz, D., Tanay, A., Rosenwasser, S., Vardi, A., 2020. A single-cell view on alga-virus interactions reveals sequential transcriptional programs and infection states. Science Advances 6 (21), eaba4137. https://doi.org/10.1126/sciadv.aba4137.

LaJeunesse, T.C., Thornhill, D.J., 2011. Improved resolution of reef-coral endosymbiont (*Symbiodinium*) species diversity, ecology, and evolution through *psbA* non-coding region genotyping. PLoS ONE 6 (12). https://doi.org/10.1371/journal.pone.0029013.

Le Bescot, N., Mahé, F., Audic, S., Dimier, C., Garet, M.J., Poulain, J., Wincker, P., de Vargas, C., Siano, R., 2016. Global patterns of pelagic dinoflagellate diversity across protist size classes unveiled by metabarcoding. Environmental Microbiology 18 (2), 609–626. https://doi.org/10.1111/1462-2920.13039.

Lenaers, G., Maroteaux, L., Michot, B., Herzog, M., 1989. Dinoflagellates in evolution. A molecular phylogenetic analysis of large subunit ribosomal RNA. Journal of Molecular Evolution 29 (1), 40–51. https://doi.org/10.1007/BF02106180.

Lenaers, G., Scholin, C., Bhaud, Y., Saint-Hilaire, D., Herzog, M., 1991. A molecular phylogeny of dinoflagellate protists (Pyrrhophyta) inferred from the sequence of 24S rRNA divergent domains D1 and D8. Journal of Molecular Evolution 32 (1), 53–63. https://doi.org/10.1007/BF02099929.

Lewis, J., Harris, A.S.D., Jones, K.J., Edmonds, R.L., 1999. Long-term survival of marine planktonic diatoms and dinofiagellates in stored sediment samples. Journal of Plankton Research 21 (2), 343–354.

Lin, S., Zhang, H., Hou, Y., Miranda, L., Bhattacharya, D., 2006. Development of a dinoflagellate-oriented PCR primer set leads to detection of picoplanktonic dinoflagellates from Long Island Sound. Applied and Environmental Microbiology 72 (8), 5626–5630. https://doi.org/10.1128/AEM.00586-06.

Lin, S., Zhang, H., Hou, Y., Zhuang, Y., Miranda, L., 2009. High-level diversity of dinoflagellates in the natural environment, revealed by assessment of mitochondrial cox1 and cob genes for dinoflagellate DNA barcoding. Applied and Environmental Microbiology 75 (5), 1279–1290. https://doi.org/10.1128/AEM.01578-08.

Lindberg, K., Moestrup, Ø., Daugbjerg, N., 2005. Studies on woloszynskioid dinoflagellates I: *Woloszynskia coronata* re-examined using light and electron microscopy and partial LSU rDNA sequences, with description of *Tovellia* gen. nov. and *Jadwigia* gen. nov. (Tovelliaceae fam. nov.). Phycologia 44 (4), 416–440. https://doi.org/10.2216/0031-8884(2005)44[416:SOWDIW]2.0.CO;2.

Litaker, R.W., Vandersea, M.W., Faust, M.A., Kibler, S.R., Chinain, M., Holmes, M.J., Holland, W.C., Tester, P.A., 2009. Taxonomy of *Gambierdiscus* including four new species, *Gambierdiscus caribaeus*, *Gambierdiscus carolinianus*, *Gambierdiscus carpenteri* and *Gambierdiscus ruetzleri* (Gonyaulacales, Dinophyceae). Phycologia 48 (5), 344–390. https://doi.org/10.2216/07-15.1.

Lundholm, N., Ribeiro, S., Andersen, T.J., Koch, T., Godhe, A., Ekelund, F., Ellegaard, M., 2011. Buried alive - germination of up to a century-old marine protist resting stages. Phycologia 50 (6), 629–640. https://doi.org/10.2216/11-16.1.

Manning, T., Thilagaraj, A.V., Mouradov, D., Piola, R., Grandison, C., Gordon, M., Shimeta, J., Mouradov, A., 2021. Diversity of dinoflagellate assemblages in coastal temperate and offshore tropical waters of Australia. BMC Ecology and Evolution 21 (1). https://doi.org/10.1186/s12862-021-01745-5.

Mantell, G.A., 1850. A Pictorial Atlas of Fossil Remains Consisting of Coloured Illustrations Selected from Parkinson's "Organic Remains of a Former World", and Artis's "Antediluvian Phytology". Henry G. Bohn, London, UK. 207 pp.

Marín, I., Aguilera, A., Reguera, B., Abad, J.P., 2001. Preparation of DNA suitable for PCR amplification from fresh or fixed single dinoflagellate cells. BioTechniques 30 (1), 88–93. https://doi.org/10.2144/01301st05.

Maroteaux, L., Herzog, M., Soyer-Gobillard, M.O., 1985. Molecular organization of dinoflagellate ribosomal DNA: evolutionary implications of the deduced 5.8S rRNA secondary structure. Biosystems 18 (3–4), 307–319. https://doi.org/10.1016/0303-2647(85)90031-0.

Mayr, E., 1981. Biological classification: toward a synthesis of opposing methodologies. Science 214 (4520), 510–516. https://doi.org/10.1126/science.214.4520.510.

Mertens, K.N., Adachi, M., Anderson, D.M., Band-Schmidt, C.J., Bravo, I., Brosnahan, M.L., Bolch, C.J.S., Calado, A.J., Carbonell-Moore, M.C., Chomérat, N., Elbrächter, M., Figueroa, R.I., Fraga, S., Gárate-Lizárraga, I., Garcés, E., Gu, H., Hallegraeff, G., Hess, P., Hoppenrath, M., Horiguchi, T., Iwataki, M., John, U., Kremp, A., Larsen, J., Leaw, C.P., Li, Z., Lim, P.T., Litaker, W., MacKenzie, L., Masseret, E., Matsuoka, K., Moestrup, Ø., Montresor, M., Nagai, S., Nézan, E., Nishimura, T., Okolodkov, Y.B., Orlova, T.Y., Reñé, A., Sampedro, N., Satta, C.T., Shin, H.H., Siano, R., Smith, K.F., Steidinger, K., Takano, Y., Tillmann, U., Wolny, J., Yamaguchi, A., Murray, S., 2020. Morphological and phylogenetic data do not support the split of *Alexandrium* into four genera. Harmful Algae 98, 101902. https://doi.org/10.1016/j.hal.2020.101902.

Moestrup, O., Hakanen, P., Hansen, G., Daugbjerg, N., Ellegaard, M., 2014. On *Levanderina fissa* gen. & comb. nov. (Dinophyceae) (syn. *Gymnodinium fissum*, *Gyrodinium instriatum*, *Gyr. uncatenum*), a dinoflagellate with a very unusual sulcus. Phycologia 53 (3), 265–292. https://doi.org/10.2216/13-254.1.

Moestrup, Ø., Hansen, G., Daugbjerg, N., 2008. Studies on woloszynskioid dinoflagellates III: on the ultrastructure and phylogeny of *Borghiella dodgei* gen. et sp. nov., a cold-water species from Lake Tovel, N. Italy, and on *B. tenuissima* comb. nov. (syn. *Woloszynskia tenuissima*). Phycologia 47 (1), 54–78. https://doi.org/10.2216/07-32.1.

Moestrup, Ø., Hansen, G., Daugbjerg, N., Flaim, G., D'Andrea, M., 2006. Studies on woloszynskioid dinoflagellates II: on *Tovellia sanguinea* sp. nov., the dinoflagellate responsible for the reddening of Lake Tovel, N. Italy. European Journal of Phycology 41 (1), 47–65. https://doi.org/10.1080/09670260600556682.

Moestrup, Ø., Lindberg, K., Daugbjerg, N., 2009a. Studies on woloszynskioid dinoflagellates IV: the genus *Biecheleria* gen. nov. Phycological Research 57 (4), 221–237. https://doi.org/10.1111/j.1440-1835.2009.00564.x.

Moestrup, Ø., Lindberg, K., Daugbjerg, N., 2009b. Studies on woloszynskioid dinoflagellates V. Ultrastructure of *Biecheleriopsis* gen. nov., with description of *Biecheleriopsis adriatica* sp. nov. Phycological Research 57 (4), 221–237. https://doi.org/10.1111/j.1440-1835.2009.00564.x.

Moon-Van Der Staay, S.Y., De Wachter, R., Vaulot, D., 2001. Oceanic 18S rDNA sequences from picoplankton reveal unsuspected eukaryotic diversity. Nature 409 (6820), 607–610. https://doi.org/10.1038/35054541.

Mordret, S., Piredda, R., Vaulot, D., Montresor, M., Kooistra, W.H.C.F., Sarno, D., 2018. DINOREF: a curated dinoflagellate (Dinophyceae) reference database for the 18S rRNA gene. Molecular Ecology Resources 18 (5), 974–987. https://doi.org/10.1111/1755-0998.12781.

Murray, S., Jørgensen, M.F., Daugbjerg, N., Rhodes, L., 2004. *Amphidinium* revisited. II. Resolving species boundaries in the *Amphidinium operculatum* species complex (Dinophyceae), including the descriptions of *Amphidinium trulla* sp nov. and *Amphidinium gibbosum* comb. nov. Journal of Phycology 40 (2), 366–382. https://doi.org/10.1046/j.1529-8817.2004.03132.x.

Murray, S.A., Garby, T., Hoppenrath, M., Neilan, B.A., 2012. Genetic diversity, morphological uniformity and polyketide production in dinoflagellates (*Amphidinium*, Dinoflagellata). PLoS ONE 7 (6), e38253. https://doi.org/10.1371/journal.pone.0038253.

Orr, R.J., Murray, S.A., Stüken, A., Rhodes, L., Jakobsen, K.S., 2012. When naked became armored: an eight-gene phylogeny reveals monophyletic origin of theca in dinoflagellates. PLoS ONE 7 (11), e50004. https://doi.org/10.1371/journal.pone.0050004.

Pascher, A., 1914. Über Flagellaten und Algen. Deutsche Botanische Gesellschaft, Berichte 32, 136–160.

Price, D.C., Bhattacharya, D., 2017. Robust Dinoflagellata phylogeny inferred from public transcriptome databases. Journal of Phycology 53 (3), 725–729. https://doi.org/10.1111/jpy.12529.

Rinke, C., Schwientek, P., Sczyrba, A., Ivanova, N.N., Anderson, I.J., Cheng, J.-F., Darling, A., Malfatti, S., Swan, B.K., Gies, E.A., Dodsworth, J.A., Hedlund, B.P., Tsiamis, G., Sievert, S.M., Liu, W.-T., Eisen, J.A., Hallam, S.J., Kyrpides, N.C., Stepanauskas, R., Rubin, E.M., Hugenholtz, P., Woyke, T., 2013. Insights into the phylogeny and coding potential of microbial dark matter. Nature 499 (7459), 431–437. https://doi.org/10.1038/nature12352.

Saldarriaga, J.F., McEwan, M.L., Fast, N.M., Taylor, F.J.R., Keeling, P.J., 2003. Multiple protein phylogenies show that *Oxyrrhis marina* and *Perkinsus marinus* are early branches of the dinoflagellate lineage. International Journal of Systematic and Evolutionary Microbiology 53 (1), 355–365. https://doi.org/10.1099/ijs.0.02328-0.

Saldarriaga, J.F., Taylor, F.J.R., Keeling, P.J., Cavalier-Smith, T., 2001. Dinoflagellate nuclear SSU rRNA phylogeny suggests multiple plastid losses and replacements. Journal of Molecular Evolution 53 (3), 204–213. https://doi.org/10.1007/s002390010210.

Saunders, G.W., Hill, D.R.A., Sexton, J.P., Andersen, R.A., 1997. Small-subunit ribosomal RNA sequences from selected dinoflagellates: testing classical evolutionary hypotheses with molecular systematic methods. In: Bhattacharya, D. (Ed.), Origins of Algae and Their Plastids. Springer Vienna, Vienna, pp. 237–259.

Scholin, C.A., Anderson, D.M., 1994. Identification of group- and strain-specific genetic markers for globally distributed *Alexandrium* (Dinophyceae): I. RFLP analysis of SSU-rRNA genes. Journal of Phycology 30 (4), 744–754. https://doi.org/10.1111/j.0022-3646.1994.00744.x.

Sieracki, M.E., Poulton, N.J., Jaillon, O., Wincker, P., de Vargas, C., Rubinat-Ripoll, L., Stepanauskas, R., Logares, R., Massana, R., 2019. Single cell genomics yields a wide diversity of small planktonic protists across major ocean ecosystems. Scientific Reports 9 (1), 6025. https://doi.org/10.1038/s41598-019-42487-1.

Smith, K.F., Biessy, L., Argyle, P.A., Trnski, T., Halafihi, T., Rhodes, L.L., 2017a. Molecular identification of *Gambierdiscus* and *Fukuyoa* (Dinophyceae) from environmental samples. Mar Drugs 15 (8). https://doi.org/10.3390/md15080243.

Smith, K.F., Kohli, G.S., Murray, S.A., Rhodes, L.L., 2017b. Assessment of the metabarcoding approach for community analysis of benthic-epiphytic dinoflagellates using mock communities. New Zealand Journal of Marine and Freshwater Research 51 (4), 555–576. https://doi.org/10.1080/00288330.2017.1298632.

Sogin, M.L., 1989. Evolution of eukaryotic microorganisms and their small subunit ribosomal RNAs. Integrative and Comparative Biology 29 (2), 487–499. https://doi.org/10.1093/icb/29.2.487.

Stepanauskas, R., Fergusson, E.A., Brown, J., Poulton, N.J., Tupper, B., Labonté, J.M., Becraft, E.D., Brown, J.M., Pachiadaki, M.G., Povilaitis, T., Thompson, B.P., Mascena, C.J., Bellows, W.K., Lubys, A., 2017. Improved genome recovery and integrated cell-size analyses of individual uncultured microbial cells and viral particles. Nature Communications 8 (1), 84. https://doi.org/10.1038/s41467-017-00128-z.

Stepanauskas, R., Sieracki, M.E., 2007. Matching phylogeny and metabolism in the uncultured marine bacteria, one cell at a time. Proceedings of the National Academy of Sciences 104 (21), 9052–9057. https://doi.org/10.1073/pnas.0700496104.

Stern, R.F., Andersen, R.A., Jameson, I., Küpper, F.C., Coffroth, M.A., Vaulot, D., Le Gall, F., Véron, B., Brand, J.J., Skelton, H., Kasai, F., Lilly, E.L., Keeling, P.J., 2012. Evaluating the ribosomal internal transcribed spacer (its) as a candidate dinoflagellate barcode marker. PLoS ONE 7 (8). https://doi.org/10.1371/journal.pone.0042780.

Stern, R.F., Horak, A., Andrew, R.L., Coffroth, M.A., Andersen, R.A., Küpper, F.C., Jameson, I., Hoppenrath, M., Véron, B., Kasai, F., Brand, J., James, E.R., Keeling, P.J., 2010. Environmental barcoding reveals massive dinoflagellate diversity in marine environments. PLoS ONE 5 (11). https://doi.org/10.1371/journal.pone.0013991.

Sze, Y., Miranda, L.N., Sin, T.M., Huang, D., 2018. Characterising planktonic dinoflagellate diversity in Singapore using DNA metabarcoding. Metabarcoding and Metagenomics 2. https://doi.org/10.3897/mbmg.2.25136.

Tang, F., Barbacioru, C., Wang, Y., Nordman, E., Lee, C., Xu, N., Wang, X., Bodeau, J., Tuch, B.B., Siddiqui, A., Lao, K., Surani, M.A., 2009. mRNA-Seq whole-transcriptome analysis of a single cell. Nature Methods 6 (5), 377–382. https://doi.org/10.1038/nmeth.1315.

Taylor, F.J.R., 1976. Flagellate phylogeny: a study in conflicts. Journal of Protozoology 23, 28–40.

Taylor, F.J.R., Hoppenrath, M., Saldarriaga, J.F., 2008. Dinoflagellate diversity and distribution. Biodiversity and Conservation 17 (2), 407–418. https://doi.org/10.1007/s10531-007-9258-3.

Thorsen, T.A., Dale, B., Nordberg, K., 1995. 'Blooms' of the toxic dinoflagellate *Gymnodinium catenatum* as evidence of climatic fluctuations in the late Holocene of southwestern Scandinavia. Holocene 5 (4), 435–446. https://doi.org/10.1177/095968369500500406.

Wetzel, O., 1933. Die in organischer Substanz erhaltenen Mikrofossilien des baltischen Kreide-Feuersteins mit einem sediment-petrographischen und stratigraphischen Anhang. Palaeontographica. Abteilung A 78. 110 pp.

Wheeler, Q., 2014. Are reports of the death of taxonomy an exaggeration? New Phytologist 201 (2), 370–371. https://doi.org/10.1111/nph.12612.
Wolters, J., 1991. The troublesome parasites - molecular and morphological evidence that Apicomplexa belong to the dinoflagellate-ciliate clade. Biosystems 25 (1–2), 75–83. https://doi.org/10.1016/0303-2647(91)90014.
Zardoya, R., Costas, E., López-Rodas, V., Garrido-Pertierra, A., Bautista, J.M., 1995. Revised dinoflagellate phylogeny inferred from molecular analysis of large-subunit ribosomal RNA gene sequences. Journal of Molecular Evolution 41 (5), 637–645. https://doi.org/10.1007/BF00175822.
Zhang, H., Bhattacharya, D., Lin, S., 2007. A three-gene dinoflagellate phylogeny suggests monophyly of Prorocentrales and a basal position for *Amphidinium* and *Heterocapsa*. Journal of Molecular Evolution 65 (4), 463–474. https://doi.org/10.1007/s00239-007-9038-4.

CHAPTER 4

From molecules to ecosystem functioning: insight into new approaches to taxonomy to monitor harmful algae diversity in Chile

Jorge I. Mardones[a,b,c], **Bernd Krock**[d], **Lara Marcus**[c,e], **Catharina Alves-de-Souza**[f], **Satoshi Nagai**[g], **Kyoko Yarimizu**[h], **Alejandro Clément**[i], **Nicole Correa**[c,i], **Sebastian Silva**[j], **Javier Paredes-Mella**[a,c], **and Peter Von Dassow**[j,k]

[a]Center for Harmful Algae Studies (CREAN), Fisheries Development Institute (IFOP), Puerto Montt, Los Lagos, Chile
[b]FONDAP Research Center Dynamics of High Latitude Marine Ecosystem (IDEAL), Valdivia, Chile
[c]CAICAI Foundation, Puerto Varas, Los Lagos, Chile
[d]Alfred Wegener Institut-Helmholtz Zentrum für Polar- und Meeresforschung, Bremerhaven, Germany
[e]Department of Zoology, Faculty of Natural and Oceanographic Sciences, University of Concepción, Concepción, Chile
[f]Algal Resources Collection, MARBIONC, Center for Marine Science, University of North Carolina Wilmington, Wilmington, NC, United States
[g]Japan Fisheries Research and Education Agency, Fisheries Stock Assessment Center, Bioinformatics and Biosciences Division, Genome Structure Analysis Group, Yokohama, Kanagawa, Japan
[h]Office of Industry-Academia-Government and Community Collaboration, Hiroshima University, Higashi-Hiroshima, Hiroshima, Japan
[i]Plancton Andino Spa., Puerto Varas Laboratory, Puerto Varas, Chile
[j]Department of Ecology, The Pontifical Catholic University of Chile, Santiago, Chile
[k]Millenium Institute of Oceanography, Universidad de Concepción, Concepción, Chile

4.1 Introduction

The Patagonian fjords in the south of Chile (41.5° S to 56° S) cover a large geographic area (240,000 km^2), and are characterized by an intricate geography and poor accessibility (Pantoja et al., 2011). These systems offer optimal areas for fishery and aquaculture activities of both shellfish and finfish (Niklitschek et al., 2013). For instance, ca. 953,296 tons of salmon and 332,725 tons of blue mussel were harvested in 2019, with annual revenues of exported products of US$ 4.400 millions and ~US$ 188m, respectively, placing Chile among the world leaders of aquaculture production (Mardones et al., 2020a). When harmful algal blooms (HABs) occur in these systems, their toxic effects can be observed at several trophic levels, leading to illness or death of aquatic organisms and/or human seafood consumers. The first HAB monitoring program led by the Chilean "Instituto de Fomento Pesquero" (IFOP) was implemented in 2006 to secure the sanitary quality of exported shellfish. Since then, public and private monitoring programs have been established to secure human health and the sustainability of the growing aquaculture activities. There is a significant effort to monitor the vast Chilean coast, which includes ten programs with an annual cost of ca. US$ 7m (Mardones et al., 2020a).

Phycotoxins have been associated with several illness syndromes in Chile, including paralytic shellfish poisoning (PSP), diarrhetic shellfish poisoning (DSP), and amnesic shellfish poisoning (ASP) (Trefault et al., 2011). Despite the detection of azaspiracids toxins, along with their producing dinoflagellate species *Azadinium* sp. and *Amphidoma* sp. (Álvarez et al., 2010), the human illness associated with azaspiracid shellfish poisoning (AZP) has not yet been reported. Similarly, there is no record of neurotoxic shellfish poisoning (NSP), although *Karenia* species have been detected in the area (Mardones et al., 2020b). Severe HAB events in Chile frequently extend over large geographic areas (hundreds of kms) and can last for days and even months, resulting in the closure of shellfish harvesting. In 2016, a massive *Alexandrium catenella* bloom resulted in a loss of employment for 1700 people, due to the inactivity of the salmon processing plants and the ban of harvesting of mussels, with estimated economic losses of US$ 2m as a result of the decline in exports (Trainer et al., 2020). Two years later, during another *A. catenella* bloom, shellfish PSP toxicity reached the extraordinary concentration of 143,130 μg STX eq. 100 g–1, a value currently considered the world record for these toxins (Chilean Ministry of Health). Regarding the effects on human health, intoxications as a result of contaminated shellfish consumption (mainly PSP and DSP) have dramatically increased in recent years, with economic costs of hospitalizations reaching values of US$ 93,119 in 2019 (Mardones et al., 2020a). These high medical expenses showed a direct relationship with highly toxic PSP events. Since HABs are increasing in severity, frequency, and biogeographical range, these events are a cause of great concern in Chile (Mardones et al., 2020a).

HAB species can also produce ichthyotoxins with a great impact on the salmon industry. Bloom events associated with finfish mortality have been recorded in Chilean fjords. For example, the phytoflagellate *Pseudochatonella verruculosa* was responsible for the largest farmed fish mortality ever recorded globally, with losses estimated at 100,000 metric tons, and export economic loss of US$ 800m in 2016 (Mardones et al., 2020c). Results from cytological bioassays using *Oncorhynchus mykiss* RTgill-W1 cell line have pointed out the potential ichthyotoxic effect of several microalgal species isolated from Chilean waters (Mardones, 2020). Despite the causative metabolites being mostly unknown, the effect of lipid peroxidation of polyunsaturated fatty acids (PUFAs), which sometimes can act synergistically with reactive oxygen species (ROS), may explain some mortalities (Mardones et al., 2015, 2018, 2020b).

The dinoflagellate genera *Alexandrium, Dinophysis, Protoceratium*, and the diatom genus *Pseudo-nitzschia* are historically known as causing HABs in Chilean Patagonian fjords. Increasing monitoring efforts in recent years have, however, revealed a higher diversity of HAB species in this area. The comparison between the IFOP phytoplankton monitoring database with the "taxonomic reference list of harmful microalgae" (Moestrup et al., 2009) identified approximately 26 species in Chilean fjords that can produce harmful effects (Mardones et al., unpublished data). This estimate is likely underestimated, as a mismatch is often observed between the toxin detection in shellfish and the presence of the putative causative species in the plankton. For instance, the detection of gymodimines in shellfish in central Chile (Mardones et al., 2020b) suggested the presence of *Karenia selliformis* and/or *Alexandrium ostenfeldii*, although gymnodimines toxin production has not yet been detected on strains of these two species isolated from Chilean waters (Mardones et al., 2020b).

Despite the decades of HAB monitoring on the vast Chilean coast, the taxonomy of many dinoflagellate and diatom species is still a challenge. Historically, microscope counting of HAB species has been the conventional monitoring methodology for identification. However, the inclusion of molecular and flow cytometry techniques, cytological bioassays, chemical identification of phycotoxins, and citizen science monitoring have proven to be relevant tools for species identification and taxonomy. In this chapter, we review the progressive inclusion of these techniques in the Chile HAB monitoring activities and evaluate their advantages and disadvantages.

4.2 Classical taxonomy approach

The precise species identification of harmful algal species is critical as it allows for accessing previous knowledge on their toxicology, physiology, and ecology (Hallegraeff et al., 2004). Although the relevance of the classical taxonomy for HAB research is indisputable, less consensus exists among the scientific community for the criterion used for species delimitation (Taylor, 1992). The biological species concept, based on interbreeding natural populations that are reproductively isolated (Mayr, 1996), is challenging to apply to protists, as they reproduce predominantly asexually (Von Dassow and Montresor, 2011). Thus the traditional morphospecies concept still constitutes the option of choice for microalgae.

The democratization of molecular tools in the last two decades has increasingly revealed distinct genetic clades within widespread HAB morphospecies. Many of these taxa were revealed to be pseudocryptic species (i.e., differentiated based on very subtle morphological or ultrastructural features), and formally described as new species after detailed morphological reexamination combined with phylogenetic analysis and, frequently, other non-morphological characters, such as pigment composition and toxin profiles (e.g., Karafas et al., 2015; Verma et al., 2016; Nascimento et al., 2017). The delimitation of true cryptic species (i.e., genetically different but with the same morphology), on the other hand, is still a tricky issue (Amato et al., 2007; Brosnahan et al., 2010). Due to exigencies in the current nomenclatural codes, many taxonomists are resistant to accept newly described species when they are not supported by morphological characters (Anderson et al., 2012), even when they present distinct non-morphological features, such as specific biogeographical origins, ecological niches, and toxin profiles (Zingone and Wyatt, 2005; Trainer et al., 2020). Except for the five formally described (and accepted) species in the *Alexandrium tamarense* complex, based on LSU rDNA clades with discrete biogeographic distribution (John et al., 2014), cryptic HAB species are mostly referred to as genetic clades (i.e., ribotypes) within morphospecies complexes, waiting for formal description (e.g., Binzer et al., 2019; Wang et al., 2019). Regardless of the "cryptic species conundrum," the proper taxonomic characterization of HAB species constitutes a very laborious task based on the so-called polyphasic approach that demands the combinations of several analyses, including culture isolation, optical and electron microscopy, and non-morphological characterization, such as genetic sequencing and toxin profiles (Fig. 4.1). Such exigencies represent an ideal that is hard to achieve during routine monitoring analyses.

HAB ecological studies and/or monitoring activities are mostly based on optical microscopy analyses alone, which makes it nearly impossible to discriminate cryptic species, or those for which morphological identi-

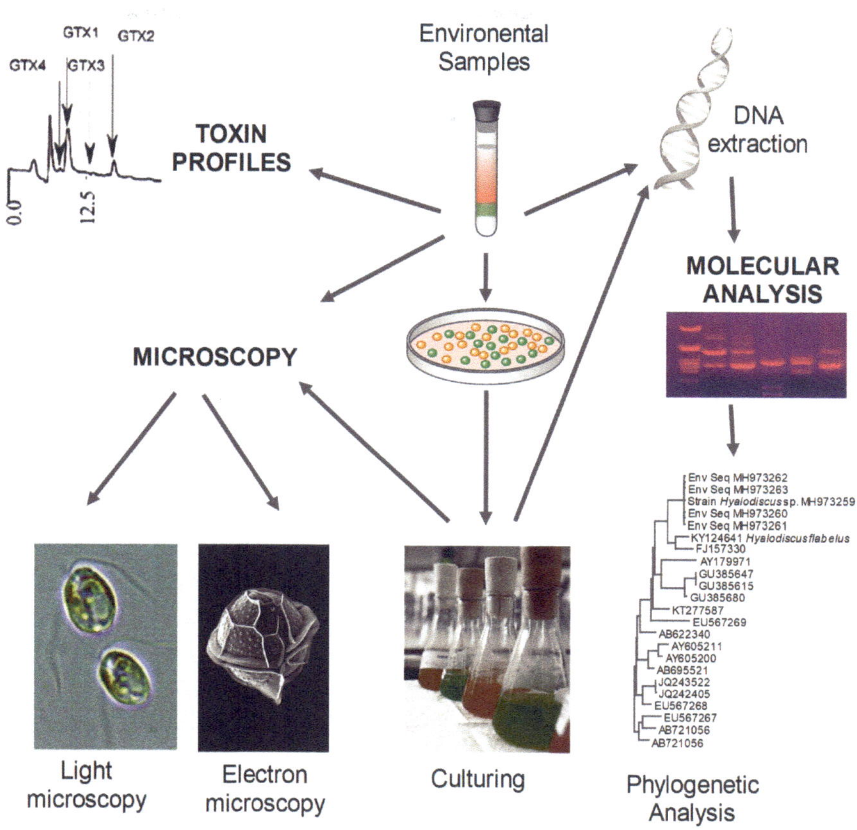

Figure 4.1 Polyphasic approach for the taxonomic characterization of HAB species.

fication is based on difficult to observe features, such as plate pattern. As preservatives frequently cause the deformation (and even the destruction) of delicate cells, HAB flagellates are frequently underestimated during counts. Moreover, due to the high number of samples and the limited trained personnel in species identification, combined with the limited quantitative sensitivity of microscopic methods, the early detection of HAB events is rarely achieved. Such a problem is especially evident in the Chilean fjords, where HAB species are still poorly taxonomically characterized. The only exceptions are the PSP producer *Alexandrium catenella* (Jedlicki et al., 2012; Varela et al., 2012; Paredes-Mella and Varela, 2020) and, more recently, the fish-killers *Pseudochatonella verruculosa* (Mardones et al., 2019) and the *Karenia selliformis* (Mardones et al., 2020b), both characterized during unexpected massive HAB events. Due to the huge geographic extension of

Chilean fjords, monitoring activities are still focused on the detection of previously reported HAB taxa for this area (e.g., *A. catenella*, *Dinophysis* spp.), with cryptic species (e.g., *Pseudo-nitzschia* spp.), flagellates (e.g., *Pseudochattonella* spp.), and small-sized species (e.g., *Azadinium* spp.) frequently overlooked. Given the frequency of emergent HAB events in Chilean fjords observed in the last years (León-Muñoz et al., 2018; Mardones et al., 2020b,c), the application of alternative approaches for HAB monitoring in this geographic area is urgent. Some examples of such methods are described in the following sections.

4.3 Molecular approach

Chile has been monitoring harmful algae for decades using microscopy, toxin analysis, and observation of water's physical properties. Though HAB mechanisms are still unclear, these approaches have provided information on the local HAB dynamics and related environmental factors. However, the current Chilean HAB monitoring programs lack molecular methodologies, a powerful tool capable of overcoming limitations given by the traditional monitoring approaches and delivering broader insights of local HAB species. For instance, the genera *Pseudochattonella* and *Alexandrium* were not detected by microscopy in samples from Metri (Los Lagos, Chile), but were identified by 18S rRNA metabarcoding analysis with less than 1% relative abundance (Yarimizu et al., 2020). John et al. (2014) analyzed the phylogeny of what was reported as *Alexandrium tamarense*, and it was found to consist of five different morphologically similar *Alexandrium* species. Thus molecular technologies have a much lower detection limit with precise identification ability (John et al., 2005; Zingone et al., 2006; Reguera and Pizarro, 2008). The addition of molecular methodology will undoubtedly improve the current Chilean monitoring systems in many ways. The following briefly introduces useful molecular techniques for HAB monitoring.

Quantitative real-time PCR (qPCR) provides gene copy numbers by amplifying target genes, allowing identification and quantification of targeted HAB species (Galluzzi et al., 2004; Penna and Galluzzi, 2013). Its high sensitivity and enumeration application advanced HAB research, improving understanding of relationships between species abundance and possible factors associated with the blooms, such as physical and chemical water properties (Bowers et al., 2000; Gray et al., 2003; Galluzzi et al., 2004). However, environmental samples often contain natural inhibitors sensitive to qPCR, which may cause variability in amplification efficiency (Ellison et al., 2011; Coyne et al., 2005). Digital PCR (dPCR) is an alternative for a quantitative HAB monitoring tool (Medlin and Orozco, 2017). Like qPCR, dPCR provides gene copy numbers, but with higher tolerance to preexisting inhibitors in field samples (Te et al., 2015; Lee et al., 2017). In addition, dPCR does not require generating a standard curve, because dPCR calculates not relative but absolute copy numbers in the DNA fragments applied to a partition (Sanders et al., 2011; Quan et al., 2018). The drawbacks of dPCR are the prohibitive cost and requirement of more concentrated target molecules, due to dPCR's smaller reaction volumes (Te et al., 2015; Quan et al., 2018). Loop-mediated isothermal amplification (LAMP) is useful for fast qualitative HAB species detection by recognizing target DNA or RNA using a reverse transcription-LAMP reaction (Tomita et al., 2008; Nagai and Itakura, 2012). Its transportability and rapid analysis time have been introduced as a part of a mobile laboratory used for HAB monitoring at Repollal and García Island, Chile (Fujiyoshi et al., 2020). However, this tool is limited to qualitative use. Fluorescence in-situ hybridization (FISH) is a technique that uses an oligo probe bound to a fluorescent marker to penetrate the cell and hybridizes to the target ribosomes, which can then be identified by a light microscope. Rather than crushing cells to extract DNA from samples, FISH maintains the whole-cell intact without rupturing cells, often a problem in using other fixatives (John et al., 2005: Medlin and Orozco, 2017). FISH uses species-specific probes that can be applied to any communities to detect target species. Applying FISH to phytoplankton was difficult, because of the autofluorescence nature of plants; however, it has been successfully applied to detect HAB species, such as *Alexandrium ostenfeldii* and *Alexandrium minutum* (John et al., 2003; Tang et al., 2012; Medlin and Orozco, 2017).

Next-generation sequencing (NGS) technologies for metabarcoding analyses provide millions of sequence reads in a single experiment. Such massively parallel sequencing (MPS) is revolutionary for environmental DNA (eDNA) surveys, because it can detect several hundreds of operational taxonomic units (OTUs) of bacteria and eukaryotes from seawater samples, even with a complex and a low-abundance population community. Additionally, the MPS-based HAB monitoring can provide precise species identification based on genetic information, even for the species that cannot be distinguished by conventional microscopy (Sogin et al., 2006; Caron et al., 2012; Tanabe et al., 2015; Nagai et al., 2016a,b,c, 2019). For example, the Sea of Okhotsk (Hokkaido, Japan) was monitored for six years weekly, and the metabarcoding analysis based on the amplification of 18S V7–V9 and 28S D1 rRNA gene regions detected 43 unique OTUs associated with toxic HAB species, with 26 of those previously not reported from the area (Sildever et al., 2019). For the time-series monitoring of Osaka Bay (Osaka, Japan), the yearly abundance pattern of *Alexandrium catenella* cells obtained by morphology-based counting coincided with those obtained by metabarcoding (Nagai et al., 2019). In Chile, the Science and Technology Research Partnership for Sustainable Development project (SATREPS) is currently monitoring 14 coastal stations using metabarcoding analysis (Yarimizu et al., 2020). So far 38 toxic HAB species have been detected from nearly 300 eDNA samples (Fig. 4.2, unpublished data). These results suggest that metabarcoding data, when used with other environmental data, such as physical and chemical properties of water, can provide HAB monitoring information on species ecology. The downfalls of metabarcoding analysis are its time-consumption and costly instrumentation and reagents requirement. Nevertheless, the introduction of molecular approaches, such as metabarcoding technologies would help gain deep insights into the microbial biodiversity and characterize local ecosystems, including HAB distributions and dynamics in Chilean coastal waters.

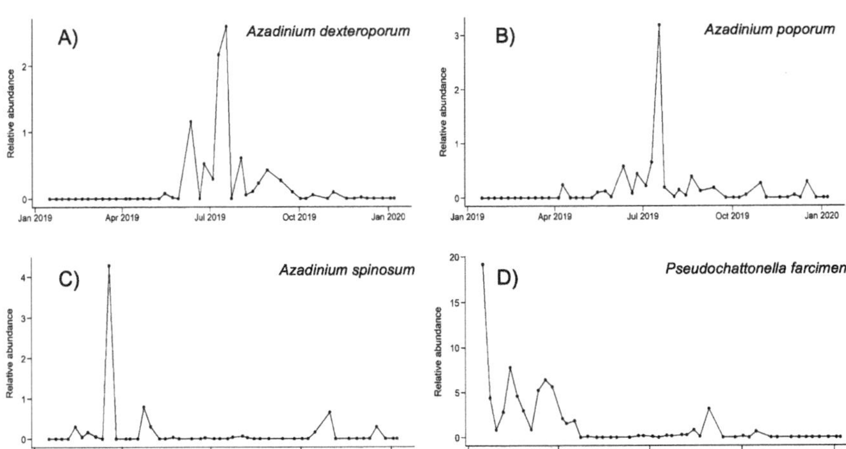

Figure 4.2 Metabarcoding results from SATREPS (Nagai et al., unpublished): The project SATREPS uses the metabarcoding technique to monitor 14 different Chilean coastal stations, starting from December 2018. In 2019, the analysis for the project showed 38 toxic HAB species from nearly 300 eDNA samples. The plots show an example of newly detected toxic HAB species in southern Chilean coastal waters: A) *Azadinium dexteroporum*, B) *Azadinium poporum*, C) *Azadinium spinosum*, and D) *Pseudochattonella farcimen*.

4.4 Chemotaxonomy

All organisms, including microalgae, produce a huge variety of chemical compounds. Those compounds that have an essential role for survival of the producing organism are called primary metabolites, and because of their fundamental functional roles, they are usually produced at higher rates and are structurally more conserved. In addition to primary metabolites, organisms produce a large suite of secondary metabolites, which in most cases are not entirely necessary for survival, but through evolutionary processes have gained beneficial properties through a vast array of functions, such as antigrazing or defense compounds, deterrents, signaling compounds, allelochemicals, sexual cues, etc. Due to their high specificity, these compounds bear an almost unlimited structural variability, and, in many cases, are highly specific for a group of organisms, a genus, and, sometimes, even a single species. This specificity allows the use of chemical constituents among other characteristics, such as morphology or genetics, to classify taxa based on these compounds, which is known as chemotaxonomy (Singh, 2016).

The degree of specificity of genetic markers and morphology is much higher than for secondary metabolites and for this reason, the aforementioned methods are the gold standard for species identification. However, chemotaxonomy can be applied for the determination of taxonomic groups instead of single species. A common example in marine biology is the characterization of phytoplanktonic groups based on pigment analysis (e.g., Carreto et al., 2016). Whereas species identification in the laboratory nowadays can be easily resolved with a suite of state-of-the-art methods and techniques, it still can be challenging in the field, as samples need to be preserved for later laboratory analysis, such as fixation of samples for microscopy, stabilization of DNA by deep-freezing or extraction with buffers or light protection of samples for pigment analysis.

Phycotoxins are secondary metabolites produced by some microalgal species that are noxious, because they can be accumulated in filter-feeding bivalves and, in some cases, in finfish and cause intoxications after consumption of contaminated fish. Even though in some cases there is evidence that these phycotoxins are used by the producing organisms for prey capture or grazing defense (Selander et al., 2006; Harðardóttir et al., 2015), in general, their biological role remains unknown. The chemical variability of phycotoxins is extremely large in terms of characteristics, such as molecular size or polarity. Based on their chemical structures, phycotoxins are categorized into different groups and within each group there can be many structural variants of the same base compound. In the case of yessotoxins (YTX) and prymnesins (PRM), up to a hundred variants have been reported (Miles et al., 2005; Binzer et al., 2019). Phycotoxins include relatively small molecules, such as saxitoxin, with a molecular weight of 300 Da up to very large molecules, such as maitotoxin, with a molecular weight over 3400 Da (Murata et al., 1993). On the other hand, phycotoxins display a wide range of polarities from ionic and highly hydrophilic in the case of many paralytic shellfish poisoning (PSP) toxin variants to relatively lipophilic in the case of high molecular weight toxins, such as ciguatoxins and azaspiracids. Despite their huge structural and chemical variability, most phycotoxins share a common feature, which is high stability. This characteristic causes issues in management since these toxins cannot be routinely destroyed by cooking; however, it makes them suitable candidates as proxies for taxonomic characterization of their producing organisms, which is particularly interesting if preservation of samples is difficult or cannot be achieved.

The use of phycotoxins as chemotaxonomic markers can of course only be applied to toxigenic species that produce known phycotoxins, which is a limitation with respect to monitoring total plankton communities. However, since HAB species are of special concern due to their threat to human health, and the exploitation of marine resources and the marine environment, there is a specific need for detection and monitoring of HAB species and their associated phycotoxins. The use of phycotoxins as taxonomic markers in some cases is not straightforward. For instance, some toxin classes, such as PSP toxins, are produced by several genera of the marine dinoflagellates *Alexandrium*, *Gymnodinium* and *Pyrodinium* as well as by several genera of limnic cyanobacteria. Nevertheless, PSP toxins still can be used as chemotaxonomic markers. For example, *Gymnodinium catenatum* (the only PSP-producing species of this genus) can be distinguished from *Alexandrium* by the presence of benzoyl derivatives (GC-toxins) that are only produced by *G. catenatum* (Vale, 2008), but not by *Alexandrium*. Even within *Alexandrium*, toxin profiles may point towards a certain species. For example, PSP toxin profiles of *A. minutum* to date have only been reported to consist of gonyautoxins (GTX), but not of nonsulfated decarbamoyl toxins (saxitoxin, neosaxitoxin) or *N*-sulfocarbamoyl toxins (B- and C-toxins), which are common to most PSP-producing *Alexandrium* (Chang et al., 1997; Chou et al., 2004; Touzet et al., 2007). However, as mentioned before, the knowledge of toxin profiles is steadily increasing, and recently only GTX-producing strains of *Alexandrium catenella* (reported as *A. tamarense*) displaying the typical *A. minutum* profile were reported from the Arctic around Greenland (Baggesen et al., 2012). But in this special case, PSP toxin profiles still can be used to identify the presence of both species in a certain area, because *A. minutum* as a subtropical-temperate species geographically does not co-occur with only GTX-producing *A. catenella* from the Atlantic Arctic.

In addition to several species producing the same toxin class, there is also the contrary case, where a single species produces different toxin classes, which, from a chemotaxonomic perspective, can be used to differentiate different populations within such a species. Such an example is *Alexandrium ostenfeldii*, which is capable of producing three different toxin classes, namely PSP toxins, spirolides (SPX), and gymnodimines (GYM) (Van de Waal et al., 2015). Interestingly, there is only one population in the Scheldt area of the Netherlands known that produces all three classes at the same time, whereas most populations either only produce SPX, or SPX and PSP, or GYM (Table 4.1) (Krohn, 2016). In contrast to the toxin variability of PSP toxins within *Alexandrium*, which is quite limited (only 12 reported variants), the structural variability of GYM (Van Wagoner et al., 2011; Harju et al., 2016; Zurhelle et al., 2018) and, especially, SPX is much higher and most likely not even fully explored yet (Tillmann et al., 2014; Martens et al., 2017; Nieva et al., 2020).

Table 4.1 Toxin class of several *Alexandrium ostenfeldii* strains.

Strain	Salinity	Origin	Toxin Class		
			PSP	SPX	GYM
AOSH2	>30	Canada	–	x	–
NX 56-07	21.1–33.3	Norway	–	x	x
NX 56-10	21.1–33.4	Norway	–	x	x
NX 56-12	21.1–33.5	Norway	–	x	x
Field Sample	33.3	North Sea	–	x	–
Field Sample	26.1	Limfjord, DK	–	x	x
Field Sample	24.5	Kattegat	–	x	–
OKNL 42	8–21	Netherlands	x	x	x
AOICW	32.5	Chile	x	x	–
AOA32-2	28.2	Chile	x	x	–
H1-G8	33.6	Argentina	x	x	–
H3-D4	33.6	Argentina	x	x	–
H2-A4	33.6	Argentina	x	x	–

In contrast to PSP toxins, which are widely distributed across species, other toxin classes are only known to be produced by a single species, and thus are much more specific in terms of chemotaxonomy. For example, spirolides (SPX) have only been reported to be produced by *Alexandrium ostenfeldii*, and not by any other species, despite the fact that *A. ostenfeldii* is capable of producing other toxins. Another example are pinnatoxins that are only produced by *Vulcanodinium rugosum* (Rhodes et al., 2010). However, it has to be taken into account that the knowledge of phycotoxins and their producing organisms is still limited, and new toxigenic species as well as new toxins are discovered and reported continuously. For this reason, it is critical to regularly review and update inventories of toxigenic plankton and their toxin profiles for chemotaxonomic approaches. For example, in 1996 (Seki et al., 1996) GYM was reported to be produced by *Gymnodinium* sp. (later renamed as *Karenia selliformis*), and only in 2011 it became evident that GYM is also produced by some strains of *A. ostenfeldii* (reported as *A. peruvianum*; Van Wagoner et al., 2011), and the number of reported GYM variants has constantly grown. A similar case is azaspiracids (AZA). AZA-1 was first described in 1998 (Satake et al., 1998), and since then four producing organisms and 26 structural AZA variants produced by Amphidomataceae have been reported to date (Krock et al., 2019). Some of the AZA are species-specific, whereas others are more common and produced by several species. Interestingly some of the specific AZA correlate with different ribotypes of a single species. For example, there are three reported ribotypes of *Azadinium spinosum* (A, B, and C), and ribotype A only produces AZA-1, -2, and -33; ribotype B produces only AZA-11 and -51, whereas ribotypes C is non-toxic and does not produce any AZA at all (Tillmann et al., 2021).

The high structural variability of currently known phycotoxins results in a large number of these compounds, which in turn reflect the high taxonomic diversity of the producing organisms. Accordingly, phycotoxins can be used to get an overview of toxigenic genera, species, and even different populations of a single species in a given water body. In remote areas with little scientific infrastructure, phycotoxins are particularly useful as chemotaxonomic markers, because of their high stability and few requirements for sample preservation. Furthermore, phycotoxins can be sampled in two different ways that complement each other. On one hand, phycotoxins can be extracted from planktonic samples, which allows for a high temporal and spatial resolution. On the other hand, dissolved phycotoxins can be sampled directly from seawater (MacKenzie et al., 2004), which results in integrated samples, due to longer sampling times. Whereas phycotoxins of plankton samples are directly linked to their producing organisms, dissolved phycotoxins in seawater samples may indicate a previous presence of toxigenic species and are not limited to a discrete sampling point.

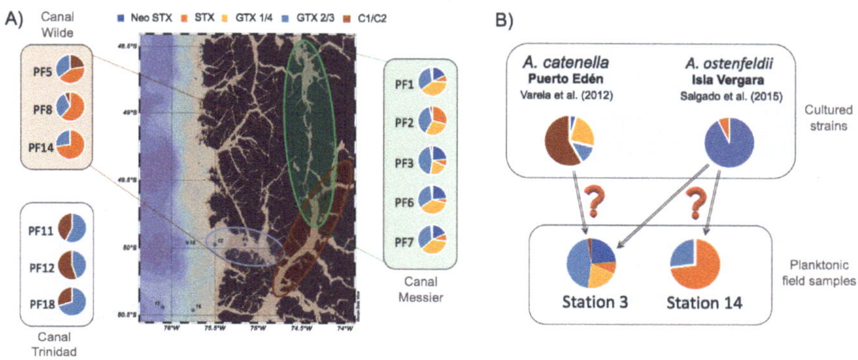

Figure 4.3 Toxin profiles of planktonic water samples during an oceanographic survey conducted in the southern Chilean fjords in November 2019. A) The presence of PSP toxins was used as a chemotaxonomic marker for the identification of *Alexandrium* populations in the study area. B) PSP toxins profiles differences between water field samples and cultured *Alexandrium* clones.

In November 2019, an oceanographic survey was conducted in the fjords and channels of the Última Esperanza province (Magallanes region) of southern Chile, which is a pristine region with poor access. The analysis of PSP toxins in surface planktonic samples revealed three different toxin profiles in three distinct water bodies (Fig. 4.3A). This clearly indicated the presence of several *Alexandrium* populations in the area. In southern Chile, two PSP producers are known: *A. catenella* and *A. ostenfeldii*. Interestingly, none of the field profiles is consistent with the known toxin profiles of each species (Fig. 4.3B). The only known toxin profile of *A. catenella* isolated in this area (Pto. Edén) is dominated by C1/2 (Varela et al., 2012), whereas field samples from this region (Canal Messier) hardly displayed any C1/2 if at all. The only known PSP profiles of *A. ostenfeldii* from the Southeast Pacific consist of saxitoxin and GTX2/3 (Salgado et al., 2015), which was isolated further North in the Moraleda Channel of the Aysén region, which mostly resembled toxin profiles from Canal Wilde. Due to the very limited number of known toxin profiles of this region, a differentiation between *A. catenella* and *A. ostenfeldii* based on chemotaxonomic markers at this stage is not possible. However, it is clear that at least one species, if not both, are present in the area and that there are three geographically distinct *Alexandrium* populations that may reflect different environmental conditions.

4.5 RTgill-W1 fish gill bioassay for ichthyotoxins detection

Most HAB species are known for producing phycotoxins (i.e., paralytic, diarrhetic and amnesic) that can accumulate in the food web with significant consequences for the aquaculture industry, tourism, and human health. Some HAB species, known as fish-killing algae (FKA), can also produce secondary metabolites (hereafter "ichthyotoxins") that can result in fish and other marine fauna kills (Dorantes-Aranda et al., 2015). However, it has been suggested that the primary advantage in releasing these toxic secondary metabolites in the water is to cease grazing by a motile predator, to produce an antibiotic effect to reduce the colonization of algal cells by bacteria or viruses, and/or to inhibit the growth of other competing phytoplankton species (allelopathy). It has thus been hypothesized that fish mortality caused by FKA species may be a secondary side effect of the release of allelochemicals (Granéli and Salomon, 2010). Some complicated issues for monitoring FKA bloom events are the following: (1) there are only few described toxins that appear related to fish kills (i.e., karlotoxins, karmitoxins, prymnesins A, B, and C-type and *K. brevisulcata* toxins), so there are limited standards available for analytical quantification; (2) most of the FKA species are mainly small flagellates of the classes Dictyochophyceae (*Pseudochattonella, Vicicitus*), Haptophyceae (i.e., *Prymnesium*), and Raphydophyceae (*Heterosigma, Chattonella*) and/or pleiomorphic dinoflagellates of the Family Kareniaceae (i.e., *Karenia, Karlodinium*) and Gymnodiniaceae (i.e., *Cochlodinium*), which are extremely fragile and difficult to identify in regular monitoring under light microscopy, and (3) many FKA species probably produce more than one toxin, therefore animal mortalities associated with blooms of FKA species cannot always be attributed to a particular toxin.

In Chile, FKA bloom events have increased in causative species diversity, duration, and intensity since the first report in 1988 (Mardones, 2020). In 2016, the most catastrophic FKA bloom event recorded in the world killed more than USD 800m in salmon, due to a massive outbreak of the toxic Dictyochophyte *Pseudochattonella* in Reloncaví Sound (Mardones et al., 2020c). Mitigation strategies attempted at the salmon farms failed to avoid mass salmon mortality, and even the analysis of more than 14,000 water samples for microalgae detection under light microscopy during the bloom event did not help salmon farmers with an early warning. A recent hypothesis, for explaining this mass salmon mortality, is that the complex life cycle exhibited by *Pseudochattonella* and its fragility against common fixatives produced an underestimation of the real cell abundance of this species and its related ichthyotoxins in 2016 (Mardones et al., 2020c). This fact makes an important call for new strategies for FKA cell and toxin monitoring. As previously stated in this chapter (Section 4.3.), molecular approaches might improve the detection of FKA species in the Chilean fjords, but the lack of FKA toxin standards and chemical analytical methods to quantify and characterize ichthyotoxins has to be pursued using new tools.

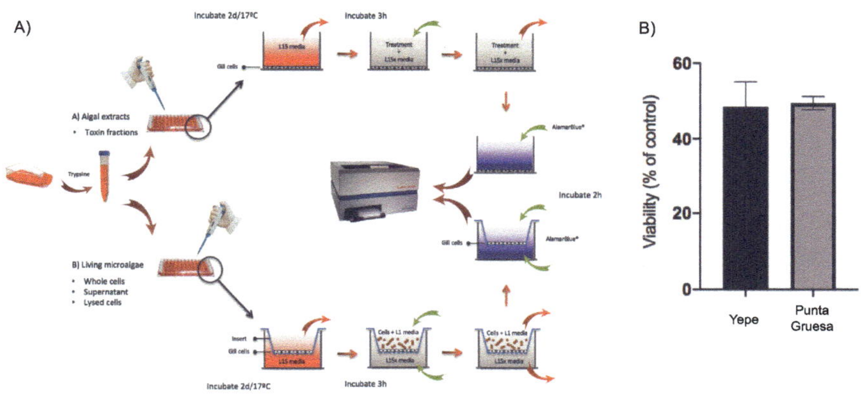

Figure 4.4 A) Conceptual diagram showing the RTgill-W1 cell line assay, and B) gill cell toxicity by two natural seawater samples from the *A. sanguinea* bloom at the localities of Yepe (122 cells mL^{-1}) and Punta Gruesa (145 cells mL^{-1}), northern Chile. Data are the mean and error bars, the standard deviation of cell viability from quadruplicate measurements.

Scientific progress for a complete understanding of the behavior of these ichthyotoxins has been mainly hindered by the use of a wide variety of different bioassay systems, such as a wide range of juvenile or adult fish (i.e., zebra, salmon, sea bass, damselfish, sheepshead minnow), *Artemia* or *Daphnia* assays, and the use of mammalian or fish erythrocytes (Hallegraeff et al., 2017). The innovative use of a standardized, reproducible, and sensitive rainbow trout RTgill-W1 cell line assay (Fig. 4.4) (Fischer et al., 2019) has allowed significant progress in characterizing ichthyotoxins produced by FKA species in Danish (Rasmussen et al., 2016), Australian (Dorantes-Aranda et al., 2015), US (Cochlan et al., 2014), and Chilean laboratories (Mardones et al., 2015, 2018, 2019). The fish cell line assay (Fig. 4.4A) has also been successfully applied to screen freshly collected seawater samples from Australian and Korean fish-killing events (Seger et al., 2017; Hallegraeff et al., 2019). In January 2020, at the localities of Yepe and Punta Gruesa, Northern Chile, an outbreak of the dinoflagellate *Akashiwo sanguinea* killed thousands of sea urchins (*Loxechinus albus* and *Tetrapygus niger*). The presence of cytotoxic metabolites in natural water samples from the toxic *A. sanguinea* event was confirmed by using the fish cell line assay (Fig. 4.4B), representing the first attempt for marine algae ichthyotoxins detection in freshly collected Chilean waters.

The use of the RTgill-W1 cell line assay on natural seawater samples for ichthyotoxins monitoring is a promising approach for an early warning alert for the salmon industry. For instance, toxins produced by newly appearing ichthyotoxic algae species, which might be taxonomically unfamiliar for phytoplankton analysts, could be detected early by the gill assay. In the last 10 years, salmon mortalities in Chilean waters have often been produced by uncommon FKA species for the region (i.e., *Karenia, Karlodinium, Margalefidinium* and *Pseudochattonella*), which has hindered the possibilities of applying successful early warning alerts or mitigation strategies on time (Mardones, 2020). On the other hand, unfixed sea water samples from salmon farms for microscopic analysis have to be transported long distances for a few days, which makes the rupture of the fragile FKA cells during the transport under high temperature, sun-light exposure, and strong shaking among other factors, unavoidable. This may contribute to an underestimation of real cell abundance in the sample. Under this scenario, persistent ichthyotoxins, such as toxic aldehydes (60 to 200 h) (Bartual and Ortega, 2013), could still be detected by the RTgill-W1cell assay, even when FKA cells are not present.

Further laboratory experiments are still needed for (i) calibration between RTgill-W1 and *in vivo* salmon responses against different microalgae ichthyotoxins, and, (ii) alternative fluorometric assays that could provide allelochemicals/ ichthyotoxins detection requiring easier and fewer technical skills (Long et al., 2018).

4.6 Imaging flow cytometry for detection of Harmful Algae

Traditional flow cytometry, where cells pass a laser in a focused sample stream and are classified by total or maximum signals of light scatter and fluorescence, has become almost routine for studying communities of pico- and nanophytoplankton since 1982 (Trask et al., 1982; Marie et al., 2005; Lomas et al., 2011). However, this technology provides almost no taxonomic resolution, with the exceptions of being able to distinguish *Synechococcus*-like from *Prochlorococcus*-like pico-cyanobacteria (both types are cells with low light scatter and red fluorescence from chlorophyll, whereas

Synechococcus-like cells also have yellow/orange fluorescence from phycoerythrin), or calcified coccolithophores, such as *Emiliania huxleyi* and its close relatives (based on specialized detectors of scatter polarization state (Olson et al., 1989; Von Dassow et al., 2012; Bendif et al., 2016)). Also, it is only useful for cells more abundant than 10^2 cells ml^{-1}, due to the low flow rates at which samples can be processed (10–100 μl min^{-1}), and single cells smaller than the laser beam (typically focused to 5–20 μm along the sample flow direction; Shapiro, 2003). In Chile, the most costly HAB problems to date are caused by the chain-forming dinoflagellate, *A. catenella* (the only marine phytoplankton species in the country to be classed by law as a "plague" species) (Mardones et al., 2015). The Chilean salmon industry uses a threshold of 20 cells ml^{-1} for *A. catenella*, whereas abundance thresholds for *Alexandrium* for regulatory actions for shellfish species in other countries can be <1 cell ml^{-1} (Anderson et al., 2001), and detectable toxins can appear in shellfish at much lower thresholds (Swan and Davidson, 2012). As mentioned above, another HAB species that resulted in disastrous fish kills in Chile is the ichthyotoxic nanoflagellate, *P. verruculosa* (Clément et al., 2017; León-Muñoz et al., 2018; Mardones et al., 2020c), which would be difficult to distinguish from other flagellates based on light scatter and fluorescence alone, and for which abundance thresholds for action range from 10–50 cells ml^{-1}. Thus instead of traditional flow cytometry, imaging flow cytometry (IFC) technologies, specifically designed for providing more taxonomic resolution for research and monitoring of larger (mostly > 10 μm) phytoplankton cells, have begun to be implemented and explored (Poulton, 2016; Dashkova et al., 2017).

One commercially available instrument designed for phytoplankton is the FlowCAM, a flow microscope, for which automatically captured images are the base data (although in some modes of operation a limited set of cytometric signals can also be recorded). It has been commercially available and utilized in phytoplankton for over 15 years and has demonstrated effectiveness in detection of HAB cells (Buskey and Hyatt, 2006), and can be used also for general characterization of nano- and microplankton (e.g., Álvarez et al., 2014; Kydd et al., 2018). A caution is that some studies have reported that cell counts did not correlate well with traditional Utermöhl counts (Garmendia et al., 2013; Kydd et al., 2018), though others report better correlations, which vary among taxonomic groups (Hrycik et al., 2019).

This instrument has been implemented for 4 years in routine private-sector monitoring operations in Chile (Clément et al., 2017, 2020). The

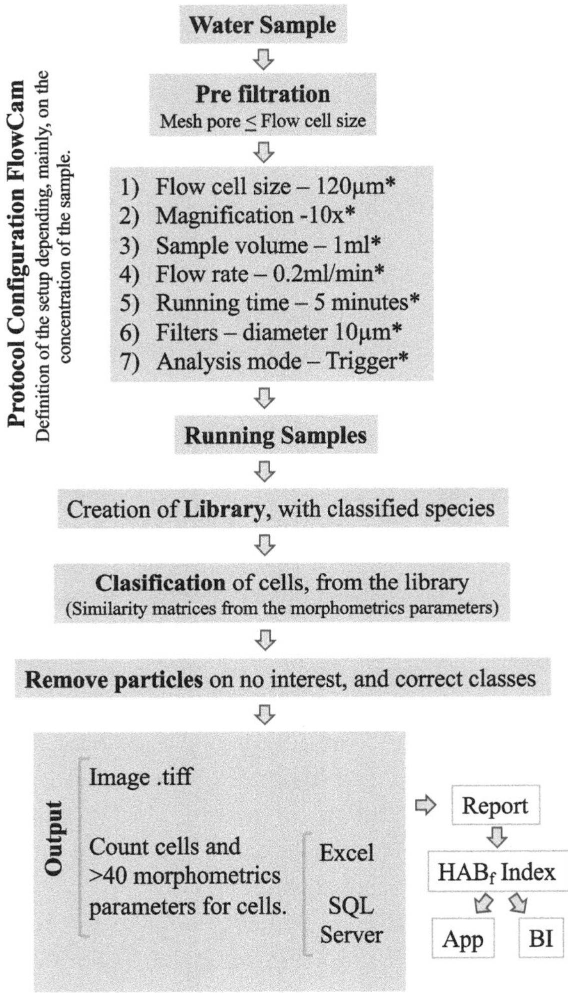

Figure 4.5 Diagram showing sample processing by FlowCAM (gray boxes) and data analysis (white boxes). Parameters with an asterisk (*) in the FlowCAM protocol correspond to a standard setup for the marine phytoplankton for Southern Chile. This protocol has been useful for the marine fjord ecosystem of Chile. Legend: App = Application; BI = Business Intelligence; SQL = Structured Query Language; HAB_f Index – Harmful Algae Bloom Index for fish farms developed by Clément et al. (2020).

routine workflow illustrated in Fig. 4.5 allows the collection of data that can be highly useful when complemented with traditional sample examination or other data. Certain trade-offs are immediately clear. The microscope objective, flow cell, and flow rate are chosen to match the size classes of interest (cells as small as 2 μm can be recorded, although in practice, as in this protocol, the focus is usually on larger cells). Another trade-off is run time: lowering the detection limits of particular HAB species would require longer times, and also increase image data processing and storage requirements from nontarget species. The taxonomic resolution provided is limited, but users find the large number of images collected helpful, especially if selected samples can be re-evaluated by other methods.

Surprisingly, chain-forming organisms present an important technical challenge for the FlowCAM (Table 4.2) and IFC more generally. The FlowCAM software counts each chain as a single entity, rather than counting the number of cells in the chain. In the case of *A. catenella*, this can be corrected in FlowCAM data using a factor for average number of cells/chain. However, this procedure does not work well for diatoms, where chain length is highly variable both between species and within species, so chain-forming diatoms are significantly under-counted.

Chain-forming organisms are of particular importance in Chile, both as important HAB species and as a prominent component of the phytoplankton community in productive waters, such as the Chilean coast (Anabalón et al., 2016) and inner seas (Alves-de-Souza et al., 2008). Chain formation is an important functional categorization, and the relative importance of such organisms is well known to correlate dynamically with oceanographic and ecological state in Chilean coastal and inner seas (Spilling et al., 2019; González et al., 2011), so adequate counting of cells in chains is of high value.

IFCs include two very different categories of instruments. The FlowCAM (as well as potential holographic IFCs that could be even more powerful, Göröcs et al., 2018), focuses completely on images. Other instruments arise from the tradition of conventional flow cytometry, and are essentially cytometers that capture complex cytometric signals (pulse-shaped recording) as base data, as well as some images. A commercial option that epitomizes this latter category is the CytoSense (Cytobuoy, Inc.), which, like the Imaging Flow Cytobot (McLane Labs), also records the shapes of cytometric signals from cells larger than the laser beam and chains of cells. The CytoSense relies principally on characterization of shapes of scatter and fluorescence signals for detecting and discriminating phytoplankton into optical categories (which do not necessarily correlate well to phylogenetic categories) (Haraguchi et al., 2017). Images can also be collected on subsets of cells classified first by the pulse-shape parameters in 2-dimensional gates. After data collection, automatic or guided classification of particles by cytometric parameters can be guided by images (Dugenne et al., 2014). The CytoSense can cytometrically detect even pico-cyanobacteria, but in practice, the trade-off is that increasing the detector gains (voltages) to detect smaller cells causes saturation of signals of the largest cells, which decreases the ability to discriminate against them.

Tests of the CytoSense suggested that it can potentially count cells in chains by the repeating cytometric signatures (Fig. 4.6A–B). In fact, when we tested several cultured *Alexandrium* species, the cytometric signatures that could most reliably distinguish *A. catenella* from its congeners related to chain formation (data not shown). Nevertheless, in practice, the ability to detect and characterize chains depends on chains passing oriented with flow (Fig. 4.6C). In the case of pennate diatoms, the fluorescence signal exhibited twice the maxima as cells in the chain, as the fluorescence is interrupted by the nucleus (Fig. 4.6D). Particularly for diatoms, we found that the counting ability to detect chains and the ability to count chains was strongly dependent on the signal used for "triggering" data acquisition. The side scatter signal did not track cells in chains well, with a high signal at the leading edge of the first cell in the chain, but then very low signals (if any) at the boundaries between subsequent cells. In contrast, forward scatter and fluorescence both provided signals that repeated faithfully with the cells in chains. Chains were often severely cut short if side scatter was used as the trigger channel, even if the gain was adequate to detect the first cell of the chain, as the signal would drop off so much from the leading edge of the cell. This resulted in both undercounting cells and eliminating a useful distinguishing characteristic from nonchain-forming organisms (Fig. 4.6E). Finally, *Chaetoceros* chains usually did not provide clear signals, due either to orientation not parallel to flow, spiral chains, or heavy spines. Thus, being able to effectively utilize pulse-shape recording requires careful optimization by experienced personnel.

While IFCs and flow microscopes represent powerful tools, it is important to recognize the challenges more generally to their implementation, particularly for countries with similar economic or geographic challenges as Chile. First of all, the cytometric profiles or images captured will rarely provide classification to better than genus-level (Camoying and Yñiguez, 2016), so must be complemented by other information (expert local knowledge, parallel analyses). Clear barriers for IFC implementation relate to their direct costs (typically USD 100,000–200,000; even if costs dropped by half they would be inaccessible to most labs and services in Chile) as well as the requirement for dedicated highly trained and experienced personnel

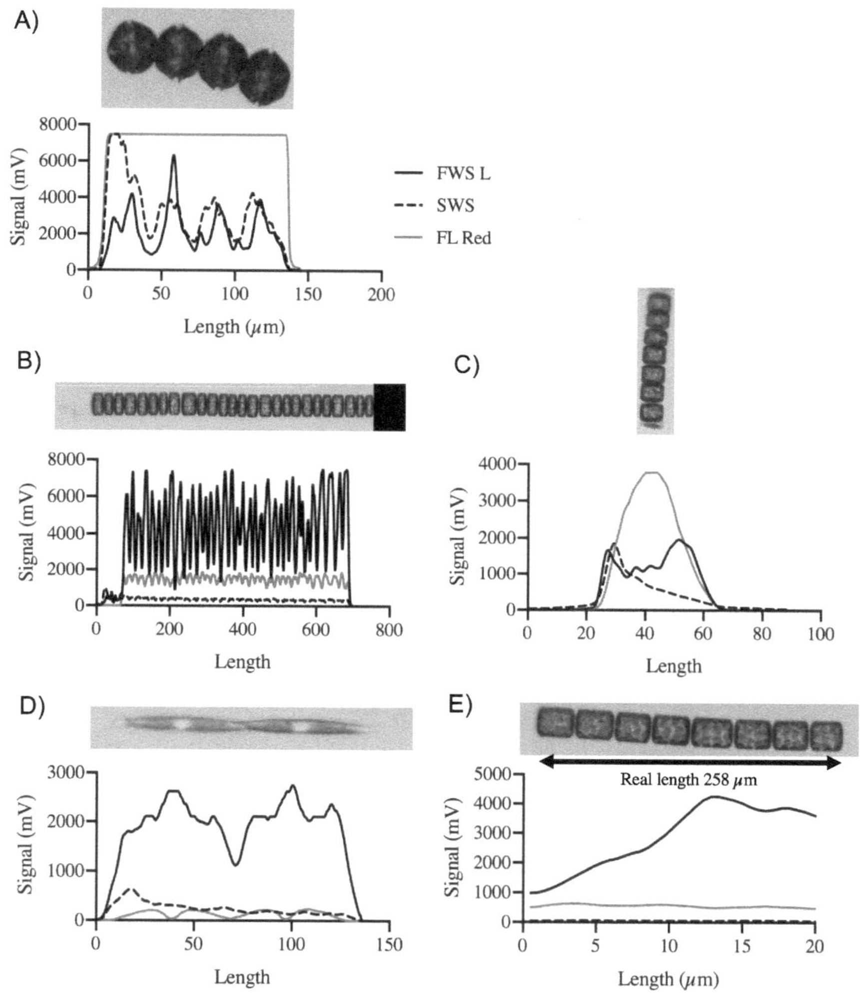

Figure 4.6 Example CytoSense cytometric profiles of phytoplankton chains. A) *Alexandrium catenella* chain in natural sample. The red fluorescence signal (FL red) is saturated, but the forward scatter (FWS L) and side scatter (SWS) have adequate gains and scan the chain as it passes. B) A long centric diatom chain, with adequate settings. Both FWS and FL red signals track well the cells in the chain, but SWS does not. C) A centric diatom chain oriented perpendicular to the flow. The cytometric profile does not capture the chain. D) 2-cell chain of a pennate diatom (*Pseudo-nitzschia*). FWS tracks the cell borders well, but SWS does not. FL red shows four, rather than two, well-separated maxima, because fluorescence is interrupted by the nucleus of each cell. E) An example of a properly oriented chain recorded with inadequate settings. In this case, SWS was used as the trigger for event acquisition, but the gain was set too low and the threshold too high, so the signal was cut off before even the first cell had passed.

to operate and maintain them effectively. In addition to trained operators, users may need to invest in acquiring separately or developing in-house external artificial intelligence or machine-learning needed to take full advantage of the data generated, both for better taxonomic discrimination and solving problems, such as counting cells in chains. A country such as Chile, with a vast marine territory, where the most costly HAB issues occur in the complex and highly heterogeneous Patagonian fjord region, covering almost 240,000 km^2 (Pantoja et al., 2011), might require a larger number of IFC instruments in distinct locations for adequate spatial coverage, with comparable investment in human operators.

An alternative is to collect samples over a larger spatial areas and process them centrally. This approach is limited in Chile, as distances are very large and transit times are correspondingly slow, and sample preservation with standard fixatives is either incompatible with IFC use or greatly reduces their efficacy. Lugol's iodine destroys fluorescence, an essential parameter for flow cytometers, and the stain would accumulate in the recirculating sheath fluid (used for hydrodynamic focusing of sample streams) in instruments such as the CytoSense. On the other hand, formaldehyde- or glutaraldehyde-fixed cells retain some fluorescence, but even thecate dinoflagellates lose distinguishing features, and unarmored flagellates, such as *Heterosigma* sp. or *P. verruculosa*, become clumped or can be completely lost upon fixation. Thus although fixed samples can be transported, even concentrated, and still provide some data, live samples give much better discrimination among phytoplankton (though their usability decreases dramatically after about a day after collection). Another potential solution could be to deploy a robust IFC with autonomous operation on ships-of-opportunity, for example, incorporating IFCs into the FerryBox instrument package (Petersen, 2014).

Finally, having access to cultures of phytoplankton species representing typical local communities as well as target HAB organisms can greatly facilitate training and optimization of protocols. Though many marine microbes are still unculturable, an important portion of HAB species can be cultured, and such cultured strains can be especially useful for species that are not consistently present or present at low abundances. However, this would require established and certified phytoplankton culture collections, and strains for training would need to be frequently replaced with new strains, as phytoplankton in long-term culture often lose important characteristics, such as chain-formation, or undergo transitions in life stage.

4.7 MicroToxMap: citizen science for the identification and monitoring of harmful algal blooms

For decades, citizen involvement in scientific projects has become a means by which individuals in the community assist scientists in investigating important research questions. Citizen science is thus a way to both gather large amounts of scientifically valuable data as well as to engage with the wider community encouraging curiosity and a greater understanding of science. After the catastrophic blooms in 2016 and 2018 in the south of Chile (Mardones et al., 2020a,b), harmful algae bloom (HAB) events have gained much public attention among the affected coastal communities. Involving these communities in the monitoring of HAB eases overcoming the logistic challenges faced by scientists and national agencies of sampling the intricate and vast regions of the Chilean fjords where HAB is recurrent. Such synergy between scientists and local communities benefits both groups by providing much needed information on HAB recurrence at a narrower spatial and temporal scale and by educating the public in complex marine phenomena that directly affects them. Several volunteer HAB monitoring programs based on the identification of HAB species or the detection of color sea changes have been developed worldwide (e.g., HABscope, https://habscope.gcoos.org; Plankton Planet, https://planktonplanet.org; and Phenomer, http://www.phenomer.org). These initiatives have proven to be low-cost and effective as a complementary approach to address this major environmental issue (e.g., Timmons et al., 2018; Hardison et al., 2019; Siano et al., 2020).

Since 2017, the MicroToxMap (MTM, https://microtoxmap.com) project represents the first monitoring program of HAB identification by the public in Chile. Overall, the MTM program aims to train volunteers in the identification of HAB species by using low-cost friendly technology. In particular, MTM is directed to those communities in the south of Chile that are more concerned by HAB events, including shellfish farmers, fish farmers, artisanal fishers, and the wider community. The information gained through MTM complements the current monitoring efforts carried out by national scientific institutions increasing the sampling frequency and extending the monitoring area. Also, education and outreach is an integral part of the MTM project filling three main aspects: (i) educate people to identify HAB species, (ii) raise awareness of the potential health and environmental risks of HAB events, and (iii) direct people on the procedures if HAB species are detected in the area (Timmons et al., 2018).

Embracing the spirit of "frugal science" (Whitesides, 2012), the MTM program provides the volunteers with an ultra-low cost (USD 3) origami-based paper microscope, the foldscope, for the identification of toxic microalgae species. The foldscope was designed at the Prakash lab (Stanford University) to be extremely portable, durable, and to give optical quality similar to conventional research microscopes (Cybulski et al., 2014). These features make the foldscope the perfect alternative tool to be used in volunteer HAB identification and monitoring programs. Volunteers are provided with a tool kit to concentrate phytoplankton, and then invited to take pictures of sampled water coupling the foldscope to their phones and report potential occurrence of HAB species by using a free smartphone application (MTM IOS and Android) (Fig. 4.7A and B). Geolocalized pictures are sent directly to taxonomists for verification via the application, thus providing real-time information from the field.

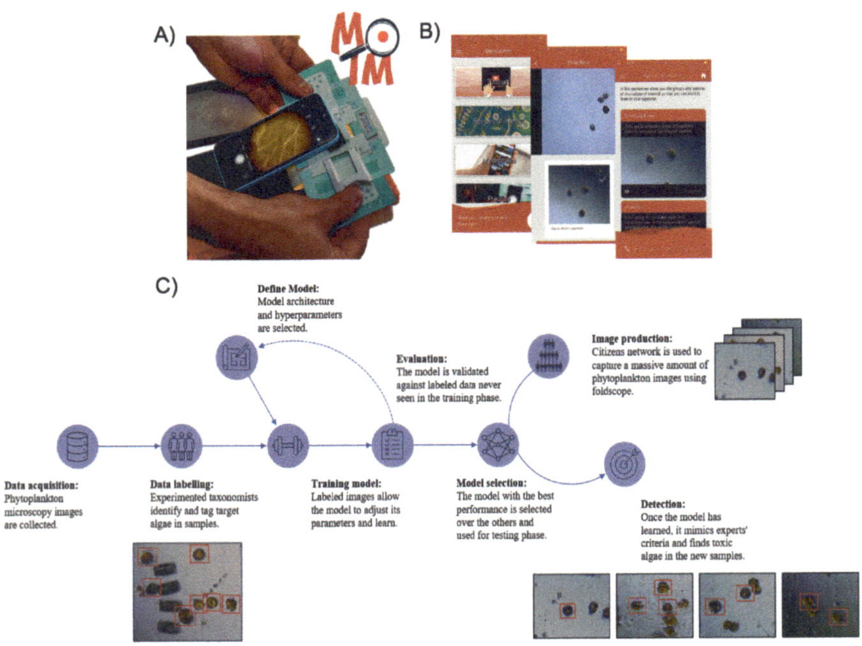

Figure 4.7 Overview of the MicroToxMap (MTM) project. A) Sample observation on a smartphone coupled to the foldscope; B) MTM smartphone application, and C) Object detection workflow.

To ensure collection of scientifically sound data, the MTM project offers quality training to volunteers on a regular basis through workshops and distance learning methods. First, a small group of selected volunteers (e.g., aquaculture farmers and teachers) in different coastal communities along the south of Chile are personally trained by the scientific team. During these workshops volunteers gain hands-on experience on collecting samples, assembling the foldscope, and identifying and reporting HAB species via the application. Thereafter, these highly trained volunteers can in turn pass their skills to other designated participants, hence creating a wide citizen network of collaborators. At all times, volunteers have access to training material (e.g., identification guidelines, video tutorials and fact-sheets) through the program's digital platforms, including the application and the web-site (Fig. 4.7B). Keeping the long-term engagement of volunteers in the project is critical for its success. For that purpose, citizen contributions are presented in the MTM project website regularly delivering a snapshot of the current HAB condition in the south of Chile. Communicating science to the general public is not always an easy task. In that sense the foldscope represents an invaluable tool for outreach. From the MTM project efforts are directed to deliver foldscopes not only to collaborators, but to the wider community for them to explore and better understand the HAB issue.

The amount of data generated by the citizen network is practically unmanageable for a small group of taxonomists. Since 2014, efficient object detection techniques on video or real-time imaging have been developed (Girshick et al., 2014; Redmon et al., 2016; Lin et al., 2017; Bochkovskiy et al., 2020), enhancing accuracy and time detection every year. The object detection problem consists mainly on finding and classifying target objects (e.g., human faces, traffic signs, or weapons) in unlabeled images based on labeled training samples. These techniques present high potential to be used in the automatized detection of toxic algae in microscopy images.

A usual object detection workflow is illustrated in Fig. 4.7C. The most essential phases for an appropriate object detection task are data acquisition and labeling (Fig. 4.7C, 1 and 2). Though data acquisition can be done either by scientists or citizens, the labeling of the objects should be made, or at least validated, by taxonomists, whose classification criteria is professional in quality and identical to those the model should have. Conversely, if data is labeled by inexperienced users, the model would mimic these mistaken examples, and its results would not be adequate. The following training phases (Fig. 4.7C, 3, 4, and 5) consists in passing the data through the model to adjust its parameters and maximize its detection performance. These steps should be repeated until the performance of the model is satisfactory (i.e., acceptably high metrics; precision, recall, and bounding box placing). Lastly, the testing (or production) phase is defined by the detection of target objects in unlabeled microscopy images (Fig. 4.6C, 6, 7, and 8). In this particular case, the model output shows targeted toxic algae with a bounding box drawn around complemented by its certainty index. Naturally, this final phase should also be confirmed by taxonomists, so the complete method is validated by experts' judgement. In the near the future, this approach could not only be used to detect toxic algae, but all kinds of phytoplankton particles to bring support to taxonomists and scientists.

4.8 Outlook/overview

As new methods continue to develop and current techniques become more advanced, scientific progress on methodological alternatives to monitor HAB species will continue to expand. The decision on which technique to use will basically depend on the scope of the monitoring program and the resources and facilities available, considering their respective pros and cons (Table 4.2).

Currently, Chilean monitoring programs for HAB species detection and enumeration relies on the use of light microscopy (LM). After several decades of experience, accurate identification of some species using this method has proven extremely difficult. For example, the ichthyotoxic flagellates *Pseudochattonella verruculosa, P. farcimen*, and *Vicicitus globosus* are frequently misclassified and/or grouped together, due to their fragility to common fixatives for cell enumeration under LM. Another example is the occurrence of the toxic dinoflagellate *Azadinium*, whose presence in southern Chilean waters was primarily possible by detecting its corresponding toxins. Generally, regular phytoplankton monitoring overlooks the presence of this genus, due to its small size or confuses it with harmless dinoflagellates of the genus *Heterocapsa*. Another issue is the extremely difficult detection of fish-killing algal (FKA) toxins or "ichthyotoxins." The lack of toxin standards and chemical analytical methods to quantify and characterize ichthyotoxins has been a nightmare for the Chilean salmon industry in the last two decades. Therefore the currently used laboratory techniques for monitoring HAB species and their toxins are not sufficiently effective to provide up-to-date data required for early warnings. However, an auspicious means to monitor HABs and marine toxins with a more efficient and rapid turnaround time is in the form of integrating new technology. Such systems combine biorecognition tools with a physical or chemical measurement component to allow for detection and quantification of a target

Table 4.2 Advantages and disadvantages of tools that can be beneficial to Chilean HAB monitoring: note that those are a general description of the pros and cons, and there are always exceptions.

Approach	Technique	Advantages	Disadvantages
Molecular	dPCR	(1) Highly quantitative, (2) highly sensitive, (3) performs species identification, (4) can estimate targe cell abundance, (5) tolerant to inhibitors in environment samples, (6) standard curve is not required.	(1) Less sensitive than qPCR, (2) cost is still expensive, (3) more concentrated target molecule is required.
	qPCR	(1) Quantitative, (2) highly sensitive, (3) perform species identification, (4) can estimate targe cell abundance.	(1) Sensitive to matrix of environmental samples causing PCR interference, (2) expensive technology, (3) a standard curve must be generated for each assay run (labor-intensive), (4) costly.
	NGS metabarcoding	(1) Can detect millions of sequence reads in a single experiment, (2) can detect species even from a complex community and a low-abundance population community.	(1) Qualitative use only, (2) expensive, (3) labor-intensive, (4) different DNA extraction, sequencing, and bioinformatic processing methods may introduce viability in results.
	FISH	(1) The whole cell remains intact, (2) use of probe instead of extraction prevent cells from rupturing.	(1) Autofluorescense of phytoplankton may interfere with fluorescent probes, (2) costly, (3) labor-intensive.
	LAMP	(1) Transportability, (2) rapid analysis time, (3) highly precise, (4) relatively cost-effective.	(1) Qualificative use only, (2) can detect a maximum of only four target species at a time.
Toxin detection	RTgill-W1 cell line bioassay	Standardized in vitro bio-assay Small sample volume (100–200 μL) High number of replicates (3–8) Time saving (2–4 h several replicates) Better control conditions	High technical skill required Strict aseptic conditions for cell line maintenance are required Nonspecific ichtyotoxin detection
	Chemotaxonomy (HPLC)	Easy sampling No need of sample preservation In some cases specificity to ribotype level Phycotoxins can be used as intra- and extracellular chemotaxonomic markers	Availability of expensive instrumentation (LC-MS/MS) Limitation to species with known phycotoxins Incomplete knowledge of toxin profiles
Imaging	Flow microscopy and Imaging Flow Cytometry	Ability to examine cells unaltered and without preservation More rapid sample analysis than traditional methods, enhancing the sample processing rate Wide size range of particles analyzed Ability to deploy in autonomous operation in the field, on ships, docks, or buoys (only CytoBuoy or Imaging Flow CytoBot at present) Permits detection of both target and novel or non-target organisms	Restriction of analyses of the phytoplankton community in a size range, depending on the size of the flow cell (FlowCAM) or detector gains (CytoSense/CytoBuoy) Taxonomic resolution may be relatively low compared to that of the conventional microscopy. For example, difficulties for the clear distinction of cells <20 μm. Counting of chains as one particle, not separate cells, thus underestimating true cell abundances Limitation of the image or particle classification process, due to the limited development and commercial availability of artificial intelligence. Lack of specificity to identify/classify a species, and in this way the classification is not so specific and is confused with "morpho-similar" species. Shortcomings in the delivery of results, with high reliance still on manual data processing.
	Citizen Science (Foldscope)	Low-cost friendly technology Increased monitoring frequency Real-time information Increase monitoring geographical area Engages the wider community in HAB phenomenon Increases environmental awareness of the local human community	Need for continuous training through workshops and distance learning methods Need for long-term commitment of volunteers Need to validate date produced by citizens

HAB species or phycotoxic analyte. Molecular toolbox, chemo-taxonomy, imaging flow cytometry, gill cell-based assays, and citizen science using low-cost microscopy improved with artificial intelligence are being progressively considered as a powerful suite of tools for HAB species and their toxins monitoring in Chilean coastal waters.

References

Álvarez, E., Moyano, M., Lopez-Urrutia, A., Nogueira, E., Scharek, R., 2014. Routine determination of plankton community composition and size structure: a comparison between FlowCam and light microscopy. Journal of Plankton Research 36 (1), 170–184.

Álvarez, G., Uribe, E., Ávalos, P., Mariño, C., Blanco, J., 2010. First identification of azaspiracid and spirolides in *Mesodesma donacium* and *Mulinia edulis* from Northern Chile. Toxicon 55, 638–641.

Alves-de-Souza, C., Gonzalez, M.T., Iriarte, J.L., 2008. Functional groups in marine phytoplankton assemblages dominated by diatoms in fjords of Southern Chile. Journal of Plankton Research 30 (11), 1233–1243.

Amato, A., Kooistra, W.H.C.F., Ghiron, J.H.L., Mann, D.G., Pröschold, T., Montresor, M., 2007. Reproductive isolation among sympatric cryptic species in marine diatoms. Protist 158 (2), 193–207.

Anabalón, V., Morales, C.E., González, H.E., Menschel, E., Schneider, W., Hormazabal, S., Valencia, L., Escribano, R., 2016. Micro-phytoplankton community structure in the coastal upwelling zone off Concepción (central Chile): annual and inter-annual fluctuations in a highly dynamic environment. Progress in Oceanography 149, 174–188.

Anderson, D.M., Andersen, P., Bricelj, V.M., Cullen, J.J., Rensel, J.E., 2001. Monitoring and management strategies for harmful algal blooms in coastal waters, APEC #201-MR-01.1. Asia Pacific Economic Program, Singapore, and Intergovernmental Oceanographic Commission Technical Series No. 59, Paris.

Anderson, D.M., Cembella, A.D., Hallegraeff, G.M., 2012. Progress in understanding harmful algal blooms: paradigm shifts and new technologies for research, monitoring, and management. Annual Review of Marine Science 4, 143–176.

Baggesen, C., Moestrup, Ø., Daugbjerg, N., Krock, B., Cembella, A.D., Madsen, S., 2012. Molecular phylogeny and toxin profiles of Alexandrium tamarense (Lebour) Balech (Dinophyceae) from the west coast of Greenland. Harmful Algae 19, 108–116.

Bartual, A., Ortega, M.J., 2013. Temperature differentially affects the persistence of polyunsaturated aldehydes in seawater. Environmental Chemistry 10, 403–408.

Bendif, E.M., Probert, I., Díaz-Rosas, F., Thomas, D., Van den Engh, G., Young, J.R., Von Dassow, P., 2016. Recent reticulate evolution in the ecologically dominant lineage of coccolithophores. Frontiers in Microbiology 7 (784).

Binzer, S.B., Svenssen, D.K., Daugbjerg, N., Alves-De-Souza, C., Pinto, E., Hansen, P.J., Larsen, T.O., Varga, E., 2019. A-, B- and C-type prymnesins are clade specific compounds and chemotaxonomic markers in *Prymnesium parvum*. Harmful Algae 81, 10–17.

Bochkovskiy, A., Wang, C.Y., Liao, H.Y.M., 2020. YOLOv4: optimal speed and accuracy of object detection. arXiv preprint. arXiv:2004.10934.

Bowers, H.A., Tengs, T., Glasgow, H.B., Burkholder, J.M., Rublee, P.A., Oldach, D.W., 2000. Development of real-time PCR assays for rapid detection of *Pfiesteria piscicida* and related Dinoflagellates. Applied and Environmental Microbiology 66 (11), 4641–4648.

Brosnahan, M.L., Kulis, D.M., Solow, A.R., Erdner, D.L., Percy, L., Lewis, J., Anderson, D.M., 2010. Outbreeding lethality between toxic Group I and nontoxic Group III *Alexandrium tamarense* spp. isolates: predominance of heterotypic encystment and implications for mating interactions and biogeography. Deep-Sea Research II 57, 175–189.

Buskey, E.J., Hyatt, C.J., 2006. Use of the FlowCam for semi-automated recognition and enumeration of red tide cells (*Karenia brevis*) in natural plankton samples. Harmful Algae 5, 685–692.
Camoying, M.G., Yñiguez, A.T., 2016. FlowCAM optimization: attaining good quality images for higher taxonomic classification resolution of natural phytoplankton samples. Limnology and Oceanography: Methods 14, 305–314.
Caron, D.A., Countway, P.D., Jones, A.C., Kim, D.Y., Schnetzer, A., 2012. Marine protistan diversity. Annual Review of Marine Science 4, 467–493.
Carreto, J.I., Montoya, N.G., Carignan, M.O., Akselman, R., Acha, E.M., Derisio, C., 2016. Environmental and biological factors controlling the spring phytoplankton bloom at the Patagonian shelf-break front – degraded fucoxanthin pigments and the importance of microzooplankton grazing. Progress in Oceanography 146, 1–21.
Chang, F.H., Anderson, D.M., Kulis, D.M., Till, D.G., 1997. Toxin production of Alexandrium minutum (Dinophyceae) from the Bay of Plenty, New Zealand. Toxicon 35 (3), 393–409.
Chou, H.N., Chen, Y.M., Chen, C.Y., 2004. Variety of PSP toxins in four culture strains of *Alexandrium minutum* collected from southern Taiwan. Toxicon 43 (3), 337–340.
Clément, A., Muñoz, F., Brito, C.G., Correa, N., Saldivia, M., Fernandez, C., Perez, F., Maluje, C., Contreras, G., Egenau, O., 2017. Climatic anomalies and harmful flagellates blooms in Southern Chile. In: Proença, L.A.O., Hallegraeff, G.M. (Eds.), Marine and Fresh-Water Harmful Algae, Proceedings of the 17th International Conference on Harmful Algae. International Society for the Study of Harmful Algae, pp. 34–37.
Clément, A., Husak, T., Clément, S., Muñoz, F., Saldivia, M., Brito, C.G., Crescini, R., Correa, N., Teiguel, K., Fernández, C., Contreras, G., Saez, S., 2020. Contribution of the HABf INDEX for fish farm's risk analysis. In: Harmful Algae 2018 – from Ecosystems to Socioecosystems, Proceedings of the 18th International Conference on Harmful Algae. Nantes. International Society for the Study of Harmful Algae, pp. 150–153.
Cochlan, W.P., Trainer, V.L., Trick, C.G., Wells, M.L., Bill, B.D., Eberhart, B.T.L., 2014. *Heterosigma akashiwo* in the Salish Sea: defining growth and toxicity leading to fish kills. In: Harmful Algae 2012, Proceedings of the 15th International Conference on Harmful Algae. International Society for the Study of Harmful Algae. 245 pp.
Coyne, K.J., Handy, S.M., Demir, E., Whereat, E.B., Hutchins, D.A., Portune, K.J., Cary, S.C., 2005. Improved quantitative real-time PCR assays for enumeration of harmful algal species in field samples using an exogenous DNA reference standard. Limnology and Oceanography: Methods 3 (9), 381–391.
Cybulski, J.S., Clements, J., Prakash, M., 2014. Foldscope: origami-based paper microscope. PLoS ONE 9 (6), e98781.
Dashkova, V., Malashenkov, D., Poulton, N., Vorobjev, I., Barteneva, N., 2017. Imaging flow cytometry for phytoplankton analysis. Methods 112, 188–200.
Dorantes-Aranda, J.J., Seger, A., Mardones, J.I., Nichols, P.D., Hallegraeff, G.M., 2015. Progress in understanding algal bloom-mediated fish kills: the role of superoxide radicals, phycotoxins and fatty acids. PLoS ONE 10 (7), e0133549.
Dugenne, M., Thyssen, M., Nerini, D., Mante, C., Poggiale, J.C., Garcia, N., Garcia, F., Grégori, G.J., 2014. Consequence of a sudden wind event on the dynamics of a coastal phytoplankton community: an insight into specific population growth rates using a single cell high frequency approach. Frontiers in Microbiology 15 (5), 485.
Ellison, S.L.R., Emslie, K.R., Kassir, Z., 2011. A standard additions method reduces inhibitor-induced bias in quantitative real-time PCR. Analytical and Bioanalytical Chemistry 401 (10), 3221–3227.
Fischer, M., Belanger, S.E., Berckmans, P., Bernhard, M.J., Bláha, L., Coman Schmid, D.E., Dyer, S.D., Haupt, T., Hermens, J., Hultman, M.T., Laue, H., Lillicrap, A., Mlnaříková, M., Natsch, A., Novák, J., Sinnige, T.L., Tollefsen, K.E., von Niederhäusern, V., Witters,

H., Županič, A., Schirmer, K., 2019. Repeatability and reproducibility of the RTgill-W1 cell line assay for predicting fish acute toxicity. Toxicological Sciences 169 (2), 353–364.
Fujiyoshi, S., Yarimizu, K., Miyashita, Y., Rilling, J., Acuña, J.J., Ueki, S., Gajardo, G., Espinoza-González, O., Guzmán, L., Jorquera, M.A., 2020. Suitcase Lab: new, portable, and deployable equipment for rapid detection of specific harmful algae in Chilean coastal waters. Environmental Science and Pollution Research, 1–12.
Galluzzi, L., Penna, A., Bertozzini, E., Vila, M., Garcés, E., Magnani, M., 2004. Development of a real-time PCR assay for rapid detection and quantification of *Alexandrium minutum* (a Dinoflagellate). Applied and Environmental Microbiology 70, 1199–1206.
Garmendia, M., Revilla, M., Zarauz, L., 2013. Testing the usefulness of a simple automatic method for particles abundance and size determination to derive cost-effective biological indicators in large monitoring networks. Hydrobiologia 704, 231–252.
Girshick, R., Donahue, J., Darrell, T., Malik, J., 2014. Rich feature hierarchies for accurate object detection and semantic segmentation. In: Proceedings of the IEEE Conference on Computer Vision and Pattern Recognition, pp. 580–587.
González, H.E., Castro, L., Daneri, G., Iriarte, J.L., Silva, N., Vargas, C.A., Giesecke, R., Sanchez, N., 2011. Seasonal plankton variability in Chilean Patagonia fjords: carbon flow through the pelagic food web of Aysen Fjord and plankton dynamics in the Moraleda Channel basin. Continental Shelf Research 31 (3–4), 225–243.
Göröcs, Z., Tamamitsu, M., Bianco, V., 2018. A deep learning-enable portable imaging flow cytometer for cost-effective, high-throughput, and label-free analysis of natural water samples. Light: Science & Applications 7 (66).
Granéli, E., Salomon, P.S., 2010. Factors influencing allelopathy and toxicity in *Prymnesium parvum*. Journal of the American Water Resources Association 46, 108–120.
Gray, M., Wawrik, B., Paul, J., Casper, E., 2003. Molecular detection and quantitation of the red tide Dinoflagellate *Karenia brevis* in the marine environment. Applied and Environmental Microbiology 69 (9), 5726–5730.
Hallegraeff, G.M., Anderson, D.M., Cembella, A.D., Enevoldsen, H., 2004. Manual on Harmful Marine Microalgae. Monographs on Oceanographic Methodology, vol. 11. UNESCO, Paris. 793 pp.
Hallegraeff, G.M., Dorantes-Aranda, J.J., Mardones, J.I., Seger, A., 2017. Review of progress in our understanding of fish-killing microalgae: implications for management and mitigation. In: Proença, L.A.O., Hallegraeff, G.M. (Eds.), Marine and Fresh-Water Harmful Algae. Proceedings of the 17th International Conference on Harmful Algae. International Society for the Study of Harmful Algae and Intergovernmental Oceanographic Commission of UNESCO, Paris, pp. 148–153.
Hallegraeff, G.M., Albinsson, M.E., Dowdney, J., Holmes, A.K., Mansour, M.P., Seger, A., 2019. Prey preference, environmental tolerances and ichthyotoxicity by the red-tide dinoflagellate *Noctiluca scintillans* cultured from Tasmanian waters. Journal of Plankton Research 41 (4), 407–418.
Haraguchi, L., Jakobsen, H.H., Lundholm, N., Carstensen, J., 2017. Monitoring natural phytoplankton communities: a comparison between traditional methods and pulse-shape recording flow cytometry. Aquatic Microbioly Ecology 80, 77–92.
Harðardóttir, S., Pančić, M., Tammilehto, A., Krock, B., Møller, E., Nielsen, T., Lundholm, N., 2015. Dangerous relations in the Arctic marine food web: interactions between toxin producing *Pseudo-nitzschia* Diatoms and Calanus Copepodites. Marine Drugs 13 (6), 3809–3835.
Hardison, D.R., Holland, W.C., Currier, R.D., Kirkpatrick, B., Stumpf, R., Fanara, T., Burris, D., Reich, A., Kirkpatrick, G.J., Litaker, W., 2019. HABscope: a tool for use by citizen scientists to facilitate early warning of respiratory irritation caused by toxic blooms of *Karenia brevis*. PLoS ONE 14 (6), e0218489.
Harju, K., Koskela, H., Kremp, A., Suikkanen, S., de la Iglesia, P., Miles, C.O., Krock, Vanninen, P., 2016. Identification of gymnodimine D and presence of gymnodimine

variants in the dinoflagellate *Alexandrium ostenfeldii* from the Baltic Sea. Toxicon 112, 68–76.

Hrycik, A.R., Shambaugh, A., Stockwell, J.D., 2019. Comparison of FlowCam and microscope biovolume measurements for a diverse freshwater phytoplankton community. Journal of Plankton Research 41 (6), 849–864.

Jedlicki, A., Fernández, G., Astorga, M., Oyarzún, P., Toro, J.E., Navarro, J.M., Martínez, V., 2012. Molecular detection and species identification of *Alexandrium* (Dinophyceae) causing harmful algal blooms along the Chilean coastline. AoB Plants 2012.

John, U., Cembella, A., Hummert, C., Elbrächter, M., Groben, R., Medlin, L., 2003. Discrimination of the toxigenic dinoflagellates *Alexandrium tamarense* and *A. Ostenfeldii* in co-occurring natural populations from Scottish coastal waters. European Journal of Phycology 38 (1), 25–40.

John, U., Medlin, L.K., Groben, R., 2005. Development of specific rRNA probes to distinguish between geographic clades of the *Alexandrium tamarense* species complex. Journal of Plankton Research 27 (2), 199–204.

John, U., Litaker, R.W., Montresor, M., Murray, S., Brosnahan, M.L., Anderson, D.M., 2014. Formal revision of the *Alexandrium tamarense* species complex (Dinophyceae) taxonomy: the introduction of five species with emphasis on molecular-based (rDNA) classification. Protist 165, 779–804.

Karafas, S., York, R., Tomas, C., 2015. Morphological and genetic analysis of the Coolia monotis species complex with the introduction of two new species, *Coolia santacroce* sp. nov. and *Coolia palmyrensis* sp. nov. (Dinophyceae). Harmful Algae 46, 18–33.

Krock, B., Tillmann, U., Tebben, J., Trefault, N., Gu, H., 2019. Two novel azaspiracids from *Azadinium poporum*, and a comprehensive compilation of azaspiracids produced by Amphidomataceae, (Dinophyceae). Harmful Algae 82, 1–8.

Krohn, T., 2016. A. ostenfeldiiund Salinität: Untersuchung von Toxinprofilen zur Bewertung der Hypothese der Salinitätsabhängigkeit von Toxinprofilen. Bachelor thesis. University of Brmerhaven. https://epic.awi.de/id/eprint/41869/ (in German).

Kydd, J., Raiakaruma, H., Briski, E., Bailey, S., 2018. Examination of a high-resolution laser optical plankton counter and FlowCam for measuring plankton concentration and size. Journal of Sea Research 133, 2–10.

Lee, H.G., Kim, H.M., Min, J., Kim, K., Park, M.G., Jeong, H.J., Kim, K.Y., 2017. An advanced tool, droplet digital PCR (ddPCR), for absolute quantification of the red-tide dinoflagellate, *Cochlodinium polykrikoides* Margalef (*Dinophyceae*). Algae 32 (3), 189–197.

León-Muñoz, J., Urbina, M.A., Garreaud, R., Iriarte, J.L., 2018. Hydroclimatic conditions trigger record harmful algal bloom in western Patagonia (summer 2016). Scientific Reports 8, 1–10.

Lin, T.Y., Goyal, P., Girshick, R., He, K., Dollár, P., 2017. Focal loss for dense object detection. arXiv preprint. arXiv:1708.02002.

Lomas, M.W., Bronk, D.A., Van den Engh, G., 2011. Use of Flow Cytometry to measure biogeochemical rates and processes in the Ocean. Annual Review of Marine Science 3 (1), 537–566.

Long, M., Tallec, K., Soudant, P., Lambert, C., Le Grand, F., Sarthou, G., Jolley, D., Hégaret, H., 2018. A rapid quantitative fluorescence-based bioassay to study allelochemical interactions from *Alexandrium minutum*. Environmental Pollution 242, 1598–1605.

MacKenzie, L., Beuzenberg, V., Holland, P., McNabb, P., Selwood, A., 2004. Solid phase adsorption toxin tracking (SPATT): a new monitoring tool that simulates the biotoxin contamination of filter feeding bivalves. Toxicon 44 (8), 901–918.

Mardones, J.I., 2020. Screening of Chilean fish-killing microalgae using a gill cell-based assay. Latin American Journal of Aquatic Research 48 (2), 329–335.

Mardones, J.I., Dorantes-Aranda, J.J., Nichols, P.D., Hallegraeff, G.M., 2015. Fish gill damage by the dinoflagellate *Alexandrium catenella* from Chilean fjords: synergistic action of ROS and PUFA. Harmful Algae 49, 40–49.

Mardones, J.I., Shabala, L., Shabala, S., Dorantes-Aranda, J.J., Seger, A., Hallegraeff, G.M., 2018. Fish gill damage by harmful microalgae newly explored by microelectrode ion flux estimation techniques. Harmful Algae 80, 55–63.

Mardones, J.I., Fuenzalida, G., Zenteno, K., Alves-De-Souza, C., Astuya, A., Dorantes-Aranda, J.J., 2019. Salinity-growth response and ichthyotoxic potency of the Chilean *Pseudochattonella verruculosa*. Frontiers in Marine Science 6, 24.

Mardones, J.I., Holland, D.S., Anderson, L., LeBihan, V., Gianella, F., Clément, A., Davidson, K., Sakamoto, S., Yoshida, T., Trainer, V.L., 2020a. Estimating and mitigating the economic costs of harmful algal blooms on commercial and recreational shellfish harvesters. In: Trainer, V.L. (Ed.), GlobalHAB. Evaluating, Reducing and Mitigating the Cost of Harmful Algal Blooms: a Compendium of Case Studies. PICES Sci. Rep. No. 59, pp. 66–83.

Mardones, J.I., Norambuena, L., Paredes, J., Fuenzalida, G., Dorantes-Aranda, J.J., Chang, K.J.L., Guzmán, L., Krock, B., Hallegraeff, G., 2020b. Unraveling the *Karenia selliformis* complex with the description of a non-gymnodimine producing Patagonian phylotype. Harmful Algae 98, 101892.

Mardones, J.I., Paredes, J., Godoy, M., Suarez, R., Norambuena, L., Vargas, V., Fuenzalida, G., Pinilla, E., Artal, O., Rojas, X., 2020c. Disentangling the environmental processes responsible for the world's largest farmed fish-killing harmful algal bloom: Chile, 2016. Science of the Total Environment, 144383.

Marie, D., Simon, N., Vaulot, D., 2005. Phytoplankton cell counting by flow cytometry. In: Andersen, R. (Ed.), Algal Culturing Techniques. Elsevier Academic Press, pp. 253–267 (Chapter 17).

Martens, H., Tillmann, U., Harju, K., Dell'Aversano, C., Tartaglione, L., Krock, B., 2017. Toxin variability estimations of 68 *Alexandrium ostenfeldii* (Dinophyceae) strains from The Netherlands reveal a novel abundant Gymnodimine. Microorganisms 5 (2).

Mayr, E., 1996. What is a species, and what is not? Philosophy of Science 63, 262–277.

Medlin, L.K., Orozco, J., 2017. Molecular techniques for the detection of organisms in aquatic environments, with emphasis on harmful algal bloom species. Sensors (Basel, Switzerland) 17 (5), 1184. https://doi.org/10.3390/s17051184.

Miles, C.O., Samdal, I.A., Aasen, J.A.B., Jensen, D.J., Quilliam, M.A., Petersen, D., Briggs, L.M., Wilkins, A.L., Rise, F., Cooney, J.M., MacKenzie, A.L., 2005. Evidence for numerous analogues of yessotoxin in *Protoceratium reticulatum*. Harmful Algae 4 (6), 1075–1091.

Moestrup, Ø., Akselmann-Cardella, R., Churro, C., Fraga, S., Hoppenrath, M., Iwataki, M., Larsen, J., Lundholm, N., Zingone, A. (Eds.), 2009. IOC-UNESCO Taxonomic Reference List of Harmful Micro Algae. http://www.marinespecies.org/hab. (Accessed 7 June 2021).

Murata, M., Naoki, H., Iwashita, T., Matsunaga, S., Sasaki, M., Yokoyama, A., Yasumoto, T., 1993. Structure of maitotoxin. Journal of the American Chemical Society 115 (5), 2060–2062.

Nagai, S., Itakura, S., 2012. Specific detection of the toxic dinoflagellates *Alexandrium tamarense* and *Alexandrium catenella* from single vegetative cells by a loop-mediated isothermal amplification method. Marine Genomics 7, 43–49.

Nagai, S., Hida, K., Urushizaki, S., Onitsuka, G., Yasuike, M., Nakamura, U., Fujiwara, A., Tajimi, S., Kimoto, K., Kobayashi, T., Gojobori, T., Ototake, M., 2016a. Influences of diurnal sampling bias on fixed-point monitoring of plankton biodiversity determined using a massively parallel sequencing-based technique. Gene 576, 667–675.

Nagai, S., Hida, K., Urusizaki, S., Takano, Y., Hongo, Y., Kameda, T., Abe, K., 2016b. Massively parallel sequencing-based survey of eukaryotic community structures in Hiroshima Bay and Ishigaki Island. Gene 576, 681–689.

Nagai, S., Miyamoto, S., Ino, K., Tajimi, S., Nishi, H., Tomono, J., 2016c. Easy detection of multiple *Alexandrium* species using DNA chromatography chip. Harmful Algae 51, 97–106.
Nagai, S., Chen, S., Kawakami, Y., Yamamoto, K., Sildever, S., Kanno, N., Oikawa, H., Yasuike, M., Nakamura, Y., Hongo, Y., Fujiwara, A., Kobayashi, T., Gojobori, T., 2019. Monitoring of the toxic dinoflagellate *Alexandrium catenella* in Osaka Bay, Japan using a massively parallel sequencing (MPS)-based technique. Harmful Algae 89, 101660.
Nascimento, S.M., Mendes, M.C.Q., Menezes, M., Rodríguez, F., Alves-De-Souza, C., Branco, S., Riobó, P., Franco, J., Nunes, J.M.C., Huk, M., 2017. Morphology and phylogeny of *Prorocentrum caipirignum* sp. nov. (Dinophyceae), a new tropical toxic benthic dinoflagellate. Harmful Algae 70, 73–89.
Nieva, J.A., Tebben, J., Tillmann, U., Wohlrab, S., Krock, B., 2020. Mass spectrometry-based characterization of new Spirolides from *Alexandrium ostenfeldii* (Dinophyceae). Marine Drugs 18 (10), 505.
Niklitschek, E.J., Soto, D., Lafon, A., Molinet, C., Toledo, P., 2013. Southward expansion of the Chilean salmon industry in the Patagonian Fjords: main environmental challenges. Reviews in Aquaculture 5 (3), 172–195.
Olson, R.J., Zettler, E.R., Anderson, O.K., 1989. Discrimination of eukaryotic phytoplankton cell types from light scatter and autofluorescence properties measured by Flow Cytometry. Cytometry 10 (63), 6–643.
Pantoja, S., Iriarte, J.L., Daneri, G., 2011. Oceanography of the Chilean Patagonia. Continental Shelf Research 31, 149–153.
Paredes-Mella, J., Varela, D., 2020. Genetic population structure of the toxigenic dinoflagellate *Alexandrium catenella* in the Patagonian Fjords System, southern Chile. Revista de Biología Marina y Oceanografia 55, 26–36.
Penna, A., Galluzzi, L., 2013. The quantitative real-time PCR applications in the monitoring of marine harmful algal bloom (HAB) species. Environmental Science and Pollution Research 20, 6851–6862.
Petersen, W., 2014. FerryBox systems: state-of-the-art in Europe and future development.
Poulton, N., 2016. FlowCam: quantification and classification of phytoplankton by imaging flow cytometry. In: Barteneva, N.S., Vorobjev, I.A. (Eds.), Imaging Flow Cytometry: Methods and Protocols. In: Methods in Molecular Biology, vol. 1389. Chapter 17.
Quan, P.L., Sauzade, M., Brouzes, E., 2018. DPCR: a technology review. Sensors (Basel, Switzerland) 18. https://doi.org/10.3390/s18041271.
Rasmussen, S.A., Meier, S., Andersen, N.G., Blossom, H.E., Duus, J.Ø., Nielsen, K.F., Hansen, P.J., Larsen, T.O., 2016. Chemodiversity of ladder-frame Prymnesin polyesters in *Prymnesium parvum*. Journal of Natural Products 79, 2250–2256.
Redmon, J., Divvala, S., Girshick, R., Farhadi, A., 2016. You only look once: unified, real-time object detection. In: Proceedings of the IEEE Conference on Computer Vision and Pattern Recognition, pp. 779–788.
Reguera, B., Pizarro, G., 2008. Planktonic dinoflagellates that contain polyether toxins of the old “DSP complex”. In: Botana, L.M. (Ed.), Seafood and Freshwater Toxins: Pharmacology, Physiology, and Detection. CRC Press Taylor & Francis Group, New York, pp. 257–284.
Rhodes, L., Smith, K., Selwood, A.I., McNabb, P., van Ginkel, R., Holland, P.T., Munday, R., 2010. Production of pinnatoxins by a peridinoid dinoflagellate isolated from Northland, New Zealand. Harmful Algae 9 (4), 384–389.
Salgado, P., Riobó, P., Rodríguez, F., Franco, J.M., Bravo, I., 2015. Differences in the toxin profiles of *Alexandrium ostenfeldii* (Dinophyceae) strains isolated from different geographic origins: evidence of paralytic toxin, spirolide, and gymnodimine. Toxicon 103, 85–98.
Sanders, R., Huggett, J., Bushell, C., Cowen, S., Scott, D., Foy, C., 2011. Evaluation of digital PCR for absolute DNA quantification. Analytical Chemistry 83, 6474–6484.

Satake, M., Ofuji, K., Naoki, H., James, K.J., Furey, A., McMahon, T., Silke, J., Yasumoto, T., 1998. Azaspiracid, a new marine toxin having unique spiro ring assemblies, isolated from Irish mussels, *Mytilus edulis*. Journal of the American Chemical Society 120 (38), 9967–9968.

Seger, A., Park, T.G., Hallegraeff, G.M., 2017. Assessment of the efficacy of clay flocculation in Korean fish farm water: *Cochlodinium* cell removal and mitigation of ichthyotoxicity. Harmful Algae 61, 46–55.

Seki, T., Satake, M., MacKenzie, A.L., Kaspar, H.F., Yasumoto, T., 1996. Gymnodimine, a novel toxic imine isolated from the Foveaux Strait oysters and *Gymnodinium* sp. Harmful and Toxic Algal Blooms. In: Yasumoto, T., Oshima, Y., Fukuyo, Y. (Eds.). Intergovernmental Oceanographic Commission of UNESCO, pp. 495–498.

Selander, E., Thor, P., Toth, G., Pavia, G., 2006. Copepods induce paralytic shellfish toxin production in marine dinoflagellates. Proceedings of the Royal Society of London. Series B, Biological Sciences 273 (1594), 1673–1680.

Shapiro, H., 2003. Practical Flow Cytometry, fourth edition. John Wiley & Sons, Hoboken, NJ.

Siano, R., Chapelle, A., Antoine, V., Michel-Guillou, E., Rigaut-Jalabert, F., Guillou, L., Hégaret, H., Leynaert, A., Curd, A., 2020. Citizen participation in monitoring phytoplankton seawater discolorations. Marine Policy 117, 103039.

Sildever, S., Kawakami, Y., Kanno, N., Kasai, H., Shiomoto, A., Katakura, S., Nagai, S., 2019. Toxic HAB species from the Sea of Okhotsk detected by a metagenetic approach, seasonality and environmental drivers. Harmful Algae 87, 101631.

Singh, R., 2016. Chemotaxonomy: a tool for plant classification. Journal of Medicinal Plants Studies 4, 90–93.

Sogin, M.L., Morrison, H.G., Huber, J.A., Welch, D.M., Huse, S.M., Neal, P.R., Arrieta, J.M., Herndl, G.J., 2006. Microbial diversity in the deep sea and the underexplored "rare biosphere". Proceedings of the National Academy of Sciences of the United States of America 103, 12115–12120.

Spilling, K., Camarena-Gómez, M.T., Lipsewers, T., Martinez-Varela, A., Díaz-Rosas, F., Eronen-Rasimus, E., Silva, N., Von Dassow, P., Montecino, V., 2019. Impacts of reduced inorganic N: P ratio on three distinct plankton communities in the Humboldt upwelling system. Marine Biology 166, 114.

Swan, S.C., Davidson, K., 2012. Monitoring Programme for the Presence of Toxin Producing Plankton in Shellfish Production Areas in Scotland. Scottish Association for Marine Science, Scottish Marine Institute, Scotland.

Tanabe, A.S., Nagai, S., Hida, H., Yasuike, M., Fujiwara, A., Nakamura, Y., Takano, Y., Katakura, S., 2015. Comparative study of the validity of three regions of 18S-rRNA gene for massively parallel sequencing-based monitoring of the planktonic eukaryote community. Molecular Ecology Resources 16 (2), 402–414.

Tang, X., Yu, R., Zhou, M., Yu, Z., 2012. Application of rRNA probes and fluorescence in situ hybridization for rapid detection of the toxic dinoflagellate *Alexandrium minutum*. Chinese Journal of Oceanology and Limnology 30 (2), 256–263.

Taylor, F.J.R., 1992. The species problem and its impact on harmful phytoplankton studies, with emphasis on dinoflagellate morphology. In: Smayda, T.J., Shimizu, Y. (Eds.), Toxic Phytoplankton Blooms in the Sea. Elsevier, New York, pp. 81–86.

Te, S.H., Chen, E.Y., Gin, K.Y.H., 2015. Comparison of quantitative PCR and droplet digital PCR multiplex assays for two genera of bloom-forming Cyanobacteria, *Cylindro spermopsis* and *Microcystis*. Applied and Environmental Microbiology 81 (15).

Tillmann, U., Kremp, A., Tahvanainen, P., Krock, B., 2014. Characterization of spirolide producing *Alexandrium ostenfeldii* (Dinophyceae) from the western Arctic. Harmful Algae 39, 259–270.

Tillmann, U., Wietkamp, S., Gu, H., Krock, B., Salas, R., Clarke, D., 2021. Multiple new strains of Amphidomataceae (Dinophyceae) from the North Atlantic revealed a high toxin profile variability of *Azadinium spinosum* and a new non-toxigenic *Azadinium* cf. *spinosum*. Microorganisms 9 (1), 134.

Timmons, M., Sweeney-Reeves, M., Morton, S.L., 2018. Harmful algal bloom education and outreach. In: Shumway, S.E., Burkholder, J.M., Morton, S.L. (Eds.), Harmful Algal Blooms: a Compendium Desk Reference. Hoboken, New Jersey, pp. 419–433.

Tomita, N., Mori, Y., Kanda, H., Notomi, T., 2008. Loop-mediated isothermal amplification (LAMP) of gene sequences and simple visual detection of products. Nature Protocols 3 (5), 877–882.

Touzet, N., Franco, J.M., Raine, R., 2007. Influence of inorganic nutrition on growth and PSP toxin production of Alexandrium minutum (Dinophyceae) from Cork Harbour, Ireland. Toxicon 50 (1), 106–119.

Trainer, V.L., Moore, S.K., Hallegraeff, G., Kudela, R.M., Clément, A., Mardones, J.I., Cochlan, W.P., 2020. Pelagic harmful algal blooms and climate change: lessons from nature's experiments with extremes. Harmful Algae 91, 101591.

Trask, B.J., Van den Engh, G.J., Elgershuizen, J.H.B.W., 1982. Analysis of phytoplankton by flow cytometry. Cytometry 2 (4), 258–264.

Trefault, N., Krock, B., Delherbe, N., Cemeblla, A., Vásquez, M., 2011. Latitudinal transects in the southeastern Pacific Ocean reveal a diverse but patchy distribution of phycotoxins. Toxicon 58 (5), 389–397.

Vale, P., 2008. Complex profiles of hydrophobic paralytic shellfish poisoning compounds in *Gymnodinium catenatum* identified by liquid chromatography with fluorescence detection and mass spectrometry. Journal of Chromatography A 1195 (1–2), 85–93.

Van de Waal, D.B., Tillmann, U., Martens, H., Krock, B., van Scheppingen, Y., John, U., 2015. Characterization of multiple isolates from an *Alexandrium ostenfeldii* bloom in The Netherlands. Harmful Algae 49, 94–104.

Van Wagoner, R.M., Misner, I., Tomas, C., Wright, J.L.C., 2011. Occurrence of 12-methylgymnodimine in a spirolide-producing dinoflagellate *Alexandrium peruvianum* and the biogenetic implications. Tetrahedron Letters 52 (33), 4243–4246.

Varela, D., Paredes, J., Alves-De-Souza, C., Seguel, M., Sfair, A., Frangópulos, M., 2012. Intraregional variation among *Alexandrium catenella* (Dinophyceae) strains from southern Chile: morphological, toxicological and genetic diversity. Harmful Algae 15, 8–18.

Verma, A., Hoppenrath, M., Dorantes-Aranda, J.J., Harwood, D.T., Murray, S.A., 2016. Molecular and phylogenetic characterization of *Ostreopsis* (Dinophyceae) and the description of a new species, *Ostreopsis rhodesae*sp. nov., from a subtropical Australian lagoon. Harmful Algae 60, 116–130.

Von Dassow, P., Montresor, M., 2011. Unveiling the mysteries of phytoplankton life cycles: patterns and opportunities behind complexity. Journal of Plankton Research 33 (1), 3–12.

Von Dassow, P., Van den Engh, G., Rodriguez, D., Gittins, J.R., 2012. Calcification state of coccolithophores can be assessed by light scatter depolarization measurements with flow cytometry. Journal of Plankton Research 34 (12), 1011–1027.

Wang, N., Mertens, K.N., Krock, B., Luo, Z., Derrien, A., Pospelova, V., Liang, Y., Bilien, G., Smith, K.F., De Schepper, S., 2019. Cryptic speciation in *Protoceratium reticulatum* (Dinophyceae): evidence from morphological, molecular and ecophysiological data. Harmful Algae 88, 101610.

Whitesides, G., 2012. The frugal way. The Economist – the World in 2012.

Yarimizu, K., Fujiyoshi, S., Kawai, M., Norambuena-Subiabre, L., Cascales, E.K., Rilling, J.I., Vilugrón, J., Cameron, H., Vergara, K., Morón-López, J., 2020. Protocols for monitoring harmful algal blooms for sustainable aquaculture and coastal fisheries in Chile. International Journal of Environmental Research and Public Health 17, 7642.

Zingone, A., Wyatt, T., 2005. Harmful algal blooms: keys to the understanding of phytoplankton ecology. In: Robinson, A.R., McCarthy, J., Rothschild, B.J. (Eds.), The Sea. Harvard University Press, Harvard, pp. 867–926.

Zingone, A., Siano, R., D'Alelio, D., Sarno, D., 2006. Potentially toxic and harmful microalgae from coastal waters of the Campania region (Tyrrhenian Sea, Mediterranean Sea). Harmful Algae 5 (3), 321–337.
Zurhelle, C., Nieva, J., Tillmann, U., Harder, T., Krock, B., Tebben, J., 2018. Identification of novel Gymnodimines and Spirolides from the marine Dinoflagellate *Alexandrium ostenfeldii*. Marine Drugs 16 (11), 446.

PART 2

Monitoring and sensing systems

Lesley A. Clementson

Phytoplankton have been a part of my research for many years, and yet I never cease to be amazed by two aspects. The first is the intricate structure of these microscopic plants, as shown in the photos on the cover of this book. The second is the ability of satellite sensors, some 700 Km above the earth's surface, to see these microscopic cells on mass. Only 10% of the signal received by the sensor is from the ocean or water body, and yet, with algorithms, we can determine with relative accuracy, the productivity of the phytoplankton community on a global scale.

Microscopic identification and taxonomic description of different species of phytoplankton provide valuable information, but is time-consuming, consequently expensive and limited to phytoplankton bigger than 8–10 μm. The technique is of little use for samples collected from oligotrophic water bodies, where picoplankton (<2 μm) dominate. Scanning electron microscopy (SEM) or fluorescence microscopy is needed to identify the smallest of the phytoplankton. Both are specialized instruments, and, again, are time-consuming and expensive to use. As microscopy alone cannot routinely resolve the complete phytoplankton community, new approaches to quantification at the species and group level have been developed and found broad application.

Pigment analysis has provided a quicker alternative to identify phytoplankton community composition at an algal group level. Several of the algal groups have a biomarker or diagnostic pigment, which is a pigment only found in one specific type of phytoplankton. For example, the pigment peridinin is found only in dinoflagellates, lutein only in chlorophytes. However, some individual species do not follow these general rules. For example, peridinin is found only in dinoflagellates, but a few species do not contain peridinin, such as the highly toxic *Karenia brevis* (formerly *Gymnodinium breve*). This species has fucoxanthin as its dominant pigment, which is a pigment generally associated with diatoms. In general, pigment analysis allows for identification of all the size classes, including the picoplankton, but it cannot identify to the species level. Pigment analysis and taxonomy combine to provide a powerful tool in describing the complete phytoplankton community composition.

Molecular techniques, such as DNA barcoding and next generation sequencing, are the most recent techniques developed to distinguish phytoplankton species. Initially the cost of these techniques was prohibitive, but with time the cost of analysis per sample has reduced and most techniques can be used routinely.

In addition to these three primarily lab-based techniques, instrument development has progressed over the past 30 years so that water bodies can be monitored remotely from moored instruments in real-time. Sophisticated instruments can also trigger the collection of a sample when a specific event occurs, such as a change in the salinity or temperature signal or an increase in the fluorescence signal. A series of moored instruments allows greater temporal and spatial coverage than the "hit and miss" approach traditional manual sampling has provided in the past. Although not mentioned in Part 2, one can imagine that monitoring and sample collection by drones, equipped with cameras and sensors, will be a regular feature of some field work in the not-too-distant future. Advances in machine learning and artificial intelligence have allowed instruments to be developed, such as the Flowcam, Imaging Flow Cytobot, and the Imaging Cytosense, which through image training, determine phytoplankton species composition and cell abundance in a host of environments and applications.

Other instrumentation has been developed to filter phytoplankton and other microbial matter from the water *in situ* and preserve the sample for molecular analysis. The environmental sample processor (ESP) developed by Monterey Bay Aquarium Research Institute is the most established and has been commercialized.

Satellite remote sensing and the development of ocean-color sensors have, over the past 25 years, enabled studies of phytoplankton distribution and dynamics at a global scale. Understanding the global distribution of phytoplankton by region and season has significantly impacted ecological studies and our understanding of oceanic and coastal processes. Recent development of phytoplankton functional type (PFT) algorithms, using satellite data, have shown an ability to determine the dominant algal group or phytoplankton size-class in an oceanic region. With the constant improvement in the number of wavelength bands and band resolution in new ocean-color sensors, the accuracy of these parameters can only improve.

These techniques all provide information, which individually or combined, add to our knowledge of phytoplankton and its role in ecological studies. However, in a world of changing climate, the information from each technique or instrument can be an important input variable in environmental models to predict the future effects on phytoplankton community distribution, harmful algal bloom occurrences, and other water quality issues.

Part 2 provides examples of how researchers have taken the traditional sampling and taxonomic techniques to study phytoplankton in the 21st century. By developing automated remote sampling devices and combining them with artificial intelligence generated image analysis, new molecular techniques, and satellite remote sensing, researchers today are able to discover—more than could ever have been imagined 50 years ago—about the various types of phytoplankton and their place in our world.

Campbell et al. in Chapter 5 discuss how a series of automated Imaging Flow Cytobots (IFCBs) deployed in the Gulf of Mexico have provided early warning of harmful algal blooms to local health authorities, as well as providing a time-series of data to understand the phytoplankton dynamics of the region. Case studies are also provided, where IFCB data in parallel with molecular approaches, such as HTS-metabarcoding and metatranscriptomics, enabled the discovery of previously unknown species and their functional activity in the phytoplankton community.

Smith et al. in Chapter 6 provides a comprehensive review of automated sampling devices and techniques. The most advanced of these is the environmental sample processor (ESP) developed by Monterey Bay Aquarium Research Institute, and a case study of how this instrument has been used in the early warning of a harmful algal bloom is provided.

Bracher et al. in Chapter 7 discuss the use of satellite remote sensing to investigate and understand phytoplankton distribution on a large scale. With the development of constantly improving ocean-color sensors in both number and resolution of bandwidths, the ability to determine, not only phytoplankton biomass or abundance, but also phytoplankton group or size-class is possible. Applications using these satellite-based products are provided.

Baird et al. in Chapter 8 review the development of models to predict phytoplankton processes, from the earliest simple statistical models to the complex and sophisticated models used today. Discussion of how the different models can be applied to harmful algal bloom prediction, community changes due to climate change, and phytoplankton dynamics within a community is provided.

Steven in Chapter 9, as the last chapter in Part 2, discusses how the techniques presented in the book can be used as tools in the management of water quality, aquaculture, and fisheries industries, as well as the emerging phytoplankton-based biotechnology industries, for societal benefit. Several case studies, where the use of taxonomic, remote monitoring, and omics techniques have been used in the management of a specific industry, are provided.

CHAPTER 5

Integrating imaging and molecular approaches to assess phytoplankton diversity

Lisa Campbell, Chetan C. Gaonkar, and Darren W. Henrichs
Department of Oceanography, Texas A&M University, College Station, TX, United States

Introduction

Phytoplankton comprise a wide spectrum of species, functional groups, sizes, morphologies, and life histories. Understanding phytoplankton diversity and its relationship with biotic and abiotic factors can help in deciphering ecosystem function and resilience. Investigations of phytoplankton diversity and dynamics historically have relied on light microscopy to identify and enumerate cells based on morphological features (Elder and Elbrachter, 2010; Hasle, 1978). Studies of the eukaryotic microplankton by conventional light microscopy are labor-intensive, time-consuming, and require a substantial level of expertise. Moreover, limited resolution may fail to distinguish cryptic species, recognize the smaller (<10 μm) cells, or detect rare species in the sample volumes typically examined. Pigment analysis (Jeffrey and Humphrey, 1975) and CHEMTAX (Mackey et al., 1996), though useful at a higher taxonomic level, cannot provide the detailed resolution at the genus or species level often required for monitoring programs or investigations into phytoplankton diversity. Electron microscopy has led to improved identification by deciphering relevant morphological traits, but it still may not be possible to elucidate the presence of cryptic species without employing molecular/genetic techniques (e.g., *Pseudo-nitzschia* species; Lundholm et al., 2006).

Here we highlight the Imaging Flow Cytobot (IFCB) as a tool that can provide continuous automated operation for monitoring the microplankton community and provide early warning of harmful algal blooms (Case study 5.1). From this time series, quantitative estimates of abundance and biomass can be used to determine the relative importance and dynamics of phytoplankton at the genus and, in some cases, species-level (Case study 5.2). Metabarcoding provides improved taxonomic resolution, uncovers cryptic species for more complete estimates of diversity (Case studies 5.3 and 5.4), but provides relative abundance based on copy number, which may not reflect actual abundance, nor does it indicate active segments of a community. Metatranscriptomics can provide a view of functional activity within the phytoplankton community. Integrating these approaches can provide a more detailed analysis of the phytoplankton responses to environmental perturbations (Case study 5.5).

Continuous automated live imaging: Imaging Flow Cytobot

The details of phytoplankton community dynamics and trends over time remain poorly described for the most part owing to a lack of methods suitable for high frequency and sustained sampling. To address this challenge, the Imaging Flow Cytobot (IFCB) was developed to allow continuous automated operation (Olson and Sosik, 2007). The IFCB is a submersible instrument designed to capture images of individual microplankton cells. The IFCB represents a significant advance over traditional flow cytometry for this component of the phytoplankton community. Using a combination of flow cytometric and video technology, IFCB collects a 5-mL sample every ~20 minutes, and then records high quality images of each phytoplankton cell in the sample (for cells ranging in size from ~10 to 150 μm). To provide estimates of cell abundances, machine-learning technology is applied for automated classification so that near real-time reporting of the abundance of individual phytoplankton taxa and community composition is possible (Sosik and Olson, 2007; Henrichs et al., 2021). Long-term deployments are achievable, because antifouling procedures and analysis of standards to monitor instrument performance are run daily as part of the automated operation.

Given the vast amount of data collected during continuous monitoring, an automated system for image classification is essential (Sosik and Olson, 2007). A set of MATLAB® routines for analysis of IFCB data were developed by H. Sosik and are freely available (https://github.com/hsosik/ifcb-analysis/wiki). These include image processing, feature extraction, construction of training sets and classifiers, and classification. Initial machine learning approaches demonstrated accuracy of the classification to be comparable to human experts (Sosik and Olson, 2007). Recent advancements, such as convolutional neural networks, have expanded the capabilities and accuracy of classifications, often at the genus- or even species-level (e.g., González et al., 2019; Henrichs et al., 2021).

The features produced by image processing include simple geometric parameters and an estimate of biovolume. Using a distance map algorithm developed by Moberg and Sosik (2012), the biovolume feature is useful for estimating biomass. Using carbon conversion factors (e.g., Menden-Deuer and Lessard, 2000) biomass per cell can be calculated for individual cells or summed to estimate the total for classes in a specific group, e.g., diatoms.

Among the challenges for implementing IFCB in studies of diversity, however, is having an adequate reference image dataset, in particular for rare species. Additional difficulties may also include cryptic species (e.g., *Chaetoceros*), species with similar morphologies (e.g., ciliates), or species that lack distinctive morphological features (e.g., some athecate dinoflagellates). Lastly, small sized cells <10 μm are difficult to resolve by IFCB.

Chain-forming species of diatoms, such as *Asterionellopsis, Chaetoceros, Pseudo-nitzschia*, or *Skeletonema*, can present a challenge for determining cell abundance from IFCB images. One approach for calibration of cells per chain was accomplished using a set of neural network models developed using precalculated image features as input to predict the number of cells in the image. Using manual counts for a set of images, both single cells and chains of varying lengths, to represent the size range for each class, an ensemble of neural networks was trained and has been employed to estimate cell abundance (e.g., Fiorendino et al., 2021).

An important advantage of the IFCB time series is that all images are recorded, so that data can be reprocessed as new automated classifiers are developed or for additional analyses. Examples of additional analyses include predator-prey interactions (Harred and Campbell, 2014), impacts of parasites (Peacock et al., 2014), life cycle stages and in situ growth rates (Brosnahan et al., 2015; Campbell et al., 2010), different seasonal patterns among individual microzooplankton species (Brownlee et al., 2016), and trends in community structure (see Case study 5.2).

Continuous automated operation of the IFCB in the Gulf of Mexico has produced a decade-long microplankton time series at the Texas observatory for algal succession time series (TOAST). Monitoring at this site has provided a wealth of data that has been utilized to examine phytoplankton dynamics. In addition to providing early warning for harmful algal blooms (Case study 5.1), in situ growth rates estimated from the frequency of dividing cells (McDuff and Chisholm, 1982) have shown that growth rates were near maximal at the initial stages of a *Dinophysis* bloom (Campbell et al., 2010), and decreased when maximum cell abundance was observed. Similar observations have been made from blooms of the dinoflagellates

Brachidinium and *Karenia* (Campbell et al., 2017; Henrichs et al., 2011). In a study of *Alexandrium fundyense* (now *A. catenella*) in Salt Pond, MA, USA, the observed in situ growth rates exceeded those from laboratory-based measurements conducted at the identical temperature, which stressed the value of being able to determine in situ growth rates from IFCB data (Brosnahan et al., 2015).

The time series of cell abundance has also been used in the testing and validation of individual-based models of phytoplankton transport (Henrichs et al., 2015). By starting cells near the TOAST location at a time and date given by the IFCB data, and running the model backwards in time, a potential region of origin for blooms of *K. brevis* was identified (Henrichs et al., 2015). Alternatively, when running the model forward in time, the IFCB cell counts can be used for validation of bloom transport models and timing of arrival (Henrichs et al., 2020). Since the IFCB captures images of all microplankton, models for multiple species can be created and validated against the timing of the appearance of cells in the IFCBs at TOAST.

Data from TOAST has provided new insights into the dynamics of phytoplankton community structure in response to environmental forcing (Anglès et al., 2015, 2019). In addition, near real-time data permits targeted sampling of ongoing phytoplankton blooms. Examples of targeted metatranscriptomics to assess phytoplankton community responses (Case study 5.5), demonstrate the benefit of combining IFCB and molecular approaches to assess the variability and diversity of the phytoplankton community.

Case Study 5.1: Imaging FlowCytobot is a proven technology for Harmful Algal Bloom early warning

Harmful algal blooms (HABs) are a global concern for both human and ecosystem health. During blooms, toxin-producing algae are concentrated in filter-feeding shellfish which, if consumed by humans, can lead to illness or fatality. Given that bloom occurrence is not easily predicted, the most effective mitigation has been early warning of blooms as they develop. Continuous, automated operation makes the IFCB a useful tool for HAB detection. Images are processed in near real-time for immediate reporting at a frequency much greater than is possible by manual sampling. In some cases, hourly resolution has allowed detection of blooms that would have otherwise been missed due to tidal variations (e.g., Campbell et al., 2010), patchiness of the bloom, or by intermittent sampling (e.g., Campbell et al., 2017).

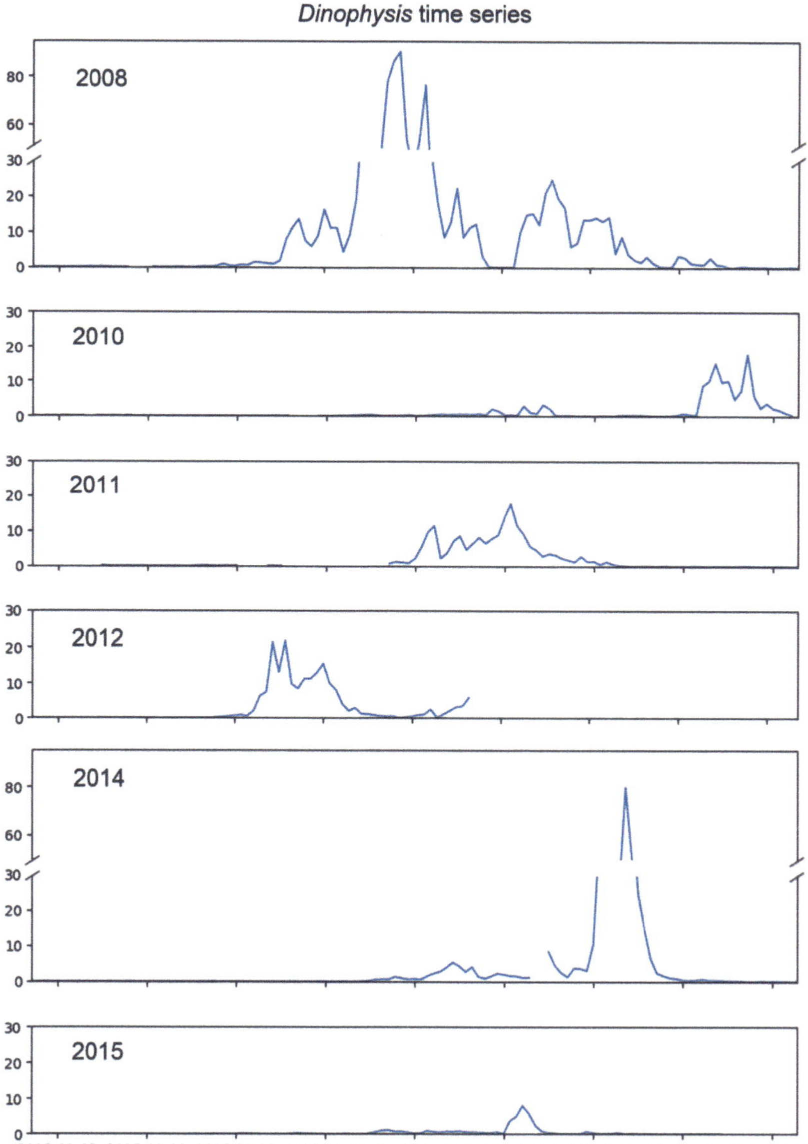

Figure 5.1 Time series of *Dinophysis* abundance in years of HAB events. In most cases, early warning of a potential bloom was indicated by an increase in cell abundance 2 weeks to 1 month in advance of the peak abundance.

In just the first decade of its deployment in the Gulf of Mexico, the high temporal-resolution imaging by the IFCB made it possible to provide early warning for six *Dinophysis* HAB events (Fig. 5.1). The 2008 *Dinophysis* bloom was the first toxic bloom of this species ever observed in Gulf of Mexico waters (Campbell et al., 2010). Although the actual target of the early warning program was *Karenia brevis* (the typical HAB species in the Gulf of Mexico; Steidinger, 2009), *Dinophysis* was recognized, because the IFCB records images of all cells in a sample. The unusually high numbers of *Dinophysis* images prompted testing for the presence of toxin in oysters, and, subsequently, shellfisheries were closed, and no human illnesses were reported (Campbell et al., 2010; Deeds et al., 2010). Since 2011 an automated email system has delivered early warning to state and local health authorities well in advance of cell concentrations that exceed action levels (Campbell et al., 2013). Most significantly, even when detection was missed by routine weekly manual sampling due to patchiness of the bloom, IFCB continuous sampling provided timely detection (Campbell et al., 2017). *Dinophysis* blooms occur in early spring, but the timing has shown considerable interannual variability, ranging from early February to mid-April (Fig. 5.1). In most cases, initiation of a potential bloom, indicated by the first observation of cell abundance above the limit of detection, occurred from 2 weeks (e.g., 2014, 2015) to 1 month (e.g., 2008, 2010) in advance of the peak abundance. Early warning has given state managers sufficient time to establish monitoring and, if necessary, shellfish closures to protect human health.

The continuous monitoring ability of the IFCB, coupled with the quantity of data collected, necessitated the development of an automated system for identifying and quantifying HAB species and, if necessary, the subsequent notification of elevated cell concentrations to state resource managers. Though the original automated notification system was designed to focus on *K. brevis*, the updated image classifier now utilizes a convolutional neural network approach that takes advantage of recent advances in the field of image recognition (Henrichs et al., 2021). The neural network approach also allows for simultaneous detection of multiple HAB species and has resulted in improved detection of *Dinophysis* cells, with fewer false positive images.

Case Study 5.2: Imaging Flow Cytobot for assessing interannual variation in microplankton community composition: the diatoms

The high temporal resolution captured by IFCB sampling enables interannual patterns in the dynamics of phytoplankton taxa to be identified. From the decade-long IFCB time series at TOAST, we observed that the contribution of diatoms to the total microplankton biomass followed a general succession of winter, spring, and summer blooms (Fig. 5.2A). Considerable interannual variability exists in the timing, duration, and magnitude of these seasonal blooms. Much of this variability can be attributed to environmental forcing factors; however, interannual variations in the composition of the diatom community may also contribute (Fig. 5.2B). Of the 27 diatom classes identified in the IFCB time series, repeated patterns of community dominance are evident for some, such as *Ditylum* each winter, or chain-forming species, e.g. *Dactyliosolen fragilissimus and Cerataulina pelagica*, each spring/summer ("DactFragCerat" in Fig. 5.2.B). Decreasing trends in abundance were observed for some groups over the past decade. For example, *Guinardia* was a large fraction of the community each winter from 2008 to 2012; however, it contributed a much smaller fraction in subsequent years. The genus *Chaetoceros* exhibited the reverse pattern and was a relatively minor component of the diatom community 2008–2012, but increased as a fraction of the total diatom community after 2012. Low abundance of *Chaetoceros* species observed in 2011 coincided with one of the warmer-than-average temperature periods and the longest and most severe drought periods reported since 2000 (https://www.drought.gov/states/texas). Though reduced runoff and, consequently, the reduced nutrient inputs may have led to low abundance in 2011, it is unclear what environmental factors led to the increase in abundance for *Chaetoceros* after 2012. After 2012, *Chaetoceros* dominated the community at times, e.g., during summer–fall of 2013 when ~half of the diatom community biomass was contributed by *Chaetoceros*.

Six classes were established in the training set to identify the key species within the *Chaetoceros* category. An expanded view of the TOAST time series for the summer–fall period (July to September) to examine the diversity of *Chaetoceros* taxa revealed the dominance of chain-forming species of *Chaetoceros* and only a minor contribution from single-cell species in most cases (Fig. 5.2C).

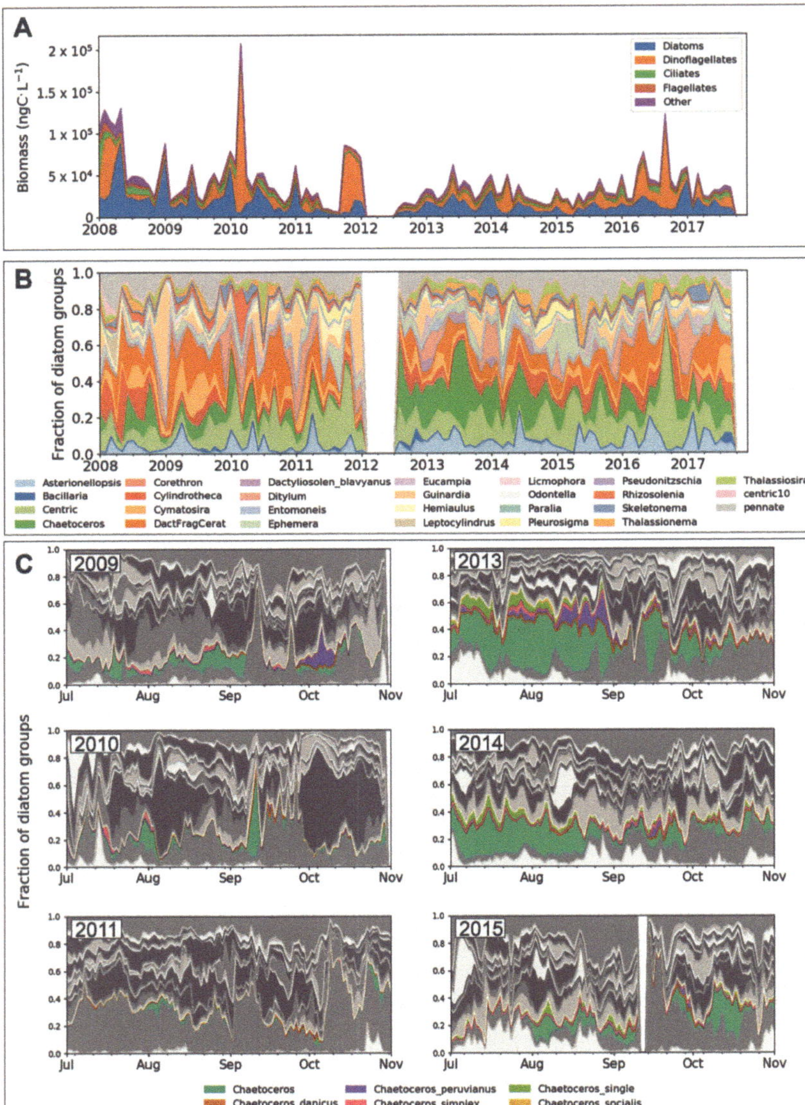

Figure 5.2 (A) Time series of microplankton community composition in the Gulf of Mexico from IFCB deployed at Port Aransas, TX based on carbon estimates (monthly resolution); (B) Time series of diatom community composition (monthly resolution); (C) Interannual variation of *Chaetoceros* species in the summer–fall period 2009–2015 (daily resolution).

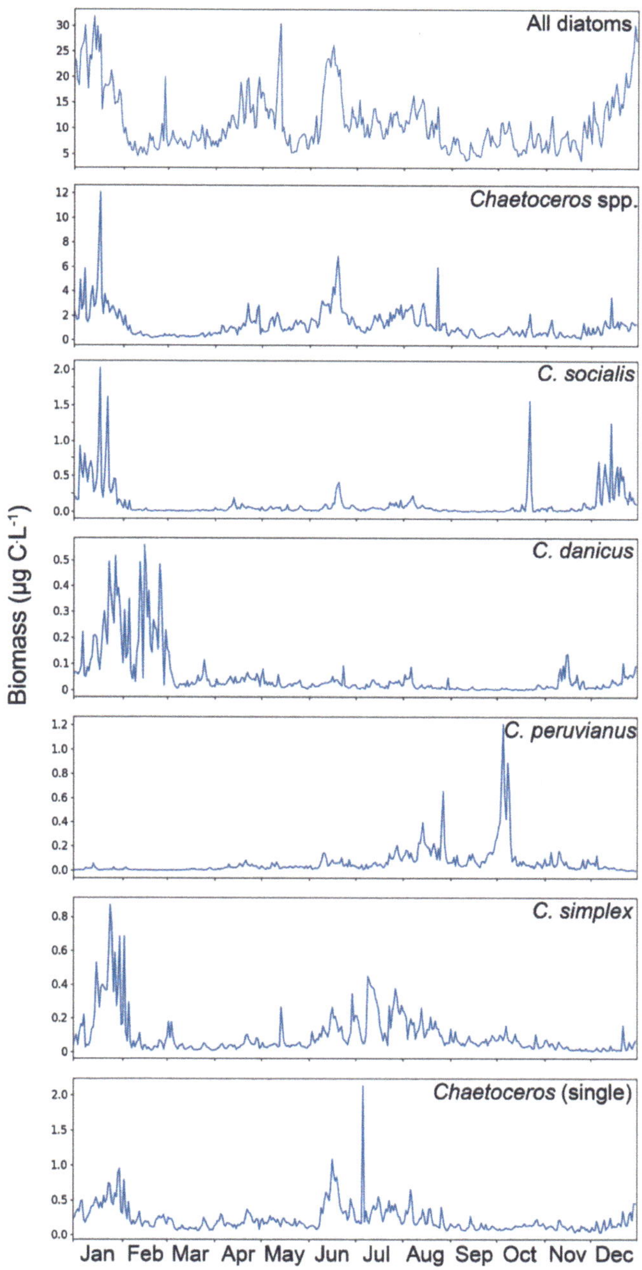

Figure 5.3 Phenology of diatom and individual *Chaetoceros* class biomass, based on IFCB automated classification of images from 2007 to 2018. (a) All diatoms; chain-forming species (b) *Chaetoceros* spp. and (c) *C. socialis*, (d) *C. danicus*, (e) *C. peruvianus*, (f) *C. simplex*, and (g) other single cells.

Chaetoceros socialis is a cosmopolitan chain-forming species that is abundant in coastal regions from the cold waters of the Arctic to warm waters of the Mediterranean Sea or the Gulf of California (Hasle and Syvertsen, 1997). It is considered to be one of the dominant species in the global ocean for its contribution to total biomass (Leblanc et al., 2012). In the Gulf of Naples, Italy, *C. socialis* is a dominant species and present virtually in every month of the year. The geography of the Gulf of Naples and Gulf of Mexico are quite similar, hence it was expected that *C. socialis* would be a dominant contributor to total diatom biomass at TOAST, but this was not the case (Fig. 5.2C).

From the decade-long time series at TOAST, the average timing of diatom blooms shows biomass peaks in winter, spring, and summer (Fig. 5.3). The chain-forming *Chaetoceros* spp. category contributes to the winter and summer blooms, which is not consistent with observations in other parts of the world (e.g., in northeastern Adriatic Sea, Mediterranean coastal zone, and Changjiang River Estuary, China), where *Chaetoceros* is a predominant component of the community during autumn and summer (Bosak et al., 2016; Yang et al., 2014). In the Gulf of Mexico, the chain-forming *Chaetoceros* spp. appear to be less important contributors during the spring diatom bloom. *Chaetoceros socialis*, although only a minor component of the diatom community, appears primarily in winter months (Dec–Jan), along with the single-cell species *C. danicus*. The other single-cell classes of *Chaetoceros*, including *C. simplex*, bloomed in winter and summer (Fig. 5.3). *Chaetoceros peruvianus* was the only species observed in late summer to early fall in most years. In 2013, the timing of the bloom shifted from late summer to early autumn (October). Gómez and Souissi (2007) have reported *C. peruvianus* to be an indicator species of warmer water conditions. Although *C. peruvianus* is a circumglobal species and is observed frequently, it never appears to dominate a bloom (Tanković et al., 2018).

Molecular approaches

The automated sequencing of DNA has made the analyses of genetic material accessible to a wide community of researchers from a variety of fields. Early sequencing technologies focused on the sequencing of a single sequence fragment (i.e., one gene); in this regard, the advancement of sequencing technology over the last decade has been extremely beneficial. The rise in output from high-throughput sequencing technologies (HTS) from ~15 megabases (454 pyrosequencing) to upwards of 3 terabases (Illumina NovaSeq) has been coincident with a dramatic lowering of costs (Reuter et al., 2015). This pairing has resulted in the use of HTS becoming rather commonplace for a wide variety of scientific fields, ranging from medicine (e.g., Lightbody et al., 2019) to soil ecosystems (e.g., Roelofs et al., 2008). HTS applications have also become important in the marine realm, with numerous studies focused on both the prokaryotic and eukaryotic domains (e.g., Alberti et al., 2017; Johnson and Martiny, 2015).

Metabarcoding

DNA "barcoding" has been one of the revolutionary steps in the advancement of microplankton ecology since the 1990s. DNA barcoding focuses on the identification of individual taxa (ideally at the species level) using a standardized genetic marker (Hebert et al., 2003). The methodology includes amplifying a taxonomic marker gene from a species-based genomic DNA sample, and then sequencing. Metabarcoding is a modification that combines community-based barcoding and HTS. Instead of a single sequence, a group of targeted sequences from an environmental sample are amplified using a universal primer, sequenced using a high-throughput DNA sequencing platform, and annotated using a taxonomic reference database (e.g., PR^2; Guillou et al., 2013). The resulting set of sequences and frequencies acts as a proxy of the microplankton community composition, which enables in-depth identification and characterization of the microplankton diversity.

The value of HTS-metabarcoding is the ability to unveil the species diversity, which includes cryptic species or plastic morphologies and rare species that may have been missed or confused using microscopic approaches. HTS-metabarcoding has reshaped the view on how to address the microplankton community and what we know about the diversity. Revealing the species diversity at high resolution helps to understand the spatial and temporal changes occurring in the microplankton community (de Vargas et al., 2015; Piredda et al., 2018; Gaonkar et al., 2020).

For surveys focusing on the environmental microplankton, "universal" eukaryotic primers targeting the hypervariable regions of 18S rDNA (e.g., V4, V8–V9 regions) have been used. The primers are modified with adaptors and indexes, which bind to the flow cell of the high throughput sequencer. Following PCR amplification, the products are sequenced, processed to produce contigs, and pre-clustered and annotated using a reference database (e.g., Gaonkar et al., 2020).

Even though HTS-metabarcoding can unveil microplankton diversity, there are still limitations to the approach. These include choice of marker region, efficacy of the universal primers, bias in PCR amplification, presence of introns, copy numbers, taxonomically validated reference sequences, and clustering approaches. The first step in metabarcoding includes the main question: *which are species of interest?* This sounds like a simple question before knowing the pros and cons of the molecular resolution of the targeted marker region. Although a region of the mitochondrial gene *COI* (cytochrome c oxidase subunit I) has been used successfully for animals, there is no universally accepted gene for barcoding algae, due to inconsistent amplification (Hall et al., 2010). Although both the V4 and V9 regions of the 18S rDNA have been extensively used to assess protistan diversity, not all protistan groups are addressed without bias. Additionally, not all groups have similar level of resolution for one given gene marker. At the phylum level, the V4 region has been used extensively for diatoms, whereas, for most dinoflagellates, the V9 region provides more resolution. For a phylogenetic study at the genus level, the improved resolution of HTS-metabarcoding may change the understanding of the ecology of the groups, e.g., Chaetocerotacean diversity (Gaonkar et al., 2020). Different levels of resolution can be seen among genera using different marker regions (e.g., *Heterocapsa* species, see Case study 5.4). Hence, the need for multiple markers may be essential to assess species diversity, especially ecological indicator species.

The second major limitation of HTS-metabarcoding is the choice of universal primers capable of amplifying all microplankton species in the community without bias. Unfortunately, this goal is simple in theory, but complicated in practice. Some of the universal primers used do not amplify certain groups (e.g., due to primer mismatch for species in the ciliate genus *Mesodinium*; Johnson et al., 2016). Such omissions can alter our understanding of the community structure, species interactions, or species responses when looking at the overall microplankton community. Also, this bias is not limited to selectivity of the primers, but also based on PCR amplification, where a dominant species is easily targeted compared to its counterpart with lower abundance or a rare species. Furthermore, longer PCR cycles can add amplification errors to the PCR products; hence, the use of good quality thermonuclease enzyme is needed for reliable PCR products.

The presence of introns may not be a major limitation. Introns were observed in only a few species in the genus *Chaetoceros* (e.g., Gaonkar et al., 2018). These introns varied in size from ~100–150+ bp, which can be captured if using a 2 × 300 bp sequencing kit. But lack of this knowledge can hinder our understanding of the protistan community. Another limitation is the variation of copy number. The marker region used in most studies are multicopy genes of rDNA, which are present in "*n*" copies in different species. This number can vary among species within the same genus. This variability can produce not only bias in read abundance, but also add to PCR and sequencing noise. Taxonomically validated reference datasets help in appropriate identification of the sequences processed from the HTS-metabarcoding approach (e.g., del Campo et al., 2018; Gaonkar et al., 2018). Lack of reference sequences can hinder identification or characterization of a species, but will not limit the genus diversity, as it would still provide information regarding the presence of additional species. The last limitation involves the clustering approach used to assess the community structure. Although clustering approaches are a simple way to denoise the metabarcoding data, the similarity threshold used can drastically affect the outcome. Given the lack of knowledge of the resolution of the marker employed, e.g., in one genus a species may be differentiated at 98%, whereas in another genus it may not be possible to differentiate at 100% similarity, clustering approaches can really hinder elucidating the community structure.

Case Study 5.3: High-throughput sequencing-metabarcoding to assess diatom species diversity

HTS-metabarcoding has become an essential tool for assessing the diversity of the microplankton community. This approach also provides assistance in exploring and revealing the hidden as well as the rare species diversity. Examination of global patterns of diatom diversity using metabarcoding has found the genus *Chaetoceros* to be among the most dominant planktonic diatom genera and most diverse with more than 200 taxonomically validated species (De Luca et al., 2019; Malviya et al., 2016). The genus is common in coastal waters and, at times, dominates the phytoplankton community, where multiple species can be present simultaneously. The lack of distinctive morphological traits makes species identification challenging. The presence of cryptic diversity also exacerbates the difficulty in identification. Molecular markers targeting ribosomal DNA have been extensively used to differentiate among the closely associated *Chaetoceros* species and for understanding their evolution (Kooistra et al., 2010; Gaonkar et al., 2018; De Luca et al., 2019). HTS-metabarcoding has revealed a higher diversity in *Chaetoceros* than previously known in the Gulf of Naples, Italy, with three years of seasonal diversity showing presence of multiple species (n = 70; Gaonkar et al., 2020). Almost half of the species found were neither taxonomically characterized nor had any reference sequence, which suggested the presence of a large component of rare and unknown species. In terms of geography and hydrography, both the Gulf of Naples and Gulf of Mexico have similar structure. Hence a similar study was conducted to assess the overall diversity within *Chaetoceros* in the Gulf of Mexico.

A maximum likelihood (ML) phylogeny was generated using raxmlGUI v2.0 (Edler et al., 2019) with 26 outgroups and 150 Chaetocerotaceae reference sequences. A total of 208 HTS-metabarcode haplotypes obtained from samples collected on a recent cruise along the Texas coast in the Gulf of Mexico (Campbell and Henrichs, 2020) were also included to generate a ML tree (Fig. 5.4). The resulting ML tree resolved into a series of well-supported clades. Each clade is represented by a dominant OTU, and some associated minor haplotypes. Most of the solitary haplotypes were treated as taxa, whereas those in the polytomies were defined as taxa depending on the number of dominant haplotypes or associated references. A total of 49 Chaetocerotacean species (48 *Chaetoceros* and one *Bacteriastrum*) were identified from the Gulf of Mexico dataset, where 22 of the haplotypes were associated with the taxonomically identified Chaetocerotacean species; two of these matched cultured *Chaetoceros*, but were not taxonomically identified (seven of these were from environmental clones), and 18 haplotypes were unassigned to any of the known Chaetocerotacean species. Among the 49 Chaetocerotacean haplotypes, 14 haplotypes had reads numbering >1000. The *Bacteriastrum* species found in this study did not match any of the available reference sequences. The most significant outcome of this HTS-metabarcoding study was the ability to identify new species. One such example is in the *C. curvisetus* clade, where the *C. curvisetus* species identified in Gulf of Mexico is distinct from the other known *C. curvisetus* species (Fig. 5.4).

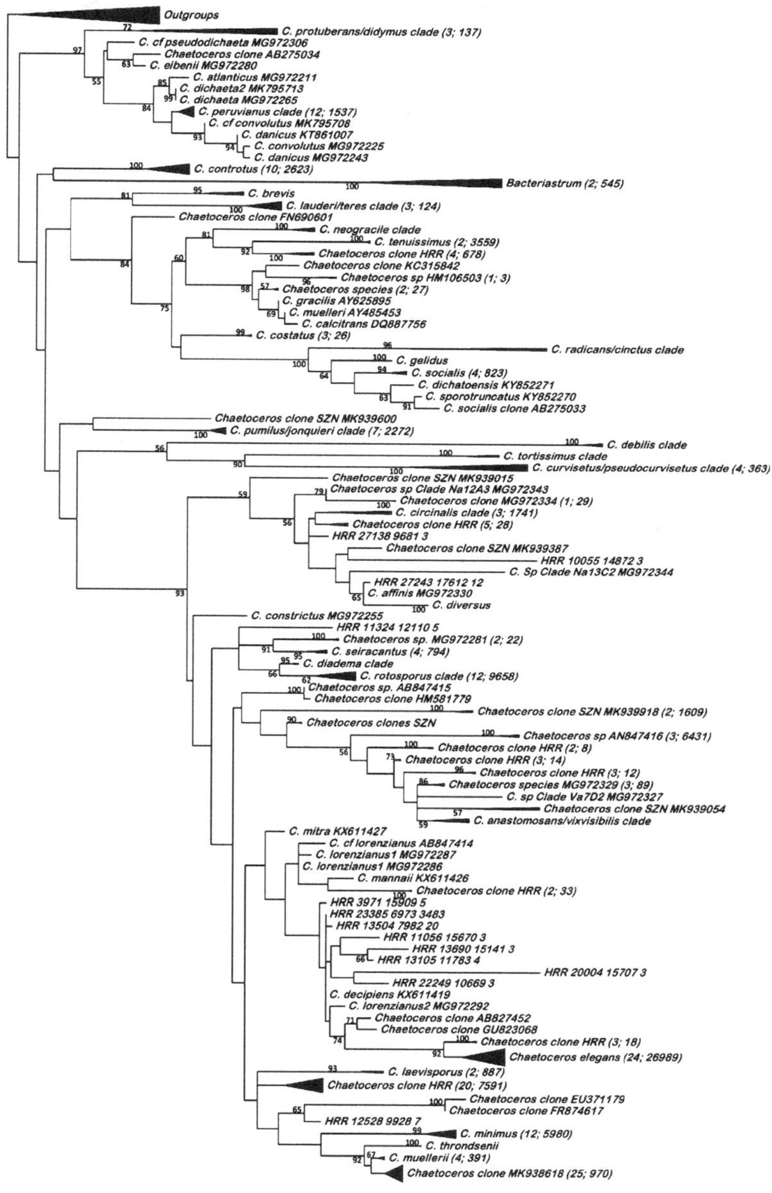

Figure 5.4 Maximum likelihood tree inferred using raxmlGUI v2.0 (Edler et al., 2019) from the 208 validated Chaetocerotacean metabarcode haplotypes from the Gulf of Mexico, 150 Chaetocerotacean references, and 26 outgroup sequences. The OTU ID includes dataset name, unique OTU number, and the last block represents the read number. Haplotypes with well-supported clades are collapsed, and IDs are shown in parentheses with a clade name, total haplotype counts, and total read numbers. Numbers on the internodes indicate the bootstrap values (1000 replicates) if $\geq$50. Complete phylogenetic tree is available at https://datadocs.bco-dmo.org/docs/HRR/data_docs/HRR_Chaetocerotacean_phylogeny.pdf.

The V9 region is a second hypervariable region that can be used in metabarcoding studies. This barcode was used in the Tara expedition for assessing the global protistan diversity (de Vargas et al., 2015). Previous reports have shown that most of the terminal taxa recovered in the 18S rDNA tree were recovered also in the V9 region (Gaonkar et al., 2018). The reason for not selecting the V9 region for this case study of *Chaetoceros* was due to the poor phylogenetic resolution, which affects placement of unknowns (Gaonkar et al., 2018). In addition, the reference sequences for the Chaetocerotacean species in the V9 region are often incomplete, as the region is at the very end of the 18S rDNA sequence.

The Chaetocerotacean diversity observed in both the IFCB images and HTS-metabarcoding data were similar (Fig. 5.5). Species in this family that were observed with both techniques included *Bacteriastrum* sp., *C. affinis*, *C. contortus*, *C. curvisetus*, *C. diadema*, *C. elegans*, *C. holsaticus*, *C. minimus*, *C. muelleri*, *C. pseudocurvisetus*, *C. peruvianus*, *C. rotosporus*, *C. simplex*, *C. socialis* and *C. tenuissimus*. Apart from these, two of the taxonomically unrepresented species of *C.* sp clade Na28A1 and *C.* sp clade Na17B2, the latter being a single-cell species, were also found.

Diversity within the single-cell *Chaetoceros* is intriguing, as these cells form a major component of the *Chaetoceros* biomass, especially during the winter and summer (Fig. 5.3). Most of these species are difficult to characterize solely based on a single IFCB image, because of their small size (< 10 μm). Coupled with the lack of distinctive morphological traits, cells in this category could not be classified to species; however, by HTS-metabarcoding, the single-cell *Chaetoceros* were identified as *C. jonquieri*, *C. minimus*, *C. muellerii*, *C. pumilus*, and *C. tenuissimus*. From the HTS-metabarcoding data it was possible to characterize only five out of fifteen species, as most of the OTUs were annotated as unknown *Chaetoceros* or uncultured clones (Fig. 5.5), which suggested the lack of appropriate knowledge of these species. This opens up the opportunity to explore these *Chaetoceros* single cells more in-depth.

Images from the IFCB for chain-forming species also included cryptic diversity. Parallel HTS-metabarcoding data identified cryptic species within the clade representing *C. protuberans* and *C. didymus, with one OTU each*; three dominant OTUs in the *C. contortus* clade;

four dominant OTUs in the *C. lorenzianus* clade; four in the clade representing *C. holsaticus* and *C. seiracantus*, which are morphologically similar to *C. diadema*. This case study demonstrated that the integrated approach of IFCB and HTS-metabarcoding enabled the discovery of multiple species that are new to science.

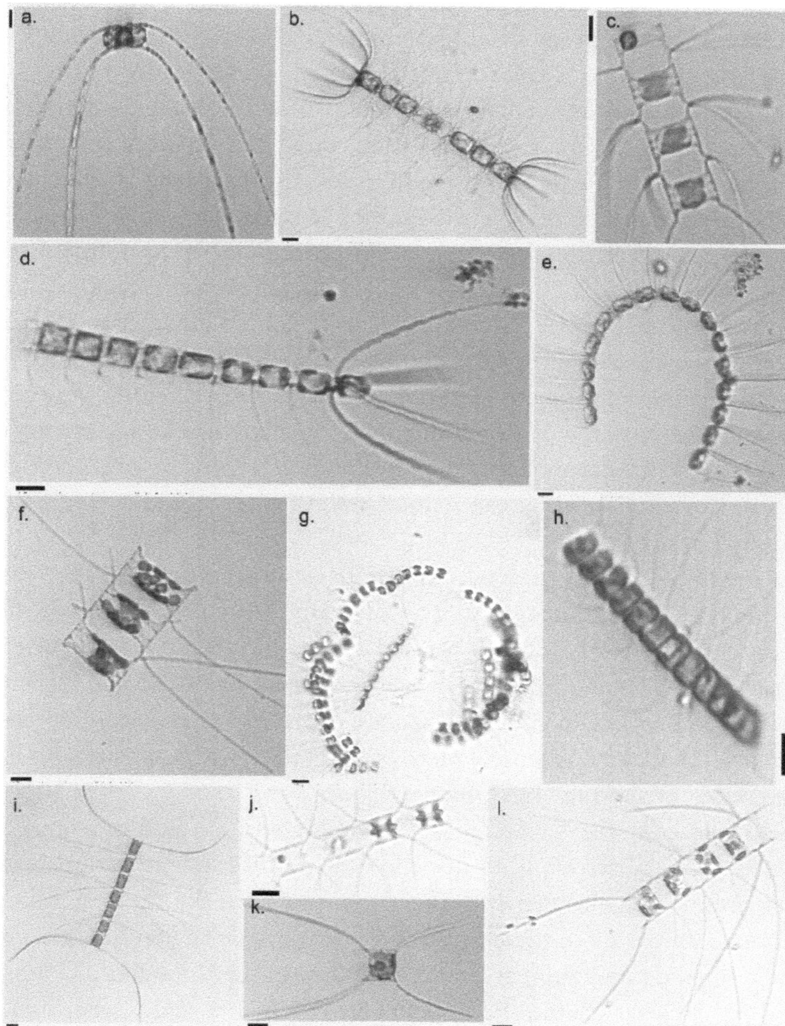

Figure 5.5 Chaetocerotacean images corresponding to species identified by both IFCB and HTS-metabarcoding. Illustration depicts (a) *C. peruvianus*, (b) *Bacteriastrum* sp., (c) *C. seiracanthus*, (d) *C. contortus*, (e) *C. curvisetus*, (f) *C. elegans*, (g) *C. socialis*, (h) *C. costatus*, (i) *C. affinis*, (j) *C. protuberans*, (k) *Chaetoceros* sp., (l) *C. deciptiens*. Scale bar = 10 μm.

Case Study 5.4: Selection of appropriate genetic marker: example for *Karenia*

Selection of the appropriate marker for metabarcoding analysis is critical for studies of microplankton diversity. Although there is no universally accepted barcode for algae, owing to a limited availability of reference data, at present the best curated database is PR^2, which focuses on eukaryotic small subunit (SSU) ribosomal RNA and DNA sequences that are annotated using a multi-level taxonomic assignment (Guillou et al., 2013). In this case study, the use of different molecular marker regions (V4 vs. V8–V9 region) to identify species is examined for a dinoflagellate bloom in the Gulf of Mexico.

Cryptic speciation has been reported among the dinoflagellates (e.g., Anglès et al., 2017; Cai et al., 2020; Daugbjerg et al., 2019; Gottschling et al., 2005; Montresor et al., 2003; Prabowo et al., 2016). The discovery of such species complexes indicates that morphology-based species determinations should be considered carefully, and underlines the importance of linking a type specimen to DNA sequence data since species cannot always be distinguished based on morphology alone; however, imaging can help to resolve cases where barcoding is limited.

In fall of 2017, a large bloom of *Karenia* was observed in the Gulf of Mexico, off the Texas coast. The fall season (August–October) is the typical period for blooms of the brevetoxin-producing HAB species, *Karenia brevis*. Not all *Karenia* species produce brevetoxin, and blooms are not monospecific (multiple *Karenia* species are present). Appropriate and accurate identification of this species is necessary should barcoding be used for HAB identification.

In this case, the V4 and V8–V9 barcode regions used individually for HTS-metabarcoding do not allow delineation of the members of the genus. The phylogenetic trees produced from the metabarcoding sequences for each of these markers do not provide an identical picture for species identification and differentiation. In the V4 region, *K. brevis* and *K. mikimotoi* cannot be resolved separately as a clade, whereas *K. selliformis* forms a clade alongside. *K. bicuneiformis* and *K. papilionacea* form another sister clade (Fig. 5.6A). In contrast, in the V8–V9 region, the two species, *K. mikimotoi* and *K. brevis* can be differentiated, but *K. selliformis* is identical to *K. mikimotoi* (Fig. 5.6B). Given the low resolution of this marker region, it becomes difficult for appropriate identification of species. Morphological identification based on IFCB images confirmed the dominant species in the bloom as *Karenia selliformis*. By integrating IFCB imaging with HTS-metabarcoding, the species could be identified and characterized appropriately.

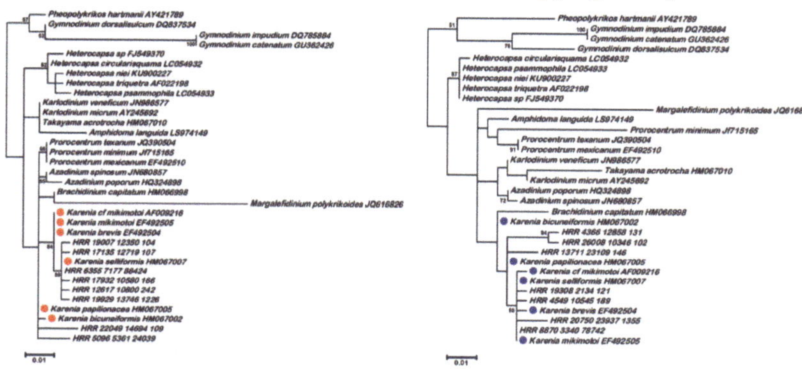

Figure 5.6 Selection of appropriate HTS-metabarcoding markers for characterizing species. (A & B) Maximum likelihood trees inferred using raxmlGUI v2.0 (Edler et al., 2019) for *Karenia* OTUs in samples collected from the Gulf of Mexico. Phylogenies based on (A) V4 region and (B) V8–V9 regions of 18S rDNA gene revealed different results. The OTU ID includes dataset name, unique OTU number, and the last block represents the read number. Numbers on the internodes indicate the bootstrap values (1000 replicates) if ≥50. Only full length 18S rDNA sequences were used for phylogeny. *Karenia* species are indicated with a dot. Only OTUs with read abundance more than 100 were included for clarity.

Metatranscriptomics

Metagenomics entails the sequencing of DNA from all organisms present within a sample. Metagenomic data provide not only a glimpse of the genetic diversity of the organisms present within a sample, but also provide an inventory of the specific genes present in that sample that are, at least theoretically, available to respond to changes in the environment. However, this inventory provides little to no information about what genes are actively being used at any moment in time. Metatranscriptomics is a natural complement to the field of metagenomics. In contrast to metagenomics, metatranscriptomics is focused on the sequencing of RNA (specifically mRNA) present within a sample. Samples are often collected from a spatial or temporal design and used to conduct an analysis of differential gene expression (Shakya et al., 2019). By comparing the number of transcripts present among samples, and combining them with differences in environmental parameters, the community response to these environmental changes can be estimated (Moran et al., 2013).

To assess the transcriptional response of a community, the mRNA from that community must first be enumerated. Community samples can be collected by filtering seawater through a filter to select different size classes of organisms or by centrifugation of the sample (with or without size selection) and subsequent removal of seawater. The RNA is then extracted from the filter (or pellet, depending on collection method) for purification. The majority of this extracted RNA is likely to be ribosomal RNA and uninformative for looking at gene expression. However, when the organisms of interest are eukaryotes, the mRNA from the sample can be enriched by selecting for the poly-A tail of mature transcripts. The enriched sample is then sequenced using a HTS approach. The output from a HTS run is a collection of short reads, where each individual read is assumed to be representative of one copy of a transcript. This approach can produce a wealth of data (e.g., ~20 billion paired end reads for a NovaSeq 6000 S4 run; http://www.illumina.com/); however, the resulting read data must be cleaned and mapped before the quantification of relative gene expression can be attempted. There is difficulty in this approach. Many species of plankton lack a reference transcriptome to which reads can be mapped (Shakya et al., 2019). To account for this, reads from a sample are often assembled into transcripts first. These assembled transcripts can then be functionally annotated based on their similarity to transcripts from known genes. The downside to the assembly approach is the lack of taxonomic assignment of transcripts. It can be difficult to identify from which organism a transcript originated. This can be overcome, to a degree, by use of reference transcriptomes available as part of the Gordon and Betty Moore Foundation's Marine Microbial Eukaryotic Transcriptome Sequencing Project (MMETSP; Keeling et al., 2014). The MMETSP provides over 650 assembled reference transcriptomes from eukaryotic marine microbes. This database of transcriptomes can be used to provide a base taxonomic assignment for the transcripts assembled from a sample, but it must be noted that the available reference transcriptomes only represent a fraction of the phytoplankton diversity present in the field. The next step in annotation is the functional annotation of the transcript. This can be accomplished in a number of ways, and different authors often use different methods and databases. Several common approaches include the following: a BLAST search against the non-redundant database hosted by the National Center for Biotechnology Information (NCBI; https://blast.ncbi.nlm.nih.gov/Blast.cgi), mapping to the Kyoto encyclopedia of genes and genomes database (KEGG; https://www.genome.jp/kegg/), and InterProScan (https://www.ebi.ac.uk/interpro/search/sequence/). The latter two of the techniques require the determination of putative amino acid sequences from the transcript before searching for similarities (e.g., TransDecoder; https://github.com/TransDecoder/TransDecoder/).

Once the transcripts have been assembled and annotated, the reads obtained from the samples are mapped back to the transcripts. The number of reads mapped to each transcript is then counted and used as a relative expression level for that transcript in that sample. Though a metatranscriptome obtained from a sample can reveal what genes were being expressed in that sample at the time of collection, it is often more informative to compare the expression of genes from one sample to the expression of the same genes in another sample or samples. These other samples could be from completely different environments or from the same environment at different time points. By mapping the reads from all samples to the same set of reference transcripts, the changes in community gene expression in response to changes in the environment can be identified.

Case Study 5.5: Targeted metatranscriptomic analysis: Karenia blooms

The ability of the IFCB to provide cell concentration information in near-real time can be further leveraged to help guide field sampling of blooms over time. In this case study, data from two bloom events are presented to illustrate the benefit of having continuous cell concentration data regarding the phytoplankton community over time and how that information can be used to help guide sampling and interpretation of results. Highlighted here are some of the significant changes in gene expression from a time series of samples collected during a 2015 bloom of *Karenia brevis*, and a time series of samples collected after Hurricane Harvey in 2017. The two, time series sets of samples were each independently sequenced, assembled, annotated, and analyzed for differential expression using the techniques mentioned above.

In September 2015, a bloom of *K. brevis* first appeared near the TOAST station at Port Aransas, Texas and persisted in the region for over a month. A total of 11 samples (3 timepoints in triplicate; 1 timepoint in duplicate) were collected from this bloom. Bloom development was monitored from cell concentrations estimated from IFCB imagery for the first three timepoints. See Table 5.1.

Table 5.1 Metatranscriptomic analysis of a 2015 bloom of *Karenia brevis*.

Sample No.	Date	Location (Lat/ Lon)	*K. brevis* concentration (cells·ml^{-1})
1	9/15/2015	27.838° N, 97.050° W	127
2	9/29/2015 (ebb tide)	27.838° N, 97.050° W	34
3	9/29/2015 (flood tide)	27.838° N, 97.050° W	100
4	10/13/2015	27.776° N, 97.391° W	~22,000[a]

[a] Sample manually counted by microscopy. Data provided D. Erdner, University of Texas Marine Science Institute.

The 11 samples collected from the bloom of *K. brevis* in 2015 produced a total of 993,652 assembled transcripts with 539,236 (54.3%) transcripts able to be taxonomically annotated using the MMETSP database. Of these taxonomically annotated transcripts, 402,702 (74.7%) were identified as originating from *K. brevis*. The high percentage of annotated transcripts identified as *K. brevis* was not surprising, given that the sampling protocol targeted a bloom of *K. brevis* based on IFCB cell concentrations. The diversity of species represented in the metatranscriptome was backed up by imagery captured by the IFCB. A phylogeny of taxonomically annotated transcripts showed a diversity of groups present, many of which were also observed in IFCB imagery (Fig. 5.7). For three of the timepoints, the fraction of mapped reads assigned to *K. brevis* transcripts showed a relative increase in number that tracked the cell concentration of *K. brevis* calculated from IFCB counts (Fig. 5.8A). However, the first timepoint showed a high fraction of reads mapped to *K. brevis* for a cell concentration that was similar to the third timepoint. This result was not completely unexpected. Though the first and third timepoints displayed similar concentrations of *K. brevis*, the relative abundance of other organisms differed between the two timepoints. The first timepoint was treated as the starting condition for comparisons of changes in gene expression in *K. brevis* for subsequent timepoints. Relative to the first timepoint, the second (2v1) and third (3v1) timepoints showed similar patterns of changes in gene expression. There was a general increase in the expression of proliferating cell nuclear antigen (PCNA) and cell division cycle and control genes (Fig. 5.9). This increased expression of genes involved in cell replication suggests that the cells were in an environment conducive to growth of the bloom, and could be an indicator that the bloom will continue to persist (Deng et al., 2019).

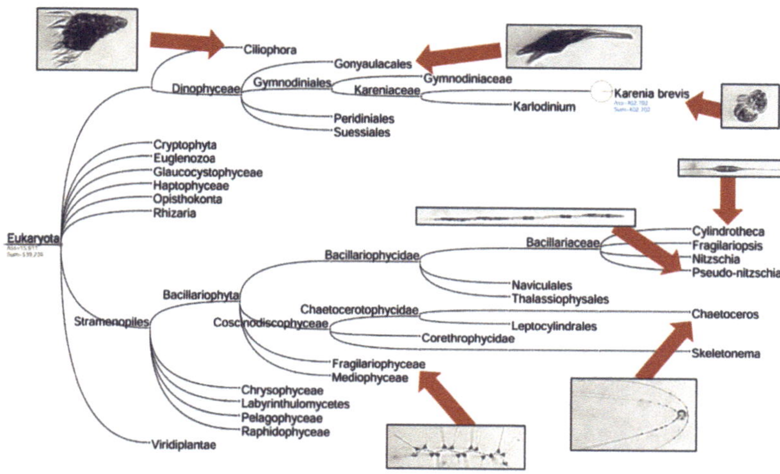

Figure 5.7 Phylogeny of taxonomically annotated transcripts from the 2015 metatranscriptome collected during the bloom of *Karenia brevis* at Port Aransas, Texas. Representative images of several taxa identified from the IFCB images and the metatranscriptome transcripts. Branches have been collapsed for clarity. Size of circle near *Karenia brevis* reflects the proportion of total transcripts (74.7%) that were annotated as *K. brevis*.

There are two possibilities for the observed increase in expression of genes involved in the acquisition of nitrogen (nitrate reductase, NR; nitrate transporters, NRT): (1) the cellular requirement for nitrogen could be increased due to growth and replication and/or (2) the extracellular concentration of nitrogen is low. Increased expression of nitrogen transporters has been observed under low nitrogen concentration conditions (Morey et al., 2011). The increase in transporters may help to scavenge what nitrate is available in the environment. The similarity between the 2v1 and 3v1 comparisons is likely due to the fact that they were both collected on the same day, several hours apart, but during different parts of the tidal cycle (timepoint 2 was an ebb tide; timepoint 3 was a flood tide). In contrast, the fourth timepoint showed decreased expression of genes involved in cell replication and growth. Though the cell concentration at this timepoint was two orders of magnitude greater than the other three, this increase was in part due to physical concentration of the cells at the point of

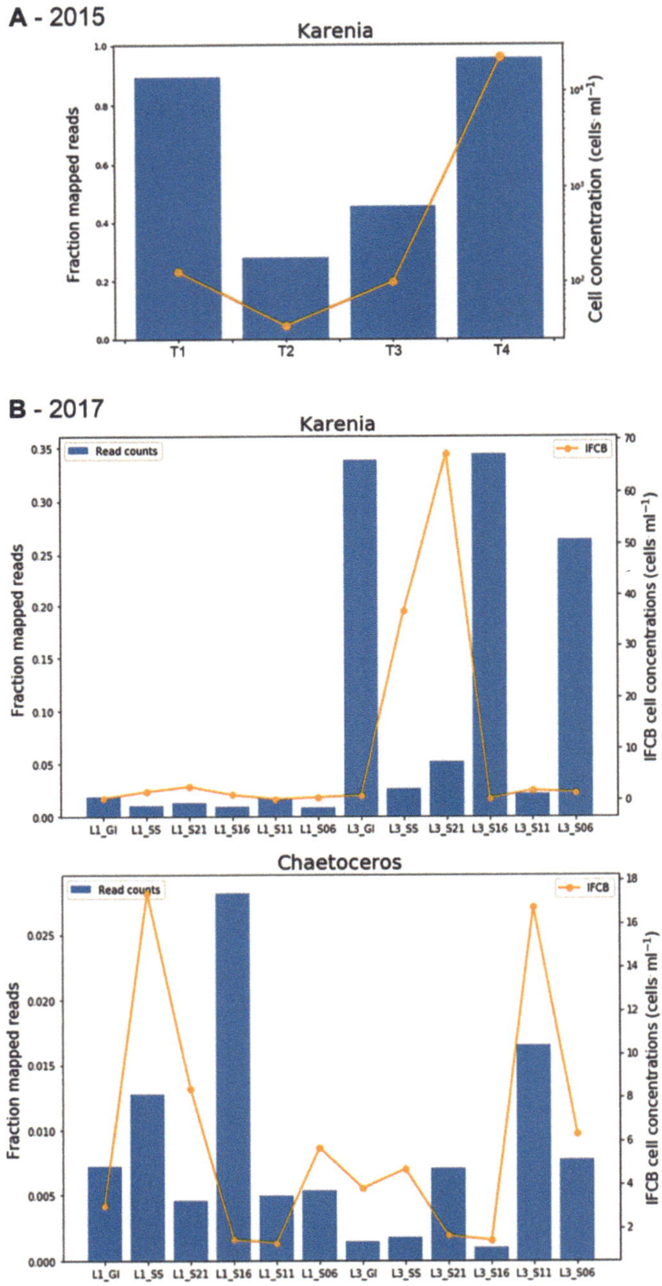

Figure 5.8 Comparison of the fraction of mapped reads (blue bars) and the cell concentrations (orange lines) of *Karenia* and *Chaetoceros* from the (A) 2015 metatranscriptome samples and (B) the 2017 metatranscriptome samples. X-axis shows different samples collected (A) over time and (B) at different stations.

collection. A concentration of cells this dense is likely to exhaust available nutrients and subsequently limit their ability to replicate. This could be an indication that the bloom is no longer growing, but provides limited information about its ability to persist in that environment.

Table 5.2 Locations for post-Hurricane Harvey samples.

Sample No.	Date	Location (Lat/ Lon)
1	9/8/2017	28.976° N, 94.915° W
2	9/15/2017	28.942° N, 95.302° W
3	9/23/2017	28.959° N, 95.094° W
4	9/30/2017	28.959° N, 95.094° W
5	10/27/2017	28.942° N, 95.302° W

In 2017, Hurricane Harvey made landfall in Texas and released a record amount of rainfall in the greater Houston region (Fiorendino et al., 2021). In the aftermath of this 1 in 1000-yr hurricane, a large pulse of freshwater flowed out of Galveston Bay and southward down the coast of Texas. A total of 20 samples (4 time points in triplicate; 1 timepoint had 8 replicates) spanning a two-month period were collected from near the TOAST station at Surfside Beach, Texas (see Table 5.2). A total of 1,242,721 transcripts were assembled from all metatranscriptome samples. Of these, a total of 61,939 transcripts were taxonomically identified to be from the genus *Karenia*. Though it was noted above that transcripts can only be taxonomically annotated if a reference transcriptome is available, in some cases a phylogenetically close relative can provide a near enough match to the target organism for sequences to be putatively identified. In this case, metabarcoding and IFCB imagery were used to identify the *Karenia* species as *K. selliformis*. There was not a reference transcriptome for *K. selliformis* available, so the reference transcriptomes for *K. brevis*, *K. mikimotoi*, and *K. papilionacea* were used to taxonomically annotate the assembled transcripts from the Hurricane Harvey metatranscriptome (Ryan and Campbell, 2016). Although the presence of low numbers of other *Karenia* species cannot be ruled out, IFCB imagery provided strong evidence that the transcripts identified as *Karenia* spp. originated from *K. selliformis* and not *K. brevis*. The fraction of mapped reads remained low at stations where IFCB cell abundance for a group was also low (e.g., *Karenia* L1, Fig. 5.8B). However, at stations where the IFCB showed an increase in cell abundance, the fraction of mapped reads did not always track the relative cell abundance well (e.g., Karenia L3, Fig. 5.8B). The first timepoint was collected approximately two weeks after the initial landfall of Harvey and consists of data from three separate stations in close proximity to each other. The remaining four timepoints were compared relative to this first timepoint. The time series of gene expression for *K. selliformis* exhibited decreased expression of PCNA and cell division genes (Fig. 5.9). There was also a decrease in the expression of nitrate transporters and nitrate reductase genes. One possibility for this result is that the effects of the large freshwater pulse were stressing cells and limiting their growth. A similar pattern of expression was also observed in timepoint 3, with the exception of PCNA genes. Timepoint 3 showed increased expression of PCNA genes and may indicate that cells were beginning to replicate. Timepoint 4 exhibited a very different pattern that is likely due to the cell concentrations observed at this timepoint. A bloom of *K. selliformis* was observed in this region during this time (Fig. 5.8B). Increased nitrogen uptake as a result of the increased cell abundance may have reduced available nitrogen within the bloom and contributed to the increase in expression of some nitrate transporters and nitrate reductase. There was also a decrease in expression of several PCNA and cell division transcripts that suggests that growth was limited at this timepoint, a growth pattern consistent with observations of cell division during blooms of *Dinophysis* and *Brachidinium* (Campbell et al., 2010; Henrichs et al., 2011). The last timepoint was collected almost two months after Hurricane Harvey landfall and showed an increase in expression of PCNA and cell division genes. By this time, the impacts of the storm appear to have been reduced and suggest that the *K. selliformis* cells had resumed growth and cell division.

Comparison of these two metatranscriptome datasets highlights the observed differences and similarities in the response to changing environmental conditions, but is also confounded by two factors. First, the target species in each dataset is different. They both belong to the genus *Karenia* and may provide important insights into how different, yet closely related, species respond to environmental changes. The second factor is the type of environmental changes experienced by the cells. In 2015, there was no massive rainfall event that contributed to a rapidly changing environment. Instead, a bloom of cells was exposed to a coastal environment and the response of the bloom to this exposure was highlighted. In 2017, the large volume of rainfall from Hurricane Harvey and resulting plume of freshwater had the potential to dramatically impact the phytoplankton community.

The benefits of combining the three different approaches (IFCB, metabarcoding, and metatranscriptomics) were observed in these two datasets. The 2015 bloom of *K. brevis* took advantage of the IFCB time series for enumerating the cell concentration at each timepoint. The 2017 dataset benefitted most from the combination of these techniques. The IFCB images showed that the *Karenia* species present was not *K. brevis*, but was predominantly *K. selliformis*, which was confirmed by the metabarcoding results. This allowed us to putatively confirm the transcripts annotated as *K. brevis* originated from *K. selliformis*.

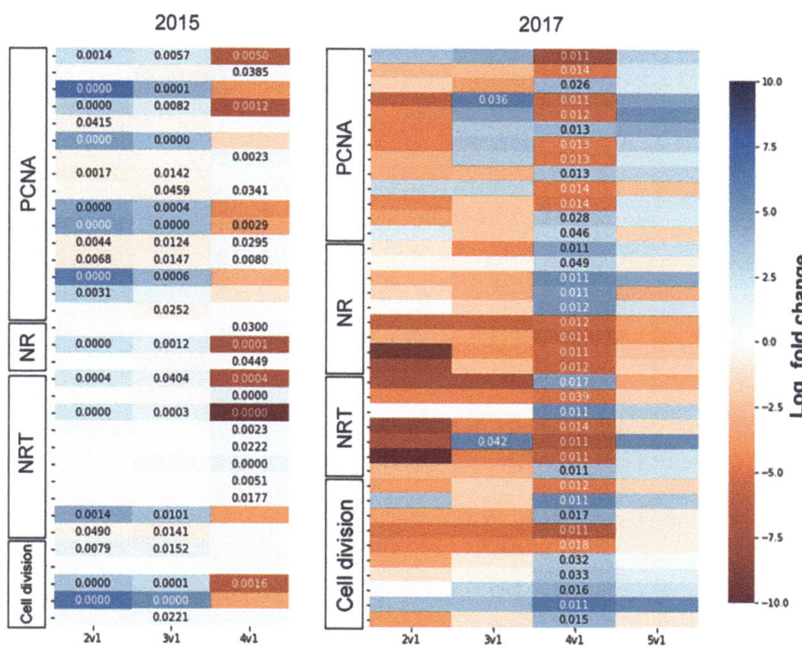

Figure 5.9 Heatmap of differential gene expression for the 2015 and 2017 metatranscriptome samples. Columns represent a comparison between a timepoint and the first sampled timepoint for that year. Rows represent individual transcripts. Numbers in boxes indicate significantly different expression between the sample relative to the first timepoint with a false discovery rate of <0.05. Color of box indicates log2 fold change difference in abundance. NR, nitrate reductase; NRT, nitrate transporter; PCNA, proliferating cell nuclear antigen.

Conclusion

As ecologists, our foremost tasks are to discover the composition of the microplankton community and to reveal the role of diversity in ecosystem functioning. This chapter has introduced some of the emerging technologies for investigating the marine phytoplankton and provided examples of how integrating these approaches can improve our understanding of the diversity and dynamics of the microplankton. By solely using one approach, we can only address one dimension of a question. The case studies presented here have addressed the advantages of integrating the data-rich and high throughput systems for investigating phytoplankton. High-frequency monitoring with the IFCB provides a near real-time series of images that together with automated image analysis can provide early warning of harmful algal blooms to protect human health. The decade-long time series of the phytoplankton abundance and biomass at TOAST has also addressed the challenge of identifying trends in the diversity of the microplankton community. HTS-metabarcoding provides improved taxonomic resolution, uncovers cryptic species and new species, but provides relative abundance based on copy number, which may not reflect actual abundance, nor indicate active segments of community. Metatranscriptomics can provide a view of functional activity within the phytoplankton community. By integrating these approaches, a detailed analysis of the phytoplankton community structure, composition, and dynamics can be obtained.

References

Alberti, A., Poulain, J., Engelen, S., Labadie, K., Romac, S., Ferrera, I., et al., 2017. Viral to metazoan marine plankton nucleotide sequences from the Tara Oceans expedition. Sci. Data 4, 170093.

Anglès, S., Jordi, A., Campbell, L., 2015. Responses of the coastal phytoplankton community to tropical cyclones revealed by high-frequency imaging flow cytometry. Limnol. Oceanogr. 60 (5), 1562–1576.

Anglès, S., Jordi, A., Henrichs, D.W., Campbell, L., 2019. Influence of coastal upwelling and river discharge on the phytoplankton community composition in the northwestern Gulf of Mexico. Prog. Oceanogr. 173, 26–36.

Anglès, S., Reñé, A., Garcés, E., Lugliè, A., Sechi, N., Camp, J., Satta, C.T., 2017. Morphological and molecular characterization of *Bysmatrum subsalsum* (Dinophyceae) from the western Mediterranean Sea reveals the existence of cryptic species. J. Phycol. 53 (4), 833–847.

Bosak, S., Godrijan, J., Šilović, T., 2016. Dynamics of the marine planktonic diatom family Chaetocerotaceae in a Mediterranean coastal zone. Estuar. Coast. Shelf Sci. 180 (5), 69–81.

Brosnahan, M.L., Velo-Suarez, L., Ralston, D.K., Fox, S.E., Sehein, T.R., Shalapyonok, A., et al., 2015. Rapid growth and concerted sexual transitions by a bloom of the

harmful dinoflagellate Alexandrium fundyense (Dinophyceae). Limnol. Oceanogr. 60 (6), 2059–2078.
Brownlee, E.F., Olson, R.J., Sosik, H.M., 2016. Microzooplankton community structure investigated with imaging flow cytometry and automated live-cell staining. Mar. Ecol. Prog. Ser. 550, 65–81.
Cai, R.B., Kayal, E., Alves-de-Souza, C., Bigeard, E., Corre, E., Jeanthon, C., et al., 2020. Cryptic species in the parasitic *Amoebophrya* species complex revealed by a polyphasic approach. Sci. Rep. 10 (1), 1–11.
Campbell, L., Henrichs, D.W., 2020. Metabarcoding samples collected from surface and chlorophyll maximum depths from R/V Pt. Sur PS 18-09 Legs 01 and 03, Hurricane Harvey RAPID Response cruise (western Gulf of Mexico) September-October 2017. Version Date 2020-09-14 [HRR-Metabarcoding]. Biological and Chemical Oceanography Data Management Office (BCO-DMO) [Feb 8, 2021].
Campbell, L., Henrichs, D.W., Olson, R.J., Sosik, H.M., 2013. Continuous automated imaging-in-flow cytometry for detection and early warning of *Karenia brevis* blooms in the Gulf of Mexico. Environ. Sci. Pollut. Res. 20 (10), 6896–6902.
Campbell, L., Henrichs, D.W., Peacock, E.E., Futrelle, J., Sosik, H.M., 2017. Imaging FlowCytobot provides novel insights on phytoplankton community dynamics. In: Proenca, L.A.O., Hallegraeff, G. (Eds.), Proceedings of the 17th International Conference on Harmful Algae. International Society for the Study of Harmful Algae and Intergovernmental Oceanographic Commission of UNESCO, pp. 74–79.
Campbell, L., Olson, R.J., Sosik, H.M., Abraham, A., Henrichs, D.W., Hyatt, C.J., et al., 2010. First harmful *Dinophysis* (DINOPHYCEAE, DINOPHYSIALES) bloom in the US is revealed by automated imaging flow cytometry. J. Phycol. 46 (1), 66–75.
Daugbjerg, N., Hansen, S.A., Richardson, K., 2019. Cryptic diversity of small-sized species of *Phalacroma* (Dinophysales, Dinophyceae) from Denmark Strait (Eastern Arctic Greenland). Phycol. Res. 67 (3), 244–249.
De Luca, D., Kooistra, W.H.C.F., Sarno, D., Gaonkar, C.C., Piredda, R., 2019. Global distribution and diversity of *Chaetoceros* (Bacillariophyta, Mediophyceae): integration of classical and novel strategies. PeerJ 7, e7410.
de Vargas, C., Audic, S., Henry, N., Decelle, J., Mahé, F., Logares, R., et al., 2015. Eukaryotic plankton diversity in the sunlit ocean. Science 348, 1261605.
Deeds, J.R., Wiles, K., Heideman, G.B., White, K.D., Abraham, A., 2010. First US report of shellfish harvesting closures due to confirmed okadaic acid in Texas Gulf coast oysters. Toxicon 55 (6), 1138–1146.
del Campo, J., Kolisko, M., Boscaro, V., Santoferrara, L.F., Nenarokov, S., Massana, R., et al., 2018. EukRef: phylogenetic curation of ribosomal RNA to enhance understanding of eukaryotic diversity and distribution. PLoS Biol. 16 (9), e2005849.
Deng, Y.Y., Hu, Z.X., Chai, Z.Y., Tang, Y.Z., 2019. Cloning and comparative studies of proliferating cell nuclear antigen (PCNA) genes for nine dinoflagellates. J. Appl. Phycol. 31, 2969–2979.
Edler, D., Klein, J., Antonelli, A., Silvestro, D., 2019. raxmlGUI 2.0 beta: a graphical interface and toolkit for phylogenetic analyses using RAxML. bioRxiv. https://doi.org/10.1101/800912.
Elder, L., Elbrachter, M., 2010. The Utermohl method for quantitative phytoplankton analysis. In: Karlson, B., Cusack, C., Bresnan, E. (Eds.), Microscopic and Molecular Methods for Quantitative Phytoplankton Analysis. UNESCO, Paris, pp. 12–20.
Fiorendino, J.M., Gaonkar, C.C., Henrichs, D.W., Campbell, L., 2021. Drivers of microplankton community assemblage following tropical cyclones. J. Plankton Res. https://doi.org/10.1093/plankt/fbab073.
Gaonkar, C.C., Piredda, R., Minucci, C., Mann, D.G., Montresor, M., Sarno, D., et al., 2018. Annotated 18S and 28S rDNA reference sequences of taxa in the planktonic diatom family Chaetocerotaceae. PLoS ONE 13 (12), e0208929.

Gaonkar, C.C., Piredda, R., Sarno, D., Zingone, A., Montresor, M., Kooistra, W.H.C.F., 2020. Species detection and delineation in the marine planktonic diatoms *Chaetoceros* and *Bacteriastrum* through metabarcoding: making biological sense of haplotype diversity. Environ. Microbiol. 22 (5), 1917–1929.

Gómez, F., Souissi, S., 2007. Unusual diatoms linked to climatic events in the northeastern English Channel. J. Sea Res. 58 (4), 283–290.

González, P., Castaño, A., Peacock, E.E., Díez, J., Del Coz, J.J., Sosik, H.M., 2019. Automatic plankton quantification using deep features. J. Plankton Res. 41 (4), 449–463.

Gottschling, M., Keupp, H., Plötner, J., Knop, R., Willems, H., Kirsch, M., 2005. Phylogeny of calcareous dinoflagellates as inferred from ITS and ribosomal sequence data. Mol. Phylogenet. Evol. 36 (3), 444–455.

Guillou, L., Bachar, D., Audic, S., Bass, D., Berney, C., Bittner, L., et al., 2013. The Protist Ribosomal Reference database (PR^2): a catalog of unicellular eukaryote Small Sub-Unit rRNA sequences with curated taxonomy. Nucleic Acids Res. 41 (D1), D597–D604.

Hall, J., Fucikova, K., Lo, C., Lewis, L., Karol, K., 2010. An assessment of proposed DNA barcodes in freshwater green algae. Algologie 31, 529–555.

Harred, L.B., Campbell, L., 2014. Predicting harmful algal blooms: a case study with *Dinophysis ovum* in the Gulf of Mexico. J. Plankton Res. 36 (6), 1434–1445.

Hasle, G.R., 1978. Using the inverted microscope. In: Sournia, A. (Ed.), Phytoplankton Manual. UNESCO, Paris, pp. 191–196.

Hasle, G.R., Syvertsen, E.E., 1997. Marine diatoms. In: Tomas, C.R. (Ed.), Identifying Marine Phytoplankton. Academic Press, San Diego, pp. 5–385.

Hebert, P.D.N., Cywinska, A., Ball, S.L., deWaard, J.R., 2003. Biological identifications through DNA barcodes. Proc. R. Soc. Lond. B 270 (1512), 313–321.

Henrichs, D.W., Anglès, S., Gaonkar, C.C., Campbell, L., 2021. Application of a convolutional neural network to improve automated early warning of harmful algal blooms. Environ. Sci. Pollut. Res. 28, 28544–28555. https://doi.org/10.1007/s11356-021-12471-2.

Henrichs, D.W., Hetland, R.D., Campbell, L., 2015. Identifying bloom origins of the toxic dinoflagellate Karenia brevis in the western Gulf of Mexico using a spatially explicit individual-based model. Ecol. Model. 313, 251–258.

Henrichs, D.W., Sosik, H.M., Olson, R.J., Campbell, L., 2011. Phylogenetic analysis of *Brachidinium capitatum* (DINOPHYCEAE) from the Gulf of Mexico indicates membership in the Kareniaceae. J. Phycol. 47 (2), 366–374.

Henrichs, D.W., Tomlinson, M.C., Campbell, L., 2020. Predicting bloom initiation on the Texas (USA) coast: combining satellite imagery with an individual-based model. In: Hess, P. (Ed.), Proceedings of the 18th International Conference on Harmful Algae. International Society for the Study of Harmful Algae, pp. 55–57.

Jeffrey, S.W., Humphrey, G.F., 1975. New spectrophotometric equations for determining chlorophyll a, b, c1, and c2 in higher plants, algae, and natural phytoplankton. Biochem. Physiol. Pflanz. 168, 191–194.

Johnson, M.D., Beaudoin, D.J., Laza-Martinez, A., Dyhrman, S.T., Fensin, E., Lin, S.J., et al., 2016. The genetic diversity of *Mesodinium* and associated cryptophytes. Front. Microbiol. 7. https://doi.org/10.3389/fmicb.2016.02017.

Johnson, Z.I., Martiny, A.C., 2015. Techniques for quantifying phytoplankton biodiversity. Annu. Rev. Mar. Sci. 7 (7), 299–324.

Keeling, P.J., Burki, F., Wilcox, H.M., Allam, B., Allen, E.E., Amaral Zettler, L.A., et al., 2014. The marine microbial eukaryote transcriptome sequencing project (MMETSP): illuminating the functional diversity of eukaryotic life in the oceans through transcriptome sequencing. PLoS Biol. 12 (6), e1001889.

Kooistra, W.H.C.F., Sarno, D., Hernández-Becerril, D.U., Assmy, P., Di Prisco, C., Montresor, M., 2010. Comparative molecular and morphological phylogenetic analyses of taxa in the Chaetocerotaceae (Bacillariophyta). Phycologia 5, 471–500.

Leblanc, K., Arístegui, J., Armand, L., Assmy, P., Beker, B., Bode, A., et al., 2012. A global diatom database–abundance, biovolume and biomass in the world ocean. Earth Syst. Sci. Data 4 (1), 149–165.
Lightbody, G., Haberland, V., Browne, F., Taggart, L., Zheng, H.R., Parkes, E., Blayney, J.K., 2019. Review of applications of high-throughput sequencing in personalized medicine: barriers and facilitators of future progress in research and clinical application. Brief. Bioinform. 20 (5), 1795–1811.
Lundholm, N., Moestrup, Ø., Kotaki, Y., Hoef-Emden, K., Scholin, C., Miller, P., 2006. Inter- and intraspecific variation of the *Pseudo-nitzschia delicatissima* complex (BACILLARIOPHYCEAE) illustrated by rRNA probes, morphological data and phylogenetetic analyses. J. Phycol. 42 (2), 464–481.
Mackey, M.D., Mackey, D.J., Higgins, H.W., Wright, S.W., 1996. CHEMTAX - a program for estimating class abundances from chemical markers: application to HPLC measurements of phytoplankton. Mar. Ecol. Prog. Ser. 144 (1–3), 265–283.
Malviya, S., Scalco, E., Audic, S., Vincenta, F., Veluchamy, A., Poulain, J., et al., 2016. Insights into global diatom distribution and diversity in the world's ocean. Proc. Natl. Acad. Sci. 113 (11), E1516–E1525.
McDuff, R.E., Chisholm, S.W., 1982. The calculation of in situ growth rates of phytoplankton populations from fractions of cells undergoing mitosis: a clarification. Limnol. Oceanogr. 27, 783–788.
Menden-Deuer, S., Lessard, E.J., 2000. Carbon to volume relationships for dinoflagellates, diatoms, and other protist plankton. Limnol. Oceanogr. 45 (3), 569–579.
Moberg, E.A., Sosik, H.M., 2012. Distance maps to estimate cell volume from two-dimensional plankton images. Limnol. Oceanogr., Methods 10, 278–288.
Montresor, M., Sgrosso, S., Procaccini, G., Kooistra, W.H., 2003. Intraspecific diversity in *Scrippsiella trochoidea* (Dinophyceae): evidence for cryptic species. Phycologia 42 (1), 56–70.
Moran, M.A., Satinsky, B., Gifford, S.M., Luo, H.W., Rivers, A., Chan, L.K., et al., 2013. Sizing up metatranscriptomics. ISME J. 7 (2), 237–243.
Morey, J.S., Monroe, E.A., Kinney, A.L., Beal, M., Johnson, J.G., Hitchcock, G.L., et al., 2011. Transcriptomic response of the red tide dinoflagellate, *Karenia brevis*, to nitrogen and phosphorus depletion and addition. BMC Genomics 12, 346.
Olson, R.J., Sosik, H.M., 2007. A submersible imaging-in-flow instrument to analyze nano-and microplankton: imaging FlowCytobot. Limnol. Oceanogr., Methods 5, 195–203.
Peacock, E.E., Olson, R.J., Sosik, H.M., 2014. Parasitic infection of the diatom *Guinardia delicatula*, a recurrent and ecologically important phenomenon on the New England Shelf. Mar. Ecol. Prog. Ser. 503, 1–10.
Piredda, R., Claverie, J.M., Decelle, J., de Vargas, C., Dunthorn, M., Edvardsen, B., et al., 2018. Diatom diversity through HTS-metabarcoding in coastal European seas. Sci. Rep. 8, 18059.
Prabowo, D.A., Shah, M.M.R., Horiguchi, T., Suda, S., 2016. Genetic diversity of *Moestrupia oblonga* (Dinophyceae) from coastal areas of Okinawa Island, Japan. Mar. Biodivers. 46 (1), 197–209.
Reuter, J.A., Spacek, D.V., Snyder, M.P., 2015. High-throughput sequencing technologies. Mol. Cell 58, 586–597.
Roelofs, D., Aarts, M.G.M., Schat, H., van Straalen, N.M., 2008. Functional ecological genomics to demonstrate general and specific responses to abiotic stress. Funct. Ecol. 22, 8–18.
Ryan, D.E., Campbell, L., 2016. Identification and phylogeny of putative PEPC genes in three toxin-producing *Karenia* (Dinophyta) species. J. Phycol. 52 (4), 618–625.
Shakya, M., Lo, C., Chain, P.S.G., 2019. Advances and challenges in metatranscriptomic analysis. Front. Genet. 10, 904. https://doi.org/10.3389/fgene.2019.00904.

Sosik, H.M., Olson, R.J., 2007. Automated taxonomic classification of phytoplankton sampled with imaging-in-flow cytometry. Limnol. Oceanogr., Methods 5, 204–216.
Steidinger, K.A., 2009. Historical perspective on *Karenia brevis* red tide research in the Gulf of Mexico. Harmful Algae 8 (4), 549–561.
Tanković, M.S., Baričević, A., Ivančić, I., Kužat, N., Medić, N., Pustijanac, E., et al., 2018. Insights into the life strategy of the common marine diatom *Chaetoceros peruvianus* Brightwell. PLoS ONE 13 (9), e0203634.
Yang, S., Han, X., Zhang, C., Sun, B., Wang, X., Shi, X., 2014. Seasonal changes in phytoplankton biomass and dominant species in the Changjiang River Estuary and adjacent seas: general trends based on field survey data 1959–2009. J. Ocean Univ. China 13, 926–934.

CHAPTER 6

Advances in *in situ* molecular systems for phytoplankton research and monitoring

Matthew C. Smith[a], **Levente Bodrossy**[b], **and Pascal Craw**[b]
[a]Sandy Bay, TAS, Australia
[b]CSIRO Oceans and Atmosphere, Hobart, TAS, Australia

Introduction

The need for high temporal and spatial resolution data with fine taxonomic detail for marine research has given rise to the idea of "ecogenomic sensors." Ecogenomic sensors are autonomous sensors, which apply molecular techniques on an in situ platform that allows for remote sample collection, processing, and molecular analyses (Preston et al., 2011). The goal of ecogenomic sensors is a deployable system that autonomously performs in situ molecular analyses and provide near real-time data on the results.

In situ, automated systems for biological sensing offer several important advantages. They significantly cut the costs associated with manual sampling and have the potential to minimize human error and contamination. Automated in situ systems enable access to samples that cannot be collected due to safety and logistical reasons. They enable a temporal and spatial resolution that cannot be easily attained by manual sampling. Automated sensing using in situ systems ultimately enables a near real-time detection and subsequent early warning of biological threats (for example, the onset or the approach of a harmful algal bloom (HAB) via ocean currents). When using molecular methods for detection, the same system can be used for different molecular targets (conceptually, molecular assays can be replaced in a plug and play manner, even if in practice this flexibility is somewhat limited). Molecular approaches (see Chapters 10, 12) enable the fine-tuning of the target. This includes targeting the assay to a narrower or broader taxonomic group, from strains to species to genera, depending on the specific monitoring needs (Kon et al., 2017; Johansson et al., 2016; Eckford-Soper and Daugbjerg, 2015; Zamor et al., 2012; Manning and La Claire, 2010). It is also possible to target functional genes, such as those associated with toxin production (Dyhrman et al., 2006; Antonella and Luca, 2013).

Medical and defence applications, particularly pathogen diagnostics and biological agent detection, have driven the development of automated nucleic acid sensing systems and the fields of microfluidics; isothermal nucleic acid amplification and lab-on-a-chip technology continue to deliver advances that bring in situ ecogenomic sensors closer to being part of a regular sensor line up.

Lab-on-a-chip technology offers many opportunities for in situ marine sensing. A number of laboratory studies have proven successful in demonstrating the possibility of lab-on-a-chip devices for HAB research and monitoring, including handheld detection devices (Casper et al., 2007; Lagier et al., 2007; Toldrà et al., 2019a), dipstick-based tests (Toldrà et al., 2019b; Zhu et al., 2019; Huang et al., 2017), and membrane-based DNA arrays (Zhang et al., 2018), but all of these systems require some level of manual handling and user supervision, making them unsuitable without further development for automated in situ operation. Unfortunately, there have been few effective demonstrations of lab-on-a-chip technology being used successfully for DNA-based detection in situ. The numerous and varied challenges associated with taking miniaturized technology from the lab to the field or commercial applications has been documented (Mohammed et al., 2015), and includes a high dependence on ancillary equipment, complexities integrating multiple coordinated operations, limitations in manufacturing, and the economics of implementation (Mohammed et al., 2015), (Volpatti and Yetisen, 2014). For ecogenomic sensors, the sample collection and DNA isolation steps must also be miniaturized (Toldrà et al., 2019b).

A large amount of basic research in the broad field of in situ phytoplankton detection exists. However, for the purpose of this chapter, we have focused on summarizing technologies and approaches that have met high technology readiness levels (Mankins, 2009) (levels 6 to 9) with peer reviewed publication and/or successful commercialization.

Current state of the art

In situ sampling

Though automated in situ samplers do not carry out molecular detection on board, they are the crucial stepping-stone towards in situ phytoplankton analysis. The notion of automated sampling instrumentation is not new. Equipment for the collection of bulk water samples, such as the Niskin bottle and van Dorn samplers, are commonplace in modern oceanography, and reflect design elements of the Nansen bottle developed in the late 19th century by Fridtjof Nansen. Development of the plankton sampling instruments in the 1920s by Alister Hardy led to the development of the continuous plankton recorder (CPR) (Reid et al., 2003). Monthly North Atlantic basin scale surveys using the CPR were initiated in 1931, and continue to this day, making it the world's longest running oceanographic survey (Batten et al., 2019). Unlike bulk water samplers, such as the Niskin bottle, the CPR represents an integrative sampling strategy. Towed behind a vessel, the CPR collects biomass from the water column using an incrementing ribbon of silk mesh (nominal size 270 μm); collected biomass is preserved in an onboard formaldehyde bath for subsequent microscopic examination in the laboratory (Richardson et al., 2006). Although tow speeds and the mesh size used by the CPR can influence the composition of the biomass collected, these and other disadvantages of sample acquisition using the CPR have been well documented (Richardson et al., 2006). However, the unprecedented temporal library of planktonic organisms, collected over large spatial scales using methods that are essentially unchanged since 1931 have had profound impact of our understanding of changes occurring in pelagic ecosystems (Batten et al., 2019; Richardson et al., 2006).

It would have been difficult for Alister Hardy to foresee the major technological advances that have been seen in molecular detection technologies over the last few decades since the CPR initiative was conceived. However, despite challenges involved in extracting DNA suitable for molecular analysis from formalin-fixed samples, a number of studies have applied molecular techniques to the analysis of phytoplankton communities from CPR silks that have been archived for over 10 years (Ripley et al., 2008; Stern et al., 2018). More recent development of in situ sampling instruments has focused on developing instruments that are capable of sample collection with flexible preservation methods that are better suited to a range of downstream molecular analyses. Many of these systems are available commercially and offer a range of sampling techniques. Samplers such as Teledyne ISCO, YSI ProSample, McLane remote access sampler (RAS) offer the ability to capture whole water samples in collection bottles or bags, and thus may be beneficial in the investigation of live phytoplankton or their dissolved metabolites. Some of these systems have the ability to collect both a continuous pumped sample or as composite samplers, where small volumes are collected and integrated over longer temporal periods, which may prove beneficial for targets that have patchy distributions due to physiological (e.g., photo taxis) or physical (e.g., tidal) influence. In many cases, these systems can have preservatives or buffers added to the collection chamber prior to deployment (e.g., saturated mercuric chloride (Winslow et al., 2014)) or have the ability to be cooled by the addition of ice to the instrument to help maintain sample integrity. In contrast to, preservative addition to sample containers prior to deployment, Robidart et al. (2013) modified an osmotic pumping collection system, initially developed for chemical sampling (Jannasch et al., 2004), to co-dispense nucleic acid preservatives (RNaLater or UmFix) as samples were collected from hydrothermal vents. Similarly, a preservative injection system was used to introduce preservation solutions into a "Niskin style" grab bottle upon actuation of the end caps, with the resulting samples used for transcriptomics analysis in the laboratory (Feike et al., 2012). Though these studies did not specifically investigate phytoplankton, they highlight alternative approaches that may be undertaken to obtain bulk water samples and amend them with preservative solutions.

Several in situ platforms are emerging that have been designed specifically to filter samples obtained from a variety of aquatic environments and preserve biomass for downstream molecular analysis. A hybrid approach (Breier et al., 2014) enabled a sampler system capable of collecting multiple sample types from deep-sea hydrothermal vent environments, including filtered and preserved biomass, as well as filtered and bulk water samples. One of the first commercial filtration-based samplers designed specifically for the collection of phytoplankton was the McLane phytoplankton sampler (PPS). This system used standard 47 mm filter holders and enabled sequential sampling of 24 samples. The PPS also offered the ability to add a sample preservative to the filtered biomass, typically to preserve the samples for microscopic examination. Although the PPS was not specifically designed for remote sampling with preservation for subsequent downstream molecular analysis, other researchers have reported modifications to the instrument to enable downstream use. For example, multiple deployments of both PPS and RAS systems in Antarctic lakes over a calendar year were performed (Winslow et al., 2014). During these deployments, sucrose lysis buffer was used to preserve DNA in the samples collected, using the PPS, for later sequencing in the laboratory. However, to our knowledge, no subsequent analysis of these samples has been published. McGillicuddy and coworkers used the PPS in conjunction with molecular probe-based cell enumeration in a study that investigated the dynamics of a bloom of the toxic dinoflagellate *Alexandrium fundyense* in the western Gulf of Maine USA (Mcgillicuddy et al., 2014).

More recent developments of filtration-based systems have been described with the aim of sampling planktonic communities. A modular system capable of capturing 8 sequential samples, each of ~2–3 L onto 0.2 μm Sterivex cartridge filters has been developed (Ribeiro et al., 2019). Nonselective sampling of heterotrophic and phototrophic plankton was preformed using a micro gear pump with each sample directed to individual Sterivex filters via the actuation of a series of solenoid valves. Sample preservation was achieved by the addition of RNAlater to the filters following sample filtration. The system achieved comparable results to samples collected using more traditional laboratory filtration techniques for both DNA recovery and for 16S- and 18S-rDNA analysis of marine microbial communities.

Molecular assays: point-of-use

Point-of-use assays have seen increased application in biomedical, food, agriculture, environmental and defence applications as they enable users to rapidly analyze discrete samples without the need for transport to laboratory infrastructure. These systems vary in the degree of complexity that is required by the user to perform the analysis. The development of effective point-of-use assays should consider sample preparation, analytical detection, and reporting. Undoubtedly, point-of-use testing has numerous benefits over traditional laboratory analysis, including faster analysis and decision making, but brings with it considerations of quality control which may encompass wide ranging aspects of data storage/transmission, privacy, reagent storage/transport, user training and data interpretation

Lateral flow assays

One of the most common approaches for point-of-use analysis for the presence of phytoplankton or their metabolites is the use of lateral flow assays. These assays provide fast, low cost and user-friendly analysis methods that can be performed by nonspecialized personnel. The analytical methodology used in various types of lateral flow assays can vary considerably and can encompass a wide range of target analytes (e.g., nucleic acids, whole cells, proteins, or other cell metabolites) resulting in various forms of qualitative or quantitative analytical outputs that can involve fluorescent, colorimetric, or chemiluminescent detection strategies. Though the analytical methodologies employed by lateral flow assays are diverse, their basic operation are similar and their operation as well as strengths and weaknesses have been well reviewed in the literature (e.g., Anfossi et al., 2012; Posthuma-Trumpie et al., 2009; Sajid et al., 2015; Carrell et al., 2019). To summarize their operation, first a raw or purified sample is applied to a sample application pad. Chromatographic separation of the sample enables transport of the sample, preimmobilized reagents involved in molecular detection, and reporting of the analyte. The conjugation area contains elements responsible for biorecognition of the analyte of interest. Strategies used at the conjugation area can include competitive or sandwich hybridization techniques and can employ various biorecognition elements with affinity to the target molecule/organism, such as antibodies, DNA/RNA aptamers, nucleic acid probes, or molecularly imprinted polymers. The detection area typically consists of both test and control lines of reagents that report the presence of the analyte(s). Common detection materials and strategies that have been employed include gold nanoparticles, colored latex beads, quantum dots, carbon nanoparticles.

Much of the research, development, and commercial application of lateral flow assays for phytoplankton detection has been aimed at the detection of toxins produced from harmful algal blooms. For example, qualitative assays (yes/no) lateral flow tests targeting toxins causing paralytic shellfish poisoning (PSP) (formally known as "mist alert" and "Jellett") (Akter et al., 2019; Jellett et al., 2006) have been commercially available from Scotia Rapid Testing (Canada) for over 10 years. These test kits provide users the ability to test shellfish flesh or concentrated algal cells from ambient water samples following some simple sample preparation steps. Kits are also available from the same manufacturer for other algal toxins, amnesic shellfish poisoning (ASP) and diarrhetic shellfish poisoning (DSP). More recently, other commercial examples of semiquantitative lateral flow assays for testing drinking and recreation waters for microcystins and nodularins have been sold by Abraxis (USA). Efficacy of commercially available lateral flow assays has been shown to produce results that are comparable as a screening tool to traditional "gold standard" laboratory analyses. For example, results from commercially available Jellett immunochromatographic lateral flow test strips for PSP were in close agreement to toxicity results derived from the mouse bioassay, with false positive results being more commonly attributed to the lateral flow assay than false negatives (Laycock et al., 2010). Similarly, outcomes from other studies (Costa et al., 2009; Turner et al., 2020) found similar results when comparing qualitative PSP lateral flow assays to more specialized laboratory-based analytical detection methodologies. Another study (Turner et al., 2020) included the

Scotia immunochromatographic lateral flow test strip with five other analytical methods—the EU reference method for PSP using pre-column oxidation liquid chromatography with fluorescence detection (LC-FLD), a post column oxidation LC-FLD method, liquid chromatography with tandem mass spectrometry, a receptor binding assay, the mouse bioassay—to detect PSP toxins in their study of 349 shellfish samples from Argentina, Mexico, Chile, and Uruguay. The lateral flow assay used in this study returned 2 false negative samples when all other methods returned positive results and had a 30% false positive rate. The occurrence of a relatively high false positive results from the PSP lateral flow assay reflected similar findings in other studies (Costa et al., 2009; Laycock et al., 2010). Similar studies have investigated commercially available lateral flow assay for the freshwater cyanotoxin microcystin. The Abraxis semi-quantitative test strips for microcystin are available to measure at concentration ranges relevant to the water source compliance requirements (drinking or recreational). The reliability of the microcystin test strips have been shown to decrease towards the lower limits of their response range, due to the increased occurrences of both false positive and negative results (Humpage et al., 2012). However, their use as a fast screening tool was supported, provided that further analytical assessment was performed when reported concentrations obtained approached compliance limits (Humpage et al., 2012).

Nucleic acid analysis

The development of the polymerase chain reaction (PCR) by Kary Mullis and colleagues in the mid-1980s (Saiki et al., 1985) has subsequently led to a diverse range of different nucleic acid amplification techniques that has enabled a wide range of analytical methods to assess organisms at the genetic level. With the advent of gene cloning, DNA sequencing techniques and real-time amplification methodologies, laboratory-based characterization of phytoplankton from cultured isolates and environmental samples has become common place. Due to their sensitivity, specificity, relative low-cost-and high throughput, laboratory-based nucleic acid analysis methods, such as quantitative real-time PCR (qPCR), digital droplet PCR, loop mediated amplification (LAMP), rolling circle amplification and nucleic sequence-based amplification (NASBA) are just a few amplification methods that have become common approaches for identifying, characterizing, and monitoring harmful algal blooms and phytoplankton from environmental and cultured samples. To perform this type of analysis, typical workflows require repeated fluidic manipulations, manual sample handling, and the need for costly laboratory equipment and facilities and the requirement of highly trained personnel. This results in a spatial and temporal separation between the collection of samples and their analysis, increasing both analysis cost and turn-around time. The ability to decentralize these types of analysis by developing portable, miniaturized systems or methods capable of being performed at the sampling site by untrained or semiskilled personnel is desirable as it will potentially enable a more cost-effective analysis and faster decision-making. For phytoplankton analysis, portable systems with varying degrees of user input have been developed. These range from modifications to traditional laboratory protocols that can be performed in the field by semitrained personnel to total analytical systems that aim to enable sample-in–answer-out processing.

In situ sample preparation

As with lab-based quantitative nucleic acid methods, sample preparation is a critical factor in the downstream analytical performance of an amplification assay. Effective cell lysis of the target organism(s), losses, or degradation of nucleic acid during purification and the copurification of various inhibitory substances all influence the overall efficiency of the analytical workflow. Similarly, reproducible extraction of high-quality DNA/RNA is essential, where quantitative interpretation of phytoplankton gene copies is being compared over spatial or temporal time scales or used to satisfy regulatory compliance.

Effective cell lysis is a critical step for reproducible and efficient nucleic acid extraction. The wide variety of cell coverings among different phytoplankton present varying challenges for a single prescribed cell lysis method to be applied in the field, as such methods described in the literature are tailored for specific organism(s) of interest. Cell lysis methods traditionally performed in the laboratory typically range from mechanical disruption (e.g., bead beating, ultrasonic disruption), chemical lysis or thermal disruption or combinations thereof. As analysis methods move from the laboratory setting to the field, modifications to these methods has become the norm as the need for specialized equipment, or reliance on toxic reagents becomes more restrictive. Many reports of point-of-use phytoplankton analysis have applied chemical lysis techniques to disrupt cells captured on filters. For example, two early examples of field amenable extraction methods for phytoplankton (Casper et al., 2007; Fries et al., 2007), both effectively used reagents containing high concentrations of guanidine isothiocycanate to lyse cells of *Karenia brevis* captured on filters. Another study (Bahi et al., 2011) described a chip-based cell capture and cell lysis solution for *K. brevis* that overcame the need for both cell capture by filtration and chemical lysis. The system first used an electric field to concentrate cells at an electrode with subsequent cell lysis using a high-strength electric field. Demonstration of this system used a relatively high concentration of cultured cells contained in a small volume of artificial media (80 000 cells in 250 μl), it was unclear how the system would perform using real-world matrices. Similarly, a microfluidic chip that coupled 3 sequential micro-cages designed for the collection and concentration of *Pseudo-nitzschia* cells was constructed (Wu et al., 2011). An integrated piezoelectric element performed cell lysis by ultrasonication. In this system cells were forced into the micro-cages and trapped by hydrodynamic focusing through combination of differing buffer and sample flow rates, with each cage capable of capturing ~6 *Pseudo-nitzschia* cells. Cell capture was performed using cultured cells in artificial media, and it is unclear on the chip performance when, in the presence of biomass and particulate matter found in real-world samples.

Purification of nucleic acids following cell lysis is most commonly performed using modifications of the Boom method (Boom et al., 1990), whereby nucleic acids are captured on a solid silica matrix in the presence of a chaotropic salt. Subsequent high-percentage alcohol washes remove impurities; a final wash with an aqueous solution elutes the purified DNA/RNA. A large number of commercially available sample purification kits are currently commercially available based on this overarching method, with numerous modifications for extraction from various sample types (e.g., soil, water, plant, bacteria, etc.,). One study (Casper et al., 2007) developed a modified protocol for the Qiagen RNeasy kit to enable its use as a field-based method for the detection of *K. brevis*. Here fluidic manipulations that required the use of a centrifuge to force reagents through the purification column were performed using a syringe coupled to a custom adapter. One concern with Boom style purification protocols, which can influence the efficiency of downstream analytical methods, is the carry-over of ethanol wash solutions into the aqueous solution containing eluted DNA/RNA. To overcome this, the ChargeSwitch gDNA mini bacteria kit (Invitrogen was modified to perform infield RNA extraction and purification from filtered *K. brevis* cells. The ChargeSwitch method relies on the binding of nucleic acid in aqueous solution to polyhistidine-coated magnetic beads under low pH conditions (<6.5); bound nucleic acids are washed under low pH conditions, and then eluted under high pH (>8.5). Following capture of biomass onto a 0.45 μm filter and chemical cell lysis, the infield

ChargeSwitch purification method could be performed using only a magnetic rack and adjustable pipettes and produced equivalent or better results to laboratory-based extraction as determined by real-time NASBA (RT-NASBA) analysis (Casper et al., 2007).

In situ nucleic acid amplification testing

A variety of instruments have been developed and demonstrated for the point-of-use detection of phytoplankton using different molecular techniques. Several instruments for performing and reporting isothermal RT-NASBA have been developed for the Florida red-tide-forming dinoflagellate *K. brevis*. Casper et al. (2007) developed a manually operated system capable of performing a single NASBA reaction contained in a standard 200 μl PCR tube. The system was capable of detecting 2 different fluorophores (excitation/emission: 470/530 and 585/635 nm), enabling the use of both target specific and internal control molecular probes (Patterson et al., 2005). The portable system was powered via mains supply using an AC to DC converter, with reaction results being reported to an attached laptop computer via serial connection. The components forming the heated fluorometer were housed in a custom 3-D printed housing which enabled LEDs, optical bandpass filters, and photodiodes to be orientated for fluorescence detection, whereas temperature control was performed using an infra-red heating element and infra-red thermometer module. In contrast to this approach, a portable standalone handheld RT-NASBA for the detection of *K. brevis* was designed and demonstrated (Smith et al., 2007). The compact battery powered system used a series of LEDs to indicate the status of the RT-NASBA reaction with positive reactions indicated if reaction fluorescence passed a user-defined threshold. As all data was saved to flashmemory, if desired, the handheld could also be connected via USB to a computer for data retrieval or for real-time data acquisition. The handheld was limited to performing a single reaction at a time, with single fluorophore detection. The core heated fluorometer portion of the system was constructed using planar fabrication methods by etching circuits for a resistive heating element, LED, and inexpensive photodiode light sensor with integrated optical color filter (567 nm) onto a flexible printed circuit board substrate. Following attachment of the LED and photodetector modules, the construct was folded to form a cylindrical heater detection module. Development of an alternative microfluidic RT-NASBA reaction strategy for the detection of *K. brevis* (Tsaloglou et al., 2011), moved away from the use of laboratory-based reaction tubes (Casper et al., 2007; Smith et al., 2007). Here, microfluidic chips containing a reaction channel for RT-NASBA was constructed from micro-milled and bonded poly-methyl methacrylate. Chips were placed in aluminium housings with temperature regulation achieved by proportional-integrative-derivative control using power resistors and thermistors. Fluorescence detection was achieved using a 20X objective lens, which served to focus excitation light from a lazer diode onto the sample. A combination of the same objective lens, two dichroic mirrors, band pass filter, and photomultiplier tube served to detect fluorescence emitted from the reaction. The microfluidic module was developed with the aim of integration with the sample collection and lysis chip (Bahi et al., 2011); however, to our knowledge no literature describing the combination of these modules has been published. Despite the differing approaches and complexities of the heating and optical designs used to construct the instrumentation described above, all three studies benchmarked their operation against the same model of laboratory-based RT-NASBA instrumentation, with each instrument producing equivalent results to their commercial counterpart. LaGier and coworkers (Lagier et al., 2007) investigated the use of a commercially available handheld electrochemical detection system to detect electrochemically labeled PCR products of *K. brevis* (and selected fecal bacterial targets). In this study no field-specific sample preparation and PCR protocol was presented with all sample manipulations prior to analyte detection being performed on standard laboratory equipment. However, in addition to traditional sample extraction and purification methods, the authors also investigated the use of crude extracts as a template for the handheld derived directly from bead-beating lysate (as described by Haugland et al., 2005) or from direct PCR of filtered cells (as described by Kirchman et al., 2001). In these cases, the crude lysates showed some reaction inhibition toward samples with higher target cell concentrations; however, the methods showed potential to streamline field-based analysis.

An alternative to systems that require a user to perform fluidic manipulations required for sample and reaction preparation prior to loading into a detector for analysis was developed (Diercks-Horn et al., 2011). The ALGADEC device is a semi-automated system to perform sandwich hybridization of ribosomal RNA's of several harmful algal bloom forming species. The assay system built on previously described assays (Diercks et al., 2008a,b) and employed species specific DNA capture and reporter probes and electrochemical detection. The system requires the user to perform cell concentration (filtration) and total RNA extraction before loading the device. The device automates temperature regulation of reaction processes with fluidic manipulations of wash and reaction components being performed using valve actuations and a peristaltic pump. Electrochemical detection was performed using a chip containing 16 working electrodes, enabling the detection of up to 14 target phytoplankton with 2 electrodes reserved for positive and negative controls. Though best results from the system were obtained using extracted and purified RNA from a commercial extraction kit, the authors also developed an extraction method that could be readily performed in the field by minimally trained personnel. The field extraction method required the user to perform sample filtration followed by addition of, and incubation with, lysis and hybridization buffers. Following incubation, the user injects the lysate into the sample inlet of the device, where the remainder of the automated analysis protocol is performed. However, simplification of the sample preparation procedure resulted in an approximate 4-fold decrease in instrument sensitivity.

Automated in situ molecular detection of phytoplankton

Fully automated, autonomous in situ analysis of phytoplankton communities remains a challenging area of research and development. Currently, the only commercially available, fully automated system for molecular detection of phytoplankton is the environmental sample processor (ESP) (Scholin et al., 2009b). The ESP (Fig. 6.1) has been through numerous generations of development and has been well reviewed (e.g., Paul et al., 2007; Scholin et al., 2017).

At the core of its operation, the ESP uses a rotating carousel to store several filters and reaction chambers (termed "pucks"), with pucks moved robotically to various positions in the instrument, depending on the process being undertaken. The reaction site in each puck may be a DNA array that is responsible for capturing rRNA from a crude lysate and enabling detection through a sandwich hybridization assay (Scholin et al., 2009a), or a protein array that uses a competitive ELISA assay (Doucette et al., 2009). Over its tenure, functionality of the ESP has been extended by the development of fluidic modules for real-time PCR for the detection of marine cyanobacteria (and other microbes) (Preston et al., 2011; Robidart et al., 2012). Similarly, its use has been demonstrated as a remote in situ sampler with subsequent laboratory-based analysis (Ottesen et al., 2011).

Though the ESP represents a commercially available system for autonomous in situ phytoplankton analysis, to our knowledge, the only other system that has been demonstrated to perform a similar operation is the

Figure 6.1 The environmental sample processor (ESP) is a self-contained, autonomous, robotic water sampler capable of performing several types of molecular biological assay in situ. (a) Instrument attached to pressure housing lid. (b) Deploying the ESP in a New Zealand shellfish farm. (c) ESP in surface mooring configuration with diver support.

autonomous microbial genosensor (AMG) (Fries et al., 2007; Paul et al., 2007). The AMG system was designed for subsurface deployment, with the prototype instrument being designed to detect *K. brevis*. The AMG used a combination of 2 rotating carousels to perform sample collection, RNA purification/concentration (using extraction on a solid silica matrix) (Boom et al., 1990), followed by detection of the Florida red tide dinoflagellate *K. brevis* (using an RT-NASBA assay targeting ribulose-1,5-bisphosphate carboxylase/oxygenase (rbcL) large subunit mRNA). The targeting of relatively short-lived, labile RNA provides a stronger line of evidence for live cells compared to the DNA-based assay, which also detects DNA persisting from dead cells. Though the AMG has not reached commercial maturity, its development reached technology readiness level of approximately 6–7 and highlights the complex nature of moving traditional laboratory-based nucleic acid detection to in situ instruments in aquatic environments.

A third instrument that successfully took molecular analysis into the marine environment was the IISA-Gene. Even though this instrument has not been used to detect phytoplankton, we mention it here as it could have easily been applied to phytoplankton work as well. The project was led by Teruo Fujii at the University of Tokyo, who used advances in lab-on-a-chip technology to develop a microfluidic PCR system (Fukuba et al., 2004). The ability to perform a PCR assay on a single flow through device that could be repeatedly used was a significant advance, but remained dependent on a user for DNA extraction, pumping, and the preparation of the reaction mixture. This technology demonstration was subsequently advanced into a standalone system called the IISA-Gene. The IISA-Gene integrated the PCR system demonstrated earlier with sample extraction, lysis, and pumping to perform a full remote analysis, without the need for an operator (Fukuba and Fujii, 2012). With a now fully automated device, in 2008 and 2009, the IISA-Gene device was integrated with a deep-sea ROV and deployed at two different hydrothermal sites in the Okinawa Trough off Japan at 1471 and 1004 m depth (Fukuba et al., 2011). On board analysis at one site consisted of one PCR analysis for the 16S rRNA gene and one PCR analysis for the pmoA gene. At a second site, time limitation meant that only a PCR for the 16S rRNA gene was performed. The results of these analyses determined that more than 1×10^4 cells of eubacteria (16s rRNA) were present in the analyzed hydrothermal fluid, whereas methane-oxidizing bacteria (pmoA) were either absent or present in numbers below the detection limit of the device. These were very limited results and lacked no template control (NTC), positive controls, and quantification standards, making the results less robust. The publication authors suggest further optimization of the process described is required as the amplified PCR products recovered appeared to contain nonspecific amplicons. Despite several shortcomings, the work described was the first description of PCR being performed on an underwater vehicle and served as a demonstration of the potential for miniaturized technology being performed in situ in the marine environment.

Miniaturized sequencing technologies

The advent of genomic sequencing has had a significant impact on phytoplankton research. Though most of this work has been restricted to the laboratory, recent advances have seen the emergence of portable sequencing technology (Van Dijk et al., 2018). Leading amongst these is Oxford Nanopore Technologies (ONT) who offer a pocket sized, portable device called the MinION for long read sequencing (www.nanopore.com/products). In the laboratory, the ONT MinION was used to sequence long-range PCR amplicons, using primers to detect to a wide range of selected dinoflagellates, meaning that although the study was primarily focused on *Alexandrium*, the applicability to three additional genera of toxic algae, namely, *Gonyaulax*, *Prorocentrum*, and *Lingulodinium*, was also possible (Hatfield et al., 2020). Recently the ONT MinION has been highlighted to assess polar microbial diversity (including phytoplankton) as a point of use shipboard analytical resource by the Earlham Institute and the British Antarctic Survey https://www.earlham.ac.uk/ship-seq-nanopore-sequencing-polar-microbes-on-board-icebreakers (last accessed: 30 May 2021) https://nanoporetech.com/resource-centre/ship-seq-nanopore-sequencing-polar-microbes-onboard-research-vessels (last accessed: 30 May 2021). Though this work remains unpublished, at this time, the application of the ONT MinION enabled better informed sampling strategies by providing near real-time analysis of microbial community structures. As the technology advances and the associated manual handling steps, particularly sample preparation, for portable sequencing are further optimized, there appears to be a very real possibility that this technology could be a useful component in in situ systems for phytoplankton research.

Case Study: The environmental sample processor

History

In the field of in situ instruments for molecular analysis the environmental sample processor (ESP) developed by Monterey Bay Aquarium Research Institute (MBARI) has led the way, developed over the past two decades (Scholin et al., 2017). The concept for the ESP started in the 1990s and by 2001 the first autonomous, sub surface HAB detections had been demonstrated, and by 2008 the system had been refined and demonstrated to work at temperatures ranging from 4 to 30°C for many months (Babin et al., 2005; Greenfield et al., 2008; Scholin et al., 2008; Yamahara et al., 2015). Current versions of the ESP now allow for these HAB species and toxin data to be reported in near-real-time and is now implemented in several monitoring activities (Bowers et al., 2016; Ryan et al., 2011, 2017; Seegers et al., 2015).

Modular approach

The ESP is a modular device that fully automates sample filtration from the environment and can perform preservation of samples or extract DNA and perform additional analyses on board. As discussed above, the ESP employs robotic handling of interchangeable "pucks" housing filter membranes, upon which different assay could be performed and imaged with a high-resolution epifluorescence camera to generate the assay results. This modular approach has allowed the evolution of numerous iterations with different assays and deployment strategies, including as a moored sampler, free drifting buoy (Aylward et al., 2015; Ottesen et al., 2013), early warning device, and an AU-based instrument (Yamahara et al., 2019). Though this modular approach has permitted great versatility, it has led to an instrument that is relatively complex and expensive, and when fitted for PCR, it requires an expert-level operator (Scholin et al., 2017).

Detection methods

The early versions employed colorimetric detection with chemiluminescence, which was replaced with a modified sandwich hybridization assay to perform sequential protein-based molecular detection of *Pseudo-nitzschia* and domoic acid (Scholin et al., 2009a; Babin et al., 2005; Scholin et al., 2008). In 2011, the additional capability to perform qPCR in situ was presented (Preston et al., 2011; Robidart et al., 2012, 2014). In this work, a reusable solid phase nucleic acid extraction system and 2-channel real-time PCR module were implemented on the ESP and used to prove the feasibility of using qPCR on a coastal mooring by determining abundance changes in SAR11, marine crenarchaea, and *Synechococcus* sp.. Later the same system was employed for near real-time, daily notification of *Pseudo-nitzschia* spp. cell count equivalents over a 45-day deployment from a wharf (Yamahara et al., 2015), demonstrating the capability of this system to provide actional data on HAB species.

Moored deployments

Subsequently, to gain deeper insights into *Pseudo-nitzschia* bloom dynamics, two moored ESP instruments were used over two bloom years (2013 & 2015) for real-time analysis and reporting on the study of interannual bloom dynamics of *Pseudo-nitzschia* ecology and to advance prediction in the Monterey Bay region, which is subject to regular blooms and associated impact on the economy. Existing sampling programs using shore and vessel-based sampling revealed highly variable *Pseudo-nitzschia* dynamics. This study capitalized on the capability of the ESP to add to these existing data and monitoring strategies with high temporal resolution sampling to thoroughly investigate the dynamics of bloom initiation, persistence, and decline (Bowers et al., 2018). The ESP samplers were programmed to collect daily samples, but were also able to collect additional samples if the attached thermometers detected the water temperature drop associated with upwelling events. In this case, additional sample collection would be triggered.

AUV deployments

In addition to the moored ESP collecting and analyzing samples, a Dorado AUV with water sample collection capability was also used to provide greater spatial coverage in addition to manual samples collected from vessels and wharfs during the two bloom events. Samples collected were subject to SHA, FISH (fluorescence in situ hybridization) and ARISA (automated ribosomal intergenic spacer analysis) for molecular taxonomy and abundance, cell culture and LC-MS for domoic acid quantification, microscopy for cell concentration. At the two ESP mooring sites, several other sensing systems were co-deployed to understand the changing biological, chemical, and physical status of the water. CTD profilers fitted with optical sensors were important in determining the movement of phytoplankton in the water column and what layers of the water column were being analyzed by the ESPs; this helped derive more meaning from the ESP samples themselves and also identify the presence of upwelling events. By combining these different physical, chemical, and biological sensing modalities, the authors were able to address both the complex species dynamics throughout the evolving bloom with high taxonomic resolution and the physicochemical drivers of the bloom with a high spatial resolution.

Real-time data

The value of the real-time capability of the ESP was demonstrated in May 2015, when high levels of domoic acid and *P.australis* were reported in real-time by one of the deployed instruments (Bowers et al., 2018). This real-time notification allowed for additional investigations to be initiated, and an AUV mission was commenced allowing much higher resolution studies to be undertaken. Without this real-time capability, the elevated toxin and *P.australis* levels would not have been detected until the completion of the deployment period and the opportunity to perform valuable additional studies would have been missed. The deployed AUV was employed to collect samples in the chlorophyll maximum layer to map the spatial distribution; this AUV data was able to map the extent of the chlorophyll maximum and collect samples to determine that it was dominated by *P.australis* patches with domoic acid concentrations that varied up to an order of magnitude.

ESPs advancing the field

Results showed that despite higher cell counts in 2013, the greatest toxin levels were seen in 2015. Though both of the bloom events, 2013 and 2015, discussed in the study showed a dominance of larger (seriata) size class *Pseudo-nitzschia* species, the molecular analyses revealed that the dominating species in the bloom was different, resulting in a low domoic acid-producing bloom of *P.fraudulenta* in 2013 compared to the much more toxic 2015 bloom of *P.australis* (Bowers et al., 2018). By using high-resolution sampling methods, the authors were able to reveal previously unknown spatial and temporal patterns. Most notably, the less toxic 2013 bloom was characterized by a dominant *P.fraudulenta* population within the bay, and the more toxic *P.australis* remained offshore. The high-resolution sampling, particularly responsive AUV-based sampling, following real-time notifications from the ESP was central in determining the existence of these two ecologically discrete populations existing in close proximity. Mapping performed by AUV was able to identify upwelling filament of cold water that existed as part of an offshore eddy system and kept the two populations of *Pseudo-nitschia* separate; the authors suggest this may be the source of cells entering Monterey Bay. During the notably more toxic 2015 bloom, the use of high temporal resolution sampling was able to demonstrate a rapid shift in the *Pseudo-nitzschia* community, from a diverse assemblage in samples from early spring to a dominance of the more toxic *P.australis* species in late spring, coinciding with a warm water anomaly detected offshore and intermittent upwelling, which the authors suggest led to the anomalous low silicate to nitrate ratio that is thought to have driven the high toxicity of this *P.australis* bloom, a finding that supports previous reports.

The capabilities of the ESP to perform frequent molecular analyses for both species ID and domoic acid toxin quantification allowed the researchers to produce large amounts of data to augment the weekly shore-based sampling performed manually. These data provided robust support for the conclusions of this study, and better inform onward prediction efforts. Without this capability it is quite possible that the finer details, including the bloom community shifts in late spring, the spatial mapping and role of upwelling from off shore eddy systems driving the nutrient changes, and the different spatial and ecological effects of *P.australis* and *P.fraudulenta*, would have eluded researchers.

Next generation ESPs

Though the initial versions of the ESPs that have been used aboard dedicated moorings and drifters, as well as installations on ships and shore-based stations (Scholin et al., 2017), they were rather large and cumbersome instruments. A third-generation ESP (3G-ESP) (Pargett et al., 2015) that is much smaller and of lower cost is in development along with handheld versions (Scholin et al., 2017). In 2019, the ESP team demonstrated the use of the third-generation ESP, coupled to a Tethys-class long-range AUV (LRAUV-ESP) to allow collection and preservation of samples for molecular analysis (Yamahara et al., 2019). Over a 12-hour period, 45 environmental samples of 1L were collected in Monterey Bay, USA. Using integrated onboard optical sensors, the samples were collected at the peak chlorophyll layer using a previously described control system (Zhang et al., 2015). This work demonstrated the capabilities and validity of using AUV samplers to increase the scale of sampling campaigns and the integration of additional sensor to perform adaptive sampling to provide informative high-resolution data.

Unmet needs and recommendations

Affordability

The single most important bottleneck to the widespread use of ecogenomic sensors for phytoplankton monitoring is affordability. Despite a clear and demonstrated utility for ecogenomic sensors in a wide range of fields, including phytoplankton research, they have yet to see widespread use. The ESP remains the only commercially available option and is expensive to purchase and operate. This high cost and operational complexity are hindrances to wider adoptions. The prospect of the lower cost, smaller third-generation ESP currently under development and the advancements seen in lab-on-a-chip and microfluidics in medical diagnostics hold promise that in the near future small, low-cost automated devices may take ecogenomic sensors from research tools to widely used devices in aquatic management, industry, and surveillance.

Real-time reporting and operational control

It is important to note that the great potential of ecogenomic sensors will not be realized with their data alone, but by integrating the analytical results they produce with existing data streams from other in situ sensors, such as CTDs and optical sensors, as was demonstrated with great effect with the ESP studies of *Pseudo-nitzschia* in Monterey Bay. The ability to operate, control, and monitor these systems with real-time communications networks will be essential, as will the ability to share and merge the data generated with external data streams, including satellite, meteorological, and environmental data to further enhance the utility of these systems in surveillance, monitoring, and modeling.

References

Akter, S., Kustila, T., Leivo, J., Muralitharan, G., Vehniäinen, M., Lamminmäki, U., 2019. Noncompetitive chromogenic lateral-flow immunoassay for simultaneous detection of microcystins and nodularin. Biosensors (Basel) 9, 79.

Anfossi, L., Baggiani, C., Giovannoli, C., D'arco, G., Giraudi, G., 2012. Lateral-flow immunoassays for mycotoxins and phycotoxins: a review. Analytical and Bioanalytical Chemistry 405, 467–480.

Antonella, P., Luca, G., 2013. The quantitative real-time PCR applications in the monitoring of marine harmful algal bloom (HAB) species. Environmental Science and Pollution Research 20, 6851–6862.

Aylward, F.O., Eppley, J.M., Smith, J.M., Chavez, F.P., Scholin, C.A., Delong, E.F., 2015. Microbial community transcriptional networks are conserved in three domains at ocean basin scales. Proceedings of the National Academy of Sciences of the United States of America 112, 5443–5448.

Babin, M., Cullen, J., Roesler, C., Donaghay, P., Doucette, G., Kahru, M., Lewis, M., Scholin, C., Sieracki, M., Sosik, H., 2005. New approaches and technologies for observing Harmful Algal Blooms. Oceanography 18, 210–227.

Bahi, M.M., Tsaloglou, M.-N., Mowlem, M., Morgan, H., 2011. Electroporation and lysis of marine microalga Karenia brevis for RNA extraction and amplification. Journal of the Royal Society Interface 8, 601–608.

Batten, S.D., Abu-Alhaija, R., Chiba, S., Edwards, M., Graham, G., Jyothibabu, R., Kitchener, J.A., Koubbi, P., Quatters-Gollop, A.M., Muxagata, E., Ostle, C., Richardson, A.J., Robinson, K.V., Takahashi, K.T., Verheye, H.M., Wilson, W., 2019. A global plankton diversity monitoring program. Frontiers in Marine Science 6, 1–14.

Boom, R., Sol, C.J.A., Salimans, M.M.M., Jansen, C.L., Wertheim-Van Dillen, P.M.E., Van Der Noordaa, J., 1990. Rapid and simple method for purification of nucleic acids. Journal of Clinical Microbiology 28, 495–503.

Bowers, H.A., Marin 3rd, R., Birch, J.M., Scholin, C.A., Doucette, G.J., 2016. Recovery and identification of Pseudo-nitzschia (Bacillariophyceae) frustules from natural samples acquired using the environmental sample processor. Journal of Phycology 52, 135–140.

Bowers, H.A., Ryan, J.P., Hayashi, K., Woods, A.L., Marin 3rd, R., Smith, G.J., Hubbard, K.A., Doucette, G.J., Mikulski, C.M., Gellene, A.G., Zhang, Y., Kudela, R.M., Caron, D.A., Birch, J.M., Scholin, C.A., 2018. Diversity and toxicity of Pseudo-nitzschia species in Monterey Bay: perspectives from targeted and adaptive sampling. Harmful Algae 78, 129–141.

Breier, J.A., Sheik, C.S., Gomez-Ibanez, D., Sayre-Mc Cord, R.T., Sanger, R., Rauch, C., Coleman, M., Bennett, S.A., Cron, B.R., Li, M., German, C.R., Toner, B.M., Dick, G.J., 2014. A large volume particulate and water multi-sampler with in situ preservation for microbial and biogeochemical studies. Deep-Sea Research 94, 195–206.

Carrell, C., Kava, A., Nguyen, M., Menger, R., Munshi, Z., Call, Z., Nussbaum, M., Henry, C., 2019. Beyond the lateral flow assay: a review of paper-based microfluidics. Microelectronic Engineering 206, 45–54.

Casper, E.T., Patterson, S.S., Bhanushali, P., Farmer, A., Smith, M., Fries, D.P., Paul, J.H., 2007. A handheld NASBA analyzer for the field detection and quantification of Karenia brevis. Harmful Algae 6, 112–118.

Costa, P.R., Baugh, K.A., Wright, B., Ralonde, R., Nance, S.L., Tatarenkova, N., Etheridge, S.M., Lefebvre, K.A., 2009. Comparative determination of paralytic shellfish toxins (PSTs) using five different toxin detection methods in shellfish species collected in the Aleutian Islands, Alaska. Toxicon 54, 313–320.

Diercks, S., Metfies, K., Medlin, L.K., 2008a. Development and adaptation of a multi-probe biosensor for the use in a semi-automated device for the detection of toxic algae. Biosensors & Bioelectronics 23, 1527–1533.

Diercks, S., Metfies, K., Medlin, L.K., 2008b. Molecular probe sets for the detection of toxic algae for use in sandwich hybridization formats. Journal of Plankton Research 30, 439–448.

Diercks-Horn, S., Metfies, K., Jackel, S., Medlin, L.K., 2011. The ALGADEC device: a semi-automated rRNA biosensor for the detection of toxic algae. Harmful Algae 10, 395–401.

Doucette, G.J., Mikulski, C.M., Jones, K.L., King, K.L., Greenfield, D.I., Marin III, R., Jensen, S., Roman, B., Elliott, C.T., Scholin, C.A., 2009. Remote, subsurface detection of the algal toxin domoic acid onboard the Environmental Sample Processor: assay development and field trials. Harmful Algae 8, 880–888.

Dyhrman, S.T., Erdner, D., La Du, J., Galac, M., Anderson, D.M., 2006. Molecular quantification of toxic Alexandrium fundyense in the Gulf of Maine using real-time PCR. Harmful Algae 5, 242–250.

Eckford-Soper, L.K., Daugbjerg, N., 2015. Development of a multiplex real-time qPCR assay for simultaneous enumeration of up to four marine toxic bloom-forming microalgal species. Harmful Algae 48, 37–43.

Feike, J., Jürgens, K., Hollibaugh, J.T., Krüger, S., Jost, G., Labrenz, M., 2012. Measuring unbiased metatranscriptomics in suboxic waters of the central Baltic Sea using a new in situ fixation system. The ISME Journal 6, 461–470.

Fries, D., Paul, J., Smith, M., Farmer, A., Casper, E., Wilson, J., 2007. The autonomous microbial genosensor, an in situ sensor for marine microbe detection. Microscopy and Microanalysis 13, 514–515.

Fukuba, T., Fujii, T., 2012. Microfabricated Flow-Through Device for In Situ Gene Analysis. Springer Protocols Handbooks, pp. 59–72.

Fukuba, T., Miyaji, A., Okamoto, T., Yamamoto, T., Kaneda, S., Fujii, T., 2011. Integrated in situ genetic analyzer for microbiology in extreme environments. RSC Advances 1 (8), 1567–1573.

Fukuba, T., Yamamoto, T., Naganuma, T., Fujii, T., 2004. Microfabricated flow-through device for DNA amplification - towards in situ gene analysis. Chemical Engineering Journal 101 (1–3), 151–156.

Greenfield, D.I., Marin, R., Doucette, G.J., Mikulski, C., Jones, K., Jensen, S., Roman, B., Alvarado, N., Feldman, J., Scholin, C., 2008. Field applications of the second-generation Environmental Sample Processor (ESP) for remote detection of harmful algae: 2006-2007. Limnology and Oceanography: Methods 6, 667–679.

Hatfield, R.G., Batista, F.M., Bean, T.P., Fonseca, V.G., Santos, A., Turner, A.D., Lewis, A., Dean, K.J., Martinez-Urtaza, J., 2020. The application of nanopore sequencing technology to the study of dinoflagellates: a proof of concept study for rapid sequence-based discrimination of potentially Harmful Algae. Frontiers in Microbiology 11, 844.

Haugland, R.A., Siefring, S.C., Wymer, L.J., Brenner, K.P., Dufour, A.P., 2005. Comparison of Enterococcus measurements in freshwater at two recreational beaches by quantitative polymerase chain reaction and membrane filter culture analysis. Water Research 39, 559–568.

Huang, H.L., Zhu, P., Zhou, C.X., He, S., Yan, X.J., 2017. The development of loop-mediated isothermal amplification combined with lateral flow dipstick for detection of Karlodinium veneficum. Harmful Algae 62, 20–29.

Humpage, A.R., Froscio, S.M., Lau, H.-M., Murphy, D., Blackbeard, J., 2012. Evaluation of the Abraxis Strip Test for Microcystins™ for use with wastewater effluent and reservoir water. Water Research 46, 1556–1565.

Jannasch, H.W., Wheat, C.G., Plant, J.N., Kastner, M., Stakes, D.S., 2004. Continuous chemical monitoring with osmotically pumped water samplers: OsmoSampler design and applications. Limnology and Oceanography: Methods 2, 102–113.

Jellett, J.F., Trainer, V., Bates, S., Somerset, I., 2006. Characteristics of the jellett rapid tests. In: Proceedings of 5th International Conference on Molluscan Shellfish Safety.

Johansson, K.S.L., Luhrig, K., Klaminder, J., Rengefors, K., 2016. Development of a quantitative PCR method to explore the historical occurrence of a nuisance microalga under expansion. Harmful Algae 56, 67–76.

Kirchman, D.L., Yu, L., Fuchs, B.M., Amann, R., 2001. Structure of bacterial communities in aquatic systems as revealed by filter PCR. Aquatic Microbial Ecology 26, 13–22.

Kon, N.F., Lau, W.L.S., Hii, K.S., Law, I.K., Teng, S.T., Lim, H.C., Takahashi, K., Gu, H.F., Lim, P.T., Leaw, C.P., 2017. Quantitative real-time PCR detection of a harmful unarmoured dinoflagellate, Karlodinium australe (Dinophyceae). Phycological Research 65, 291–298.

Lagier, M.J., Fell, J.W., Goodwin, K.D., 2007. Electrochemical detection of harmful algae and other microbial contaminants in coastal waters using hand-held biosensors. Marine Pollution Bulletin 54, 757–770.

Laycock, M.V., Donovan, M.A., Easy, D.J., 2010. Sensitivity of lateral flow tests to mixtures of saxitoxins and applications to shellfish and phytoplankton monitoring. Toxicon 55, 597–605.

Mankins, J.C., 2009. Technology readiness assessments: a retrospective. Acta Astronautica 65, 1216–1223.

Manning, S.R., La Claire, J.W., 2010. Multiplex PCR methods for the species-specific detection and quantification of Prymnesium parvum (Haptophyta). Journal of Applied Phycology 22, 587–597.

Mcgillicuddy, D.J., Brosnahan, M.L., Couture, D.A., He, R., Keafer, B.A., Manning, J.P., Martin, J.L., Pilskaln, C.H., Townsend, D.W., Anderson, D.M., 2014. A red tide of Alexandrium fundyense in the Gulf of Maine. Deep-Sea Research. Part 2. Topical Studies in Oceanography 103, 174–184.

Mohammed, M.I., Haswell, S., Gibson, I., 2015. Lab-on-a-chip or chip-in-a-lab: challenges of commercialization lost in translation. Procedia Technology 20, 54–59.

Ottesen, E.A., Marin 3rd, R., Preston, C.M., Young, C.R., Ryan, J.P., Scholin, C.A., Delong, E.F., 2011. Metatranscriptomic analysis of autonomously collected and preserved marine bacterioplankton. The ISME Journal 5, 1881–1895.

Ottesen, E.A., Young, C.R., Eppley, J.M., Ryan, J.P., Chavez, F.P., Scholin, C.A., Delong, E.F., 2013. Pattern and synchrony of gene expression among sympatric marine microbial populations. Proceedings of the National Academy of Sciences of the United States of America 110, E488–E497.

Pargett, D.M., Birch, J.M., Preston, C.M., Ryan, J.P., Zhang, Y., Scholin, C.A., 2015. Development of a mobile ecogenomic sensor. In: OCEANS 2015 - MTS/IEEE Washington. Institute of Electrical and Electronics Engineers Inc., pp. 1–6.

Patterson, S.S., Casper, E.T., Garcia-Rubio, L., Smith, M.C., Paul, J.H., 2005. Increased precision of microbial RNA quantification using NASBA with an internal control. Journal of Microbiological Methods 60, 343–352.

Paul, J., Scholin, C., Van Den Engh, G., Perry, M.J., 2007. In situ instrumentation. Oceanography 20, 70–78.

Posthuma-Trumpie, G.A., Korf, J., Van Amerongen, A., 2009. Lateral flow (immuno)assay: its strengths, weaknesses, opportunities and threats. A literature survey. Analytical and Bioanalytical Chemistry 393, 569–582.

Preston, C.M., Harris, A., Ryan, J.P., Roman, B., Marin 3rd, R., Jensen, S., Everlove, C., Birch, J., Dzenitis, J.M., Pargett, D., Adachi, M., Turk, K., Zehr, J.P., Scholin, C.A., 2011. Underwater application of quantitative PCR on an ocean mooring. PLoS ONE 6, e22522.

Reid, P.C., Colebrook, J.M., Matthews, J.B.L., Aiken, J., Team, C.P.R., 2003. The Continuous Plankton Recorder: concepts and history, from plankton indicator to undulating recorders. Progress in Oceanography 58, 117–173.

Ribeiro, H., Martins, A., Goncalves, M., Guedes, M., Tomasino, M.P., Dias, N., Dias, A., Mucha, A.P., Carvalho, M.F., Almeida, C.M.R., Ramos, S., Almeida, J.M., Silva, E., Magalhaes, C., 2019. Development of an autonomous biosampler to capture in situ aquatic microbiomes. PLoS ONE 14, e0216882.

Richardson, A.J., Walne, A.W., John, A.W.G., Jonas, T.D., Lindley, J.A., Sims, D.W., Stevens, D., Witt, M., 2006. Using continuous plankton recorder data. Progress in Oceanography 68, 27–74.

Ripley, S.J., Baker, A.C., Miller, P.I., Walne, A.W., Schroeder, D.C., 2008. Development and validation of a molecular technique for the analysis of archived formalin-preserved phytoplankton samples permits retrospective assessment of Emiliania huxleyi communities. Journal of Microbiological Methods 73, 118–124.

Robidart, J., Callister, S.J., Song, P., Nicora, C.D., Wheat, C.G., Girguis, P.R., 2013. Characterizing microbial community and geochemical dynamics at hydrothermal vents using osmotically driven continuous fluid samplers. Environmental Science & Technology 47, 4399–4407.

Robidart, J.C., Church, M.J., Ryan, J.P., Ascani, F., Wilson, S.T., Bombar, D., Marin, R., Richards, K.J., Karl, D.M., Scholin, C.A., Zehr, J.P., 2014. Ecogenomic sensor reveals controls on N2-fixing microorganisms in the North Pacific Ocean. The ISME Journal 8, 1175–1185.

Robidart, J.C., Preston, C.M., Paerl, R.W., Turk, K.A., Mosier, A.C., Francis, C.A., Scholin, C.A., Zehr, J.P., 2012. Seasonal Synechococcus and Thaumarchaeal population dynamics examined with high resolution with remote in situ instrumentation. The ISME Journal 6, 513–523.

Ryan, J., Greenfield, D., Marin, R., Preston III, C., Roman, B., Jensen, S., Pargett, D., Birch, J., Mikulski, C., Doucette, G., Scholin, C., 2011. Harmful phytoplankton ecology studies using an autonomous molecular analytical and ocean observing network. Limnology and Oceanography 56, 1255–1272.

Ryan, J.P., Kudela, R.M., Birch, J.M., Blum, M., Bowers, H.A., Chavez, F.P., Doucette, G.J., Hayashi, K., Marin, R., Mikulski, C.M., Pennington, J.T., Scholin, C.A., Smith, G.J., Woods, A., Zhang, Y., 2017. Causality of an extreme harmful algal bloom in

Monterey Bay, California, during the 2014-2016 northeast Pacific warm anomaly. Geophysical Research Letters 44, 5571–5579.

Saiki, R.K., Scharf, S., Faloona, F., Mullis, K.B., Horn, G.T., Erlich, H.A., Arnheim, N., 1985. Enzymatic amplification of b-globin genomic sequences and restriction site analysis for diagnosis of sickle cell anemia. Science 230, 1350–1354.

Sajid, M., Kawde, A.-N., Daud, M., 2015. Designs, formats and applications of lateral flow assay: a literature review. Journal of Saudi Chemical Society 19, 689–705.

Scholin, C., Doucette, G., Jensen, S., Roman, B., Pargett, D., Marin III, R., Preston, C., Jones, W., Feldman, J., Everlove, C., Harris, A., Alvarado, N., Massion, E., Birch, J., Greenfield, D., Wheeler, K., Vrijenhoek, R., Mikulski, C., Jones, K., 2009a. Remote detection of marine microbes, small invertebrates, harmful algae, and biotoxins using the Environmental Sample Processor (ESP). Oceanography 22, 158–167.

Scholin, C., Doucette, G., Jensen, S., Roman, B., Pargett, D., Marin, R., Preston, C., Jones, W., Feldman, J., Everlove, C., Harris, A., Alvarado, N., Massion, E., Birch, J., Greenfield, D., Vrijenhoek, R., Mikulski, C., Jones, K., 2009b. Remote detection of marine microbes, small invertebrates, harmful algae, and biotoxins using the Environmental Sample Processor (ESP). Oceanography 22, 158–167.

Scholin, C.A., Birch, J., Jensen, S., Marin III, R., Massion, E., Pargett, D., Preston, C., Roman, B., Ussler III, W., 2017. The quest to develop ecogenomic sensors: a 25-year history of the Environmental Sample Processor (ESP) as a case study. Oceanography 30, 100–113.

Scholin, C.A., Doucette, G.J., Cembella, A., 2008. Prospects for developing automated systems for in situ detection of harmful algae and their toxins. In: Babin, M., Roesler, C.S., Cullen, J.J. (Eds.), EPIC3In: Real-Time Coastal Observing Systems for Marine Ecosystem Dynamics and Harmful Algal Blooms. In: Oceanographic Methodology Series. UNESCO Publications, Paris, pp. 413–462.

Seegers, B.N., Birch, J.M., Marin, R., Scholin, C.A., Caron, D.A., Seubert, E.L., Howard, M.D.A., Robertson, G.L., Jones, B.H., 2015. Subsurface seeding of surface harmful algal blooms observed through the integration of autonomous gliders, moored environmental sample processors, and satellite remote sensing in southern California. Limnology and Oceanography 60, 754–764.

Smith, M.C., Steimle, G., Ivanov, S., Holly, M., Fries, D.P., 2007. An integrated portable hand-held analyser for real-time isothermal nucleic acid amplification. Analytica Chimica Acta 598, 286–294.

Stern, R., Moore, S.K., Trainer, V.L., Bill, B.D., Fischer, A., Batten, S., 2018. Spatial and temporal patterns of Pseudo-nitzschia genetic diversity in the North Pacific Ocean from Continuous Plankton Recorder surveys. Marine Ecology. Progress Series 606, 7–82.

Toldrà, A., Alcaraz, C., Diogène, J., O'sullivan, C.K., Campàs, M., 2019a. Detection of Ostreopsis cf. ovata in environmental samples using an electrochemical DNA-based biosensor. Science of the Total Environment 689, 655–661.

Toldrà, A., O'sullivan, C.K., Campàs, M., 2019b. Detecting Harmful Algal Blooms with Isothermal Molecular Strategies. Elsevier Ltd.

Tsaloglou, M.-N., Bahi, M.M., Waugh, E.M., Morgan, H., Mowlem, M., 2011. On-chip real-time nucleic acid sequence-based amplification for RNA detection and amplification. Analytical Methods 3, 2127–2133.

Turner, A.D., Tarnovius, S., Hatfield, R.G., Teixeira-Alves, M., Broadwater, M., Dolah, F.V., Garcia-Mendoza, E., Medina, D., Salhi, M., Goya, A.B., Barrera, F., Carrasco, D., Rubilar, I., Suarez-Isla, B.A., 2020. Application of six detection methods for analysis of paralytic shellfish toxins in shellfish from four regions within Latin America. Mar Drugs 18.

Van Dijk, E.L., Jaszczyszyn, Y., Naquin, D., Thermes, C., 2018. The third revolution in sequencing technology. Trends in Genetics 34, 666–681.

Volpatti, L.R., Yetisen, A.K., 2014. Commercialization of microfluidic devices. Trends in Biotechnology 32, 347–350.

Winslow, L.A., Dugan, H.A., Buelow, H.N., Cronin, K.D., Priscu, J.C., Takacs-Vesbach, C., Doran, P.T., 2014. Autonomous year-round sampling and sensing to explore the physical and biological habitability of permanently ice-covered Antarctic lakes. Marine Technology Society Journal 48, 8–17.

Wu, C., Lillehoj, P.B., Sabet, L., Wang, P., Ho, C.-M., 2011. Ultrasonication on a microfluidic chip to lyse single and multiple Pseudo-nitzschia for marine biotoxin analysis. Biotechnology Journal 6, 150–155.

Yamahara, K.M., Demir-Hilton, E., Preston, C.M., Marin, R., Pargett, D., Roman, B., Jensen, S., Birch, J.M., Boehm, A.B., Scholin, C.A., 2015. Simultaneous monitoring of faecal indicators and harmful algae using an in-situ autonomous sensor. Letters in Applied Microbiology 61, 130–138.

Yamahara, K.M., Preston, C.M., Birch, J.M., Walz, K.R., Marin III, R., Jensen, S., Pargett, D., Roman, B., Zhang, Y., Ryan, J., Ussler, B., Hobson, B., Kieft, B., Raanan, B.Y., Goodwin, K.D., Chavez, F.P., Scholin, C.A., 2019. In-situ autonomous acquisition and preservation of marine environmental DNA using an autonomous underwater vehicle. Frontiers in Marine Science 6, 373.

Zamor, R.M., Glenn, K.L., Hambright, K.D., 2012. Incorporating molecular tools into routine HAB monitoring programs: using qPCR to track invasive Prymnesium. Harmful Algae 15, 1–7.

Zhang, C., Chen, G., Wang, Y., Sun, R., Nie, X., Zhou, J., 2018. MHBMDAA: membrane-based DNA array with high resolution and sensitivity for toxic microalgae monitoring. Harmful Algae 80, 107–116.

Zhang, Y., Kieft, B., Mcewen, R., Stanway, J., Bellingham, J., Ryan, J., Hobson, B., Pargett, D., Birch, J., Scholin, C., 2015. Tracking and sampling of a phytoplankton patch by an autonomous underwater vehicle in drifting mode. In: OCEANS 2015 - MTS/IEEE Washington. Institute of Electrical and Electronics Engineers Inc., pp. 1–5.

Zhu, P., Huang, H.L., Zhou, C.X., Xu, J., Qiao, L.L., Dang, C.Y., Pang, J.H., Gao, W.F., Yan, X.J., 2019. Sensitive and rapid detection of Prymnesium parvum (Haptophyceae) by loop-mediated isothermal amplification combined with a lateral flow dipstick. Aquaculture 505, 199–205.

CHAPTER 7

Applications of satellite remote sensing technology to the analysis of phytoplankton community structure on large scales

Astrid Bracher[a,b], Robert J.W. Brewin[c], Aurea M. Ciotti[d], Lesley A. Clementson[e], Takafumi Hirata[f,g], Tihomir S. Kostadinov[h], Colleen B. Mouw[i], and Emanuele Organelli[j]

[a]Physical Oceanography of Polar Seas, Climate Sciences, Alfred Wegener Institute, Helmholtz Centre for Polar and Marine Research, Bremerhaven, Germany
[b]Institute of Environmental Physics, Faculty of Physics and Electrical Engineering, University Bremen, Bremen, Germany
[c]Centre for Geography and Environmental Science, College of Life and Environmental Sciences, University of Exeter, Cornwall, United Kingdom
[d]Center for Marine Biology, University of São Paulo, São Sebastião, SP, Brazil
[e]CSIRO Oceans and Atmosphere, Hobart, TAS, Australia
[f]Arctic Research Center, Hokkaido University, Sapporo, Japan
[g]Graduate School of Environmental Science, Hokkaido University, Sapporo, Japan
[h]Department of Liberal Studies, California State University, San Marcos, CA, United States
[i]University of Rhode Island, Graduate School of Oceanography, Narragansett, RI, United States
[j]National Research Council (CNR), Institute of Marine Sciences (ISMAR), Rome, Italy

7.1 Challenges of sensors and methods

7.1.1 Sensor characteristics

For more than 15 years, algorithms have exploited multispectral ocean-color satellite data to derive metrics on phytoplankton composition (e.g., see IOCCG, 2014, and Section 7.1.2). Images from passive ocean-color sensors (see overview in https://ioccg.org/resources/missions-instruments/) typically encompass pixel size ranges from 300 m to 1000 m, with the footprint of the satellite sensor covering the whole globe within 1 to 3 days, acquiring information only during daytime, cloud- and sea-ice-free conditions, and only for the upper layer of the oceans (at maximum around 50 m). Current ocean-color sensors have six to ten spectral bands in the visible wavelength range, with a 10 to 20 nm band resolution and a signal to noise ratio (SNR) of higher than 1000. Such spectral and sensor characteristics can limit the discrimination of multiple groups simultaneously (IOCCG, 2014). Indeed, phytoplankton groups (PG; following the classification of Sieburth et al., 1978), despite having specific marker pigments, contain a similar suite of pigments, or pigments with similar absorptive properties that can be distinguished only when several bands are taken into consideration (Ciotti et al., 2002; Organelli et al., 2013; Wolanin et al., 2016; Organelli et al., 2017). To partly overcome limitations in PG and phytoplankton size class (PSC) detection, the European Space Agency (ESA) launched in 2016 a new ocean-color sensor (Ocean and Land Color Instrument, OLCI) on Sentinel-3a, which provides two more bands in the visible range than its predecessor the MEdium Resolution Imaging Spectrometer (MERIS), that had 8 bands. It is anticipated that the number of bands will increase further for future multispectral ocean-color sensors, and many space agencies' efforts have been, and are, directed towards the development of high spectral resolution ocean-color sensors, the so-called hyperspectral ocean-color remote sensing.

Hyperspectral satellite sensors, with fine spectral resolution (1 nm and better) and high signal-to-noise ratio (>1000), and which were originally designed for atmospheric applications, provide additional opportunities for distinguishing multiple PG (e.g., the PhytoDOAS method developed by Bracher et al., 2009 and Sadeghi et al., 2012a). However, the exploitation of hyperspectral satellite data for ocean-color applications has been limited, because hyperspectral sensors, like the SCanning Imaging Absorption spectroMeter for Atmospheric CartographY (SCIAMACHY, spectral resolution <0.5 nm), do not provide operational water-leaving radiance products. Most of these data are characterized by low spatial resolution (~0.5°) and temporal coverage (taking several days to build up a global composite) and were not designed for phytoplankton observations. Moreover, SCIAMACHY data acquisition ended when contact with the ENVISAT satellite was lost (April 2012). Nonetheless, PhytoDOAS has been recently adapted for the ozone monitoring instrument (OMI) sensor (operational since 2004) and for the Sentinel-5-Precursor (S-5-P) satellite (launched in 2017) that carries the TROPOspheric Monitoring Instrument (TROPOMI). First results of these attempts are very promising (Oelker et al., 2019; ESA S5POC project) and suggest future success to obtain spectrally derived PG estimates with improved global coverage (daily) and smaller spatial resolution (18 × 24 km and 3.5 × 7 km, respectively). TROPOMI is also a precursor to the ultraviolet/visible/near-infrared instrument (UVN). This instrument has a similar spatial resolution to TROPOMI and is due to be launched in 2023 on both Sentinel-4 and Sentinel-5 (https://sentinel.esa.int/web/sentinel/missions/). The hyperspectral Ocean Color Instrument (OCI) on NASA's Plankton, Aerosol, Cloud, ocean Ecosystem (PACE) mission (expected to launch in 2022) will revolutionize mainstream ocean-color science. The OCI instrument will have continuous spectral coverage from 340 nm to 890 nm in 5 nm steps and 5 nm bandwidth (with additional SWIR bands), a 2-day global temporal coverage and 1 km spatial resolution (Werdell et al., 2019, see https://pace.oceansciences.org/). The hyperspectral data of PACE, with an optimal combination of SNR, spatial and temporal coverage and resolution, is expected to fill the gap for improved detection of PG and PSCs in the open ocean (for more details, see Vandermeulen et al., 2017).

Due to the characteristics of available ocean-color multispectral sensors and hyperspectral atmospheric sensors, the estimations of phytoplankton composition using satellite data yield time-series data at coarse spatial resolution (>300 m pixel size). For resolving smaller spatial scales (10–100 m), which are important for understanding community dynamics in coastal and inland waters and at some oceanic fronts, several high spatial resolution sensors have been used, including the Hyperspectral Imager for the Coastal Ocean (HICO), which provided data from 2009 to 2014 at high spectral (6 nm, between 400–800 nm) resolution, with SNR above 200, and a spatial resolution of 100 m. Since HICO was hosted on the International Space Station (ISS), its data coverage was limited to a few selected coastal areas within the tropical and temperate oceans. Mostly, the lack of robust atmospheric correction for HICO and the low SNR has prevented the exploitation of the full spectrum and consistency between scenes. Only standard phytoplankton biomass metrics (chlorophyll-a concentration (Chl-a), fluorescence line height), which can also be derived from multispectral data, have been produced using HICO (Braga et al., 2013; Ryan et al., 2014). More recently, machine learning techniques have been developed to enable the retrieval of the hyperspectral phytoplankton absorption spectrum (Pahlevan et al., 2021) that is essential for improved PG and PSC detection (e.g., Ciotti et al., 2002).

Optical sensors designed for terrestrial applications have proven attractive for inland and coastal water-color applications. Multispectral imaging sensors, such as the MultiSpectral Instrument (MSI) on Sentinel-2 and Ocean Land Imager (OLI) on Landsat-8 can be used to track turbidity as well as Chl-a, though the limited spectral information makes PG and PSC detection unlikely. More recently, efforts have been targeted on retrieving PGs from hyperspectral terrestrial satellite sensors, including the Hyperion spectrometer onboard NASA's earth observing one, which acquired from 2001–2017 hyperspectral images from 400 to 2500 nm with a 30-m pixel size, and the Compact High-Resolution Imaging Spectrometer (CHRIS) on ESA's Proba-1 microsatellite, which has been operating since 2001. Two more hyperspectral terrestrial sensors have recently started operating in space: Deutsche Luft- und Raumfahrt's (DLR's) Earth Sensing Imaging Spectrometer (DESIS) launched in 2018 from Germany, and the PRecursore IperSpettrale della Missione Applicativa (PRISMA) launched in 2019 from Italy. Similar sensors are due to be launched soon, including the Environmental MAPping and analysis program (EnMAP) from Germany, the Hyperspectral Imager SUIte (HISUI) from Japan, and the Surface Biology and Geology (SBG) from the United States.

For these optical-terrestrial operating and upcoming sensors, the technical characteristics are more suited for inland and coastal water applications (e.g., Alonso et al., 2019; Giardino et al., 2020). These sensors will provide data over the complete visible range, at 2.5 nm (DESIS), 6 nm (EnMAP), and between 8 to 12 nm (SBG, HISUI, PRISMA) spectral resolution, with parts of the SWIR also covered. SNR in the visible region of the spectrum for these sensors is up to 500 for DESIS, ENMAP, and HISUI; 400 for SBG, and 200 for PRISMA, and the spatial resolution is $\leq$30 m. Despite the potential of these new sensors, there are challenges to their use for PG and PSC detection. It must be demonstrated whether quantitative detection of various phytoplankton groups is achievable with these sensor characteristics. Furthermore, these sensors only sample a few coastal and inland waters, with many regions not covered. Some coastal and inland waters are particularly challenging to monitor, for example fjords, over which images typically show a low SNR in the blue spectral range (very dark waters), owing to high concentrations of colored dissolved organic matter that absorb light strongly at blue wavelengths (CEOS, 2018; Muller-Karger et al., 2018).

To understand and monitor how current and future anthropogenic stressors, and those derived from climate change, influence trends in phytoplankton community structure and consequently the upper trophic chain and functioning of marine ecosystems, long-term time-series data are needed. Considering the typical life-span of ocean-color sensors of 5 to 10 years, the construction of long-term time-series data require merged ocean-color products (e.g., Maritorena et al., 2010; Sathyendranath et al., 2019), such as those distributed operationally through the European Copernicus program (https://marine.copernicus.eu/), containing data from NASA's SeaWiFS (Sea-viewing Wide Field-of-view Sensor) and MODIS (MODerate resolution Imaging Spectroradiometer), NOAA's VIIRS (Visible Infrared Imaging Radiometer Suite), and ESA's MERIS and OLCI ocean-color sensors. Discontinuities (e.g., due to inter-sensor biases) and gaps in the data records can significantly compromise the ability to detect trends and need to be taken into account (e.g., Weatherhead et al., 1998; Beaulieu et al., 2013). The synergistic use of multiple sensor types opens the opportunity to improve products performances. For instance, the SynSenPFT algorithm of Losa et al. (2017) is an algorithm that combines data on PG Chl-a derived from SCIAMACHY-PhytoDOAS and Chl-a-OC-PFT from the Ocean Color Climate Change Initiative (OC-CCI) to improve PG products. This algorithm can obtain high spatially and temporally resolved Chl-a (see details in Section 7.1.2) PG data, using the spectral imprints retrieved from hyperspectral satellite data (SCIAMACHY), and the high resolution available through multispectral PG products (daily data available from 2002 to 2012 on 4 × 4 km pixel resolution). Further efforts will be necessary to enable the continuation of these global PG data sets, from 1997 until today and beyond.

7.1.2 Satellite algorithms

Several overviews describing and evaluating global and regional algorithms on phytoplankton composition are available in the literature (e.g., Nair et al., 2008; IOCCG, 2014; Mouw et al., 2017). A summary of the main principles behind these algorithms can be categorized into three groups (Brewin et al., 2011): (1) abundance-based approaches, (2) ecological-based approaches, and (3) spectral-response based approaches.

Abundance-based approaches exploit relationships between the trophic status of the environment (*sensu* Margalef, 1967, corresponding to the biomass or abundance of phytoplankton) and expected contributions from different PSCs or PGs (see Nair et al., 2008). Generally, microphytoplankton (i.e., cell or colony size >20 μm) are dominated by diatoms and dinoflagellates, and their relative abundance is expected to increase as a function of total biomass or abundance. On the other hand, picophytoplankton dominate low Chl-a waters, and as the total biomass or abundance increases their proportion is expected to decrease (Yentsch and Phinney, 1989). The models developed to capture these relationships have been statistical (e.g., Uitz et al., 2006), empirical (e.g., Hirata et al., 2011), or conceptual (e.g., Brewin et al., 2010). The simplicity of the abundance-based approach makes this family of algorithms fairly versatile and easy to implement. An abundance-based algorithm can be applied to any Chl-a product, not only from satellite data, but also from other ocean robotic platforms (e.g., Sauzede et al., 2015; Organelli and Claustre, 2019). These relationships can be easily tuned for regional applications (e.g., Brotas et al., 2013; Sammartino et al., 2015; Lamont et al., 2018; Sun et al., 2018; Corredor-Acosta et al., 2018), by using *in-situ* data, typically marker pigments detected by high-performance liquid chromatography (HPLC), or physical sequential size fractionated pigment extractions. However, for these same reasons, abundance-based algorithms are not capable of capturing scenarios when PG or PSC deviate from the established relationships, and care needs to be taken, for example, when interpreting changes in the relationships, for example, that could be related to climate change (Sathyendranath et al., 2017).

Ecological-based approaches use satellite ocean-color derived products with additional satellite-derived environmental variables (e.g., sea surface temperature, light availability, wind stress) to identify ecological niches and the specific phytoplankton communities associated with them (Raitsos et al., 2008; Palacz et al., 2013; Stock and Subramaniam, 2020). These methods use advanced statistical tools (e.g., artificial neural networks) to link PG data to environmental, mostly satellite-based, data sets. Efforts have also been made to combine abundance- and ecological-based approaches (e.g., Brewin et al., 2015; Moore and Brown, 2020). As with the abundance-based approaches, care must also be taken when interpreting time-series data using the ecological-based approaches, given the potential for natural or anthropogenically driven deviations from the tuned ecological relationships.

Spectral-response-based algorithms analyze spectral variations in optical properties to detect PG or PSCs, without making assumptions on their relationship with the phytoplankton biomass. The rationale is that spectral optical properties change with phytoplankton pigment composition and cell size (i.e., packaging effect), as well as photophysiology (Morel and Bricaud, 1981; Stramski and Kiefer, 1991). Some spectral algorithms directly link satellite-measured remote-sensing reflectance or normalized water-leaving radiance to PG or PSC (e.g., Alvain et al., 2005; Ben Mustapha et al., 2014; Li et al., 2013; Zhu et al., 2019; Xi et al., 2020) by applying statistical methods, such as principal component analysis and more recently, machine learning approaches. Other algorithms require first inverting satellite-measured radiance to obtain spectral light absorption and/or optical backscattering coefficients (the namely inherent optical properties (IOPs) of near-surface seawater), either via direct radiative inversion or based on look-up tables (LUTs) from forward modeling (see details below), the spectral characteristics of which are used to infer PG or PSCs (e.g., Sathyendranath et al., 2004; Ciotti and Bricaud, 2006; Bracher et al., 2009; Kostadinov et al., 2009; Mouw and Yoder, 2010; Devred et al., 2011; Sadeghi et al., 2012a; Roy et al., 2013; Shang et al., 2014; Kramer et al., 2018). They are based on physical principles established by radiative transfer equations and approximations (*see* comprehensive overview Mouw et al., 2017) and minimize the use of empirical relationships. They are thus thought to be better suited for monitoring long-term shifts in phytoplankton composition (Sathyendranath et al., 2017). Despite the intrinsic advantages of these mechanistic approaches, their performances are still subject to uncertainties, such as difficulties in distinguishing among PGs with the same or similar optical signatures (Organelli et al., 2017). Furthermore, as the interactions between natural oceanic particle assemblages and the light field are too complex to be modeled entirely by first principles, a certain degree of empiricism persists in semi-analytical algorithms.

Radiative transfer models (RTM) are tools used to both develop and assess the sensitivity of these spectral algorithms, as well as to apply corrections and develop LUTs. Evers-King et al. (2014) and Xi et al. (2015) used the ocean RTM software HydroLight (Sequoia Scientific) to model variations in the composition of PSC or certain (dominant) PG, respectively, assessing the potential of retrievals in different water types. Werdell et al. (2014) and Wolanin et al. (2016) have used the generalized inherent optical property (GIOP) model software (Werdell et al., 2013) to invert reflectance spectra (either water-leaving or top of atmosphere), and Wolanin et al. (2015) used the coupled ocean-atmosphere RTM Sciatran (Rozanov et al., 2014) to test the sensitivity of a PG retrieval (PhytoDOAS). Werdell et al. (2014) optimized the GIOP inversion scheme to retrieve the absence or presence of the dinoflagellate *Noctiluca miliaris* from MODIS data, whereas Wolanin et al. (2016) used the same method to identify optimal band placements for multi- and hyper-spectral satellite data, for successful retrievals of certain PG. Their results indicated that the inclusion of 381, 473, 532, and 594 nm bands for the Sentinel-3 ocean and land color instrument (OLCI) would potentially enable absorption-based quantitative retrievals of diatoms, cyanobacteria, and coccolithophores. Spectral-response-based approaches to retrieve PGs and PSCs will capitalize on analysis of hyperspectral optical properties (Ciotti et al., 2002; Organelli et al., 2013; Bracher et al., 2015), and thus hyperspectral satellites. Hyperspectral-based approaches may also help further reduce PG and PSC detection uncertainties by quantifying the influence of phytoplankton physiology on spectral characteristics of light absorption (Organelli et al., 2017).

Data sets on the composition of phytoplankton present either the dominance of a particular group, their fraction (mostly based on their size), or their biomass, with the latter given typically as the PG or PSC contribution to estimated or observed biomass in Chl-a concentration. However, biogeochemical models designed to study the carbon cycle typically simulate phytoplankton groups in carbon biomass, prompting recent algorithm efforts in that direction (e.g., see ESA BICEP project). For example, backscattering-based and absorption-based algorithms of Kostadinov et al. (2016) and Roy et al. (2013), which derive the particle size distribution (PSD) and carbon-based PSCs, and photoacclimation models (e.g., Sathyendranath et al., 2020).

Finally, we stress that there are large uncertainties in the *in situ* PG and PSC data used for product validation (Brewin et al., 2014; Kramer and Siegel, 2019; Chase et al., 2020). Over the past decade, there have been some efforts to intercompare these different satellite approaches (Brewin et al., 2011; Hirata et al., 2012; Kostadinov et al., 2017; Stock and Subramaniam, 2020), and to identify the best algorithms for a particular application. However, undoubtedly, more work is needed in this direction (IOCCG, 2014) as well a better definition of PG and PSC products with comparable metrics that are also compliant with user's needs and with *in situ* observations (Bracher et al., 2017a).

7.2 Application of satellite-derived phytoplankton diversity for fulfilling user needs

7.2.1 Tracing global phytoplankton diversity

Satellite remote sensing of phytoplankton composition allows us to determine where, in the ocean, certain types of phytoplankton are blooming, or their relative dominance in low phytoplankton biomass waters. These data, essential for following PG or PSC changes in both time and space, can be useful to the fishing industry (e.g., by identifying phytoplankton types that are important food sources for certain species of fish, see next section), for a better understanding of the ocean carbon cycle by improving primary production estimates (Uitz et al., 2010; Brewin et al., 2017a) and for assessment and modeling of the biological carbon pump (e.g., Siegel et al., 2014; Dutkiewicz et al., 2015). Considering the availability of over two decades of continuous satellite observations, phytoplankton composition from space is likely to become a key variable to estimate future effects of climate change on marine plankton. Indeed, satellite products of PGs are assimilated into ocean biogeochemistry models to improve hindcast and forecast simulations (e.g., Xiao and Friedrichs, 2014; Ciavatta et al., 2018; Skákala et al., 2018; Pradhan et al., 2020; IOCCG, 2020).

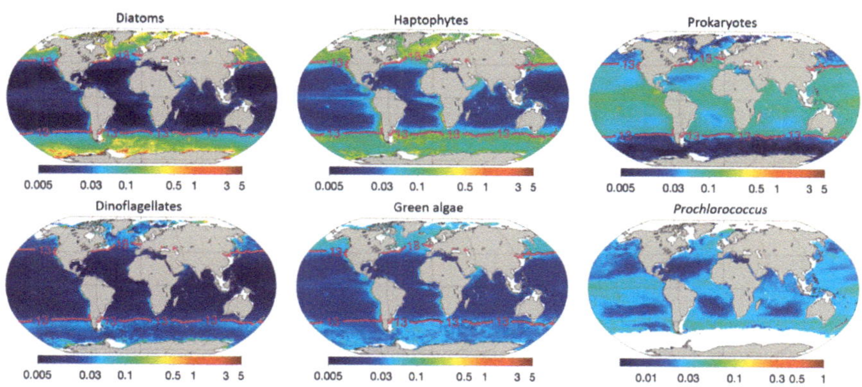

Figure 7.1 The Chl-a [mg m^{-3}] of various phytoplankton groups from GlobColour merged products (from left to right: (upper panel), diatoms, haptophytes, all prokaryotic phytoplankton, (lower panel), dinoflagellates, green algae, *Prochlorococcus*) as mean for 2002–2012 (OLCI-PFT algorithm v3.0 from Xi et al., 2021). Note that the color scale is logarithmic, and *Prochlorococcus* (bottom right panel) is given on a different scale from the rest. In pink, the mean isotherm of 13°C is shown. Figure courtesy of Hongyan Xi, AWI.

In this section, we illustrate synoptic global information on the biomass of all major groups using the recent algorithm of Xi et al. (2021), which is an update from Xi et al. (2020), and which integrates satellite ocean-color and sea surface temperature (SST) data to estimate Chl-a of the major PGs, namely diatoms, haptophytes, dinoflagellates, green algae, and prokaryotes, with per-pixel uncertainties, at global scale. These data span from 2002 until the present day, at 4 km resolution (for data products, see Table 7.1). The long-term means for Chl-a over a 10-year time series for each phytoplankton group from this algorithm (Fig. 7.1) show diatom Chl-a concentration is generally greater in high latitudes, marginal seas, and coastal upwelling regions than in tropical regions and subtropical gyres. The average diatom Chl-a over the globe is ~0.08 mg m^{-3}, with higher concentrations north of 40 °N (North Atlantic, Bering Sea, and Labrador Sea, up to the Arctic

Ocean), the Patagonian Shelf, and most parts of the Southern Ocean. Dinoflagellate Chl-a are low over the majority of the globe ($\sim$0.02 mg m^{-3}), but higher in the Arctic Ocean and the Patagonian Shelf region. Haptophytes, with a global average Chl-a concentration of 0.09 mg m^{-3}, follow the distributions of the diatoms, but show slightly greater values over wider spatial areas in the high latitudes, waters near the coasts, and equatorial regions, such as the west coast of Africa. Global Chl-a estimates of green algae are 0.03 mg m^{-3} on average, but higher in the Arctic and the coastal oceans around the south-eastern part of South America. Prokaryotic phytoplankton (also further partitioned into *Prochlorococcus*, appearing mainly in the tropical and subtropical waters) have a global average Chl-a of 0.05 mg m^{-3}, and are more abundant in the subtropical regions, where they dominate the biomass, particularly in the oligotrophic areas. Higher abundances of prokaryotic phytoplankton are also observed at the three Atlantic upwelling regions off the African coast and waters south of Panama. They contribute slightly ($\sim$5–20% of total Chl-a) to the overall community in the Arctic Ocean.

7.2.2 Satellite products on phytoplankton composition for tackling scientific objectives

7.2.2.1 Food web dynamics

7.2.2.1.1 Case study on energy content and transfer

Satellite observations of the ocean are highly relevant to fisheries and aquaculture, and this relevance will only increase as new satellite ocean products become available (IOCCG, 2009). It is well established that the community structure of phytoplankton plays a pivotal role in how energy is transferred through the marine food web (Michaels and Silver, 1988), as the energy content and elemental composition varies with phytoplankton group (Roy, 2018). Thus, satellite observations of phytoplankton composition offer much potential for understanding and monitoring food-web dynamics, fisheries, and aquaculture.

In addition to the rate of carbon fixation in photosynthesis, the rate of synthesis of essential fatty acids (FAs) is critical to the growth of marine organisms. Certain phytoplankton taxonomic groups, such as diatoms, are known to be enriched in FAs, such as eicosapentaenoic acids (EPAs), a long chain omega-3 polyunsaturated fatty acid. These FAs are essential to the health of most organisms, including humans. Using *in situ* observations, Budge et al. (2014) quantified the relationships between FAs produced predominantly by diatoms, and then applied the satellite remote-sensing algorithm of Sathyendranath et al. (2004) to estimate the fraction of total Chl-a attributed to diatoms in the Northwest Atlantic. These estimates were then used to map the distribution of EPAs relative to diatom carbon concentrations (C) in the Northwest Atlantic, using methods for converting the Chl-a to C (Sathyendranath et al., 2009). Using published estimates of global primary production, predictions of the contributions from large phytoplankton to global primary production (Uitz et al., 2010) and using the mean mass ratio of EPA:diatom carbon from the analysis in the Northwest Atlantic, Budge et al. (2014) provided a first global estimate of EPA of around 240 Mt y^{-1}. Additional considerations of the energy transfer through the food web suggested that the current supply of EPA was barely sufficient to meet the nutritional demand of the world population in the present day. The work of Budge et al. (2014) demonstrates how satellite PG products can be used to monitor fluctuations of an essential fatty acid.

7.2.2.1.2 Satellite products for studying phytoplankton phenology

In addition to the rates of production of organic carbon and fatty acids, the phenology of phytoplankton plays a critical role in food availability for higher trophic levels and affects energy transfer through the food web (Cushing, 1990; Platt et al., 2003; Koeller et al., 2009). Phenology refers to the seasonal cycle of growth and senescence/mortality, driven by the astronomical seasons on Earth. The North Atlantic spring bloom is a well-studied example that attracted the attention of early western oceanographers, who set out to explain it mechanistically, namely through the Sverdrup critical depth hypothesis (Sverdrup, 1953); the exact mechanism of bloom initiation and termination, as well as the importance of controlling variables, such as grazing, is sufficiently complex that it is still a hot topic in modern biological oceanography (Behrenfeld, 2010).

Phenological analysis is not only useful for studying the seasonal cycle of total phytoplankton biomass (given as Chl-a or C, for example), but also the biomass of PGs and PSCs, which are expected to change as the annual cycle progresses (Chivers et al., 2020). In the North Atlantic, diatoms can dominate during the spring bloom period, and are typically followed by dinoflagellates in summer, with picophytoplankton dominating during the oligotrophic stage of the season. This seasonal change in composition impacts the particle size distribution (Kostadinov et al., 2010) and biogeochemical cycles. Studying the details of PG succession is possible from satellite by using PG and PSC algorithms (e.g., see Mouw et al., 2017). Considering the variety of algorithms available, it becomes important to intercompare them, to assess their consistency and accuracy, and limitations to guide users choices on what algorithm is best suited for a specific application. However, considering that the various PG or PSC algorithms are based on different principles, retrieve different variables, with differing units and scales (e.g., Chl-a due to diatoms vs. microphytoplankton C, IOCCG, 2014; Mouw et al., 2017; Kostadinov et al., 2017), direct comparisons are often nontrivial. Generally, comparison of phenological parameters can be more meaningful than directly comparing disparate remote-sensing variables. In addition, comparing individual time series of variables, even if they are on the same scales, can lead to meaningless or incomplete results if there is a phase shift (i.e., phenological difference). By comparing phenological metrics instead, these issues can be avoided to a certain extent.

Kostadinov et al. (2017) compared phenology metrics among 10 different PSC and PG algorithms using a 5-year (2003–2007) monthly satellite dataset based mainly from SeaWiFS, and SCIAMACHY for PhytoDOAS (Bracher et al., 2009; Sadeghi et al., 2012a), and seven coupled model intercomparison project earth system models (CMIP5), as part of the International Satellite PFT Algorithm Inter-Comparison Project (Hirata et al., 2012). The phenological inter-comparisons were based on the fraction of microphytoplankton, or a similar variable for the satellite PG algorithms, and on diatom C for the CMIP5 models. Phenology metrics derived per-pixel globally for each PFT algorithm and CMIP5 model (using a Fourier transform-based model of the seasonal cycle) were seasonal amplitude, the month of maximum, bloom duration, and percent seasonal variance for the primary and a secondary bloom (if present). Results of this study indicate that although algorithms agree on a global scale and to first order (e.g., they have similar global patterns of bloom amplitude: low amplitude in the subtropical gyres and higher amplitude in coastal and high latitude regions), there were significant differences among them, with some algorithms exhibiting primary bloom peaks several months apart from each other at the two regions chosen for analysis: the North Atlantic drift region (NADR) and the North Atlantic subtropical gyre west (NASW), two of the provinces defined by Longhurst (2006). Overall, the majority of algorithms agree on a May peak of the primary bloom at NADR, and a March or April bloom peak at NASW. The ensemble-mean month of maximum bloom in the North Atlantic exhibits a clear progression northward from peaks in February at ~30°N to peaks in May or later at ~50°N. Differences were found between satellite data and the CMIP5 models. For example, seasonal cycles explain the most variance in satellite algorithms along the zonal bands around 30° latitude in both hemispheres, but not in CMIP5 model simulations. Bloom duration was higher in the mid-latitudes for the satellite algorithms when compared with CMIP5 models. Most satellite algorithms exhibited a zonally alternating pattern of single and double bloom peaks (i.e., a secondary bloom) in the mid- and high latitudes, whereas the CMIP5 model simulations did not.

Moving forward, it will be important to investigate to what extent observed inter-algorithm differences may be attributed to: (1) the different algorithms actually capturing genuine ecosystem deviations, for example, differences in Chl-a and bio-volume, or alterations in diatoms and microphytoplankton phenology, (2) limitations of multispectral satellite data, for example, noise in the signal, (3) actual algorithm limitations/shortcomings, and (4) limitations or artifacts of the inter-comparison methodology (e.g., chosen thresholds for bloom detection). These comparison exercises between satellite algorithms and Earth system models (ESMs), are critical for understanding the accuracy of current ESM simulations and for informing future development of the ocean biogeochemistry modules of ESMs. Ultimately, this will help improve future predictions of the response of the ocean and Earth system to climate change.

In a study by Gittings et al. (2021), the phenology of phytoplankton larger than 2 μm in diameter was studied in the northern and central Red Sea, using 20 years of ESA OC-CCI merged ocean-color data (Sathyendranath et al., 2019). Using a regionally tuned satellite algorithm (Gittings et al., 2019), they investigated interannual variability in phenology metrics for large phytoplankton (bloom initiation and duration), which were related to fish landings of two commercially important fisheries (*Sardinella spp.* and *Teuthida*). The maximum catches of the two species were found to coincide with the earliest bloom initiation and longest bloom duration of the time series (e.g., see Fig. 7.2). This work illustrates the potential for using satellite phytoplankton group phenology products as a tool for regional fisheries management.

Using satellite-derived diatom Chl-a from September to April each year from 1997 to 2012, the mean patterns and the interannual variability of the diatom bloom phenology in the Southern ocean was examined (Soppa et al., 2016). Diatoms are the major primary producer and a key component of the carbon and silicate biogeochemical cycles of this region. It was shown that the mean spatial patterns of timing and duration of diatom blooms are generally associated with the position of the Southern Antarctic Circumpolar Current front and of the maximum sea-ice extent. In several

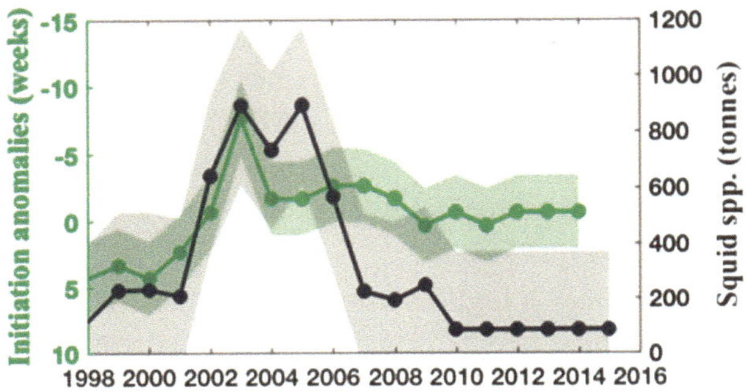

Figure 7.2 Interannual variability in the landings of *Squid spp.* (*Teuthida*, in tonnes) and the anomalies of phytoplankton cells larger than 2 μm in diameter, bloom initiation (weeks) from 1998 to 2014. One year was subtracted from the time series of *Squid* landings, and the plotted time series corresponds to one year after the phytoplankton bloom (i.e., the interannual variability in the timing of food availability is evident in the catch of the following year). Figure taken from Gittings et al. (2021).

areas, the anomalies of the phenological indices were found to be correlated with El Niño Southern Oscillation (ENSO) and Southern Annual Mode (SAM) indices. Composite maps of the anomalies revealed distinct spatial patterns and opposite events of ENSO and SAM had similar effects on the diatom phenology. For example, in the Ross Sea, a later start of the bloom and lower diatom biomass were associated with El Niño and negative SAM events, likely influenced by an increase in sea-ice extent during these events. This study illustrates a close coupling between climate variability and PG phenology, which could prove useful for improving future predictions of the response of phytoplankton to climate change.

7.2.2.2 Impact of phytoplankton size on export flux in the global ocean: a case study

The marine biological carbon pump is a key process in Earth's carbon cycle, in which fixed organic carbon is exported from the near surface ocean to depth (e.g., Siegel et al., 2014; DeVries and Weber, 2017). Here, we describe an example of a study, namely Mouw et al. (2016), that used satellite total Chl-a and size class data to assess the export of particulate organic carbon. As known before, the efficiency of the biological carbon pump is governed largely by the biologically mediated export of carbon from the surface ocean to the deep ocean, and its remineralization as a function of depth. However, the efficiency of the biological pump depends not only on phytoplankton abundance and production rate, but also on phytoplankton composition, as different cell sizes and morphologies impact food web functioning and sinking rates. In the past, global studies have incorporated satellite Chl-a, net primary production (NPP) and satellite physical variables (e.g., SST) to understand the role of phytoplankton in these processes (e.g., Laws et al., 2000; Henson et al., 2012; Vicente et al., 2021). The study by Mouw et al. (2016) exemplifies how incorporating satellite retrievals of phytoplankton composition (fraction of microphytoplankton) with NPP and Chl-a products, can improve estimates of particulate organic carbon (POC) vertical flux, remineralization, and transfer, using discrete field datasets of POC from sediment traps and ^{234}Th measurements. Their findings support the idea of high export flux, but low transfer efficiency, in productive regions and the inverse for oligotrophic regions. However, for some regions dominated by small cells, and at certain periods of the year, they unexpectedly observed greater vertical export flux efficiency and lower transfer efficiencies (see Fig. 7.3), emphasizing that phytoplankton size alone does not capture all the variability of the system, and that a taxonomic discrimination maybe necessary to include, due to differential impacts on silica (e.g., diatoms) and carbonate fractions (e.g., coccolithophores) of the flux. This was not included in this study, but other satellite products (see Section 7.3 for available products) could be used to investigate further.

In a separate study, Siegel et al. (2014) estimated that the direct phytoplankton contribution to total export flux was on average only 12.7%, with values greater than 20% found only in high-latitude and upwelling regions, which suggested a significant portion of the carbon exported is controlled by the flux of fecal matter from zooplankton grazing. Other biological and physical processes have been shown to play important roles in POC flux (Dall'Olmo et al., 2016; Boyd et al., 2019; Le Moigne, 2019; Briggs et al., 2020). By partitioning their analysis into provinces, Mouw et al. (2016) assumed similar food web structures and similar impacts of higher trophic levels on phytoplankton composition, particle aggregation, zooplankton grazing, and fecal production. In the natural world, where all processes are aggregated in the resulting POC flux, the absolute impact of phytoplankton alone on export flux is challenging to quantify. However, it is clear from the Mouw et al. (2016) analysis that the absolute magnitude of carbon flux, and the efficiency with which it is exported, remineralized, and transferred, does have a clear dependency on the size of the phytoplankton cell, but also on its coating and ballasting upon which the ecosystem is acting.

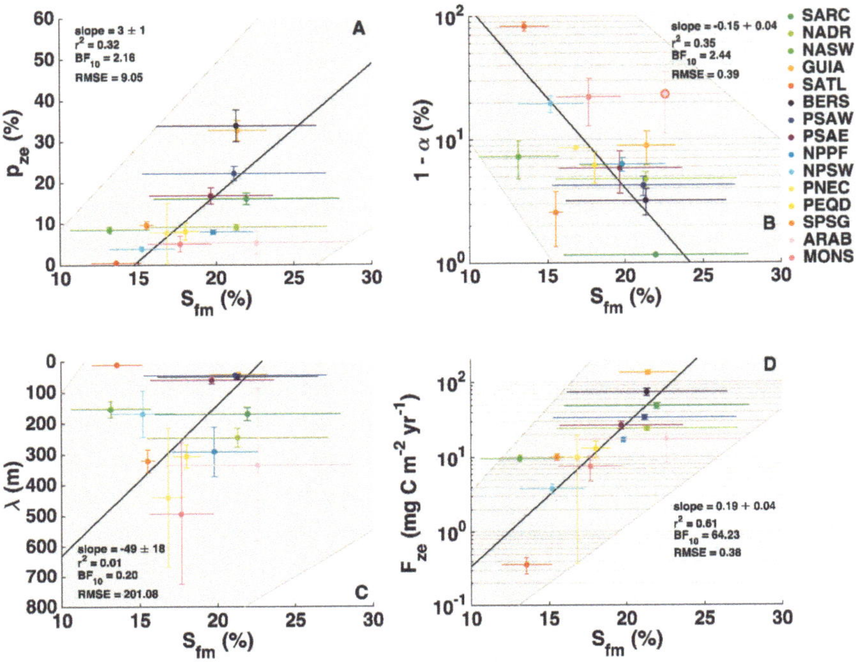

Figure 7.3 Relationships between mean percent microplankton (S_{fm}) and (a) export flux efficiency (P_{ze}), (b) refractory fraction ($1 - \alpha$), (c) remineralization length scale (λ), and (d) POC flux at the depth of export (F_{ze}) for fitted provinces. Regions are colored to indicate Longhurst (2006) provinces. Regressions are Type II, with data circled in red excluded as outliers. Shaded regions represent the 95% confidence interval of the regression fit. Increases in microplankton fraction were associated with greater export efficiency, lower refractory fractions, and greater absolute exported POC. Figure taken from Mouw et al. (2016).

7.2.2.3 Other applications

Satellite time series of PG or PSC data have been used in many more applications to assess regional and global variability, trends in phytoplankton community composition (e.g., Kostadinov et al., 2010; Brewin et al., 2012; Sadeghi et al., 2012b; Sharma et al., 2019), and PG phenology (Alvain et al., 2013; Soppa et al., 2016). The benefits of applying satellite phytoplankton diversity products for assessing the radiation feedback from aerosols and clouds for the global climate system have been illustrated by Arnold et al. (2009) and Booge et al. (2016). Globally, natural aerosols, principally from marine biogenic primary and secondary aerosol sources, explain more than half of the spatiotemporal variability in aerosol distributions. Budgeting of marine aerosols in the Arctic Ocean is usually not quantified, due to insufficient reliable data on producers and knowledge on formation processes. In Booge et al. (2016), in-water measurements of the precursors of cloud condensation nuclei (isoprene and related substances such as dimethyl-sulfide-phosphate) from three different oceanic regions, together with satellite Chl-a concentration and PG specific Chl-a data (based on Hirata et al., 2011), were investigated to improve estimates of global marine isoprene emissions. It was shown that the use of satellite-derived Chl-a concentration alone, to estimate isoprene, underestimates the measured oceanic isoprene concentration by a mean factor of 20 (19 ± 12). The model was improved by using (satellite) PG specific Chl-a, and PG-dependent production values based on the relationship between major PGs and isoprene production, as determined from *in situ* data, resulting in only a slight underestimation (factor of 1.7 ± 1.2).

Other applications of satellite PG data include the detection of regional harmful algal bloom (HAB) events (e.g., Kurekin et al., 2014; Blondeau-Patissier et al., 2014; Hill et al., 2020; IOCCG, 2021) and the estimation of recruitment of juvenile fish (Trzcinski et al., 2013). Roy (2018) used satellite PSC data to improve estimations of the distributions of phytoplankton carbohydrates, proteins, and lipids in the world oceans. It is envisaged that the applications of satellite PG and PSC products will continue to increase in the future. The development of hyperspectral missions, which are envisioned to expand the number of PGs detected and decrease the uncertainties of PGs and PSCs detection (see Section 7.1.1), will inevitably play a significant role in expanding the applications of PGs and PSC products.

7.3 Societal benefits of satellite PG and PSC products: integrating into services

Making satellite products of phytoplankton composition available through marine ecosystem services (ideally multi-platform merged data sets) ensures ease of access by research and operational communities, maximizing impact, and ultimately leading to greater societal benefits. Coasts are especially vulnerable to threats of harmful algal blooms, eutrophication, hypoxia, and other processes that deteriorate water quality. High resolution data on phytoplankton diversity are urgently needed for important socio-economic management efforts (e.g., fisheries, aquaculture, and coastal management, see IOCCG, 2009).

Satellite PG and PSC products have been implemented into several international and national marine service platforms. A major objective of the EU Copernicus Marine Environment Monitoring Service (CMEMS, https://resources.marine.copernicus.eu/) is to deliver and maintain a competitive and state-of-the-art European service that responds to public and private intermediate user needs, and thus transparently involves users in the service delivery definition. In CMEMS, a diversity of satellite products have been available operationally since 2015. The catalogue offers time-series of satellite ocean-color products (e.g., remote sensing reflectance, Chl-a, phytoplankton absorption) for the global ocean as well as tailored for European seas (e.g., Mediterranean Sea, Eastern Atlantic Ocean, Arctic Ocean, Baltic Sea and Black Sea), including outputs of biogeochemical models (e.g., Chl-a, dissolved nutrients and other environmental variables) useful for forecasting. In 2020, CMEMS has been enriched by satellite datasets encompassing regional and globally tuned products on Chl-a of major PGs and PSCs, as derived from abundance-based approaches (Brewin et al., 2017b; Di Cicco et al., 2017) and spectral-based approaches (Xi et al., 2020, updated in 2021 to Xi et al., 2021). These datasets, by discriminating among PGs, may allow improved monitoring of carbon export at sea, as well as a better understanding of primary production and subsequent interactions across marine trophic levels.

The UK Natural Environmental Research Council (NERC) supports an operational remote sensing service called the NERC Earth Observation Data Acquisition and Analysis Service, NEODAAS (https://www.neodaas.ac.uk/Home). It is hosted at Plymouth Marine Laboratory and overseen by the UK National Centre for Earth Observation. The service provides operational satellite data processing for the scientific community and provides near-real-time support (and rapid response), useful for guiding field-based studies in the ocean. Satellite PSC products (Brewin et al., 2010, 2015, 2017b) have been integrated into the NEODAAS product stream and are available on request. These have been used for international student training activities and supported research (e.g., Brewin et al., 2019).

The U.S. National Oceanic and Atmospheric Administration (NOAA) has a mandate to ensure the productivity and sustainability of fisheries and fishing communities in the United States through science-based decision-making. To do this, they have adopted a holistic Ecosystem-Based Fisheries Management (EBFM) policy to maintain ecosystems in a healthy, productive, and sustainable condition so they can provide the services humans want and need, by taking into account the entire ecosystem of the species being managed. Given the critical role of phytoplankton in the food web and the diverse function different PGs play in the ecosystem, improved understanding of phytoplankton composition from satellite estimates will help advance fisheries modeling and better inform management decisions. More specifically, understanding how phytoplankton composition directly influences the transfer efficiency of photosynthetically derived energy to higher trophic levels, will help set constraints on the total amount of energy available to the ecosystem (Fogarty et al., 2016). Furthermore, in 2020, the NOAA Climate and Fisheries Initiative (CFI) was established to meet requirements for information on past, current, and future climate and ocean conditions and ensure informed, climate-ready fisheries management and protected species conservation in the face of rapid environmental change. Considering the PG data available from satellite products, and by feeding this information into numerical models to forecast fisheries production, critical information could be made available to understand how climate and oceanic conditions are affecting phytoplankton composition and how these changes could impact marine species and ecosystems. It is also envisaged that other US ocean-color service providers, such as the NASA Ocean Biological Processing Group (https://oceancolor.gsfc.nasa.gov/), will expand their product stream to include PSC and PG products in the future.

Table 7.1 A selection of phytoplankton size class (PSC), phytoplankton group (PG), and particle size distribution (PSD) products available to the community (updated from Mouw et al., 2017). Note links may become outdated.

Type	Reference	Link
PG	Xi et al. (2020, 2021)	https://marine.copernicus.eu/
PG	Bracher et al. (2009, 2017b) & Sadeghi et al. (2012a)	https://doi.org/10.1594/PANGAEA.870486
PG	Losa et al. (2017)	https://doi.pangaea.de/10.1594/PANGAEA.875873
PG	Ben Mustapha et al. (2014)	https://hermes.acri.fr/index.php?class=demonstration_products
PG/PSC	Di Cicco et al. (2017)	https://marine.copernicus.eu/
PSC	Brewin et al. (2010, 2015, 2017b)	https://catalogue.ceda.ac.uk/uuid/7201151d40683a4420da90c30640d4fa https://marine.copernicus.eu/ https://www.neodaas.ac.uk/Home
PSC	Mouw and Yoder (2010)	https://doi.pangaea.de/10.1594/PANGAEA.860474
PSD	Kostadinov et al. (2009, 2016)	https://hermes.acri.fr/index.php?class=demonstration_products https://doi.pangaea.de/10.1594/PANGAEA.859005

Table 7.1 provides information to users on where some of these satellite PG and PSC products can be found, as of early 2021.

References

Alonso, K., Bachmann, M., Burch, K., Carmona, E., Cerra, D., de los Reyes, R., Dietrich, D., Heiden, U., Holderlin, A., Ickes, J., 2019. Data products, quality and validation of the DLR Earth Sensing Imaging Spectrometer (DESIS). Sensors 19, 4471.

Alvain, S., Moulin, C., Dandonneau, Y., Breon, F., 2005. Remote sensing of phytoplankton groups in case 1 waters for global SeaWiFS imagery. Deep-Sea Research. Part 1 52, 1989–2004.

Alvain, S., Le Quéré, C., Bopp, L., Racault, M.F., et al., 2013. Rapid climatic driven shifts of diatoms at high latitudes. Remote Sensing of Environment 132, 195–201.

Arnold, S.R., Spracklen, D.V., Williams, J., Yassaa, N., et al., 2009. Evaluation of the global oceanic isoprene source and its impacts on marine organic carbon aerosol. Atmospheric Chemistry and Physics 9, 1253–1262.

Beaulieu, C., Henson, S.A., Sarmiento, J.L., Dunne, J.P., Doney, S.C., Rykaczewski, R.R., Bopp, L., 2013. Factors challenging our ability to detect long-term trends in ocean chlorophyll. Biogeosciences 10 (4), 2711–2724.

Behrenfeld, M.J., 2010. Abandoning Sverdrup's Critical Depth Hypothesis on phytoplankton blooms. Ecology 91 (4), 977–989. https://doi.org/10.1890/09-1207.1.

Ben Mustapha, Z., Alvain, S., Jamet, C., Loisel, H., et al., 2014. Automatic classification of water-leaving radiance anomalies from global SeaWiFS imagery: application to the detection of phytoplankton groups in open ocean waters. Remote Sensing of Environment 146, 97–112. https://doi.org/10.1016/j.rse.2013.08.046.

Blondeau-Patissier, D., Gower, J.F.R., Dekker, A.G., Phionn, S.R., Brando, V.E., 2014. A review of ocean color remote sensing methods and statistical techniques for the detection, mapping and analysis of phytoplankton blooms in coastal and open oceans. Progress in Oceanography 123, 123–144.

Booge, D., Marandino, C.A., Schlundt, C., Palmer, P.I., Schlundt, M., Atlas, E.L., Bracher, A., Saltzman, E.S., Wallace, D.W.R., 2016. Can simple models predict large scale surface ocean isoprene concentrations? Atmospheric Chemistry and Physics 16, 11807–11821. https://doi.org/10.5194/acp-16-11807-2016.

Boyd, P.W., Claustre, H., Levy, M., et al., 2019. Multi-faceted particle pumps drive carbon sequestration in the ocean. Nature 568, 327–335. https://doi.org/10.1038/s41586-019-1098-2.

Bracher, A., Vountas, M., Dinter, T., Burrows, J.P., Röttgers, R., Peeken, I., 2009. Quantitative observation of cyanobacteria and diatoms from space using PhytoDOAS on SCIAMACHY data. Biogeosciences 6, 751–764.

Bracher, A., Taylor, B.B., Taylor, M., Dinter, T., Röttgers, R., Steinmetz, F., 2015. Using empirical orthogonal functions derived from remote sensing reflectance for the prediction of concentrations of phytoplankton pigments. Ocean Science 11, 139–158.

Bracher, A., Bouman, H.A., Bricaud, A., Brewin, R.W.J., Brotas, V., Ciotti, A.M., Clementson, L., Devred, E., Di Cicco, A.M., Dutkiewicz, S., Hardman-Mountford, N.J., Hickman, A.E., Hieronymi, M., Hirata, T., Losa, S.N., Mouw, C.B., Organelli, E., Raitsos, D.E., Uitz, J., Vogt, M., Wolanin, A., 2017a. Obtaining Phytoplankton Diversity from Ocean Color: A Scientific Roadmap for Future Development. Frontiers in Marine Science 4, 55. https://doi.org/10.3389/fmars.2017.00055.

Bracher, A., Dinter, T., Wolanin, A., Rozanov, V.V., Losa, S., Soppa, M.A., 2017b. Global monthly mean chlorophyll a surface concentrations from August 2002 to April 2012 for diatoms, coccolithophores and cyanobacteria from PhytoDOAS algorithm version

3.3 applied to SCIAMACHY Data, link to NetCDF files in ZIP archive. PANGAEA. https://doi.org/10.1594/PANGAEA.870486.
Braga, F., Giardino, C., Bassani, C., Matta, E., Candiani, G., Strömbeck, N., Adamo, M., Bresciani, M., 2013. Assessing water quality in the northern Adriatic Sea from HICO™ data. Remote Sensing Letters 4, 1028–1037.
Brewin, R.J.W., Sathyendranath, S., Hirata, T., Lavender, S.J., Barciela, R.M., Hardman-Mountford, N.J., 2010. A three-component model of phytoplankton size class for the Atlantic Ocean. Ecological Modelling 221 (11), 1472–1483.
Brewin, R.J.W., Hardman-Mountford, N.J., Lavender, S.J., Raitsos, D.E., Hirata, T., Uitz, J., Devred, E., Bricaud, A., Ciotti, A., Gentili, B., 2011. An intercomparison of bio-optical techniques for detecting dominant phytoplankton size class from satellite remote sensing. Remote Sensing of Environment 115 (2), 325–339.
Brewin, R.J.W., Hirata, T., Hardman-Mountford, N.J., Lavender, S.J., et al., 2012. The influence of the Indian Ocean Dipole on interannual variations in phytoplankton size structure as revealed by Earth Observation. Deep-Sea Research. Part 2 77–80, 117–127. https://doi.org/10.1016/j.dsr2.2012.04.009.
Brewin, R.J.W., Sathyendranath, S., Lange, P.K., Tilstone, G., 2014. Comparison of two methods to derive the size-structure of natural populations of phytoplankton. Deep-Sea Research. Part 1. Oceanographic Research Papers 85, 72–79.
Brewin, R.J.W., Sathyendranath, S., Jackson, T., Barlow, R., et al., 2015. Influence of light in the mixed-layer on the parameters of a three-component model of phytoplankton size class. Remote Sensing of Environment 168, 437–450.
Brewin, R.J.W., Ciavatta, S., Sathyendranath, S., Jackson, T., Tilstone, G., Curran, K., Airs, R.L., Cummings, D., Brotas, V., Organelli, E., Dall'Olmo, G., Raitsos, D.E., 2017b. Uncertainty in ocean-color estimates of chlorophyll for phytoplankton groups. Frontiers in Marine Science 4, 104. https://doi.org/10.3389/fmars.2017.00104.
Brewin, R.J.W., Tilstone, G.H., Jackson, T., Cain, T., Miller, P.I., Lange, P.K., Misra, A., Airs, R.L., 2017a. Modelling size-fractionated primary production in the Atlantic Ocean from remote sensing. Progress in Oceanography 158, 130–149.
Brewin, R.J.W., Ciavatta, S., Sathyendranath, S., Skákala, J., Bruggeman, J., Ford, D., Platt, T., 2019. The influence of temperature and community structure on light absorption by phytoplankton in the North Atlantic. Sensors 19, 4182. https://doi.org/10.3390/s19194182.
Briggs, N., Dall'Olmo, G., Claustre, H., 2020. Major role of particle fragmentation in regulating biological sequestration of CO2 by the oceans. Science 367 (6479), 791–793.
Brotas, V., Brewin, R.J.W., Sá, C., Brito, A.C., Silva, A., Mendes, C.R., Diniz, T., Kaufmann, M., Tarran, G., Groom, S.B., Platt, T., 2013. Deriving phytoplankton size classes from satellite data: validation along a trophic gradient in the eastern Atlantic Ocean. Remote Sensing of Environment 134, 66–77.
Budge, S.M., Devred, E., Forget, M.H., Stuart, V., Trzcinski, M.K., Sathyendranath, S., Platt, T., 2014. Estimating concentrations of essential omega-3 fatty acids in the ocean: supply and demand. ICES Journal of Marine Science 71 (7), 1885–1893.
CEOS, 2018. Feasibility study for an aquatic ecosystem Earth observing system. Report v. 1.1. In: Dekker, A.G., Pinnel, N. (Eds.), Committee on Earth Observation Satellites (CEOS) and Commonwealth Scientific and Industrial Research Organization (CSIRO). CSIRO, Canberra, ACT, Australia.
Chase, A.P., Kramer, S.J., Haëntjens, N., Boss, E.S., Karp-Boss, L., Edmondson, M., Graff, J.R., 2020. Evaluation of diagnostic pigments to estimate phytoplankton size classes. Limnology and Oceanography: Methods 18 (10), 570–584.
Chivers, W.J., Edwards, M., Hays, G.C., 2020. Phenological shuffling of major marine phytoplankton groups over the last six decades. Diversity and Distributions 26 (5), 536–548.
Ciavatta, S., Brewin, R.J.W., Skákala, J., Polimene, L., de Mora, L., Artioli, Y., Allen, J.I., 2018. Assimilation of ocean-color plankton functional types to improve marine

ecosystem simulations. Journal of Geophysical Research: Oceans 123, 834–854. https://doi.org/10.1002/2017JC013490.

Ciotti, A.M., Bricaud, A., 2006. Retrievals of a size parameter for phytoplankton and spectral light absorption by colored detrital matter from water-leaving radiances at SeaWiFS channels in a continental shelf region off Brazil. Limnology and Oceanography: Methods 4, 237–253.

Ciotti, A.M., Lewis, M.R., Cullen, J.J., 2002. Assessment of the relationships between dominant cell size in natural phytoplankton communities and the spectral shape of the absorption coefficient. Limnology and Oceanography 47 (2), 404–417.

Corredor-Acosta, A., Morales, C.E., Brewin, R.J., Auger, P.A., Pizarro, O., Hormazabal, S., Anabalón, V., 2018. Phytoplankton size structure in association with mesoscale eddies off central-southern Chile: the satellite application of a phytoplankton size-class model. Remote Sensing 10 (6), 834.

Cushing, D.H., 1990. Plankton production and year-class strength in fish populations: an update of the match/mismatch hypothesis. Advances in Marine Biology 26, 249–293. https://doi.org/10.1016/S0065-2881(08)60202-3.

Dall'Olmo, G., Dingle, J., Polimene, L., et al., 2016. Substantial energy input to the mesopelagic ecosystem from the seasonal mixed-layer pump. Nature Geoscience 9, 820–823. https://doi.org/10.1038/ngeo28.

Devred, E., Sathyendranath, S., Stuart, V., Platt, T., 2011. A three component classification of phytoplankton absorption spectra: application to ocean-color data. Remote Sensing of Environment 115 (9), 2255–2266.

DeVries, T., Weber, T., 2017. The export and fate of organic matter in the ocean: new constraints from combining satellite and oceanographic tracer observations. Global Biogeochemical Cycles 31 (3), 535–555.

Di Cicco, A., Sammartino, M., Marullo, S., Santoleri, R., 2017. Regional empirical algorithms for an improved identification of phytoplankton functional types and size classes in the Mediterranean Sea using satellite data. Frontiers in Marine Science 4, 126. https://doi.org/10.3389/fmars.2017.00126.

Dutkiewicz, S., Hickman, A.E., Jahn, O., Gregg, W.W., Mouw, C.B., Follows, M.J., 2015. Capturing optically important constituents and properties in a marine biogeochemical and ecosystem model. Biogeosciences 12, 4447–4481. https://doi.org/10.5194/bg-12-4447-2015.

Evers-King, H., Bernard, S., Lain, L.R., Probyn, T.A., 2014. Sensitivity in reflectance attributed to phytoplankton cell size: forward and inverse modelling approaches. Optics Express 22 (10), 11536–11551.

Fogarty, M.J., Rosenberg, A.A., Cooper, A.B., Dickey-Collas, M., Fulton, E.A., Gutiérrez, N.L., Hyde, K.J.W., Kleisner, K.M., Kristiansen, T., Longo, C., Minte-Vera, C.V., Minto, C., Mosqueira, I., Osio, G.C., Ovando, D., Selig, E.R., Thorson, J.T., Ye, Y., 2016. Fishery production potential of large marine ecosystems: a prototype analysis. Environmental Development 17 (Supplement 1), 211–219.

Giardino, C., Bresciani, M., Braga, F., Fabbretto, A., Ghirardi, N., Pepe, M., Gianinetto, M., Colombo, R., Cogliati, S., Ghebrehiwot, S., Laanen, M., Peters, S., Schroeder, T., Concha, J.A., Brando, V.E., 2020. First evaluation of PRISMA Level 1 data for water applications. Sensors 20, 4553.

Gittings, J.A., Brewin, R.J.W., Raitsos, D.E., Kheireddine, M., Ouhssain, M., Jones, B.H., Hoteit, I., 2019. Remotely sensing phytoplankton size structure in the Red Sea. Remote Sensing of Environment 234, 111387.

Gittings, J.A., Raitsos, D., Brewin, R.J.W., Hoteit, I., 2021. Links between phenology of large phytoplankton and fisheries in the northern and central Red Sea. Remote Sensing 13 (2), 231.

Henson, S.A., Sanders, R., Madsen, E., 2012. Global patterns in efficiency of particulate organic carbon export and transfer to the deep ocean. Global Biogeochemical Cycles 26, GB1028. https://doi.org/10.1029/2011GB004099.
Hill, P.R., Kumar, A., Temimi, M., Bull, D.R., 2020. HABNet: machine learning, remote sensing-based detection of Harmful Algal Blooms. IEEE Journal of Selected Topics in Applied Earth Observations and Remote Sensing 13, 3229–3239. https://doi.org/10.1109/JSTARS.2020.3001445.
Hirata, T., Hardman-Mountford, N.J., Brewin, R.J.W., Aiken, J., et al., 2011. Synoptic relationships between surface Chlorophyll-a and diagnostic pigments specific to phytoplankton functional types. Biogeosciences 8, 311–327. https://doi.org/10.5194/bg-8-311-2011.
Hirata, T., Hardman-Mountford, N., Brewin, R.J.W., 2012. Comparing satellite-based phytoplankton classification methods. Eos Transactions American Geophysical Union 93 (6), 59.
IOCCG, 2009. Remote sensing in fisheries and aquaculture. In: Forget, M.-H., Stuart, V., Platt, T. (Eds.), Reports of the International Ocean-Colour Coordinating Group, No. 8. IOCCG, Dartmouth, Canada.
IOCCG, 2014. Phytoplankton functional types from space. In: Sathyendranath, S., Stuart, V. (Eds.), Reports of the International Ocean Color Coordinating Group, No. 15. IOCCG, Dartmouth, Canada.
IOCCG, 2020. Synergy between ocean colour and biogeochemical/ecosystem models. In: Dutkiewicz, S. (Ed.), IOCCG Report Series, No. 19. International Ocean Colour Coordinating Group, Dartmouth, Canada.
IOCCG, 2021. Observation of Harmful Algal Blooms with ocean colour radiometry. In: Bernard, S., Kudela, R., Robertson Lain, L., Pitcher, G.C. (Eds.), IOCCG Report Series, No. 20. International Ocean Colour Coordinating Group, Dartmouth, Canada.
Koeller, P., et al., 2009. Basin-scale coherence in phenology of shrimps and phytoplankton in the North Atlantic Ocean. Science 324 (5928), 791–793. https://doi.org/10.1126/science.1170987.
Kostadinov, T., Milutinovic, S., Marinov, I., Cabré, A., 2016. Carbon-based phytoplankton size classes retrieved via ocean color estimates of the particle size distribution. Ocean Science 12 (2), 561–575. https://doi.org/10.5194/os-12-561-2016.
Kostadinov, T.S., Siegel, D.A., Maritorena, S., 2009. Retrieval of the particle size distribution from satellite ocean color observations. Journal of Geophysical Research 114, C09015. https://doi.org/10.1029/2009JC005303.
Kostadinov, T.S., Siegel, D.A., Maritorena, S., 2010. Global variability of phytoplankton functional types from space: assessment via the particle size distribution. Biogeosciences 7, 3239–3257. https://doi.org/10.5194/bg-7-3239-2010.
Kostadinov, T.S., Cabré, A., Vedantham, H., Marinov, I., Bracher, A., Brewin, R.J.W., Bricaud, A., Hirata, T., Hirawake, T., Hardman-Mountford, N.J., Mouw, C., 2017. Inter-comparison of phytoplankton functional type phenology metrics derived from ocean color algorithms and Earth System Models. Remote Sensing of Environment 190, 162–177.
Kramer, S.J., Siegel, D.A., 2019. How can phytoplankton pigments be best used to characterize surface ocean phytoplankton groups for ocean color remote sensing algorithms? Journal of Geophysical Research: Oceans 124 (11), 7557–7574.
Kramer, S.J., Roesler, C.S., Sosik, H.M., 2018. Bio-optical discrimination of diatoms from other phytoplankton in the surface ocean: evaluation and refinement of a model for the Northwest Atlantic. Remote Sensing of Environment 217, 126–143.
Kurekin, A.A., Miller, P.I., Van der Woerd, H.J., 2014. Satellite discrimination of Karenia mikimotoi and Phaeocystis harmful algal blooms in European coastal waters: merged classification of ocean color data. Harmful Algae 31, 163–176. https://doi.org/10.1016/j.hal.2013.11.003.

Lamont, T., Brewin, R.J.W., Barlow, R.G., 2018. Seasonal variation in remotely-sensed phytoplankton size structure around southern Africa. Remote Sensing of Environment 204, 617–631.
Laws, E.A., Falkowski, P.G., Smith Jr, W.O., Ducklow, H., McCarthy, J.J., 2000. Temperature effects on export production in the open ocean. Global Biogeochemical Cycles 14 (4), 1231–1246.
Le Moigne, F., 2019. Pathways of organic carbon downward transport by the oceanic biological carbon pump. Frontiers in Marine Science 6, 634. https://doi.org/10.3389/fmars.2019.00634.
Li, Z., Li, L., Song, K., Cassar, N., 2013. Estimation of phytoplankton size fractions based on spectral features of remote sensing ocean color data. Journal of Geophysical Research: Oceans 118 (3), 1445–1458. https://doi.org/10.1002/jgrc.20137.
Longhurst, A.R., 2006. Ecological Geography of the Sea, 2nd ed. Academic Press, San Diego. 560 pp.
Losa, S., Soppa, M.A., Dinter, T., Wolanin, A., Brewin, R.J.W., Bricaud, A., Oelker, J., Peeken, I., Gentili, B., Rozanov, V.V., Bracher, A., 2017. Synergistic exploitation of hyper- and multispectral precursor Sentinel measurements to determine Phytoplankton Functional Types at best spatial and temporal resolution (SynSenPFT). Frontiers in Marine Science 4, 203. https://doi.org/10.3389/fmars.2017.00203.
Margalef, R., 1967. Some concepts relative to the organization of plankton. Oceanography and Marine Biology Annual Review 5, 257–289.
Maritorena, S., Fanton d'Andon, O.H., Mangin, A., Siegel, D.A., 2010. Merged satellite ocean color data products using a bio-optical model: characteristics, benefits and issues. Remote Sensing of Environment 114, 1791–1804.
Michaels, A.F., Silver, M.W., 1988. Primary production, sinking fluxes and the microbial food web. Deep-Sea Research. Part A, Oceanographic Research Papers 35 (4), 473–490.
Moore, T.S., Brown, C.W., 2020. Incorporating environmental data in abundance-based algorithms for deriving phytoplankton size classes in the Atlantic Ocean. Remote Sensing of Environment 240, 111689.
Morel, A., Bricaud, A., 1981. Theoretical results concerning light absorption in a discrete medium, and application to specific absorption of phytoplankton. Deep-Sea Research 28, 1375–1393.
Mouw, C.B., Yoder, J.A., 2010. Optical determination of phytoplankton size composition from global SeaWiFS imagery. Journal of Geophysical Research 115, C12018. https://doi.org/10.1029/2010JC006337.
Mouw, C.B., Barnett, A., McKinley, G.A., Gloege, L., Pilcher, D., 2016. Phytoplankton size impact on export flux in the global ocean. Global Biogeochemical Cycles 30, 1542–1562. https://doi.org/10.1002/2015GB005355.
Mouw, C.B., Hardman-Montford, N., Alvain, S., Bracher, A., et al., 2017. A consumer's guide to satellite remote sensing of multiple phytoplankton groups in the global ocean. Frontiers in Marine Science 4, 41. https://doi.org/10.3389/fmars.2017.00041.
Muller-Karger, F.E., Hestir, E., Ade, C., Turpie, K., Roberts, D., Siegel, D., Miller, R., Humm, D., Izenberg, N., Keller, M., Morgan, F., Frouin, R., Dekker, A., Gardner, R., Goodman, J., Schaeffer, B., Franz, B., Pahlevan, N., Mannino, A.G., Concha, J.A., Ackleson, S.G., Najjar, R., Cavanaugh, K., Romanou, A., Tzortziou, M., Boss, E., Pavlick, R., Schimel, D., Freeman, A., Rousseaux, C.S., Dunne, J., Long, M., Klein, E., McKinley, G., Letelier, R., Roffer, M., Goes, J., Bracher, A., Arrigo, K.R., Dierssen, H., Zhang, X., Davis, F., Best, B., Guralnick, R., Moisan, J., Sosik, H.M., Kavanaugh, M., Kudela, R., Mouw, C.B., Barnard, A., Palacios, S., Roessler, C., Drakou, E., Appeltans, W., 2018. Satellite sensor requirements for monitoring essential biodiversity variables of coastal ecosystems. Ecological Applications 28 (3), 749–760. https://doi.org/10.1002/eap.1682.

Nair, A., Sathyendranath, S., Platt, T., Morales, J., Stuart, V., Forget, M.H., Devred, E., Bouman, H., 2008. Remote sensing of phytoplankton functional types. Remote Sensing of Environment 112 (8), 3366–3375.

Oelker, J., Richter, A., Dinter, T., Rozanov, V.V., Burrows, J.P., Bracher, A., 2019. Global diffuse attenuation coefficient derived from vibrational Raman scattering detected in hyperspectral backscattered satellite spectra. Optics Express 27 (12), A829–A855. https://doi.org/10.1364/OE.27.00A829.

Organelli, E., Claustre, H., 2019. Small phytoplankton shapes colored dissolved organic matter dynamics in the North Atlantic subtropical gyre. Geophysical Research Letters 46, 12183–12191. https://doi.org/10.1029/2019GL084699.

Organelli, E., Bricaud, A., Antoine, D., Uitz, J., 2013. Multivariate approach for the retrieval of phytoplankton size structure from measured light absorption spectra in the Mediterranean Sea (BOUSSOLE site). Applied Optics 52 (11), 2257–2273. https://doi.org/10.1364/AO.52.002257.

Organelli, E., Nuccio, C., Lazzara, L., Uitz, J., Bricaud, A., Massi, L., 2017. On the discrimination of multiple phytoplankton groups from light absorption spectra of assemblages with mixed taxonomic composition and variable light conditions. Applied Optics 56 (14), 3952–3968. https://doi.org/10.1364/AO.56.003952.

Pahlevan, N., Smith, B., Binding, C., Li, L., Bresciani, M., Giardino, C., 2021. Hyperspectral retrievals of phytoplankton absorption and chlorophyll-a in inland and nearshore coastal waters. Remote Sensing of Environment 2021 (253), 112200.

Palacz, A.P., St. John, M.A., Brewin, R.J.W., Hirata, T., et al., 2013. Distribution of phytoplankton functional types in high-nitrate low-chlorophyll waters in a new diagnostic ecological indicator model. Biogeosciences 10, 7553–7574. https://doi.org/10.5194/bg-10-7553-2013.

Platt, T., Fuentes-Yaco, C., Frank, K.T., 2003. Spring algal bloom and larval fish survival. Nature 423 (6938), 398. https://doi.org/10.1038/423398b.

Pradhan, H.K., Völker, C., Losa, S.N., Bracher, A., Nerger, L., 2020. Global assimilation of ocean-color data of phytoplankton functional types: impact of different datasets. Journal of Geophysical Research: Oceans 125, e2019JC015586. https://doi.org/10.1029/2019JC015586.

Raitsos, D.E., Lavender, S.J., Maravelias, C.D., Haralambous, J., et al., 2008. Identifying phytoplankton functional groups from space: an ecological approach. Limnology and Oceanography 53 (2), 605–613.

Roy, S., 2018. Distributions of phytoplankton carbohydrate, protein and lipid in the world oceans from satellite ocean colour. The ISME Journal (ISSN 1751-7370) 12 (6), 1457–1472. https://doi.org/10.1038/s41396-018-0054-8.

Roy, S., Sathyendranath, S., Bouman, H., Platt, T., 2013. The global distribution of phytoplankton size spectrum and size classes from their light-absorption spectra derived from satellite data. Remote Sensing of Environment 139, 185–197.

Rozanov, V.V., Rozanov, A., Kokhanovsky, A., Burrows, J.P., 2014. Radiative transfer through atmosphere and ocean: software package SCIATRAN. Journal of Quantitative Spectroscopy & Radiative Transfer 133, 13–71. https://doi.org/10.1016/j.jqsrt.2013.7.004.

Ryan, J.P., Davis, C.O., Tufillaro, N.B., Kudela, R.M., et al., 2014. Application of the hyperspectral imager for the coastal ocean to phytoplankton ecology studies in Monterey Bay, CA, USA. Remote Sensing 6, 1007–1025.

Sadeghi, A., Dinter, T., Vountas, M., Taylor, B., et al., 2012b. Remote sensing of coccolithophore blooms in selected oceanic regions using the PhytoDOAS method applied to hyper-spectral satellite data. Biogeosciences 9, 2127–2143.

Sadeghi, A., Dinter, T., Vountas, M., Taylor, B.B., et al., 2012a. Improvement to the PhytoDOAS method for identification of coccolithophores using hyper-spectral satellite data. Ocean Science 8, 1055–1070.

Sammartino, M., Di Cicco, A., Marullo, S., Santoleri, R., 2015. Spatio-temporal variability of micro-, nano-and pico-phytoplankton in the Mediterranean Sea from satellite ocean colour data of SeaWiFS. Ocean Science 11 (5), 759–778.

Sathyendranath, S., Watts, L., Devred, E., Platt, T., Caverhill, C., Maass, H., 2004. Discrimination of diatoms from other phytoplankton using ocean-colour data. Marine Ecology. Progress Series 272, 59–68.

Sathyendranath, S., Stuart, V., Nair, A., Oka, K., Nakane, T., Bouman, H., Forget, M.H., Maass, H., Platt, T., 2009. Carbon-to-chlorophyll ratio and growth rate of phytoplankton in the sea. Marine Ecology. Progress Series 383, 73–84.

Sathyendranath, S., Brewin, R.J.W., Jackson, T., Mélin, F., Platt, T., 2017. Ocean-colour products for climate-change studies: what are their ideal characteristics? Remote Sensing of Environment 203, 125–138.

Sathyendranath, S., Brewin, R.J.W., Brockmann, C., Brotas, V., et al., 2019. An ocean-colour time series for use in climate studies: the experience of the ocean-colour climate change initiative (OC-CCI). Sensors 19 (19), 4285. https://doi.org/10.3390/s19194285.

Sathyendranath, S., Platt, T., Kovač, Ž., Dingle, J., Jackson, T., Brewin, R.J.W., Franks, P., Marañón, E., Kulk, G., Bouman, H., 2020. Reconciling models of primary production and photoacclimation. Applied Optics 59, C100–C114.

Sauzede, R., Claustre, H., Jamet, C., Uitz, J., Ras, J., Mignot, A., et al., 2015. Retrieving the vertical distribution of chlorophyll a concentration and phytoplankton community composition from in situ fluorescence profiles: a method based on a neural network with potential for global-scale applications. Journal of Geophysical Research 120 (1), 451–470. https://doi.org/10.1002/2014JC010355.

Shang, S., Wu, J., Huang, B., Lin, G., Lee, Z., Liu, J., Shang, S., 2014. A new approach to discriminate dinoflagellate from diatom blooms from space in the East China Sea. Journal of Geophysical Research: Oceans 119 (7), 4653–4668.

Sharma, P., Marinov, I., Cabre, A., Kostadinov, T., Singh, A., 2019. Increasing biomass in the warm oceans: unexpected new insights from SeaWiFS. Geophysical Research Letters 46, 3900–3910. https://doi.org/10.1029/2018GL079684.

Sieburth, J.M., Smetacek, V., Lenz, J., 1978. Pelagic ecosystem structure: heterotrophic compartments of the plankton and their relationship to plankton size fractions 1. Limnology and Oceanography 23 (6), 1256–1263.

Siegel, D.A., Buesseler, K.O., Doney, S.C., Sailley, S.F., Behrenfeld, M.J., Boyd, P.W., 2014. Global assessment of ocean carbon export by combining satellite observations and food-web models. Global Biogeochemical Cycles 28 (3), 181–196.

Skákala, J., Ford, D.A., Brewin, R.J.W., McEwan, R., Kay, S., Taylor, B.H., et al., 2018. The assimilation of phytoplankton functional types for operational forecasting in the northwest European Shelf. Journal of Geophysical Research: Oceans 123. https://doi.org/10.1029/2018JC014153.

Soppa, M.A., Völker, C., Bracher, A., 2016. Diatom phenology in the Southern Ocean: mean patterns, trends and the role of climate oscillations. Remote Sensing 8, 420. https://doi.org/10.3390/rs8050420.

Stock, A., Subramaniam, A., 2020. Accuracy of empirical satellite algorithms for mapping phytoplankton diagnostic pigments in the open ocean: a supervised learning perspective. Frontiers in Marine Science 7, 599. https://doi.org/10.3389/fmars.2020.00599.

Stramski, D., Kiefer, D.A., 1991. Light scattering by microorganisms in the open ocean. Progress in Oceanography 28 (4), 343–383.

Sun, X., Shen, F., Liu, D., Bellerby, R.G., Liu, Y., Tang, R., 2018. In situ and satellite observations of phytoplankton size classes in the entire continental shelf sea, China. Journal of Geophysical Research: Oceans 123 (5), 3523–3544.

Sverdrup, H.U., 1953. On conditions for the vernal blooming of phytoplankton. ICES Journal of Marine Science 18 (3), 287–295.

Trzcinski, M.K., Devred, E., Platt, T., Sathyendranath, S., 2013. Variation in ocean color may help predict cod and haddock recruitment. Marine Ecology. Progress Series 491, 187–197.

Uitz, J., Claustre, H., Morel, A., Hooker, S.B., 2006. Vertical distribution of phytoplankton communities in open ocean: an assessment based on surface chlorophyll. Journal of Geophysical Research 111, C08005. https://doi.org/10.1029/2005JC003207.

Uitz, J., Claustre, H., Gentili, B., Stramski, D., 2010. Phytoplankton class-specific primary production in the world's oceans: seasonal and interannual variability from satellite observations. Global Biogeochemical Cycles 24 (3). https://doi.org/10.1029/2009GB003680.

Vandermeulen, R.A., Mannino, A., Neeley, A., Werdell, J., Arnone, R., 2017. Determining the optimal spectral sampling frequency and uncertainty thresholds for hyperspectral remote sensing of ocean color. Optics Express 25, A785–A797.

Vicente, T.M., Yamashita, C., de Mello e Sousa, S.H., Ciotti, A.M., 2021. Evaluation of the relationship between biomass of living (stained) benthic foraminifera and particulate organic matter vertical flux in an oligotrophic region, Campos Basin, southeastern Brazilian continental margin. Journal of Sea Research (ISSN 1385-1101) 176, 102110. https://doi.org/10.1016/j.seares.2021.102110.

Weatherhead, E.C., Reinsel, G.C., Tiao, G.C., Meng, X.L., Choi, D., Cheang, W.K., et al., 1998. Factors affecting the detection of trends: statistical considerations and applications to environmental data. Journal of Geophysical Research 103 (D14), 17,149–17,161. https://doi.org/10.1029/98JD00995.

Werdell, P.J., Franz, B.A., Bailey, S.W., Feldman, G.C., et al., 2013. Generalized ocean color inversion model for retrieving marine inherent optical properties. Applied Optics 52, 2019.

Werdell, P.J., Roesler, C.S., Goes, J.I., 2014. Discrimination of phytoplankton functional groups using an ocean reflectance inversion model. Applied Optics 53, 4833.

Werdell, P.J., Behrenfeld, M.J., Bontempi, P.S., Boss, E., Cairns, B., et al., 2019. The Plankton, Aerosol, Cloud, ocean Ecosystem (PACE) mission: status, science, advances. Bulletin of the American Meteorological Society 100 (9), 1775–1794. https://doi.org/10.1175/bams-d-18-0056.1.

Wolanin, A., Dinter, T., Bracher, A., 2015. SY-4Sci Synergy R & D Study 4: Phytoplankton Functional Types (SynSenPFT), D2.1 - Report on using radiative transfer modelling to develop a correction scheme and investigate the sensitivity of the improved PhytoDOAS (version 3.0) algorithm. Version 1, 30 Oct 2015, ESRIN Contract No.: 4000112410/14/I-NB. https://www.awi.de/fileadmin/user_upload/AWI/Forschung/Klimawissenschaft/Physikalische_Ozeanographie_der_Polarmeere/SEOM-SynSenPFT-D2.1_v1.pdf.

Wolanin, A., Soppa, M.A., Bracher, A., 2016. Investigation of spectral band requirements for improving retrievals of phytoplankton functional types. Remote Sensing 8, 871. https://doi.org/10.3390/rs8100871.

Xi, H., Hieronymi, M., Röttgers, R., Krasemann, H., et al., 2015. Hyperspectral differentiation of phytoplankton taxonomic groups: a comparison between using remote sensing reflectance and absorption spectra. Remote Sensing 7, 14781–14805. https://doi.org/10.3390/rs71114781.

Xi, H., Losa, S., Mangin, A., Soppa, M.A., Garnesson, P., Demaria, J., Liu, Y., Hembise Fanton d'Andon, O., Bracher, A., 2020. Global retrieval of phytoplankton functional types based on empirical orthogonal functions using CMEMS GlobColour merged products and further extension to OLCI data. Remote Sensing of Environment 240, 111704. https://doi.org/10.1016/j.rse.2020.111704.

Xi, H., Losa, S.N., Mangin, A., Garnesson, P., Bretagnon, M., Demaria, J., Soppa, M.A., Hembise Fanton d'Andon, O., Bracher, A., 2021. Global chlorophyll a concentrations of phytoplankton functional types with detailed uncertainty assessment using multi-sensor

ocean color and sea surface temperature satellite products. Journal of Geophysical Research: Oceans 126, e2020JC017127. https://doi.org/10.1029/2020JC017127.
Xiao, Y., Friedrichs, M.A.M., 2014. The assimilation of satellite-derived data into a one-dimensional lower trophic level marine ecosystem model. Journal of Geophysical Research 119, 2691–2712. https://doi.org/10.1002/2013JC009433.
Yentsch, C.S., Phinney, D.A., 1989. A bridge between ocean optics and microbial ecology. Limnology and Oceanography 34 (8), 1694–1705.
Zhu, Q., Shen, F., Shang, P., Pan, Y., Li, M., 2019. Hyperspectral remote sensing of phytoplankton species composition based on transfer learning. Remote Sensing 11 (17), 200.

CHAPTER 8

Modeling phytoplankton processes in multiple functional types

Mark Baird[a], Stephanie Dutkiewicz[b], Anna Hickman[c], Mathieu Mongin[a], Monika Soja-Wozniak[a], Jennifer Skerratt[a], and Karen Wild-Allen[a]

[a]CSIRO Oceans and Atmosphere, Hobart, TAS, Australia
[b]Department of Earth, Atmospheric and Planetary Sciences, Massachusetts Institute of Technology, Cambridge, MA, United States
[c]University of Southampton, National Oceanography Centre Southampton, Southampton, United Kingdom

8.1 Introduction

The existence of phytoplankton, and an understanding of their importance in natural water bodies, can be traced back to the early use of microscopes in the 1670s. By the mid-1800s it was recognized that some microscopic life-forms were photosynthetic (see Karlusich et al., 2020, for a history of early phytoplankton research). The first appreciation of the process of photosynthesis began to evolve with the identification of photosynthetic oxygen production by Joseph Priestly in 1774 (Shevela and Björn, 2018). The first "models" of photosynthetic processes, which are still components of more sophisticated phytoplankton models used today, were Liebig's law of the minimum (von Liebig, 1840), which represents the most limiting resource in biological growth; and the representation of photosynthesis as the stoichiometric combination of water, carbon dioxide, and energy to create organic matter (Ingen-Housz, 1796).

A variety of mathematical models have been applied to phytoplankton physiology and their roles in biogeochemical cycles and ecosystem functioning. An example of a simple statistical model is the tight relationship between total phosphorus and chlorophyll concentration found in freshwater lakes (Dillon and Rigler, 1974). More sophisticated statistical techniques, such as machine learning, can provide spatially resolved predictions of phytoplankton population diversity (Flombaum et al., 2020). Due to a need to understand the impact of individual processes on plankton dynamics, the most common approach has been to develop process-based models.

In process-based (or mechanistic) models, the most significant processes are individually parameterized, and their impact on plankton dynamics summed. Such models usually consider the flow of matter (and energy) from one "pool" such as inorganic nutrient to a second pool such as phytoplankton biomass, and then to a grazer pool, and a detrital pool (Fig. 8.1A). This is the classic NPZD configuration. Process-based modeling is the most common approach taken to quantifying phytoplankton dynamics and is the focus of this chapter.

The first process-based model to explain phytoplankton blooms in the natural environment was developed more than 70 years ago (Riley, 1947). Riley described the plankton with a single phytoplankton component. The Riley study was followed by models that considered more components of the plankton (Steele and Henderson, 1981). Subsequent studies have included more complete descriptions of phytoplankton growth (Jassby and Platt, 1976; Geider, 1987), and more complex biogeochemical cycling (Fasham et al., 1990). Since the beginning of the 21st century, modelers have become yet more ambitious, both in their representation of phytoplankton and in the many functional roles phytoplankton play in aquatic environments.

This ambition saw three, initially distinct, directions taken (Fig. 8.1). The first direction saw modelers significantly increase the number of functional roles of phytoplankton (i.e., additional pools in the NPZD format, Quere et al., 2005), with a goal, partly, to align with similar approaches for the terrestrial system and with global Earth system and climate models in mind (e.g., Haxeltine and Prentice, 1996). They began to consider additional elements (such as iron and silica), more complex processes (such as mixotrophy and nitrogen fixation), and sometimes with large numbers of plankton types (Follows et al., 2007). A second direction recognized the key role organism size played in rates of planktonic processes (Platt and Denman, 1977), with modelers developing multiple size-class (Moloney and Field, 1989), and even size-resolving (Baird and Suthers, 2007) plankton models. A third direction, coined "trait-based modeling" and common to models of other ecological systems, involved representing varying rates due to different taxa as a changing parameter value for the same mathematical description of a phytoplankton process (Wirtz and Eckhardt, 1996; Jones et al., 2010; Bruggeman, 2011).

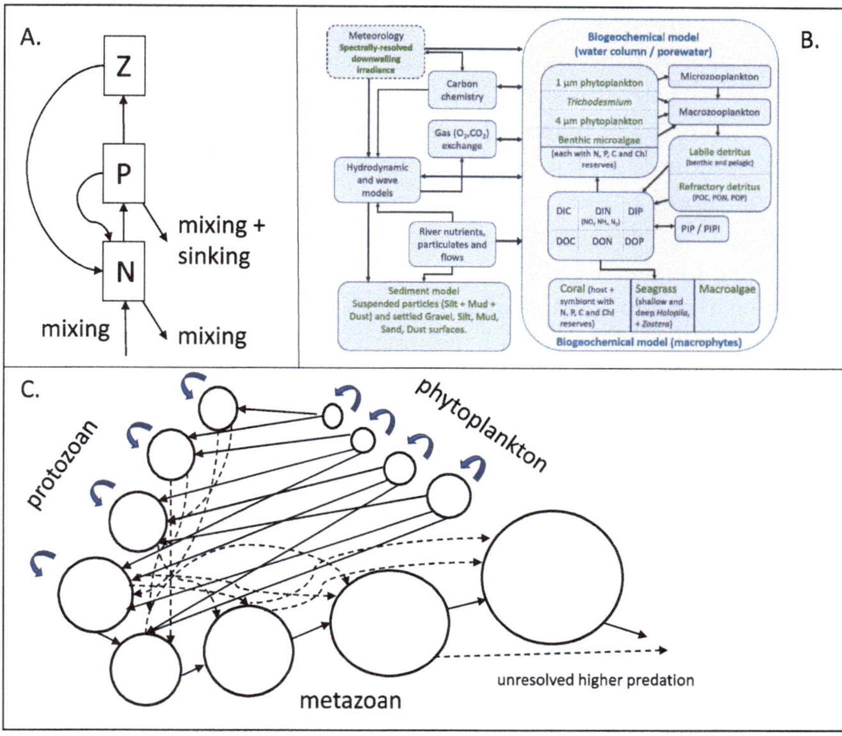

Figure 8.1 Schematic of biogeochemical models with phytoplankton types. A: Simple NPZ model (Edwards and Brindley, 1999); B: Functional biogeochemical eReefs model (Skerratt et al., 2019); C: Size-resolved model (Baird and Suthers, 2007).

The distinction between functional-type, size-based, and trait-based modeling quickly broke down as researchers became motivated by a new range of academic challenges and practical applications, such as harmful algal bloom dynamics, resolving more ecological properties such as biodiversity, or to capture subtle changes in algal communities due to climate change. In these new configurations, functional-type models were inevitably capturing different size-classes (large blooming diatom species with fast sinking rates vs. tiny nonsinking cyanobacteria); trait-based models looked to size-relationships for changing parameter values, and size-based models needed functional complexity to represent biogeochemical processes accurately (Baird et al., 2020). Thus models of plankton diversity often contain elements of all three approaches (Dutkiewicz et al., 2020).

A further motivating factor in the evolution of phytoplankton models has been new observational techniques, including spectrally resolved absorption measurements (Clementson and Wojtasiewicz, 2019a,b), "omics" (i.e., genomics, metagenomics; Karlusich et al., 2020), and hyperspectral satellite observations. To take advantage of these new observations, modelers are providing more bespoke parameterizations that are informed by the new observations (e.g., particle scattering properties; Soja-Woźniak et al., 2019), and model outputs for direct comparison with observations, (e.g., spectrally resolved apparent optical properties; Fujii et al., 2007; Dutkiewicz et al., 2015a; Jones et al., 2016).

This chapter looks across the range of phytoplankton modeling approaches described above to provide insights into the process descriptions that have been used to parameterize phytoplankton processes, with an emphasis when these descriptions vary between functional types.

8.2 Photosynthesis and chlorophyll synthesis

Phytoplankton absorption. Phytoplankton absorption represents one of the most obvious and achievable processes to distinguish between phytoplankton types. Absorption spectra are available for both the most important individual pigments for absorption (Clementson and Wojtasiewicz, 2019a) and for a variety of phytoplankton species (Clementson and Wojtasiewicz, 2019b). With the use of spectrally resolved optical models, plankton modelers have been able to calculate phytoplankton absorption on either functional group-specific spectrally resolved absorption coefficients (Dutkiewicz et al., 2015a; Gregg and Rousseaux, 2017) or functional-group-specific pigment assemblages with cell sizes (Baird et al., 2020). The importance of these studies is that phytoplankton absorption coefficients are such fundamental parameters (being the initial slope of the photosynthesis vs. irradiance curve and contributing to attenuation of light) as well as being a sensitive parameter in biogeochemical models. Additionally, phytoplankton absorption influences water-leaving radiances, allowing model simulations to be directly assessed against satellite products (Laiolo et al., 2021).

Photosynthesis. The processes of photon absorption, leading to carbon fixation, is a complex quantum/molecular chain of chemical reactions (Falkowski and Raven, 2013). Models of photosynthesis range from simple photosynthesis vs. irradiance curves with an initial slope of the absorption cross section and saturating at a photosynthesis maximum rate, to those with photoinhibition (Jassby and Platt, 1976), to even more complex models that consider reactive oxygen build-up dynamics (Baird et al.,

2018) or flash kinetics (Rubio et al., 2003). Interestingly, though important differences in photophysiology between phytoplankton functional groups exists, these differences have rarely been distinguished in multiple functional type models. For example, dinoflagellates contain a form of the Rubisco (Ribulose-1,5-bisphosphate carboxylase-oxygenase) enzyme used in photosynthesis that has a significantly different temperature tolerance to other forms, as seen in the photoinhibition of zooxanthellae that drives coral bleaching (Baird et al., 2018), yet biogeochemical models of dinoflagellates commonly have the same photophysiological representation as other phytoplankton groups (Wild-Allen et al., 2010).

Chlorophyll synthesis. The pigment content of a phytoplankton cell is important for both the calculation of the absorption cross section and often, because of the use of optical sensors, the quantification of phytoplankton biomass. As a result, models have been developed that consider pigment synthesis so as to represent a dynamical chlorophyll content, and therefore ratio of chlorophyll to carbon of cells (Geider et al., 1997), and these are commonly applied in multiple functional type models. Typically, these models are quantified based on the chlorophyll-*a* (Chl-a) molecule, allowing a changing Chl-a pigment content, in combination with a Chl-a specific absorption spectra, to determine phytoplankton absorption. More niche components of chlorophyll synthesis models include a cost of biosynthesis (Geider et al., 1997) and the dependence of the pigment synthesis on the opaqueness of the cell (Baird et al., 2020).

8.3 Impact of phytoplankton on ocean optics

Phytoplankton affect the underwater light field, and the water-leaving radiance, as a result of their absorption, scattering and, in particular cases, their fluorescence and luminescence (Mobley, 1994). As noted above, the differing phytoplankton absorption spectra is one of the distinguishing features that has been used to model different phytoplankton functional types. The scattering spectra varies even more among phytoplankton types. This is in part because scattering is strongly dependent on particle size and some phytoplankton cell walls, especially coccolithophorids, backscatter strongly. These differing optical properties result in differing water-leaving radiance, introducing the possibility of observing phytoplankton functional groups remotely (Soja-Woźniak et al., 2018).

8.4 Macronutrient uptake and stoichiometry

Nutrient uptake. Models of nutrient uptake in phytoplankton range from essentially nutrient-limited growth functions, such as typified by the Monod growth equation (Fasham et al., 1990), through to models that consider transport of nutrients to the cell surface and uptake into the cell interior (Pasciak and Gavis, 1975), as well as internal biochemical processing (John and Flynn, 2000). The most important distinction is between those models with dynamical stores (or reserves or quotas) of nutrients, which therefore have nutrient uptake timescales faster than cell growth (often called surge uptake), and those models such as the Monod equation that consider growth and uptake in one process (Monod, 1949; Heliweger, 2017).

Stoichiometry. Perhaps the most remarkable marine biogeochemical phenomenon is the consistency of the elemental ratio of dissolved inorganic nitrogen to phosphorus of the world's oceans (N:P = 16:1 mol N:mol P), and a corresponding similar ratio of C:N:P of 106:16:1 of phytoplankton, coined the Redfield ratio. It was Redfield's assertion that by combining elements at a set ratio, and their subsequent remineralization at depth, phytoplankton production was controlling the ocean nutrient ratios (Redfield et al., 1963). Furthermore, he noted that the elemental ratio of phytoplankton growing at their maximum growth rate is the Redfield ratio, that is, when phytoplankton are unlimited by the supply of C, N, or P, the sum of all their components is also at the Redfield ratio.

There are two important implications for phytoplankton modeling of Redfield's framework. Firstly, that when a limiting nutrient is taken up and is converted into biomass (quantified by C, N, or P), a predictable quantity of phytoplankton biomass is generated. This constraint explains the success of aquatic biogeochemical models in quantifying phytoplankton blooms associated with nutrient injections, such as water column overturning or coastal upwelling. Secondly, derivations from the Redfield ratio can be considered a physiological adjustment to environmental conditions (e.g., greater internal storage of a replete nutrient compared to a deplete nutrient) with the turnover of elements through phytoplankton growth remaining at Redfield, thus maintaining the observed ocean dissolved nutrient ratios.

Variable stoichiometry. Although the elemental ratio of dissolved macronutrients in the ocean interior may be at Redfield, surface waters can become deplete in one nutrient, driving phytoplankton to non-Redfield ratios.

Early models with variable stoichiometry considered a steady-state between environmental conditions and cellular elemental ratios (Laws and Chalup, 1990). By assuming steady-state the phytoplankton state could still be represented by one variable (i.e., biomass), but the effect of variable stoichiometry could be included in phytoplankton growth rates and phytoplankton uptake, export, and remineralization. Though this was a significant improvement at little (computational) cost, it could not resolve the mismatched timing of uptake of nutrients and photosynthesis (i.e., nitrate uptake at night or at depth). This led to the need to develop dynamic models of variable stoichiometry.

Dynamic energy budgets (DEB) models. DEB theory is a broad description of physiological models of organisms that partitions biomass into one or more reserves and one or more structures (Kooijman and Kooijman, 2010). Though the term DEB is mostly used for complex animal models, the DEB approach has been used to consider the interaction of nutrient uptake, light capture, and growth for phytoplankton (Kooijman and Kooijman, 2010; Baird et al., 2001). The key attribute of DEB models, compared to say the Monod growth or steady-state (Laws and Chalup, 1990) models mentioned above, is that the supply of nutrients and light can be decoupled from the growth, allowing cells to have a metabolic memory that smooths out fluctuations in supply of nutrients, and that individual cells can have different physiological states.

The quota-style models and the DEB models are similar in the sense that they allow for non-Redfield uptake and release of resources and must be formulated to conserve mass. However, by representing the non-Redfield component in different state variables (quotas vs. reserves), they encourage slightly different functional forms for uptake and growth, with quota models generally following Droop-style models (Droop, 1968).

Macromolecular models. The next step in phytoplankton physiological sophistication is to split the "reserves" or "quota" into multiple molecular pools. For example, Inomura et al. (2020) considered internal stores of elements within proteins, pigments, nucleic acids, carbohydrates, and lipids. As mentioned in the Introduction, this breakdown allows a more meaningful comparison with biochemical and "omic" observations. Furthermore, it allows for a distinct use of each element beyond simple stoichiometric combination, such as the separate combinations of N and P in RNA and protein molecules that leads to different growth versus nutrient status of N and P (Inomura et al., 2020). The penalty of having 5 stores of each element (rather than one) for each phytoplankton type has limited the use of these models in multifunctional-type three-dimensional environmental models.

8.5 Micronutrients and silica

Early marine biogeochemical models were generally based on nitrogen (Fasham et al., 1990). To incorporate multiple phytoplankton functional groups it has been essential to include more limiting nutrients, as often the most important differentiating factor is their elemental requirements and means of obtaining other nutrients. The simplest example is the silica required by diatoms. The need to represent phosphorus limitation of N-fixers has resulted in inclusion of phosphorus (Aumont et al., 2015). Iron limitation has been recognized as a limiting nutrient in the Southern Ocean, impacting on the photosystems of all phytoplankton. The split between ammonium and nitrate uptake also provides a differentiating factor in algal groups.

8.6 Nitrogen fixation, calcification, mixotrophy

Biogeochemical models have also incorporated phytoplankton with additional biogeochemical functions. Organisms that are capable of fixing nitrogen (using N_2 gas to create organic matter) are ubiquitous in the marine environment, becoming most important when N is the limiting macronutrient. A number of parameterizations for N-fixing organisms have been used in marine models (Robson et al., 2013; Baird et al., 2020; Dutkiewicz et al., 2015a; Dutheil et al., 2018). It remains unclear what controls the distributions of nitrogen-fixing organisms. Modeling studies that explicitly include both N fixers and the necessary biogeochemical cycles have been used to explore these drivers (e.g., Dutkiewicz et al., 2012; Landolfi et al., 2015), finding the importance of phosphorous and iron cycling.

Another important phytoplankton type is the calcifiers (Yool et al., 2013). Though calcifiers have not become standard in models of phytoplankton production, their impact on carbon chemistry has meant they have been included in many of the global biogeochemical models (Aumont et al., 2015; Dutkiewicz et al., 2015b). The modeling study of Monteiro et al. (2016) has been used to analyze the cost and benefits of calcifying, suggesting that coccoliths may protect from grazing, viruses, and light.

Mixotrophs have been strongly advocated by some modelers (Mitra and Flynn, 2010). The importance of mixotrophs lies not only in the different function they play among the phytoplankton, but also through grazing, how they structure the rest of the phytoplankton groups (Section 8.7). Model work has suggested that mixotrophy might also be important in controlling the size distribution within a phytoplankton community, with important feedback to marine biogeochemistry (Ward and Follows, 2016).

8.7 Phytoplankton mortality

Phytoplankton mortality due to grazing, sinking, viral lysis, and other ecosystem processes are the balance to the growth processes described above. Though many of these processes are assumed the same for all phytoplankton types, mortality by grazing is often resolved according to phytoplankton defence (e.g., cell walls). Although it is not the purpose of this chapter to consider the multiple functional zooplankton types that are involved in grazing phytoplankton, it is important to note that when mortality terms, such as zooplankton grazing, selectively impact on phytoplankton functional types, this can have as great a role in determining the phytoplankton community composition as the determinants of phytoplankton growth. The paragraph that follows describes a set of numerical experiments to illustrate the role of grazing selection based on size.

In Baird (2010), a size-resolved pelagic ecosystem was set-up, in which all parameters values were determined by the size of the plankton. Thus functional groups in the study were essentially size-classes. Critically for the discussion here, one parameter in the model was the size-range of prey for a predator. This essentially determined who ate whom. The study then undertook a sensitivity analysis to determine the most sensitive biological parameter. The results clearly showed that the model was more sensitive to the parameter determining size range of diets than the other parameters typically associated with phytoplankton functional diversity, such as maximum growth rates or light absorption. From this, it can be concluded that model representation of multiple phytoplankton classes must be considered in the context of the single or multiple zooplankton functional types in the model ecosystem.

8.8 Summary of process descriptions

Models have incorporated the different phytoplankton processes in several ways, using different structures and assumptions, depending on their specific purpose. Various frameworks exist to define how phytoplankton grow in different conditions, for example, energy and/or resource allocation within the cells (e.g., Geider, DEB theory). Many models describe a number of phytoplankton types (or traits) and the dominant ones emerge based on competition and selection in the modeled environmental conditions. Some models have required (and have shed light on) our understanding of how phytoplankton traits are related to one another (for example, which traits are related to phytoplankton size), and the importance of trade-offs, or evolutionary constraints that dictate trait combinations (Follows et al., 2007). Some of these approaches have been embedded within three-dimensional physical models and can follow regional or global biogeochemical cycles, including air-sea fluxes of carbon.

8.9 Marine applications

The development of diverse approaches has dramatically improved our fundamental understanding of phytoplankton ecology as well as provided model capability to tackle a myriad of pressing scientific problems. It also means that the practitioner who is assembling these models must balance the advantage of including each innovation with the cost that complexity can introduce in terms of computing time as well as the greater variability in model behavior. In this section, we give some case studies of the applications of these phytoplankton models.

Case Study 8.1: Global change

Anthropogenic-driven global change is and will continue to impact the oceans. Warming and associated changes in mixing and circulation, as well as increased carbon of the waters (ocean acidification) will affect phytoplankton. Process-based models, as described in this chapter, provide a laboratory to explore these potential changes. Recent climate assessment analysis (e.g., IPCC) have included results from Earth system models (which include atmosphere, land, and ocean) that have included phytoplankton explicitly within a global context (e.g., Bopp et al., 2013). These models suggest a global decrease in primary production, though with regionally variable patterns of change, including some high latitude regions having increased productivity. The variable patterns arise as drivers of change, such as increased phytoplankton growth rates with warmer water, are balanced against reduced nutrient supply as a result of less mixing and altered circulation (Dutkiewicz et al., 2013). Models can additionally suggest how phytoplankton community structure might shift in the future ocean. For instance, models show that smaller phytoplankton might become more dominant, as they are better adapted to low nutrient conditions (Bopp et al., 2005; Dutkiewicz et al., 2013). Moreover, model studies can also show that the differing physiological response to ocean acidification between phytoplankton species will likely also lead to substantial rearrangement of phytoplankton communities (Dutkiewicz et al., 2015b, 2019).

Case Study 8.2: Impact of phytoplankton community structure on carbon export fluxes

The biological carbon pump is responsible for the downward transfer of organic carbon produced in the surface ocean to the deep layers via the sinking of organic particles, and its eventual sequestration for geological timescales (Lam et al., 2011). Estimations suggest that the biological carbon pump accounts for about two-thirds of the global oceanic carbon pump. The two main pathways of carbon export are physical aggregation of phytoplankton cells and their subsequent sinking in the form of large aggregates, or through biological aggregation of grazed phytoplankton cells and their sinking in the form of dense zooplankton fecal pellets.

Parameterizations of sinking velocity in most biogeochemical models used to estimate carbon fluxes are either constant values or scaled on size and/or depth based on empirical relationships (Kriest and Oschlies, 2013; Pasquer et al., 2005; Baird and Emsley, 1999).

The natural variation in the carbon export efficiency is large (order of magnitude in some cases) and occurs over a wide range of spatiotemporal scales (Buesseler, 1998). The role of planktonic communities in driving these export variations has been suspected since the 1990s, and extensively studied since then (Boyd and Newton, 1995). Phytoplankton morphological types, through their influence on particle formation and structure, may indirectly control large particle sinking velocities (Laurenceau-Cornec et al., 2015). The changes in the composition of detritus, closely related to seasonal shifts of surface plankton communities, leads to significant modification of detritus sinking velocities and subsequent alteration of the efficiency, amplitude, and timing of carbon export.

Modifying the sinking velocity parameterization in BGC models with parameterizations based on detritus origin is a way to account for the influence of planktonic communities on carbon export. It has been shown that using such a parameterization can resolve the large errors in estimations of carbon export fluxes calculated from BGC models using a constant sinking velocity (Laurenceau-Cornec et al., 2015). A variable sinking velocity of detritus calculated from the relative contributions of different plankton sources could improve the accuracy of carbon flux estimations made by BGC models.

Case Study 8.3: Harmful algal blooms (HABs)

The modeling of HABs is a good example of where the prediction of the abundance of multiple algal types is highly desirable, and where it is often only one or a few rare species that are of prime interest. Approaches to prediction of HABs have included remote sensing (Blondeau-Patissier et al., 2014), statistical modeling, and multiple functional group processes modeling (Ralston and Moore, 2020). Ralston and Moore (2020) argue that for changing environmental conditions, it is process-based models of phytoplankton functional types that provide the most robust predictions. They identify 15 different studies using process-based HAB models, with the processes that are important in their parameterizations including rates of growth, mortality, mobility, and toxin production. Interestingly, HABs are often considered using individual-based models (IBMs) in preference to Eulerian (concentration-based) models (Gillibrand et al., 2016). One significant advantage of IBMs is that life-cycle processes, such as the age of passive and mobile phases, which are near-impossible to include in Eulerian models, are naturally represented in IBMs.

Case Study 8.4: Coastal water quality

In the aquatic environment, absorption and scattering by phytoplankton cells reduces the vertical penetration of light. Often anthropogenic inputs of nutrients increase phytoplankton concentrations, adding to the rate of light attenuation and degrading the natural environment. On the Great Barrier Reef (GBR) off northeast Australia, where seabed habitats of seagrass and corals are critical for ecosystem function, changes in optical properties are important metrics of ecosystem health (Baird et al., 2018).

To represent the impact of phytoplankton on water quality, the eReefs marine biogeochemical model of the GBR contains 4 phytoplankton functional types (small and large phytoplankton, benthic microalgae, and *Trichodesmium*) as well as a dinoflagellate coral symbiont (Baird et al., 2020). The differences between small and large phytoplankton and benthic microalgae lie in their size, sinking rates, and absorption spectra (Fig. 8.2). Another phytoplankton functional type, nitrogen-fixer *Trichodesmium*, has the additional processes of obtaining nitrogen from dissolved N_2 gas and a variable sinking rate, due to physiological state.

The functional diversity plays two important roles in predicting water quality. Firstly, the plankton populations are represented with differing roles in biogeochemical cycling: small phytoplankton maintain a low population under tight grazing control; large phytoplankton escape grazing control and bloom, due to catchment inputs and upwelling events; *Trichodesmium* provides a source of nitrogen in deplete open ocean waters; and benthic microalgae provide a seabed reservoir that is intermittently resuspended during high bottom-stress events. Without this functional diversity, biogeochemical models do not capture the nitrogen cycling on the GBR.

On the GBR, where the bleaching of corals due to the expulsion of algal symbionts (zooxanthellae) has become an important anthropogenic impact, the modeling of the zooxanthellae has been developed to include xanthophyll cycling and photosynthetic reaction center dynamics (Baird et al., 2018).

The second significant role of diversity of plankton types is to capture the differing optical properties of the algal types on the spectrally resolved light field. This is achieved in the eReefs model through specifying a unique absorption spectra for each functional type (Fig. 8.2), and then applying this rate of absorption per unit length for the different sizes of phytoplankton, and their time-varying pigment content. This approach captures the impact of the package effect, whereby highly pigmented cells have a reduced absorption of light per unit of pigment (Soja-Woźniak et al., 2020). This sophistication becomes important for considering the spectral quality of light reaching the seabed. For benthic plants, light reaching the bottom at the chlorophyll peaks is absorbed better, and therefore more important than downwelling light at other wavelengths.

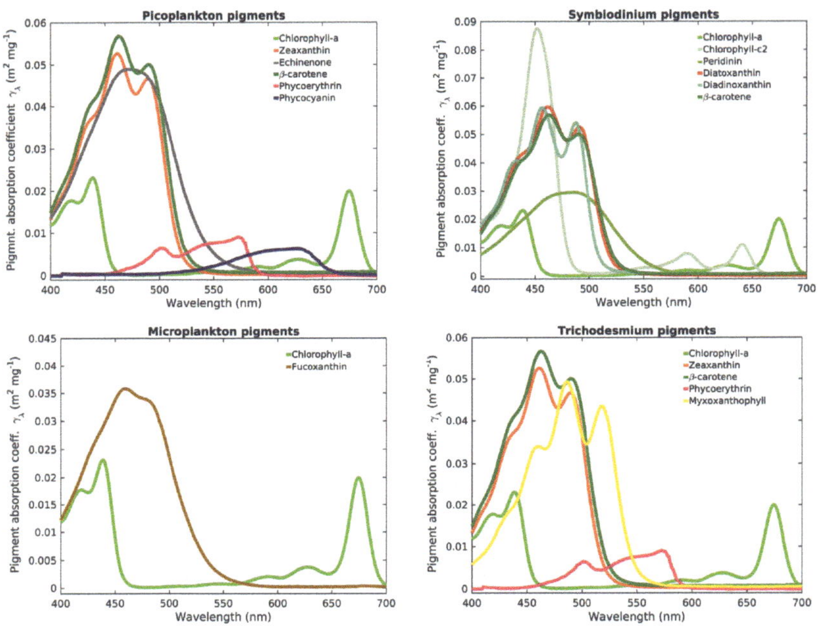

Figure 8.2 Pigment-specific absorption spectra of 4 phytoplankton functional types used in the eReefs marine biogeochemical model of the Great Barrier Reef. A: Picoplankton spectra used for small phytoplankton based on Synechococcus; B: Spectra from the dinoflagellate symbiont that inhabits coral polyps and includes a photoprotective pigment (red line) that is important in the simulation of photoinhibition/photoacclimation; C: Microplankton spectra based on a diatom used for large phytoplankton and benthic microalgae; D: Spectra from the nitrogen-fixer Trichodesmium. Original data sourced from Clementson and Wojtasiewicz (2019a,b).

Case Study 8.5: Operational models

A number of modeling teams have developed biogeochemical models that are now deployed in operational environments (Fennel et al., 2019). Operational modeling systems are typically well-tested, post-development versions that are less likely to contain large numbers of phytoplankton types. For example, the NEMO-MEDUSA model (Yool et al., 2013) deployed by the United Kingdom Meteorological Office runs as an operational global physical forecasting system containing only two phytoplankton groups (diatom and non-diatom) with the main distinction between the two being the silica requirement of diatoms.

Mercator Ocean International runs an operational global model for short-term forecasts and reanalyses with the biogeochemical model PISCES (Aumont et al., 2015). Though PISCES contains only two phytoplankton classes, it interestingly contains additional physiological processes compared to MEDUSA. PISCES is a mixed Monod–quota model. Thus nitrogen and phosphorus limitations are based on a Monod parameterization, where growth depends on the external nutrient concentrations, whereas iron limitation is modeled according to a classical quota approach (Section 8.6).

Some more complex models have been deployed in regional operational settings. The eReefs model used in the Great Barrier Reef/Coral Sea region contains four phytoplankton types (small and large phytoplankton, benthic microalgae and *Trichodesmium*), as well as the dinoflagellate *Symbiodinium* inside a coral holobiont (Jones et al., 2016; Baird et al., 2020). In the eReefs case, regional biological oceanographic properties have determined the phytoplankton function types: the coastal waters off northeast Australia are highly nitrogen-limiting with nitrogen fixation providing an important source of nitrogen to surface waters, and therefore requiring a nitrogen fixing phytoplankton, but is rarely silica limited, thus negating the need for a dedicated diatom functional type.

Ocean-color and biogeochemical data assimilation. Biogeochemical models commonly assimilate observations by comparing the mismatch of modeled and observed phytoplankton properties. With biogeochemical data assimilation most commonly using remotely-sensed observations, the ability to interpret the presence of phytoplankton by a change in water-leaving radiance is critical. Ocean models that resolve multiple phytoplankton size-classes can add more variability in model predictions of the distribution of chlorophyll, and therefore ocean-color: a challenge when assimilating ocean-color data (IOCCG, 2020).

In terms of multiple phytoplankton functional types, the smaller phytoplankton classes contribute disproportionality more to surface ocean-color signals, due to higher carbon to chlorophyll ratios and less packaging of pigments (Soja-Woźniak et al., 2020; Laiolo et al., 2021). The inferred relationship between chlorophyll and reflectance ratios is driven by the change of chlorophyll a concentration, the change in the phytoplankton size-class structure, and on the covariance of other optically significant constituents with chlorophyll a (Sathyendranath et al., 2017). However, these relationships are based on global mean relationships between phytoplankton chlorophyll and reflectance: thus a timing mismatch in modeled phytoplankton groups of, say, a progression from small to large phytoplankton could cause a large change in ocean-color, and therefore impose a data assimilation shock on the model.

At the same time, new hyperspectral satellite sensors, such as the NASA plankton, aerosol, cloud, ocean ecosystem (PACE) mission will provide additional data (Gregg and Rousseaux, 2017), and it is hoped that the identification of phytoplankton functional types will aid in the assessment of these more complex aspects of biogeochemical models.

8.10 Summary

The observations, parameterizations, and ambitions of scientists to represent multiple phytoplankton types in aquatic models has increased greatly over the past decade. Realizing the benefits of this work will require asking new questions of the models, undertaking more targeted simulations (with a likely focus on impacts of climate change), and to undertake more sophisticated comparison of simulated and observed phytoplankton diversity.

References

Aumont, O., Éthé, C., Tagliabue, A., Bopp, L., Gehlen, M., 2015. PISCES-v2: an ocean biogeochemical model for carbon and ecosystem studies. Geoscientific Model Development 8 (8), 2465–2513.

Baird, M.E., 2010. Limits to prediction in a size-resolved pelagic ecosystem model. Journal of Plankton Research 32 (8), 1131–1146.

Baird, M.E., Emsley, S.M., 1999. Towards a mechanistic model of plankton population dynamics. Journal of Plankton Research 21 (1).

Baird, M.E., Emsley, S.M., McGlade, J.M., 2001. Modelling the interacting effects of nutrient uptake, light capture and temperature on phytoplankton growth. Journal of Plankton Research 23 (8), 829–840.

Baird, M.E., Mongin, M., Rizwi, F., Bay, L.K., Cantin, N.E., Soja-Woźniak, M., Skerratt, J., 2018. A mechanistic model of coral bleaching due to temperature-mediated light-driven reactive oxygen build-up in zooxanthellae. Ecological Modelling 386, 20–37.

Baird, M.E., Suthers, I.M., 2007. A size-resolved pelagic ecosystem model. Ecological Modelling 203 (3–4), 185–203.

Baird, M.E., Wild-Allen, K.A., Parslow, J., Mongin, M., Robson, B., Skerratt, J., Rizwi, F., Soja-Woźniak, M., Jones, E., Herzfeld, M., Margvelashvili, N., 2020. CSIRO Environmental Modelling Suite (EMS): scientific description of the optical and biogeochemical models (vB3p0). Geoscientific Model Development 13 (9), 4503–4553.
Blondeau-Patissier, D., Gower, J.F., Dekker, A.G., Phinn, S.R., Brando, V.E., 2014. A review of ocean color remote sensing methods and statistical techniques for the detection, mapping and analysis of phytoplankton blooms in coastal and open oceans. Progress in Oceanography 123, 123–144.
Bopp, L., et al., 2013. Multiple stressors of ocean ecosystems in the 21st century: projections with CMIP5 models. Biogeosciences 10, 6225–6245.
Bopp, L., Aumont, O., Cadule, P., Alvain, S., Gehlen, M., 2005. Response of diatoms distribution to global warming and potential implications: a global model study. Geophysical Research Letters 32, L19606. https://doi.org/10.1029/2005GL023653.
Boyd, P., Newton, P., 1995. Evidence of the potential influence of planktonic community structure on the interannual variability of particulate organic carbon flux. Deep-Sea Research. Part 1. Oceanographic Research Papers 42 (5), 619–639.
Bruggeman, J., 2011. A phylogenetic approach to the estimation of phytoplankton traits 1. Journal of Phycology 47 (1), 52–65.
Buesseler, K.O., 1998. The decoupling of production and particulate export in the surface ocean. Global Biogeochemical Cycles 12 (2), 297–310.
Clementson, L.A., Wojtasiewicz, B., 2019a. Dataset on the absorption characteristics of extracted phytoplankton pigments. Data in Brief 24, 103875.
Clementson, L.A., Wojtasiewicz, B., 2019b. Dataset on the in vivo absorption characteristics and pigment composition of various phytoplankton species. Data in Brief 25, 104020.
Dillon, P.J., Rigler, F.H., 1974. The phosphorus-chlorophyll relationship in lakes 1, 2. Limnology and Oceanography 19 (5), 767–773.
Droop, M.R., 1968. Vitamin B12 and marine ecology. IV. The kinetics of uptake, growth and inhibition in Monochrysis lutheri. Journal of the Marine Biological Association of the UK 48 (3), 689–733.
Dutheil, C., Aumont, O., Gorgues, T., Lorrain, A., Bonnet, S., Rodier, M., Dupouy, C., Shiozaki, T., Menkes, C., 2018. Modelling N_2 fixation related to Trichodesmium sp.: driving processes and impacts on primary production in the tropical Pacific Ocean. Biogeosciences 15 (14), 4333–4352.
Dutkiewicz, S., Cermeno, P., Jahn, O., Follows, M.J., Hickman, A.E., Taniguchi, D.A., Ward, B.A., 2020. Dimensions of marine phytoplankton diversity. Biogeosciences 17 (3), 609–634.
Dutkiewicz, S., Hickman, A.E., Jahn, O., Gregg, W.W., Mouw, C.B., Follows, M.J., 2015a. Capturing optically important constituents and properties in a marine biogeochemical and ecosystem model. Biogeosciences 12 (14), 4447–4481.
Dutkiewicz, S., Hickman, A.E., Jahn, O., Henson, S., Beaulieu, C., Monier, E., 2019. Ocean colour signature of climate change. Nature Communications 10 (1), 1–13.
Dutkiewicz, S., Morris, J., Follows, M., Scott, J., Levitan, O., Dyhrman, S., Berman-Frank, I., 2015b. Impact of ocean acidification on the structure of future phytoplankton communities. Nature Climate Change 5, 1002–1006. https://doi.org/10.1038/nclimate2722.
Dutkiewicz, S., Scott, J., Follows, M.J., 2013. Winners and losers: phytoplankton biogeochemical and ecological changes in a warmer world. Global Biogeochemical Cycles 27, 463–477. https://doi.org/10.1002/gbc.20042.
Dutkiewicz, S., Ward, B.A., Monteiro, F., Follows, M.J., 2012. Interconnection between nitrogen fixers and iron in the Pacific Ocean: theory and numerical model. Global Biogeochemical Cycles 26, GB1012. https://doi.org/10.1029/2011GB004039.
Edwards, A.M., Brindley, J., 1999. Zooplankton mortality and the dynamical behaviour of plankton population models. Bulletin of Mathematical Biology 61 (2), 303–339.

Falkowski, P.G., Raven, J.A., 2013. Aquatic Photosynthesis. Princeton University Press.
Fasham, M.J., Ducklow, H.W., McKelvie, S.M., 1990. A nitrogen-based model of plankton dynamics in the oceanic mixed layer. Journal of Marine Research 48 (3), 591–639.
Fennel, K., Gehlen, M., Brasseur, P., Brown, C.W., Ciavatta, S., Cossarini, G., Crise, A., Edwards, C.A., Ford, D., Friedrichs, M.A., Gregoire, M., 2019. Advancing marine biogeochemical and ecosystem reanalyses and forecasts as tools for monitoring and managing ecosystem health. Frontiers in Marine Science 6, 89.
Flombaum, P., Wang, W.L., Primeau, F.W., Martiny, A.C., 2020. Global picophytoplankton niche partitioning predicts overall positive response to ocean warming. Nature Geoscience 13 (2), 116–120.
Follows, M.J., Dutkiewicz, S., Grant, S., Chisholm, S.W., 2007. Emergent biogeography of microbial communities in a model ocean. Science 315 (5820), 1843–1846.
Fujii, M., Boss, E., Chai, F., 2007. The value of adding optics to ecosystem models: a case study. Biogeosciences 4 (5), 817–835.
Geider, R.J., 1987. Light and temperature dependence of the carbon to chlorophyll a ratio in microalgae and cyanobacteria: implications for physiology and growth of phytoplankton. New Phytologist 106 (1), 1–34.
Geider, R.J., MacIntyre, H.L., Kana, T.M., 1997. Dynamic model of phytoplankton growth and acclimation: responses of the balanced growth rate and the chlorophyll a: carbon ratio to light, nutrient-limitation and temperature. Marine Ecology. Progress Series 148, 187–200.
Gillibrand, P.A., Siemering, B., Miller, P.I., Davidson, K., 2016. Individual-based modelling of the development and transport of a Karenia mikimotoi bloom on the North-West European continental shelf. Harmful Algae 53, 118–134.
Gregg, W.W., Rousseaux, C.S., 2017. Simulating PACE global ocean radiances. Frontiers in Marine Science 4, 60.
Haxeltine, A., Prentice, I.C., 1996. BIOME3: an equilibrium terrestrial biosphere model based on ecophysiological constraints, resource availability, and competition among plant functional types. Global Biogeochemical Cycles 10 (4), 693–709.
Heliweger, F.L., 2017. 75 years since Monod: it is time to increase the complexity of our predictive ecosystem models (opinion). Ecological Modelling 346, 77–87.
Ingen-Housz, J., 1796. Essay on the Food of Plants and the Renovation of Soils. Additional Appendix (no. III) to the Outlines of the Fifteenth Chapter of the Proposed General Report from the Board of Agriculture/on the subject of Manures. Printed by W. Bulmer and Co, London.
Inomura, K., Omta, A.W., Talmy, D., Bragg, J., Deutsch, C., Follows, M.J., 2020. A mechanistic model of macromolecular allocation, elemental stoichiometry, and growth rate in phytoplankton. Frontiers in Microbiology 11, 86.
IOCCG, 2020. Synergy between ocean colour and biogeochemical/ecosystem models. In: Dutkiewicz, S. (Ed.), Reports of the International Ocean-Colour Coordinating Group, No. 19. International Ocean-Colour Coordinating Group (IOCCG), Dartmouth, NS, Canada. 184 pp.
Jassby, A.D., Platt, T., 1976. Mathematical formulation of the relationship between photosynthesis and light for phytoplankton. Limnology and Oceanography 21 (4), 540–547.
John, E.H., Flynn, K.J., 2000. Modelling phosphate transport and assimilation in microalgae; how much complexity is warranted? Ecological Modelling 125 (2–3), 145–157.
Jones, E., Parslow, J., Murray, L., 2010. A Bayesian approach to state and parameter estimation in a Phytoplankton-Zooplankton model. Australian Meteorological and Oceanographic Journal 59 (SP), 7–16.
Jones, E.M., Baird, M.E., Mongin, M., Parslow, J., Skerratt, J., Lovell, J., Margvelashvili, N., Matear, R.J., Wild-Allen, K., Robson, B., Rizwi, F., 2016. Use of remote-sensing reflectance to constrain a data assimilating marine biogeochemical model of the Great Barrier Reef. Biogeosciences 13 (23), 6441–6469.

Karlusich, J.J.P., Ibarbalz, F.M., Bowler, C., 2020. Exploration of marine phytoplankton: from their historical appreciation to the omics era. Journal of Plankton Research 42 (6), 595–612.
Kooijman, B., Kooijman, S.A.L.M., 2010. Dynamic Energy Budget Theory for Metabolic Organisation. Cambridge University Press.
Kriest, I., Oschlies, A., 2013. Swept under the carpet: organic matter burial decreases global ocean biogeochemical model sensitivity to remineralization length scale. Biogeosciences 10 (12), 8401–8422.
Laiolo, L., Matear, R., Soja-Woźniak, M., Suggett, D.J., Hughes, D.J., Baird, M.E., Doblin, M.A., 2021. Modelling the impact of phytoplankton cell size and abundance on inherent optical properties (IOPs) and a remotely sensed chlorophyll-a product. Journal of Marine Systems 213, 103460.
Lam, P.J., Doney, S.C., Bishop, J.K., 2011. The dynamic ocean biological pump: insights from a global compilation of particulate organic carbon, CaCO3, and opal concentration profiles from the mesopelagic. Global Biogeochemical Cycles 25 (3).
Landolfi, A., Koeve, W., Dietze, H., Kähler, P., Oschlies, A., 2015. A new perspective on environmental controls of marine nitrogen fixation. Geophysical Research Letters 42 (11), 4482–4489.
Laurenceau-Cornec, E.C., Trull, T.W., Davies, D.M., Bray, S.G., Doran, J., Planchon, F., Carlotti, F., Jouandet, M.P., Cavagna, A.J., Waite, A.M., Blain, S., 2015. The relative importance of phytoplankton aggregates and zooplankton fecal pellets to carbon export: insights from free-drifting sediment trap deployments in naturally iron-fertilised waters near the Kerguelen Plateau. Biogeosciences 12 (4), 1007–1027.
Laws, E.A., Chalup, M.S., 1990. A microalgal growth model. Limnology and Oceanography 35 (3), 597–608.
Mitra, A., Flynn, K.J., 2010. Modelling mixotrophy in harmful algal blooms: more or less the sum of the parts? Journal of Marine Systems 83 (3–4), 158–169.
Mobley, C.D., 1994. Light and Water: Radiative Transfer in Natural Waters. Academic Press.
Moloney, C.L., Field, J.G., 1989. General allometric equations for rates of nutrient uptake, ingestion, and respiration in plankton organisms. Limnology and Oceanography 34 (7), 1290–1299.
Monod, J., 1949. The growth of bacterial cultures. Annual Review of Microbiology 3 (1), 371–394.
Monteiro, F.M., Bach, L.T., Brownlee, C., Brown, P., Rickaby, R.E.M., Tyrrell, T., Beaufort, L., Dutkiewicz, S., Gibbs, S., Gutowska, M.A., Lee, R., Poulton, A.J., Riebesell, U., Young, J., Ridgwell, A., 2016. Why marine phytoplankton calcify? Science Advances 2. https://doi.org/10.1126/sciadv.1501822.
Pasciak, W.J., Gavis, J., 1975. Transport limited nutrient uptake rates in Ditylum brightwellii 1. Limnology and Oceanography 20 (4), 604–617.
Pasquer, B., Laruelle, G., Becquevort, S., Schoemann, V., Goosse, H., Lancelot, C., 2005. Linking ocean biogeochemical cycles and ecosystem structure and function: results of the complex SWAMCO-4 model. Journal of Sea Research 53 (1–2), 93–108.
Platt, T., Denman, K., 1977. Organisation in the pelagic ecosystem. Helgoländer Wissenschaftliche Meeresuntersuchungen 30 (1), 575–581.
Quere, C.L., Harrison, S.P., Colin Prentice, I., Buitenhuis, E.T., Aumont, O., Bopp, L., Claustre, H., Cotrim Da Cunha, L., Geider, R., Giraud, X., Klaas, C., 2005. Ecosystem dynamics based on plankton functional types for global ocean biogeochemistry models. Global Change Biology 11 (11), 2016–2040.
Ralston, D.K., Moore, S.K., 2020. Modeling harmful algal blooms in a changing climate. Harmful Algae 91, 101729.
Redfield, A.C., Ketchum, B.H., Richards, F.A., 1963. The influence of organisms on the composition of seawater. The Sea 2, 26–77.

Riley, G.A., 1947. A theoretical analysis of the zooplankton population of Georges Bank. Journal of Marine Research 6, 104–113.
Robson, B.J., Baird, M., Wild-Allen, K., 2013. A physiological model for the marine cyanobacteria, Trichodesmium. In: MODSIM2013, 20th International Congress on Modelling and Simulation. Modelling and Simulation Society of Australia and New Zealand.
Rubio, F.C., Camacho, F.G., Sevilla, J.F., Chisti, Y., Grima, E.M., 2003. A mechanistic model of photosynthesis in microalgae. Biotechnology and Bioengineering 81 (4), 459–473.
Sathyendranath, S., Brewin, R.J., Jackson, T., Mélin, F., Platt, T., 2017. Ocean-colour products for climate-change studies: what are their ideal characteristics? Remote Sensing of Environment 203, 125–138.
Shevela, D., Björn, L.O., 2018. Photosynthesis: Solar Energy for Life. World Scientific Publishing.
Skerratt, J.H., Mongin, M., Baird, M.E., Wild-Allen, K.A., Robson, B.J., Schaffelke, B., Davies, C.H., Richardson, A.J., Margvelashvili, N., Soja-Wozniak, M., Steven, A.D.L., 2019. Simulated nutrient and plankton dynamics in the Great Barrier Reef (2011–2016). Journal of Marine Systems 192, 51–74.
Soja-Woźniak, M., Baird, M., Schroeder, T., Qin, Y., Clementson, L., Baker, B., Boadle, D., Brando, V., Steven, A.D., 2019. Particulate backscattering ratio as an indicator of changing particle composition in coastal waters: observations from Great Barrier Reef waters. Journal of Geophysical Research: Oceans 124 (8), 5485–5502.
Soja-Woźniak, M., Darecki, M., Wojtasiewicz, B., Bradtke, K., 2018. Laboratory measurements of remote sensing reflectance of selected phytoplankton species from the Baltic Sea. Oceanologia 60 (1), 86–96.
Soja-Woźniak, M., Laiolo, L., Baird, M.E., Matear, R., Clementson, L., Schroeder, T., Doblin, M.A., Suthers, I.M., 2020. Effect of phytoplankton community size structure on remote-sensing reflectance and chlorophyll a products. Journal of Marine Systems 211, 103400.
Steele, J.H., Henderson, E.W., 1981. A simple plankton model. The American Naturalist 117 (5), 676–691.
von Liebig, J., 1840. Chemistry in Its Application to Agriculture and Physiology. Taylor and Walton, London.
Ward, B.A., Follows, M.J., 2016. Marine mixotrophy increases trophic transfer efficiency, mean organism size and vertical carbon flux. Proceedings of the National Academy of Sciences 113 (11), 2958–2963. https://doi.org/10.1073/pnas.1517118113.
Wild-Allen, K., Herzfeld, M., Thompson, P.A., Rosebrock, U., Parslow, J., Volkman, J.K., 2010. Applied coastal biogeochemical modelling to quantify the environmental impact of fish farm nutrients and inform managers. Journal of Marine Systems 81 (1–2), 134–147.
Wirtz, K.W., Eckhardt, B., 1996. Effective variables in ecosystem models with an application to phytoplankton succession. Ecological Modelling 92 (1), 33–53.
Yool, A., Popova, E.E., Anderson, T.R., 2013. MEDUSA-2.0: an intermediate complexity biogeochemical model of the marine carbon cycle for climate change and ocean acidification studies. Geoscientific Model Development 6 (5), 1767–1811.

CHAPTER 9

Managing the societal uses of phytoplankton: technology applications and needs

Andrew D.L. Steven
CSIRO Oceans and Atmosphere, Queensland Biosciences Precinct, University of Queensland, Brisbane, QLD, Australia

9.1 Introduction

In this chapter, we consider how many of the technologies that are discussed in this book can be applied to address the multifaceted and evolving societal uses of phytoplankton, including both their uses and their consequences, to ensure better, more responsive management.

Today, management agencies, confronted by a growing number of complex issues affecting phytoplankton dynamics, are increasingly looking to science and technology to better understand, monitor, forecast, and find solutions to issues that range from: the consequences of changing climate conditions on phytoplankton and the many ecosystems dependent upon them; mitigating the pervasive increase in frequency and severity of harmful algal blooms (HABs) and their impacts on industries and the safety of food and drinking water upon which we rely; how to regulate established industries, such as fishing, to ensure they are both sustainable, yet remain profitable; and emerging industries seeking to harness the unique capabilities of phytoplankton to help address some of the world's most pressing challenges, including alternative energies, food production, waste remediation, and the production of high-value products.

This chapter takes a necessarily broad view of phytoplankton management needs, as it does in considering which emerging technologies might be of benefit, and what gaps remain to be addressed. The technical description of these technologies is addressed by many of the other chapters in this volume, notably for in situ monitoring (Chapters 5 & 6) remote sensing (Chapter 7), and modeling (Chapter 8). Here, this chapter focuses on the following: (a) summarizing current and emerging societal applications of phytoplankton and the issues that arise and require management; (b) considering the range of emerging technologies in terms of their potential management application, and (c) illustrating six societal benefit areas that rely on phytoplankton, with case studies from around the world that highlight how these technologies are being used to address specific management issues.

9.2 Societal uses of phytoplankton and their management

Societal uses of phytoplankton and management context

The many spectacular and biodiverse ecosystems that occur globally across open ocean and coastal marine waters to freshwaters in rivers and lakes, all depend upon the productivity of taxonomically diverse phytoplankton communities at the base of their food chains. This productivity is also the basis for the many goods and services phytoplankton provide within the framework of nature's contribution to people (Díaz et al., 2015). Though goods such as fisheries and clean water are well-known, there is growing recognition of the key role phytoplankton play in the provision of regulating services such as oxygen production, carbon sequestration, and biogeochemical cycling upon which we—and planet earth—rely. As humanity grapples with finding sustainable solutions to addressing growing global demand for food, potable water, and clean energy, and addressing the climate crisis, novel uses of phytoplankton, including the development of alternative feedstocks for aquaculture and energy production, bioremediation, and the production of high-value pharma- and nutraceutical products, are being developed and scaled-up (Sutherland et al., 2021).

Though the great majority of human interactions with phytoplankton are positive and beneficial, the occurrence of HABs—although they can occur naturally—are increasing globally and require dedicated prevention, mitigation, and control management strategies to ensure safe food, drinking and bathing waters, and minimize economic and public health consequences.

Although the management of HABs has been the primary focus of responsible authorities over at least the last forty years, the modern management context today is far more complex, as it contends with managing *inter alia* the consequences of accelerating coastal and offshore development and resource use, increasing eutrophication (not only from adjacent landuses, but a growing aquaculture industry), shipping ballast water introductions of phytoplankton species, and climate change pressures (leading to warmer and more acidic waters that are changing the distribution and diversity of phytoplankton).

These issues can often interact in complex and hard-to-predict ways, resulting in a higher likelihood of negative impacts, which can span local to regional to global scales, and require technological innovation across a range of fields to inform the development of effective policies and sustainable, socially acceptable operations. Key challenges faced by management are to discriminate between natural phytoplankton dynamics, and against a backdrop of a changing climate, and those resulting from human impacts, situations which made more difficult where there are time-lags or reciprocal (i.e., hysteresis) feedbacks (Hoagland et al., 2014).

Management needs and approaches

A number of international conventions directly relate to the management of phytoplankton, including for the management of ballast water (*International Convention for the Control and Management of Ships Ballast Water and Sediments*) and food standards, such as the *Codex Alimentarius*. Indirectly, conventions such as the *United Nations Framework Convention on Climate Change* (UNFCC), the *Convention on Biological Diversity* (CBD), and the *Nagoya Protocol on Access and Benfit Sharing*, also shape the way in which the various uses of phytoplankton and their consequences are managed.

At national and subnational levels, a bewildering array of legislation can be enacted, for example, to protect the environment or human health, or to promote or protect certain industrial sectors, but these vary greatly in terms of the standards set, how they are enforced, and who pays. For example, in the European Union, the *Water* and *Marine Strategy Framework* Directives (WFD Directive 2000/60/EC, and MSFD Directive 2008/56/EC, respectively) legislate environmental standards for fish and marine waters and compel member states to undertake monitoring of specified "harmful" marine phytoplankton species. In the United States, federal government acts, such as the *National Environmental Policy Act* (NEPA, 1969), the *Clean Water Act* (CWA, 1972) the *Marine Protection, Research, and Sanctuaries Act* (MPRSA, 1972), and the *Ocean Dumping Act* (1988), are enacted to protect and maintain the marine environment from a range of human activities.

As it is beyond the scope of this chapter to detail specific management actions taken by individual countries, below is an overview of the general principles and approaches taken by management, and how this influences their preference for, and adoption of, particular technologies.

Responsibility for the management of these legislative requirements can reside with a number of agencies, including—most commonly—environmental protection agencies, fisheries, water management and health authorities. These agencies exercise responsibilities according to their specific mandate and jurisdiction, and are sometimes overlapping or contradictory. These responsibilities might include ensuring the condition of a designated waterbody by managing activities that occur in and adjacent to it, or to sustainably manage a specific resource (e.g., fisheries, water impoundments), or to set environmental quality standards, and are implemented most commonly through the development of policies and management plans, and the regular conduct of monitoring, reporting, and evaluation protocols to assess the effectiveness of management interventions (see next section) and to assess operational performance by various industries with designated compliance standards. Often, responsibility for the day-to-day management of water quality and responses to HABs and their consequences, such as beach closures and fish-kills, is further devolved to local government bodies.

Though many government agencies have responsibility only for a single issue, sector, or species, increasingly integrated ecosystem-based management (EBM) approaches are being adopted to coordinate multisector management, address cumulative impacts, and deliver triple bottom-line outcomes. EBM provides a best-practice approach to managing natural resources by including humans and their sustainable use of the environment as a central part of the ecosystem and by linking management objectives and actions to ecosystem health through appropriate indicators (Borja et al., 2013; Smit et al., 2021).

Although effective management strategies aim to minimize the related socioeconomic costs by being accurate, appropriate for the affected area, and fast-acting, they still need to understand the underlying cause of particular phenomena to deliver effective management strategies. They are also seeking to develop better forecasting of future conditions on likely impacts to be able to respond proactively to prevent or mitigate more effectively any potential impacts.

A risk-based management approach is often taken to managing hazards, and there are several points at which management actions may be implemented. For example, in the generic case of responding to algal blooms, three types of management intervention can be progressively implemented. Firstly, *prevention* refers to actions taken to keep HABs from happening, or from directly impacting a particular resource. *Mitigation* activities are undertaken to deal with an existing or ongoing bloom, and to reduce negative impacts, and doing so can include monitoring for, and forecasting of, the presence of blooms and alerting the public to potential hazards. Finally, *control* strategies seek to apply technological solutions to suppress or eliminate blooms, using mechanical, biological, chemical, genetic, and environmental control methods, but they can be the most challenging and costly aspect of HAB management (Yu et al., 2018).

Though increasing levels of response can lead to ever-greater benefits, there may be diminishing returns in terms of reducing any hazard-related damages. Thus it is important to quantify the economic impacts of HABs to identify and implement effective and cost-efficient mitigation strategies in the management of these coastal hazards. Such information helps managers (and politicians) understand the benefits of these programs and can also inform private investment decision-making in different sectors by improving the characterization of relevant financial risks.

The growing number of novel uses of phytoplankton for production of algal feedstocks, products for human consumption and use, bioremediation and energy production also require the development of policies that on the one had provide the right techno-economic incentives to grow these industries to scale, but need to be balanced by effective safeguards that protect the surrounding environment from inappropriate resource exploitation, release of contaminants, or genetically modified organisms, and to ensure the rights of affected human populations, including first nations people.

Monitoring programs, indicators, targets, thresholds and standards

The maintenance of, or recovery to a specified environment quality standard (EQS) is designated in environmental policies of many nations, and all broadly seek to protect or mitigate activities that diminish the integrity of aquatic ecosystems and the services they provide. These conditions are most commonly referred to as either good environmental (ecological is also used in some policies) status (GES), or condition (GEC), although other terms such as "health", "state", "quality" can be used synonymously (Smit et al., 2021). Assessing the current environmental status usually takes an integrative ecosystem-based approach, employing a number of indicators diagnostic of both pressures and condition and requires a comparison between current conditions and an established reference/baseline/threshold/trigger, with any difference attributed to anthropogenic activities (Borja et al., 2012).

A condition of many of these regulations is the requirement to monitor and report on the condition of designated areas relative to set environmental standards, or for industries to report on their operational performance against compliance criteria. The extent of these monitoring programs varies greatly in relation to their objective(s), how comprehensive they are, what technologies they use, and what standards they benchmark or report against. The important design elements of a monitoring program will include the following: (a) determining the relevant indicators and standards; (b) defining the relevant geographic scales and seasonal trends that need to be accounted for; (c) the number of sites and frequency of monitoring to be undertaken, and (d) the analyses reporting and communication protocols that need to be implemented. The cost and amount of effort involved for sample collection, analysis, coordination of sampling, management of data, and other associated activities are significant practicalities that also influence the design and implementation of monitoring programs. Whatever the monitoring and management strategy, a key component will always be the timely communication of observations and results to relevant stakeholders.

Indicators are quantifiable metrics that are used to monitor the status of ecosystems and their response to environmental perturbations and most commonly include measures of diversity, abundance or concentration, and extent of various taxa, including phytoplankton (Smit et al., 2021; Borja et al., 2012; McQuatters-Gollop et al., 2019). For ecosystems requiring remediation, targets (aspirational timebound endpoints that represent attainment of GES) are also being increasingly set to drive coordinated management actions.

Some indicators trigger management action when a threshold is reached, whereas others play an interpretive, or surveillance, role in informing management and are effective tools for summarizing and communicating key aspects of ecosystem condition. Often indicators can be aggregated into an index which usually consists of multiple components and are expressed as a single numerical value, for example, the ocean health index (Halpern, 2020; Halpern et al., 2012).

Standards, triggers, and thresholds are values that define the upper limit of what are considered natural levels for indicators and are typically developed for individual classes of water bodies. They are usually designated within policies for compliance and attainment purposes, and exceedances may trigger further monitoring or management action. Changes in these indicators beyond set standards are diagnostic of phytoplankton blooms and can initiate additional monitoring and research, or trigger management interventions.

Phytoplankton biomass, as well as the occurrence of particular sentinel taxa or species, are commonly used indicators of environmental condition, and their monitoring is a key component of most routine monitoring programs. Chlorophyll concentration is the most common metric of phytoplankton biomass and is often used as a proxy of nutrient, or trophic status, and as part of broader ecosystem condition assessment. This is in large part due to established relationships between chlorophyll concentration and concentrations of the macronutrients' phosphorus – limiting to phytoplankton growth in freshwaters –, and nitrogen – limiting mainly in marine waters. Thus the trophic status of waters is often classified according to their chlorophyll average concentration ($\mu g\ L^{-1}$), which for fresh and marine, respectively, being either oligo- (8, or <0.1), meso- (8–25, 0.1–1), or eu-trophic (>5, >1). Chlorophyll has a number of other advantages over the direct measurement of nutrients, including requiring minimal processing and storage in the field, relatively cheap analysis costs, and the availability of in situ fluorescence detections methods, or satellites, that can directly measure chlorophyll concentrations. The detection and quantification of specific phytoplankton taxa and species are also measured as an indicator of diversity, or for surveillance of HABs. The fluorescence spectra from specific accessory pigments that are diagnostic of particular phytoplankton taxa, (e.g., Phycocyanin in Cyanophytes) and the derivation of phytoplankton functional types (PFTs) from satellite data have enabled routine measurement of these parameters. Less commonly measured is primary production, a first-order index of local ecosystem carrying capacity, which is logistically difficult to measure in the field, though a number of satellite products indirectly estimate phytoplankton productivity from established relationships with chlorophyll.

Of course, the challenge of determining what is good environmental condition is now very much more complicated by the significant shifts in species distributions and ecosystems that are occurring in response to climate change. Understanding how phytoplankton will respond in a future warmer climate, typified by increased sea surface temperatures and water column stratification, and being able to partition these effects from those that can be attributed to other putative anthropogenic impacts require long-term consistent time-series (Wells et al., 2020; Griffith and Gobler, 2020; Glibert, 2020; Townhill et al., 2019). Here, models are being used to make future mid- and end-of-century projections of oceanographic and phytoplankton parameters, and are often based on Intergovernmental Panel on Climate Change (IPCC) representative concentration pathways (RCP) scenarios.

9.3 Emerging technologies for management applications

The emerging technologies—technologies whose development and practical applications, are still largely unrealized—used to advance the understanding of phytoplankton biology and ecology and improve their detection and enumeration spans the use of omic-based technologies through to the development of satellite remote-sensing products and models. In some cases, these technologies are still very much emerging, such as many transcript- and metabolomics, whereas others (e.g., satellites) are more evolutionary as they improve in resolution, accuracy, scalability, performance, and cost.

Many of the technologies being successfully applied to phytoplankton detection, identification, and quantification have, like nearly all fields of environment science, leveraged rapid advances in platform technologies, including notably metagenomic high throughput sequencing (HTS) for genomic characterization and novel marker identification. Likewise, information and communications technologies (ICT) technologies, from sensor automation and communication to analysis of large volume of data with machine learning (ML) approaches, have revolutionized the ability to communicate information in near real-time, to "mine" relationships from massive datasets, and to distil and deliver information that is readily interpretable by management. The most powerful approaches have come from the synergy of methodologies applied to issues including the detection of harmful algal blooms and the development of improved microalgal strains for biotechnology applications.

The needs of management to better detect, prevent, and mitigate HABs has been a primary impetus over the last thirty or so years for instrumentation development for monitoring applications (Anderson et al., 2012). Major global research programs on HABs, such as the Global Ecology and Oceanography of Harmful Algal Blooms, GEOHAB (Kudela et al., 2017) and its successor, GlobalHAB (Berdalet et al., 2018), have been very successful in progressing our understanding of these complex phenomena and developing new approaches and technologies for research and diffusion to management applications. Likewise, coordination of regional and global ocean observing systems, compilations of data into publicly accessible databases and access to free, full, and open satellite archives have greatly advanced adoption by management agencies.

The remainder of this section summarizes the criteria considered by management agencies in their adoption of technologies and overview issues relating to three classes of technology used in the detection and characterization of phytoplankton and the antecedent environmental conditions that may lead to their proliferation or spread: in situ sensors, earth observing technologies, and management models. The technical details of these technologies are described in Chapters 5 to 8 of this book, whereas detailed reviews of emerging technologies relevant to HABs are provided by Doucette et al. (2018) and Glibert et al. (2018). Importantly, how the combinations of these technologies are being used to better inform operational decision-making are considered.

Management criteria for monitoring and technology application

The adoption of emerging technologies by management involves consideration of many factors, many of which are summarized below:

Fit for purpose: Does the technology provide relevant information and meet defined performance requirements that align with the desired outcome or purpose? Does the instrument make measurements at the spatial and temporal resolution required to correctly characterize the dynamics, and to not lead to erroneous conclusions resulting from type I (false positives) or II (false negative) errors?

Maturity: Does the technology meet the criteria for operational deployment identified through a technology readiness assessment which grades the technology on a technology readiness level (TRL) scale from 1 to 9, with 9 being the highest and indicating that the technology is proven in an operational environment?

Operationally deployable: Is the technology able to be deployed operationally and safely by management personnel? Can it operate in remote or sometimes extreme environments? What are the power and telecommunications requirements? Is it likely to drift or easily foul, or is it prone to damage or theft?

Information interpretable and translatable to decisions: Is the information provided readily interpretable by management? Is it a direct measurement, or if it is a proxy, is the relationship with the direct parameter of interest well quantified for the region of interest?

Timeliness: Can the information be communicated in near real-time (NRT) to allow decision-makers to make timely decisions?

Cost-effectiveness: Does the technology provide improved information for decision-making that outweighs both the capital cost and the ongoing operational costs of deployments and analysis? Can it be integrated with other existing technology?

In situ monitoring

Although many management agencies still often rely on manual sampling and microscopy for identification and enumeration of individual phytoplankton species, in situ sensor technologies are increasingly being adopted, as the benefits to management become apparent. Such instruments can provide high-frequency monitoring that cannot be easily attained by manual sampling, minimize human error and contamination, and provide automated detection and near real-time reporting that improves the responsiveness of management agencies.

The use of single- and multi-parameter sondes are now routine in many monitoring programs. In addition to salinity and temperature, most sondes will provide a fluorescence-based detection of chlorophyll, and in some cases a phycocyanin sensor for the detection of cyanobacteria concentrations.

In the last decade or so, development of in situ molecular and imaging instrumentation have been used mainly in research contexts to better understand plankton dynamics, from food web characterization to harmful algal bloom dynamics (Spanbauer et al., 2020). Despite the demonstrated utility of commercially available automated imagery instruments, such as Imaging Flow Cytobot (IFCB) and ecogenomic sensors, such as environmental sample processor (ESP), which are respectively discussed in detail in Chapters 5 (Campbell et al., 2022) and 6 (Smith et al., 2022)—with the exception of some US agencies—there has been only limited adoption by management agencies. Their relatively high cost and operational complexity are the key impediments to wider adoption into monitoring programs for in situ detection of phytoplankton and characterization of their and dynamics. However, future advances—leveraging medical diagnostic improvements in microfluidics and the development of lab-on-a-chip technologies—as well as the integration of analytical results with existing data streams from other in situ sensors will hopefully lead to affordable, readily deployable sensors that can be adopted for aquatic management applications.

Satellite and drone remote sensing

The need for synoptic, systematic, and cost-effective monitoring of surface waters has led to notable advances in satellite technology, with a spatial coverage impossible to achieve from in situ sampling. Today, satellites are routinely applied to the detection, mapping, and monitoring of chlorophyll, as well as important ancillary parameters, such as sea surface temperature (SST) and sea surface height (SSH). Improvements in spatial, temporal, and spectral resolution of many satellite sensors, the development of algorithms that can more accurately quantify chlorophyll in coastal waters and significant new capabilities for the discrimination of phytoplankton groups (PG), and composition using phytoplankton size classes (PSC), have greatly improved routine monitoring and forecasting of phytoplankton dynamics. With near to subdaily revisit times now a reality, satellite information can allow for more nuanced decision-making; for example, showing where blooms begin to form and the trajectories and patterns in which are they are likely to spread (Stroming et al., 2020). The use of satellite information for validation of, or assimilation into, process models has also greatly improved their forecasting skill (see next section).

The uptake of satellite data by management agencies and the provision of bespoke services have been enabled by public availability of multiple sources of satellite imagery with phytoplankton data and products now available through platform services, such as the European Commission's Copernicus Marine Environment Monitoring Service (CMEMS; https://resources.marine.copernicus.eu/) and the National Ocean and Atmospheric Administration (NOAA; https://www.ncdc.noaa.gov/data-access/marineocean-data).

The availability of space-based hyperspectral sensors, such as National Aeronautics and Space Administration (NASA) Plankton, Aerosol, Cloud, Ecosystem (PACE) mission in 2022, will further improve the capability to make broad taxonomic distinctions at lower phytoplankton biomass than with multispectral sensors.

Despite these advances, satellites still have a number of limitations that hinder their application to a broader range of management situations. Insufficient spatial resolution remains a major limitation for monitoring smaller inland and estuarine waterbodies. Further improvements in determining phytoplankton parameters in optically complex coastal waters are required, while interference from cloud cover remains a significant constraint for ocean-color sensors (Coffer et al., 2018; Schaeffer and Myer, 2020). Ocean-color also has little application when HAB abundance is low relative to the broader phytoplankton assemblage, or when HABs are in thin layers or other subsurface aggregations (Raine et al., 2018).

The advent of autonomous vehicles (drones), mounted with high-resolution multi- and hyperspectral cameras, has greatly enabled targeted aerial reconnaissance and detection of HABs, which is not possible with satellites. Given their affordability, drones are now being utilized by management authorities for water quality and algal bloom monitoring (Becker et al., 2019; Wu et al., 2019).

Management models

A variety of process and increasingly sophisticated statistical models (discussed in the next section) are being developed and applied operationally, both globally and regionally to assess environmental status, test management scenarios, provide HAB and fisheries forecasts, and make projections of how future climate scenarios may induce shifts in planktonic communities and their productivity.

Baird et al. (2022) (Chapter 8) provide a detailed review of advances in modeling key phytoplankton processes that are used to differentiate plankton and provide a number of applications. Here, the focus is on management models (operational systems that are used to guide management decision-making) and may operate in several modalities, to provide near real-time information, short-term (2–3 days typically) forecasts of environmental conditions, or to undertake scenario assessments.

Typically, such models are run by operational agencies, and are built upon a spatially explicit hydrodynamic model coupled with a biogeochemical model that can range in complexity from simple to sophisticated multicompartment representations of nutrient cycling and phytoplankton production. Such models rely on information from a number of observational sources to force and validate the model; both in situ and satellite-derived data can be assimilated to improve model predictions and forecasts. Sometimes, ensembles of different models, or cases, are run to improve the statistical robustness of the predictions or forecasts.

Progress in forecasting has been advanced in recent years by the development of region-specific models, as well as short-term (once or twice weekly) and seasonal forecasts to predict the severity of blooms and to help minimize any negative economic and health consequences. Through the HAB operational forecasting system (HABOFS), NOAA supports a number of regional HAB forecasting systems and issues regular forecasts (see Case study 9.2).

Models may also be used to interrogate how a system may respond to future changes, whether it be climate change, a nutrient reduction action, or evaluation of the risk to a developing aquaculture region. Scenario evaluations to test the effect(s) of specific management actions (e.g., targeted nutrient reduction) can be quantified in terms of effects on the plankton community biomass (as chlorophyll) or composition (usually size classes and sometimes specific species. Such models can be used to provide scenario outcomes to such questions as the following: (a) If nutrients are reduced by x amount, what will be the phytoplankton response?; (b) What will be the comparative effect of reducing N versus P? (Franks, 2018). Case study 9.1 summarizes an information system, *eReefs*, used for the Great Barrier Reef in Australia to routinely test scenarios in relation to proposed nutrient reductions, and to produce model output that is incorporated into annual ecosystem condition assessment reports.

As the sophistication of such models has increased and the datasets used to evaluate them have expanded, the metrics by which their skill can be assessed have received more attention. Future advances in modeling will occur at this junction of models and data, where integrating data from traditional and emerging techniques into process models has the potential to rapidly improve knowledge of processes and enable more accurate risk assessment and mitigation strategies Franks, 2018; Glibert et al., 2010; Ralston and Moore, 2020; Flynn and McGillicuddy, 2018.

ICT applications

The utilization of various platform information and communications technologies (ICT) continue to enable the ongoing development, operation, and integration of many in situ and satellite sensors, and models that are currently in use, allowing the transmission and management of massive environmental datasets, and providing timely information to managers needing to make operational decisions. Such "smart" monitoring is now in common use, for example, for the management of drinking-water-supply processes, HAB forecasting, detection of infrastructure failure, and public health protection (Park et al., 2020). The ability to share and merge the data generated with data streams from other meteorological and environmental sensors (the Internet of Things (IOT)) further enhances the utility these systems in surveillance, monitoring, and modeling. More generally, open data principles, such as FAIR (findability, accessibility, interoperability, and reusability) have democratized access to much of this data and is also transforming the way scientists and managers can combine data and undertake global analyses.

Advanced data analysis, notably machine learning techniques that include support vector machines and artificial neural networks (including convolutional neural networks), are now widely used to analyze and "mine" relationships from the large data-volumes available from in situ and satellite sensors and to develop more accurate and efficient algorithms (Yu et al., 2020; Hill et al., 2020). Such statistical models are finding wide application in freshwater operational HAB forecasting in China (Hu et al., 2020), Korea (Lee and Lee, 2018), and the USA (Liu et al., 2020).

The development of global databases such, as the harmful algal event database (HAEDAT, http://haedat.iode.org) that record HAB events and their associated management actions and economic/ecological impact, and HABMAP-OBIS (a database on the geographic range of harmful algal species, http://www.iobis.org) have enabled the comprehensive regional- and national-scale analyses of regional timeseries of HAB occurrences and their consequences, and have elucidated the underlying causes.

The technical challenges confronting managers, in accessing satellite information have largely been overcome with the implementation of data cubes and analysis ready data (ARD) that allow rapid analysis through timeseries of imagery (Malthus et al., 2019). Likewise, information systems and visualization portals are increasingly common, allowing users to access relevant information, products, and services without having to "peer under the hood" of the models, sensors, and interoperable data protocols servicing these requests. The delivery of public and commercial services, such as fishing or HAB forecasts, to portable devices, such as smart phones, has done much to communicate information to those industry sectors making day-to-day operational decisions, and the provision of health and safety information bulletins to the general public, such as whether recreational water areas are safe for swimming (Mishra et al., 2020; Hill et al., 2020).

9.4 Management applications and technology needs

This section reviews six societal benefits we derive from phytoplankton, and considers the management context, uptake of emerging technologies to date, and future needs related to each: (1) phytoplankton as indicators of environmental status; (2) the management of HABs and their food and human health consequences; (3) management of shipping ballast water; (4) achieving more sustainable fisheries and aquaculture; (5) industries, and

(6) biotechnology applications of phytoplankton as sustainable alternatives for energy, aquaculture feeds, and waste remediation. Case studies from Australia, Chile, India, and the United States of America provide relevant examples of management approaches.

Managing ecosystem conditions

Phytoplankton are regarded as a key ecosystem condition indicator and are most commonly related to measuring changes in the nutrient-status of water bodies. There is also a growing recognition of the ecosystem provisioning and regulating services phytoplankton provide by supporting production of higher trophic levels and as a CO_2 sink, and that consequently, management needs to give greater consideration to the potential impact of human activities on these services (Tweddle et al., 2018).

Chlorophyll concentrations, phytoplankton diversity, and the occurrence of designated nuisance species are the primary indicators (Section 9.2) measured in many monitoring programs with varying levels of sophistication. Though many monitoring programs still use manual sampling methods, there is increasing adoption of fluorescence-based in situ sensors, and the use of satellites to measure chlorophyll and some phytoplankton functional types. This data can be used to establish initial baseline conditions for the ecosystem, and subsequently to routinely assess and report changes in condition. Management models are also now being used to assess and report ecosystem condition and to test the efficacy of potential management interventions, or progress in attaining management targets, such as whether nutrient reduction strategies are resulting in improved environmental status (see Case study 9.1).

Given the broader set of management issues to be addressed, and greater emphasis now on assessment and reporting, there is a need to move beyond simplistic and token monitoring programs to ones that can adequately discern cumulative impacts and differentiate climate variability and change from those directly related to use of the aquatic environment.

The challenge of determining what is good environmental condition is of course complicated now by observed climate change-induced shifts in phytoplankton distribution and abundance patterns, which confounds attribution to singular putative human development and resource use pressures (Glibert, 2020). Modeling phytoplankton responses to climate change will likely expand as resource managers and policy makers increasingly demand projections at both near-term and longer time scales, to inform the development of strategies to reduce socioeconomic and public health impacts, and to increase resilience of socioecological systems. Extending models to decadal time scales or longer, extrapolating to forcing regimes that are outside historical observations, representing potential regime shifts in the dominant processes controlling HAB development, and incorporating uncertainty and variability in physical climate model projections are challenging but feasible tasks (Ralston and Moore, 2020).

Case Study 9.1: Great Barrier Reef: assessing ecosystem conditions and managing nutrient reductions

The UNESCO world heritage listed Great Barrier Reef (GBR) is the world's largest coral reef ecosystem, extending over 2300 km along the Australian coast, and is Australia's most prized environmental asset and a key contributor to its economy. However, the GBR is faced with a number of potentially cumulative and hard-to-predict threats that challenge its long-term resilience, including climate change, agricultural pollutants, and other forms of reduced water quality. At a policy level, the *Reef 2050 Plan* and *Reef Water Quality Protection Plan* set ambitious end-of-catchment load reduction targets for nitrogen and phosphorus (~80%), and sediments (~50%), to be achieved by 2025, through improvements in on-farm management practices, and assessed through reporting of end-of-catchment annual loads.

eReefs is a comprehensive interoperable information platform (Steven et al., 2019b), built upon an integrated system of data, catchment, and marine models, visualization, reporting and decision support tools, that has been developed and operated for the GBR region to provide users with access via a portal to improved environmental intelligence (http://portal.ereefs.info). *eReefs* is used to provide yearly assessments of water quality state of the GBR during the preceding reporting period. Scenario modeling is also undertaken to assess if nutrient loads reduction targets are likely to reduce chlorophyll concentration in GBR waters. The most recent analysis over the 2011–2018 period showed that chlorophyll concentration in 16 river plumes was reduced by 0.01–0.10 mg m^{-3} demonstrating, quantitatively, the benefits of improved catchment management on GBR health (Baird et al., 2021). *eReefs* has also been used to investigate planktonic and nutrient responses to climatic variation, showing that during La Niña episodes, when greater than average winds, rainfall, and river discharge occur, dissolved inorganic nitrogen, primary production, and phytoplankton biomass concentrations increase in inshore and mid-shelf waters (Skerratt et al., 2018).

Management of harmful algal blooms

Harmful algal blooms occur in both fresh and marine waters, and as the name implies, can have negative consequences for both the associated ecosystems, marine vertebrates (e.g., fish, birds, reptiles, and marine mammals), and the socioeconomic benefits we derive. Thus HABs can be considered a societal disbenefit with impacts ranging from loss of amenity and aesthetics, loss of income for economic sectors (including commercial fisheries and aquaculture, recreational fisheries and coastal tourism), to significant public health impacts, such as food and water poisoning. These direct effects can propagate through regional economies, causing negative impacts to other sectors (Adams et al., 2018).

Management of HABs is a challenging task in changing, often unpredictable coastal, marine and freshwater environments. Though HABs have always been a natural phenomenon, over the last thirty years, regional analyses demonstrate an increased prevalence of HABs in nearly all coastal waters of the world (see *Harmful Algae* special issue on global harmful algal blooms status reporting, volume 102, 2021). Similar analyses of freshwater lakes and reservoirs over a 30-year period, using earth observation, demonstrate that the frequency and severity of blooms have also increased, and also highlights the changes wrought by a changing climate and the ambiguity that arises in interpreting long-term trends and attributing them to putative and manageable human activities (Ho et al., 2019).

Rapid urbanization, growth of the aquaculture industry, and increased global shipping traffic (see following sections for more detailed discussion) are some of reasons that have led to more widespread and frequent HABs and species introductions. But monitoring and surveillance has also significantly increased over this time, most notably in the aquaculture industry, and a recent analysis of the HAEDAT database found that after normalizing data for monitoring effort, there were no discernible global trends, and contrasting regional trends resulted from differences in bloom species, type, and emergent impacts (Hallegraeff et al., 2021).

From a management perspective, HABs are a hazard, the risk of which depends upon their (1) proximity to human populations, (2) duration and frequency of occurrence, (3) cell densities and toxicities, and (4) weather conditions (Adams et al., 2018). As discussed in Section 9.2, preventative, mitigation, and control methods can be employed. A common prevention strategy is to control the supply and availability of nutrients, achieved through and the designation of nutrient reduction measures and the setting of environmental compliance standards. The ability to prevent or mitigate the adverse impacts of harmful algal blooms (HAB) on humans, wildlife, fisheries, and ecosystems, as well as to identify the environmental factors driving HAB population growth and toxicity, is based largely on early detection of causative organisms and their toxins (Doucette and Kudela, 2017). However, once a bloom is established, adequate control and mitigation technologies are needed to stop its rapid proliferation using a range of physical, chemical, or biological methods, and to mitigate potential harmful effects, especially in critical locations, (e.g., freshwater reservoirs, aquaculture sites, recreational bathing waters).

To better anticipate and respond to HAB events, management efforts across many countries have implemented operational HAB monitoring and forecasting systems, for both fresh and marine waters, including in China (Hu et al., 2020), Korea (Lee and Lee, 2018), Australia (Malthus et al., 2019), South Africa (Smith and Bernard, 2020), Chile (Sandoval et al., 2018), and the USA (Anderson et al., 2019). Individual forecasting systems differ according to their regional needs, but usually combine information from in situ monitoring, satellites, and management models to provide forecasts that can be used to inform management and issue public information bulletins. Case study 9.2 provides a summary of some of the regionally based operational HAB monitoring and modeling in the USA.

Though there have been significant improvements in the technologies for the detection and management of HABs, a lack of epidemiological studies, a scarcity of studies of chronic exposure, and the need for a greater understanding of the likely consequences of climate change are cited as key gaps in the evidence base needed to better inform management of HABs and human-health outcomes (Young et al., 2020).

Case Study 9.2: Operational HAB monitoring and modeling in the USA

The National Oceanic and Atmospheric Administration (NOAA) supports a HAB operational forecast system for several regions of the conterminous USA (https://tidesandcurrents.noaa.gov/hab/faq.html). Different types of HABs have different suites of models and measuring/monitoring tools and data that populate these models. In the Gulf of Mexico, nowcasts and short-term forecasts of *Karenia brevis* blooms are based on satellite imagery, field observations, models, public health reports, and buoy data to provide the large spatial scale and high frequency of observations. Information is conveyed regarding potential development, intensification, transport, and associated impacts of *K. brevis* blooms. Similar operational forecasting models have been developed or are a priority for other regions, including Lake Erie (cyanobacteria), Gulf of Maine (*Alexandrium fundyense*), Washington Coast (*Pseudo-nitzschia*), California coast, and Chesapeake Bay. A challenge for many of these regions and the development of operational models is that they routinely experience multiple different types of HABs.

Managing ballast water introductions

Ships' ballast water and sediment in ballast tanks are an important vector for the transportation of nonindigenous species (NIS) of phytoplankton and other marine pest taxa between different regions of the world, and once established can result in significant ecological, societal, and economic impacts (Hulme, 2021; Anton et al., 2019). Being small, ubiquitous, and in some cases, capable of surviving in the dark, phytoplankton can be successfully transported via ballast tanks to other ports, where their introduction poses significant risks to the health of ecosystems and industries, such as fisheries, aquaculture, and tourism (Khandeparker et al., 2020; Casas-Monroy et al., 2015). Though the potential of ballast water to act as a major introduction vector for toxic phytoplankton is beyond doubt, the evidence linking suspected introductions with source populations is less convincing, with only a relatively small number of ballast water introductions being recognized to date (Gerhard and Gunsch, 2019).

Ballast water exchange (BWE) has been the key strategy used to minimize such introductions, but its efficacy is highly variable with studies showing that depending on ship type, method, and location of exchange, and type of organism, BWE purges only 70–100% of planktonic organisms, but is less successful for smaller phytoplankton (nominally 10 $<$ 50 μm) phytoplankton. BWE is also considered most effective in freshwater, due to osmotic shock, and least effective for vessels conducting coastal exchange (50–20 nautical miles offshore), where plankton densities can be higher than offshore (Ardura et al., 2020; Shang et al., 2019; Gerhard and Gunsch, 2019).

The regulatory landscape for ballast water management changed significantly in 2017 when the International Ballast Water Management Convention (IBWMC) entered in force, compelling ships engaged in international transits to implement a ship-specific ballast water management plan, and to manage their ballast water and sediments to a numeric discharge standard (the D-2 standards), which limits the density of organisms released to recipient ecosystems to less than 10 viable organisms. Full global compliance with these discharge standards are required by 2024, when vessels are to have installed on-board multistage ballast water treatment systems, such as filtration and electro-chlorination, to remove large particles and chemically inactivate smaller particles (https://www.imo.org/, last accessed 13th June 2021). These standards have been similarly adopted by legal instruments governing ballast water transport at many national levels.

The monitoring methods employed to check compliance with IBWMC requirements requires substantial further development and operationalization if they are to harness the rapid diagnostic advantages of molecular and metabarcoding methods. Microarray and multiplex PCR techniques that allow for rapid detection of multiple microorganisms have been developed and trialed, but their application in regulatory contexts (e.g., testing compliance with numerical discharge standards) remains a developing field. Phytoplankton diversity studies employing such methods have demonstrated, particularly in freshwater, greater diversity and heterogeneity of phytoplankton than identified with traditional microscopy methods (Chen et al., 2016). But to date, most studies have been limited to relatively small sample sizes; the largest recently published effort utilized 41 ballast samples from 4 international ports. Quality assurance steps are required to establish confidence in such detections in Asia, Africa, and North America (Gerhard and Gunsch, 2019). As ballast water transport is likely to continue as a vector for the translocation of aquatic species in the foreseeable future, the adoption of HTS approaches will likely grow with improvements in reference databases, sequencing technology, bioinformatic processing, and understanding of the ecology of environmental DNA and the development of standard operating procedures and quality assurance methods (Gollasch et al., 2019). Further studies are also necessary to determine the viability of phytoplankton after experiencing ballast tank conditions, and how effective approved treatment systems will be under the broad range of conditions to which they might be exposed to (Casas-Monroy et al., 2018).

Managing aquaculture

Aquaculture is the fastest growing food sector, increasing 16-fold between 1985 and 2018, and surpassing in 2014 wild capture fisheries production (FAO, 2020). The continued expansion of aquaculture globally intersects with phytoplankton in a number of ways. The impacts of concentrated finfish farming, often in protected embayments, can have a detrimental impact on the surrounding environment. Increased organic matter from animal feeding and waste, leading to deoxygenated water, can give rise to increased frequency and severity of HABs, with some species persisting in the underlying sediments as cysts until "bloom" conditions occur. However, the industry is also vulnerable to region-wide HAB events caused by large-scale antecedent climate and oceanographic conditions, such as those occurring in the austral summer of 2016 in Chile when blooms of *Pseudochattonella verruculosa* led to massive losses for the industry and their insurers (see Case study 9.3). Thus to manage aquaculture expansion and long-term sustainability and profitability, it is essential to evaluate how climate change will affect HAB dynamics (Brown et al., 2020).

Faced with increasing aquaculture expansion, the requirement to protect human health, and negative social perceptions, many countries have implemented management strategies that encompass both strategic planning for new regions, as well as for tactical operating requirements. To minimize the risk of HABs, strategies that can be used include (i) spatial and temporal planning of operations; (ii) direct interventions for controlling (e.g., physical, chemical, biological methods) the presence or abundance of HAB species, and (iii) holistic environmental management, such as integrated multitrophic aquaculture (IMTA), which employs extractive bivalve shellfish and macroalgae alongside finfish and crustaceans, to recycle nutrients, thus maximizing productivity and water quality simultaneously. In the absence of better information, responsible management authorities will usually take a precautionary approach and can sometimes direct the closure of farms and the killing of entire production cycles. With entrenched negative social perceptions of aquaculture in many countries, it is also recognized that there is a need to "go beyond compliance" by exceeding environmental quality standards and to recognize aquaculture needs to be managed within a socioecological framework that explicitly recognizes ecosystem services and EBM (Willot et al., 2019; McGhee et al., 2019).

Monitoring approaches for aquaculture differ among countries in terms of their regulatory requirements, the methodologies, indicators, and standards imposed, and the degree to which the costs are met by industry and government (Yarimizu et al., 2020). Nevertheless, compliance programmes usually comprise a system of monitoring that spans field monitoring of environmental and operational conditions to food-safety testing, including for the presence of toxins. Many of these compliance monitoring programs still rely on traditional methods of cell counts and toxin testing and have been slow to adopt many of the "omic" technologies for detection and quantification. Thus there is considerable scope to capitalize on advances in automation and eco-sensor technologies to inform operations and meet compliance requirements. Remote sensing is a well-established technology used in the siting of new aquaculture leases, although spatial resolution remains an issue when assessing small-scale ventures. The development of new capabilities for the discrimination of phytoplankton groups and size classes has enabled the development of better detection and forecasting of HABs, and, operationally, these are in use in several countries, most notably in the USA, China, Korea, Japan, Norway, and Chile. A number of modeling platforms and information systems have been developed specifically for the industry, including *SIMA* in Chile (Steven et al., 2019a), *Manolin* (https://manolinaqua.com) in Norway, and the FAO *BlueCloud Aquaculture Monitor* (https://www.blue-cloud.org/demonstrators/aquaculture-monitor). These systems, though not explicitly concerned with phytoplankton, do capture the broader socio-ecological impacts and facilitate tactical and strategic decision-making (Steven et al., 2019a). However, in developing countries there are significant challenges to adoption of technologies, including secondary education, diversified on-farm activities, farm size, production levels, attendance of extension training, ease of understanding, and ease of handling technologies (Obiero et al., 2019).

In response to coastal areas becoming more crowded and subject to poorer environmental and growing conditions, there is a growing interest in establishing offshore aquaculture (Buck et al., 2018). Regulatory policy for sustainable offshore aquaculture has recently been developed in the United States, but is still to be formulated in other countries (Lester et al., 2018). However, HABs can often originate naturally offshore, and some species may present an even greater risk here than inshore (Gobler et al., 2017).

Finally, to address sustainability issues, and remain profitable, there are significant opportunities for phytoplankton biotechnology to be used in sustainable growth of aquaculture to reduce its environmental footprint, through for example, IMTA and green-pond approaches; provide sustainable production of whole-cell and formulated feeds as a replacement to fishmeal, and reducing HAB and disease outbreaks with animal health and growth supplements (Fabris et al., 2020). These are discussed in greater detail in the section on phytoplankton biotechnology.

Case Study 9.3: Chilean aquaculture and the impacts of Harmful Algal Blooms

Chile is one of the largest producers of farmed salmon, and has a rapidly expanding shellfish culture industry. HAB events first affected the industry in 1988 and have subsequently increased in intensity, frequency, and the diversity of harmful species, spreading progressively from Patagonian Chile, where salmon farming is concentrated, northwards along the coast to areas where shellfish are now mainly cultured. This has caused significant harm to the industry and other economic sectors, as well as social unrest (Díaz et al., 2019; Soto et al., 2019).

HAB impacts culminated in the austral summer of 2016 when massive blooms of *Pseudochattonella verruculosa* in the south and *Alexandrium catenella* in the north, dubbed the "Godzilla-red tide event," occurred with losses of 100,000 metric tons of salmon and trout, equivalent to an export loss of USD$800 million, or ~15% of Chile's yearly production. These blooms were linked to large-scale climatic anomalies, resulting from a strong El Niño event and a positive phase of the southern annular mode (SAM), resulting in strong vertical water column stratification and low water flushing rates. These conditions stimulated the development of a dynamic *P. verruculosa* thin layer bloom that persisted for several weeks with chlorophyll-a concentrations of up to 38 $\mu g\ L^{-1}$, and algal cell densities of 7000–20,000 cells mL^{-1} (Mardones et al., 2021; Trainer et al., 2020). Subsequently, in the 2020 austral summer, blooms of dinoflagellate *Cochlodinium* spp. occurred and, though not as large as the 2016 events, caused massive fish mortalities and economic impacts for the industry and its insurers.

The degree to which these HAB events and their expansion northward are the result of regional climatic events and climate change, or the compounded impacts of local-scale coastal development, agriculture, and farm operations that can cause eutrophication, and the possible translocation of the cysts of dinoflagellate species in nets and well-boats (Armijo et al., 2020; Rodriguez-Villegas et al., 2020). Both the industry and the Chilean government have invested considerably in developing better monitoring and forecasting capabilities of such events, particularly in satellite remote sensing, most recently the use of Sentinel 2 and 3 (Rodriguez-Benito et al., 2020), and the better assessment of data and the use of biophysical and socioecological models in an information system, such as SIMA (Steven et al., 2019a). Despite significant investment in the development of molecular methods, including infrastructure and training, they have not yet been adopted in aquaculture monitoring programs, and traditional microscopic and toxin analyses are still most common (Yarimizu et al., 2020). It is clear that there is an urgent need for investing in integrated approaches to investigate, forecast, prevent, and, if feasible, mitigate the emergence of HABs to address the negative impacts on industry and human health.

Fisheries

Effective management of the world's fish stocks, many of which are in significant decline, today adopt holistic ecosystems approaches to fisheries management (EAFM) and, at the same time, find ways to balance the costs of fishing to remain profitable. Operational remote sensing and modeling are now commonly used by many countries, and in international waters, to better target fishing efforts and to provide forecasts of oceanographic conditions that can be used to estimate seasonal and interannual variability in productivity and its consequences for fisheries production (FAO, 2020; Stuart et al., 2011).

Satellite remote sensing has for several decades been routinely used to help manage fisheries at sustainable levels, including for stock assessment, fish harvesting, and reductions in bycatch. Remote sensing is also used to guide fishing fleets to locate fish schools more efficiently by identifying potential fishing zones (PFZs), and relies upon established relationships between target species and environmental factors and features, such as oceanographic fronts, where fish tend to aggregate. The latest remote-sensing techniques have progressed from single to multiple parameter algorithms that incorporate sea surface temperature (SST) and height (SSH), and chlorophyll, and phytoplankton functional types, and are further improved when in situ measurements are included and used to validate synoptic conditions. Monitoring these parameters together with phytoplankton phenology and size structure provide important information to help understand the response of marine ecosystems to environmental change and will likely become increasingly important as climate change alters the geographical distribution and the timing of phytoplankton blooms (Gittings et al., 2021).

To assist the fishing industry to become more sustainable and more economically viable, many countries have developed marine fisheries forecasting service systems. The most advanced include Japan, Australia, Canada, Chile, France, Norway, Portugal, and Russia. The remotely-sensed data are provided in near real-time to help fishermen save fuel—and reduce greenhouse gas emissions—and ship time during their search for fish, and potentially reduces bycatch. Though concerns have been raised that such systems can lead to overexploitation of fish stocks, when implemented alongside other conservation-based management tools, such as quota systems, they are effective (Klemas, 2013). Case study 9.4 showcases the Indian fisheries services.

Case Study 9.4: Indian fishing services

In India ~7 million people live along the coastline and are dependent on fishing for their livelihood. Since 1990 India, has progressively developed the use of satellite data to identify PFZs and provide information to assist fishermen. Today, the Indian National Centre for Ocean Information Services (INCOIS) at Hyderabad generates and disseminates daily, multilingual PFZ forecasts in several media formats to the Indian fishing community working in coastal and pelagic waters (http://www.incois.gov.in/Incois/advisory_pfz_main.jsp). PFZ advisories are issued for 14 sectors along the Indian coast and are generated using satellite-derived SST and chlorophyll data, and wind speed. They are available in native language and contain both maps of the location of PFZ and text on the location, depth, and proximity to landmarks. These bulletins are reported to have helped artisanal fisheries reduce search time by up to 70% and increase the catch per unit effort (Solanki et al., 2003). Models are also now being increasingly used to improve forecasts of the probable location of fish aggregations and to monitor environmental conditions (Giri et al., 2015).

Phytoplankton biotechnology applications

The application of phytoplankton (or microalgae as it is commonly referred to) biotechnology is rapidly developing, to address growing demands for more sustainable forms of industrial and agricultural products, including as a renewable feedstock for aquaculture, biofuel production, high-value products, pharmaceuticals, therapeutics, health supplements and personal care products, and environmental clean technology applications in wastewater treatment and atmospheric carbon dioxide (CO_2) mitigation (Sutherland et al., 2021; Mandotra et al., 2021; Fabris et al., 2020; Dębowski et al., 2020; Yarnold et al., 2019; Souza et al., 2019).

This potential is due to the high photosynthetic efficiencies of phytoplankton, which can rapidly build biomass that is rich in carbohydrates lipids, pigments, and other bioactives. Phytoplankton also possess other intrinsic advantages over conventional agriculture, such as low production costs, no requirement for arable land, and the capacity to be rapidly grown in either large-scale outdoor systems or self-contained photobioreactors (Fabris et al., 2020; Souza et al., 2019). Processes such as bioenergy with carbon capture and sequestration (BECCS) can be used for ethanol production and at the same time achieve substantial carbon removal (Choi et al., 2019).

The development of these biotechnology applications has been bolstered by decades of foundational research on algal biochemistry and physiology, and leveraged functional genetic and molecular tools, for improved strain selection and process engineering.

Much effort has gone into optimizing and scaling-up production facilities, but with the exception of a relatively few commercial scale operations producing high-value products, other biotechnology applications are still mostly at pilot-scale implementation (Barsanti and Gualtieri, 2018). The cultivation systems used, range from low technology, low capital, and operational cost systems, such as high rate algal ponds (HRAPS, which that are often used in wastewater treatment), to high technology, capital and operational cost systems, such as photobioreactors. Which system is appropriate for cultivating microalgae will depend greatly on the final product being produced. For example, food, animal feed, nutritional supplements, and biofuels made from microalgae are considered low-value products and as such favor systems where high volumes of microalgae can be cultured at low cost. In contrast, high-value products, such as medicines, require culture systems, such as enclosed photobioreactors or fermenters, to meet regulatory requirements for quality and purity and/or the containment of genetically modified organisms (Sutherland et al., 2021). The development of efficient algal biorefineries that allow cosourcing of different products, minimize waste and maximize productivity, are widely considered as one of the most tractable solutions (Fabris et al., 2020).

To further expand large-scale operations for the production and use of microalgal biomass, a number of economic, technological, and legal difficulties must be overcome. Technologically, expanded algal collections for breeding purposes and genome sequencing, identification of a broader suite of phytoplankton species than the few currently being used, and the development of systems that can support multitrophic and multilayered production are needed. Cost is a key roadblock, with a high initial investment required to develop biorefineries capable of at-scale production. More cost-effective and efficient methods for cultivation, harvesting and drying of microalgae biomass, and separation and concentration of refined products, are also required to be competitive with conventional production techniques. Substantial cost decreases can potentially be achieved through the development of biorefinery approaches to maximize value from microalgal biomass (Fabris et al., 2020).

Though many governments, most notably the EU, engage in the techno-economic development of these industries, legislation and policy is also required to safeguard the environment and the rights and livelihoods of peoples who may be affected. Environmental concerns, such as the high use of water in cultivation and the potential impact of releases to the adjacent environment, also require new technologies and methods. It is notable that the majority of marine gene patents (98%) are held by a few large international companies based in 10 countries, while 165 countries are unrepresented (Blasiak et al., 2018). Instruments such as the *Nagoya Protocol on Access and Benefit Sharing* seeks to protect indigenous rights in this regard.

9.5 Conclusions

This chapter has briefly reviewed the evolving societal uses of phytoplankton and highlighted where emerging technologies are being used to assist in their sustainable management. For ecosystem and industry-specific monitoring and surveillance applications, in situ sensors and satellites have been increasingly adopted by management agencies with improvements in cost, access, and interpretability of data. Some of the most relevant applications for management rely on the integration of several technologies, for example, the integration of HTS and ML with in situ technologies are being used to conduct faster, more comprehensive identification of phytoplankton species and toxins, and satellite and drone data are being integrated and assimilated in models that are operationally used to provide managers with nowcasts and forecasts of phytoplankton and their associated environmental conditions. Management needs of course continue to evolve as new uses of marine and freshwater environments require approvals and performance conditions for their operation, and emerging issues, such as the consequences of climate change, must be considered and addressed. To ensure these technologies are fit-for-purpose, management should be specific on the criteria they use to evaluate whether a technology is suited for their needs. With renewed political will and widespread commitment to addressing global climate, biodiversity, sustainability, and equity challenges, industry-scale applications of microalgal biotechnologies with proactive techno-economic development have the potential to deliver solutions to a diversity of societal needs, including many of the UN sustainable development goals.

References

Adams, C.M., Larkin, S.L., Hoagland, P., Sancewich, B., 2018. Assessing the economic consequences of Harmful Algal Blooms: a summary of existing literature, research methods, data, and information gaps. In: Shumway, S.E., Burkholder, J.M., Morton, S.L. (Eds.), Harmful Algal Blooms: a Compendium Desk Reference, pp. 337–354.

Anderson, C.R., Berdalet, E., Kudela, R.M., Cusack, C.K., Silke, J., O'Rourke, E., et al., 2019. Scaling up from regional case studies to a global Harmful Algal Bloom observing system. Frontiers in Marine Science 6, 1–24.

Anderson, D.M., Cembella, A.D., Hallegraeff, G.M., 2012. Progress in understanding Harmful Algal Blooms: paradigm shifts and new technologies for research, monitoring, and management. In: Carlson, C.A., Giovannoni, S.J. (Eds.), Annual Review of Marine Science, pp. 143–176.

Anton, A., Geraldi, N.R., Lovelock, C.E., Apostolaki, E.T., Bennett, S., Cebrian, J., et al., 2019. Global ecological impacts of marine exotic species. Nature Ecology & Evolution 3, 787–800.

Ardura, A., Borrell, Y.J., Fernandez, S., Arenales, M.G., Martinez, J.L., Garcia-Vazquez, E., 2020. Nuisance algae in ballast water facing international conventions. Insights from DNA metabarcoding in ships arriving in Bay of Biscay. Water 12, 2168.

Armijo, J., Oerder, V., Auger, P.A., Bravo, A., Molina, E., 2020. The 2016 red tide crisis in southern Chile: possible influence of the mass oceanic dumping of dead salmons. Marine Pollution Bulletin 150, 110603.

Baird, M.E., Dutkiewicz, S., Hickman, A., Mongin, M., Soja-Wozniak, M., Skerratt, J., Wild-Allen, K., 2022. Modelling phytoplankton processes in multiple functional types. In: Clementson, L.A., Eriksen, R.S., Willis, A. (Eds.), Advances in Phytoplankton Ecology: Applications of Emerging Technology. Elsevier. Chapter 8 (in this book).

Baird, M.E., Mongin, M., Skerratt, J., Margvelashvili, N., Tickell, S., Steven, A.D.L., et al., 2021. Impact of catchment-derived nutrients and sediments on marine water quality on the Great Barrier Reef: an application of the eReefs marine modelling system. Marine Pollution Bulletin 167, 112297.

Barsanti, L., Gualtieri, P., 2018. Is exploitation of microalgae economically and energetically sustainable? Algal Research 31, 107–115.

Becker, R.H., Sayers, M., Dehm, D., Shuchman, R., Quintero, K., Bosse, K., Sawtell, R., 2019. Unmanned aerial system based spectroradiometer for monitoring harmful algal blooms: a new paradigm in water quality monitoring. Journal of Great Lakes Research 45, 444–453.

Berdalet, E., Kudela, R., Banas, N.S., Al, E., 2018. GlobalHAB: fostering international co-ordination on harmful algal bloom research in aquatic systems. In: Glibert, P.M., Berdalet, E., Burford, M.A., Pitcher, G.C., Zhou, M. (Eds.), Global Ecology and Oceanography of Harmful Algal Blooms. Springer, Cham, Switzerland, pp. 425–447.
Blasiak, R., Jouffray, J.-B., Wabnitz, C.C.C., Sundström, E., Österblom, H., 2018. Corporate control and global governance of marine genetic resources. Science Advances 4, eear5237.
Borja, Á., Dauer, D.M., Grémare, A., 2012. The importance of setting targets and reference conditions in assessing marine ecosystem quality. Ecological Indicators 12, 1–7.
Borja, A., Elliott, M., Andersen, J.H., Cardoso, A.C., Carstensen, J., Ferreira, J.G., et al., 2013. Good Environmental Status of marine ecosystems: what is it and how do we know when we have attained it? Marine Pollution Bulletin 76, 16–27.
Brown, A.R., Lilley, M., Shutler, J., Lowe, C., Artioli, Y., Torres, R., et al., 2020. Assessing risks and mitigating impacts of harmful algal blooms on mariculture and marine fisheries. Reviews in Aquaculture 12, 1663–1688.
Buck, B.H., Troell, M.F., Krause, G., Angel, D.L., Grote, B., Chopin, T., 2018. State of the art and challenges for offshore integrated multi-trophic aquaculture (IMTA). Frontiers in Marine Science 5, 165. https://doi.org/10.3389/fmars.2018.00165.
Campbell, L., Gaonkar, C., Henrichs, D., 2022. Integrating imaging and molecular approaches to assess phytoplankton diversity. In: Clementson, L.A., Eriksen, R.S., Willis, A. (Eds.), Advances in Phytoplankton Ecology: Applications of Emerging Technology. Elsevier. Chapter 5 (in this book).
Casas-Monroy, O., Linley, R.D., Adams, J.K., Chan, F.T., Drake, D.A., Bailey, S.A., 2015. Relative invasion risk for plankton across marine and freshwater systems: examining efficacy of proposed international ballast water discharge standards. PLoS ONE 10, e0118267.
Casas-Monroy, Oscar, Linley, Robert D., Chan, Po-Shun, Kydd, Jocelyn, Byllaardt, Julie Vanden, Bailey, Sarah, 2018. Evaluating efficacy of filtration + UV-C radiation for ballast water treatment at different temperatures. Journal of Sea Research 133, 20–28.
Chen, X.F., Zhou, Q.J., Duan, W.J., Zhou, C.X., Duan, L.J., Zhang, H.L., et al., 2016. Development and evaluation of a DNA microarray assay for the simultaneous detection of nine harmful algal species in ship ballast and seaport waters. Chinese Journal of Oceanology and Limnology 34, 86–101.
Choi, Y.Y., Patel, A.K., Hong, M.E., Chang, W.S., Sim, S.J., 2019. Microalgae Bioenergy with Carbon Capture and Storage (BECCS): an emerging sustainable bioprocess for reduced CO2 emission and biofuel production. Bioresource Technology Reports 7, 100270.
Coffer, M.M.A., Schaeffer, B.A., Urquhart, E.A., Darling, J.A., Salls, W.B., 2018. A method for quantifying the number of US lakes with cyanobacterial harmful algal blooms using satellite remote sensing. In: Gao, W., Cang, N.B., Wang, J. (Eds.), Remote Sensing and Modeling of Ecosystems for Sustainability XV.
Dębowski, M., Zieliński, M., Kazimierowicz, J., Kujawska, N., Talbierz, S., 2020. Microalgae cultivation technologies as an opportunity for bioenergetic system development—advantages and limitations. Sustainability 12, 9980.
Díaz, P.A., Álvarez, G., Varela, D., Pérez-Santos, I., Díaz, M., Molinet, C., et al., 2019. Impacts of harmful algal blooms on the aquaculture industry: Chile as a case study. Perspectives in Phycology 6, 39–50.
Díaz, S., Demissew, S., Carabias, J., Joly, C., Lonsdale, M., Ash, N., et al., 2015. The IPBES Conceptual Framework — connecting nature and people. Current Opinion in Environmental Sustainability 14, 1–16.
Doucette, G.J., Kudela, R.M., 2017. In situ and real-time identification of toxins and toxin-producing microorganisms in the environment. In: Recent Advances in the Analysis of Marine Toxins, pp. 411–443.

Doucette, G.J., Medlin, L.K., McCarron, P., Hess, P., 2018. Detection and surveillance of harmful algal bloom species and toxins. In: Shumway, S.E., Burkholder, J.M., Morton, S.L. (Eds.), Harmful Algal Blooms: a Compendium Desk Reference. John Wiley & Sons, Hoboken, NJ, pp. 39–113.

Fabris, M., Abbriano, R.M., Pernice, M., Sutherland, D.L., Commault, A.S., Hall, C.C., et al., 2020. Emerging technologies in algal biotechnology: toward the establishment of a sustainable, algae-based bioeconomy. Frontiers of Plant Science 11, 279.

FAO, 2020. The State of World Fisheries and Aquaculture 2020: Sustainability in Action. Rome.

Flynn, K.J., McGillicuddy, D.J., 2018. Modeling marine harmful algal blooms: current status and future prospects. In: Shumway, S.E., Burkholder, J.M., Morton, S.L. (Eds.), Harmful Algal Blooms: a Compendium Desk Reference. John Wiley & Sons, Hoboken, NJ, pp. 115–133.

Franks, P.J.S., 2018. Recent advances in modelling of harmful algal blooms. In: Glibert, P.M., Berdalet, E., Burford, M.A., Pitcher, G.C., Zhou, M. (Eds.), Global Ecology and Oceanography of Harmful Algal Blooms, pp. 359–377.

Gerhard, W.A., Gunsch, C.K., 2019. Metabarcoding and machine learning analysis of environmental DNA in ballast water arriving to hub ports. Environment International 124, 312–319.

Giri, S., Manna, S., Chanda, A., Chowdhury, A., Mukhopadhyay, A., Chakraborty, S., Hazra, S., 2015. Implementing a spatial model to derive potential fishing zones in the Northern Bay of Bengal lying adjacent to West Bengal Coast, India. Journal of the Indian Society of Remote Sensing 44, 59–66.

Gittings, J.A., Raitsos, D.E., Brewin, R.J.W., Hoteit, I., 2021. Links between phenology of large phytoplankton and fisheries in the northern and central Red Sea. Remote Sensing 13 (2), 231.

Glibert, P.M., 2020. Harmful algae at the complex nexus of eutrophication and climate change. Harmful Algae 91, 101583.

Glibert, P.M., Allen, J.I., Bouwman, A.F., Brown, C.W., Flynn, K.J., Lewitus, A.J., Madden, C.J., 2010. Modeling of HABs and eutrophication Status, advances, challenges. Journal of Marine Systems 83, 262–275.

Glibert, P.M., Pitcher, G.C., Bernard, S., Li, M., 2018. Advancements and continuing challenges of emerging technologies and tools for detecting harmful algal blooms, their antecedent conditions and toxins, and applications in predictive models. In: Glibert, P.M., Berdalet, E., Burford, M.A., Pitcher, G.C., Zhou, M. (Eds.), Global Ecology and Oceanography of Harmful Algal Blooms. Springer, Cham, Switzerland, pp. 339–357.

Gobler, C.J., Doherty, O.M., Hattenrath-Lehmann, T.K., Griffith, A.W., Kang, Y., Litaker, R.W., 2017. Ocean warming since 1982 has expanded the niche of toxic algal blooms in the North Atlantic and North Pacific oceans. Proceedings of the National Academy of Sciences of the United States of America 114, 4975–4980.

Gollasch, S., Hewitt, C.L., Bailey, S., David, M., 2019. Introductions and transfers of species by ballast water in the Adriatic Sea. Marine Pollution Bulletin 147, 8–15.

Griffith, A.W., Gobler, C.J., 2020. Harmful algal blooms: a climate change co-stressor in marine and freshwater ecosystems. Harmful Algae 91, 101590.

Hallegraeff, G.M., Anderson, D.M., Belin, C., Bottein, M.-Y.D., Bresnan, E., Chinain, M., et al., 2021. Perceived global increase in algal blooms is attributable to intensified monitoring and emerging bloom impacts. Communications Earth & Environment 2, 117.

Halpern, B.S., 2020. Building on a decade of the ocean health index. One Earth 2, 30–33.

Halpern, B.S., Longo, C., Hardy, D., McLeod, K.L., Samhouri, J.F., Katona, S.K., et al., 2012. An index to assess the health and benefits of the global ocean. Nature 488, 615–620.

Hill, P.R., Kumar, A., Temimi, M., Bull, D.R., 2020. HABNet: machine learning, remote sensing-based detection of harmful algal blooms. IEEE Journal of Selected Topics in Applied Earth Observations and Remote Sensing 13, 3229–3239.
Ho, J.C., Michalak, A.M., Pahlevan, N., 2019. Widespread global increase in intense lake phytoplankton blooms since the 1980s. Nature 574, 667–670.
Hoagland, P., Jin, D., Beet, A., Kirkpatrick, B., Reich, A., Ullmann, S., et al., 2014. The human health effects of Florida red tide (FRT) blooms: an expanded analysis. Environment International 68, 144–153.
Hu, M., Wang, Y., Sun, Z., Su, Y., Li, S., Bao, Y., Wen, J., 2020. Performance of ensemble-learning models for predicting eutrophication in Zhuyi Bay, Three Gorges Reservoir. River Research and Applications. https://doi.org/10.1002/rra.3739.
Hulme, P.E., 2021. Unwelcome exchange: international trade as a direct and indirect driver of biological invasions worldwide. One Earth 4, 666–679.
Khandeparker, L., Kuchi, N., Desai, D.V., Anil, A.C., 2020. Changes in the ballast water tank bacterial community during a trans-sea voyage: elucidation through next generation DNA sequencing. Journal of Environmental Management 273, 111018.
Klemas, V., 2013. Fisheries applications of remote sensing: an overview. Fisheries Research 148, 124–136.
Kudela, R., Berdalet, E., Enevoldsen, H., Pitcher, G., Raine, R., Urban, E., 2017. GEOHAB–the global ecology and oceanography of harmful algal blooms program: motivation, goals, and legacy. Oceanography 30, 12–21.
Lee, S., Lee, D., 2018. Improved prediction of harmful algal blooms in four major South Korea's rivers using deep learning models. International Journal of Environmental Research and Public Health 15, 1322.
Lester, S.E., Gentry, R.R., Kappel, C.V., White, C., Gaines, S.D., 2018. Opinion: offshore aquaculture in the United States: untapped potential in need of smart policy. Proceedings of the National Academy of Sciences of the United States of America 115, 7162–7165.
Liu, Q., Rowe, M.D., Anderson, E.J., Stow, C.A., Stumpf, R.P., Johengen, T.H., 2020. Probabilistic forecast of microcystin toxin using satellite remote sensing, in situ observations and numerical modeling. Environmental Modelling & Software 128, 104703.
Malthus, T.J., Lehmann, E., Ho, X., Botha, E., Anstee, J., 2019. Implementation of a satellite based inland water algal bloom alerting system using analysis ready data. Remote Sensing 11, 2954.
Mandotra, S.K., Sharma, C., Srivastava, N., Ahluwalia, A.S., Ramteke, P.W., 2021. Current prospects and future developments in algal bio-hydrogen production: a review. Biomass Conversion and Biorefinery. https://doi.org/10.1007/s13399-021-01414-z.
Mardones, J.I., Paredes, J., Godoy, M., Suarez, R., Norambuena, L., Vargas, V., et al., 2021. Disentangling the environmental processes responsible for the world's largest farmed fish-killing harmful algal bloom: Chile 2016. Science of the Total Environment 766, 144383.
McGhee, C., Falconer, L., Telfer, T., 2019. What does 'beyond compliance' look like for the Scottish salmon aquaculture industry? Marine Policy 109. https://doi.org/10.1016/j.marpol.2019.103668.
McQuatters-Gollop, A., Mitchell, I., Vina-Herbon, C., Bedford, J., Addison, P.F.E., Lynam, C.P., et al., 2019. From science to evidence – how biodiversity indicators can be used for effective marine conservation policy and management. Frontiers in Marine Science 6. https://doi.org/10.3389/fmars.2019.00109.
Mishra, D.R., Kumar, A., Ramaswamy, L., Boddula, V.K., Das, M.C., Page, B.P., Weber, S.J., 2020. CyanoTRACKER: a cloud-based integrated multi-platform architecture for global observation of cyanobacterial harmful algal blooms. Harmful Algae 96. https://doi.org/10.1016/j.hal.2020.101828.
Obiero, K.O., Waidbacher, H., Nyawanda, B.O., Munguti, J.M., Manyala, J.O., Kaunda-Arara, B., 2019. Predicting uptake of aquaculture technologies among smallholder fish farmers in Kenya. Aquaculture International 27, 1689–1707.

Park, J., Kim, K.T., Lee, W.H., 2020. Recent advances in information and communications technology (ICT) and sensor technology for monitoring water quality. Water 12. https://doi.org/10.3390/w12020510.

Raine, R., Berdalet, E., Yamazaki, H., Jenkinson, I., Reguera, B., 2018. Key questions and recent research advances in harmful algal blooms in stratified systems. In: Glibert, P.M., Berdalet, E., Burford, M.A., Pitcher, G.C., Zhou, M. (Eds.), Global Ecology and Oceanography of Harmful Algal Blooms. Springer, pp. 165–198.

Ralston, D.K., Moore, S.K., 2020. Modeling harmful algal blooms in a changing climate. Harmful Algae 91, 101729.

Rodriguez-Benito, C.V., Navarro, G., Caballero, I., 2020. Using Copernicus Sentinel-2 and Sentinel-3 data to monitor harmful algal blooms in Southern Chile during the COVID-19 lockdown. Marine Pollution Bulletin 161, 111722.

Rodriguez-Villegas, C., Diaz, P.A., Pizarro, G., Salgado, P., Perez-Santos, I., Diaz, M., et al., 2020. Alexandrium catenella cyst accumulation by passive and active dispersal agents: implications for the potential spreading risk in Chilean Patagonian fjords. Harmful Algae 96, 101832.

Sandoval, M.S., Parada, C., Torres, R., 2018. Proposal of an integrated system for forecasting Harmful Algal Blooms (HAB) in Chile. Latin American Journal of Aquatic Research 46, 424–451.

Schaeffer, B.A., Myer, M.H., 2020. Resolvable estuaries for satellite derived water quality within the continental United States. Remote Sensing Letters 11, 535–544.

Shang, L., Hu, Z., Deng, Y., Liu, Y., Zhai, X., Chai, Z., et al., 2019. Metagenomic sequencing identifies highly diverse assemblages of dinoflagellate cysts in sediments from ships' ballast tanks. Microorganisms 7. https://doi.org/10.3390/microorganisms7080250.

Skerratt, J.H., Mongin, M., Baird, M.E., Wild-Allen, K.A., Robson, B.J., Schaffelke, B., Davies, C.H., Richardson, A.J., Margvelashvili, N., Soja-Wozniak, M., Steven, A.D.L., 2018. Simulated nutrient and plankton dynamics in the Great Barrier Reef (2011–2016). Journal of Marine Systems. https://doi.org/10.1016/j.jmarsys.2018.12.006.

Smit, K.P., Bernard, A.T.F., Lombard, A.T., Sink, K.J., 2021. Assessing marine ecosystem condition: a review to support indicator choice and framework development. Ecological Indicators 121, 107148.

Smith, M.C., Bodrossy, L., Pascal, C., 2022. Advances in *in situ* systems for phytoplankton research and monitoring. In: Clementson, L.A., Eriksen, R.S., Willis, A. (Eds.), Advances in Phytoplankton Ecology: Applications of Emerging Technology. Elsevier. Chapter 6 (in this book).

Smith, M.E., Bernard, S., 2020. Satellite ocean color based harmful algal bloom indicators for aquaculture decision support in the Southern Benguela. Frontiers in Marine Science 7. https://doi.org/10.3389/fmars.2020.00061.

Solanki, H.U., Dwivedi, R.M., Nayak, S., Gulati, D.K., John, M.E., Somvanshi, V.S., 2003. Potential fishing zones (PFZ) forecast using satellite data derived biological and physical processes. Journal of the Indian Society of Remote Sensing 31, 67–69.

Soto, D., León-Muñoz, J., Dresdner, J., Luengo, C., Tapia, F.J., Garreaud, R., 2019. Salmon farming vulnerability to climate change in southern Chile: understanding the biophysical, socioeconomic and governance links. Reviews in Aquaculture 11, 354–374.

Souza, M.P., Hoeltz, M., Brittes Benitez, L., Machado, Ê.L., De Souza Schneider, R.D.C., 2019. Microalgae and clean technologies: a review. CLEAN – Soil, Air, Water 47, 1800380.

Spanbauer, T.L., Briseno-Avena, C., Pitz, K.J., Suter, E., 2020. Salty sensors, fresh ideas: the use of molecular and imaging sensors in understanding plankton dynamics across marine and freshwater ecosystems. Limnology and Oceanography Letters 5, 169–184.

Steven, A.D.L., Aryal, S., Bernal, P., Bravo, F., Bustamante, R.H., Condie, S., et al., 2019a. SIMA Austral: an operational information system for managing the Chilean aquacul-

ture industry with international application. Journal of Operational Oceanography 12, S29–S46.
Steven, A.D.L., Baird, M.E., Brinkman, R., Car, N.J., Cox, S.J., Herzfeld, M., et al., 2019b. eReefs: an operational information system for managing the Great Barrier Reef. Journal of Operational Oceanography, 1–17. https://doi.org/10.1080/1755876x.2019.1650589.
Stroming, S., Robertson, M., Mabee, B., Kuwayama, Y., Schaeffer, B., 2020. Quantifying the human health benefits of using satellite information to detect cyanobacterial harmful algal blooms and manage recreational advisories in US lakes. Geohealth 4, e2020GH000254.
Stuart, V., Platt, T., Sathyendranath, S., 2011. The future of fisheries science in management: a remote-sensing perspective. ICES Journal of Marine Science 68, 644–650.
Sutherland, D.L., McCauley, J., Labeeuw, L., Ray, P., Kuzhiumparambil, U., Hall, C., et al., 2021. How microalgal biotechnology can assist with the UN Sustainable Development Goals for natural resource management. Current Research in Environmental Sustainability 3, 100050.
Townhill, B.L., Tinker, J., Jones, M., Pitois, S., Creach, V., Simpson, S.D., et al., 2019. Harmful algal blooms and climate change: exploring future distribution changes. ICES Journal of Marine Science 76, 353.
Trainer, V.L., Moore, S.K., Hallegraeff, G., Kudela, R.M., Clement, A., Mardones, J.I., Cochlan, W.P., 2020. Pelagic harmful algal blooms and climate change: lessons from nature's experiments with extremes. Harmful Algae 91, 101591.
Tweddle, J.F., Gubbins, M., Scott, B.E., 2018. Should phytoplankton be a key consideration for marine management? Marine Policy 97, 1–9.
Wells, M.L., Karlson, B., Wulff, A., Kudela, R., Trick, C., Asnaghi, V., et al., 2020. Future HAB science: directions and challenges in a changing climate. Harmful Algae 91. https://doi.org/10.1016/j.hal.2019.101632.
Willot, P.-A., Aubin, J., Salles, J.-M., Wilfart, A., 2019. Ecosystem service framework and typology for an ecosystem approach to aquaculture. Aquaculture 512. https://doi.org/10.1016/j.aquaculture.2019.734260.
Wu, D., Li, R., Zhang, F., Liu, J., 2019. A review on drone-based harmful algae blooms monitoring. Environmental Monitoring and Assessment 191. https://doi.org/10.1007/s10661-019-7365-8.
Yarimizu, K., Fujiyoshi, S., Kawai, M., Norambuena-Subiabre, L., Cascales, E.-K., Rilling, J.-I., et al., 2020. Protocols for monitoring harmful algal blooms for sustainable aquaculture and coastal fisheries in Chile. International Journal of Environmental Research and Public Health 17. https://doi.org/10.3390/ijerph17207642.
Yarnold, J., Karan, H., Oey, M., Hankamer, B., 2019. Microalgal aquafeeds as part of a circular bioeconomy. Trends in Plant Science 24, 959–970.
Young, N., Sharpe, R.A., Barciela, R., Nichols, G., Davidson, K., Berdalet, E., Fleming, L.E., 2020. Marine harmful algal blooms and human health: a systematic scoping review. Harmful Algae 98, 101901.
Yu, B., Xu, L., Peng, J., Hu, Z., Wong, A., 2020. Global chlorophyll-a concentration estimation from moderate resolution imaging spectroradiometer using convolutional neural networks. Journal of Applied Remote Sensing 14 (3), 034520.
Yu, Z., Song, X., Cao, X., Liu, Y., 2018. Mitigation and control of harmful algal blooms. In: Glibert, P.M., Berdalet, E., Burford, M.A., Pitcher, G.C., Zhou, M. (Eds.), Global Ecology and Oceanography of Harmful Algal Blooms, pp. 403–423.

PART 3

Omics in aquatic ecology

Anusuya Willis

Next generation sequencing has instigated one of the most dramatic changes in methods to study phytoplankton ecology. The development of omics techniques, which utilize DNA and RNA sequences, have created new ways of investigating enduring questions of phytoplankton identification, evolutionary relationships, and population structure and dynamics.

Throughout my own research I have used omics techniques to link cell biology and physiology with genome content and gene expression. My interest in genes began when direct isolation of extracellular proteins involved in adhesion was unsuccessful, but subsequent investigation of the genome revealed many potential genes of interest. These genes could then be investigated through different avenues, in this case I used reverse genetic engineering to increase gene expression, thereby providing new opportunities for uncovering gene function. Omics techniques utilize mass sequence data, either DNA or RNA, and can offer different methods of enquiry and reveal details that might otherwise be overlooked.

The third part of this book focusses on techniques using next generation sequencing; where large amounts of data are generated comprising either short sequences for OTUs (operational taxonomic units), targeted to single genes, or long sequences (spanning chromosome sections, genes and intergenic regions). These sequences are then used to examine species identification or abundance, gene presence or expression, or whole genome content.

As omics technologies continue to improve and the analytical costs of obtaining sequence decrease, the volume of sequence data has increased dramatically. The bottleneck, and cost, to using omics technology is currently the computing power and bioinformatic expertise that is required to analyze the huge amounts of data and tease out the relevant details. Analysis of sequence data, from an organisms with no reference genome is like assembling a multi-million piece puzzle from which many or most of the pieces are missing, and with no picture on the box. Then comparing this incomplete image with another incomplete image to spot the differences. However, as quantities of data increase, these images become easier to resolve and see. Lack of reference databases for DNA are another of the limitations to omics data analysis, for both species identification and functional analysis. For example, in some metabarcoding samples, only a third of OTUs can be identified, and full genome sequences contain a high proportion, as much as 50% of "hypothetical proteins." Future advances in reference databases will allow archived data to be reanalyzed for new insights; this highlights the value of studies that identify new species or investigate the function of single genes or proteins. Archived sequence data will continue to be a valuable resource for many years.

The increase in our understanding of phytoplankton in aquatic ecology through omics techniques is vast and far-reaching. In Part 1 of this book, the impact of sequencing for species identification on taxonomy was described, and a wealth of evolutionary relationships defined through single-gene comparisons. This phylogenetic information does not require functional understanding of the genes, but comes directly from evolutionary markers, single nucleotide polymorphisms, in the DNA sequence. Vastly more information is acquired through functional annotation of genomes: understanding the variation in gene content between organisms and species, and their expression under different environments, enables a functional characterization of species dynamics and their relation to environmental factors.

The third part of this books contains six chapters that explore major omics techniques and their application for exploring phytoplankton ecology. Chapters cover real-time PCR (10), metabarcoding (11 and 15), transcriptomes (12), comparative genomics (13), linking genes to ecological processes (14), and global ocean phytoplankton populations (15).

Coyne et al., in Chapter 10, describe the use of qPCR to investigate factors impacting phytoplankton distributions and activities. Explaining techniques for species detection, including cryptic species, intraspecific diversity, harmful algae bloom (HAB) species, and gene expression. An overview of technology gives insight into the practical applications of qPCR.

Lopes dos Santos et al., in Chapter 11, use metabarcoding or high-throughput sequencing to analyze biodiversity across environments and taxonomic groups, thereby establishing biogeographic patterns and uncovering new lineages.

Willis et al., in Chapter 12, provide an overview of the use of comparative genomics to investigate the intraspecific genome diversity of a cyanobacterial species to gain insights into strain variation within populations.

Kolody et al., in Chapter 13, describe how transcriptomics offers the ability to view the immediate cellular response of algae and algal communities to varying environmental conditions, as well as an alternative to the challenge of sequencing the whole genome of protists.

Berges et al., in Chapter 14 use a combination of omics approaches: genomic, transcriptomic, and proteomics, to link gene content to ecosystem response and ecological processes. This chapter is an example of how these techniques and careful experimentation can be used to link subcellular processes to ecosystem responses.

The final Chapter 15, by *Vincent et al.*, describes the global sampling effort undertaken by the *Tara* Oceans expedition, which currently represents the largest DNA sequencing effort ever undertaken by ocean scientists. These data provide the scientific community with unprecedented resources that are transforming how we study the oceans and assess climate change, and is an example of the power of big sequencing data.

These six chapters provide useful information on omics techniques and their application, with case studies demonstrating how omics can contribute to the advancement of understanding of phytoplankton ecology.

CHAPTER 10

Current applications and technological advances in quantitative real-time PCR (qPCR): a versatile tool for the study of phytoplankton ecology

Kathryn J. Coyne[a], Yanfei Wang[a], Susanna A. Wood[b], Peter D. Countway[c], and Sydney M. Greenlee[d]
[a]University of Delaware, Lewes, DE, United States
[b]Cawthron Institute, Nelson, New Zealand
[c]Bigelow Laboratory for Ocean Sciences, East Boothbay, ME, United States
[d]University of Maine, Orono, ME, United States

Introduction

An understanding of phytoplankton ecology and its contribution to ecosystem functioning requires the accurate assessment of the abundance and distribution of algal species in complex environments. Though microscopy is often considered the "gold standard" for enumeration of microbial species, this approach is both time-intensive and requires a high level of expertise to accurately identify species. Even then, many species cannot be reliably identified, due to small size or cryptic morphological features, and the determination of cell abundance is often limited to large species or those in high abundance. Advances in molecular technology offer efficient and powerful alternatives to microscopic methods for detection and enumeration of phytoplankton species in natural assemblages (reviewed by Ebenezer et al., 2012; Humbert et al., 2010; Kudela et al., 2010; Metfies et al., 2016; Tan et al., 2015; Toldra et al., 2020; Wood et al., 2013). In particular, the shift from microscopy to molecular methods for phytoplankton monitoring and research has been driven by the requirement for expediency, and for increased accuracy and sensitivity. It should be noted here that molecular methods can be "open," meaning that knowledge of species composition or nucleic acid sequences is not a prerequisite, or "closed," in which advanced knowledge of the target species and nucleic acid sequence is necessary for assay development and implementation. In this chapter, we discuss a closed method, quantitative real-time PCR (qPCR), and its application to advance our understanding of phytoplankton ecology, with a focus on harmful algal bloom species (HABs). A case study is presented illustrating the use of quantitative reverse-transcriptase PCR (RT-qPCR) to investigate transcriptional responses of HABs to nitrogen input. Recent advances in qPCR technology are described, including the development of portable, field-ready qPCR instrumentation that allows rapid identification and enumeration of algal species in the field. As this technology becomes more widely adopted, its use will alleviate bottlenecks in monitoring for species of interest. A case study comparing unreplicated samples, run using a portable qPCR system with traditional lab-based technology, demonstrates the utility of the instrument. This is followed by a description of droplet digital PCR (ddPCR), illustrating improvements in instrument technology that circumvent many of the challenges associated with PCR amplification of field samples, such as the presence of PCR inhibitors or small sample sizes. Finally, the adaptation of qPCR for community-based monitoring programs and potential contributions to the field of HAB ecology are discussed.

Method overview

DNA amplification using the polymerase chain reaction (PCR) is perhaps the most frequently applied molecular technique in environmental research. Its application to environmental research was first reported in 1993, when it was used to investigate the presence and activity of sediment bacteria (e.g., Erb and Wagnerdobler, 1993; Herrick et al., 1993). PCR has since become an indispensable tool in microbial ecology for both monitoring and research, allowing detection of a single cell within a complex community (Panicker et al., 2004; Popels et al., 2003), and has enhanced investigations of microbial community dynamics (e.g., Beneduce et al., 2014; Muyzer et al., 1993). This method makes use of a thermally stable DNA polymerase to extend a set of oligonucleotide primers, resulting in a copy of the DNA fragment delineated by the primer pair. The DNA fragment is then replicated during each cycle of the reaction, generating copies of the fragment in an exponential fashion.

Conventional PCR requires end-point analysis and is limited in its ability to accurately determine the abundance of target species. In contrast, quantitative real-time PCR (qPCR) uses fluorescent probes or dyes to measure the change in product concentration during each PCR cycle (Heid et al., 1996). qPCR is often used to measure differences in the relative amount of the target DNA in a sample, and with careful design, can also be used to quantify absolute numbers of gene copies or cells (e.g., Bowers et al., 2000; Coyne et al., 2005; Godhe et al., 2008). Selected reviews of qPCR are provided in Table 10.1.

Table 10.1 Select reviews of quantitative real-time PCR (qPCR) in the literature and key topics covered.

Topic of review	Reference
Advantages and limitations of qPCR for ecological studies	Smith and Osborn (2009)
Molecular methods for detection of microalgae	Ebenezer et al. (2012)
Overview of qPCR and quantitative reverse-transcriptase PCR	Dymond (2013)
Application of qPCR for detection of HABs	Antonella and Luca (2013)
Molecular technology for monitoring aquatic habitats	Wood et al. (2013)
qPCR assays for monitoring HABs	Penna and Galluzzi (2013)
Molecular tools for HAB monitoring	Bruce et al. (2015)
Retrospective review of qPCR to assess cyanotoxin risk	Pacheco et al. (2016)
Molecular methods for detecting HABs	Toldra et al. (2020)
Sample extraction and preventing PCR inhibition	Sidstedt et al. (2020)

For qPCR, the increase in PCR product concentration is measured as an increase in fluorescence emission (ΔRn), generated either by a fluorescent reporter dye conjugated to a template-specific oligonucleotide probe, or by a nonspecific fluorescent dye that intercalates within the double-stranded DNA product, as described below. More details about sample preparation, primer and probe design, and assessment can be found elsewhere (e.g., Antonella and Luca, 2013; Bustin et al., 2009; Coyne et al., 2005, 2001; Dymond, 2013; Eckford-Soper and Daugbjerg, 2015; Eland et al., 2012; Gaget et al., 2017; Rueckert and Morgan, 2007; Sidstedt et al., 2020; Smith and Osborn, 2009; Stoeckel et al., 2009; Tomiokai et al., 2008; Yuan et al., 2015).

Fluorescent reporters

The simplest and most cost-efficient approach to qPCR uses a minor groove-binding dye, such as SYBR Green or SYTO9, as the fluorescent reporter (Andersen et al., 2006; Gudnason et al., 2007; Monis et al., 2005).

These intercalating dyes fluoresce only when bound to double-stranded DNA, and their fluorescence is proportional to the concentration of PCR product produced during the elongation step of the reaction (Fig. 10.1A). This approach is convenient, cost-effective, and can be applied directly to any PCR assay. However, intercalating dyes, such as SYBR Green, bind nonspecifically to all double-stranded DNA, including primer-dimers and nonspecific products, potentially resulting in a fluorescence signal that is no longer proportional to the target amplicon. For this reason, the specificity of qPCR product formation is assessed at the end of reaction using dissociation curve, or melt point, analysis (Giglio et al., 2003), which discriminates PCR products from spurious amplification products according to their melting temperature.

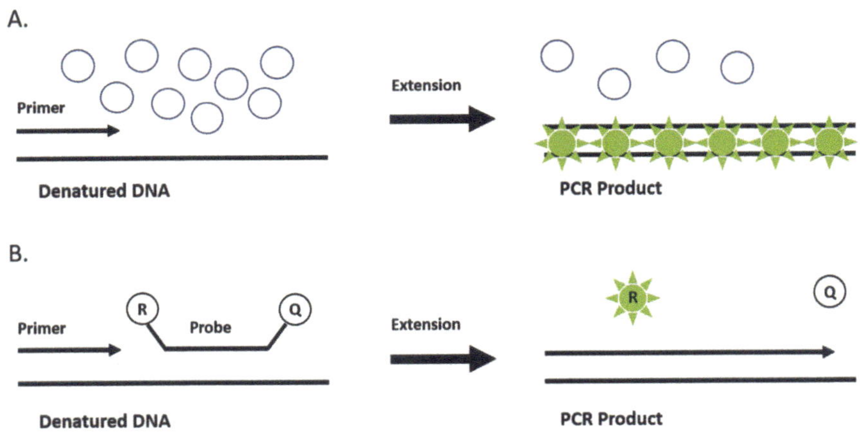

Figure 10.1 Cartoon illustrating SYBR Green (A) and Taqman Probe (B) approaches to quantitative PCR. See text for details.

Hydrolysis probes, such as TaqMan (ThermoFisher Scientific), provide an additional level of specificity for the reaction (Andersen et al., 2006; Suzuki et al., 2000). This method employs a template-specific oligonucleotide probe designed to hybridize to the target DNA sequence between priming sites (Fig. 10.1B). TaqMan probes contain a fluorescent reporter dye at the 5'-end and a quencher at the 3'-end. The quencher prevents the emission of fluorescence while the probe is intact due to Förster energy resonance transfer (Livak et al., 1995). During PCR elongation, the 5' nuclease activity of Taq polymerase cleaves the probe, separating the reporter and quencher so that fluorescence quenching no longer occurs.

The increase in fluorescence during each cycle is then proportional to the product.

Calculating target abundance

The threshold or quantification cycle number (referred to as Ct or Cq) for each reaction is the cycle number at which the fluorescence emission changes significantly over baseline levels. By plotting the Ct values versus the log of known starting concentrations of standards, the relative concentrations of target DNA for each unknown sample can be determined by linear regression analysis. Comparison to a "calibrator" sample with a known number of cells for the target species also provides a means to calculate the cell density of phytoplankton in water or sediment samples (Coyne et al., 2005; Demir et al., 2008; Handy et al., 2005, 2008).

Strengths and limitations

Quantitative real-time PCR has a range of detection over seven orders of magnitude (e.g., Coyne et al., 2005) and can detect a single cell (Panicker et al., 2004) extracted from a complex matrix, or as few as two copies of the target gene (Perini et al., 2011). PCR and qPCR also have a high level of species or strain specificity, a quality that drove development of some of the earliest qPCR applications for cryptic species. The specificity is dependent primarily on primer design, however, and requires extensive validation (and sometimes revalidation) for use in environmental research and monitoring. A recent study by Tillmann et al. (2019), for example, revealed the genetic diversity of *Azodinium spinosum* in Argentinean coastal waters, including the presence of a strain with mismatches to qPCR primers currently used by monitoring programs. The authors suggest that the assay may underestimate abundance of *A. spinosum* in this region. Determining cell density by qPCR analysis can also be affected by nutrient limitation (Main et al., 2014), or by growth stage. Populations that include a large number of cells rapidly dividing (such as during the early stages of a bloom) will have a greater average DNA content per cell, and may result in an overestimate of cell abundance (but see Main et al., 2014). Calibrators consisting of samples with known cell counts derived from the same environment as unknowns may provide for more accurate determination of cell abundance (Coyne et al., 2005).

A significant source of error in the application of molecular methods to environmental samples is the presence of inhibitory compounds, such as humic acids, heavy metals, phenolic compounds, or polysaccharides, that can affect amplification efficiency (Sidstedt et al., 2020). Commercially available DNA extraction kits and qPCR master mixes have been developed specifically for molecular analysis of environmental samples, with the aim of reducing the effects of inhibitory substances on molecular analysis or preventing coextraction of these compounds. Alternative strategies have been designed to correct for the presence of inhibitors by including internal amplification controls (IACs), or by examining the kinetics of amplification for each reaction (reviewed by Sidstedt et al., 2020). Coyne et al. (2005), for example, introduced a known concentration of exogenous plasmid DNA into the DNA extraction buffer as an IAC. The concentration of target DNA was then measured by qPCR using target-specific primers, and results were normalized to the internal standard in the same extract. Handy et al. (2006) investigated the accuracy and sensitivity of single-tube detection and quantification of both the internal standard and target microalgal species, showing that results were comparable to enumeration of each in individual qPCR reactions, essentially reducing the number of qPCR reactions required by 50%. van Doorn et al. (2009) expanded upon this approach by incorporating a series of IACs of decreasing concentrations in each reaction, resulting in a standard curve being generated for each individual reaction.

Comparisons to other methods

Since microscopy is often cited as the gold standard for enumeration of cells from complex samples, the development of qPCR assays often includes a comparison to microscopic cell counts. An early application of qPCR to investigate the abundance of *Alexandrium fundyense* in transect samples from the Gulf of Maine, for example, used both microscopy and an oligonucleotide probe assay for cross confirmation of qPCR (Dyhrman et al., 2006). Although cell counts were low in all samples, qPCR results exhibited a high degree of agreement with the other two methods. Rueckert et al. (2007) developed a qPCR assay for the noxious filamentous cyanobacteria, *Dolichospermum planctonicum* (formerly *Anabaena planktonica*), and directly compared the assay to microscopy using samples collected from Lower Karori Reservoir in New Zealand. qPCR results were significantly correlated to microscopic cell counts over a wide range, from less than 10 cells mL^{-1} to over 17,000 cells mL^{-1}, with greater agreement at higher cell densities. In a more recent example, Vassalli et al. (2018) compared qPCR assays for *Ostreopsis* spp. in both planktonic and benthic samples to manual cell counts and an automated optical recognition and counting method

(OPR) based on microscopic analysis. Results demonstrated a strong correlation between all three methods, although the authors noted a tendency toward an overestimation of cell abundance by qPCR, in agreement with other comparative studies (e.g., Godhe et al., 2007). Comparisons between qPCR and sandwich hybridization assay (SHA), using a single cellular homogenate, highlighted differences based on the target nucleic acid. qPCR uses DNA as a template, whereas SHA measures concentrations of the target RNA (Doll et al., 2014; Main et al., 2014, 2018). Analysis of environmental samples showed a good correlation between qPCR and SHA when cell concentrations were above bloom levels (10^5 cells L^{-1}), but not at lower cell densities (Main et al., 2018), which may be related to differences in sensitivity of each method when applied to complex samples.

Quantitative PCR applications in phytoplankton ecology

Some of the earliest qPCR assays applied to the study of phytoplankton ecology were designed for species that were difficult to identify by microscopy (Bowers et al., 2000; Foulds et al., 2002; Popels et al., 2003). Bowers et al. (2000) developed a qPCR assay for two HAB dinoflagellates, *Pfiesteria piscicida* and *P. shumwayae*, which until then, could only be distinguished using scanning electron microscopy (SEM). The assays were shown to be specific to each species and allowed for detection against a complex background of nonspecific DNA. Foulds et al. (2002) and Vaitomaa et al. (2003) designed qPCR assays targeting the microsystin synthetase gene cluster in cyanobacteria. This assay permitted identification and enumeration of toxic strains in environmental water samples, and discriminated toxic from nontoxic strains that may co-occur in the same environment. Popels et al. (2003) designed a qPCR assay for *Aureococcus anaphagefferens*, a small pelagophyte that is not easily identified due to its size and lack of distinguishing morphological features. Since these early applications, qPCR assays have been developed for a broad range of phytoplankton species, including freshwater and marine HABs and invasive species (e.g., Cary et al., 2014). Table 10.2 lists select qPCR assays developed for HAB species, with more extensive lists provided in Antonella and Luca (2013), Penna and Galluzzi (2013), and Pacheco et al. (2016).

Multispecies detection

The application of qPCR to enumerate multiple species of interest presents a challenge that can delay the adoption of this technology for routine mon-

Table 10.2 Select quantitative real-time PCR assays developed for harmful algal bloom species.

Target species	Reference
Alexandrium spp.	Dyhrman et al. (2006); Erdner et al. (2010); Garneau et al. (2011); Ruvindy et al. (2018); Vandersea et al. (2017)
Dolichospermum spp. (formerly *Anabaena*)	Al-Tebrineh et al. (2010); Rueckert et al. (2007)
Aureococcus anophagefferens	Guo et al. (2015); Popels et al. (2003); Zhang et al. (2020)
Chloromorum toxicum	Handy et al. (2006)
Chattonella spp.	Coyne et al. (2005); Park et al. (2012)
Cochlodinium spp.	Howard et al. (2012); Park et al. (2012)
Raphidiopsis (formerly *Cylindrospermopsis*) *raciborskii*	Lei et al. (2019)
Dinophysis spp.	Kavanagh et al. (2010)
Didymosphenia geminata	Cary et al. (2014)
Fibrocapsa japonica	Demir et al. (2008); Engesmo et al. (2018)
Gyrodinium instriatum	Handy et al. (2008)
Hematodinium sp.	Nagle et al. (2009)
Heterosigma akashiwo	Coyne et al. (2005); Engesmo et al. (2018); Park et al. (2012)
Karenia spp.	Engesmo et al. (2018); Yuan et al. (2012a)
Karlodinium veneficum	Handy et al. (2008); Engesmo et al. (2018)
Lingulodinium polyedrum	S. Moorthi et al. (2006)
Microcystis spp.	Hotto et al. (2008); Ngwa et al. (2014b); Zuo et al. (2018)
Ostreopsis spp.	Countway and Caron (2006); Hariganeya et al. (2013)
Prorocentrum spp.	Handy et al. (2008); Yuan et al. (2012b)
Pseudo-nitzschia spp.	Yamahara et al. (2015)
Thalassiosira rotula	He et al. (2007)

itoring. Assays are typically designed for one species at a time; analysis of multiple species in the same sample requires additional time, effort, and cost. This is not a deficiency in technological capacity, as many commonly used qPCR instruments are capable of distinguishing fluorescence emission spectra from multiple fluorophores, allowing for detection of several species-specific fluorescent probes. Instead, the challenge is in the design and validation of qPCR assays for multiple species. Handy et al. (2006) described the development of "multiprobing" assays, using a single primer pair with species-specific TaqMan probes to distinguish between two related harmful raphidophyte species that co-occur in Delaware's coastal bays. The TaqMan probes incorporated different fluorescent markers, allowing for detection and discrimination between species. Multiplex assays, in which species-specific primers and probes are included in the same qPCR reaction mix, have also been developed for phytoplankton species that are commonly found together. Te et al. (2015), for example, developed and validated a multiplex qPCR assay using primers and probes targeting two cyanobacteria species, whereas Al-Tebrineh et al. (2012) designed primers and probes to simultaneously detect genes from four cyanotoxin biosynthetic pathways in addition to primers and probe for cyanobacteria 16S rRNA as a control.

Ecotype- and strain-level detection

A distinct advantage of qPCR over imaging technologies is the ability to identify phytoplankton strains or ecotypes that cannot be distinguished morphologically. A recent study, for example, applied qPCR to evaluate environmental factors driving the abundance of two strains of unicellular N_2-fixing cyanobacteria (UCYN-A1 and -A2) over time and space (Henke et al., 2018). Using qPCR primers targeting variable regions in the *nifH* gene and correlating abundance with environmental factors, results of this study suggested the two strains occupied separate ecological niches. Other studies used qPCR to identify strains based on differences in the internal transcribed spacer (ITS), a noncoding region of the rRNA operon that can be used to distinguish genetic lineages distributed over large spatial scales or distinct habitats (e.g., Ahlgren and Rocap, 2012; Ahlgren et al., 2006; Wilkinson et al., 2016). Wolf et al. (2019) also developed a qPCR assay using allele-specific primers to investigate intraspecific interactions and responses to climate change within a single population of *Thalassiosira hyalina*. In a final example, Toebe et al. (2013) developed qPCR assays to distinguish between toxic and nontoxic ribotypes of the "*Alexandrium tamarense* species complex". Application of these assays to field samples collected along the Scottish coast revealed the cooccurrence of two ribotypes, supporting the hypothesis that they are cryptic species, rather than strains of the same species (Wolf et al., 2019).

Contributions to phytoplankton ecology

Application of qPCR assays has led to a deeper understanding of phytoplankton ecology, and factors regulating the distribution and activities of phytoplankton over time and space. Early applications of qPCR in the field, for example, (1) demonstrated the seasonal distribution of *Ostreococcus* and its contribution to picoeukaryote biomass in the eastern North Pacific (Countway and Caron, 2006), (2) identified toxic populations of *Microcystis* in Lake Erie (Rinta-Kanto et al., 2005), (3) evaluated vertical migration of *Heterosigma akashiwo* and *Chattonella subsalsa* in a shallow estuary (Handy et al., 2005), and (4) quantified microzooplankton grazing on *H. akashiwo* in mixed natural assemblages (Demir et al., 2008). These investigations highlighted the benefits of qPCR for investigating natural populations of phytoplankton in terms of increased accuracy and sensitivity, as well as a decrease in time and effort required to complete the analysis compared to traditional microscopic methods of analysis. Species- or group-specific qPCR assays are now commonly used to examine environmental factors that play a role in the distribution and abundance of HABs and other phytoplankton species (Demir-Hilton et al., 2011; Gast et al., 2014; Menning et al., 2018; Qiao et al., 2017; Vandersea et al., 2018; Zhen et al., 2016). A number of studies have also used qPCR to address novel questions about phytoplankton ecology, such as the potential for plastic debris to serve as a vector for HABs dispersal (Casabianca et al., 2019), to identify correlations between particle-associated pathogens and phytoplankton species (Main et al., 2015), or to assess the effect of an algicide on target HABs and non-target phytoplankton and other protists (Tilney et al., 2014). Unexpected discrepancies between cell counts obtained by qPCR and microscopy because of mismatches in primer or probe sequences have led to discoveries of new species (Coyne et al., 2005; Demir-Hilton et al., 2012) and ecotypes (Zinser et al., 2006). For example, the application of a qPCR assay for *H. akashiwo* failed to detect the species in bloom samples where microscopic cell counts for this species exceeded 10 million cells L^{-1} (Coyne et al., 2005). Isolated cells resembling *H. akashiwo* were further examined by electron microscopy, along with pigment and sequence analysis, to support the discovery of a new genus and species, *Viridilobus marinus* (Demir-Hilton et al., 2012).

Quantitative PCR applications for cyanobacteria

In addition to applications for eukaryotic species mentioned above, qPCR has been widely used to investigate the distribution and activity of cyanobacteria and changes in the relative abundance of toxic versus non-toxic strains in laboratory culture experiments, as well as in the field. A number of reports, for example, describe the application of qPCR to address questions about the response of cyanobacteria to nutrients (e.g., Davis et al., 2009; Li et al., 2012; Pimentel and Giani, 2014; Yoshida et al., 2007), temperature (Davis et al., 2009; Joung et al., 2011; Panksep et al., 2020), light (Renaud et al., 2011; Wang et al., 2019), CO_2 (Van de Waal et al., 2011; Yu et al., 2015), and the presence of grazers (Davis and Gobler, 2011; Park et al., 2018). Quantitative PCR assays for toxic strains have also been evaluated as a proxy for cyanotoxin concentrations in water samples. Direct measurement of toxins is laborious and expensive (reviewed by Pacheco et al., 2016), whereas microscopy is not able to distinguish toxic vs. non-toxic strains of the same species, which may cooccur in field samples. To address these issues, qPCR assays were designed to target toxin synthesis genes, such as the microcystin synthetase (*mcy*) gene cluster in *Microcystis*, *Planktothrix* and *Dolichospermum* spp. (formerly *Anabeana*), sometimes used in parallel with qPCR assays targeting 16S rRNA gene sequences to determine total cyanobacteria abundance (e.g., Gagala et al., 2014; Lin et al., 2011; Muller et al., 2017; Rinta-Kanto et al., 2005). These and other studies (e.g., Al-Tebrineh et al., 2011; Rueckert et al., 2007) demonstrate the value of qPCR as an early warning monitoring tool, with increased sensitivity over microscopic methods and the ability to distinguish between toxic and nontoxic strains. Correlations between *mcy* gene abundance and toxin concentrations in field samples, however, have yielded mixed results (reviewed by Pacheco et al., 2016). In an early effort to apply qPCR in the field, Vaitomaa et al. (2003) demonstrated a significant positive correlation between *Microcystin* and *Dolichospermum mcyE* gene abundance and microcystin concentrations in several Finnish lakes, but only when all data sets were combined. That relationship deteriorated for data collected from a single lake. In a retrospective analysis, Pacheco et al. (2016) examined 33 reports evaluating correlations between qPCR assays for *mcy* gene abundance and either microcystin concentrations or cyanobacteria cell counts in field samples. Among these, only 67% of the reports demonstrated a significant correlation with toxin concentrations, whereas 84% showed a significant correlation with cell counts. Due to the lack of consistent correlation between toxin concentrations and *mcy* gene abundance in environmental water samples, Panksep et al. (2020) instead used qPCR analysis to evaluate risk. They defined the toxin quota as the microcystin concentration per unit of *mcy* gene copies as determined by qPCR. A high toxin quota, then, would signify greater risk to the public, regardless of cell density. Further analysis of environmental factors indicated a significant correlation between water temperature and toxin quota, supporting previous work by Wood et al. (2017) that demonstrated a correlation between microcystin quotas on a per biomass basis with temperature.

The lack of correlation between *mcy* gene abundance and toxin content is not unexpected, as several reports demonstrate that *mcy* expression is not constitutive (e.g., Ngwa et al., 2014a; Wood et al., 2011, 2012). The presence of mixed assemblages of toxigenic species within a population may play a role in toxin production, where *mcyE* gene expression and toxin concentrations were repressed in mixed cultures (Ngwa et al., 2014a). A better understanding of factors driving growth and toxic activity for cyanobacteria and other phytoplankton species requires analysis of cellular response to environmental stimuli, such as changes in transcript abundance discussed below.

Reverse transcriptase qPCR (RT-qPCR)

The qPCR assays described above used DNA as template in the reaction, permitting enumeration of target cells. As highlighted above for cyanobacteria, however, these assays may not indicate activity or cellular response of phytoplankton species beyond changes in growth. Quantitative reverse transcription PCR (RT-qPCR), on the other hand, uses RNA as the starting material to determine changes in mRNA transcript abundance for the target gene. For this method, RNA is first reverse transcribed into complementary DNA (cDNA), which is then used as template to conduct qPCR (reviewed by Medlin and Orozco, 2017). The RT and qPCR reactions can be conducted individually or combined as a "one-step" RT-qPCR reaction with the reverse transcription and PCR amplification occurring in the same tube. The one-step approach employs gene-specific primers for cDNA synthesis and PCR amplification. This method is convenient with minimal handling and may be more sensitive than the two-step approach (Wacker and Godard, 2005). In the two-step approach, cDNA synthesis and qPCR are conducted in separate reactions, which allows for optimization of each step, and the cDNA product of the RT reaction may then be interrogated for multiple transcripts in separate qPCR reactions.

Reverse transcriptase-qPCR has also been used as a tool to verify data generated by other technologies such as microarray (e.g., Dittami et al., 2013; Johnson et al., 2012; Ritter et al., 2014) and RNA-seq sequencing (e.g., Lauritano et al., 2019; Lv et al., 2013; Zhang et al., 2019) in studies on both laboratory monocultures and samples collected from natural communities. Although technologies such as microarray and RNA-seq sequencing can provide wide coverage of gene expression profiles, RT-qPCR is still considered the gold standard in the quantification of transcript changes (Bustin, 2010; Pinto et al., 2012; Shim et al., 2016).

Analysis of gene expression in phytoplankton

RT-qPCR has been widely applied to study algal gene expression under different environmental conditions or in response to stimuli, typically for laboratory monocultures (Colina et al., 2019; González-Hourcade et al., 2020; Lauritano et al., 2019; Wang et al., 2018). For instance, Lauritano et al. (2015) applied RT-qPCR to investigate the expression of 18 stress-related genes (e.g., antioxidant genes, heat shock proteins, and aldehyde dehydrogenases) in diatom *Skeletonema marinoi* under various conditions, including nutrient limitation/starvation, CO_2 enrichment, and the combination of these stressors. Results of this study suggested that *S. marinoi* differentially regulates transcription of stress response genes as a bioenergetic saving strategy. In another example, Yu et al. (2019) demonstrated down-regulation of genes involved in photosynthesis and microcystin synthesis in *Microcystis aeruginosa* in response to algicidal substances produced by bacterium *Streptomyces amritsarensis*, providing insights into the molecular mechanisms associated with algicidal activity.

Compared to the wide application of RT-qPCR on laboratory monocultures (Table 10.3), fewer studies have used RT-qPCR to investigate dynamics of harmful algal blooms in co-cultivation systems (e.g., Scherer et al., 2018) or natural communities (e.g., Dittami et al., 2013; Lu et al., 2020, 2019; Wood et al., 2011; Yu et al., 2019; Zhang et al., 2019). Questions remain about the relevance of unialgal laboratory culture experiments to natural field samples. Wurch et al. (2014) for example, investigated the expression of genes responding to nitrogen- (xanthine/uracil/vitamin C permease; *XUV*) and phosphorous-deficiency (inorganic phosphate transporter, *PTA3*) in *A. anophagefferens* from samples collected during a natural bloom, and compared results to laboratory culture experiments. The authors demonstrated *XUV* expression peaked at the bloom peak, but was lower than expression levels in nitrogen-deficient unialgal laboratory cultures. In contrast, *PTA3* expression peaked early during the bloom and surpassed the level indicating phosphorous-deficiency in laboratory cultures.

Table 10.3 Select references for RT-qPCR application in algal studies.

RT-qPCR Targeted Algal Species	Reference
Laboratory monoculture(s)	
Aureococcus anophagefferens (Pelagophyceae)	Wurch et al. (2014)
Bostrychia moritziana (Florideophyceae)	Shim et al. (2016)
Chlamydomonas reinhardtii (Chlorophyceae)	Colina et al. (2019); Lv et al. (2013)
Chattonella subsalsa (Raphidophyceae)	Wang et al. (2018)
Coccomyxa simplex (Trebouxiophyceae)	González-Hourcade et al. (2020)
Heterosigma akashiwo (Raphidophyceae)	Coyne (2010)
Karenia brevis (Dinophyceae)	Johnson et al. (2012)
Lyngbya aestuarii (Cyanophyceae)	Pinto et al. (2012)
Microcystis aeruginosa (Cyanophyceae)	Yu et al. (2019)
Nostoc sp. (Cyanophyceae)	Pinto et al. (2012)
Skeletonema marinoi (Bacillariophyceae)	Lauritano et al. (2015)
Synechocystis sp. (Cyanophyceae)	Pinto et al. (2012)
Tetraselmis suecica (Chlorodendrophyceae)	Lauritano et al. (2019)
Trebouxia sp. (Trebouxiophyceae)	González-Hourcade et al. (2020)
Co-cultivation system	
Microcystis aeruginosa (Cyanophyceae)	Scherer et al. (2018)
Natural communities	
Aphanizomenon flos-aquae (Cyanophyceae)	Lu et al. (2019)
Aureococcus anophagefferens (Pelagophyceae)	Wurch et al. (2014)
Dolichospermum sp. (Cyanophyceae)	Lu et al. (2019)
Microcystis sp. (Cyanophyceae)	Lu et al. (2020); Lu et al. (2019); Yu et al. (2019)
Nostocales sp. (Cyanophyceae)	Lu et al. (2019)
Planktothrix sp. (Cyanophyceae)	Lu et al. (2020)
Prorocentrum donghaiense (Dinophyceae)	Zhang et al. (2019)
Pseudochattonella sp. (Dictyochophyceae)	Dittami et al. (2013)

Case Study 10.1: Application of RT-qPCR

Here we describe the application of RT-qPCR to investigate microbial community dynamics and differential regulation of nitrate reductase genes during harmful algal blooms caused by *Chattonella subsalsa* (Raphidophyceae). *Chattonella* spp. (*C. antiqua*, *C. marina*, *C. subsalsa*, and *C. ovata*) are harmful algae that have been associated with fish mortality worldwide (Basti et al., 2016; García-Mendoza et al., 2018; Imai and Yamaguchi, 2012; Qiu et al., 2020; Viana et al., 2019). In

Delaware's Inland Bays (DIB; Delaware, US), *C. subsalsa* forms mixed blooms with other harmful raphidophytes, including *H. akashiwo* and *Fibrocapsa japonica* (Zhang et al., 2006). Factors contributing to the persistence of these blooms and the ability of cooccurring raphidophytes to avoid competitive exclusion are unclear. In DIB, nitrate is the dominant nitrogen source, and its enrichment may be an essential factor for bloom initiation (reviewed by Coyne, 2010). As one of the main nitrogen sources utilized by *C. subsalsa* (Zhang et al., 2006), nitrate and its assimilation may play a role in the competitive dynamics of this species.

In plants and phytoplankton, nitrate is reduced to ammonium by two sequential reactions, in which nitrate reductase (NR) catalyzes the first and rate-limiting step (Flores et al., 2005). *NR* gene expression and enzyme activity are highly regulated by environmental conditions, such as change of light, temperature, and nitrogen sources (reviewed by Glibert et al., 2016; Sanz-Luque et al., 2015). Previous research identified two novel nitrate reductases in *C. subsalsa*, CsNR2-2/2HbN (NR2; Stewart and Coyne, 2011) and NR3 (Wang et al., 2018). *NR2* and *NR3* transcript abundances for laboratory monocultures of *C. subsalsa* were evaluated using RT-qPCR, demonstrating that gene expression was differentially regulated by environmental conditions (Wang et al., 2018).

Here, the transcriptional response of *C. subsalsa* to a single pulse of nitrogen (nitrate, ammonium, or both nitrate and ammonium) was investigated during a bloom of this species in DIB. RT-qPCR analysis was conducted to investigate temporal changes in expression of *C. subsalsa NR2* and *NR3* and to evaluate the contribution of nitrogen source to competitive dynamics of *C. subsalsa* in a mixed natural community.

Materials and methods

A detailed description of this investigation was described in Wang (2016). Briefly, water samples were collected in July 2015 from the Inland River Bay, DE, US, during a bloom of *C. subsalsa* and divided into four treatments (N = 4). No nutrients other than nitrogen were added (ambient nitrate = 2.38 μM; ambient ammonium = 4.72 μM). Treatments included additions of 100 μM nitrate (designated NO3), 100 μM ammonium (designated NH4), or 50 μM nitrate and 50 μM ammonium (designated BOTH). Control samples did not receive additional nitrogen.

Samples were collected for analysis at the initial timepoint (T0) just prior to the nitrogen addition and again at 1 hour (T1) and 24 hours (T24) after the addition of nitrogen. Details for dissolved nutrient and chlorophyll *a* analysis, as well as RNA extraction can be found in Wang (2016). Growth rates were calculated based on cell counts for *C. subsalsa* and on chlorophyll *a* content for the entire phytoplankton community. The synthesis of cDNA from total RNA was performed as described in Wang et al. (2018). Expression of *NR2*, *NR3*, and *ß-tubulin* (as a normalizing gene) was determined in separate 10 μL reactions by qPCR using SYBR green master mix (ThermoFisher Scientific) and gene-specific primers (Table 10.4) on the ABI Prism 7500 sequence detection system (ThermoFisher Scientific), as described in Wang et al. (2018). A dissociation curve was generated after amplification to evaluate specificity of the reaction.

Table 10.4 Sequences of *Chattonella subsalsa*-specific primers for RT-qPCR analysis of nitrate reductase (*NR2* and *NR3*) and *ß-tubulin* mRNA expression.

Primers[a]	Sequence (5′-3′)	Gene Targeted
CsNR Glob 1005F	GAAGCTGCAATCCAAGCTGTAGT	*NR2*
CsNR Glob 1240R	CCAACTCAGTCAACGTGTTCA	*NR2*
CsNR3 1143F	CTGGACACAGCTGAGTTACA	*NR3*
CsNR3 1228R	GCAATGCGATGAGTTGGAG	*NR3*
Cs ß-Tubulin 695F	CACTTGCTGGAACCACTTGC	*ß-Tubulin*
Cs ß-Tubulin 878R	TGCTGTTGAGTCAGCTCAGG	*ß-Tubulin*

[a] F, forward primer; R, reverse primer.

Details for statistical analysis can be found in Wang (2016).

Results and discussion

On the day of collection, the cell density of *C. subsalsa* was 4.3×10^6 cells L^{-1}. A number of other harmful algal species were also identified in the field sample, including *Chloromorum toxicum*, *H. akashiwo*, *Levanderina fissa* (formerly *Gyrodinium instriatum*), *Heterocapsa rotundata*, *Peridinium* sp., *Prorocentrum triestinum*, *Protoperidinium quinquecorne*, and *Karlodinium veneficum*.

The growth rate of *C. subsalsa* was positive in all treatments (Fig. 10.2). In the No addition control group, the growth rate of

C. subsalsa was significantly higher than the entire assemblage ($p < 0.05$), with a positive growth for *C. subsalsa* and a negative growth for the entire assemblage, suggesting that *C. subsalsa* outcompeted other species under low nitrogen concentrations. Previous research reported the ability of *Chattonella* spp. to utilize nitrogen sources at low levels (Zhang et al., 2006), as well as their ability to maintain low cell densities as background flora, becoming dominant with the reloading of nutrients (reviewed by Imai and Yamaguchi, 2012). *Chattonella* spp. are also mixotrophic and may be able to acquire nutrients through predation to maintain their growth and population size under low nutrient conditions (Jeong, 2011).

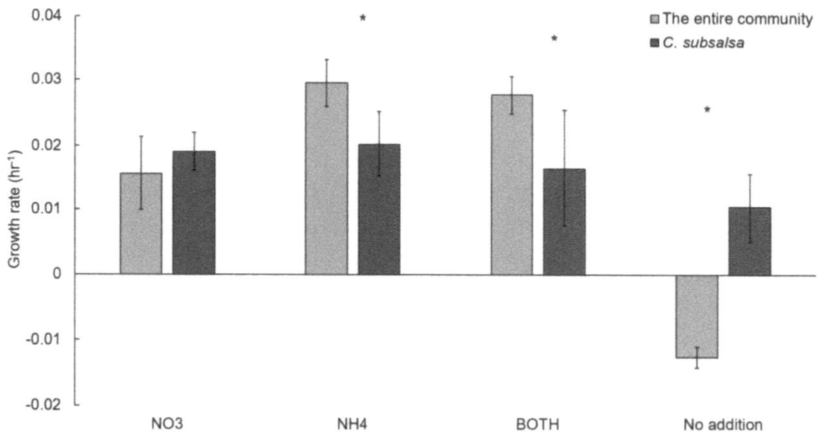

Figure 10.2 Growth rates (hr^{-1}) of *C. subsalsa* and the phytoplankton community in treatments with added nitrate (NO3), ammonium (NH4), both nitrate and ammonium (BOTH), or controls with no nutrient addition (No addition) over 24 hours. Error bars are standard deviations of four biological replicates. Asterisks above the bars indicate the significant difference in the growth rate between *C. subsalsa* and the entire community in the same treatment ($p < 0.05$).

RT-qPCR analysis revealed distinct differences in expression of *NR2* and *NR3* at T1hr for each treatment. *NR2* expression significantly increased in the NO3 treatment compared to T0, whereas it decreased significantly in the NH4 and BOTH treatments ($p < 0.05$; Fig. 10.3A). These results were consistent with previous studies on the effects of nitrate and ammonium on *NR* gene expression, in which the NR transcript abundance was up-regulated by nitrate and down-regulated by ammonium in a wide range of algal species (Coyne, 2010; de Montaigu et al., 2011; Hou et al., 2017; Imamura et al., 2010;

Song and Ward, 2004). In contrast to these results, prior research on laboratory cultures of *C. subsalsa* indicated that *NR2* expression was significantly higher in the ammonium treatment compared to those cultured with nitrate (Wang et al., 2018). This may reflect differences in the cellular nutrient status and/or the different growth stages of this alga in these two studies. The concentrations of other nutrients in cultures may also affect the response of this NR gene. For instance, Hou et al. (2017) demonstrated that *NR* gene expression in *Haematococcus pluvialis* increased significantly under nitrate deficiency, at an expression level higher than or equal to the cells cultured with nitrate. It should also be noted that *C. subsalsa* cultures in Wang et al. (2018) were acclimated to nitrate or ammonium as the sole nitrogen source, whereas in this study, additional nitrogen sources were present that may have played a role in the differential responses of *C. subsalsa NR2* in these two studies.

NR2 expression significantly decreased in the NO3 treatment as well as the No addition group at T24hr compared to T1hr ($p < 0.05$; Fig. 10.3A), whereas there was a significant increase in *NR2* expression in the BOTH treatment. This pattern is similar to *NR* expression measured in nitrogen-starved *D. tertiolecta* after nitrate addition (Song and Ward, 2004). The increase in *NR2* expression in the BOTH treatment was accompanied by an increase in nitrate to ammonium ratios from 0.5 to 1.1 in the cultures (data not shown), suggesting that ammonium is preferentially taken up by this species, and that regulation of *NR* by these nitrogen sources may depend on their relative abundance (reviewed by Glibert et al., 2016). Other research demonstrated a similar effect of nitrate to ammonium ratios on *NR* expression in *Chlamydomonas* (de Montaigu et al., 2011; Sanz-Luque et al., 2015).

In contrast to *NR2* expression, there was a significant increase in *NR3* expression in the No addition group at T1hr compared to T0 ($p < 0.05$; Fig. 10.3B), whereas addition of nitrogen had no significant effect on expression in any of the other treatments. This is consistent with previous research in unialgal laboratory cultures showing that *NR3* expression was less sensitive to nitrogen source or the balance between nitrate and ammonium than *NR2*, but more sensitive to other environmental factors, such as a change in light (Wang et al., 2018). At T24hr, *NR3* expression in the No addition group was significantly higher than the NH4 treatment ($p < 0.05$), but there were no significant differences in *NR3* expression between any other treatments.

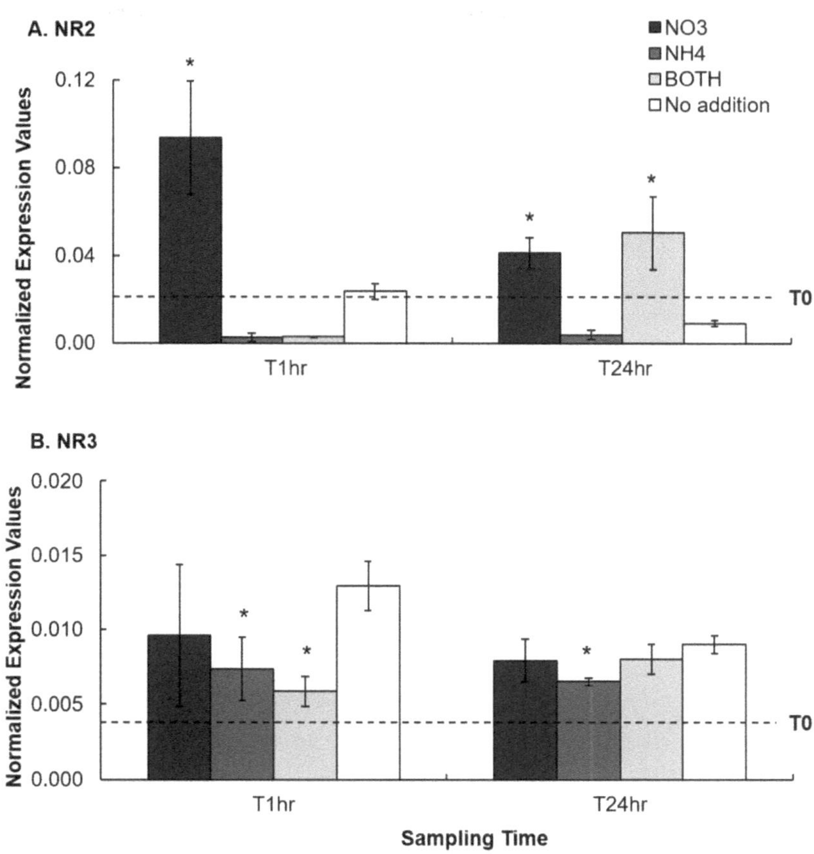

Figure 10.3 *NR2* (A) and *NR3* (B) expression, normalized to $\beta - tubulin$ at 1 hour (T1hr) and 24 hours (T24hr) after adding nitrate (NO3), ammonium (NH4), both nitrate and ammonium (BOTH), or in controls with no nutrient addition (No addition). Gene expression at T0 (the initial time point) is indicated by the dashed line. Error bars indicate standard deviations of four biological replicates. Asterisks indicate the significant difference in *NR* expression between the indicated group and the No addition group at the same time point ($p < 0.05$).

The rapid response of *NR2* to nitrate input may provide an advantage to relatively large cells of *C. subsalsa* in environments where they must compete with other smaller-celled species, such as *H. akashiwo*. Rapid down-regulation of NR2 with ammonium input is also a cost-saving measure, with lower energy required for the assimilation of ammonium over nitrate. On the other hand, regulation of NR3 may allow *C. subsalsa* to fine-tune its response to nitrogen in dynamic coastal environments where nitrate and ammonium concentrations fluctuate over small timescales.

Conclusion

This case study illustrates the power of RT-qPCR to investigate temporal changes in transcript abundance and potential mechanisms contributing to bloom initiation by algal species. Results of this study contribute to our understanding of bloom dynamics and the ability of *C. subsalsa* to outcompete other species in environments with large fluxes in nitrogen input and changes in nitrogen source, while also highlighting differences with monoculture laboratory experiments (Wang et al., 2018).

Recent advances in qPCR technology

Recent advances in qPCR technology (reviewed by Beeton-Kempen, 2020) reduce assay time and reaction volumes while increasing throughput, with benchtop instruments now able to accommodate over 5000 reactions in a single run (SmartChip Real-Time PCR System, Takara Bio, Inc.). Analysis of multiplexed samples that can be discriminated by melt point analysis (HRM; Wittwer, 2009), or by single fluorescence channel multiplexing that differentiates target amplicons by adjusting detection temperatures within each cycle (MuDT; Lee et al., 2014) opens up qPCR assays for detection of a greater number of species in a single reaction. Novel instrumentation, such as droplet-on-thermocouple silhouette real-time PCR (DOTS qPCR; Harshman et al., 2015) can provide results in as few as four cycles (three minutes and 30 seconds), whereas portable, field-ready qPCR instrumentation and droplet digital PCR (ddPCR; Hindson et al., 2011), both described more fully below, have already proven to be of value to HAB research.

Detection of harmful algal blooms via portable PCR

New technologies for the detection of HABs and other phytoplankton via DNA and RNA are becoming smaller, more affordable, and highly portable; in some cases, interfacing directly with widely available smartphones. In practice, this trend towards remote and/or *in situ* detection of genetic signatures from microbial species began more than a decade ago. Specifically, the development of the environmental sample processor

(ESP; Scholin et al., 2009; Yamahara et al., 2015) to detect harmful *Pseudo-nitzschia* (Greenfield et al., 2006) and other taxa via hybridization probes and qPCR (Preston et al., 2011; Scholin et al., 2017) prompted interest in field-based genetic detection of HABs. Initially designed to be moored at a fixed location, changes in this technology have allowed ESP deployment on gliders (Seegers et al., 2015) for qPCR detection of HAB populations in subsurface waters off Southern California. A key advantage of this platform is the accessibility of the data via cloud services or email immediately following a run, leading to rapid data analysis and interpretation of results, independent of proximity to a traditional lab. Such autonomous observatories with embedded genetic analysis systems provide high-resolution data from remote sites, but also come with high operational costs and requirements for advanced training. For these reasons, observatory-scale systems are somewhat impractical for routine HAB monitoring needs at geographically dispersed locations, prompting the need for simple, inexpensive, and portable genetic technology.

One of the first portable genetic analyzers to be designed specifically for HAB research, commonly known as the "Tricorder" among HAB researchers, has proven to be useful for the analysis of *Karenia brevis* (Casper et al., 2007). The previous approach is based on amplification of RNA by the nucleic acid sequence based amplification (NASBA) method, which is highly sensitive, generates rapid results, and relies upon isothermal amplification, therefore simplifying hardware requirements. Similar NASBA assays including internal controls (IC-NASBA) were designed for *Karenia mikimotoi* (Ulrich et al., 2010) and *P. multiseries* (Delaney et al., 2011), thereby expanding options for rapid HAB quantification. Other portable genetic systems are based on detection via loop-mediated isothermal amplification (LAMP), a technique with widespread applications for HAB research, including detection of *Alexandrium* sp. (Nagai and Itakura, 2012; Wang et al., 2020a), *Prymensium parvum* (Zhang et al., 2014), and *K. mikimotoi* (Huang et al., 2020; Wang et al., 2020b).

There are increasing options for traditional qPCR (e.g., primer-based assays that utilize Sybr Green and probe-based assays, such as TaqMan and Molecular Beacon) on portable systems that can be used virtually anywhere. For example, the Two3 system (Biomeme Inc., Philadelphia, PA) has been used to confirm the presence of a potentially harmful cyanobacteria species at a remote location in Maine, and to confirm the presence of marine bacteria with Dimethylsulfoniopropionate (DMSP) metabolism capabilities in a remote Antarctic field station (Countway, unpublished). Although the Two3 had limited throughput capability, it was still deemed useful for a number of field-based applications, including detection of aquatic invasive species (Sepulveda et al., 2018; Thomas et al., 2020), pathogenic viruses (Brown et al., 2020; Hole and Nfon, 2019; Russell et al., 2018), and pathogenic bacteria (Nguyen et al., 2018). The Two3 has been replaced by the Three9 ("Franklin") system, a reference to its 3-color, 9 sample capacity (Biomeme Inc., Philadelphia, PA), which has greatly improved sample throughput. Features of the Three9 include battery-powered capability, a user-friendly programming interface via the smartphone application "Biomeme Go," Bluetooth or USB connectivity to the smartphone, similarity to bench-top qPCR systems in terms of workflow and speed of analysis, and accessibility to both raw and processed data immediately after the run. Other portable instruments have greater capacity, such as the Open qPCR system (Chai Bio, Santa Clara, CA), which retains a traditional qPCR format with 2-color, 16 sample capacity. The Open qPCR system has been used to detect RNA from malaria parasites (Taylor et al., 2017) and to aid in the selection of aptamers that can bind to a virus that infects *Drosophila melanogaster* (Damase et al., 2018). Key features of the Open qPCR include its rapid thermal cycling—40 cycles in 20 minutes—as well as the "open source" philosophy regarding its associated software and API, its ability to perform high-resolution melt-curve analysis, and the ability to generate quantitative results from standard curves. A key feature of both portable qPCR systems is their ability to run existing qPCR assays that were developed with benchtop instruments, with little to no optimization or customization required. One caveat is that existing fluorogenic probes that are used on lab-based qPCR systems must be compatible with one of the detection channels on these portable systems.

There will likely be a proliferation of new, low-cost, portable genetic detection systems in the near future, such as the one developed by Ganguli et al. (2020), spurred by the need for rapid and affordable SARS-CoV-2 testing options in the face of the global pandemic. These new and recently developed portable platforms could be used for a wide range of diagnostic applications for HAB monitoring and phytoplankton research.

Field collections and extraction of environmental DNA

A major challenge with the application of portable DNA/RNA detection technology is extraction of DNA and/or RNA samples outside of a traditional laboratory setting. In general, the collection of samples for monitoring HABs is trivial, and is often as simple as deploying a bucket with attached line, a water bottle to parse out individual samples, and a cooler with ice packs for storage until filtration. Filtration of water samples can be accomplished at the point of collection with portable vacuum or peristaltic pumps connected to filter towers, or inline filters/filter cartridges, respectively. A manually operated vacuum hand-pump attached to a one-liter plastic side-arm flask equipped with a 500 mL magnetic filter tower is sufficient for most field applications of portable qPCR. A 47 mm GF/F filter installed in the filter tower allows for passage of relatively large volumes of water (generally 250–2000 mL, depending on water clarity and particle load) while retaining particles >0.7 μm, which should be sufficient for sampling most phytoplankton species. Filtered samples should be stored on ice, frozen at −20 °C (<1–2 weeks) or −80 °C (longer) as soon as possible, if not extracted immediately for DNA or RNA and analyzed by qPCR.

Choices for DNA/RNA extraction outside of the lab generally rely upon approaches that utilize a combination of low-tech procedures involving physical (heating), chemical (lysis buffer), and mechanical (bead-beating) cell lysis, resulting in a mixture of DNA/RNA and lysis buffer. Assuming that samples are processed minimally, to the point of generating a crude cell lysate (CCL), all of these cell lysis procedures can be achieved in the field, or outside of a traditional molecular biology lab as described below. Solutions for field-based extractions are becoming more common. These range from simple syringe-based extraction and purification protocols (Zhang et al., 2013) to a miniature sonicator/bead-beater (Lim et al., 2016). Commercially-available DNA (or RNA) extraction kits designed for use in the field have been applied to a variety of environmental targets (Brown et al., 2020; Nguyen et al., 2018; Thomas et al., 2020), including the harmful dinoflagellate *Karlodinium* (Toldrà et al., 2018). The latest version of one such commercial kit (M1 sample prep and homogenization kit, Biomeme Inc.) incorporates manual bead-beating of filtered eDNA samples (using a single ball bearing in a 5 mL tube) followed by DNA capture, purification, and elution from a syringe-based column that requires no specialized lab equipment and results in minimal chemical waste.

A simple bead-beating procedure can also be used to generate a CCL, followed by dilution with molecular grade water to obtain quick and reproducible extractions from filtered HAB samples. Minimal processing of DNA samples to generate CCLs has the primary advantage of minimizing losses of DNA, resulting in less variability among DNA replicates. The only known disadvantage of this minimal approach is the requirement for dilution of the CCL 1:100 with molecular grade water to relieve PCR inhibition from the lysis buffer and cellular constituents, which would otherwise be removed in more extensive DNA extraction. Dilution at this level might reduce rare DNA templates to less than detectable levels; however, this can sometimes be counteracted by filtering greater volumes to maximize chances for sample detection and using as much diluted sample as possible (e.g., 5 μl in a 20 μl reaction). In practice, this method has been used to detect the genomic equivalent of just 16 *Ostreococcus* sp. cells per reaction at the lowest concentration of cell-based standards, illustrating the sensitivity of the approach (Countway and Caron, 2006). Two studies that utilized the extraction procedure described above, but targeting the HABs *Alexandrium catenella* (Garneau et al., 2011) and *Lingulodinium polyedrum* (Moorthi et al., 2006), reported detection limits of 12 and 21 cells per liter, respectively. Detection limits for many dinoflagellate species are quite low, due to the high copy numbers of the gene target in each cell (Gong and Marchetti, 2019; Zhu et al., 2005), a feature that makes it easier to detect low cell concentrations of many HAB dinoflagellates compared to other phytoplankton taxa, such as *Ostreococcus* (e.g., Countway and Caron, 2006).

Case Study 10.2: Portable qPCR

In the fall of 2016, the state of Maine (USA) experienced a widespread and persistent bloom of *Pseudo-nitzschia*, resulting in high concentrations of domoic acid in shellfish and shellfish harvesting closures along much of Maine's coastline. Although the Gulf of Maine harbors multiple species of *Pseudo-nitzschia* (Fernandes et al., 2014), the 2016 bloom included the first confirmed detection of *Pn. australis* along the Maine coast and the high toxicity in shellfish was attributed to this species (Clark et al., 2019).

Methods

Scientists from Bigelow Laboratory for Ocean Sciences (East Boothbay, ME) sampled the *Pseudo-nitzschia* bloom between the midcoast and downeast sectors of the Maine coast over a period of several days in early October of 2016. Samples were collected with a hand-deployed Niskin bottle at depths of 2 and 10 m, and immediately filtered. Samples were processed to the stage of crude cell lysates (CCL), as described previously (Countway and Caron, 2006), and diluted 1:100 prior to qPCR analysis.

Quantitative PCR was conducted using both the portable Three9 qPCR device (Biomeme Inc., Philadelphia, PA), and a traditional benchtop machine, the CFX96 qPCR system (Bio-Rad, Hercules, CA) to evaluate the efficacy, cost-effectiveness, and accuracy of the Three9 system when using single samples, rather than multiple technical replicates. DNA standards to calibrate the qPCR assay were prepared from a mixture of two plasmids containing cloned 18S rRNA genes of *Pn. pungens* and *Pn. fraudulenta* and quantified by Qubit fluorometry (ThermoFisher Scientific, Waltham, MA) to estimate gene copy numbers. The linearized plasmid DNA was serially diluted to generate a series of standards to achieve gene copies per reaction, ranging over 8 orders of magnitude. Quantitative PCR was then performed using genus specific *Pseudo-nitzschia* qPCR primers (Fitzpatrick et al., 2010) and PerfeCTa Sybr Green FastMix (QuantaBio, Beverly, MA) on both Three9 and CFX96 qPCR systems. Samples and standards were run as triplicate technical replicates on the CFX96 system, and as single assays on the Three9, due to the limited number of samples that can be accommodated by this instrument. Standards and samples were also run separately on the Three9, due to the smaller sample block.

Results and discussion

Comparison of results for both qPCR platforms showed similar trends of *Pseudo-nitzschia* 18S rRNA gene copies per liter of seawater as displayed in Ocean Data View (ODV) (Schlitzer, 2020) for the 2 m depth horizon (Fig. 10.4). Results for the 10 m samples were similarly congruent between the portable Three9 system and the lab-based CFX96 system when plotted in ODV (not shown). The resulting gene abundance estimates for *Pseudo-nitzschia* spanned a large dynamic range, from 4.1×10^5–9.7×10^7 (Three9) to 5.3×10^5–9.3×10^7 (Bio-Rad) copies per liter. Both systems revealed a "hotspot" of *Pseudo-nitzschia* abundance in Machias Bay, with lower concentrations at offshore stations (Fig. 10.4).

When *Pseudo-nitzschia* gene abundance data from all depths and stations from each qPCR platform were plotted against one another, the resulting trend was highly significant ($R^2 = 0.8979$, $p < 0.001$) (Fig. 10.5). Overall, the portable Biomeme Three9 system appeared to perform well, but slightly overestimated the copy number compared to results from the lab-based Bio-Rad CFX96 system. Additionally, the diluted crude cell lysates did not appear to present a problem for detection by the Three9, reducing time and costs associated with qPCR analysis of field samples.

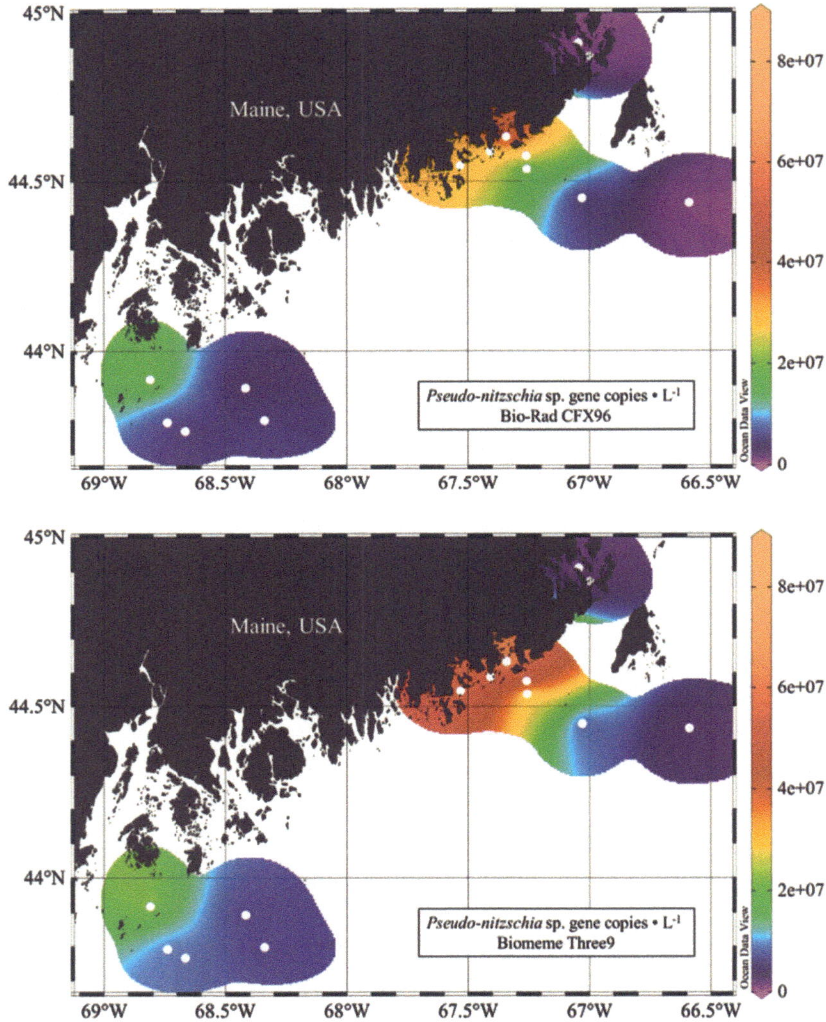

Figure 10.4 Estimates of *Pseudo-nitzschia* sp. gene copies per liter by qPCR along the coast of Maine on the lab-based Bio-Rad CFX96 (top panel) and the portable Biomeme Three9 (bottom panel). Color scaling represents gene copies per liter. Shading was determined by weighted-average gridding in Ocean Data View (v. 5.3).

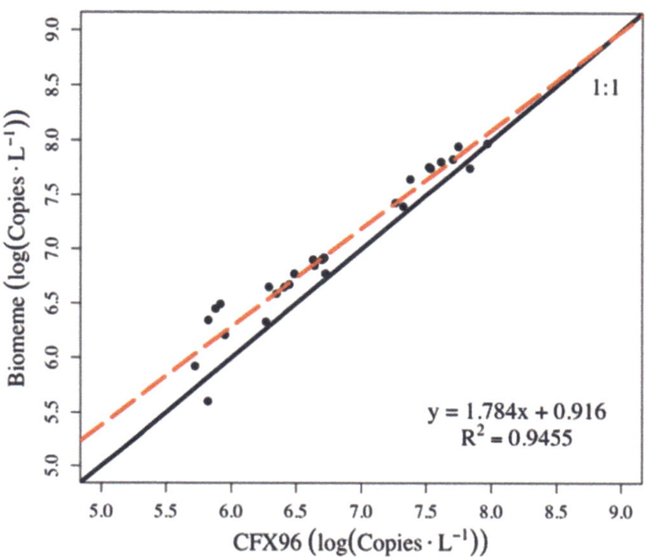

Figure 10.5 Regression analysis of *Pseudo-nitzschia* gene copies per liter as determined on the portable Three9 (Biomeme) system vs. the lab-based CFX96 (Bio-Rad). The regression (dashed red line) was highly significant ($p < 0.001$), although Biomeme estimates were slightly higher than the estimates from Bio-Rad for most of the sample comparisons.

Droplet digital PCR

Droplet digital PCR (ddPCR) is an alternative method of quantitative PCR that derives the target's abundance by utilizing a water-oil emulsion droplet system. Droplet digital PCR uses a combination of microfluidics and proprietary surfactant chemistries to divide PCR samples into approximately 20,000 individual nanoliter-sized droplets (Fig. 10.6; Hindson et al., 2011; Li et al., 2020; Salipante and Jerome, 2020). An individual PCR reaction occurs within each droplet with detection based on fluorescence that can be generated either by (i) double-stranded DNA binding dyes, or (ii) fluorescently labeled oligonucleotide probes. After thermal cycling, the droplets are analyzed one-by-one in a specialized instrument and assigned "1" if positive and "0" if negative (the basis for the term digital). The absolute number of copies of the target is calculated from the proportion of positive partitions and statistically corrected with a Poisson distribution (Huggett and Whale, 2013).

The basic principle of digital PCR is that there is a random distribution of *m* molecules into *n* partitions. The average number of targets is the

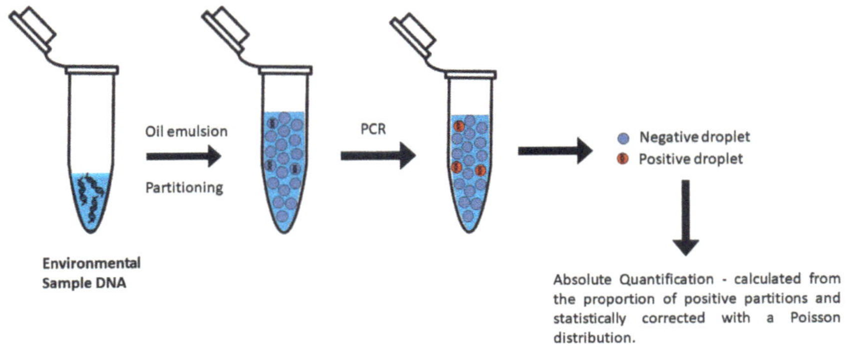

Figure 10.6 Schematic of the typical droplet digital qPCR workflow.

partition (λ) following Eq. (10.1), which is estimated based on the Poisson distribution as per Eq. (10.2). The *m* and C of an unknown sample are then calculated from the percentage of empty partitions *P* (Eqs. (10.3) and (10.4)). The confidence interval (CI) of the Poisson distribution is influenced by the probability of an empty partition *P* and the average number of targets per partition λ, and the 95% CI for the expected concentration (Eq. (10.5)).

$$\lambda = \frac{m}{n} = C \times V, \tag{10.1}$$

$$\lambda = -\ln P, \tag{10.2}$$

$$C = \frac{\lambda}{V} = \frac{-\ln P}{V}, \tag{10.3}$$

$$m = n \times \lambda = -n \ln P, \tag{10.4}$$

$$95\% CI = n * e\frac{\ln(\lambda) \pm 1.96 * \sqrt{1-P}}{-\lambda * \sqrt{\mathrm{n} * \mathrm{P}}}, \tag{10.5}$$

where λ is the average number of targets per partition, m is the number of targets in the sample, n is the number of partitions, C is the sample concentration, V is the partition volume, P is the probability of an empty partition, CI is the confidence interval, and 1.96 is a constant for a 95% confidence interval. An example of ddPCR output is presented in Fig. 10.7.

There are several key advantages of ddPCR over conventional qPCR technology. Reactions are generally not affected by PCR inhibitors in environmental samples; for example, in ddPCR, positive droplets are counted at the endpoint, lessening the impact of inhibitors that typically affect the

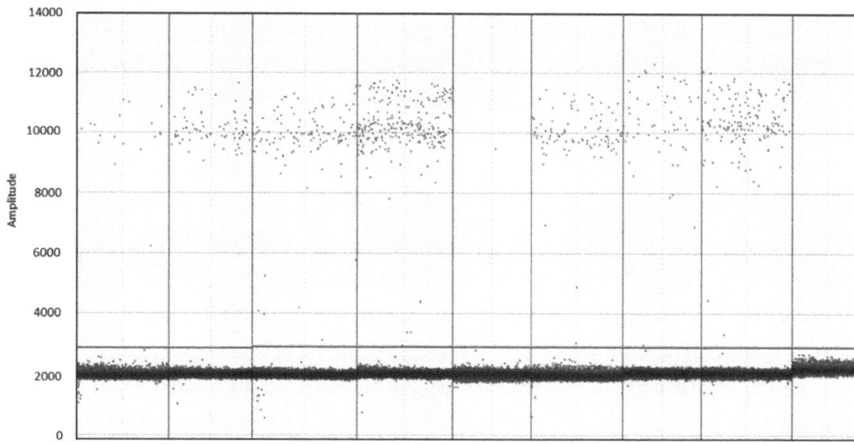

Figure 10.7 Example of droplet digital PCR output from a study investigating the copy number of *Microcoleus*-specific anaC gene (a gene involved in anatoxin production) in *Microcoleus* mats. Sample number is on the Y-axis, and each sample's amplification within each droplet is represented by a dot on the graph.

calculation of Ct values in standard qPCR methods. Other key advantages are that ddPCR does not require standard curves or reference samples (Sanders et al., 2011), thus reducing overall costs. In addition, assays do not require replicate samples and the sample size needed is smaller than conventional qPCR systems, thus preserving precious samples. In most instances, assays designed for qPCR can be readily adapted to ddPCR. Digital minimum information for publication of quantitative digital PCR experiments (dMIQE) guidelines have been developed to help standardize experimental protocols and improve the reproducibility of data (Huggett et al., 2013, 2020). Recent studies also indicate the ddPCR is more sensitive than qPCR for target detection in environmental samples (e.g., Doi et al., 2015; Wood et al., 2019). For example, Wood et al. (2019) assessed the ability of qPCR and ddPCR to detect nuisance marine species, the Mediterranean fanworm *Sabella spallanzanii* in environmental sample. The qPCR analyses detected *S. spallanzanii* in 53% of water and 85% of biofouling samples. Using ddPCR, *S. spallanzanii* was detected in 61% of water and 95% of biofouling samples. In contrast, Nathan et al. (2014) found that qPCR and ddPCR produced similar estimates of DNA concentrations for the round goby (*Neogobius melanostomus*). However, they noted that ddPCR was faster, the results more consistent, and capital expenditure aside, ddPCR was more cost-effective than qPCR.

To date, most applications of ddPCR have been designed for biomedical and food safety purposes (e.g., Cai et al., 2014; Strain et al., 2013), although an increasing number of assays have been designed to detect rare or invasive species in a range of sample types (e.g., Capo et al., 2019; Nathan et al., 2014; Zink et al., 2017). The application of ddPCR assays is also increasing in the field of environmental microbiology, with a wide range of uses, including the quantification of pathogenic bacteria (Porcellato et al., 2016; Talarico et al., 2016), viral loads (Nicot et al., 2016), and functional genes involved in processes such as nitrification and denitrification (*nirS* and *nosZ*; Dong et al., 2014) and antibiotic-resistant genes (ARGs; Cavé et al., 2016).

Compared to qPCR, studies on phytoplankton using ddPCR assays are relatively limited. Te et al. (2015) developed primers and probes to quantify two common bloom-forming cyanobacteria: *Microcystis* and *Raphidiopsis* (formerly *Cylindrospermopsis*). These assays were then tested on both qPCR and ddPCR platforms and showed high linearity and quantitative correlations. The ddPCR assays had lower variability, were less impacted by inhibition and performed better when the assays were duplexed. In contrast to most other studies that have compared ddPCR and qPCR assays, the authors showed that the qPCR assays had higher sensitivity, a wider linear dynamic range, a shorter analysis time, and were more cost-effective than ddPCR. The authors suggested a tiered approach for analysis of environmental samples. They suggest qPCR to be used for initial screening, which could then be followed by ddPCR when more accurate quantitative data is required.

Recently a ddPCR assay targeting the *anaC* gene (a gene involved in anatoxin production) was developed and used to assess copy numbers in dried samples of benthic cyanobacterial mats (Kelly et al., 2018; Wood and Puddick, 2017). In these studies, the authors determined the concentration of anatoxin congeners using liquid chromatography-mass spectrometry. Since both the gene copy number and toxin concentration were normalized to the weight of sample, an anatoxin quota could be determined. This allowed the researchers to gain valuable insights into the abundance of toxin and nontoxic strains within mats and to explore potential environmental variables that regulate anatoxin production (Kelly et al., 2018, 2019; Thomson-Laing et al., 2020; Wood and Puddick, 2017). Wood and Puddick (2017) carried out a study to determine if toxic and nontoxic genotypes cooccur in the mats of the benthic cyanobacteria *Microcoleus autumnalis*, and whether the relative abundance of the different genotypes affects the total anatoxin concentrations in a mat. Previous studies had shown that within small spatial scales (<1 m), the toxin concentrations of mats could vary markedly. They selected three mats at sites in two rivers and sampled these every 2–3 h for 24–26 h. The samples were lyophilized and ground to a fine powder. Two aliquots of known weights were analyzed for anatoxin congeners using liquid chromatography-mass spectrometry, or ddPCR with *Microcoleus*-specific *anaC* primers. Anatoxin concentrations in the mats varied 59- and 303-fold and gene copies 53- and 2828-fold. When converted to anatoxin quotas, there was markedly less variability (42- and 16-fold). The results highlighted that variability in anatoxin concentrations among mats was largely and primarily due to the abundance of toxic genotypes. Digital droplet PCR has also been applied to detect and quantify genes involved in microcystins production (and cyanobacterial abundance) in lake sediment cores, allowing insights into shifts in historical community composition and drivers of change (Dodsworth, 2015; Harris et al., 2020).

In the marine environment, ddPCR has been used for the quantification of the red-tide dinoflagellate, *Cochlodinium polykrikoides* (Lee et al., 2017) and the paralytic shellfish poison producing dinoflagellate *Alexandrium* (Lee et al., 2020). Lee et al. (2017) showed a strong correlation between the number of *C. polykrikoides* cells estimated by ddPCR and those visually counted using a microscope in samples from cultures and natural waters. The identification and quantification of *Alexandrium* species, based on morphology, is challenging. It is relatively common for different *Alexandrium* species to cooccur (e.g., *A. affine*, *A. catenella*, and *A. pacificum*), further complicating microscope-based assessments. Lee et al. (2020) designed a series of ddPCR assays that target the internal transcribed spacer region. These assays clearly distinguished between the three species and accurately estimated their abundance in environmental samples.

The number of ddPCR assays for detection of HABs or cyanoHAB species is likely to increase as more research and management institutions invest in ddPCR systems. While the capital cost of these systems is currently quite high compared to traditional qPCR technology, the cost will probaby come down over time. With the savings associated with reduced labor and the elimination of the requirement to run standard curves or run samples in triplicate, the analysis costs per sample are comparable to qPCR platforms.

Building qPCR into community-based monitoring programs

Over the past several years, there has been increased interest in DNA/RNA technology for applications, ranging from detection of HABs (both marine and freshwater) to beach water quality and microbial source tracking (MST), and seafood forensics and/or safety. The interest is not limited to other academic researchers, but includes members of state agencies, community-level environmental organizations, public and private school educators, and concerned members of the public (e.g., Bos et al., 2019; Hamburger et al., 2018; Kelly et al., 2016; Levesque et al., 2017). Implementation of low-cost, user-friendly, and robust qPCR technology will allow community scientist partners to play an increasingly larger role in routine monitoring and response to HAB events. One might imagine a future where community scientists provide diagnostic reports based on highly specific qPCR assays (e.g., Ardura and Zaiko, 2018), with detailed information about the presence and abundance of harmful algal species. Rudko et al. (2020), for example, evaluated results of qPCR analysis for toxic cyanobacteria and other pathogens by a community-based monitoring (CBM) program, and compared these results to those generated by experts using traditional laboratory-based analysis. CBM partners extracted DNA and performed qPCR in the field with prepared reagents and standards. A survey of participants found a wide range of prior experience, and all participants received the same level of training. Results showed that qPCR performed by the CBM monitoring program yielded accurate results (Rudko et al., 2020). Though there are still many challenges to adaptation of qPCR for community-based programs, the sheer number of observations has the potential to greatly enhance our knowledge of HAB dynamics, leading to improvements in predictive capabilities, and contributing more broadly to our understanding of phytoplankton ecology.

Acknowledgments

Many thanks to Dr. Ed Whereat (Citizens Monitoring Group, UD) for assistance with field sampling and to Anna Schutschkow (UD) for help processing samples for Case Study 10.1. KJC was funded by the US National Oceanic and Atmospheric Association National Center for Coastal Ocean Science (NOAA NCCOS; Grants #NA10NOS4780141, #NA15NOS4780182 and #NA04NOS4780240; MERHAB publication # 232; ECOHAB publication #990), US National Science Foundation (NSF; Grant # IOS-0745102), the Gordon and Betty Moore Foundation through Grant #2637 to the National Center for Genome Resources, and the Delaware Sea Grant College Program (Grant #R/HCE-4). PDC and SMG were supported by NSF award #OIA-1849227 to the Maine EPSCoR program at the University of Maine. PDC received additional support from Bigelow Laboratory for Ocean Sciences. Support for *Pseudo-nitzschia* sample collection was provided by the NOAA NCCOS HAB Event Response Program. SAW was funded by the New Zealand Ministry of Business, Innovation and Employment research programme: Our lakes' health; past, present and future (C05X1707).

References

Ahlgren, N., Rocap, G., 2012. Diversity and distribution of marine *Synechococcus*: multiple gene phylogenies for consensus classification and development of qPCR assays for sensitive measurement of clades in the ocean. Frontiers in Microbiology 3, 213. https://doi.org/10.3389/fmicb.2012.00213.

Ahlgren, N., Rocap, G., Chisholm, S., 2006. Measurement of *Prochlorococcus* ecotypes using real-time polymerase chain reaction reveals different abundances of genotypes with similar light physiologies. Environmental Microbiology 8 (3), 441–454. https://doi.org/10.1111/j.1462-2920.2005.00910.x.

Al-Tebrineh, J., Gehringer, M., Akcaalan, R., Neilan, B., 2011. A new quantitative PCR assay for the detection of hepatotoxigenic cyanobacteria. Toxicon 57 (4), 546–554. https://doi.org/10.1016/j.toxicon.2010.12.018.

Al-Tebrineh, J., Mihali, T., Pomati, F., Neilan, B., 2010. Detection of saxitoxin-producing cyanobacteria and *Anabaena circinalis* in environmental water blooms by quantitative PCR. Applied and Environmental Microbiology 76 (23), 7836–7842. https://doi.org/10.1128/AEM.00174-10.

Al-Tebrineh, J., Pearson, L., Yasar, S., Neilan, B., 2012. A multiplex qPCR targeting hepato- and neurotoxigenic cyanobacteria of global significance. Harmful Algae 15, 19–25. https://doi.org/10.1016/j.hal.2011.11.001.

Andersen, C.B., Holst-Jensen, A., Berdal, K.G., Thorstensen, T., Tengs, T., 2006. Equal performance of TaqMan, MGB, molecular beacon, and SYBR green-based detection assays in detection and quantification of roundup ready soybean. Journal of Agricultural and Food Chemistry 54 (26), 9658–9663. https://doi.org/10.1021/jf061987c.

Antonella, P., Luca, G., 2013. The quantitative real-time PCR applications in the monitoring of marine harmful algal bloom (HAB) species. Environmental Science and Pollution Research 20 (10), 6851–6862. https://doi.org/10.1007/s11356-012-1377-z.

Ardura, A., Zaiko, A., 2018. PCR-based assay for *Mya arenaria* detection from marine environmental samples and tracking its invasion in coastal ecosystems. Journal for Nature Conservation 43, 1–7. https://doi.org/10.1016/j.jnc.2018.02.007.

Basti, L., Nagai, K., Go, J., Okano, S., Oda, T., Tanaka, Y., Nagai, S., 2016. Lethal effects of ichthyotoxic raphidophytes, *Chattonella marina*, *C. antiqua*, and *Heterosigma akashiwo*, on post-embryonic stages of the Japanese pearl oyster, *Pinctada fucata martensii*. Harmful Algae 59, 112–122. https://doi.org/10.1016/j.hal.2016.08.003.

Beeton-Kempen, N., 2020. The Latest Advances in Quantitative PCR. Retrieved January 31, 2021 from https://www.technologynetworks.com/genomics/lists/the-latest-advances-in-quantitative-pcr-328426.

Beneduce, L., Spano, G., Lamacchia, F., Bellucci, M., Consiglio, F., Head, I.M., 2014. Correlation of seasonal nitrification failure and ammonia-oxidizing community dynamics in a wastewater treatment plant treating water from a saline thermal spa. Annals of Microbiology 64 (4), 1671–1682. https://doi.org/10.1007/s13213-014-0811-5.

Bos, J.S., Nanayakkara, L., Hurlbert, M., Finlay, K., 2019. Citizen science for Saskatchewan lakes: a pilot project. Lake and Reservoir Management 35 (1), 77–89. https://doi.org/10.1080/10402381.2018.1538172.

Bowers, H., Tengs, T., Glasgow, H., Burkholder, J., Rublee, P., Oldach, D., 2000. Development of real-time PCR assays for rapid detection of *Pfiesteria piscicida* and related

dinoflagellates. Applied and Environmental Microbiology 66 (11), 4641–4648. https://doi.org/10.1128/AEM.66.11.4641-4648.2000.
Brown, A.T., McAloose, D., Calle, P.P., Auer, A., Posautz, A., Slavinski, S., Brennan, R., Walzer, C., Seimon, T.A., 2020. Development and validation of a portable, point-of-care canine distemper virus qPCR test. PLoS ONE 15 (4), e0232044. https://doi.org/10.1371/journal.pone.0232044.
Bruce, K., Leterme, S., Ellis, A., Lenehan, C., 2015. Approaches for the detection of harmful algal blooms using oligonucleotide interactions. Analytical and Bioanalytical Chemistry 407 (1), 95–116. https://doi.org/10.1007/s00216-014-8193-x.
Bustin, S.A., 2010. Why the need for qPCR publication guidelines?-the case for MIQE. Methods 50 (4), 217–226. https://doi.org/10.1016/j.ymeth.2009.12.006.
Bustin, S.a., Benes, V., Garson, J.a., Hellemans, J., Huggett, J., Kubista, M., Mueller, R., Nolan, T., Pfaffl, M.W., Shipley, G.L., Vandesompele, J., Wittwer, C.T., 2009. The MIQE guidelines: minimum information for publication of quantitative real-time PCR experiments. Clinical Chemistry 55 (4), 611–622. https://doi.org/10.1373/clinchem.2008.112797.
Cai, Y.C., Li, X., Lv, R., Yang, J.L., Li, J., He, Y.P., Pan, L.W., 2014. Quantitative analysis of pork and chicken products by droplet digital PCR. Biomed Research International 2014, 810209. https://doi.org/10.1155/2014/810209.
Capo, E., Spong, G., Norman, S., Konigsson, H., Bartels, P., Bystrom, P., 2019. Droplet digital PCR assays for the quantification of brown trout (*Salmo trutta*) and Arctic char (*Salvelinus alpinus*) from environmental DNA collected in the water of mountain lakes. PLoS ONE 14 (12), e0226638. https://doi.org/10.1371/journal.pone.0226638.
Cary, S.C., Coyne, K.J., Rueckert, A., Wood, S.A., Kelly, S., Gemmill, C.E.C., Vieglais, C., Hicks, B.J., 2014. Development and validation of a quantitative PCR assay for the early detection and monitoring of the invasive diatom *Didymosphenia geminata*. Harmful Algae 36, 63–70. https://doi.org/10.1016/j.hal.2014.04.003.
Casabianca, S., Capellacci, S., Giacobbe, M., Dell'Aversano, C., Tartaglione, L., Varriale, F., Narizzano, R., Risso, F., Moretto, P., Dagnino, A., Bertolotto, R., Barbone, E., Ungaro, N., Penna, A., 2019. Plastic-associated harmful microalgal assemblages in marine environment. Environmental Pollution 244, 617–626. https://doi.org/10.1016/j.envpol.2018.09.110.
Casper, E.T., Patterson, S.S., Bhanushali, P., Farmer, A., Smith, M., Fries, D.P., Paul, J.H., 2007. A handheld NASBA analyzer for the field detection and quantification of *Karenia brevis*. Harmful Algae 6 (1), 112–118. https://doi.org/10.1016/j.hal.2006.11.001.
Cavé, L., Brothier, E., Abrouk, D., Bouda, P.S., Hien, E., Nazaret, S., 2016. Efficiency and sensitivity of the digital droplet PCR for the quantification of antibiotic resistance genes in soils and organic residues. Applied Microbiology and Biotechnology 100 (24), 10597–10608. https://doi.org/10.1007/s00253-016-7950-5.
Clark, S., Hubbard, K.A., Anderson, D.M., McGillicuddy Jr., D.J., Ralston, D.K., Townsend, D.W., 2019. *Pseudo-nitzschia* bloom dynamics in the Gulf of Maine: 2012–2016. Harmful Algae 88, 101656. https://doi.org/10.1016/j.hal.2019.101656.
Colina, F., Amaral, J., Carbó, M., Pinto, G., Soares, A., Cañal, M.J., Valledor, L., 2019. Genome-wide identification and characterization of CKIN/SnRK gene family in *Chlamydomonas reinhardtii*. Scientific Reports 9 (1), 1–16. https://doi.org/10.1038/s41598-018-35625-8.
Countway, P., Caron, D., 2006. Abundance and distribution of *Ostreococcus* sp. in the San Pedro Channel, California, as revealed by quantitative PCR. Applied and Environmental Microbiology 72 (4), 2496–2506. https://doi.org/10.1128/AEM.72.4.2496-2506.2006.
Coyne, K., 2010. Nitrate reductase (NR1) sequence and expression in the harmful alga *Heterosigma akashiwo* (Raphidophyceae). Journal of Phycology 46 (1), 135–142. https://doi.org/10.1111/j.1529-8817.2009.00781.x.

Coyne, K., Handy, S., Demir, E., Whereat, E., Hutchins, D., Portune, K., Doblin, M., Cary, S., 2005. Improved quantitative real-time PCR assays for enumeration of harmful algal species in field samples using an exogenous DNA reference standard. Limnology and Oceanography: Methods 3, 381–391. https://doi.org/10.4319/lom.2005.3.381.

Coyne, K.J., Hutchins, D.A., Hare, C., Cary, S.C., 2001. Assessing temporal and spatial variability in *Pfiesteria piscicida* distributions using molecular probing techniques. Aquatic Microbial Ecology 24, 275–285. https://doi.org/10.3354/ame024275.

Damase, T.R., Miura, T.A., Parent, C.E., Allen, P.B., 2018. Application of the open qPCR instrument for the in vitro selection of DNA aptamers against epidermal growth factor receptor and *Drosophila* C Virus. ACS Combinatorial Science 20 (2), 45–54. https://doi.org/10.1021/acscombsci.7b00138.

Davis, T., Berry, D., Boyer, G., Gobler, C., 2009. The effects of temperature and nutrients on the growth and dynamics of toxic and non-toxic strains of *Microcystis* during cyanobacteria blooms. Harmful Algae 8 (5), 715–725. https://doi.org/10.1016/j.hal.2009.02.004.

Davis, T., Gobler, C., 2011. Grazing by mesozooplankton and microzooplankton on toxic and non-toxic strains of *Microcystis* in the Transquaking River, a tributary of Chesapeake Bay. Journal of Plankton Research 33 (3), 415–430. https://doi.org/10.1093/plankt/fbq109.

de Montaigu, A., Sanz-Luque, E., Macias, M.I., Galvan, A., Fernandez, E., 2011. Transcriptional regulation of CDP1 and CYG56 is required for proper NH_4^+ sensing in *Chlamydomonas*. Journal of Experimental Botany 62 (4), 1425–1437. https://doi.org/10.1093/jxb/erq384.

Delaney, J.A., Ulrich, R.M., Paul, J.H., 2011. Detection of the toxic marine diatom *Pseudo-nitzschia multiseries* using the RuBisCO small subunit (rbcS) gene in two real-time RNA amplification formats. Harmful Algae 11, 54–64. https://doi.org/10.1016/j.hal.2011.07.005.

Demir, E., Coyne, K., Doblin, M., Handy, S., Hutchins, D., 2008. Assessment of microzooplankton grazing on *Heterosigma akashiwo* using a species-specific approach combining quantitative real-time PCR (QPCR) and dilution methods. Microbial Ecology 55 (4), 583–594. https://doi.org/10.1007/s00248-007-9263-9.

Demir-Hilton, E., Hutchins, D.a., Czymmek, K.J., Coyne, K.J., 2012. Description of *Viridilobus marinus* (Gen. et ep. nov.), a new raphidophye from Delaware's inland bays. Journal of Phycology 48 (5), 1220–1231. https://doi.org/10.1111/j.1529-8817.2012.01212.x.

Demir-Hilton, E., Sudek, S., Cuvelier, M., Gentemann, C., Zehr, J., Worden, A., 2011. Global distribution patterns of distinct clades of the photosynthetic picoeukaryote *Ostreococcus*. The ISME Journal 5 (7), 1095–1107. https://doi.org/10.1038/ismej.2010.209.

Dittami, S., Hostyeva, V., Egge, E., Kegel, J., Eikrem, W., Edvardsen, B., 2013. Seasonal dynamics of harmful algae in outer Oslofjorden monitored by microarray, qPCR, and microscopy. Environmental Science and Pollution Research 20 (10), 6719–6732. https://doi.org/10.1007/s11356-012-1392-0.

Dodsworth, W., 2015. Temporal trends in cyanobacteria through paleo-genetic analyses. Unpublished MSc Thesis. University of Guelph. https://ruor.uottawa.ca/bitstream/10393/40167/5/Dodsworth_William_2020_Thesis.pdf.

Doi, H., Uchii, K., Takahara, T., Matsuhashi, S., Yamanaka, H., Minamoto, T., 2015. Use of droplet digital PCR for estimation of fish abundance and biomass in environmental DNA surveys. PLoS ONE 10 (3), e0122763. https://doi.org/10.1371/journal.pone.0122763.

Doll, C., Main, C., Bianco, C., Coyne, K., Greenfield, D., 2014. Comparison of sandwich hybridization assay and quantitative PCR for the quantification of live and preserved cultures of *Heterosigma akashiwo* (Raphidophyceae). Limnology and Oceanography: Methods 12, 232–245. https://doi.org/10.4319/lom.2014.12.232.

Dong, L.H., Meng, Y., Wang, J., Liu, Y.Y., 2014. Evaluation of droplet digital PCR for characterizing plasmid reference material used for quantifying ammonia oxidizers and

denitrifiers. Analytical and Bioanalytical Chemistry 406 (6), 1701–1712. https://doi.org/10.1007/s00216-013-7546-1.
Dyhrman, S.T., Erdner, D., La Du, J., Galac, M., Anderson, D.M., 2006. Molecular quantification of toxic *Alexandrium fundyense* in the Gulf of Maine using real-time PCR. Harmful Algae 5 (3), 242–250. https://doi.org/10.1016/j.hal.2005.07.005.
Dymond, J.S., 2013. Explanatory chapter: quantitative PCR. In: Lorsch, J. (Ed.), Laboratory Methods in Enzymology: DNA, Vol. 529. Elsevier Academic Press Inc., pp. 279–289.
Ebenezer, V., Medlin, L.K., Ki, J.S., 2012. Molecular detection, quantification, and diversity evaluation of microalgae. Marine Biotechnology 14 (2), 129–142. https://doi.org/10.1007/s10126-011-9427-y.
Eckford-Soper, L., Daugbjerg, N., 2015. Examination of six commonly used laboratory fixatives in HAB monitoring programs for their use in quantitative PCR based on Taqman probe technology. Harmful Algae 42, 52–59. https://doi.org/10.1016/j.hal.2014.12.007.
Eland, L., Davenport, R., Mota, C., 2012. Evaluation of DNA extraction methods for freshwater eukaryotic microalgae. Water Research 46 (16), 5355–5364. https://doi.org/10.1016/j.watres.2012.07.023.
Engesmo, A., Strand, D., Gran-Stadniczenko, S., Edvardsen, B., Medlin, L., Eikrem, W., 2018. Development of a qPCR assay to detect and quantify ichthyotoxic flagellates along the Norwegian coast, and the first Norwegian record of *Fibrocapsa japonica* (Raphidophyceae). Harmful Algae 75, 105–117. https://doi.org/10.1016/j.hal.2018.04.007.
Erb, R.W., Wagnerdobler, I., 1993. Detection of polychlorinated biphenyl degradation genes in polluted sediments by direct DNA extraction and polymerase chain-reaction. Applied and Environmental Microbiology 59 (12), 4065–4073. https://doi.org/10.1128/AEM.59.12.4065-4073.1993.
Erdner, D.L., Percy, L., Keafer, B., Lewis, J., Anderson, D.M., 2010. A quantitative real-time PCR assay for the identification and enumeration of *Alexandrium* cysts in marine sediments. Deep-Sea Research. Part 2. Topical Studies in Oceanography 57 (3–4), 279–287. https://doi.org/10.1016/j.dsr2.2009.09.006.
Fernandes, L.F., Hubbard, K.A., Richlen, M.L., Smith, J., Bates, S.S., Ehrman, J., Léger, C., Mafra, L.L., Kulis, D., Quilliam, M., Libera, K., McCauley, L., Anderson, D.M., 2014. Diversity and toxicity of the diatom *Pseudo-nitzschia* Peragallo in the Gulf of Maine, Northwestern Atlantic Ocean. Deep-Sea Research. Part 2. Topical Studies in Oceanography 103, 139–162. https://doi.org/10.1016/j.dsr2.2013.06.022.
Fitzpatrick, E., Caron, D.A., Schnetzer, A., 2010. Development and environmental application of a genus-specific quantitative PCR approach for *Pseudo-nitzschia* species. Marine Biology 157 (5), 1161–1169. https://doi.org/10.1007/s00227-009-1383-y.
Flores, E., Frías, J.E., Rubio, L.M., Herrero, A., 2005. Photosynthetic nitrate assimilation in cyanobacteria. Photosynthesis Research 83 (2), 117–133. https://doi.org/10.1007/s11120-004-5830-9.
Foulds, I., Granacki, A., Xiao, C., Krull, U., Castle, A., Horgen, P., 2002. Quantification of microcystin-producing cyanobacteria and *E-coli* in water by 5'-nuclease PCR. Journal of Applied Microbiology 93 (5), 825–834. https://doi.org/10.1046/j.1365-2672.2002.01772.x.
Gagala, I., Izydorczyk, K., Jurczak, T., Pawelczyk, J., Dziadek, J., Wojtal-Frankiewicz, A., Jozwik, A., Jaskulska, A., Mankiewicz-Boczek, J., 2014. Role of environmental factors and toxic genotypes in the regulation of microcystins-producing cyanobacterial blooms. Microbial Ecology 67 (2), 465–479. https://doi.org/10.1007/s00248-013-0303-3.
Gaget, V., Keulen, A., Lau, M., Monis, P., Brookes, J., 2017. DNA extraction from benthic Cyanobacteria: comparative assessment and optimization. Journal of Applied Microbiology 122 (1), 294–304. https://doi.org/10.1111/jam.13332.
Ganguli, A., Mostafa, A., Berger, J., Aydin, M.Y., Sun, F., Ramirez, S.A.S.d., Valera, E., Cunningham, B.T., King, W.P., Bashir, R., 2020. Rapid isothermal amplification and

portable detection system for SARS-CoV-2. Proceedings of the National Academy of Sciences 117 (37), 22727–22735. https://doi.org/10.1073/pnas.2014739117.

García-Mendoza, E., Cáceres-Martínez, J., Rivas, D., Fimbres-Martinez, M., Sánchez-Bravo, Y., Vásquez-Yeomans, R., Medina-Elizalde, J., 2018. Mass mortality of cultivated northern bluefin tuna *Thunnus thynnus orientalis* associated with *Chattonella* species in Baja California, Mexico. Frontiers in Marine Science 5 (DEC), 454. https://doi.org/10.3389/fmars.2018.00454.

Garneau, M., Schnetzer, A., Countway, P., Jones, A., Seubert, E., Caron, D., 2011. Examination of the seasonal dynamics of the toxic dinoflagellate *Alexandrium catenella* at Redondo Beach, California, by quantitative PCR. Applied and Environmental Microbiology 77 (21), 7669–7680. https://doi.org/10.1128/AEM.06174-11.

Gast, R., McKie-Krisberg, Z., Fay, S., Rose, J., Sanders, R., 2014. Antarctic mixotrophic protist abundances by microscopy and molecular methods. FEMS Microbiology, Ecology 89 (2), 388–401. https://doi.org/10.1111/1574-6941.12334.

Giglio, S., Monis, P.T., Saint, C.P., 2003. Demonstration of preferential binding of SYBR Green I to specific DNA fragments in real-time multiplex PCR. Nucleic Acids Research 31 (22), e136. https://doi.org/10.1093/nar/gng135.

Glibert, P.M., Wilkerson, F.P., Dugdale, R.C., Raven, J.A., Dupont, C.L., Leavitt, P.R., Parker, A.E., Burkholder, J.M., Kana, T.M., 2016. Pluses and minuses of ammonium and nitrate uptake and assimilation by phytoplankton and implications for productivity and community composition, with emphasis on nitrogen-enriched conditions. Limnology and Oceanography 61 (1), 165–197. https://doi.org/10.1002/lno.10203.

Godhe, A., Asplund, M., Harnstrom, K., Saravanan, V., Tyagi, A., Karunasagar, I., 2008. Quantification of diatom and dinoflagellate biomasses in coastal marine seawater samples by real-time PCR. Applied and Environmental Microbiology 74 (23), 7174–7182. https://doi.org/10.1128/AEM.01298-08.

Godhe, A., Cusack, C., Pedersen, J., Andersen, P., Anderson, D.M., Bresnan, E., Cembella, A., Dahl, E., Diercks, S., Elbrächter, M., Edler, L., Galluzzi, L., Gescher, C., Gladstone, M., Karlson, B., Kulis, D., LeGresley, M., Lindahl, O., Marin III, R., McDermott, G., Medlin, L.K., Naustvoll, L.-J., Penna, A., Töbe, K., 2007. Intercalibration of classical and molecular techniques for identification of *Alexandrium fundyense* (Dinophyceae) and estimation of cell densities. Harmful Algae 6 (1), 56–72. https://doi.org/10.1016/j.hal.2006.06.002.

Gong, W., Marchetti, A., 2019. Estimation of 18S gene copy number in marine eukaryotic plankton using a next-generation sequencing approach. Frontiers in Marine Science 6 (APR), 219. https://doi.org/10.3389/fmars.2019.00219.

González-Hourcade, M., del Campo, E.M., Braga, M.R., Salgado, A., Casano, L.M., 2020. Disentangling the role of extracellular polysaccharides in desiccation tolerance in lichen-forming microalgae. First evidence of sulfated polysaccharides and ancient sulfotransferase genes. Environmental Microbiology 22 (8), 3096–3111. https://doi.org/10.1111/1462-2920.15043.

Greenfield, D.I., Marin III, R., Jensen, S., Massion, E., Roman, B., Feldman, J., Scholin, C.A., 2006. Application of environmental sample processor (ESP) methodology for quantifying *Pseudo-nitzschia australis* using ribosomal RNA-targeted probes in sandwich and fluorescent in situ hybridization formats. Limnology and Oceanography: Methods 4 (11), 426–435. https://doi.org/10.4319/lom.2006.4.426.

Gudnason, H., Dufva, M., Bang, D.D., Wolff, A., 2007. Comparison of multiple DNA dyes for real-time PCR: effects of dye concentration and sequence composition on DNA amplification and melting temperature. Nucleic Acids Research 35 (19), e127. https://doi.org/10.1093/nar/gkm671.

Guo, H., Liu, Y., Zhang, Q., Yuan, X., Zhang, W., Zhang, Z., 2015. A quantitative polymerase chain reaction assay for the enumeration of brown tide algae *Aureococcus anophag-*

efferens in coastal waters of Qinhuangdao. Acta Oceanologica Sinica 34 (2), 132–136. https://doi.org/10.1007/s13131-015-0615-x.

Hamburger, S., Gioeli, K.T., Berthold, D., Laughinghouse, H.D., 2018. Volunteer Algae Monitoring Program (VAMP) in the Indian River Lagoon. Marine Technology Society Journal 52 (4), 88–93. https://doi.org/10.4031/mtsj.52.4.7.

Handy, S., Coyne, K., Portune, K., Demir, E., Doblin, M., Hare, C., Cary, S., Hutchins, D., 2005. Evaluating vertical migration behavior of harmful raphidophytes in the Delaware Inland Bays utilizing quantitative real-time PCR. Aquatic Microbial Ecology 40 (2), 121–132. https://doi.org/10.3354/ame040121.

Handy, S., Demir, E., Hutchins, D., Portune, K., Whereat, E., Hare, C., Rose, J., Warner, M., Farestad, M., Cary, S., Coyne, K., 2008. Using quantitative real-time PCR to study competition and community dynamics among Delaware Inland Bays harmful algae in field and laboratory studies. Harmful Algae 7 (5), 599–613. https://doi.org/10.1016/j.hal.2007.12.018.

Handy, S., Hutchins, D., Cary, S., Coyne, K., 2006. Simultaneous enumeration of multiple raphidophyte species by quantitative real-time PCR: capabilities and limitations. Limnology and Oceanography: Methods 4, 193–204. https://doi.org/10.4319/lom.2006.4.193.

Hariganeya, N., Tanimoto, Y., Yamaguchi, H., Nishimura, T., Tawong, W., Sakanari, H., Yoshimatsu, T., Sato, S., Preston, C.M., Adachi, M., 2013. Quantitative PCR method for enumeration of cells of cryptic species of the toxic marine dinoflagellate *Ostreopsis* spp. in coastal waters of Japan. PLoS ONE 8 (3), e57627. https://doi.org/10.1371/journal.pone.0057627.

Harris, T., Yun, J-H., Baker, D.S., Kastens, J., Sturm, B., Leavitt, P., Ketterer, M., St. Amand, A., 2020. Phytoplankotn and water quality in Milford Reservoir: Results of paleolimnological sediment core and historical data analyses. Kansas Biological Survey Report No. 197. Kansas Water Office, Kansas, USA. https://kars.ku.edu/media/downloads/HABs/Milford_Feb2020/Milford_Final_Report_KBS_197.pdf.

Harshman, D.K., Rao, B.M., McLain, J.E., Watts, G.S., Yoon, J.Y., 2015. Innovative qPCR using interfacial effects to enable low threshold cycle detection and inhibition relief. Science Advances 1 (8), e1400061. https://doi.org/10.1126/sciadv.1400061.

He, S., Yu, Z., Mi, T., 2007. Development of a real-time PCR method for *Thalassiosira rotula* rapid detection. Acta Oceanologica Sinica 26 (2), 133–139.

Heid, C., Stevens, J., Livak, K., Williams, P., 1996. Real time quantitative PCR. Genome Research 6, 986–994. https://doi.org/10.1101/gr.6.10.986.

Henke, B., Turk-Kubo, K., Bonnet, S., Zehr, J., 2018. Distributions and abundances of sublineages of the N_2-fixing Cyanobacterium Candidatus *Atelocyanobacterium thalassa* (UCYN-A) in the New Caledonian Coral Lagoon. Frontiers in Microbiology 9. https://doi.org/10.3389/fmicb.2018.00554.

Herrick, J.B., Madsen, E.L., Batt, C.A., Ghiorse, W.C., 1993. Polymerase chain-reaction amplification of naphthalene-catabolic and 16S ribosomal-RNA gene-sequences from indigenous sediment bacteria. Applied and Environmental Microbiology 59 (3), 687–694. https://doi.org/10.1128/AEM.59.3.687-694.1993.

Hindson, B.J., Ness, K.D., Masquelier, D.A., Belgrader, P., Heredia, N.J., Makarewicz, A.J., Bright, I.J., Lucero, M.Y., Hiddessen, A.L., Legler, T.C., Kitano, T.K., Hodel, M.R., Petersen, J.F., Wyatt, P.W., Steenblock, E.R., Shah, P.H., Bousse, L.J., Troup, C.B., Mellen, J.C., Wittmann, D.K., Erndt, N.G., Cauley, T.H., Koehler, R.T., So, A.P., Dube, S., Rose, K.A., Montesclaros, L., Wang, S.L., Stumbo, D.P., Hodges, S.P., Romine, S., Milanovich, F.P., White, H.E., Regan, J.F., Karlin-Neumann, G.A., Hindson, C.M., Saxonov, S., Colston, B.W., 2011. High-throughput droplet digital PCR system for absolute quantitation of DNA copy number. Analytical Chemistry 83 (22), 8604–8610. https://doi.org/10.1021/ac202028g.

Hole, K., Nfon, C., 2019. Foot-and-mouth disease virus detection on a handheld real-time polymerase chain reaction platform. Transboundary and Emerging Diseases 66 (4), 1789–1795. https://doi.org/10.1111/tbed.13227.

Hotto, A., Satchwell, M., Berry, D., Gobler, C., Boyer, G., 2008. Spatial and temporal diversity of microcystins and microcystin-producing genotypes in Oneida Lake, NY. Harmful Algae 7 (5), 671–681. https://doi.org/10.1016/j.hal.2008.02.001.

Hou, L.l., Liu, F., Zang, X., Zhang, X., He, B., Ding, Y., Song, X., Xiao, D., Wang, H., 2017. Cloning and transcription analysis of the nitrate reductase gene from *Haematococcus pluvialis*. Biotechnology Letters 39 (4), 589–597. https://doi.org/10.1007/s10529-016-2283-0.

Howard, M.D.A., Jones, A.C., Schnetzer, A., Countway, P.D., Tomas, C.R., Kudela, R.M., Hayashi, K., Chia, P., Caron, D.A., 2012. Quantitative real-time polymerase chain reaction for *Cochlodinium fulvescens* (Dinophyceae), a harmful dinoflagellate from California coastal waters. Journal of Phycology 48 (2), 384–393. https://doi.org/10.1111/j.1529-8817.2012.01120.x.

Huang, H.L., Gao, W.F., Zhu, P., Zhou, C.X., Qiao, L.L., Dang, C.Y., Pang, J.H., Yan, X.J., 2020. Molecular method for rapid detection of the red tide dinoflagellate *Karenia mikimotoi* in the coastal region of Xiangshan Bay, China. Journal of Microbiological Methods 168, 105801. https://doi.org/10.1016/j.mimet.2019.105801.

Huggett, J.F., Foy, C.A., Benes, V., Emslie, K., Garson, J.A., Haynes, R., Hellemans, J., Kubista, M., Nolan, R., Pfaffl, M.W., Pfaffl, M.W., Shipley, G.L., Vandesompele, J., Wittwer, C.T., Bustin, S.A., 2013. The digital MIQE guidelines: minimum information for publication of quantitative digital PCR experiments. Clinical Chemistry 59 (6), 892–902. https://doi.org/10.1373/clinchem.2013.206375.

Huggett, J.F., Whale, A., 2013. Digital PCR as a novel technology and its potential implications for molecular diagnostics. Clinical Chemistry 59 (12), 1691–1693. https://doi.org/10.1373/clinchem.2013.214742.

Huggett, J.F., Whale, A.S., De Spiegelaere, W., Nour, A.A., Bae, Y.K., Benes, V., Burke, D., Cleveland, M., Corbisier, P., Devonshire, A.S., Dong, L.H., Drandi, D., Foy, C.A., Garson, J.A., He, H.J., Hellemans, J., Kubista, M., Lievens, A., Makrigiorgos, M.G., Milavec, M., Mueller, R.D., Nolan, T., O'Sullivan, D.M., Pfaffl, M.W., Rodiger, S., Romsos, E.L., Shipley, G.L., Taly, V., Untergasser, A., Wittwer, C.T., Bustin, S.A., Vandesompele, J., 2020. The digital MIQE guidelines update: minimum information for publication of quantitative digital PCR experiments for 2020. Clinical Chemistry 66 (8), 1012–1029. https://doi.org/10.1093/clinchem/hvaa125.

Humbert, J.F., Quiblier, C., Gugger, M., 2010. Molecular approaches for monitoring potentially toxic marine and freshwater phytoplankton species. Analytical and Bioanalytical Chemistry 397 (5), 1723–1732. https://doi.org/10.1007/s00216-010-3642-7.

Imai, I., Yamaguchi, M., 2012. Life cycle, physiology, ecology and red tide occurrences of the fish-killing raphidophyte *Chattonella*. Harmful Algae 14, 46–70. https://doi.org/10.1016/j.hal.2011.10.014.

Imamura, S., Terashita, M., Ohnuma, M., Maruyama, S., Minoda, A., Weber, A.P.M., Inouye, T., Sekine, Y., Fujita, Y., Omata, T., Tanaka, K., 2010. Nitrate assimilatory genes and their transcriptional regulation in a unicellular red alga *Cyanidioschyzon merolae*: genetic evidence for nitrite reduction by a sulfite reductase-like enzyme. Plant and Cell Physiology 51 (5), 707–717. https://doi.org/10.1093/pcp/pcq043.

Jeong, H.J., 2011. Mixotrophy in red tide algae raphidophytes. Journal of Eukaryotic Microbiology 58 (3), 215–222. https://doi.org/10.1111/j.1550-7408.2011.00550.x.

Johnson, J.G., Morey, J.S., Neely, M.G., Ryan, J.C., van Dolah, F.M., 2012. Transcriptome remodeling associated with chronological aging in the dinoflagellate, *Karenia brevis*. Marine Genomics 5, 15–25. https://doi.org/10.1016/j.margen.2011.08.005.

Joung, S., Oh, H., Ko, S., Ahn, C., 2011. Correlations between environmental factors and toxic and non-toxic *Microcystis* dynamics during bloom in Daechung Reservoir, Korea. Harmful Algae 10 (2), 188–193. https://doi.org/10.1016/j.hal.2010.09.005.

Kavanagh, S., Brennan, C., O'Connor, L., Moran, S., Salas, R., Lyons, J., Silke, J., Maher, M., 2010. Real-time PCR detection of *Dinophysis* Species in Irish coastal waters. Marine Biotechnology 12 (5), 534–542. https://doi.org/10.1007/s10126-009-9238-6.

Kelly, L., Wood, S., McAllister, T., Ryan, K., 2018. Development and application of a quantitative PCR assay to assess genotype dynamics and anatoxin content in *Microcoleus autumnalis*-dominated mats. Toxins 10 (11), 431. https://doi.org/10.3390/toxins10110431.

Kelly, L.T., Bouma-Gregson, K., Puddick, J., Fadness, R., Ryan, K.G., Davis, T.W., Wood, S.A., 2019. Multiple cyanotoxin congeners produced by sub-dominant cyanobacterial taxa in riverine cyanobacterial and algal mats. PLoS ONE 14 (12), e0220422. https://doi.org/10.1371/journal.pone.0220422.

Kelly, M.G., Krokowski, J., Harding, J.P.C., 2016. RAPPER: a new method for rapid assessment of macroalgae as a complement to diatom-based assessments of ecological status. Science of the Total Environment 568, 536–545. https://doi.org/10.1016/j.scitotenv.2015.12.068.

Kudela, R.M., Howard, M.D.A., Jenkins, B.D., Miller, P.E., Smith, G.J., 2010. Using the molecular toolbox to compare harmful algal blooms in upwelling systems. Progress in Oceanography 85 (1–2), 108–121. https://doi.org/10.1016/j.pocean.2010.02.007.

Lauritano, C., Orefice, I., Procaccini, G., Romano, G., Ianora, A., 2015. Key genes as stress indicators in the ubiquitous diatom *Skeletonema marinoi*. BMC Genomics 16. https://doi.org/10.1186/s12864-015-1574-5.

Lauritano, C., De Luca, D., Amoroso, M., Benfatto, S., Maestri, S., Racioppi, C., Esposito, F., Ianora, A., 2019. New molecular insights on the response of the green alga *Tetraselmis suecica* to nitrogen starvation. Scientific Reports 9 (1), 1–12. https://doi.org/10.1038/s41598-019-39860-5.

Lee, H.G., Kim, H.M., Min, J., Kim, K., Park, M.G., Jeong, H.J., Kim, K.Y., 2017. An advanced tool, droplet digital PCR (ddPCR), for absolute quantification of the red-tide dinoflagellate, *Cochlodinium polykrikoides* Margalef (Dinophyceae). Algae 32 (3), 189–197. https://doi.org/10.4490/algae.2017.32.9.10.

Lee, H.G., Kim, H.M., Min, J., Park, C., Jeong, H.J., Lee, K., Kim, K.Y., 2020. Quantification of the paralytic shellfish poisoning dinoflagellate *Alexandrium* species using a digital PCR. Harmful Algae 92, 101726. https://doi.org/10.1016/j.hal.2019.101726.

Lee, Y.-J., Kim, D., Lee, K., Chun, J.-Y., 2014. Single-channel multiplexing without melting curve analysis in real-time PCR. Scientific Reports 4 (1), 7439. https://doi.org/10.1038/srep07439.

Lei, L., Lei, M., Lu, Y., Peng, L., Han, B., 2019. Development of real-time PCR for quantification of *Cylindrospermopsis raciborskii* cells and potential cylindrospermopsin-producing genotypes in subtropical reservoirs of southern China. Journal of Applied Phycology 31 (6), 3749–3758. https://doi.org/10.1007/s10811-019-01898-3.

Levesque, D., Cattaneo, A., Deschamps, G., Hudon, C., 2017. In the eye of the beholder: assessing the water quality of shoreline parks around the Island of Montreal through citizen science. Science of the Total Environment 579, 978–988. https://doi.org/10.1016/j.scitotenv.2016.10.175.

Li, D., Kong, F., Shi, X., Ye, L., Yu, Y., Yang, Z., 2012. Quantification of microcystin-producing and non-microcystin producing *Microcystis* populations during the 2009 and 2010 blooms in Lake Taihu using quantitative real-time PCR. Journal of Environmental Sciences 24 (2), 284–290. https://doi.org/10.1016/S1001-0742(11)60745-6.

Li, H.T., Pan, J.Z., Fang, Q., 2020. Development and application of digital PCR technology. Progress in Chemistry 32 (5), 581–593. https://doi.org/10.7536/pc190913.

Lim, H.J., Lee, E.H., Yoon, Y., Chua, B., Son, A., 2016. Portable lysis apparatus for rapid single-step DNA extraction of *Bacillus subtilis*. Journal of Applied Microbiology 120 (2), 379–387. https://doi.org/10.1111/jam.13011.

Lin, S., Shen, J., Liu, Y., Wu, X., Liu, Q., Li, R., 2011. Molecular evaluation on the distribution, diversity, and toxicity of *Microcystis* (Cyanobacteria) species from Lake Ulungur-a mesotrophic brackish desert lake in Xinjiang, China. Environmental Monitoring and Assessment 175 (1–4), 139–150. https://doi.org/10.1007/s10661-010-1500-x.

Livak, K.J., Flood, S.J.A., Marmaro, J., Giusti, W., Deetz, K., 1995. Oligonucleotides with fluorescent dyes at opposite ends provide a quenched probe system useful for detecting PCR product and nucleic-acid hybridization. PCR-Methods and Applications 4 (6), 357–362.

Lu, J., Struewing, I., Wymer, L., Tettenhorst, D.R., Shoemaker, J., Allen, J., 2020. Use of qPCR and RT-qPCR for monitoring variations of microcystin producers and as an early warning system to predict toxin production in an Ohio inland lake. Water Research 170. https://doi.org/10.1016/j.watres.2019.115262.

Lu, J., Zhu, B., Struewing, I., Xu, N., Duan, S., 2019. Nitrogen–phosphorus-associated metabolic activities during the development of a cyanobacterial bloom revealed by metatranscriptomics. Scientific Reports 9 (1), 1–11. https://doi.org/10.1038/s41598-019-38481-2.

Lv, H., Qu, G., Qi, X., Lu, L., Tian, C., Ma, Y., 2013. Transcriptome analysis of *Chlamydomonas reinhardtii* during the process of lipid accumulation. Genomics 101 (4), 229–237. https://doi.org/10.1016/j.ygeno.2013.01.004.

Main, C., Salvitti, L., Whereat, E., Coyne, K., 2015. Community-level and species-specific associations between phytoplankton and particle-associated *Vibrio* species in Delaware's Inland Bays. Applied and Environmental Microbiology 81 (17), 5703–5713. https://doi.org/10.1128/AEM.00580-15.

Main, C.R., Doll, C., Bianco, C., Greenfield, D.I., Coyne, K.J., 2014. Effects of growth phase, diel cycle and macronutrient stress on the quantification of *Heterosigma akashiwo* using qPCR and SHA. Harmful Algae 37, 92–99. https://doi.org/10.1016/j.hal.2014.05.014.

Main, C.R., Greenfield, D.I., Doll, C., Wang, Y., Whereat, E.B., Mortensen, R., Pettay, D.T., Coyne, K.J., 2018. Critical comparison of molecular methods for detection and enumeration of the harmful algal species, *Heterosigma akashiwo*, in environmental water samples. Journal of Applied Phycology 30 (4), 2425–2434. https://doi.org/10.1007/s10811-018-1444-z.

Medlin, L.K., Orozco, J., 2017. Molecular techniques for the detection of organisms in aquatic environments, with emphasis on harmful algal bloom species. Sensors 17 (5), 1184. https://doi.org/10.3390/s17051184.

Menning, D., Carraher-Stross, W., Graham, E., Thomas, D., Phillips, A., Scharping, R., Garey, J., 2018. Aquifer discharge drives microbial community change in Karst Estuaries. Estuaries and Coasts 41 (2), 430–443. https://doi.org/10.1007/s12237-017-0281-7.

Metfies, K., Schroeder, F., Hessel, J., Wollschlager, J., Micheller, S., Wolf, C., Kilias, E., Sprong, P., Neuhaus, S., Frickenhaus, S., Petersen, W., 2016. High-resolution monitoring of marine protists based on an observation strategy integrating automated on-board filtration and molecular analyses. Ocean Science 12 (6), 1237–1247. https://doi.org/10.5194/os-12-1237-2016.

Monis, P.T., Giglio, S., Saint, C.P., 2005. Comparison of SYTO9 and SYBR Green I for real-time polymerase chain reaction and investigation of the effect of dye concentration on amplification and DNA melting curve analysis. Analytical Biochemistry 340 (1), 24–34. https://doi.org/10.1016/j.ab.2005.01.046.

Moorthi, S.D., Countway, P.D., Stauffer, B.A., Caron, D.A., 2006. Use of quantitative real-time PCR to investigate the dynamics of the red tide dinoflagellate *Lingulodinium polyedrum*. Microbial Ecology 52 (1), 136–150. https://doi.org/10.1007/s00248-006-9030-3.

Muller, K., Chhun, A., Guildford, S., Yakobowski, S., Jonlija, M., 2017. Molecular and ecological characterization of toxic cyanobacteria from the Bay of Quinte (Lake Ontario) and Maumee Bay (Lake Erie). Journal of Great Lakes Research 43 (6), 1067–1083. https://doi.org/10.1016/j.jglr.2017.03.021.

Muyzer, G., De Waal, E.C., Uitierlinden, A.G., 1993. Profiling of complex microbial populations by Denaturing Gradient Gel Electrophoresis analysis of Polymerase Chain Reaction-amplified genes coding for 16S rRNA. Applied and Environmental Microbiology 59 (3), 695–700. https://doi.org/10.1128/AEM.59.3.695-700.1993.

Nagai, S., Itakura, S., 2012. Specific detection of the toxic dinoflagellates *Alexandrium tamarense* and *Alexandrium catenella* from single vegetative cells by a loop-mediated isothermal amplification method. Marine Genomics 7, 43–49. https://doi.org/10.1016/j.margen.2012.03.001.

Nagle, L., Place, A.R., Schott, E.J., Jagus, R., Messick, G., Pitula, J.S., 2009. Real-time PCR-based assay for quantitative detection of *Hematodinium* sp. in the blue crab *Callinectes sapidus*. Diseases of Aquatic Organisms 84 (1), 79–87. https://doi.org/10.3354/dao02027.

Nathan, L.M., Simmons, M., Wegleitner, B.J., Jerde, C.L., Mahon, A.R., 2014. Quantifying environmental DNA signals for aquatic invasive species across multiple detection platforms. Environmental Science & Technology 48 (21), 12800–12806. https://doi.org/10.1021/es5034052.

Nguyen, P.L., Sudheesh, P.S., Thomas, A.C., Sinnesael, M., Haman, K., Cain, K.D., 2018. Rapid detection and monitoring of *Flavobacterium psychrophilum* in water by using a handheld, field-portable quantitative PCR system. Journal of Aquatic Animal Health 30 (4), 302–311. https://doi.org/10.1002/aah.10046.

Ngwa, F., Madramootoo, C., Jabaji, S., 2014a. Comparison of cyanobacterial microcystin synthetase (mcy) E gene transcript levels, mcy E gene copies, and biomass as indicators of microcystin risk under laboratory and field conditions. Microbiologyopen 3 (4), 411–425. https://doi.org/10.1002/mbo3.173.

Ngwa, F., Madramootoo, C., Jabaji, S., 2014b. Development and application of a multiplex qPCR technique to detect multiple microcystin-producing cyanobacterial genera in a Canadian freshwater lake. Journal of Applied Phycology 26 (4), 1675–1687. https://doi.org/10.1007/s10811-013-0199-9.

Nicot, F., Cazabat, M., Lhomme, S., Marion, O., Saune, K., Chiabrando, J., Dubois, M., Kamar, N., Abravanel, F., Izopet, J., 2016. Quantification of HEV RNA by Droplet Digital PCR. Viruses-Basel 8 (8), 233. https://doi.org/10.3390/v8080233.

Pacheco, A., Guedes, I., Azevedo, S., 2016. Is qPCR a reliable indicator of cyanotoxin risk in freshwater? Toxins 8 (6), 172. https://doi.org/10.3390/toxins8060172.

Panicker, G., Myers, M.L., Bej, A.K., 2004. Rapid detection of *Vibrio vulnificus* in shellfish and Gulf of Mexico water by real-time PCR. Applied and Environmental Microbiology 70 (1), 498–507. https://doi.org/10.1128/aem.70.1.498-507.2004.

Panksep, K., Tamm, M., Mantzouki, E., Rantala-Ylinen, A., Laugaste, R., Sivonen, K., Tammeorg, O., Kisand, V., 2020. Using microcystin gene copies to determine potentially-toxic blooms, example from a shallow eutrophic lake Peipsi. Toxins 12 (4), 211. https://doi.org/10.3390/toxins12040211.

Park, B., Baek, S., Ki, J., Cattolico, R., Han, M., 2012. Assessment of EvaGreen-based quantitative real-time PCR assay for enumeration of the microalgae *Heterosigma* and *Chattonella* (Raphidophyceae). Journal of Applied Phycology 24 (6), 1555–1567. https://doi.org/10.1007/s10811-012-9816-2.

Park, B., Li, Z., Kang, Y., Shin, H., Joo, J., Han, M., 2018. Distinct bloom dynamics of toxic and non-toxic *Microcystis* (Cyanobacteria) subpopulations in Hoedong Reservoir (Korea). Microbial Ecology 75 (1), 163–173. https://doi.org/10.1007/s00248-017-1030-y.

Penna, A., Galluzzi, L., 2013. The quantitative real-time PCR applications in the monitoring of marine harmful algal bloom (HAB) species. Environmental Science and Pollution Research 20, 6851–6862. https://doi.org/10.1007/s11356-012-1377-z.

Perini, F., Casabianca, A., Battocchi, C., Accoroni, S., Totti, C., Penna, A., 2011. New approach using real-time PCR method for estimation of the toxic marine dinoflagellate *Ostreopsis* cf. *ovata* in marine environment. PLoS ONE 6 (3), e17699. https://doi.org/10.1371/journal.pone.0017699.

Pimentel, J., Giani, A., 2014. Microcystin production and regulation under nutrient stress conditions in toxic *Microcystis* strains. Applied and Environmental Microbiology 80 (18), 5836–5843. https://doi.org/10.1128/AEM.01009-14.

Pinto, F., Pacheco, C., Ferreira, D., Moradas-Ferreira, P., Tamagnini, P., 2012. Selection of suitable reference genes for RT-qPCR analyses in cyanobacteria. PLoS ONE 7 (4), e34983. https://doi.org/10.1371/journal.pone.0034983.

Popels, L.C., Cary, S.C., Hutchins, D.A., Forbes, R., Pustizzi, F., Gobler, C.J., Coyne, K.J., 2003. The use of quantitative polymerase chain reaction for the detection and enumeration of the harmful alga *Aureococcus anophagefferens* in environmental samples along the United States East Coast. Limnology and Oceanography: Methods 1, 92–102. https://doi.org/10.4319/lom.2003.1.92.

Porcellato, D., Narvhus, J., Skeie, S.B., 2016. Detection and quantification of *Bacillus cereus* group in milk by droplet digital PCR. Journal of Microbiological Methods 127, 1–6. https://doi.org/10.1016/j.mimet.2016.05.012.

Preston, C.M., Harris, A., Ryan, J.P., Roman, B., Marin, R., Jensen, S., Everlove, C., Birch, J., Dzenitis, J.M., Pargett, D., Adachi, M., Turk, K., Zehr, J.P., Scholin, C.A., 2011. Underwater application of quantitative PCR on an Ocean Mooring. PLoS ONE 6 (8), e22522. https://doi.org/10.1371/journal.pone.0022522.

Qiao, L., Chen, Y., Mi, T., Zhen, Y., Gao, Y., Yu, Z., 2017. Quantitative PCR analysis of the spatiotemporal dynamics of *Aureococcus anophagefferens* and *Minutocellus polymorphus* and the relationship between brown tides and nutrients in the coastal waters of Qinhuangdao, China. Journal of Applied Phycology 29 (1), 297–308. https://doi.org/10.1007/s10811-016-0959-4.

Qiu, X., Matsuyama, Y., Furuse, M., Shimasaki, Y., Oshima, Y., 2020. Effects of *Chattonella antiqua* on the swimming behavior and brain monoamine metabolism of juvenile yellowtail (*Seriola quinqueradiata*). Marine Pollution Bulletin 152. https://doi.org/10.1016/j.marpolbul.2020.110896.

Renaud, S., Pick, F., Fortin, N., 2011. Effect of light intensity on the relative dominance of toxigenic and nontoxigenic strains of *Microcystis aeruginosa*. Applied and Environmental Microbiology 77 (19), 7016–7022. https://doi.org/10.1128/AEM.05246-11.

Rinta-Kanto, J., Ouellette, A., Boyer, G., Twiss, M., Bridgeman, T., Wilhelm, S., 2005. Quantification of toxic *Microcystis* spp. during the 2003 and 2004 blooms in western Lake Erie using quantitative real-time PCR. Environmental Science & Technology 39 (11), 4198–4205. https://doi.org/10.1021/es048249u.

Ritter, A., Dittami, S.M., Goulitquer, S., Correa, J.A., Boyen, C., Potin, P., Tonon, T., 2014. Transcriptomic and metabolomic analysis of copper stress acclimation in *Ectocarpus siliculosus* highlights signaling and tolerance mechanisms in brown algae. BMC Plant Biology 14 (1), 116. https://doi.org/10.1186/1471-2229-14-116.

Rudko, S.P., Reimink, R.L., Peter, B., White, J., Hanington, P.C., 2020. Democratizing water monitoring: implementation of a community-based qPCR monitoring program for recreational water hazards. PLoS ONE 15 (5), e0229701. https://doi.org/10.1371/journal.pone.0229701.

Rueckert, A., Morgan, H.W., 2007. Removal of contaminating DNA from polymerase chain reaction using ethidium monoazide. Journal of Microbiological Methods 68 (3), 596–600. https://doi.org/10.1016/j.mimet.2006.11.006.

Rueckert, A., Wood, S.A., Cary, S.C., 2007. Development and field assessment of a quantitative PCR for the detection and enumeration of the noxious bloom-former *Anabaena planktonica*. Limnology and Oceanography: Methods 5, 474–483. https://doi.org/10.4319/lom.2007.5.474.

Russell, J.A., Campos, B., Stone, J., Blosser, E.M., Burkett-Cadena, N., Jacobs, J.L., 2018. Unbiased strain-typing of Arbovirus directly from mosquitoes using nanopore sequencing: a field-forward biosurveillance protocol. Scientific Reports 8 (1), 5417. https://doi.org/10.1038/s41598-018-23641-7.

Ruvindy, R., Bolch, C., MacKenzie, L., Smith, K., Murray, S., 2018. qPCR Assays for the detection and quantification of multiple paralytic shellfish toxin-producing species of *Alexandrium*. Frontiers in Microbiology 9, 3153. https://doi.org/10.3389/fmicb.2018.03153.

Salipante, S.J., Jerome, K.R., 2020. Digital PCR-an emerging technology with broad applications in microbiology. Clinical Chemistry 66 (1), 117–123. https://doi.org/10.1373/clinchem.2019.304048.

Sanders, R., Huggett, J.F., Bushell, C.A., Cowen, S., Scott, D.J., Foy, C.A., 2011. Evaluation of digital PCR for absolute DNA quantification. Analytical Chemistry 83 (17), 6474–6484. https://doi.org/10.1021/ac103230c.

Sanz-Luque, E., Chamizo-Ampudia, A., Llamas, A., Galvan, A., Fernandez, E., 2015. Understanding nitrate assimilation and its regulation in microalgae. Frontiers of Plant Science 6. https://doi.org/10.3389/fpls.2015.00899.

Scherer, P.I., Absmeier, C., Urban, M., Raeder, U., Geist, J., Zwirglmaier, K., 2018. Influence of cyanobacteria, mixotrophic flagellates, and virioplankton size fraction on transcription of microcystin synthesis genes in the toxic cyanobacterium *Microcystis aeruginosa*. Microbiologyopen 7 (1), e00538. https://doi.org/10.1002/mbo3.538.

Schlitzer, R., 2020. Ocean Data View. https://odv.awi.de.

Scholin, C., Doucette, G., Jensen, S., Roman, B., Pargett, D., Marin III, R., Preston, C., Jones, W., Feldman, J., Everlove, C., Harris, A., Alvarado, N., Massion, E., Birch, J., Greenfield, D., Vrijenhoek, R., Mikulski, C., Jones, K., 2009. Remote detection of marine microbes, small invertebrates, harmful algae, and biotoxins using the Environmental Sample Processor (ESP). Oceanography 22 (2), 158–167. https://doi.org/10.5670/oceanog.2009.46.

Scholin, C.A., Birch, J., Jensen, S., Marin III, R., Massion, E., Pargett, D., Preston, C., Roman, B., Ussler III, W., 2017. The quest to develop ecogenomic sensors a 25-year history of the environmental sample processor (ESP) as a case study. Oceanography 30 (4), 100–113. https://doi.org/10.5670/OCEANOG.2017.427.

Seegers, B.N., Birch, J.M., Marin, R., Scholin, C.A., Caron, D.A., Seubert, E.L., Howard, M.D.A., Robertson, G.L., Jones, B.H., 2015. Subsurface seeding of surface harmful algal blooms observed through the integration of autonomous gliders, moored environmental sample processors, and satellite remote sensing in southern California. Limnology and Oceanography 60 (3), 754–764. https://doi.org/10.1002/lno.10082.

Sepulveda, A.J., Hutchins, P.R., Massengill, R.L., Dunker, K.J., 2018. Tradeoffs of a portable, field-based environmental DNA platform for detecting invasive northern pike (*Esox lucius*) in Alaska. Management of Biological Invasions 9 (3), 253–258. https://doi.org/10.3391/mbi.2018.9.3.07.

Shim, J., Shim, E., Kim, G.H., Han, J.W., Zuccarello, G.C., 2016. Keeping house: evaluation of housekeeping genes for real-time PCR in the red alga, *Bostrychia moritziana* (Florideophyceae). Algae 31 (2), 167–174. https://doi.org/10.4490/algae.2016.31.5.25.

Sidstedt, M., Radstrom, P., Hedman, J., 2020. PCR inhibition in qPCR, dPCR and MPS-mechanisms and solutions. Analytical and Bioanalytical Chemistry 412 (9), 2009–2023. https://doi.org/10.1007/s00216-020-02490-2.

Smith, C.J., Osborn, A.M., 2009. Advantages and limitations of quantitative PCR (Q-PCR)-based approaches in microbial ecology. FEMS Microbiology, Ecology 67 (1), 6–20. https://doi.org/10.1111/j.1574-6941.2008.00629.x.

Song, B., Ward, B.B., 2004. Molecular characterization of the assimilatory nitrate reductase gene and its expression in the marine green alga *Dunaliella tertiolecta* (Chlorophyceae). Journal of Phycology 40 (4), 721–731. https://doi.org/10.1111/j.1529-8817.2004.03078.x.

Stewart, J.J., Coyne, K.J., 2011. Analysis of raphidophyte assimilatory nitrate reductase reveals unique domain architecture incorporating a 2/2 hemoglobin. Plant Molecular Biology 77 (6), 565–575. https://doi.org/10.1007/s11103-011-9831-8.

Stoeckel, D.M., Stelzer, E.A., Dick, L.K., 2009. Evaluation of two spike-and-recovery controls for assessment of extraction efficiency in microbial source tracking studies. Water Research 43 (19), 4820–4827. https://doi.org/10.1016/j.watres.2009.06.028.

Strain, M.C., Lada, S.M., Luong, T., Rought, S.E., Gianella, S., Terry, V.H., Spina, C.A., Woelk, C.H., Richman, D.D., 2013. Highly precise measurement of HIV DNA by droplet digital PCR. PLoS ONE 8 (4), e55943. https://doi.org/10.1371/journal.pone.0055943.

Suzuki, M.T., Taylor, L.T., DeLong, E.F., 2000. Quantitative analysis of small-subunit rRNA genes in mixed microbial populations via 5'-nuclease assays. Applied and Environmental Microbiology 66 (11), 4605–4614. https://doi.org/10.1128/aem.66.11.4605-4614.2000.

Talarico, S., Safaeian, M., Gonzalez, P., Hildesheim, A., Herrero, R., Porras, C., Cortes, B., Larson, A., Fang, F.C., Salama, N.R., 2016. Quantitative detection and genotyping of *Helicobacter pylori* from stool using droplet digital PCR reveals variation in bacterial loads that correlates with cagA virulence gene carriage. Helicobacter 21 (4), 325–333. https://doi.org/10.1111/hel.12289.

Tan, B., Ng, C., Nshimyimana, J., Loh, L., Gin, K., Thompson, J., 2015. Next-generation sequencing (NGS) for assessment of microbial water quality: current progress, challenges, and future opportunities. Frontiers in Microbiology 6, 1027. https://doi.org/10.3389/fmicb.2015.01027.

Taylor, B.J., Lanke, K., Banman, S.L., Morlais, I., Morin, M.J., Bousema, T., Rijpma, S.R., Yanow, S.K., 2017. A direct from blood reverse transcriptase polymerase chain reaction assay for monitoring *Falciparum* malaria parasite transmission in elimination settings. The American Journal of Tropical Medicine and Hygiene 97 (2), 533–543. https://doi.org/10.4269/ajtmh.17-0039.

Te, S., Chen, E., Gin, K., 2015. Comparison of quantitative PCR and droplet digital PCR multiplex assays for two genera of bloom-forming cyanobacteria, *Cylindrospermopsis* and *Microcystis*. Applied and Environmental Microbiology 81 (15), 5203–5211. https://doi.org/10.1128/AEM.00931-15.

Thomas, A.C., Tank, S., Nguyen, P.L., Ponce, J., Sinnesael, M., Goldberg, C.S., 2020. A system for rapid eDNA detection of aquatic invasive species. Environmental DNA 2 (3), 261–270. https://doi.org/10.1002/edn3.25.

Thomson-Laing, G., Puddick, J., Laroche, O., Fulton, S., Steiner, K., Heath, M.W., Wood, S.A., 2020. Broad and fine scale variability in bacterial diversity and cyanotoxin quotas in benthic cyanobacterial mats. Frontiers in Microbiology 11, 129. https://doi.org/10.3389/fmicb.2020.00129.

Tillmann, U., Gottschling, M., Krock, B., Smith, K., Guinder, V., 2019. High abundance of *Amphidomataceae* (Dinophyceae) during the 2015 spring bloom of the Argentinean Shelf and a new, non-toxigenic ribotype of *Azadinium spinosum*. Harmful Algae 84, 244–260. https://doi.org/10.1016/j.hal.2019.01.008.

Tilney, C.L., Pokrzywinski, K.L., Coyne, K.J., Warner, M.E., 2014. Effects of a bacterial algicide, IRI-160AA, on dinoflagellates and the microbial community in microcosm experiments. Harmful Algae 39 (0), 210–222. https://doi.org/10.1016/j.hal.2014.08.001.

Toebe, K., Alpermann, T., Tillmann, U., Krock, B., Cembella, A., John, U., 2013. Molecular discrimination of toxic and non-toxic *Alexandrium* species (Dinophyta) in natural phytoplankton assemblages from the Scottish coast of the North Sea. European Journal of Phycology 48 (1), 12–26. https://doi.org/10.1080/09670262.2012.752870.

Toldrà, A., Andree, K.B., Fernández-Tejedor, M., Diogène, J., Campàs, M., 2018. Dual quantitative PCR assay for identification and enumeration of *Karlodinium veneficum* and *Karlodinium armiger* combined with a simple and rapid DNA extraction method. Journal of Applied Phycology 30 (4), 2435–2445. https://doi.org/10.1007/s10811-018-1446-x.

Toldra, A., O'Sullivan, C., Diogene, J., Campas, M., 2020. Detecting harmful algal blooms with nucleic acid amplification-based biotechnological tools. Science of the Total Environment 749, 141605. https://doi.org/10.1016/j.scitotenv.2020.141605.

Tomiokai, N., Nagai, T., Kawasaki, T., Imai, A., Matsushige, K., Kohata, K., 2008. Quantification of *Microcystis* in a eutrophic lake by simple DNA extraction and SYBR Green real-time PCR. Microbes and Environments 23 (4), 306–312. https://doi.org/10.1264/jsme2.ME08515.

Ulrich, R.M., Casper, E.T., Campbell, L., Richardson, B., Heil, C.A., Paul, J.H., 2010. Detection and quantification of *Karenia mikimotoi* using real-time nucleic acid sequence-based amplification with internal control RNA (IC-NASBA). Harmful Algae 9 (1), 116–122. https://doi.org/10.1016/j.hal.2009.08.010.

Vaitomaa, J., Rantala, A., Halinen, K., Rouhiainen, L., Tallberg, P., Mokelke, L., Sivonen, K., 2003. Quantitative real-time PCR for determination of microcystin synthetase E copy numbers for *Microcystis* and *Anabaena* in lakes. Applied and Environmental Microbiology 69 (12), 7289–7297. https://doi.org/10.1128/AEM.69.12.7289-7297.2003.

Van de Waal, D., Verspagen, J., Finke, J., Vournazou, V., Immers, A., Kardinaal, W., Tonk, L., Becker, S., Van Donk, E., Visser, P., Huisman, J., 2011. Reversal in competitive dominance of a toxic versus non-toxic cyanobacterium in response to rising CO_2. The ISME Journal 5 (9), 1438–1450. https://doi.org/10.1038/ismej.2011.28.

van Doorn, R., Klerks, M.M., van Gent-Pelzer, M.P.E., Speksnijder, A., Kowalchuk, G.A., Schoen, C.D., 2009. Accurate quantification of microorganisms in PCR-inhibiting environmental DNA extracts by a novel internal amplification control approach using Biotrove OpenArrays. Applied and Environmental Microbiology 75 (22), 7253–7260. https://doi.org/10.1128/aem.00796-09.

Vandersea, M., Kibler, S., Tester, P., Holderied, K., Hondolero, D., Powell, K., Baird, S., Doroff, A., Dugan, D., Litaker, R., 2018. Environmental factors influencing the distribution and abundance of *Alexandrium catenella* in Kachemak Bay and Lower Cook Inlet, Alaska. Harmful Algae 77, 81–92. https://doi.org/10.1016/j.hal.2018.06.008.

Vandersea, M., Kibler, S., Van Sant, S., Tester, P., Sullivan, K., Eckert, G., Cammarata, C., Reece, K., Scott, G., Place, A., Holderied, K., Hondolero, D., Litaker, R., 2017. qPCR assays for *Alexandrium fundyense* and *A. ostenfeldii* (Dinophyceae) identified from Alaskan waters and a review of species-specific *Alexandrium* molecular assays [Review]. Phycologia 56 (3), 303–320. https://doi.org/10.2216/16-41.1.

Vassalli, M., Penna, A., Sbrana, F., Casabianca, S., Gjeci, N., Capellacci, S., Asnaghi, V., Ottaviani, E., Giussani, V., Pugliese, L., Jauzein, C., Lemee, R., Hachani, M., Turki, S., Acaf, L., Saab, M., Fricke, A., Mangialajo, L., Bertolotto, R., Totti, C., Accoroni, S., Berdalet, E., Vila, M., Chiantore, M., 2018. Intercalibration of counting methods for *Ostreopsis* spp. blooms in the Mediterranean Sea. Ecological Indicators 85, 1092–1100. https://doi.org/10.1016/j.ecolind.2017.07.063.

Viana, T.V., Fistarol, G.O., Amario, M., Menezes, R.B., Carneiro, B.L.R., Chaves, D.M., Hargreaves, P.I., Silva-Lima, A.W., Valentin, J.L., Tenenbaum, D.R., Arruda, E.F., Paranhos, R., Salomon, P.S., 2019. Massive blooms of *Chattonella subsalsa* Biecheler (Raphi-

dophyceae) in a hypereutrophic, tropical Estuary-Guanabara Bay, Brazil. Frontiers in Marine Science 6 (MAR), 85. https://doi.org/10.3389/fmars.2019.00085.
Wacker, M.J., Godard, M.P., 2005. Analysis of one-step and two-step real-time RT-PCR using SuperScript III. Journal of Biomolecular Techniques 16 (3), 266–271. https://pubmed.ncbi.nlm.nih.gov/16461951.
Wang, L., Chen, G., Zhang, C., Wang, Y., Sun, R., 2020a. Application of loop-mediated isothermal amplification combined with lateral flow dipstick to rapid and sensitive detection of *Alexandrium catenella*. Environmental Science and Pollution Research 27 (4), 4246–4257. https://doi.org/10.1007/s11356-019-06889-y.
Wang, L., Zhang, C., Chen, G., Wang, Y., Fu, M., 2020b. Development of a rapid screening test for *Karenia mikimotoi* by using loop-mediated isothermal amplification and lateral flow dipstick. Journal of Applied Phycology 32 (5), 3143–3155. https://doi.org/10.1007/s10811-020-02203-3.
Wang, Y., 2016. Expression and activity of novel nitrate reductase enzymes in *Chattonella subsalsa* and implications for competitive dynamics in marine environments. Master's Thesis. University of Delaware.
Wang, Y., Bouchard, J.N., Coyne, K.J., 2018. Expression of novel nitrate reductase genes in the harmful alga, *Chattonella subsalsa*. Scientific Reports 8 (1), 1–12. https://doi.org/10.1038/s41598-018-31735-5.
Wang, Z., Zhang, Y., Huang, S., Peng, C., Hao, Z., Li, D., 2019. Nitrogen limitation significantly reduces the competitive advantage of toxic *Microcystis* at high light conditions. Chemosphere 237, 124508. https://doi.org/10.1016/j.chemosphere.2019.124508.
Wilkinson, S., Pontasch, S., Fisher, P., Davy, S., 2016. The distribution of intra-genomically variable dinoflagellate symbionts at Lord Howe Island, Australia. Coral Reefs 35 (2), 565–576. https://doi.org/10.1007/s00338-015-1357-0.
Wittwer, C.T., 2009. High-resolution DNA melting analysis: advancements and limitations. Human Mutation 30 (6), 857–859. https://doi.org/10.1002/humu.20951.
Wolf, K., Romanelli, E., Rost, B., John, U., Collins, S., Weigand, H., Hoppe, C., 2019. Company matters: the presence of other genotypes alters traits and intraspecific selection in an Arctic diatom under climate change. Global Change Biology 25 (9), 2869–2884. https://doi.org/10.1111/gcb.14675.
Wood, S., Borges, H., Puddick, J., Biessy, L., Atalah, J., Hawes, I., Dietrich, D., Hamilton, D., 2017. Contrasting cyanobacterial communities and microcystin concentrations in summers with extreme weather events: insights into potential effects of climate change. Hydrobiologia 785 (1), 71–89. https://doi.org/10.1007/s10750-016-2904-6.
Wood, S., Puddick, J., 2017. The abundance of toxic genotypes is a key contributor to anatoxin variability in *Phormidium*-dominated benthic mats. Marine Drugs 15 (10), 307. https://doi.org/10.3390/md15100307.
Wood, S., Rueckert, A., Hamilton, D., Cary, S., Dietrich, D., 2011. Switching toxin production on and off: intermittent microcystin synthesis in a *Microcystis* bloom. Environmental Microbiology Reports 3 (1), 118–124. https://doi.org/10.1111/j.1758-2229.2010.00196.x.
Wood, S., Smith, K., Banks, J., Tremblay, L., Rhodes, L., Mountfort, D., Cary, S., Pochon, X., 2013. Molecular genetic tools for environmental monitoring of New Zealand's aquatic habitats, past, present and the future. New Zealand Journal of Marine and Freshwater Research 47 (1), 90–119. https://doi.org/10.1080/00288330.2012.745885.
Wood, S.A., Dietrich, D.R., Cary, S.C., Hamilton, D.P., 2012. Increasing *Microcystis* cell density enhances microcystin synthesis: a mesocosm study. Inland Waters 2 (1), 17–22. https://doi.org/10.5268/IW-2.1.424.
Wood, S.A., Pochon, X., Laroche, O., von Ammon, U., Adamson, J., Zaiko, A., 2019. A comparison of droplet digital polymerase chain reaction (PCR), quantitative PCR and metabarcoding for species-specific detection in environmental DNA. Molecular Ecology Resources 19 (6), 1407–1419. https://doi.org/10.1111/1755-0998.13055.

Wurch, L., Gobler, C., Dyhrman, S., 2014. Expression of a xanthine permease and phosphate transporter in cultures and field populations of the harmful alga *Aureococcus anophagefferens*: tracking nutritional deficiency during brown tides. Environmental Microbiology 16 (8), 2444–2457. https://doi.org/10.1111/1462-2920.12374.

Yamahara, K.M., Demir-Hilton, E., Preston, C.M., Marin, R., Pargett, D., Roman, B., Jensen, S., Birch, J.M., Boehm, A.B., Scholin, C.A., 2015. Simultaneous monitoring of faecal indicators and harmful algae using an in-situ autonomous sensor. Letters in Applied Microbiology 61 (2), 130–138. https://doi.org/10.1111/lam.12432.

Yoshida, M., Yoshida, T., Takashima, Y., Hosoda, N., Hiroishi, S., 2007. Dynamics of microcystin-producing and non-microcystin-producing *Microcystis* populations is correlated with nitrate concentration in a Japanese lake. FEMS Microbiology Letters 266 (1), 49–53. https://doi.org/10.1111/j.1574-6968.2006.00496.x.

Yu, L., Kong, F., Shi, X., Yang, Z., Zhang, M., Yu, Y., 2015. Effects of elevated CO_2 on dynamics of microcystin-producing and non-microcystin-producing strains during *Micyocystis* blooms. Journal of Environmental Sciences 27, 251–258. https://doi.org/10.1016/j.jes.2014.05.047.

Yu, Y., Zeng, Y., Li, J., Yang, C., Zhang, X., Luo, F., Dai, X., 2019. An algicidal *Streptomyces amritsarensis* strain against *Microcystis aeruginosa* strongly inhibits microcystin synthesis simultaneously. Science of the Total Environment 650, 34–43. https://doi.org/10.1016/j.scitotenv.2018.08.433.

Yuan, J., Li, M.Z., Lin, S.J., 2015. An improved DNA extraction method for efficient and quantitative recovery of phytoplankton diversity in natural assemblages. PLoS ONE 10 (7), e0133060. https://doi.org/10.1371/journal.pone.0133060.

Yuan, J., Mi, T., Zhen, Y., Yu, Z., 2012a. Development of a rapid detection and quantification method of *Karenia mikimotoi* by real-time quantitative PCR. Harmful Algae 17, 83–91. https://doi.org/10.1016/j.hal.2012.03.004.

Yuan, J., Mi, T., Zhen, Y., Yu, Z., 2012b. Development of a real-time PCR method (Taqman) for rapid identification and quantification of *Prorocentrum donghaiense*. Journal of Ocean University of China 11 (3), 366–374. https://doi.org/10.1007/s11802-012-1911-0.

Zhang, F., Shi, Y., Jiang, K., Song, W., Ma, C., Xu, Z., Ma, L., 2014. Rapid detection and quantification of *Prorocentrum minimum* by loop-mediated isothermal amplification and real-time fluorescence quantitative PCR. Journal of Applied Phycology 26 (3), 1379–1388. https://doi.org/10.1007/s10811-013-0163-8.

Zhang, M., Liu, Y., Chen, L., Quan, S., Jiang, S., Zhang, D., Yang, L., 2013. One simple DNA extraction device and its combination with modified visual loop-mediated isothermal amplification for rapid on-field detection of genetically modified organisms. Analytical Chemistry 85 (1), 75–82. https://doi.org/10.1021/ac301640p.

Zhang, Q., Chen, Z., Zhao, J., Yu, R., Qiu, L., Kong, F., Wang, Y., Yan, T., Zhou, M., 2020. Development of a sensitive qPCR method for the detection of pelagophyte *Aureococcus anophagefferens*. Limnology and Oceanography: Methods 18 (2), 41–51. https://doi.org/10.1002/lom3.10344.

Zhang, Y., Fu, F.X., Whereat, E., Coyne, K.J., Hutchins, D.A., 2006. Bottom-up controls on a mixed-species HAB assemblage: a comparison of sympatric *Chattonella subsalsa* and *Heterosigma akashiwo* (Raphidophyceae) isolates from the Delaware Inland Bays, USA. Harmful Algae 5 (3), 310–320. https://doi.org/10.1016/j.hal.2005.09.001.

Zhang, Y., Lin, X., Shi, X., Lin, L., Luo, H., Li, L., Lin, S., 2019. Metatranscriptomic signatures associated with phytoplankton regime shift from diatom dominance to a dinoflagellate bloom. Frontiers in Microbiology 10 (MAR), 590. https://doi.org/10.3389/fmicb.2019.00590.

Zhen, Y., Qiao, L., Gu, B., Mi, T., 2016. Characteristics of eukaryotic microalgal community and its abiotic influencing factors during brown tide blooms near Qinhuangdao, China. Harmful Algae 57, 1–12. https://doi.org/10.1016/j.hal.2016.05.001.

Zhu, F., Massana, R., Not, F., Marie, D., Vaulot, D., 2005. Mapping of picoeucaryotes in marine ecosystems with quantitative PCR of the 18S rRNA gene. FEMS Microbiology, Ecology 52 (1), 79–92. https://doi.org/10.1016/j.femsec.2004.10.006.

Zink, F.A., Tembrock, L.R., Timm, A.E., Farris, R.E., Perera, O.P., Gilligan, T.M., 2017. A droplet digital PCR (ddPCR) assay to detect *Helicoverpa armigera* (Lepidoptera: Noctuidae) in bulk trap samples. PLoS ONE 12 (5), e0178704. https://doi.org/10.1371/journal.pone.0178704.

Zinser, E., Coe, A., Johnson, Z., Martiny, A., Fuller, N., Scanlan, D., Chisholm, S., 2006. *Prochlorococcus* ecotype abundances in the North Atlantic Ocean as revealed by an improved quantitative PCR method. Applied and Environmental Microbiology 72 (1), 723–732. https://doi.org/10.1128/AEM.72.1.723-732.2006.

Zuo, J., Chen, L., Shan, K., Hu, L., Song, L., Gan, N., 2018. Assessment of different mcy genes for detecting the toxic to non-toxic *Microcystis* ratio in the field by multiplex qPCR. Journal of Oceanology and Limnology 36 (4), 1132–1144. https://doi.org/10.1007/s00343-019-7186-1.

CHAPTER 11

Phytoplankton diversity and ecology through the lens of high throughput sequencing technologies

Adriana Lopes dos Santos[a], Catherine Gérikas Ribeiro[b], Denise Ong[a], Laurence Garczarek[c], Xiao Li Shi[d], Scott D. Nodder[e], Daniel Vaulot[c,a], and Andres Gutiérrez-Rodríguez[e]

[a]Asian School of the Environment, Nanyang Technological University, Singapore
[b]GEMA Center for Genomics, Ecology & Environment, Faculty of Sciences, Universidad Mayor, Santiago, Chile
[c]UMR 7144, ECOMAP, CNRS, Sorbonne Université, CNRS, Station Biologique de Roscoff, Roscoff, France
[d]State Key Laboratory of Lake Science and Environment, Nanjing Institute of Geography and Limnology, Chinese Academy of Sciences, Nanjing, China
[e]National Institute of Water and Atmospheric Research (NIWA), Wellington, New Zealand

11.1 Introduction

The study of taxonomic diversity and structure of biological communities is a cornerstone of ecology. Plankton communities are taxonomically and functionally diverse, with organisms covering a great range of cell sizes, morphologies, and trophic modes (autotrophs, heterotrophs, mixotrophs), and belonging to archaea, bacteria, protists (microbial eukaryotes), and metazoans (Caron et al., 2012). In addition, plankton communities are extremely dynamic, with high growth and mortality rates. For example, phytoplankton growth rates are typically in excess of 1.0 day^{-1}, which is faster than those of archaea and bacteria (around 0.1 day^{-1}), although some of the most abundant bacterial species (from the SAR11 clade) can grow at rates similar to phytoplankton (Kirchman, 2016). The structural complexity and dynamic nature of plankton communities has historically precluded an accurate and comprehensive assessment of community structure and species diversity at temporal and spatial scales.

During the last 30 years, the analysis of environmental diversity by molecular approaches has highlighted a massive unknown diversity, including entire lineages without any cultured representatives, for which only environmental sequences are available (Giovannoni et al., 1990). New information on the genetic diversity of small-sized phytoplankton groups is particularly important since these groups lack distinctive morphological attributes (López-García et al., 2001; Moon-van der Staay et al., 2001; Vaulot et al., 2008). Morphological analyses combined with molecular approaches has demonstrated the existence of cryptic species, even within well-known phytoplankton groups, such as diatoms (Amato et al., 2007; Kaczmarska et al., 2014), powering the debate on phytoplankton species delineation (Leliaert et al., 2014). Currently, the advent of high-throughput sequencing techniques combined with improved taxonomically annotated reference sequence databases allow rapid and cost-effective assessment of the composition of plankton communities at a global level (de Vargas et al., 2015).

11.1.1 The concept of molecular markers

"Science is impelled by technological advance and a guiding vision"
- Carl Woese (2009)

Carl Woese and George Fox in 1977 established a "third domain of life," called Archaebacteria, after examining the sequences of the 16S/18S ribosomal ribonucleic acid (rRNA) of plants, yeast, human cells, bacteria, and methanogenic archaea (Woese and Fox, 1977). Comparative analysis of biomolecules to determine evolutionary relationships was already considered a powerful approach (Fitch and Margoliash, 1967; Zuckerkandl and Pauling, 1965) and had been applied to different groups of eukaryotes (vertebrates and invertebrates) (Fitch, 1976). However, Woese and Fox were interested in determining "the relationships covering the entire spectrum of extant living systems" (Woese and Fox, 1977) and therefore they needed what is today called a molecular marker. The molecule of choice was the rRNA, and the arguments they used to select rRNA still hold today (Olsen et al., 1986). Firstly, the rRNA is functionally and evolutionary homologous in all living organisms, and therefore a common denominator by which all organisms can be compared. Secondly, rRNA sequences have both highly conserved and highly variable regions, allowing the examination of both distant and close evolutionary relationships. Thirdly, to maintain its function, the structure of the sequences does not change rapidly, allowing homologous regions within the rRNA to be properly aligned and accurately compared. Lastly, the rRNA gene is presumably less susceptible to horizontal gene transfer (HGT), as its product interacts with many molecules (also known as the complexity hypothesis, Jain et al.,

1999). Hence, one would expect only evolutionary relationships to be reflected in rRNA sequences, although there are a few reported cases today of horizontal gene transfer for rRNA (Kitahara and Miyazaki, 2013; Yabuki et al., 2014).

In 1977 David Lane and collaborators proposed that only homologous regions that were different enough to allow statistically significant comparisons should be sequenced (Lane et al., 1985). They noticed that conserved sequences were located next to less-conserved regions. Conserved regions would serve as initiation sites for primer elongation sequencing techniques, whereas less-conserved regions would allow phylogenetic evaluations. The suggested approach of primer regions flanking target gene markers is still an important concept today.

11.1.2 The advent of environmental sequencing

Medlin et al. (1988)'s landmark paper proposed the first set of primers and the polymerase chain reaction (PCR) conditions for the amplification of small subunit (SSU) rRNA eukaryotic gene (18S rRNA). This allowed determination of the nearly complete 18S rRNA sequence from the marine diatom *Skeletonema costatum*, opening the door for using SSU rRNA gene as biodiversity marker in studies targeting the diversity of natural plankton communities, published 13 years later (Díez et al., 2001; López-García et al., 2001; Moon-van der Staay et al., 2001). These early studies revealed a widely unsuspected diversity among small planktonic eukaryotes, including the discovery of new lineages within well-known phytoplankton groups, such as prasinophytes, haptophytes, dinoflagellates and stramenopiles. In subsequent years, a wide diversity of aquatic environments has been explored by the cloning approaches, including deep-sea ecosystems, anoxic environments, or oligotrophic regions (Edgcomb et al., 2002; Lovejoy et al., 2006; Not et al., 2007b; Stoeck et al., 2003), leading to the discovery of other novel uncultivated groups (e.g., picobiliphytes, now renamed Picozoa, Not et al., 2007a). Moreover, other genes besides the 18S rRNA gene have been used, for example, the 28S rRNA or the plastid 16S rRNA, *rbcL* or *psba* genes (e.g., Man-Aharonovich et al., 2010; Rodríguez-Martínez et al., 2013; Samanta and Bhadury, 2016).

11.1.3 The transition to high throughput sequencing (HTS)

About 15 years ago, Sanger sequencing was overtaken by next generation sequencing (NGS) or high throughput sequencing (HTS). HTS could provide millions of sequences of the target genetic marker, without the need to construct clone libraries, which was one of the rate limiting factors. This led to a rapid increase in the environmental studies of community structure.

The two main HTS technologies used for microbial diversity studies have been developed by two companies: 454 (Roche) and Illumina. HTS follows a sequencing by synthesis approach (Goodwin et al., 2016). As each nucleotide base complementary to the DNA fragment is added using a polymerase, a signal will identify what nucleotide was added to the elongating strand. This signal can be a fluorophore or a change in ionic concentration. Using this method, many different DNA fragments can be sequenced at one time, contributing to the massive parallelization, and hence high throughput of the process. The use of the 454 system was first suggested for metabarcoding by Kysela et al. (2005): as 454 was not optimized to sequence many samples individually, these authors proposed tagging the sequences from different samples such that they could be sequenced together and separated later. Sogin et al. (2006) were the first to apply 454 technology to sequence 16S rRNA gene amplicons to determine the bacteria and archaea diversity of marine samples. Soon after, the V9 region of the 18S rRNA gene was sequenced to study the diversity of marine protists (Amaral-Zettler et al., 2009; Stoeck et al., 2009). The 454 technology was initially more popular compared to Illumina, because it generated longer reads, although with a higher error rate (Luo et al., 2012). However, Roche stopped production of 454 sequencing machines in late 2013, and discontinued its technical support in 2016 (Pedrós-Alió et al., 2018). Meanwhile, Illumina increased the length of the fragments it could sequence (up to 600 bp), and is now the preferred HTS technology (Jurburg et al., 2020).

Metabarcoding, which consists of amplifying a marker gene with PCR and sequencing the amplicon using HTS, is the focus of this chapter. It can be also referred to as "amplicon sequencing," "tag sequencing" and "HTS". Sometimes it is mistankenly cited as "metagenomics" which refers to sequencing unamplified DNA. The methodological advantages of metabarcoding attracted an increasing number of researchers to apply this approach to microbial diversity studies (Santoferrara et al., 2020), in particular in aquatic ecosystems (Fig. 11.1).

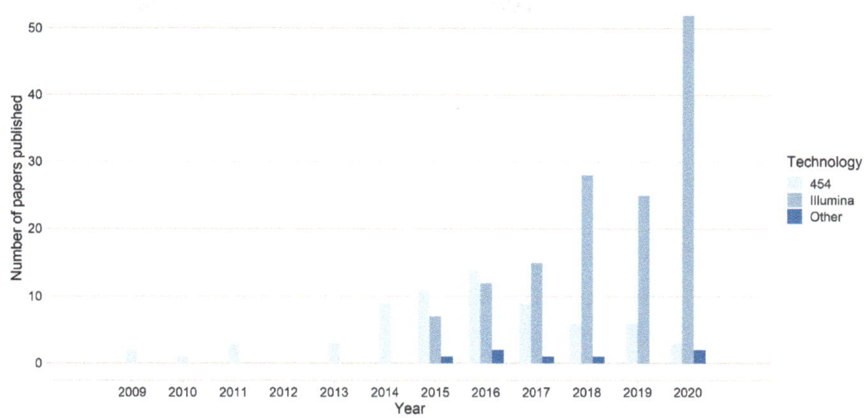

Figure 11.1 Papers using metabarcoding for aquatic protist studies published per year and technology (see Table S1 and https://github.com/vaulot/Chapter-2020-Lopes-Metabarcoding-phytoplankton).

11.2 The different steps of metabarcoding

11.2.1 Sampling and DNA extraction

Sampling of phytoplankton (or more generally of plankton) involves the collection of the cells that are free-floating in the water either by filtration, by using a plankton net, or by separating individual cells by pipetting or flow cytometry. Phytoplankton taxa can also be found on other substrates, such as ice, sediment, or even on the surface of other organisms, such as coral or seagrasses (see below Section 11.4), all of which require specialized sampling techniques. The most common approach consists collecting sea or freshwater water with bottles (e.g., Niskin bottles) or with *in situ* pumps and a series of decreasing pore size filters (for example 20 μm, 2 μm, 0.2 μm) with the aim of separating the community into the plankton size-classes proposed by Sieburth et al. (1978). The collected biomass is stored frozen or more rarely in ethanol or even formalin (Shiozaki et al., 2021).

Samples are then transported to the laboratory for acid nucleic (DNA and/or RNA) extraction. Several methods and commercial kits are available for this purpose; the choice may vary depending on the substrate considered, as well as the amount of material available. Most studies have focused on DNA. However, because phytoplankton can be found in a range of metabolic states (dormant, growing, decaying), a number of studies have sequenced the reverse-transcribed RNA (cDNA) in addition to DNA. The ratio of RNA to DNA sequences (RNA:DNA) is then used as a proxy of metabolic activity (Charvet et al., 2014; Egge et al., 2015; Hu et al., 2016; Massana et al., 2015; Giner et al., 2020). Yet, evidence indicates that the general use of rRNA as an indicator of metabolic state of microbial communities has several limitations (e.g., rRNA copy numbers vary as a factor of metabolic state or cell size) (Blazewicz et al., 2013).

11.2.2 Marker gene selection

The genetic markers used in a given study will vary, depending on the target organisms and availability of reference sequences. The 18S rRNA gene is by far the most commonly targeted one used in eukaryotic plankton studies (Table 11.1). For prokaryotic phytoplankton (cyanobacteria), however, the equivalent 16S rRNA gene is not the most used (see Section 11.5). Though sequencing of the entire SSU rRNA is possible using Sanger sequencing, the sequence is too long for HTS. Small hypervariable regions of the 18S rRNA have been tested and selected over the years, based on their ability to distinguish different taxa. Hypervariable regions 4 (V4) and 9 (V9) are the most often used to access the eukaryotic diversity in phytoplankton natural assemblages. The V4 region is located around 550 base pair (bp) from the beginning of the 18S rRNA gene (reference sequence *Saccharomyces cerevisiae*) and is ~450 bp long, whereas V9 is located at the end of the gene with a ~150 bp length (Vaulot et al., 2021). Though both regions are able to provide a similar picture of eukaryotic phytoplankton diversity at higher taxonomic levels (division, class, see Tragin and Vaulot, 2018), at lower taxonomic levels (e.g. species) specific groups may be better resolved using a specific region. For example, V9 distinguish better between phytoplankton species of the class Chloropicophyceae (Lopes dos Santos et al., 2017), whereas it has poor resolution among the order Dolichomastigales (class Mamiellophyceae, (Tragin and Vaulot, 2018)). One drawback of the V9 region is the limited number of reference sequences in public databases. In aquatic metabarcoding studies, the V4 region is clearly the most often used, probably because of the better coverage of reference databases (Table 11.1).

Other marker genes have been used (Tables 11.1 and S1) to access eukaryotic phytoplankton diversity, for example, the plastid 16S rRNA (Trefault et al., 2021). The use of plastid 16S rRNA for metabarcoding has not been popularized, as had been earlier with clone libraries (e.g., Shi et al., 2011), but 16S plastid sequences have been obtained as a by-product of bacteria 16S metabarcoding studies (e.g., Needham and Fuhrman, 2016). One issue is the few reference sequences for plastidial 16S, despite efforts to increase the number of cultures sequenced (Decelle et al., 2015). The

Table 11.1 Most frequently used gene markers and 18S rRNA region in eukaryotic metabarcoding studies of aquatic phytoplankton with the number of papers (N) where used (Table S1).

Marker gene	N	18S region	N
18S rRNA	221	V4	160
16S rRNA plastid	7	V9	34
28S rRNA	6	Other	27
ITS2	3		
ITS1	2		
rbcL plastid	2		
rRNA operon	2		
23S rRNA plastid	1		
coi mitochondrion	1		

D1–D2 region of the large subunit 28S rRNA gene has been used to determine diversity for Haptophyta (Bittner et al., 2013; Gran-Stadniczeñko et al., 2017). Although there are more reference sequences available for 18S rRNA (Edvardsen et al., 2016), the 28S rRNA has more variable regions, especially between closely related species of Haptophyta (Bittner et al., 2013; Liu et al., 2009). For example, Gran-Stadniczeñko et al. (2017) found more diversity using the D1–D2 region of 28S rRNA compared to the V4 region of 18S rRNA, but the lack of reference sequences remains a problem. Other marker genes that have been used with the clone library approach prior to the development of HTS, such as the ITS (internally transcribed spacer of the rRNA operon), *psbA* or *rbcL* genes (Man-Aharonovich et al., 2010; Rodríguez-Martínez et al., 2013; Samanta and Bhadury, 2016), have been very little used for phytoplankton metabarcoding, either due to the absence of suitable short region, lack of general primers, or absence of reference sequences.

Once a marker gene has been selected, it is necessary to select suitable primers. For 18S rRNA, a wide range of primers and primer sets have been developed (for a review see Vaulot et al., 2021). Several criteria must be met. First, the primers must have as little bias as possible, i.e., amplify all taxonomic groups with the same efficiency. Second, they must produce a fragment with a compatible size with the sequencing technology of choice. For example, a fragment should not surpass 550 bp when Illumina 2×300 kits are used, allowing for some overlap between the forward and reverse fragments sequenced. In studies of marine protists, including phytoplankton, by far the most used primer pair is composed of

Table 11.2 Eukaryotic 18S rRNA primers most often used in metabarcoding studies for aquatic phytoplankton and protists (see Table S1) with the number of papers (N) where used.

Name	Sequence	Region	Direction	Reference	DOI	N
TAReuk454FWD1	CCAGCASCYGCGGTAATTCC	V4	fwd	Stoeck et al. (2010)	10.1111/j.1365-294X.2009.04480.x	83
1380F	CCCTGCCHTTTGTACACAC	V9	fwd	Amacher et al. (2009)	10.1371/journal.pone.0006372	19
528F	GCGGTAATTCCAGCTCCAA	V4	fwd	Cheung et al. (2010)	10.1038/ismej.2010.26	18
E572F	CYGCGGTAATTCCAGCTC	V4	fwd	Comeau et al. (2011)	10.1371/journal.pone.0027492	16
1391F	GTACACACCGCCCGTC	V9	fwd	Lane (1991)		7
3NDF	GGCAAGTCTGGTGCCAG	V4	fwd	Cavalier-Smith et al. (2009)	10.1016/j.protis.2009.03.003	6
TAReukREV3	ACTTTCGTTCTTGATYRA	V4	rev	Stoeck et al. (2010)	10.1111/j.1365-294X.2009.04480.x	61
1510R	CCTTCYGCAGGTTCACCTAC	V9	rev	López-García et al. (2003)	10.1073/pnas.0235779100	24
E1009R	AYGGTATCTRATCRTCTTYG	V4	rev	Comeau et al. (2011)	10.1371/journal.pone.0027492	16
V4 18S Next.Rev	ACTTTCGTTCTTGATYRATGA	V4	rev	Piredda et al. (2017)	10.1093/femsec/fiw200	13
EukB	TGATCCTTCTGCAGGTTCACCTAC	V9	rev	Medlin et al. (1988)	10.1016/0378-1119(88)90066-2	8
964iR	ACTTTCGTTCTTGATYRR	V4	rev	Balzano et al. (2015)	10.3354/ame01740	7

TAReuk454FWD1 and TAReukREV3, as forward and reverse primers, respectively (Tables 11.2 and 11.3), which were among the first primers designed for metabarcoding (Stoeck et al., 2010). These primers target the V4 region, yielding a ~420 bp amplicon suitable for Illumina sequencing. One problem is that the reverse primer has one mismatch to Haptophyta, a very important phytoplankton group, at the 3' end of the primer, which is particularly unfavorable, resulting in datasets where Haptophyta are absent. A small modification of this primer (V4 18S Next.Rev, Table 11.2) removes this bias, allowing more complete datasets (Piredda et al., 2017). For the V9 region, two primer sets, 1380F/1510R and 1389F/1510R, are the ones most often used. Some primer sets are designed to target specific groups, enhancing, for example, the coverage of haptophytes or diatoms (Gran-Stadniczeñko et al., 2017; Rynearson et al., 2020).

Table 11.3 18S rRNA primer sets most often used in metabarcoding studies of aquatic phytoplankton and protists (see Table S1) with the number of papers (N) where used. Refer to Table 11.2 for sequence and reference of primers.

Primer fwd	Primer rev	Region	N
TAReuk454FWD1	TAReukREV3	V4	60
1380F	1510R	V9	19
E572F	E1009R	V4	15
TAReuk454FWD1	V4 18S Next.Rev	V4	13
1391F	EukB	V9	7
528F	1055R	V4	6
Other			101

11.2.3 PCR and sequencing

Once DNA is extracted and the marker gene and the relevant primers selected, the next step involves one or two rounds of PCR, depending on the material available. Low amounts of DNA, for example, from samples sorted by flow cytometry require usually 2 PCR rounds (Gérikas Ribeiro et al., 2018). As for any PCR, conditions (annealing temperature, primer concentration, etc.) need to be optimized. Another round of PCR is then needed to add sample "barcodes" that allow a large number of samples to be run together. Typically, for an Illumina MiSeq run, up to 300 samples can be multiplexed together, with each sample producing 50 to 100 thousand sequences (reads). Higher throughput sequencers, such as HiSeq, can yield 20 to 50 times more sequences per sample.

11.2.4 Data processing

In this section, we will focus mostly on 18S rRNA metabarcodes, but the process would be similar for any other marker gene. HTS produces millions of raw sequences (reads) that need to be processed to produce meaningful data. In brief, the different steps are the following: demultiplexing, quality filtering, removal of primers, assembly, clustering, removal of chimera, and taxonomic assignment. Several bioinformatics programs have been developed to take care of these different steps (for a recent review Pérez-Cobas et al., 2020). Some can perform almost all necessary steps, such as Qiime (Bolyen et al., 2019), mothur (Schloss et al., 2009), VSEARCH (Rognes et al., 2016), or dada2 (Callahan et al., 2016); whereas others perform only specific steps, such as cutadapt (Martin, 2011), which focuses on primer removal. Qiime, a suite of programs written in Python is the most used to date, despite its initial versions being complicated to install, whereas dada2 running under R is becoming increasingly popular (Table 11.4). Another new option is PEMA, which groups a number of tools, such as VSEARCH and Swarm in a ready-to-use Docker container (Zafeiropoulos et al., 2020).

Table 11.4 Software most often used to process sequence in metabarcoding studies of aquatic phytoplankton and protists (see Table S1) with the number of papers (N) where used.

Software	N
USEARCH	77
QIIME	68
mothur	43
swarm	25
dada2	19
Other	14

It is also possible to use web services to process datasets. VAMPS (https://vamps2.mbl.edu/) is one of the oldest such site and offers, besides data processing, the possibility to download and visualize public datasets (Huse et al., 2014). It is superseded by a service, such as SILVAngs (https://ngs.arb-silva.de/), which is linked to the SILVA sequence database (see below) or FROGS (http://frogs.toulouse.inra.fr/). Such services suffer from having to upload your data on a web server, which can be slow, and from offering only fixed processing options (e.g., clustering or reference database).

Some web sites also facilitate downloading processed data from large scale projects such as Tara or Malaspina, (see below) and the Ocean Barcode Atlas (http://oba.mio.osupytheas.fr/ocean-atlas/, Vernette et al., 2021).

Read processing. Many samples are usually sequenced together, and usually the demultiplexing step is performed by the sequencing software, but programs such as cutadapt can also perform this step. It is always necessary to remove the primers if this is not performed before, which can be done with cutadapt, and eventually combined with demultiplexing. The next step is to remove reads that are either too short or of bad quality. Quality can be checked using, for example, the function *plotQualityProfile* in the R dada2 package, allowing for the adjustment of parameters using the *FilterandTrim* function. When using Illumina technology, the next step is to build contigs or paired-end assemblies (Kozich et al., 2013) from the R1 (forward) and R2 (reverse) reads. Each of these steps may remove a significant number of reads, but it is always better to be more rigorous at this stage to prevent artifacts at the latter processing stages.

Clustering. Highly similar sequences are grouped together and initial strategies defined as operational taxonomic units (OTUs), where ideally each OTU would represent one microbial species (Pedrós-Alió et al., 2018). The conventional practice suggests that microbial organisms with a 97% similarity threshold (conversely a dissimilarity of 3%) are considered part of the same species (Schloss and Handelsman, 2005; Stackebrandt and Goebel, 1994), and this threshold has been widely used in metabarcoding studies (Table S1). However, this threshold is somewhat arbitrary, varying between the taxonomic groups that are being studied and the length and region of the marker gene used (Edgar, 2018). Depending on the aim of the study, 97% can be too low and should be increased to 99% or 100%, for example, when looking at species distributions (Tragin and Vaulot, 2019). The use of OTUs with a fixed similarity level has several drawbacks (Callahan et al., 2017). First, this clustering depends on the number of samples included in the analysis, and if the number of samples is changed, the analysis must be redone. Second, the sequences need to be aligned to a reference alignment, which is a very time-consuming step, as well as dependent on the quality of the reference alignment. Lastly, it is difficult to compare the results from different studies if they choose a different threshold. Therefore in recent years, more "natural" clustering algorithms have been developed, which includes swarm (Mahé et al., 2014) and dada2 (Callahan et al., 2016). These approaches are more computationally efficient and yield data that can be easily compared between datasets. The clusters produced by these methods are called either "swarms" or "ASVs" (amplicon sequence variants). The processing software assigns sample reads to each cluster, forming what is known as an ASV or OTU table, with one line per OTU/ASV and one column per sample, the cells representing the number of reads for a given combination of OTU/ASV and sample.

One important source of error is linked to chimera formation during PCR amplification. Chimeric sequences are two or more fragments from distinct species that are combined during PCR to form a sequence, which is further amplified and could be inaccurately detected as a novel species (Meyerhans et al., 1990). If undetected, the resulting diversity estimates will be falsely skewed (Porazinska et al., 2012; Wang and Wang, 1997). A variety of tools have been developed to detect and remove these sequences, such as UCHIME (Edgar et al., 2011) or the *removeBimeraDenovo* function in dada2. However, this process is often not satisfactory and chimera removal may have to be done by visual inspection of an alignment of OTU/ASV and reference sequences.

Taxonomic assignation. The taxonomic assignation of the OTU/ASV sequences is at the heart of the metabarcoding analysis. OTU/ASV representative sequences are assigned according to their similarity to those from a reference database, for which the taxonomy has been curated by experts. The database should cover the whole targeted taxonomic domain, e.g., eukaryotes when using the 18S rRNA gene. If some important groups are missing, the corresponding OTU/ASV will not be correctly annotated. The sequences from the reference database should also be correctly annotated: for example, if a reference sequence is annotated as a diatom, whereas it is in reality a green alga, this will result in errors in the final community structure. For eukaryotes, three general databases can been used. Genbank nr contains all published sequences and is therefore the most comprehensive in terms of coverage. Its main handicap is the taxonomic annotation is not curated and can be very minimal, in particular for environmental sequences (e.g., many sequences are simply annotated as "eukaryotes"). SILVA (Quast et al., 2013) is better annotated than Genbank, and is periodically updated (https://www.arb-silva.de/). For protists, however, the best choice is the Protist Ribosomal Reference database (PR^2, https://pr2-database.org), which contains fewer sequences (~180,000), but these sequences are periodically re-annotated by experts from each group using 8 taxonomic levels, from the kingdom to the species (Guillou et al., 2013). Some specialized databases have also been developed for specific taxonomic groups, such as dinoflagellates (DINOREF, Mordret et al., 2018), or specific genes, such as the plastidial 16S rRNA (PHYTOREF, Decelle et al., 2015).

Once a reference database is selected, the taxonomic assignment can be done in several different ways. The best strategy is to use alignment-independent approaches, such as the naive Bayesian classifier, also known as the RDP (Ribossomal Database Project) classifier (Wang et al., 2007). RDP is implemented in pipelines such as mothur or dada2, and provides bootstrap values for the confidence of the assignment at each taxonomic level. Similar approaches are used by the SINTAX algorithm (Edgar, 2016), as implemented in VSEARCH or the *IdTaxa* function from the R DECIPHER package (Wright Erik, 2016). Other methods rely on alignments such as a LCAClassifier of Crest (Lanzén et al., 2012), which uses the aligned SILVA database as a reference, or pplacer (Matsen et al., 2010), which places sequences on a fixed reference tree. A simple BLAST (Altschul et al., 1990) search can also be used to obtain the closest matching sequences from either Genbank or a reference database, but it is difficult to implement for a large number of OTU/ASV, and does not provide bootstrap values. If one is interested to investigate low taxonomic levels (e.g., genus or species), it is often wise to confirm OTU/ASV assignments with alignments and phylogenetic trees using closely related sequences obtained by BLAST.

Abundance normalization. Metabarcoding data are used in general to calculate changes in relative taxa abundance between different samples. A first issue is that the total number of reads vary between samples. There are roughly two ways of normalizing samples (Pérez-Cobas et al., 2020). One approach is to rarefy all samples to the least abundant sample, which may lead to the loss of rare OTU/ASV using, for example, the *rarefy* function from the R vegan package (Oksanen et al., 2016). Alternatively, one can compute the relative abundance of a given OTU/ASV in each sample, or normalize each sample to a fixed number of reads, e.g., the mean or median of all samples. More sophisticated approaches can be taken (Love et al., 2014). Another concern is that the rRNA gene copy number varies between species. Though picoplankton typically have a few copies of rRNA operon, larger eukaryotes can have several thousands of copies, especially among groups such as dinoflagellates (Gong and Marchetti, 2019; Zhu et al., 2005). As such, eukaryotes with large number of rRNA gene copies will be over-represented in metabarcoding studies. As the number of species for which the copy number is known remains very small, there are only few ways to correct for this bias.

Downstream analyses. Once an OTU/ASV table has been constructed with the taxonomy of each sequence assigned and abundance normalized, many different analyses can be performed. One very useful R package is phyloseq (McMurdie and Holmes, 2013). It combines a large number of tools in a single package, providing ways to filter (for a sample type) and group the data (e.g., at higher taxonomic level), to display the data under a variety of graphic formats (barplots, heatmaps), to perform alpha and beta diversity analyses, such as NMDS (nonmetric multidimensional scaling), or to visualize networks. Another useful tool is the DESeq2 package (Love et al., 2014), which allows detecting OTU/ASV that are specific of a given set of samples.

Making data available: a critical but overlooked step. Data availability is key for scientific reproducibility. Researchers should be able to start from the original data and reproduce all steps to reach the final conclusions, following the FAIR Data Principles (Wilkinson et al., 2016). Moreover, the primary data from any paper should be available to the community, so they can be reused in new studies. Therefore it is imperative that raw sequences are deposited into public repositories, preferably the NCBI/EBI/DDBJ SRA (Sequence Read Archive), with standardized environmental metadata and technical information (Yilmaz et al., 2011). The accession number provided should be stated in the publication. Today very few journals request explicitly that authors provide primary sequence files, and even when it is stated in the instructions to authors, this is only partially enforced. It is very revealing that only 80% of the HTS studies listed here (Table S1) have deposited raw sequences into public repositories, a situation that has been also pointed out for 16S metabarcoding data (Jurburg et al., 2020). Scripts and pipelines should be provided and clearly documented. Derived data, such as OTU/ASV tables and sequences with their assignation, should also be freely available in open repositories (e.g., GitHub, Zenodo or Figshare). Statements such as "only made available upon request" should result in immediate refusal for serious journal to process submissions (Eren, 2019).

11.3 Protist metabarcoding studies in aquatic environments

To date, more than 240 studies have used metabarcoding to look at protist communities in aquatic systems (Table S1). The number of studies appears to increase exponentially in recent years (Fig. 11.1). A few of these studies have targeted specifically phytoplankton (e.g., using plastid 16S rRNA primers), whereas most of them obtained data on the overall protist community since the most commonly used primer sets (Table 11.3) do not discriminate against heterotrophic groups, even if some of the studies focused on the sole analysis of phytoplankton groups. The majority of studies investigated marine systems, in particular oceanic and coastal (Table 11.5). Freshwater systems have been less intensively sampled, with the exception of lakes. Most aquatic studies have sampled the water and, much fewer, sediments (Table 11.5). Some key aquatic biotopes, such as coral reefs, have been scarcely sampled as have been epiphytic communities on algae or sea grasses.

Table 11.5 Ecosystems and substrates targeted by aquatic metabarcoding studies (see Table S1).

Ecosystem	N	Substrate	N
marine oceanic	106	water	206
marine coastal	91	sediments	29
freshwater lakes	23	ice	6
estuarine	11	biofilm	5
freshwater rivers	11	sorted phytoplankton	5
lagoons	6	coral	4
freshwater ponds	4	sediment trap	4
aquaculture	3	copepod gut content	2
freshwater gutters	1	periphyton	2
freshwater supply	1	ballast water	1
saltern	1	ice-algal aggregates	1
		macroalgae epiphytes	1
		rock	1
		seagrass	1
		sponge	1
		suspended and sinking particles	1
		water (experiment)	1
		zooplankton	1

In addition to these studies that targeted one or few sampling locations, three global surveys have been performed that contributed to popularize metabarcoding as a tool to study protist and phytoplankton communities. The Ocean Sampling Day network (Kopf et al., 2015) aimed at simultaneous sampling (on the day of the summer solstice in the Northern Hemisphere) coastal stations (Fig. 11.2). Metabarcoding data are available for two years (2014 and 2015), but sampling has resumed since 2018

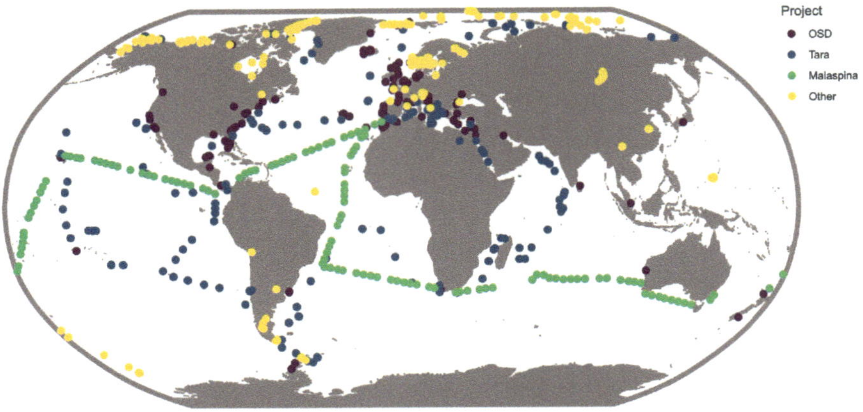

Figure 11.2 Map of stations investigated in a representative subset of 45 aquatic metabarcoding studies (see https://github.com/vaulot/Chapter-2020-Lopes-Metabarcoding-phytoplankton).

(https://www.assembleplus.eu/research/ocean-sampling-day) as part of the ASSEMBLE Plus. Although initially heavily biased towards Europe, the network of stations has gradually expanded. Numerous papers have made use of this resource (e.g., Tragin and Vaulot, 2018). The Tara *Oceans* launched in 2009 was much more ambitious, as it covered all three major oceans (Atlantic, Pacific, and Indian, Fig. 11.2), as well as the Arctic Ocean. Samples were taken at 2 depths (in general surface and deep-chlorophyll maximum) and separated into four size fractions (pico, nano, micro, meso). A large number of ancillary measurements (temperature, salinity, nutrients) was acquired, as well as samples for microscopy. With respect to metabarcoding, one of the remarkable characteristics of the Tara expedition is the depth of sequencing achieved (typically one million of reads per sample), allowing sampling to cover the "rare" biosphere, i.e., organisms in very low abundance (Logares et al., 2015). Moreover, samples were also analyzed by metagenomics and metatranscriptomics approaches, complementing the metabarcoding data. A final key element was the quick release of the primary and secondary data (raw and processed sequences), as well as the associated environmental data. This led besides the initial papers (e.g., de Vargas et al., 2015) to a very large number of papers making use of the data, for example, to look at the distribution of specific plankton groups (e.g., giant protists, Biard et al., 2016; Malviya et al., 2016), or at global diversity trends (Lopes dos Santos et al., 2017; Ibarbalz et al., 2019). For more details on Tara *Oceans* expeditions, please refer to Chapter 15. The Malaspina expedition (2010–2011, Duarte, 2015) performed round-the-world sampling (Fig. 11.2) not only of surface, but also of deep waters (Giner et al., 2020; Logares et al., 2020). Combining the data obtained from these expeditions with other more localized studies allow inference of the ocean-wide distribution of newly described species (*Mantoniella beaufortii*, Yau et al., 2020), as well as of larger taxonomic groups (diatoms, Malviya et al., 2016).

11.3.1 Arctic and Antarctic communities

The southern and northern poles are among the most environmentally susceptible regions to climate change (Haumann et al., 2020; Perovich and Richter-Menge, 2009; Turner et al., 2005) and ocean acidification (Orr et al., 2005; Shadwick et al., 2013), and the least studied ecosystems on the planet, due to logistical constraints.

Metabarcoding has revealed the community progression is driven by drastic seasonal changes on solar energy input (Joli et al., 2017; Vick-Majors et al., 2014), and has demonstrated the interconnectivity between polar and subpolar regions (Sow et al., 2020), and between the Arctic and Antarctic domains (Segawa et al., 2018; Wolf et al., 2015). Metabarcoding also allows questions that microscopy-based methods could only address with difficulty: for example, the seasonal progression during polar night of pico-sized algae, for which morphological features are hard to distinguish (Joli et al., 2017), or the composition of the rare microbial biosphere and its ecological roles and biogeography (Galand et al., 2009). Overall, many studies have collectively generated a solid knowledge base on polar protist diversity, unveiling phytoplankton community structure of the water (Duret et al., 2020; Kalenitchenko et al., 2019), ice (Comeau et al., 2013; Hardge et al., 2017; Lutz et al., 2018) and snow (Davey et al., 2019; Soto et al., 2020).

11.3.1.1 The risk of habitat loss for ice-associated communities

Polar environments are subject to complex hydrographic, atmospheric, and seasonal interactions, with prolonged periods of constant light or complete darkness, where ice dynamics plays a major role in structuring plankton communities (Janout et al., 2016; Van Leeuwe et al., 2018), from light attenuation effects to changes in water stability patterns during melting (Rozema et al., 2017). Besides providing a complex matrix, where different sympagic (ice-associated) algae communities thrive (Comeau et al., 2013), sea ice has a recruiting/seeding role (Hardge et al., 2017; Niemi et al., 2011; Olsen et al., 2017) for the phytoplankton spring bloom in the water column. The question of how phytoplankton production, diversity, and community structure will change due to climate-related loss of sea ice has been increasingly addressed, especially in the Arctic (Horvat et al., 2017; Leu et al., 2011), where this phenomenon has been steeply accentuated in the last decades (Serreze et al., 2007; Stroeve et al., 2014). Hardge et al. (2017) compared Arctic protist diversity on sea ice and in melt pond water, under-ice water, and the deep-chlorophyll maximum using metabarcoding, and demonstrated that the sea ice, and specially multiyear ice, harbors a particularly high diversity of unique taxa. This study highlighted the specificity of the sympagic community dominated by diatoms, and a potential diversity loss in both sea ice and water column as multiyear sea ice decreases drastically in the region.

Due to the on-going reduction of sea ice thickness and extent, melt ponds are increasingly becoming a common feature of Arctic's landscape, fueling a positive feedback on sea ice retreat and enabling the formation of massive sub-ice blooms (Arrigo et al., 2014; Horvat et al., 2017). Using RNA-based metabarcoding, Xu et al. (2020) reported that melt ponds harbor a very different and less diverse active protist community compared to that of the sea water, dominated by ciliates, cercozoan, and chrysophytes which are controlled by ecological drift, rather than dispersal limitation. Kilias et al. (2014) observed that melt pond and bottom-ice protist communities were also dissimilar, with melt ponds being dominated by the genera *Carteria*, *Ochromonas* and *Dinobryon*. These studies indicate that, although subjected to variable degrees of community exchange, sea ice, melt ponds, and the water column provide different community-structuring selective pressures on protists, and thus ecosystems shifts imposed by climate change will undoubtedly affect its diversity across polar environments.

11.3.1.2 Impact of protist diversity on ocean cycles and novel biogeographic patterns

As spring approaches and the light reaching the water column becomes sufficient to support the growth of autotrophic populations, a vast phytoplankton bloom is formed beneath the melting sea ice for many kilometers from the ice edge into the pack ice (Arrigo et al., 2014). Metabarcoding coupled with other methods can investigate the magnitude and structure of polar phytoplankton blooms and their impact on biogeochemical cycles. For example, Duret et al. (2020) investigated differences between eukaryotic plankton groups and the efficiency of carbon transfer via suspended/sinking particles in the Southern Ocean, using 18S rRNA metabarcoding data as a tool for identifying groups linked to a stronger biological carbon pump.

Chaetoceros-enriched communities were prone to a more efficient carbon flux to deeper water layers, in contrast to *Phaeocystis*-enriched suspended particles, which tended to go through remineralization within the upper layers of the water column (Duret et al., 2020).

Microdiversity analysis of metabarcoding data (meaning, setting a threshold for diversity using a single nucleotide difference instead of fixed percentages of similarity, see Section 11.2.4) has enabled a higher degree of precision on determining ecotype biogeographical patterns within certain species. This approach has been increasingly recognized as ecologically robust, providing insights into evolutionary factors shaping microbial dynamics (Needham et al., 2017), and linked to functional stability within microbial communities (García-García et al., 2019). For example, *Phaeocystis*, an important member of high-latitude environments, seems to display biogeographic patterns tightly linked to oceanographic boundaries, especially the polar and subantarctic fronts (Sow et al., 2020). Tragin and Vaulot (2019) analyzed microdiversity within Mamiellophyceae members and reported a previously undetected *Micromonas* subarctic clade, which is phylogenetically close to the panarctic *Micromonas polaris*, and possibly has been masked until now by studies using clone libraries and similarity percentages of 98% or less. The *Micromonas* B3 clade has an overlapping distribution with *M. polaris*, near 60°N latitude, but seems to have a markedly different distribution band towards lower latitudes down to 45°N, indicating a fairly distinct niche preference (Tragin and Vaulot, 2019) (Fig. 11.3).

11.3.1.3 Bipolarity studies as a proof of concept for microbial dispersal theories

Although opposed to each other geographically, the North and the South poles share similarities such as extreme seasonal variation of solar energy input and a community phenology tightly linked to sea-ice dynamics. The existence of bipolar taxa evokes questions regarding interconnectivity paths and rates of evolutionary divergence (Darling et al., 2000; Sul et al., 2013). The use of metabarcoding enables a refined look at planktonic protists (including photosynthetic groups) that thrive in both Arctic and Antarctic ecosystems, a comparison which was before mainly performed by morphological traits analysis (Fryxell et al., 1981; Montresor et al., 2003). Wolf et al. (2015), using the 18S rRNA V4 region and a 98% OTU similarity level, reported that more than 10% of the OTUs were shared between the two poles, most of them belonging to the rare biosphere, with the exception of two taxa: an unknown alveolate and *Micromonas*. Segawa et al.

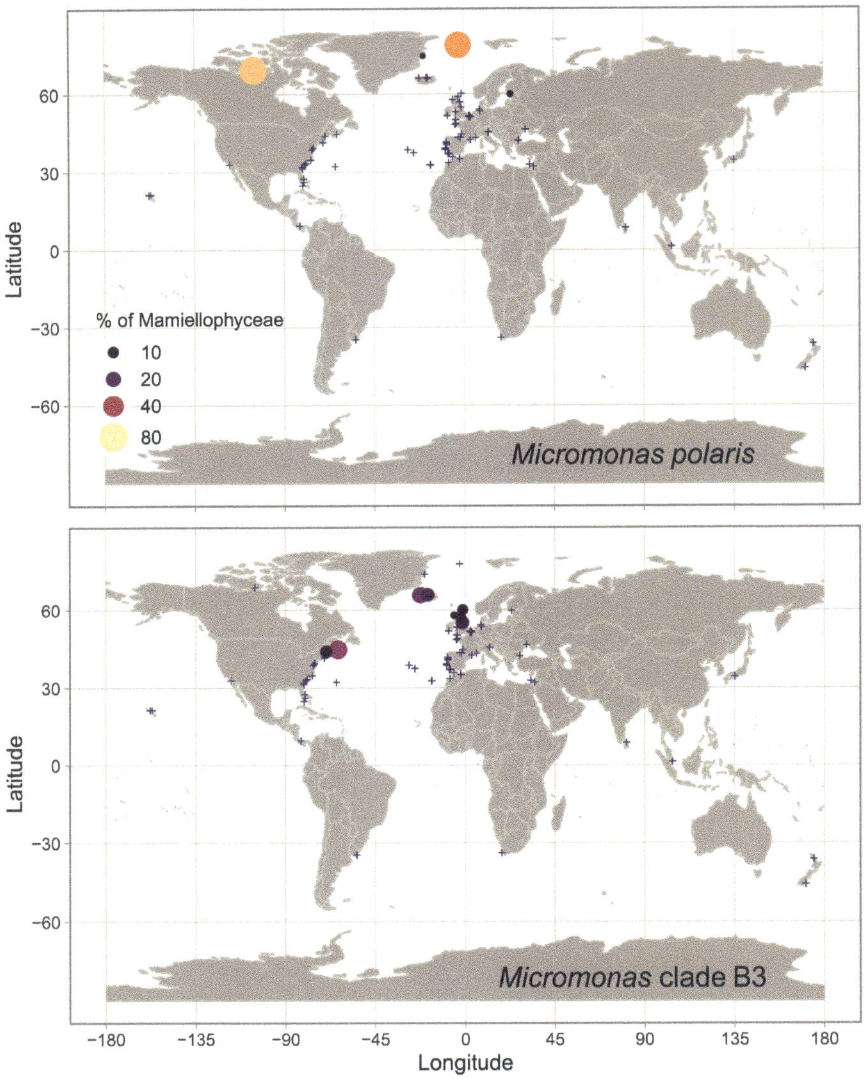

Figure 11.3 Distribution of the polar *Micromonas polaris* and the subarctic *Micromonas* B3 clade. Extracted from Tragin and Vaulot (2019).

(2018) analyzed distribution of red snow algae at both poles using metabarcoding of the ITS (faster-evolving genomic region than the 18S rRNA) and showed a high degree of bipolarity, but also a pronounced endemism between Arctic and Antarctic ribotypes. These results question the Baas-Becking dispersion hypothesis that states: "Everything is everywhere, but the environment selects".

11.3.1.4 Metabarcoding as a way to measure vulnerability of polar environments

Physicochemical changes, such as ocean warming and acidification, have already been reported to affect polar plankton diversity based on experiments testing organism response to a given set of environmental parameters (Benner et al., 2019; Beszteri et al., 2018; Hoppe et al., 2018). However, the responses of the total community may vary from the results obtained in laboratory experiments, since the complexity of natural interactions can be rarely (if ever) duplicated in controlled environments. The use of metabarcodes to investigate plankton and phytoplankton diversity is an important tool also to analyze and predict their vulnerability to the ongoing drastic changes in these regions of the planet, and is complimentary to experiments on specific organisms, by taking a broader view of the complete community structure.

Metabarcoding can be used as a tool for assessing the dynamic response of communities to a fast-changing environment and a way to monitor fluctuations on both the abundant and the rare biosphere. For example, the ecotype boundaries described by Tragin and Vaulot (2019) and Sow et al. (2020) for Arctic *Micromonas* and Antarctic *Phaeocystis*, respectively, can now be monitored as the ocean currents and their physicochemical parameters change. Moreover, metabarcoding monitoring can be useful at a much longer time scales, considering the paleo-oceanographic evolution of polar habitats. Pawłowska et al. (2020) investigated the microdiversity variability from Arctic planktonic foraminifera from the last 140,000 years using sediment ancient DNA (*seda*DNA), and reported population-level responses to deglaciation and sea-ice retreat, due to Atlantic water inflow.

11.3.2 The biological carbon pump

The biological carbon pump (BCP) refers to the process by which organic matter produced photosynthetically in the sunlit surface ocean is transported vertically downwards to depth through particle sinking, vertical mixing, and active transport by larger organisms (Boyd et al., 2019; Turner, 2015; Volk and Hoffert, 1985). The BCP is fundamental for the functioning of marine ecosystems by: (1) removing carbon from the upper ocean contributing to long-term carbon sequestration; (2) providing remineralized sources of inorganic and organic nutrients at depth, due to microbial decomposition of downward sinking particles; (3) fueling mesopelagic, bathypelagic, and benthic ecosystems that rely primarily on sinking particulate organic matter (POM) from the euphotic zone.

Despite its importance for marine ecosystem functioning and earth's climate, our understanding of the BCP, its variability, and its role in the carbon cycling is still rudimentary (Falkowski et al., 1998; Honjo et al., 2014). The imbalance between vertical export and metabolic demand in the dark ocean accentuates the knowledge gap around the functioning of the BCP (Burd et al., 2010; Herndl and Reinthaler, 2013), which is partly due to the sparsity of export measurements, the high temporal and spatial variability of the biological and physical processes driving the BCP (Boyd et al., 2019; Honjo et al., 2014), as well as the lack of information about the biological composition and ecological interactions that produce and transform sinking POM throughout the water column (Worden et al., 2015).

11.3.2.1 Molecular approaches applied to biological communities associated with sinking particles

Gravitational sinking is considered the dominant mechanism of POM vertical export, although physical mechanisms, by which POM is captured or drawn down into the deep ocean, as well as the active transport of organic materials by the vertical migration of larger animals can be equally significant (Boyd et al., 2019; Resplandy et al., 2019; Stukel et al., 2017). The importance of community composition and food web structure to the magnitude, strength, and efficiency of the BCP is widely accepted (Bach et al., 2019; Boyd and Newton, 1999; De La Rocha and Passow, 2007; Henson et al., 2019; Stukel et al., 2011). There is a growing body of evidence indicating that particle properties and processes (e.g., porosity and repackaging), which are intimately linked to plankton community composition and trophic structure, affect the sinking velocity and remineralization of particles (Bach et al., 2019; Cavan et al., 2017; Le Moigne et al., 2014; Puigcorbé et al., 2015; Siegel et al., 2014; Stemmann and Boss, 2011). By altering the abundance, size, and composition of sinking particles, phytoplankton (through the formation of marine aggregates and/or ballasting of POM), zooplankton (via consumption, transformation, and repackaging of POM), and bacteria (by the remineralization of POM) influence the temporal and spatial characteristics of vertical export.

Vertical export has been traditionally investigated with sediment traps designed to intercept and collect sinking particles (Boyd and Trull, 2007; Buesseler et al., 2007; Knauer et al., 1979). Bulk and specific geochemical methods can be used to broadly infer the dominant contributors to export flux (e.g. biogenic silica fluxes are often related to diatoms; pigment or organic compound studies are used to infer functional groups), however they

Table 11.6 Studies of vertical fluxes. na: not applicable; nd: not determined; metaB: metabarcoding; metaG: metagenomics; metaT: metatranscriptomics.

Study	Region	Surface Chl*a* range (mg/m^3)	Sampling method	Depth of traps deployment (m)	Duration of deployment	Preservation	Molecular approach	Gene marker	Main finding
Amacher et al., 2009	NASG-E - ESTOC	1.2–1.6	CTD-Niskin (euphotic) + VERTEX-type PITs	200, 500	26 h	Unpreserved brine	Clone Libraries	18S rDNA	Small rather than larger (diatom) phytoplankton taxa dominated sequences in trap material
Amacher et al., 2013	NASG-W - Bermuda Atlantic Time-series Study	0.1–0.5	CTD-Niskin (euphotic) + VERTEX-type PITs	150	72 h	Formaldehyde (0.74%) and unpreserved brine	DGGE	18S & 16S rRNA	Small phytoplankton taxa present in trap samples. *Synechococcus* overrepresented in sediment trap samples. Increased abundance of small eukaryotes (Pras, Prym) in water column following deep mixing not reflected in sediment traps - stimulated production recycled locally
LeCleir et al. (2014)	Subtropical SW Pacific - FeCycle II	0.5–2.5	CTD-Niskin (euphotic) and RESPIRE traps	100, 120	24 h–72 h	Unpreserved brine	MetaB	16S rRNA	Rapid colonization of sinking particles by bacteria in the first 24 h and little changes in the following 72 h. Sequences affiliated to Roseobacter enriched in traps relative to water samples confering them a key role in degradation of sinking particles
Fontanez et al., 2015	NPSG-HOT	nd	CTD-Niskin (euphotic) and VERTEX-type PITs	150, 200, 300, 500	12 days	RNA Later and unpreserved brine	MetaG	18S & 16S rRNA mitags	Distinctive communities in live, poison and water column samples with live traps enriched in both bacterial and eukaryotic taxa associated with particle and dissolved organic matter degradation. Depth-related partitioning of bacterial taxa in sinking particles possibly related to colonization of particles by eukaryotic taxa
Gutierrez-Rodriguez et al., 2019	California Current Ecosystem	0.08–0.69	CTD-Niskin (euphotic) + VERTEX-type PITs	60, 100, 150	72 h	Formaldehyde (0.4%) & unpreserved brine	MetaB	18S rRNA	Rhizaria dominance in preserved traps across all trophic conditions suggest important role in export. Important compositional differences between preserved and unpreserved traps with significant increase of heterotrophic protists (Stramenopiles-HNF-*Caecitellus* spp. and Phaeodaria) stress the importance of heterotrophic protist in remineralization and transformation of sinking particles
Duret et al. (2019)	Southern Ocean	0.2–2.2	CTD + Marine Snow Catcher	Bottom of mixed layer & Mesopelagic (DCM +110 m)	na	na	MetaB	16S rRNA	Distinctive prokaryotic communities associated with sinking and suspended particles. Pseudomonadales, Vibrionales and Rhodobacterales enriched in sinking particles while Flavobacteriales were enriched in suspended particles

continued on next page

Table 11.6 *(continued)*

Study	Region	Surface Chl*a* range (mg/m^3)	Sampling method	Depth of traps deployment (m)	Duration of deployment	Preservation	Molecular approach	Gene marker	Main finding
Mestre et al. (2018)	Tropical and subtropical Atlantic, Pacific, Indian Oceans	nd	CTD-Niskin (surface to 4000 m)	na	na	na	MetaB	16S rRNA	Most abundant taxa in deep ocean are present in surface waters suggesting strong connectivity between both realms driven mainly through large size particles. Higher dissimilarity among free-living communities indicates stronger vertical partitioning.
Boeuf et al. (2019)	NPSG - HOT	nd	CTD + PARFLUX Deep sediment trap	4000	12 days/bottle	RNA Later	MetaB (MetaG & MetaT)	18S & 16S rRNA	Systematic dominance of Alteromonadales and Campylobacterales across 9-month survey in all MetaG, MetaT and MetaB libraries. More diverse contribution of eukaryotic groups with Syndiniales, Ciliates and Rhizaria dominating the protistan sequences
Duret et al. (2020)	Southern Ocean	0.2–2.2	CTD + Marine Snow Catcher	Bottom of mixed layer & Mesopelagic (DCM +110 m)	na	na	MetaB	18S rRNA	Diatoms were enriched in particles sinking into the upper mesopelagic while suspended particles here were enriched in Prymnesiophytes. Suggesting higher transfer efficiency for diatom-enriched particles. Diverse community of heterotrophic protist (Choanoflagellates, Radiolaria, dinoflagellates, ciliates and fungi) associated with particles. HNF and Acantharea were more abundant in sinking than suspended particles
Preston et al. (2020)	California Current Ecosystem	nd	CTD + PARFLUX Deep sediment trap	3900, 3950	11 days/bottle	RNA Later	MetaB	18S rRNA	Diatoms associated with greater POC flux event to deep ocean, with co-occurring increases of zooplankton sequences suggesting their involvement in diatom-driven production transport. Small phytoplankton taxa were found to contribute substantially to deep export during the winter with metazoan consumers likely acting as vectors, while a diverse suit of Radiolarian taxa dominated the material reaching the traps frequently. Collodaria sequences dominated the community in POM collected between March and April (99%–65% of total reads), although their presence was not associated with increased POC fluxes

are often not particularly resolutive for individual plankton studies (Nodder, 1997). Microscopy of the material collected in sediment traps has been used to characterize the biological components and to infer ecological processes driving POM vertical export. For example, taxonomic identification of intact cells and resting spore composition in sediment trap samples has shown the importance of diatoms in sinking marine aggregates (Martin, 2011; Scharek et al., 1999; Rynearson et al., 2013; Waite et al., 1992) as well as the contribution of other plankton groups such as coccolithophores (Honjo, 1976), radiolaria (Decelle et al., 2013; Gowing and Coale, 1989; Gowing, 1993; Lampitt et al., 2009; Martin et al., 2010; Michaels et al., 1995), and foraminifera (Sautter and Thunell, 1989; Schiebel, 2002).

Taxonomic composition analysis of sinking particles based solely on morphological criteria is often challenging, due to high levels of material degradation/transformation by microbial activity. Early applications of molecular approaches demonstrated the potential of genomic tools to characterize the biological composition of particles collected by sediment traps. Table 11.6 provides a summary of the studies that have coupled particle collection methods with molecular approaches to investigate the vertical fluxes of sinking particles. Sequencing of 18S rRNA clone libraries from water column and sediment trap samples showed for the first time significant differences between the eukaryotic plankton community associated with sinking particles and general water column communities in the eastern subtropical North East Atlantic (Amacher et al., 2009). A similar sampling strategy with DNA-based molecular fingerprinting of protist and cyanobacteria communities at the Bermuda Atlantic Time-series (BATS) site in the western subtropical Atlantic showed a similar pattern of dissimilarity between water column and traps over a year-round sampling period, suggesting that not all phytoplankton groups contributed equally to vertical export (Amacher et al., 2013). Using metabarcoding, LeCleir et al. (2014) showed significant differences in free-living and particle-associated bacterial communities of water column, sinking, and incubated POM. Differences in diversity of the particle-associated bacterial community after 24 hours of *in situ* incubation suggested that rapid changes may occur in communities associated with particles sinking from the euphotic zone. Results from these pioneering studies indicated that both the contribution of primary producers and heterotrophic bacteria to export production and remineralization of sinking particles differed between taxonomic groups.

More recent studies applying HTS to sinking POM have also reported significant differences between planktonic and particle-associated communities across contrasting oceanographic regions (Boeuf et al., 2019; Duret et al., 2020; Fontanez et al., 2015; Gutierrez-Rodriguez et al., 2019; Preston et al., 2019). The confirmation of these differences and the prevalence of certain taxa in sinking particles across a broad range of oceanographic conditions and regions further supports the distinctive contributions that specific taxonomic groups make to the processes driving or resulting from vertical export, such as particle sinking, remineralization/transformation, and food web connectivity in the ocean.

11.3.2.2 Phytoplankton and vertical export

Diatoms are arguably one of the most important contributors to POC export accounting for nearly 40% globally (Jin et al., 2006; Nelson et al., 1995; Sarthou et al., 2005). Their relatively large cell size, silica-rich skeletons, and bloom-forming capacity confer diatoms with high export potential (Tréguer et al., 2018). Visual and pigment analysis of the material collected by sediment traps have provided evidence of the important contribution made by diatoms to sinking POM across contrasting trophic conditions, seasons, and latitudes, through direct sinking and aggregation, or mediated by zooplankton consumption and fecal pellet production (Boyd et al., 2005; Durkin et al., 2016; Ebersbach et al., 2014; Martin et al., 2010; Scharek et al., 1999; Smetacek et al., 2012). Yet, early molecular studies of sediment trap materials in the low productivity/oligotrophic waters of the North Atlantic subtropical gyre revealed lower than expected contributions of diatoms relative to small nanoflagellates and the cyanobacterium *Synechococcus* (Amacher et al., 2009, 2013), which is consistent with recent studies providing evidence of considerable small phytoplankton contributions to export (Waite et al., 2000; Richardson, 2019).

In more productive systems such as the California current upwelling, metabarcoding analysis of POM collected in short-term floating sediment traps deployed in the upper mesopelagic zone (<500 m) identified diatoms and dinoflagellates as the dominant phytoplankton groups contributing to export production, despite the suppressed upwelling conditions during the survey period in summer (Fig. 11.4, Gutierrez-Rodriguez et al., 2019). Similar analysis conducted on sinking particles collected in deep sediment traps (4000 m), deployed in the same upwelling region over a 9-month period, found that diatom contributions to total abundance of 18S rRNA reads was relatively low, although the highest POC flux event coincided with samples where diatom sequences in trap samples were higher (Preston et al., 2019).

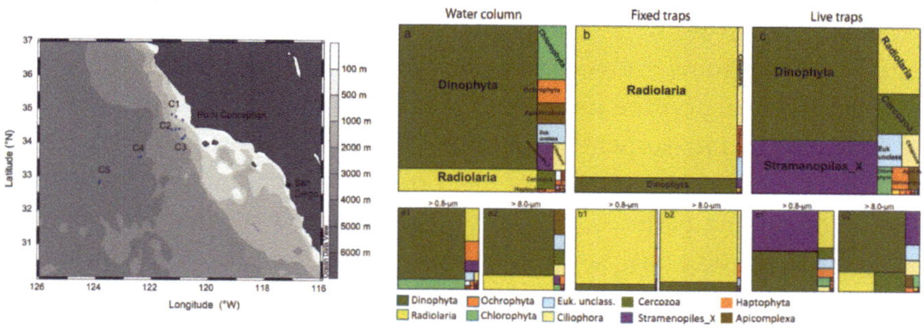

Figure 11.4 Protist community from the California current in the water column and in traps where material has been fixed or kept live (reprinted from Gutierrez-Rodriguez et al., 2019).

This study reported a succession in the diatom species associated with particles reaching the seafloor, with different species of *Chaetoceros* (*C. didymus*, *C. neogracile*, *C. socialis*) dominating in winter, followed by species belonging to *Minidiscus*, *Thalassiosira*, *Actinocyclus*, *Skeletonema*, and *Ditylum* genera in spring, whereas *Thallassiosira aestivalis*, not previously detected in the study, appeared to dominate the diatom assemblage associated with the high POC flux observed in late June (early summer, Preston et al., 2019). The analysis of photosynthetic community composition from 16S rRNA gene metabarcoding provided evidence of the relative dominance of *Synechococcus* and prasinophytes in sinking particles during early and late winter, respectively, consistent with molecular analysis of sinking particle at BATS (Amacher et al., 2013).

Although there is a general agreement that larger phytoplankton cells are considered to drive vertical carbon export (Michaels and Silver, 1988), the relative contribution of picophytoplankton to downward fluxes and efficiency of the BCP are still under debate (Henson et al., 2019; Puigcorbé et al., 2015; Richardson and a Jackson, 2007; Stukel et al., 2013; Waite et al., 2000). The formation of marine aggregates and trophic repackaging into fecal pellets, particularly by pelagic tunicates, such as salps and appendicularians (Stone and Steinberg, 2016), are considered the main mechanisms by which small phytoplankton contribute to downward fluxes (Richardson, 2019).

Duret et al. (2020) investigated the phytoplankton community composition of sinking and suspended particles sampled with the marine snow catcher at four stations with different productivity in the Scotia Sea (Southern Ocean). The study showed that diatoms and prymnesiophytes were enriched in sinking and suspended particles, respectively. These results suggest that diatoms are more efficient in carbon transfer to the upper mesopelagic than prymnesiophytes in this area, since the latter are more easily disintegrated into suspended particles, therefore supporting the paradigm of higher export potential of larger (diatom) phytoplankton. Metabarcoding analysis of sinking particles, especially if combined with size fractionation and biogeochemical analyses, can provide valuable information in this regard.

Metabarcoding of functional genes can also shed light on the contribution of specific groups to export. For example, Farnelid et al. (2019) applied 16S rRNA and nitrogenase gene (*nifH*) molecular barcodes to investigate the role of nitrogen-fixing organisms in the vertical export within the North Pacific subtropical gyre. This approach revealed for the first time the presence of a diverse suit of diazotrophs in sinking particles, indicating that new nitrogen could be directly exported vertically. In addition, the comparison between taxonomic affiliation of free-living and particle-associated N^2 fixers indicated that some taxa, assigned to larger size classes, were preferentially associated with particles, and hence present higher export potential.

Dinoflagellata and Radiolaria groups often dominate the protist community associated with sinking organic particles below the euphotic zone (Boeuf et al., 2019; Duret et al., 2020; Fontanez et al., 2015; Gutierrez-Rodriguez et al., 2019; Preston et al., 2019). Their high relative abundance revealed by molecular analysis is likely influenced, to some extent, by the high number of 18S rRNA gene copies characteristic of these groups (Biard et al., 2016; Lin, 2011). However, their tendency to sink directly and to produce fast sinking fecal pellets cannot be ignored. For instance, the siliceous or strontium skeleton of polycystine radiolaria and acantharia, the large size (mm to cm) that solitary and colonial forms of some species can reach, and the sticky nature of the pseudopodia and mucilaginous matrix that these amoeboid organisms use as feeding structures, are attributes that are also consistent with a high export potential.

Metabarcoding analysis of 18S rRNA gene of sinking material reaching the seafloor (4000 m) of the California current over extended periods of time showed the dominance of different radiolarian taxa (Collodaria, Spumellarida, Acantharea and RAD-A, Preston et al., 2019). Similar analysis obtained from POM collected with moored traps (4000 m) deployed in the North Pacific subtropical gyre reported sporadic dominance of rhizarian that were affiliated mainly to Foraminifera and Phaeodaria (Boeuf et al.,

2019). Whether the abundance of Rhizaria in trap-collected POM is due to direct sinking of mixotrophic species from the sunlit layer and enhanced vertical export, or due to consumption and remineralization of sinking particles by heterotrophic species that favor the attenuation of vertical fluxes, is not clear.

Though metabarcoding can be only be taken, at best, as a semiquantitative approach, these studies demonstrate that, when combined with targeted sampling strategies, it can provide valuable information about biological sources and ecological processes transforming sinking POM. Coupled with elemental and size composition of particles, collected with particle collection methods (e.g. sediment traps, *in situ* pumps), and indirect estimates of export (e.g., ^{238}U-^{234}Th disequilibria, oxygen, and nutrient mass balances), this molecular approach can improve our understanding of the role of different plankton groups in carbon export. Any step towards improving the quantitative nature of metabarcoding analysis (e.g., reducing uncertainty around taxon-specific copy number variability) and integration with biogeochemical metrics (sequence to carbon conversion factors) will certainly enhance the value of metabarcoding approaches, not only for biogeochemical studies, but also for diversity and ecological studies. It is also important to acknowledged that different methods commonly used to collect POM exported from the euphotic zone (e.g. sediment traps, *in situ* pumps, marine snow catcher, size fractionation of water bottle samples) are biased towards the different types of particles collected (Lee, 2019; Peterson et al., 2005) and therefore can reflect different phytoplankton community composition. VERTEX-type or Particle Interceptor Traps (PITs), for instance, are more efficient at capturing rapidly sinking particles than suspended particles and marine snow (McDonnell et al., 2015). Sampling biases between CTD-Niskin bottle and high volume *in situ* pump methods (Liu et al., 2009; Twining et al., 2015) highlighted the importance of combining different methods whenever possible, and the need to consider these biases when comparing and interpreting results from different studies (Puigcorbé et al., 2020). These methodological differences can also be used to our advantage. (Duret et al., 2020) for instance used the marine snow catcher that allows separate collection of sinking and suspended particles (Lampitt et al., 1993) in combination DNA metabarcoding analysis of different size fractions to investigate differences in their biological sources and particle transformations in the Southern Ocean. These authors observed differences in the relative abundance of phytoplankton groups between sinking particles (enriched in diatoms) and suspended particles (enriched in prymnesiophytes).

11.3.3 Predator-prey interactions and trophic connectivity

Functioning of ecosystems is maintained by the flow of material and energy through the food webs, which are a representation of trophic interactions between species (Thompson et al., 2012). In marine pelagic ecosystems, the bulk of organic material and energy is produced and consumed at the base of the food web, where the three domains of life constitute an intricate network of interactions. The microbial component of the food web comprises many trophic levels, through which over 95% of total primary and secondary production is processed before reaching larger metazoan zooplankton that are directly available as prey to higher trophic levels (Steinberg and Landry, 2017).

Despite their crucial role, our current knowledge of the structural and dynamic properties of planktonic food webs (i.e., "what eats what") remains fairly incomplete. This is due in part because of the extreme functional and taxonomic diverse nature of eukaryotic plankton (Caron et al., 2012; Mitra et al., 2016), and because of the high degree of omnivory and dynamic nature of predator-prey interactions (Banse, 2013; Calbet and Saiz, 2005; Isaacs, 1972; Steinberg and Landry, 2017; Zeldis et al., 2002). Conventional methods used for dietary and food web reconstruction (e.g., stable isotope analysis, visual inspection of stomach contents) are somewhat inadequate for such diverse and often unstructured plankton systems (Craig et al., 2013; Landry, 2002; Maloy et al., 2013). Metabarcoding represents an alternative that is rapidly gaining traction in trophic ecology, due to its high sensitivity, taxonomic resolution, and cost-effective nature (Clare, 2014; King et al., 2008; Kress et al., 2015; Majdi et al., 2018). When applied to gut, stomach, or scat contents, metabarcoding can provide taxonomically comprehensive information on what a predator eats (Deagle et al., 2019), and allow for dietary richness estimates that illuminate niche partitioning from plankton to marine fishes, birds, and mammals (Carroll et al., 2019; Casey et al., 2019; Craig et al., 2013; Kaunisto et al., 2017; Leray et al., 2015; Ray et al., 2016).

Zamora-Terol et al. (2020) investigated the species-specific trophic links and seasonality in the Baltic Sea by applying 16S/18S rRNA metabarcoding to the gut content of several species of relevant meso-zooplankton groups (copepods and cladocerans). Their comprehensive study revealed the shift from diatoms and dinoflagellate dominance in copepods diet during the spring, whereas in the summer zooplankton, including cladocerans, showed a more diverse diet, dominated by cyanobacteria and heterotrophic protists. Their results indicated that copepods change their natural diet over seasons, and overall adapt their feeding strategies to the available prey spectrum.

Generating a good quality PCR product for the specific genomic region and group of interest remains a crucial step of the DNA metabarcoding workflow. This can be particularly challenging in the case of gut content analysis, where nontargeted DNA from the predator represents the majority of the originally extracted DNA material. To minimize the contribution of the predator's material to the PCR amplicon, it is recommended to dissect out the predator gut/stomach where possible, and to adopt molecular strategies (e.g., protist-specific or blocking primers and restriction enzymes) that restrain the amplification of the predator's DNA (Bower et al., 2004; Craig et al., 2013; Maloy et al., 2013; O'Rorke et al., 2013; Vestheim and Jarman, 2008).

11.4 Marine picocyanobacteria

11.4.1 Use of the universal marker gene, the 16S rRNA

Marine picocyanobacteria of the genera *Prochlorococcus* and marine *Synechococcus* are the two most abundant photosynthetic prokaryotes in the ocean, contributing up to 8 and 16% of the net primary productivity, respectively (Flombaum et al., 2013; Scanlan et al., 2009). They likely diverged from a common ancestor around 150 million years ago and, on the basis of 16S rRNA gene sequences, most members of the two groups would be considered to be the same species (Dufresne et al., 2005). The main difference between the two genera resides in their photosynthetic apparatus. Like most cyanobacteria, *Synechococcus* main light-harvesting antenna is the phycobilisome, which comprises phycobiliproteins (for example, phycoerythrin and phycocyanin), each binding to one or several light-harvesting chromophores, such as phycoerythrobilin and phycocyanobilin (Ting et al., 2002). *Prochlorococcus* (as *Prochloron* and *Prochlorothrix*) lacks phycobilisomes and its main light-harvesting antenna complex is composed of prochlorophyte chlorophyll-binding protein (Pcb), which binds divinyl chlorophyll *a* and *b* (Ting et al., 2002). Differences in cell size and photosynthetic pigments of these genera produce distinct flow cytometry profiles, which was used to reveal their abundance and ubiquity, and highlighted some of the main ecological idiosyncrasies of each genus. When they co-occur, *Prochlorococcus* cell abundance is about one to two orders of magnitude larger than *Synechococcus*. *Prochlorococcus* is more abundant in warm (>15 °C) oligotrophic waters, whereas *Synechococcus* dominates in coastal and more temperate or mesotrophic open ocean waters (Blanchot and Rodier, 1996; Campbell and Vaulot, 1993; Olson et al., 1985; Partensky et al., 1999).

Studies using 16S rRNA gene, clone libraries, and Sanger sequencing brought the first insights into marine picocyanobacteria genetic diversity. In agreement with pigment data, these studies suggested the existence of two distinct *Prochlorococcus* ribotypes with different adaptations to light availability (Scanlan and West, 2002; Urbach et al., 1998): a "high light-adapted ecotype" (HL) containing sequences from *Prochlorococcus* cells collected in surface waters and a deeply branching lineage called "low light-adapted ecotype" (LL), including sequences from members isolated deeper in the water column. Moore et al. (1998) demonstrated that co-occurring populations, displaying discrete distribution of Chl autofluorescence and light scatter, contain physiologically and genetically distinct types of *Prochlorococcus*, with different growth responses over a range of light intensities. Despite the complexity of their environment, these broad categories of *Prochlorococcus*, low- and high-light, shift their relative abundance throughout annual cycles in a predictable manner (Malmstrom et al., 2010; Zwirglmaier et al., 2007). Although the knowledge on *Prochlorococcus* diversity has increased with the availability of sequencing data and methods (Kashtan et al., 2014), the notion of the HL and LL groups that emerged from the analysis of 16S rRNA sequences combined with phenotypic and ecological data is still valid today (see Biller et al., 2015).

In contrast to *Prochlorococcus*, *Synechococcus* can be found in both marine and freshwater and is present in almost all marine environments, including high latitude and nutrient-rich coastal waters (Partensky et al., 1999). These differences might be explained by some of *Synechococcus* physiological traits, such as the capacity to tune its phycobilisome antenna systems to changing temperatures (Mackey et al., 2013; Pittera et al., 2014) and its higher growth rates (Chen et al., 2009). Marine *Synechococcus* have long been classified into marine clusters A, B, and C (MC-A, MC-B, and MC-C), based on the composition of their major light-harvesting pigments, their swimming motility (Waterbury et al., 1985), their salt requirement for growth, and their GC content (Waterbury and Rippka, 1989). *Synechococcus* 16 rRNA gene sequences from various marine isolates led to the identification of up to 10 clades within marine cluster-A (Fuller et al., 2003; Urbach et al., 1992, 1998; West and Scanlan, 1999), which was renamed subcluster (SC) 5.1 (Herdman et al., 2015).

Different approaches have been developed to assess the structure of marine picocyanobacteria communities based on 16S rRNA sequences. For instance, clade-specific oligonucleotides indicated that the dominance of specific *Synechococcus* clades could be correlated with physical features of the water column (Fuller et al., 2006, 2003, 2005). These studies and notably the study by Zwirglmaier et al. (2008) allowed a broad description of the main characteristics of four dominant *Synechococcus* clades and of their global biogeography. Clades I and IV were predominant in both coastal and cold, high latitude waters (above c. 30°N and below 30°S); clade II was found in coastal/continental shelf zones in tropical and subtropical waters; clade III was detected in a fairly narrow window of nitrate and phosphate concentrations, suggesting members of this clade are oligotrophs, whereas other clades, such as IX and X, were only detected at low abundance in the field (Zwirglmaier et al., 2007, 2008). These early studies targeted only *Synechococcus* clades established from isolated and sequenced strains, and therefore were likely missing the uncultivated diversity.

Several studies have focused on which variable region of the 16S rRNA gene is more suitable to describe bacterial diversity in natural assemblages, and although these comparisons led to contrasting results, the V4 and/or V4–V5 fragments were generally found to be the best for establishing phylogenetic relationships (Apprill et al., 2015; Caporaso et al., 2011; Parada et al., 2016). The variable region of the 16S rRNA gene that best reflects the full-length gene phylogeny may also depend on the tax onomic groups being studied. For picocyanobacteria, the V5, V6, and V7 hypervariable regions were shown to provide the highest taxonomic and phylogenetic resolution for the genera *Synechococcus*, *Prochlorococcus*, and *Cyanobium* (Huber et al., 2019).

The number of reads recovered from the cyanobacteria phylum using general 16S rRNA gene primers usually hardly exceeds 20% of the total read abundance, even when applying HTS methods, such as metagenomics and metabarcoding (Doherty et al., 2017; Sunagawa et al., 2015). Huber et al. (2019) developed new 16S rRNA primers targeting specifically picocyanobacteria, suitable to metabarcoding. By using these cyanobacteria-specific primers, most of the sequences obtained (97%) could be assigned to picocyanobacteria and used in phylogenetic reconstructions that featured topologies comparable to those generated when using the complete 16S rRNA gene sequence. Although they likely deserve to be further tested with more environmental samples and/or mock communities, these primers appear well-suited to study the diversity, community structure, and dynamics of picocyanobacteria communities by 16S rRNA gene metabarcoding.

11.4.2 More resolutive markers

A general disadvantage of bacterial ribosome-based diversity analyses is the relatively low sequence divergence between closely related groups or species. For marine picocyanobacteria, the considerable genetic diversity observed within both *Prochlorococcus* and *Synechococcus* genera is only manifested through small-scale variations, about 2.8% and 4.5%, respectively (Doré et al., 2020). The well-known high- and low-light adapted clades of *Prochlorococcus*, for example, differ by less than 3% of their 16S rRNA sequences (Biller et al., 2015). Moreover, the number of rRNA gene copies varies among bacteria, including picocyanobacteria. For example, two to three identical copies of the rRNA operon are present in most marine *Synechococcus* genomes (Dufresne et al., 2008; Farrant et al., 2016; Fuller et al., 2003), and two in some members of LL adapted *Prochlorococcus* clades, although most *Prochlorococcus* genomes have only one (Martiny et al., 2009).

These limitations led the scientific community to use other marker genes to analyze the genetic diversity of marine picocyanobacteria (Table 11.7). Generally, these markers show more variability than the 16S rRNA gene, and thus provide more phylogenetic resolution over a short sequence stretch. For instance, the sequence identities between different picocyanobacteria clades are lower for ITS (<93%: Rocap et al., 2002), *petB* (<94%: Mazard et al., 2012); *ntcA* (86%: Penno et al., 2006), *rpoC1* (82%: Toledo and Palenik, 1997) than for 16S rRNA (<97%: Fuller et al., 2003; Moore et al., 1998). Furthermore, most protein encoding genes are single-copy, such as *rpoC1* (Bergsland and Haselkorn, 1991; Tai and Palenik, 2009), *petB* (Mazard et al., 2012; Farrant et al., 2016), *ntcA* (Lindell et al., 1998; Penno et al., 2006), *rbcL* (Scanlan et al., 2009), and *narB* (Scanlan et al., 2009). The *psbA* multigene family is an exception to the list, since it can be found in a variable number (1 to 6) of copies per genome in marine *Synechococcus* and *Cyanobium*. It was shown to be subjected to intragenome homogenization (D1 protein of photosystem II, PSII), most probably mediated by gene conversion (Garczarek et al., 2008). However, numerous genes involved in carbon and energy metabolism (e.g., *psbA* or some subunits of the F0F1 ATP synthase) have been found to be present in cyanophages, which could bias the interpretation of diversity studies based on these markers (Chénard and Suttle, 2008; Mann et al., 2003; Millard et al., 2004). Similarly, as mentioned before, a good molecular marker

Table 11.7 Genes used for metabarcoding of *Synechococcus*.

Gene	Coding for	References
ITS	16S-23S rRNA internal transcribed spacer	Rocap et al., 2002
		Ernst et al., 2003
		Chen et al., 2006
		Cai et al., 2010
		Choi and Noh, 2009
		Ahlgren and Rocap, 2012
		Huang et al., 2012
		Kashtan et al., 2014
rpoC1	gamma subunit of RNA polymerase	Palenik, 1994
		Toledo and Palenik, 1997
		Ferris and Palenik, 1998
		Toledo and Palenik, 1997
		Ma et al., 2004
		Muhling et al., 2005
		Tai and Palenik, 2009
		Gutiérrez-Rodríguez et al., 2014
petB	cytochrome b6	Mazard et al., 2012
		Farrant et al., 2016
rbcL	large subunit of the RuBisCO (Ribulose 1,5-bisphosphate carboxylase-oxygenase)	Chen et al., 2004
		Chen et al., 2006
		Paerl et al., 2012
narB	nitrate reductase	Paerl et al., 2008
		Paerl et al., 2011
		Jenkins et al., 2006
atpBE	intergenic region	Lockhart and Penny, 1992
psbA	D1 protein of photosystem II reaction center	Kishino et al., 1990
		Hess et al., 1995
		Zeidner et al., 2003
cpeB	C-phycoerythrin class I, beta subunit	Steglich et al., 2003
ntcA	N-regulation gene	Penno et al., 2006
		Post et al., 2011

should also not be the subject to horizontal gene transfer, which constitute a common, although not so frequent, gene gain mechanism in marine picocyanobacteria (Doré et al., 2020; Martiny et al., 2009). For most marker genes, lateral gene transfer has been checked by showing the congruence between gene marker and 16S rRNA phylogenies (Mazard et al., 2012; Urbach et al., 1998).

The different marker genes also differ on how easily and reliably sequences can be accessed and/or aligned. Some of them, such as *petB* or *ntcA*, are highly conserved in length and sequences, allowing sequences to be automatically aligned (Mazard et al., 2012; Penno et al., 2006). In contrast, *rpoC1* and ITS gene sequences often require manual refinement of the alignment, with the necessity to split the analysis per taxa and/or to suppress the most variable parts. For the ITS, the presence of inserts of variable size between members of *Synechococcus* SC 5.1, 5.2, and 5.3 makes it impossible to automatically align the sequences from different SC (Mella-Flores et al., 2012; Rocap et al., 2002). The higher the sequence variability of the gene marker is, the harder will be the identification of conserved primer binding areas, leading to the design of degenerated primers. Finally, the use of new (or not widely used) markers also suffers from databases with a low number of reference sequences, which can lead to primer amplification biases. For example, metabarcoding *petB* primer set is heavily biased towards marine *Synechococcus* SC 5.1 (Mazard et al., 2012), although *petB* gene is an excellent marker to reliably assign both *Prochlorococcus* and *Synechococcus* with a high taxonomical resolution (Farrant et al., 2016). For *petB*, the $_{mi}$tags approach seem to be the alternative (see Section 11.4.3).

11.4.3 $_{mi}$tags as an alternative to picocyanobacteria metabarcoding

Metagenomics approaches are particularly suitable for diversity and functional studies of marine picocyanobacteria. Their relative high abundance and ubiquity *in situ* compared to other taxa (Biller et al., 2015; Partensky et al., 1999) results in the dominance of their gene sequences in oceanic metagenomics datasets (DeLong et al., 2006; Rusch et al., 2007; Venter et al., 2004). The availability of several complete or near-complete genomes can be used as references to annotate taxonomically and functionally environmental reads (Biller et al., 2015; Doré et al., 2020; Dufresne et al., 2008; Kettler et al., 2007; Scanlan et al., 2009). These characteristics allowed the mining off *Prochlorococcus* and *Synechococcus* sequences in early environmental metagenomic libraries, such as the whole-genome shotgun datasets from the Sargasso Sea (Mühling et al., 2006; Venter et al., 2004), or the Global Ocean Sampling (GOS) (Biers et al., 2009; Huang et al., 2012; Rusch et al., 2007) projects. The $_{mi}$tags approach is a non-PCR-based strategy to retrieve 16S rRNA sequences from metagenome datasets to access the taxonomic diversity and structure of prokaryotic communities (Logares et al., 2014). The $_{mi}$tags approach recovers sufficient reads for

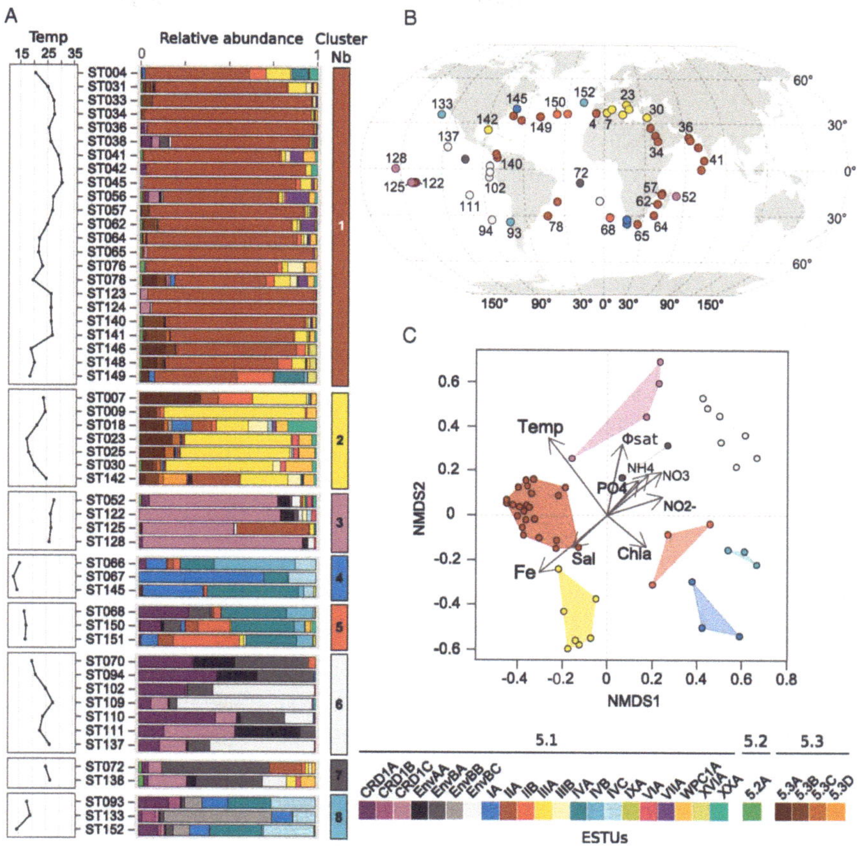

Figure 11.5 Biogeography of *Synechococcus* ESTUs in surface Tara *Oceans* metagenomes and relation to physico chemical parameters. (A) Histograms of the relative abundance of *Synechococcus* ESTUs at each station sorte by similarity. Left panels indicate seawater temperature (°C) at each station. (B) Distribution of the ESTU assemblages, color-coded as in A, along the Tara *Oceans* transect. (C) NMDS analysis of stations according to Bray–Curtis distance between *Synechococcus* assemblages, with fitted physicochemical parameters. Reprinted from Farrant et al. (2016).

community taxonomic profiling, as well as for richness, evenness, and beta diversity estimations. Using a reference database of the *petB* gene, Farrant et al. (2016) recruited reads out of 109 metagenomes from the Tara *Oceans* expedition (Fig. 11.5). This work revealed novel genetic diversity within both *Prochlorococcus* and *Synechococcus*, even among the most abundant and well-characterized clades. In addition, 136 completely new *petB* sequences were successfully assembled from metagenomic reads, enriching the reference database. By combining the genetic information with environmental parameters, "ecologically significant taxonomic units (ESTUs)" were further defined, representing organisms belonging to the same phylogenetic clade defined with *petB* and occupying a common ecological niche.

For example, within clade II, ESTU IIB was found to co-occur with clades I and IV in cold waters, whereas ESTU IIA dominated in warm, (sub)tropical waters. With respect to clade III, although the presence of the two ESTUs defined within this clade was confirmed in warm, oligotrophic waters, they seem to be particularly thriving in phosphorus-depleted waters, a previously unknown characteristic for members of this clade. This was later confirmed by comparative genomics (Doré et al., 2020). The CRD1 and EnvB *Synechococcus* genotypes were shown to dominate the Pacific stations analyzed between 33°S and 35°N, and to be locally abundant in the South and North Atlantic, as well as in the Indian Ocean, all corresponding to Fe-limited ocean regions. Moreover, each of these clades could be split into three distinct ESTUs, showing different distribution patterns: CRD1B and EnvBB thrive in cold waters, CRD1C and EnvBC in warm areas, whereas CRD1A and EnvBA were found in both types of ecological niches, suggesting that these latter populations are able to acclimate to a wide range of temperatures.

Although the metagenomic recruitment of marker genes has not been popular among studies focusing on eukaryotic phytoplankton (in particular due to the low representation of eukaryotic reads in environmental metagenome datasets), the $_{mi}$tags approach seem to constitute a powerful alternative to the metabarcoding for picocyanobacteria community, allowing to have access to the still undescribed lineages and microdiversity within existing lineages and to outline the physicochemical parameters driving the distribution of these lineages using the numerous metagenomics datasets now available. However, the recovery of sequences from low abundance components of the plankton community requires surveys with sequence depth coverage (i.e., the number of sequences obtained per sample) that are many orders of magnitude larger than those usually reported in the literature. Metagenomics thus remains complementary to metabarcoding approaches, which are still faster, less expensive, and easier to analyze, especially for long-term monitoring or studies including large numbers of samples.

11.5 Future directions

The metabarcoding approach has brought an avalanche of exciting new data on the diversity of plankton. Key scientific findings were enabled by metabarcoding, as reviewed by Santoferrara (2020). For example, it has revealed the existence of the "rare biosphere," where a large number of species are detected in very low abundance (Sogin et al., 2006). Though the "rare biosphere" is seemingly less important in ecosystem functioning when compared to abundant groups, it is hypothesized that they act as a "bank of redundant functions" that help to maintain a continuous ecosystem following shifts in environmental conditions, making the ecosystem more resilient to change (Caron and Countway, 2009; Dawson and Hagen, 2009; Dunthorn et al., 2014). Metabarcoding also revealed novel and diverse lineages among well-known taxonomic groups (e.g., de Vargas et al., 2015).

Although metabarcoding is subject to various technical biases, currently no other method can offer the analysis of a large numbers of samples with the same sensitivity, robustness, and taxonomic resolution at a similar cost. These exceptional characteristics have led metabarcoding to be suggested and applied as a promising tool in biomonitoring programs, in both freshwater (Kermarrec et al., 2014; Visco et al., 2015; Zimmermann et al., 2015) and marine ecosystems (Chariton et al., 2015; Pawlowski et al., 2014, 2016). Cordier et al. (2018) went further and showed that metabarcoding combined with supervised machine learning (SML) can accurately predict biotic indices, regardless of the taxonomic affiliation of the sequences. The predictive models were built using metabarcoding data from five different variable regions within the ribosomal small subunit rRNA gene for monitoring the impact of salmon farming activities in Norway on the benthic communities (Cordier et al., 2018).

Because of the high sensitivity of metabarcoding (e.g., trace concentrations of DNA can be PCR amplified and sequence), this method has been applied to the characterization of food webs and predator-prey interactions, as mentioned in Section 11.4.3. However, metabarcoding is a qualitative diversity tool and the nature of the sample, as well as the several complex laboratory and bioinformatics steps of this technique, can affect the final sequence abundance. The way read counts are used—i.e., frequency of occurrence of different prey taxa or their relative abundance—can influence derived dietary metrics (Deagle et al., 2019). The application of quantitative molecular techniques, such as qPCR to assess absolute abundance of targeted groups can help ground-truth metabarcoding results (Schwarz et al., 2018). Similarly, the combination of DNA sequencing with more quantitative approaches, such as stable isotopes commonly used in trophic ecology, has the potential to produce taxonomically resolved predator-prey interactions and improve dietary inferences from metabarcoding and associated material and energetic fluxes within the marine food web (Hardy et al., 2010; Nakamura et al., 2020; Whitaker et al., 2019).

The possibility of applying metabarcoding to archived samples also adds an historical angle that will allow ecosystem and biodiversity changes to be assessed through time (Metfies et al., 2017). The development of methods allowing for DNA extraction and amplification from preserved, sediment traps acquired over a time-series, for example, represents not only an affordable solution, but also a significant step towards integration of molecular and biogeochemical measurements and opens the possibility for retrospective analyses of archived samples from long-term sampling programs. Shiozaki et al. (2021) have proposed that the DNA extraction of preserved plankton samples with a lysis solution of borate-NaOH buffer (pH 11), SDS, and proteinase K can effectively remove the cross-links between acid nucleic and cellular proteins formed by formalin fixation. Studies on the DNA extraction of DNA formalin-fixed samples stored in museums and laboratories are scarce (Bucklin and Allen, 2004; Ruane and Austin, 2017; Schander and Kenneth, 2003), and given the increasing use of molecular approaches and the variety of preservation methods used for plankton samples, further research on the effects of different preservatives on molecular and biogeochemical measurements is warranted.

The concept of functional diversity is generally centered around understanding communities and ecosystems from what function organisms can perform, rather than from their evolutionary history (Petchey and Gaston, 2006). Several considerations should be addressed before applying functional diversity frameworks to the analysis of communities. For example, which functional information or traits about the organism are relevant and how different traits should be weighted according to their impact in the role of the organisms in the ecosystem. Given the complexity of plankton communities, both in terms of function and diversity, the understanding of the relationship between functional and taxonomic diversity remains a major challenge for plankton ecologists (Litchman and Klausmeier, 2008). Few studies have attempted to correlate plankton molecular diversity with functional traits (Ramond et al., 2019; Schneider et al., 2020). Their results point to a promising field in plankton ecology. A global effort to create a consensus list of plankton traits, combined with an open curated reference sequence database, is needed to advance our understanding of crucial ecosystem services provided to the biosphere by planktonic microbes.

Short amplicons produced by metabarcoding have a comparatively low phylogenetic signal. This can present a problem for identification, especially when an environmental sequence is distantly related to reference sequences. A new wave of sequencing technology is capable of long-read sequencing with high-throughput, minus the rigorous preparation required for Sanger sequencing. These technologies are known as single-molecule long-read sequencing or long-read amplicon sequencing. They might be confused with Illumina or 454, as they can also be referred to as high-throughput sequencing (HTS) or next generation sequencing (for the same reasons discussed above). Here, it will be called long-read amplicon sequencing. The two sequencing approaches that have been applied to microbes are from Pacific Biosciences (PacBio) and Oxford Nanopore Technologies (MinION), that can produce reads up to 20 Kb and 200 Kb, respectively (Goodwin et al., 2016). With such technology, longer genetic markers can be targeted, such as the ITS, large subunit (LSU) rRNA, or the entire rRNA operon (which includes the SSU, ITS, and LSU). These regions can span up to 4500 base pairs, which would greatly increase taxonomic resolution to uncover new taxonomic groups. Jamy et al. (2020) reported greater accuracy and sensitivity using a new "phylogeny-aware" approach, only possible with longer reads, compared to the similarity-based and phylogenetic placement-based methods for shorter reads. Longer reads also provide a new opportunity to directly investigate phylogenetic relationships, which would be especially beneficial for groups only known from environmental sequences and not cultures (Jamy et al., 2020). Long-read amplicon sequencing has only recently started to be applied to diversity studies of the marine microbial community (Fu et al., 2020; Thompson et al., 2020). There are still some limitations that need to be addressed. At the time of writing this chapter, the error rate of long-read amplicon sequencing can be up to 12%, much higher than HTS at 0.1% (Goodwin et al., 2016). The use of alternative long-read sequencing strategies (such as PacBio circular consensus sequencing) with specific pipelines was reported to reduce error rates near to zero (Callahan et al., 2019), and along with the continuous development of sequencing technologies, might be changing the long-read metabarcoding scenario soon. Moreover, new reference sequence databases containing longer gene fragments would need to be developed.

Acknowledgments

We thank Fabrice Not for the suggestion to write this chapter. We thank publishers for allowing the inclusion of the previous published material. ALS was supported by the

Singapore Ministry of Education, Academic Research Fund Tier 1 (RG26/19). DV was supported by ANR contract PhytoPol (ANR-15-CE02-0007). CGR was supported by FONDECYT project 3190827.

Appendix 11.6 Supplementary material

Supplementary material related to this chapter can be found online at https://doi.org/10.1016/B978-0-12-822861-6.00020-0.

References

Ahlgren, N., Rocap, G., 2012. Diversity and distribution of marine *Synechococcus*: multiple gene phylogenies for consensus classification and development of qPCR assays for sensitive measurement of clades in the ocean. Frontiers in Microbiology 3, 213.

Altschul, S.F., Gish, W., Miller, W., Myers, E.W., Lipman, D.J., 1990. Basic local alignment search tool. Journal of Molecular Biology 215, 403–410.

Amacher, J., Neuer, S., Anderson, I., Massana, R., 2009. Molecular approach to determine contributions of the protist community to particle flux. Deep-Sea Research. Part 1. Oceanographic Research Papers 56, 2206–2215.

Amacher, J., Neuer, S., Lomas, M., 2013. DNA-based molecular fingerprinting of eukaryotic protists and cyanobacteria contributing to sinking particle flux at the Bermuda Atlantic time-series study. Deep-Sea Research. Part 2. Topical Studies in Oceanography 93, 71–83.

Amaral-Zettler, L.A., McCliment, E.A., Ducklow, H.W., Huse, S.M., 2009. A method for studying protistan diversity using massively parallel sequencing of V9 hypervariable regions of small-subunit ribosomal RNA genes. PLoS ONE 4, e6372.

Amato, A., Kooistra, W.H., Levialdi Ghiron, J.H., Mann, D.G., Pröschold, T., Montresor, M., 2007. Reproductive isolation among sympatric cryptic species in marine diatoms. Protist 158, 193–207.

Apprill, A., Mcnally, S., Parsons, R., Weber, L., 2015. Minor revision to V4 region SSU rRNA 806R gene primer greatly increases detection of SAR11 bacterioplankton. Aquatic Microbial Ecology.

Arrigo, K.R., et al., 2014. Phytoplankton blooms beneath the sea ice in the Chukchi Sea. Deep-Sea Research. Part 2. Topical Studies in Oceanography 105, 1–16.

Bach, L.T., Stange, P., Taucher, J., Achterberg, E.P., Algueró-Muñiz, M., Horn, H., Esposito, M., Riebesell, U., 2019. The influence of plankton community structure on sinking velocity and remineralization rate of marine aggregates. Global Biogeochemical Cycles 33, 971–994.

Balzano, S., Abs, E., Leterme, S.C., 2015. Protist Diversity along a Salinity Gradient in a Coastal Lagoon. Aquatic Microbial Ecology 74 (3), 263–277.

Banse, K., 2013. Reflections about chance in my career, and on the top-down regulated world. Annual Review of Marine Science 5, 1–19.

Benner, I., Irwin, A.J., Finkel, Z.V., 2019. Capacity of the common Arctic picoeukaryote *Micromonas* to adapt to a warming ocean. Limnology and Oceanography Letters 5, 221–227.

Bergsland, K.J., Haselkorn, R., 1991. Evolutionary relationships among Eubacteria, Cyanobacteria, and Chloroplasts: evidence from the *rpoC1* gene of *Anabaena* sp. strain PCC 7120. Journal of Bacteriology 173, 3446–3455.

Beszteri, S., Thoms, S., Benes, V., Harms, L., Trimborn, S., 2018. The response of three Southern Ocean phytoplankton species to ocean acidification and light availability: a transcriptomic study. Protist 169, 958–975.

Biard, T., Stemmann, L., Picheral, M., Mayot, N., Vandromme, P., Hauss, H., Gorsky, G., Guidi, L., Kiko, R., Not, F., 2016. In situ imaging reveals the biomass of giant protists in the global ocean. Nature 532, 504–507.

Biers, E.J., Sun, S., Howard, E.C., 2009. Prokaryotic genomes and diversity in surface ocean waters: interrogating the global ocean sampling metagenome. Applied and Environmental Microbiology 75, 2221–2229.

Biller, S.J., Berube, P.M., Lindell, D., Chisholm, S.W., 2015. *Prochlorococcus*: the structure and function of collective diversity. Nature Review, Microbiology 13, 13–27.

Bittner, L., Gobet, A., Audic, S., Romac, S., Egge, E.S., Santini, S., Ogata, H., Probert, I., Edvardsen, B., de Vargas, C., 2013. Diversity patterns of uncultured haptophytes unravelled by pyrosequencing in Naples Bay. Molecular Ecology 22, 87–101.

Blanchot, J., Rodier, M., 1996. Picophytoplankton abundance and biomass in the western tropical Pacific Ocean during the 1992 El Niño year: results from flow cytometry. Deep-Sea Research. Part 1. Oceanographic Research Papers 43, 877–895.

Blazewicz, S.J., Barnard, R.L., Daly, R.A., Firestone, M.K., 2013. Evaluating rRNA as an indicator of microbial activity in environmental communities: limitations and uses. The ISME Journal 7, 2061–2068.

Boeuf, D., Edwards, B.R., Eppley, J.M., Hu, S.K., Poff, K.E., Romano, A.E., Caron, D.A., Karl, D.M., DeLong, E.F., 2019. Biological composition and microbial dynamics of sinking particulate organic matter at abyssal depths in the oligotrophic open ocean. Proceedings of the National Academy of Sciences of the United States of America 116, 11824–11832.

Bolyen, E., et al., 2019. Reproducible, interactive, scalable and extensible microbiome data science using QIIME 2. Nature Biotechnology 37, 852–857.

Bower, S.M., Carnegie, R.B., Goh, B., Jones, S.R., Lowe, G.J., Mak, M.W., 2004. Preferential PCR amplification of parasitic protistan small subunit rDNA from metazoan tissues. Journal of Eukaryotic Microbiology 51, 325–332.

Boyd, P.W., Claustre, H., Levy, M., Siegel, D.A., Weber, T., 2019. Multi-faceted particle pumps drive carbon sequestration in the ocean. Nature 568, 327–335.

Boyd, P.W., Newton, P.P., 1999. Does planktonic community structure determine downward particulate organic carbon flux in different oceanic provinces? Deep-Sea Research. Part 1 46, 63–91.

Boyd, P.W., Strzepek, R., Takeda, S., Jackson, G., Wong, C.S., McKay, R.M., Law, C., Kiyosawa, H., Saito, H., Sherry, N., Johnson, K., Gower, J., Ramaiah, N., 2005. The evolution and termination of an iron-induced mesoscale bloom in the Northeast Subarctic Pacific. Limnology and Oceanography 50, 1872–1886.

Boyd, P.W., Trull, T.W., 2007. Understanding the export of biogenic particles in oceanic waters: is there consensus? Progress in Oceanography 72, 276–312.

Bucklin, A., Allen, L., 2004. MtDNA sequencing from zooplankton after long-term preservation in buffered formalin. Molecular Phylogenetics and Evolution 30, 879–882.

Buesseler, K.O., Lamborg, C.H., Boyd, P.W., Lam, P.J., Trull, T.W., Bidigare, R.R., Bishop, J.K.B., Casciotti, K.L., Dehairs, F., Elskens, M., Honda, M., Karl, D.M., Siegel, D.A., Silver, M.W., Steinberg, D.K., Valdes, J., Van Mooy, B., Wilson, S., 2007. Revisiting carbon flux through the ocean's twilight zone. Science 316, 567–570.

Burd, A.B., Hansell, D.A., Steinberg, D.K., Anderson, T.R., Aristegui, J., Baltar, F., Beaupre, S.R., Buesseler, K.O., DeHairs, F., Jackson, G.A., Kadko, D.C., Koppelmann, R., Lampitt, R.S., Nagata, T., Reinthaler, T., Robinson, C., Robison, B.H., Tamburini, C., Tanaka, T., 2010. Assessing the apparent imbalance between geochemical and biochemical indicators of meso- and bathypelagic biological activity: what the @$#! is wrong

with present calculations of carbon budgets? Deep-Sea Research. Part 2. Topical Studies in Oceanography 57, 1557–1571.
Cai, H., Wang, K., Huang, S., Jiao, N., Chen, F., 2010. Distinct patterns of picocyanobacterial communities in winter and summer in the Chesapeake Bay. Applied and Environmental Microbiology 76, 2955–2960.
Calbet, A., Saiz, E., 2005. The ciliate-copepod link in marine ecosystems. Aquatic Microbial Ecology 38, 157–167.
Callahan, B.J., McMurdie, P.J., Holmes, S.P., 2017. Exact sequence variants should replace operational taxonomic units in marker-gene data analysis. The ISME Journal 11, 2639–2643.
Callahan, B.J., McMurdie, P.J., Rosen, M.J., Han, A.W., Johnson, A.J.A., Holmes, S.P., 2016. DADA2: high-resolution sample inference from illumina amplicon data. Nature Methods 13, 581–583.
Callahan, B.J., Wong, J., Heiner, C., Oh, S., Theriot, C.M., Gulati, A.S., McGill, S.K., Dougherty, M.K., 2019. High-throughput amplicon sequencing of the full-length 16S rRNA gene with single-nucleotide resolution. Nucleic Acids Research 47 (18), e103.
Campbell, L., Vaulot, D., 1993. Photosynthetic picoplankton community structure in the sub-tropical North Pacific Ocean near Hawaii (station ALOHA). Deep-Sea Research 40, 2043–2060.
Caporaso, J.G., Lauber, C.L., Walters, W.A., Berg-Lyons, D., Lozupone, C.A., Turnbaugh, P.J., Fierer, N., Knight, R., 2011. Global patterns of 16S rRNA diversity at a depth of millions of sequences per sample. Proceedings of the National Academy of Sciences of the United States of America.
Caron, D., Countway, P., 2009. Hypotheses on the role of the protistan rare biosphere in a changing world. Aquatic Microbial Ecology 57, 227–238.
Caron, D.A., Countway, P.D., Jones, A.C., Kim, D.Y., Schnetzer, A., 2012. Marine protistan diversity. Annual Review of Marine Science 4, 467–493.
Carroll, E.L., Gallego, R., Sewell, M.A., Zeldis, J., Ranjard, L., Ross, H.A., Tooman, L.K., O'Rorke, R., Newcomb, R.D., Constantine, R., 2019. Multi-locus DNA metabarcoding of zooplankton communities and scat reveal trophic interactions of a generalist predator. Scientific Reports 9, 281.
Casey, J.M., Meyer, C.P., Morat, F., Brandl, S.J., Planes, S., Parravicini, V., 2019. Reconstructing hyperdiverse food webs: gut content metabarcoding as a tool to disentangle trophic interactions on coral reefs. Methods in Ecology and Evolution. https://doi.org/10.1111/2041-210X.13206.
Cavalier-Smith, T., Lewis, R., Chao, E.E., Oates, B., Bass, D., 2009. *Helkesimastix Marina* n. Sp. (Cercozoa: Sainouroidea Superfam. n.) a Gliding Zooflagellate of Novel Ultrastructure and Unusual Ciliary Behaviour. Protist 160 (3), 452–479.
Cavan, E.L., Henson, S.A., Belcher, A., Sanders, R., 2017. Role of zooplankton in determining the efficiency of the biological carbon pump. Biogeosciences 14, 177–186.
Chariton, A.A., Stephenson, S., Morgan, M.J., Steven, A.D., Colloff, M.J., Court, L.N., Hardy, C.M., 2015. Metabarcoding of benthic eukaryote communities predicts the ecological condition of estuaries. Environmental Pollution 203, 165–174.
Charvet, S., Vincent, W.F., Lovejoy, C., 2014. Effects of light and prey availability on Arctic freshwater protist communities examined by high-throughput DNA and RNA sequencing. FEMS Microbiology, Ecology 88, 550–564.
Chen, B., Liu, H., Landry, M.R., Chen, M., Sun, J., Shek, L., Chen, X., Harrison, P.J., 2009. Estuarine nutrient loading affects phytoplankton growth and microzooplankton grazing at two contrasting sites in Hong Kong coastal waters. Marine Ecology. Progress Series 379.
Chen, F., Wang, K., Kan, J., Bachoon, D.S., Lu, J., Lau, S., Campbell, L., 2004. Phylogenetic diversity of *Synechococcus* in the Chesapeake Bay revealed by Rilbulose-1, 5-Bisphosphate

carboxylase-oxygenase (RuBisCO) large subunit gene (rbcL) sequences. Aquatic Microbial Ecology 36, 153–164.

Chen, F., Wang, K., Kan, J., Suzuki, M.T., Wommack, K.E., 2006. Diverse and unique picocyanobacteria in Chesapeake Bay, revealed by 16S-23S rRNA internal transcribed spacer sequences. Applied and Environmental Microbiology 72, 2239–2243.

Chénard, C., Suttle, C.A., 2008. Phylogenetic diversity of sequences of cyanophage photosynthetic gene *psbA* in marine and freshwaters. Applied and Environmental Microbiology 74, 5317–5324.

Cheung, M.K., Au, C.H., Chu, K.H., Kwan, H.S., Wong, C.K., 2010. Composition and genetic diversity of Picoeukaryotes in subtropical coastal waters as revealed by 454 Pyrosequencing. The ISME Journal 4 (8), 1053–1059.

Choi, D.H., Noh, J.H., 2009. Phylogenetic diversity of synechococcus strains isolated from the East China Sea and the East Sea. FEMS Microbiology, Ecology 69, 439–448.

Clare, E.L., 2014. Molecular detection of trophic interactions: emerging trends, distinct advantages, significant considerations and conservation applications. Evolutionary Applications 7, 1144–1157.

Comeau, A.M., Li, W.K.W., Tremblay, J.-E., Carmack, E.C., Lovejoy, C., 2011. Arctic Ocean Microbial Community Structure before and after the 2007 Record Sea Ice Minimum. PLoS ONE 6 (11), 1–12. https://doi.org/10.1371/journal.pone.0027492.

Comeau, A.M., Philippe, B., Thaler, M., Gosselin, M., Poulin, M., Lovejoy, C., 2013. Protists in Arctic drift and land-fast sea ice. Journal of Phycology 49, 229–240.

Cordier, T., Forster, D., Dufresne, Y., Martins, C.I., Stoeck, T., Pawlowski, J., 2018. Supervised machine learning outperforms taxonomy-based environmental DNA metabarcoding applied to biomonitoring. Molecular Ecology Resources 18, 1381–1391.

Craig, C., Kimmerer, W.J., Cohen, C.S., 2013. A DNA-based method for investigating feeding by copepod nauplii. Journal of Plankton Research 36, 271–275.

Darling, K.F., Wade, C.M., Stewart, I.A., Kroon, D., Dingle, R., Brown, A.J.L., 2000. Molecular evidence for genetic mixing of Arctic and Antarctic subpolar populations of planktonic foraminifers. Nature 405, 43–47.

Davey, M.P., Norman, L., Sterk, P., Huete-Ortega, M., Bunbury, F., Loh, B.K.W., Stockton, S., Peck, L.S., Convey, P., Newsham, K.K., Smith, A.G., 2019. Snow algae communities in Antarctica: metabolic and taxonomic composition. New Phytologist 222, 1242–1255.

Dawson, S.C., Hagen, K.D., 2009. Mapping the protistan 'rare biosphere'. Journal of Biology.

De La Rocha, C.L., Passow, U., 2007. Factors influencing the sinking of POC and the efficiency of the biological carbon pump. Deep-Sea Research. Part 2. Topical Studies in Oceanography 54, 639–658.

de Vargas, C., et al., 2015. Eukaryotic plankton diversity in the sunlit ocean. Science 348, 1261605.

Deagle, B.E., Thomas, A.C., McInnes, J.C., Clarke, L.J., Vesterinen, E.J., Clare, E.L., Kartzinel, T.R., Eveson, J.P., 2019. Counting with DNA in metabarcoding studies: how should we convert sequence reads to dietary data? Molecular Ecology 28, 391–406.

Decelle, J., Martin, P., Paborstava, K., Pond, D.W., Tarling, G., Mahé, F., de Vargas, C., Lampitt, R., Not, F., 2013. Diversity, ecology and biogeochemistry of cyst-forming acantharia (radiolaria) in the oceans. PLoS ONE 8, e53598.

Decelle, J., Romac, S., Stern, R.F., Bendif, E.M., Zingone, A., Audic, S., Guiry, M.D., Guillou, L., Tessier, D., Le Gall, F., Gourvil, P., Lopes dos Santos, A., Probert, I., Vaulot, D., de Vargas, C., Christen, R., 2015. PhytoREF: a reference database of the plastidial 16S rRNA gene of photosynthetic eukaryotes with curated taxonomy. Molecular Ecology Resources 15, 1435–1445.

DeLong, E.F., Preston, C.M., Mincer, T., Rich, V., Hallam, S.J., Frigaard, N.U., Martinez, A., Sullivan, M.B., Edwards, R., Brito, B.R., Chisholm, S.W., Karl, D.M., 2006.

Community genomics among stratified microbial assemblages in the ocean's interior. Science 311, 496–503.
Díez, B., Pedrós-Alió, C., Massana, R., 2001. Study of genetic diversity of eukaryotic picoplankton in different oceanic regions by small-subunit rRNA gene cloning and sequencing. Applied and Environmental Microbiology 67, 2932–2941.
Doherty, M., Yager, P.L., Moran, M.A., Coles, V.J., Fortunato, C.S., Krusche, A.V., Medeiros, P.M., Payet, J.P., Richey, J.E., Satinsky, B.M., Sawakuchi, H.O., Ward, N.D., Crump, B.C., 2017. Bacterial biogeography across the Amazon River-ocean continuum. Frontiers in Microbiology.
Doré, H., et al., 2020. Evolutionary mechanisms of long-term genome diversification associated with niche partitioning in marine picocyanobacteria. Frontiers in Microbiology.
Duarte, C.M., 2015. Seafaring in the 21st century: the Malaspina 2010 circumnavigation expedition. Limnology and Oceanography Bulletin 24, 11–14.
Dufresne, A., Garczarek, L., Partensky, F., 2005. Accelerated evolution associated with genome reduction in a free-living prokaryote. Genome Biology 6, R14.
Dufresne, A., Ostrowski, M., Scanlan, D.J., Garczarek, L., Mazard, S., Palenik, B.P., Paulsen, I.T., de Marsac, N.T., Wincker, P., Dossat, C., Ferriera, S., Johnson, J., Post, A.F., Hess, W.R., Partensky, F., 2008. Unraveling the genomic mosaic of a ubiquitous genus of marine cyanobacteria. Genome Biology 9, R90.
Dunthorn, M., Stoeck, T., Clamp, J., Warren, A., Mahé, F., 2014. Ciliates and the rare biosphere: a review. Journal of Eukaryotic Microbiology 61, 404–409.
Duret, M.T., Lampitt, R.S., Lam, P., 2019. Prokaryotic niche partitioning between suspended and sinking marine particles. Environmental Microbiology Reports 11 (3), 386–400.
Duret, M.T., Lampitt, R.S., Lam, P., 2020. Eukaryotic influence on the oceanic biological carbon pump in the Scotia Sea as revealed by 18S rRNA gene sequencing of suspended and sinking particles. Limnology and Oceanography 65, S49–S70. https://doi.org/10.1002/lno.11319.
Durkin, C.A., Van Mooy, B.A.S., Dyhrman, S.T., Buesseler, K.O., 2016. Sinking phytoplankton associated with carbon flux in the Atlantic Ocean. Limnology and Oceanography 61, 1172–1187.
Ebersbach, F., Assmy, P., Martin, P., Schulz, I., Wolzenburg, S., Nöthig, E.-M., 2014. Particle flux characterisation and sedimentation patterns of protistan plankton during the iron fertilisation experiment LOHAFEX in the Southern Ocean. Deep-Sea Research. Part 1. Oceanographic Research Papers 89, 94–103.
Edgar, R., 2016. SINTAX: a simple non-Bayesian taxonomy classifier for 16S and ITS sequences. bioRxiv 74161.
Edgar, R.C., 2018. Updating the 97% identity threshold for 16S ribosomal RNA OTUs. Bioinformatics 34, 2371–2375.
Edgar, R.C., Haas, B.J., Clemente, J.C., Quince, C., Knight, R., 2011. UCHIME improves sensitivity and speed of chimera detection. Bioinformatics 27, 2194–2200.
Edgcomb, V.P., Kysela, D.T., Teske, A., Gomez, A.d.V., Sogin, M.L., 2002. Benthic eukaryotic diversity in the Guaymas Basin hydrothermal vent environment. Proceedings of the National Academy of Sciences of the United States of America 99, 7658–7662.
Edvardsen, B., Egge, E.S., Vaulot, D., 2016. Diversity and distribution of haptophytes revealed by environmental sequencing and metabarcoding – a review. Perspectives in Phycology 3, 77–91.
Egge, E.S., Johannessen, T.V., Andersen, T., Eikrem, W., Bittner, L., Larsen, A., Sandaa, R.A., Edvardsen, B., 2015. Seasonal diversity and dynamics of haptophytes in the Skagerrak, Norway, explored by high-throughput sequencing. Molecular Ecology 24, 3026–3042.
Eren, A.M., 2019. Fantastic data and how to share them: a plea to journal editors and reviewers. Meren Lab.. https://merenlab.org/2019/02/24/fantastic-data/.

Ernst, A., Becker, S., Wollenzien, U.I., Postius, C., 2003. Ecosystem-dependent adaptive radiations of picocyanobacteria inferred from 16S rRNA and ITS-1 sequence analysis. Microbiology 149, 217–228.

Falkowski, P.G., et al., 1998. Biogeochemical controls and feedbacks on ocean primary production. Science 281, 200–206.

Farnelid, H., Turk-Kubo, K., Ploug, H., Ossolinski, J.E., Collins, J.R., Van Mooy, B.A.S., Zehr, J.P., 2019. Diverse diazotrophs are present on sinking particles in the North Pacific Subtropical Gyre. The ISME Journal 13, 170–182.

Farrant, G.K., Doré, H., Cornejo-Castillo, F.M., Partensky, F., Ratin, M., Ostrowski, M., Pitt, F.D., Wincker, P., Scanlan, D.J., Iudicone, D., Acinas, S.G., Garczarek, L., 2016. Delineating ecologically significant taxonomic units from global patterns of marine picocyanobacteria. Proceedings of the National Academy of Sciences of the United States of America 113, E3365–E3374.

Ferris, M.J., Palenik, B., 1998. Niche adaptation in ocean cyanobacteria. Nature 396, 226–228.

Fitch, W.M., 1976. The molecular evolution of cytochrome c in eukaryotes. Journal of Molecular Evolution 8, 13–40.

Fitch, W.M., Margoliash, E., 1967. Construction of phylogenetic trees. Science 155, 279–284.

Flombaum, P., Gallegos, J.L., Gordillo, R.A., Rincón, J., Zabala, L.L., Jiao, N., Karl, D.M., Li, W.K., Lomas, M.W., Veneziano, D., Vera, C.S., Vrugt, J.A., Martiny, A.C., 2013. Present and future global distributions of the marine cyanobacteria *Prochlorococcus* and *Synechococcus*. Proceedings of the National Academy of Sciences of the United States of America 110, 9824–9829.

Fontanez, K.M., Eppley, J.M., Samo, T.J., Karl, D.M., Delong, E.F., 2015. Microbial community structure and function on sinking particles in the North Pacific Subtropical Gyre. Frontiers in Microbiology 6, 1–14.

Fryxell, G.A., Doucette, G.J., Hubbard, G.F., 1981. The genus *Thalassiosira*: the bipolar diatom *T. antarctica* comber. Botanica Marina 24, 321–336.

Fu, Y., Zheng, P., Zhang, X., Zhang, Q., Ji, D., 2020. Protist interactions and seasonal dynamics in the coast of Yantai, Northern Yellow Sea of China as revealed by metabarcoding. Journal of Ocean University of China 19, 961–974.

Fuller, N.J., Campbell, C., Allen, D.J., Pitt, F.D., Le Gall, F., Vaulot, D., Scanlan, D.J., 2006. Analysis of photosynthetic picoeukaryote diversity at open ocean sites in the Arabian Sea using a PCR biased towards marine algal plastids. Aquatic Microbial Ecology 43, 79–93.

Fuller, N.J., Marie, D., Partensky, F., Vaulot, D., Post, A.F., Scanlan, D.J., 2003. Clade-specific 16S ribosomal DNA oligonucleotides reveal the predominance of a single marine *Synechococcus* clade throughout a stratified water column in the Red Sea. Applied and Environmental Microbiology 69, 2430–2443.

Fuller, N.J., West, N.J., Marie, D., Yallop, M., Rivlin, T., Post, A.F., Scanlan, D.J., 2005. Dynamics of community structure and phosphate status of picocyanobacterial populations in the Gulf of Aqaba, Red Sea. Limnology and Oceanography 50, 363–375.

Galand, P.E., Casamayor, E.O., Kirchman, D.L., Lovejoy, C., 2009. Ecology of the rare microbial biosphere of the Arctic Ocean. Proceedings of the National Academy of Sciences of the United States of America 106, 22427–22432.

García-García, N., Tamames, J., Linz, A.M., Pedrós-Alió, C., Puente-Sánchez, F., 2019. Microdiversity ensures the maintenance of functional microbial communities under changing environmental conditions. The ISME Journal 13, 2969–2983.

Garczarek, L., Dufresne, A., Blot, N., Cockshutt, A.M., Peyrat, A., Campbell, D.A., Joubin, L., Six, C., 2008. Function and evolution of the *psbA* gene family in marine *Synechococcus*: *Synechococcus* sp. WH7803 as a case study. The ISME Journal.

Gérikas Ribeiro, C., Lopes dos Santos, A., Marie, D., Pereira Brandini, F., Vaulot, D., 2018. Small eukaryotic phytoplankton communities in tropical waters off Brazil are dominated by symbioses between haptophyta and nitrogen-fixing cyanobacteria. The ISME Journal 12, 1360–1374.
Giner, C.R., Pernice, M.C., Balagué, V., Duarte, C.M., Gasol, J.M., Logares, R., Massana, R., 2020. Marked changes in diversity and relative activity of picoeukaryotes with depth in the world ocean. The ISME Journal 14, 437–449.
Giovannoni, S.J., Britschgi, T.B., Moyer, C.L., Field, K.G., 1990. Genetic diversity in Sargasso Sea bacterioplankton. Nature 345, 60–63.
Gong, W., Marchetti, A., 2019. Estimation of 18S gene copy number in marine eukaryotic plankton using a next-generation sequencing approach. Frontiers in Marine Science 6, 219.
Goodwin, S., McPherson, J.D., Richard McCombie, W., 2016. Coming of age: ten years of next-generation sequencing technologies. Nature Reviews. Genetics 17, 333–351.
Gowing, M.M., 1993. Seasonal radiolarian flux at the VERTEX North Pacific time-series site. Deep-Sea Research. Part 1. Oceanographic Research Papers 40, 517–545.
Gowing, M.M., Coale, S.L., 1989. Fluxes of living radiolarians and their skeletons along a northeast Pacific transect from coastal upwelling to open ocean waters. Deep-Sea Research. Part A, Oceanographic Research Papers 36, 561–576.
Gran-Stadniczeñko, S., Šupraha, L., Egge, E.D., Edvardsen, B., 2017. Haptophyte diversity and vertical distribution explored by 18S and 28S ribosomal RNA gene metabarcoding and scanning electron microscopy. Journal of Eukaryotic Microbiology 64, 514–532.
Guillou, L., et al., 2013. The protist ribosomal reference database (PR^2): a catalog of unicellular eukaryote small sub-unit rRNA sequences with curated taxonomy. Nucleic Acids Research 41, D597–D604.
Gutiérrez-Rodríguez, A., Slack, G., Daniels, E.F., Selph, K.E., Palenik, B., Landry, M.R., 2014. Fine spatial structure of genetically distinct picocyanobacterial populations across environmental gradients in the Costa Rica Dome. Limnology and Oceanography 59, 705–723.
Gutierrez-Rodriguez, A., Stukel, M.R., Lopes dos Santos, A., Biard, T., Scharek, R., Vaulot, D., Landry, M.R., Not, F., 2019. High contribution of rhizaria (radiolaria) to vertical export in the California current ecosystem revealed by DNA metabarcoding. The ISME Journal 13, 964–976.
Hardge, K., Peeken, I., Neuhaus, S., Lange, B.A., Stock, A., Stoeck, T., Weinisch, L., Metfies, K., 2017. The importance of sea ice for exchange of habitat-specific protist communities in the central Arctic Ocean. Journal of Marine Systems 165, 124–138.
Hardy, C.M., Krull, E.S., Hartley, D.M., Oliver, R.L., 2010. Carbon source accounting for fish using combined DNA and stable isotope analyses in a regulated lowland River Weir Pool. Molecular Ecology 19, 197–212.
Haumann, F.A., Gruber, N., Münnich, M., 2020. Sea-ice induced Southern Ocean subsurface warming and surface cooling in a warming climate. AGU Advances 1, e2019AV000132.
Henson, S., Le Moigne, F., Giering, S., 2019. Drivers of carbon export efficiency in the global ocean. Global Biogeochemical Cycles 33, 891–903.
Herdman, M., Castenholz, R.W., Waterbury, J.B., Rippka, R., 2015. Form-Synechococcus. Bergey's Manual of Systematics of Archaea and Bacteria, pp. 1–7.
Herndl, G.J., Reinthaler, T., 2013. Microbial control of the dark end of the biological pump. Nature Geoscience 6, 718–724.
Hess, W.R., Weihe, A., Loiseaux-de Goër, S., Partensky, F., Vaulot, D., 1995. Characterization of the single *psbA* gene of *Prochlorococcus marinus* CCMP 1375 (Prochlorophyta). Plant Molecular Biology 27, 1189–1196.
Honjo, S., 1976. Coccoliths: production, transportation and sedimentation. Marine Micropaleontology 1, 65–79.

Honjo, S., Eglinton, T., Taylor, C., Ulmer, K., Sievert, S., Bracher, A., German, C., Edgcomb, V., Francois, R., Iglesias-Rodriguez, M.D., Van Mooy, B., Rapeta, D., 2014. Understanding the role of the biological pump in the global carbon cycle: an imperative for ocean science. Oceanography 27, 10–16.

Hoppe, C.J.M., Flintrop, C.M., Rost, B., 2018. The Arctic picoeukaryote *Micromonas pusilla* benefits synergistically from warming and ocean acidification. Biogeosciences 15, 4353–4365.

Horvat, C., Jones, D.R., Iams, S., Schroeder, D., Flocco, D., Feltham, D., 2017. The frequency and extent of sub-ice phytoplankton blooms in the Arctic Ocean. Science Advances 3, e1601191.

Hu, S.K., Campbell, V., Connell, P., Gellene, A.G., Liu, Z., Terrado, R., Caron, D.A., 2016. Protistan diversity and activity inferred from RNA and DNA at a coastal ocean site in the eastern North Pacific. FEMS Microbiology, Ecology 92, fiw050.

Huang, S., Wilhelm, S.W., Harvey, H.R., Taylor, K., Jiao, N., Chen, F., 2012. Novel lineages of *Prochlorococcus* and *Synechococcus* in the global oceans. The ISME Journal 6, 285–297.

Huber, P., Cornejo-Castillo, F.M., Ferrera, I., Sánchez, P., Logares, R., Metz, S., Balagué, V., Acinas, S.G., Gasol, J.M., Unrein, F., 2019. Primer design for an accurate view of picocyanobacterial community structure by using high-throughput sequencing. Applied and Environmental Microbiology 85, e02659.

Huse, S.M., Mark Welch, D.B., Voorhis, A., Shipunova, A., Morrison, H.G., Eren, A.M., Sogin, M.L., 2014. VAMPS: a website for visualization and analysis of microbial population structures. BMC Bioinformatics 15, 41.

Ibarbalz, F.M., et al., 2019. Global trends in marine plankton diversity across kingdoms of life. Cell 179. 1084–1097.e21.

Isaacs, J.D., 1972. Unstructured marine food webs and "pollutant analogues". United States National Marine Fisheries Service. Fishery Bulletin 70, 1053–1059.

Jain, R., Rivera, M.C., Lake, J.A., 1999. Horizontal gene transfer among genomes: the complexity hypothesis. Proceedings of the National Academy of Sciences of the United States of America 96, 3801–3806.

Jamy, M., Foster, R., Barbera, P., Czech, L., Kozlov, A., Stamatakis, A., Bending, G., Hilton, S., Bass, D., Burki, F., 2020. Long-read metabarcoding of the eukaryotic rDNA operon to phylogenetically and taxonomically resolve environmental diversity. Molecular Ecology Resources 20, 429–443.

Janout, M.A., Hölemann, J., Waite, A.M., Krumpen, T., von Appen, W.J., Martynov, F., 2016. Sea-ice retreat controls timing of summer plankton blooms in the eastern Arctic Ocean. Geophysical Research Letters 43, 12,493–12,501.

Jenkins, B.D., Zehr, J.P., Gibson, A., Campbell, L., 2006. Cyanobacterial assimilatory nitrate reductase gene diversity in coastal and oligotrophic marine environments. Environmental Microbiology 8, 2083–2095.

Jin, X., Gruber, N., Dunne, J.P., Sarmiento, J.L., Armstrong, R.A., 2006. Diagnosing the contribution of phytoplankton functional groups to the production and export of particulate organic carbon, $CaCO_3$, and opal from global nutrient and alkalinity distributions. Global Biogeochemical Cycles 20.

Joli, N., Monier, A., Logares, R., Lovejoy, C., 2017. Seasonal patterns in Arctic prasinophytes and inferred ecology of *Bathycoccus* unveiled in an Arctic winter metagenome. The ISME Journal 11, 1372–1385.

Jurburg, S.D., Konzack, M., Eisenhauer, N., Heintz-Buschart, A., 2020. The archives are half-empty: an assessment of the availability of microbial community sequencing data. Communications Biology 3, 474.

Kaczmarska, I., Mather, L., Aluddington, I., Muise, F., Ehrman, J.M., 2014. Cryptic diversity in a cosmopolitan diatom known as *Asterionellopsis glacialis* (Fragilariaceae): im-

plications for ecology, biogeography, and taxonomy. American Journal of Botany 101, 267–286.
Kalenitchenko, D., Joli, N., Potvin, M., Tremblay, J., Lovejoy, C., 2019. Biodiversity and species change in the Arctic Ocean: a view through the lens of nares strait. Frontiers in Marine Science 6, 1–17.
Kashtan, N., Roggensack, S.E., Rodrigue, S., Thompson, J.W., Biller, S.J., Coe, A., Ding, H., Marttinen, P., Malmstrom, R.R., Stocker, R., Follows, M.J., Stepanauskas, R., Chisholm, S.W., 2014. Single-cell genomics reveals hundreds of coexisting subpopulations in wild *Prochlorococcus*. Science 344, 416–420.
Kaunisto, K.M., Roslin, T., Sääksjärvi, I.E., Vesterinen, E.J., 2017. Pellets of proof: first glimpse of the dietary composition of adult odonates as revealed by metabarcoding of feces. Ecology and Evolution 7, 8588–8598.
Kermarrec, L., Franc, A., Rimet, F., Chaumeil, P., Frigerio, J.-M., Humbert, J.-F., Bouchez, A., 2014. A next-generation sequencing approach to river biomonitoring using benthic diatoms. Freshwater Science 33, 349–363.
Kettler, G.C., Martiny, A.C., Huang, K., Zucker, J., Coleman, M.L., Rodrigue, S., Chen, F., Lapidus, A., Ferriera, S., Johnson, J., Steglich, C., Church, G.M., Richardson, P., Chisholm, S.W., 2007. Patterns and implications of gene gain and loss in the evolution of *Prochlorococcus*. PLoS Genetics 3, e231.
Kilias, E.S., Peeken, I., Metfies, K., 2014. Insight into protist diversity in Arctic sea ice and melt-pond aggregate obtained by pyrosequencing. Polar Research 33, 23466.
King, R.A., Read, D.S., Traugott, M., Symondson, W.O.C., 2008. Molecular analysis of predation: a review of best practice for DNA-based approaches. Molecular Ecology 17, 947–963.
Kirchman, D.L., 2016. Growth rates of microbes in the oceans. Annual Review of Marine Science 8, 285–309.
Kishino, H., Miyata, T., Hasegawa, M., 1990. Maximum likelihood inference of protein phylogeny and the origin of chloroplasts. Journal of Molecular Evolution 31, 151–160.
Kitahara, K., Miyazaki, K., 2013. Revisiting bacterial phylogeny. Mobile Genetic Elements 3, e24210.
Knauer, G.A., Martin, J.H., Bruland, K.W., 1979. Fluxes of particulate carbon, nitrogen, and phosphorus in the upper water column of the northeast Pacific. Deep-Sea Research 26, 97–108.
Kopf, A., et al., 2015. The ocean sampling day consortium. GigaScience 4, 27.
Kozich, J.J., Westcott, S.L., Baxter, N.T., Highlander, S.K., Schloss, P.D., 2013. Development of a dual-index sequencing strategy and curation pipeline for analyzing amplicon sequence data on the miseq illumina sequencing platform. Applied and Environmental Microbiology 79, 5112–5120.
Kress, W.J., Garcia-Robledo, C., Uriarte, M., Erickson, D.L., 2015. DNA barcodes for ecology, evolution, and conservation. Trends in Ecology & Evolution 30, 25–35.
Kysela, D.T., Palacios, C., Sogin, M.L., 2005. Serial analysis of V6 ribosomal sequence tags (SARST-V6): a method for efficient, high-throughput analysis of microbial community composition. Environmental Microbiology 7, 356–364.
Lampitt, R.S., Salter, I., Johns, D., 2009. Radiolaria: major exporters of organic carbon to the deep ocean. Global Biogeochemical Cycles 23, 1–9.
Lampitt, R.S., Wishner, K.F., Turley, C.M., Angel, M.V., 1993. Marine snow studies in the Northeast Atlantic-Ocean - distribution, composition and role as a food source for migrating plankton. Marine Biology 116 (4), 689–702.
Landry, M.R., 2002. Integrating classical and microbial food web concepts: evolving views from the open-ocean tropical Pacific. Hydrobiologia 480, 29–39.
Lane, D.J., 1991. 16S/23S rRNA sequencing. In: Stackebrandt, E., Goodfellow, M. (Eds.), Nucleic Acid Techniques in Bacterial Systematics. John Wiley and Sons, New York, N.Y., pp. 115–175.

Lane, D.J., Pace, B., Olsen, G.J., Stahl, D.A., Sogin, M.L., Pace, N.R., 1985. Rapid determination of 16S ribosomal RNA sequences for phylogenetic analyses. Proceedings of the National Academy of Sciences of the United States of America 82, 6955–6959.
Lanzén, A., Jørgensen, S.L., Huson, D.H., Gorfer, M., Grindhaug, S.H., Jonassen, I., Øvreås, L., Urich, T., 2012. CREST - classification resources for environmental sequence tags. PLoS ONE 7.
Le Moigne, F.A.C., Pabortsava, K., Marcinko, C.L.J., Martin, P., Sanders, R.J., 2014. Where is mineral ballast important for surface export of particulate organic carbon in the ocean? Geophysical Research Letters 41, 8460–8468.
LeCleir, G.R., DeBruyn, J.M., Maas, E.W., Boyd, P.W., Wilhelm, S.W., 2014. Temporal changes in particle-associated microbial communities after interception by nonlethal sediment traps. FEMS Microbiology, Ecology 87, 153–163.
Lee, C., 2019. Passing the baton to the next generation: a few problems that need solving. Annual Review of Marine Science 11, 1–13.
Leliaert, F., Verbruggen, H., Vanormelingen, P., Steen, F., López-Bautista, J.M., Zuccarello, G.C., Clerck, O.D., 2014. DNA-based species delimitation in algae DNA. European Journal of Phycology 49, 179–196.
Leray, M., Meyer, C.P., Mills, S.C., 2015. Metabarcoding dietary analysis of coral dwelling predatory fish demonstrates the minor contribution of coral mutualists to their highly partitioned, generalist diet. PeerJ 3, e1047.
Leu, E., Søreide, J.E., Hessen, D.O., Falk-Petersen, S., Berge, J., 2011. Consequences of changing sea-ice cover for primary and secondary producers in the European Arctic Shelf seas: timing, quantity, and quality. Progress in Oceanography 90, 18–32.
Lin, S., 2011. Genomic understanding of dinoflagellates. Research in Microbiology 162, 551–569.
Lindell, D., Padan, E., Post, A.F., 1998. Regulation of ntcA expression and nitrite uptake in the marine *Synechococcus* sp. strain WH7803. Journal of Bacteriology 180, 1878–1886. https://doi.org/10.1128/JB.180.7.1878-1886.1998.
Litchman, E., Klausmeier, C.A., 2008. Trait-based community ecology of phytoplankton. Annual Review of Ecology, Evolution, and Systematics 39, 615–639.
Liu, Z., Cochran, J.K., Lee, C., Gasser, B., Miquel, J.C., Wakeham, S.G., 2009. Further investigations on why POC concentrations differ in samples collected by Niskin Bottle and in situ pump. Deep-Sea Research. Part 2. Topical Studies in Oceanography 56, 1558–1567.
Lockhart, P., Penny, D., 1992. The problem of GC content, evolutionary trees and the origins of Chl-a/b photosynthetic organelles: are the prochlorophytes a eubacterial model for higher plant photosynthesis. Research in Photosynthesis III, 499–505.
Logares, R., et al., 2014. Patterns of rare and abundant marine microbial eukaryotes. Current Biology 24, 813–821.
Logares, R., Deutschmann, I.M., Junger, P.C., Giner, C.R., Krabberød, A.K., Schmidt, T.S.B., Rubinat-Ripoll, L., Mestre, M., Salazar, G., Ruiz-González, C., Sebastián, M., de Vargas, C., Acinas, S.G., Duarte, C.M., Gasol, J.M., Massana, R., 2020. Disentangling the mechanisms shaping the surface ocean microbiota. Microbiome 8, 55.
Logares, R., Mangot, J.-F., Massana, R., 2015. Rarity in aquatic microbes: placing protists on the map. Research in Microbiology 166, 831–841.
Lopes dos Santos, A., Pollina, T., Gourvil, P., Corre, E., Marie, D., Garrido, J.L., Rodríguez, F., Noël, M.-H., Vaulot, D., Eikrem, W., 2017. Chloropicophyceae, a new class of picophytoplanktonic prasinophytes. Scientific Reports 7, 14019.
López-García, P., Philippe, H., Gail, F., Moreira, D., 2003. Autochthonous eukaryotic diversity in hydrothermal sediment and experimental microcolonizers at the Mid-Atlantic Ridge. Proceedings of the National Academy of Sciences of the United States of America 100 (2), 697–702.

López-García, P., Rodriguez-Valera, F., Pedrós-Alió, C., Moreira, D., 2001. Unexpected diversity of small eukaryotes in deep-sea Antarctic plankton. Nature 409, 603–607.
Love, M.I., Huber, W., Anders, S., 2014. Moderated estimation of fold change and dispersion for RNA-seq data with DESeq2. Genome Biology 15, 550.
Lovejoy, C., Massana, R., Pedros-Alio, C., 2006. Diversity and distribution of marine microbial eukaryotes in the Arctic Ocean and adjacent seas. Applied and Environmental Microbiology 72, 3085–3095.
Luo, C., Tsementzi, D., Kyrpides, N., Read, T., Konstantinidis, K.T., 2012. Direct comparisons of illumina vs. Roche 454 sequencing technologies on the same microbial community DNA sample. PLoS ONE 7, e30087.
Lutz, S., McCutcheon, J., McQuaid, J.B., Benning, L.G., 2018. The diversity of ice algal communities on the Greenland ice sheet as revealed by oligotyping. Microbial Genomics 4, 1–10.
Ma, Y., Jiao, N.Z., Zeng, Y.H., 2004. Natural community structure of cyanobacteria in the South China Sea as revealed by rpoC1 gene sequence analysis. Letters in Applied Microbiology 39, 353–358.
Mackey, K.R., Paytan, A., Caldeira, K., Grossman, A.R., Moran, D., Mcilvin, M., Saito, M.A., 2013. Effect of temperature on photosynthesis and growth in marine *Synechococcus* spp. Plant Physiology 163, 815–829.
Mahé, F., Rognes, T., Quince, C., de Vargas, C., Dunthorn, M., 2014. Swarm: robust and fast clustering method for amplicon-based studies. PeerJ 2, e593.
Majdi, N., et al., 2018. There's no harm in having too much: a comprehensive toolbox of methods in trophic ecology. Food Webs 17, e00100.
Malmstrom, R.R., Coe, A., Kettler, G.C., Martiny, A.C., Frias-Lopez, J., Zinser, E.R., Chisholm, S.W., 2010. Temporal dynamics of *Prochlorococcus* ecotypes in the Atlantic and Pacific Oceans. The ISME Journal 4, 1252–1264.
Maloy, A.P., Culloty, S.C., Slater, J.W., 2013. Dietary analysis of small planktonic consumers: a case study with marine bivalve larvae. Journal of Plankton Research 35, 866–876.
Malviya, S., Scalco, E., Audic, S., Vincent, F., Veluchamy, A., Poulain, J., Wincker, P., Iudicone, D., de Vargas, C., Bittner, L., Zingone, A., Bowler, C., 2016. Insights into global diatom distribution and diversity in the world's ocean. Proceedings of the National Academy of Sciences of the United States of America 113, E1516–E1525.
Man-Aharonovich, D., Philosof, A., Kirkup, B.C., Le Gall, F., Yogev, T., Berman-Frank, I., Polz, M.F., Vaulot, D., Béjà, O., 2010. Diversity of active marine picoeukaryotes in the eastern Mediterranean Sea unveiled using photosystem-II *psbA* transcripts. The ISME Journal 4, 1044–1052.
Mann, N.H., Cook, A., Millard, A., Bailey, S., Clokie, M., 2003. Marine ecosystems: bacterial photosynthesis genes in a virus. Nature 424, 741.
Martin, M., 2011. Cutadapt removes adapter sequences from high-throughput sequencing reads. EMBnet.journal 17, 10.
Martin, P., Allen, J.T., Cooper, M.J., Johns, D.G., Lampitt, R.S., Sanders, R., Teagle, D.A.H., 2010. Sedimentation of acantharian cysts in the Iceland basin: strontium as a ballast for deep ocean particle flux, and implications for acantharian reproductive strategies. Limnology and Oceanography 55, 604–614.
Martiny, A.C., Tai, A.P., Veneziano, D., Primeau, F., Chisholm, S.W., 2009. Taxonomic resolution, ecotypes and the biogeography of *Prochlorococcus*. Environmental Microbiology 11, 823–832.
Massana, R., et al., 2015. Marine protist diversity in European coastal waters and sediments as revealed by high-throughput sequencing. Environmental Microbiology 17, 4035–4049.
Matsen, F.A., Kodner, R.B., Armbrust, E.V., 2010. Pplacer: linear time maximum-likelihood and Bayesian phylogenetic placement of sequences onto a fixed reference tree. BMC Bioinformatics 11, 538.

Mazard, S., Ostrowski, M., Partensky, F., Scanlan, D.J., 2012. Multi-locus sequence analysis, taxonomic resolution and biogeography of marine *Synechococcus*. Environmental Microbiology 14, 372–386.

McDonnell, A.M., Boyd, P.W., Buesseler, K.O., 2015. Effects of sinking velocities and microbial respiration rates on the attenuation of particulate carbon fluxes through the mesopelagic zone. Global Biogeochemical Cycles 29, 175–193.

McMurdie, P.J., Holmes, S., 2013. Phyloseq: an r package for reproducible interactive analysis and graphics of microbiome census data. PLoS ONE 8, e61217.

Medlin, L., Elwood, H.J., Stickel, S., Sogin, M.L., 1988. The characterization of enzymatically amplified eukaryotic 16S-like rRNA-coding regions. Gene 71, 491–499.

Mella-Flores, D., Six, C., Ratin, M., Partensky, F., Boutte, C., Le Corguillé, G., Marie, D., Blot, N., Gourvil, P., Kolowrat, C., Garczarek, L., 2012. *Prochlorococcus* and *Synechococcus* have evolved different adaptive mechanisms to cope with light and Uv stress. Frontiers in Microbiology 3, 285.

Mestre, M., Ruiz-gonzález, C., Logares, R., Duarte, C.M., Gasol, J.M., 2018. Sinking particles promote vertical connectivity in the ocean microbiome. Proceedings of the National Academy of Sciences of the United States of America 115 (29), 6799–6807.

Metfies, K., Bauerfeind, E., Wolf, C., Sprong, P., Frickenhaus, S., Kaleschke, L., Nicolaus, A., Nöthig, E.-M., 2017. Protist communities in moored long-term sediment traps (Fram Strait, Arctic)–preservation with Mercury chloride allows for PCR-based molecular genetic analyses. Frontiers in Marine Science 4, 1–13.

Meyerhans, A., Vartanian, J.P., Wain-Hobson, S., 1990. DNA recombination during PCR. Nucleic Acids Research 18, 1687–1691.

Michaels, A.F., Caron, D.A., Swanberg, N.R., Howse, F.A., Michaels, C.M., 1995. Planktonic sarcodines (acantaria, radiolaria, foraminifera) in surface waters near Bermuda: abundance, biomass and vertical flux. Journal of Plankton Research 17, 131–163.

Michaels, A.F., Silver, M.W., 1988. Primary production, sinking fluxes and the microbial food web. Deep-Sea Research. Part A, Oceanographic Research Papers 35, 473–490.

Millard, A., Clokie, M.R., Shub, D.A., Mann, N.H., 2004. Genetic organization of the *psbAD* region in phages infecting marine *Synechococcus* strains. Proceedings of the National Academy of Sciences of the United States of America 101, 11007–11012.

Mitra, A., et al., 2016. Defining planktonic protist functional groups on mechanisms for energy and nutrient acquisition; incorporation of diverse mixotrophic strategies. Protist 167, 106–120.

Montresor, M., Lovejoy, C., Orsini, L., Procaccini, G., Roy, S., 2003. Bipolar distribution of the cyst-forming dinoflagellate polarella glacialis. Polar Biology 26, 186–194.

Moon-van der Staay, S.Y., De Wachter, R., Vaulot, D., 2001. Oceanic 18S rDNA sequences from picoplankton reveal unsuspected eukaryotic diversity. Nature 409, 607–610.

Moore, L.R., Rocap, G., Chisholm, S.W., 1998. Physiology and molecular phytogeny of coexisting *Prochlorococcus* ecotypes. Nature 393, 464–467.

Mordret, S., Piredda, R., Vaulot, D., Montresor, M., Kooistra, W.H.C.F., Sarno, D., 2018. Dinoref: a curated dinoflagellate (dinophyceae) reference database for the 18S rRNA gene. Molecular Ecology Resources 18, 974–987.

Muhling, M., Fuller, N.J., Millard, A., Somerfield, P.J., Marie, D., Wilson, W.H., Scanlan, D.J., Post, A.F., Joint, I., Mann, N.H., 2005. Genetic diversity of marine *Synechococcus* and co-occurring cyanophage communities: evidence for viral control of phytoplankton. Environmental Microbiology 7, 499–508.

Mühling, M., Fuller, N.J., Somerfield, P.J., Post, A.F., Wilson, W.H., Scanlan, D.J., Joint, I., Mann, N.H., 2006. High resolution genetic diversity studies of marine *Synechococcus* isolates using *rpoC1*-based restriction fragment length polymorphism. Aquatic Microbial Ecology 45, 263–275.

Nakamura, Y., Tuji, A., Makino, W., Matsuzaki, S.-i.S., Nagata, N., Nakagawa, M., Noriko, Takamura, 2020. Feeding ecology of a mysid species, neomysis awatschensis in the Lake Kasumigaura: combining approach with microscopy, stable isotope analysis and DNA metabarcoding. Plankton and Benthos Research 15, 44–54.

Needham, D.M., Fuhrman, J.A., 2016. Pronounced daily succession of phytoplankton, archaea and bacteria following a spring bloom. Nature Microbiology 1, 16005.

Needham, D.M., Sachdeva, R., Fuhrman, J.A., 2017. Ecological dynamics and co-occurrence among marine phytoplankton, bacteria and myoviruses shows microdiversity matters. The ISME Journal 11, 1614–1629.

Nelson, D.M., Tréguer, P., Brzezinski, M.A., Leynaert, A., Quéguiner, B., 1995. Production and dissolution of biogenic silica in the ocean: revised global estimates, comparison with regional data and relationship to biogenic sedimentation. Global Biogeochemical Cycles 9, 359–372.

Niemi, A., Michel, C., Hille, K., Poulin, M., 2011. Protist assemblages in winter sea ice: setting the stage for the spring ice algal bloom. Polar Biology 34, 1803–1817.

Nodder, S.D., 1997. Short-term sediment trap fluxes from Chatham Rise, Southwest Pacific Ocean. Limnology and Oceanography 42, 777–783.

Not, F., Gausling, R., Azam, F., Heidelberg, J.F., Worden, A.Z., 2007b. Vertical distribution of picoeukaryotic diversity in the Sargasso Sea. Environmental Microbiology 9, 1233–1252.

Not, F., Valentin, K., Romari, K., Lovejoy, C., Massana, R., Töbe, K., Vaulot, D., Medlin, L.K., 2007a. Picobiliphytes, a new marine picoplanktonic algal group with unknown affinities to other eukaryotes. Science 315, 252–254.

Oksanen, A.J., Blanchet, F.G., Friendly, M., Kindt, R., Legendre, P., Mcglinn, D., Minchin, P.R., Hara, R.B.O., Simpson, G.L., Solymos, P., Stevens, M.H.H., Szoecs, E., 2016. Package 'Vegan' (Version 2.4-0).

Olsen, G.J., Lane, D.J., Giovannoni, S.J., Pace, N.R., Stahl, D.A., 1986. Microbial ecology and evolution: a ribosomal RNA approach. Annual Review of Microbiology 40, 337–365.

Olsen, L.M., Laney, S.R., Duarte, P., Kauko, H.M., Fernández-méndez, M., Mundy, C.J., Rösel, A., Meyer, A., Itkin, P., Cohen, L., Peeken, I., Tatarek, A., Róźańska-Pluta, M., Wiktor, J., Taskjelle, T., Pavlov, A.K., Hudson, S.R., Granskog, M.A., Hop, H., Assmy, P., 2017. The seeding of ice algal blooms in Arctic pack ice: the multiyear ice seed repository hypothesis. Journal of Geophysical Research: Biogeosciences 122, 1529–1548.

Olson, R.J., Vaulot, D., Chisholm, S.W., 1985. Marine phytoplankton distributions measured using shipboard flow cytometry. Deep-Sea Research. Part A, Oceanographic Research Papers 32, 1273–1280.

O'Rorke, R., Lavery, S.D., Wang, M., Nodder, S.D., Jeffs, A.G., 2013. Determining the diet of larvae of the red rock lobster (*Jasus edwardsii*) using high-throughput DNA sequencing techniques. Marine Biology 161, 551–563.

Orr, J.C., et al., 2005. Anthropogenic ocean acidification over the twenty-first century and its impact on calcifying organisms. Nature 437, 681–686.

Paerl, R.W., Foster, R.A., Jenkins, B.D., Montoya, J.P., Zehr, J.P., 2008. Phylogenetic diversity of cyanobacterial *narB* genes from various marine habitats. Environmental Microbiology 10, 3377–3387.

Paerl, R.W., Johnson, K.S., Welsh, R.M., Worden, A.Z., Chavez, F.P., Zehr, J.P., 2011. Differential distributions of *Synechococcus* subgroups across the California current system. Frontiers in Microbiology 2, 1–22.

Paerl, R.W., Turk, K.A., Beinart, R.A., Chavez, F.P., Zehr, J.P., 2012. Seasonal change in the abundance of *Synechococcus* and multiple distinct phylotypes in Monterey Bay determined by *rbcL* and *narB* quantitative PCR. Environmental Microbiology 14, 580–593.

Palenik, B., 1994. Cyanobacterial community structure as seen from RNA polymerase gene sequence analysis. Applied and Environmental Microbiology 60, 3212–3219.

Parada, A.E., Needham, D.M., Fuhrman, J.A., 2016. Every base matters: assessing small subunit rRNA primers for marine microbiomes with mock communities. Time series and global field samples. Environmental Microbiology.

Partensky, F., Blanchot, J., Vaulot, D., 1999. Differential distribution and ecology of *Prochlorococcus* and *Synechococcus* in oceanic waters: a review. Marine Cyanobacteria. Monaco.

Pawłowska, J., Wollenburg, J.E., Zajaczkowski, M., Pawlowski, J., 2020. Planktonic foraminifera genomic variations reflect paleoceanographic changes in the Arctic: evidence from sedimentary ancient DNA. Scientific Reports 10, 15102.

Pawlowski, J., Esling, P., Lejzerowicz, F., Cedhagen, T., Wilding, T.A., 2014. Environmental monitoring through protist next-generation sequencing metabarcoding: assessing the impact of fish farming on benthic foraminifera communities. Molecular Ecology Resources 14, 1129–1140.

Pawlowski, J., Esling, P., Lejzerowicz, F., Cordier, T., Visco, J., Martins, C., Kvalvik, A., Staven, K., Cedhagen, T., 2016. Benthic monitoring of salmon farms in Norway using foraminiferal metabarcoding. Aquaculture Environment Interactions 8, 371–386.

Pedrós-Alió, C., Acinas, S.G., Logares, R., Massana, R., 2018. Marine microbial diversity as seen by high-throughput sequencing. Microbial Ecology of the Oceans, 47–97.

Penno, S., Lindell, D., Post, A.F., 2006. Diversity of *Synechococcus* and *Prochlorococcus* populations determined from DNA sequences of the N-regulatory gene *ntcA*. Environmental Microbiology.

Pérez-Cobas, A.E., Gomez-Valero, L., Buchrieser, C., 2020. Metagenomic approaches in microbial ecology: an update on whole-genome and marker gene sequencing analyses. Microbial Genomics, mgen000409.

Perovich, D.K., Richter-Menge, J.A., 2009. Loss of sea ice in the Arctic. Annual Review of Marine Science 1, 417–441.

Petchey, O.L., Gaston, K.J., 2006. Functional diversity: back to basics and looking forward. Ecology Letters 9, 741–758.

Peterson, M.L., Wakeham, S.G., Lee, C., Askea, M.A., Miquel, J.C., 2005. Novel techniques for collection of sinking particles in the ocean and determining their settling rates. Limnology and Oceanography: Methods 3, 520–532.

Piredda, R., Tomasino, M.P., D'Erchia, A.M., Manzari, C., Pesole, G., Montresor, M., Kooistra, W.H.C.F., Sarno, D., Zingone, A., 2017. Diversity and temporal patterns of planktonic protist assemblages at a Mediterranean long term ecological research site. FEMS Microbiology, Ecology 93, fiw200.

Pittera, J., Humily, F., Thorel, M., Grulois, D., Garczarek, L., Six, C., 2014. Connecting thermal physiology and latitudinal niche partitioning in marine *Synechococcus*. The ISME Journal.

Porazinska, D.L., Giblin-Davis, R.M., Sung, W., Thomas, W.K., 2012. The nature and frequency of chimeras in eukaryotic metagenetic samples. Journal of Nematology 44, 18–25.

Post, A.F., Penno, S., Zandbank, K., Paytan, A., Huse, S.M., Welch, D.M., 2011. Long term seasonal dynamics of *Synechococcus* population structure in the Gulf of Aqaba, Northern Red Sea. Frontiers in Microbiology 2, 131.

Preston, C.M., Durkin, C.A., Yamahara, K.M., 2019. DNA metabarcoding reveals organisms contributing to particulate matter flux to abyssal depths in the North East Pacific Ocean. Deep-Sea Research. Part 2. Topical Studies in Oceanography 173, 104708.

Preston, C.M., Durkin, C.A., Yamahara, K.M., 2020. DNA metabarcoding reveals organisms contributing to particulate matter flux to abyssal depths in the North East Pacific ocean. Deep-Sea Research. Part 2. Topical Studies in Oceanography 173, 104708.

Puigcorbé, V., Benitez-Nelson, C.R., Masqué, P., Verdeny, E., White, A.E., Popp, B.N., Prahl, F.G., Lam, P.J., 2015. Small phytoplankton drive high summertime carbon and nutrient export in the Gulf of California and eastern tropical North Pacific. Global Biogeochemical Cycles 29, 1309–1332.

Puigcorbé, V., Ruiz-González, C., Masqué, P., Gasol, J.M., 2020. Sampling device-dependence of prokaryotic community structure on marine particles: higher diversity recovered by *in situ* pumps than by oceanographic bottles. Frontiers in Microbiology. https://doi.org/10.3389/fmicb.2020.01645.

Quast, C., Pruesse, E., Yilmaz, P., Gerken, J., Schweer, T., Yarza, P., Peplies, J., Glöckner, F.O., 2013. The SILVA ribosomal RNA gene database project: improved data processing and web-based tools. Nucleic Acids Research 41, D590.

Ramond, P., Sourisseau, M., Simon, N., Romac, S., Schmitt, S., Rigaut-Jalabert, F., Henry, N., de Vargas, C., Siano, R., 2019. Coupling between taxonomic and functional diversity in protistan coastal communities. Environmental Microbiology.

Ray, J.L., Skaar, K.S., Simonelli, P., Larsen, A., Sazhin, A., Jakobsen, H.H., Nejstgaard, J.C., Troedsson, C., 2016. Molecular gut content analysis demonstrates that calanus grazing on *Phaeocystis pouchetii* and *Phaeocystis pouchetii* is sensitive to bloom phase but not prey density. Marine Ecology. Progress Series 542, 63–77.

Resplandy, L., Lévy, M., McGillicuddy Jr., D.J., 2019. Effects of eddy-driven subduction on ocean biological carbon pump. Global Biogeochemical Cycles 33, 1071–1084.

Richardson, T.L., 2019. Mechanisms and pathways of small-phytoplankton export from the surface ocean. Annual Review of Marine Science 11, 57–74.

Richardson, T.L., a Jackson, G., 2007. Small phytoplankton and carbon export from the surface ocean. Science 315, 838–840.

Rocap, G., Distel, D.L., Waterbury, J.B., Chisholm, S.W., 2002. Resolution of *Prochlorococcus* and *Synechococcus* ecotypes by using 16S-23S ribosomal DNA internal transcribed spacer sequences. Applied and Environmental Microbiology 68, 1180–1191.

Rodríguez-Martínez, R., Rocap, G., Salazar, G., Massana, R., 2013. Biogeography of the uncultured marine picoeukaryote MAST-4: temperature-driven distribution patterns. The ISME Journal 7, 1531–1543.

Rognes, T., Flouri, T., Nichols, B., Quince, C., Mahé, F., 2016. VSEARCH: a versatile open source tool for metagenomics. PeerJ 4, e2584.

Rozema, P.D., Venables, H.J., van de Poll, W.H., Clarke, A., Meredith, M.P., Buma, A.G.J., 2017. Interannual variability in phytoplankton biomass and species composition in northern Marguerite Bay (West Antarctic Peninsula) is governed by both winter sea ice cover and summer stratification. Limnology and Oceanography 62, 235–252.

Ruane, S., Austin, C.C., 2017. Phylogenomics using formalin-fixed and 100+ year-old intractable natural history specimens. Molecular Ecology Resources 17, 1003–1008.

Rusch, D.B., et al., 2007. The sorcerer II global ocean sampling expedition: northwest Atlantic through eastern tropical Pacific. PLoS Biology 5, e77.

Rynearson, T.A., Flickinger, S.A., Fontaine, D.N., 2020. Metabarcoding reveals temporal patterns of community composition and realized thermal niches of *Thalassiosira* spp. (Bacillariophyceae) from the narragansett bay long-term plankton time series. Biology 9, 19.

Rynearson, T.A., Richardson, K., Lampitt, R.S., Sieracki, M.E., Poulton, A.J., Lyngsgaard, M.M., Perry, M.J., 2013. Major contribution of diatom resting spores to vertical flux in the sub-polar North Atlantic. Deep-Sea Research. Part 1. Oceanographic Research Papers 82, 60–71.

Samanta, B., Bhadury, P., 2016. A comprehensive framework for functional diversity patterns of marine chromophytic phytoplankton using rbcL phylogeny. Scientific Reports 6, 20783.

Santoferrara, L., Burki, F., Filker, S., Logares, R., Dunthorn, M., McManus, G.B., 2020. Perspectives from ten years of protist studies by high-throughput metabarcoding. Journal of Eukaryotic Microbiology 67, 612–622.

Sarthou, G., Timmermans, K.R., Blain, S., Tréguer, P., 2005. Growth physiology and fate of diatoms in the ocean: a review. Journal of Sea Research 53, 25–42.
Sautter, L.R., Thunell, R.C., 1989. Seasonal succession of planktonic foraminifera; results from a four-year time-series sediment trap experiment in the northeast Pacific. Journal of Foraminiferal Research 19, 253–267.
Scanlan, D.J., Ostrowski, M., Mazard, S., Dufresne, A., Garczarek, L., Hess, W.R., Post, A.F., Hagemann, M., Paulsen, I., Partensky, F., 2009. Ecological genomics of marine picocyanobacteria. Microbiology and Molecular Biology Reviews 73, 249–299.
Scanlan, D.J., West, N.J., 2002. Molecular ecology of the marine cyanobacterial genera *Prochlorococcus* and *Synechococcus*. FEMS Microbiology, Ecology 40, 1–12.
Schander, C., Kenneth, H.M., 2003. DNA, PCR and formalinized animal tissue – a short review and protocols. Organisms Diversity & Evolution 3, 195–205.
Scharek, R., Tupas, L.M., Karl, D.M., 1999. Diatom fluxes to the deep sea in the oligotrophic North Pacific gyre at station ALOHA. Marine Ecology. Progress Series 182, 55–67.
Schiebel, R., 2002. Planktic foraminiferal sedimentation and the marine calcite budget. Global Biogeochemical Cycles 16, 3-1–3-21.
Schloss, P.D., Handelsman, J., 2005. Introducing DOTUR, a computer program for defining operational taxonomic units and estimating species richness. Applied and Environmental Microbiology 71, 1501–1506.
Schloss, P.D., Westcott, S.L., Ryabin, T., Hall, J.R., Hartmann, M., Hollister, E.B., Lesniewski, R.A., Oakley, B.B., Parks, D.H., Robinson, C.J., Sahl, J.W., Stres, B., Thallinger, G.G., Van Horn, D.J., Weber, C.F., 2009. Introducing mothur: open-source, platform-independent, community-supported software for describing and comparing microbial communities. Applied and Environmental Microbiology 75, 7537–7541.
Schneider, L., Anestis, K., Mansour, J., Anschütz, A., Gypens, N., Hansen, P., John, U., Klemm, K., Martin, J., Medic, N., Not, F., Stolte, W., 2020. A dataset on trophic modes of aquatic protists. Biodiversity Data Journal 8.
Schwarz, D., Spitzer, S.M., Thomas, A.C., Kohnert, C.M., Keates, T.R., Acevedo-Gutierrez, A., 2018. Large-scale molecular diet analysis in a generalist marine mammal reveals male preference for prey of conservation concern. Ecology and Evolution 8, 9889–9905.
Segawa, T., Matsuzaki, R., Takeuchi, N., Akiyoshi, A., Navarro, F., Sugiyama, S., Yonezawa, T., Mori, H., 2018. Bipolar dispersal of red-snow algae. Nature Communications 9.
Serreze, M.C., Holland, M.M., Stroeve, J., 2007. Perspectives on the Arctic's shrinking sea-ice cover. Science 315, 1533–1536.
Shadwick, E.H., Trull, T.W., Thomas, H., Gibson, J.A.E., 2013. Vulnerability of polar oceans to anthropogenic acidification: comparison of Arctic and Antarctic seasonal cycles. Scientific Reports 3, 2339.
Shi, X.L., Lepère, C., Scanlan, D.J., Vaulot, D., 2011. Plastid 16S rRNA gene diversity among eukaryotic picophytoplankton sorted by flow cytometry from the South Pacific Ocean. PLoS ONE 6, e18979.
Shiozaki, T., Itoh, F., Hirose, Y., Onodera, J., Kuwata, A., Harada, N., 2021. A DNA metabarcoding approach for recovering plankton communities from archived samples fixed in formalin. PLoS ONE 16, e0245936.
Sieburth, J.M., Smetacek, V., Lenz, J., 1978. Pelagic ecosystem structure: heterotrophic compartments of the plankton and their relationship to plankton size fractions 1. Limnology and Oceanography 23, 1256–1263.
Siegel, D.A., Buesseler, K.O., Doney, S.C., Sailley, S.F., Behrenfeld, M.J., Boyd, P.W., 2014. Global assessment of ocean carbon export by combining satellite observations and food-web models. Global Biogeochemical Cycles 28, 181–196.
Smetacek, V., et al., 2012. Deep carbon export from a Southern Ocean iron-fertilized diatom bloom. Nature 487, 313–319.

Sogin, M.L., Morrison, H.G., Huber, J.A., Welch, D.M., Huse, S.M., Neal, P.R., Arrieta, J.M., Herndl, G.J., 2006. Microbial diversity in the deep sea and the underexplored "rare biosphere". Proceedings of the National Academy of Sciences of the United States of America 103, 12115–12120.
Soto, D.F., Fuentes, R., Huovinen, P., Gómez, I., 2020. Microbial composition and photosynthesis in Antarctic snow algae communities: integrating metabarcoding and pulse amplitude modulation fluorometry. Algal Research 45, 101738.
Sow, L.S.S., Trull, T.W., Bodrossy, L., 2020. Oceanographic fronts shape *Phaeocystis* assemblages: a high-resolution 18S rRNA gene survey from the ice-edge to the equator of the South Pacific. Frontiers in Microbiology 11, 1847.
Stackebrandt, E., Goebel, B.M., 1994. Taxonomic note: a place for DNA-DNA reassociation and 16S rRNA sequence analysis in the present species definition in bacteriology. International Journal of Systematic and Evolutionary Microbiology 44, 846–849.
Steglich, C., Post, A.F., Hess, W.R., 2003. Analysis of natural populations of *Prochlorococcus* spp. in the northern Red Sea using phycoerythrin gene sequences. Environmental Microbiology 5, 681–690.
Steinberg, D.K., Landry, M.R., 2017. Zooplankton and the ocean carbon cycle. Annual Review of Marine Science 9, 413–444.
Stemmann, L., Boss, E., 2011. Plankton and particle size and packaging: from determining optical properties to driving the biological pump. Annual Review of Marine Science 4, 263–290.
Stoeck, T., Bass, D., Nebel, M., Christen, R., Jones, M.D.M., Breiner, H.W., Richards, T.A., 2010. Multiple marker parallel tag environmental DNA sequencing reveals a highly complex eukaryotic community in marine anoxic water. Molecular Ecology 19, 21–31.
Stoeck, T., Behnke, A., Christen, R., Amaral-Zettler, L., Rodriguez-Mora, M.J., Chistoserdov, A., Orsi, W., Edgcomb, V.P., 2009. Massively parallel tag sequencing reveals the complexity of anaerobic marine protistan communities. BMC Biology 7, 72.
Stoeck, T., Taylor, G.T., Epstein, S.S., 2003. Novel eukaryotes from the permanently anoxic cariaco basin (Caribbean Sea). Applied and Environmental Microbiology 69, 5656–5663.
Stone, J.P., Steinberg, D.K., 2016. Salp contributions to vertical carbon flux in the Sargasso Sea. Deep-Sea Research. Part 1. Oceanographic Research Papers 113, 90–100.
Stroeve, J.C., Markus, T., Boisvert, L., Miller, J., Barrett, A., 2014. Changes in Arctic melt season and implications for sea ice loss. Geophysical Research Letters 41, 1216–1225.
Stukel, M.R., Aluwihare, L.I., Barbeau, K.A., Chekalyuk, A.M., Goericke, R., Miller, A.J., Ohman, M.D., Ruacho, A., Song, H., Stephens, B.M., Landry, M.R., 2017. Mesoscale ocean fronts enhance carbon export due to gravitational sinking and subduction. Proceedings of the National Academy of Sciences of the United States of America 114, 1252–1257.
Stukel, M.R., Décima, M., Selph, K.E., Taniguchi, D.A.A., Landry, M.R., 2013. The role of *Synechococcus* in vertical flux in the Costa Rica upwelling dome. Progress in Oceanography 112-113, 49–59.
Stukel, M.R., Landry, M.R., Benitez-Nelson, C.R., Goericke, R., 2011. Trophic cycling and carbon export relationships in the California current ecosystem. Limnology and Oceanography 56, 1866–1878.
Sul, W.J., Oliver, T.A., Ducklow, H.W., Amaral-Zettler, L.A., Sogin, M.L., 2013. Marine bacteria exhibit a bipolar distribution. Proceedings of the National Academy of Sciences of the United States of America 110, 2342–2347.
Sunagawa, S., et al., 2015. Structure and function of the global ocean microbiome. Science.
Tai, V., Palenik, B., 2009. Temporal variation of *Synechococcus* clades at a coastal Pacific Ocean monitoring site. The ISME Journal.
Thompson, J.R., Hatfield, R.G., Batista, F.M., Bean, T.P., Fonseca, V.G., 2020. The application of nanopore sequencing technology to the study of dinoflagellates: a proof

of concept study for rapid sequence-based discrimination of potentially harmful algae. Frontiers in Microbiology 11, 1–16.

Thompson, R.M., Brose, U., Dunne, J.A., Hall, R.O., Hladyz, S., Kitching, R.L., Martinez, N.D., Rantala, H., Romanuk, T.N., Stouffer, D.B., Tylianakis, J.M., 2012. Food webs: reconciling the structure and function of biodiversity. Trends in Ecology & Evolution 27, 689–697.

Ting, C.S., Rocap, G., King, J., Chisholm, S.W., 2002. Cyanobacterial photosynthesis in the oceans: the origins and significance of divergent light-harvesting strategies. Trends in Microbiology 10, 134–142.

Toledo, G., Palenik, B., 1997. *Synechococcus* diversity in the California current as seen by RNA polymerase (*rpoC1*) gene sequences of isolated strains. Applied and Environmental Microbiology 63, 4298–4303.

Tragin, M., Vaulot, D., 2018. Green microalgae in marine coastal waters: the ocean sampling day (OSD) dataset. Scientific Reports 8, 14020.

Tragin, M., Vaulot, D., 2019. Novel diversity within marine Mamiellophyceae (Chlorophyta) unveiled by metabarcoding. Scientific Reports 9, 5190.

Trefault, N., De la Iglesia, R., Moreno-Pino, M., Lopes dos Santos, A., Gérikas Ribeiro, C., Parada-Pozo, G., Cristi, A., Marie, D., Vaulot, D., 2021. Annual phytoplankton dynamics in coastal waters from Fildes Bay, Western Antarctic Peninsula. Scientific Reports 11, 1368.

Tréguer, P., Bowler, C., Moriceau, B., Dutkiewicz, S., Gehlen, M., Aumont, O., Bittner, L., Dugdale, R., Finkel, Z., Iudicone, D., Jahn, O., Guidi, L., Lasbleiz, M., Leblanc, K., Levy, M., Pondaven, P., 2018. Influence of diatom diversity on the ocean biological carbon pump. Nature Geoscience 11, 27–37.

Turner, J., Colwell, S.R., Marshall, G.J., Lachlan-Cope, T.A., Carleton, A.M., Jones, P.D., Lagun, V., Reid, P.A., Iagovkina, S., 2005. Antarctic climate change during the last 50 years. International Journal of Climatology 25, 279–294.

Turner, J.T., 2015. Zooplankton fecal pellets, marine snow, phytodetritus and the ocean's biological pump. Progress in Oceanography 130, 205–248.

Twining, B.S., Rauschenberg, S., Morton, P.L., Ohnemus, D.C., Lam, P.J., 2015. Comparison of particulate trace element concentrations in the North Atlantic Ocean as determined with discrete bottle sampling and in situ pumping. Deep-Sea Research. Part 2. Topical Studies in Oceanography 116, 273–282.

Urbach, E., Robertson, D.L., Chisholm, S.W., 1992. Multiple evolutionary origins of prochlorophytes within the cyanobacterial radiation. Nature 355, 267–270.

Urbach, E., Scanlan, D.J., Distel, D.L., Waterbury, J.B., Chisholm, S.W., 1998. Rapid diversification of marine picophytoplankton with dissimilar light-harvesting structures inferred from sequences of *Prochlorococcus* and *Synechococcus* (cyanobacteria). Journal of Molecular Evolution 46, 188–201.

Van Leeuwe, M.A., Tedesco, L., Arrigo, K.R., Assmy, P., Campbell, K., Meiners, K.M., Rintala, J.-M., Selz, V., Thomas, D.N., Stefels, J., Deming, J.W., Leeuwe, M.A.V., Tedesco, L., Arrigo, K.R., Assmy, P., Meiners, K.M., Rintala, J.-M., Selz, V., David, N., Stefels, J., 2018. Microalgal community structure and primary production in Arctic and Antarctic sea ice: a synthesis. Elementa: Science of the Anthropocene 6, 1–25.

Vaulot, D., Eikrem, W., Viprey, M., Moreau, H., 2008. The diversity of small eukaryotic phytoplankton (<3 μm) in marine ecosystems. FEMS Microbiology Reviews 32, 795–820.

Vaulot, D., Geisen, S., Mahé, F., Bass, D., 2021. Pr2-Primers: an 18S rRNA Primer Database for Protists. Molecular Ecology Resources. https://doi.org/10.1111/1755-0998.13465. In press.

Venter, J.C., et al., 2004. Environmental genome shotgun sequencing of the Sargasso Sea. Science 304, 66–74.

Vernette, C., Henry, N., Lecubin, J., de Vargas, C., Hingamp, P., Lescot, M., 2021. The ocean barcode atlas: a web service to explore the biodiversity and biogeography of marine organisms. Molecular Ecology Resources 21 (4), 1347–1358.

Vestheim, H., Jarman, S.N., 2008. Blocking primers to enhance PCR amplification of rare sequences in mixed samples - a case study on prey DNA in Antarctic krill stomachs. Frontiers in Zoology 5, 12.

Vick-Majors, T.J., Priscu, J.C., Amaral-Zettler, L.A., 2014. Modular community structure suggests metabolic plasticity during the transition to polar night in ice-covered Antarctic lakes. The ISME Journal 8, 778–789.

Visco, J.A., Apothéloz-Perret-Gentil, L., Cordonier, A., Esling, P., Pillet, L., Pawlowski, J., 2015. Environmental monitoring: inferring the diatom index from next-generation sequencing data. Environmental Science & Technology 49, 7597–7605.

Volk, T., Hoffert, M.I., 1985. Ocean carbon pumps: analysis of relative strengths and efficiencies in ocean-driven atmospheric CO_2 changes. In: The Carbon Cycle and Atmospheric CO_2: Natural Variations Archean to Present. In: Geophysical Monograph Series, vol. 32, pp. 99–110.

Waite, A., Bienfang, P.K., Harrison, P.J., 1992. Spring bloom sedimentation in a subarctic ecosystem. Marine Biology 114, 131–138.

Waite, A.M., Safi, K.A., Hall, J.A., Nodder, S.D., 2000. Mass sedimentation of picoplankton embedded in organic aggregates. Limnology and Oceanography.

Wang, G.C.Y., Wang, Y., 1997. Frequency of formation of chimeric molecules is a consequence of PCR coamplification of 16S rRNA genes from mixed bacterial genomes. Applied and Environmental Microbiology 63, 4645–4650.

Wang, Q., Garrity, G.M., Tiedje, J.M., Cole, J.R., 2007. Naive Bayesian classifier for rapid assignment of rRNA sequences into the new bacterial taxonomy. Applied and Environmental Microbiology 73, 5261–5267.

Waterbury, J., Rippka, R., 1989. The order Chroococcales. Bergey's Manual of Systematic Bacteriology 3, 1728–1746.

Waterbury, J.B., Willey, J.M., Franks, D.G., Valois, F.W., Watson, S.W., 1985. A cyanobacterium capable of swimming motility. Science 230, 74–76.

West, N.J., Scanlan, D.J., 1999. Niche-partitioning of *Prochlorococcus* populations in a stratified water column in the eastern North Atlantic Ocean? Applied and Environmental Microbiology.

Whitaker, M.R.L., Baker, C.C.M., Salzman, S.M., Martins, D.J., Pierce, N.E., 2019. Combining stable isotope analysis with DNA metabarcoding improves inferences of trophic ecology. PLoS ONE 14, e0219070.

Wilkinson, M.D., et al., 2016. The FAIR guiding principles for scientific data management and stewardship. Scientific Data 3, 1–9.

Woese, C.R., 2009. A new biology for the 21st century. Frontiers in Ecology and the Environment 7, 455.

Woese, C.R., Fox, G.E., 1977. Phylogenetic structure of the prokaryotic domain: the primary kingdoms. Proceedings of the National Academy of Sciences of the United States of America 74, 5088–5090.

Wolf, C., Kilias, E., Metfies, K., 2015. Protists in the polar regions: comparing occurrence in the Arctic and Southern Oceans using pyrosequencing. Polar Research 34, 23225.

Worden, A.Z., Follows, M.J., Giovannoni, S.J., Wilken, S., Zimmerman, A.E., Keeling, P.J., 2015. Rethinking the marine carbon cycle: factoring in the multifarious lifestyles of microbes. Science 347, 1257594.

Wright Erik, S., 2016. Using DECIPHER v2.0 to analyze big biological sequence data in r. The R Journal 8, 352.

Xu, D., Kong, H., Yang, E.-j., Li, X., Jiao, N., Warren, A., Wang, Y., Lee, Y., Jung, J., Kang, S.-h., 2020. Contrasting community composition of active microbial eukaryotes in melt ponds and sea water of the Arctic Ocean revealed by high throughput sequencing. Frontiers in Microbiology 11, 1–15.

Yabuki, A., Toyofuku, T., Takishita, K., 2014. Lateral transfer of eukaryotic ribosomal RNA genes: an emerging concern for molecular ecology of microbial eukaryotes. The ISME Journal 8, 1544–1547.

Yau, S., Lopes dos Santos, A., Eikrem, W., Gérikas Ribeiro, C., Gourvil, P., Balzano, S., Escande, M.-L., Moreau, H., Vaulot, D., 2020. *Mantoniella beaufortii* and *Mantoniella baffinensis* sp. nov. (Mamiellales, Mamiellophyceae), two new green algal species from the high Arctic. Journal of Phycology 56, 37–51.

Yilmaz, P., et al., 2011. Minimum information about a marker gene sequence (MIMARKS) and minimum information about any (x) sequence (MIxS) specifications. Nature Biotechnology 29, 415–420.

Zafeiropoulos, H., Viet, H.Q., Vasileiadou, K., Potirakis, A., Arvanitidis, C., Topalis, P., Pavloudi, C., Pafilis, E., 2020. PEMA: a flexible pipeline for environmental DNA metabarcoding analysis of the 16S/18S ribosomal RNA, ITS, and COI marker genes. GigaScience 9, 1–12.

Zamora-Terol, S., Novotny, A., Winder, M., 2020. Reconstructing marine plankton food web interactions using DNA metabarcoding. Molecular Ecology 29, 3380–3395.

Zeidner, G., Preston, C.M., Delong, E.F., Massana, R., Post, A.F., Scanlan, D.J., Beja, O., 2003. Molecular diversity among marine picophytoplankton as revealed by *psbA* analyses. Environmental Microbiology 5, 212–216.

Zeldis, J., James, M.R., Grieve, J., Richards, L., 2002. Omnivory by copepods in the New Zealand subtropical frontal zone. Journal of Plankton Research 24, 9–23.

Zhu, F., Massana, R., Not, F., Marie, D., Vaulot, D., 2005. Mapping of picoeucaryotes in marine ecosystems with quantitative PCR of the 18S rRNA gene. FEMS Microbiology, Ecology 52, 79–92.

Zimmermann, J., Glöckner, G., Jahn, R., Enke, N., Gemeinholzer, B., 2015. Metabarcoding vs. morphological identification to assess diatom diversity in environmental studies. Molecular Ecology Resources 15, 526–542.

Zuckerkandl, E., Pauling, L., 1965. Molecules as documents of evolutionary history. Journal of Theoretical Biology 8, 357–366.

Zwirglmaier, K., Heywood, J.L., Chamberlain, K., Woodward, E.M., Zubkov, M.V., Scanlan, D.J., 2007. Basin-scale distribution patterns of picocyanobacterial lineages in the Atlantic Ocean. Environmental Microbiology 9, 1278–1290.

Zwirglmaier, K., Jardillier, L., Ostrowski, M., Mazard, S., Garczareck, L., Vaulot, D., Not, F., Massana, R., Ulloa, O., Scanlan, D.J., 2008. Global phylogeography of marine *Synechococcus* and *Prochlorococcus* reveals a distinct partitioning of lineages amongst oceanic biomes. Environmental Microbiology 10, 147–161.

CHAPTER 12

Comparative genomics for understanding intraspecific diversity: a case study of the cyanobacterium *Raphidiopsis raciborskii*

Anusuya Willis[a], Jason N. Woodhouse[b], Brett A. Neilan[c], and Michele A. Burford[d]

[a]Australian National Algae Culture Collection, CSIRO National Collections and Marine Infrastructure, Hobart, TAS, Australia
[b]Department of Experimental Limnology, Leibniz-Institute of Freshwater Ecology and Inland Fisheries, Berlin, Germany
[c]School of Environmental and Life Sciences, The University of Newcastle, Newcastle, NSW, Australia
[d]Australian Rivers Institute, Griffith University, Brisbane, QLD, Australia

Glossary

Average nucleotide identity (ANI) A measure of nucleotide-level genomic similarity between the coding regions of two genomes.

Core genome The genes present in all strains of a species.

CRISPR Clustered Regularly Interspaced Short Palindromic Repeats, which are the hallmark of a bacterial defence system that forms the basis for CRISPR-Cas9 genome editing technology.

Ecotype A distinct organism form occupying a particular habitat.

Genome strategy Refers to the evolutionary changes of a genome, either expanding or reducing.

Genotype An individual's collection of genes.

Homologous recombination (HR) Genetic recombination in which nucleotide sequences are exchanged between two similar or identical molecules of DNA.

Horizontal gene transfer (HGT) The acquisition by an organism of genetic information by transfer (e.g., via plasmid or virus).

Internal transcribed spacer (ITS) The spacer DNA situated between the small-subunit ribosomal RNA (rRNA) and large-subunit rRNA genes in the chromosome or the corresponding transcribed region in the polycistronic rRNA precursor transcript.

Interspecific distance (ISD) Distance between two species, e.g., genetic distance determined by average nucleotide identity.

Intraspecific Diversity Differences between strains of the same species.

Pangenome the entire gene set of all strains of a species, including the genes in the core and variable genomes.

Strain A low-level taxonomic rank used at the intraspecific level.
Variable shell genome The genes present in some but not all strains of a species.

Introduction

Phytoplankton, comprised of photosynthetic microscopic organisms, are a mix of many thousands of species. There are five main groups of phytoplankton: stramenopiles (e.g., diatoms); alveolates (e.g., dinoflagellates); chlorophytes (green algae); haptophytes (e.g., coccolithophores); and cyanobacteria (i.e., blue-green algae). Cyanobacteria, despite being commonly termed "blue-green algae" are bacteria, with asexual reproduction, circular chromosomes, and without membrane-bound organelles. Characterization and identification of phytoplankton species is important for understanding ecosystem health, food-web or trophic interactions, detecting invasive or harmful species, and a myriad of other application. Traditional characterization of algae has been based on morphology (light microscopy, scanning electron microscopy (SEM) and transmission electron microscopy (TEM)), physiology via lab-based experiments of isolated cultures, and ecology using field observations or mesocosm experiments.

The rise of genome sequencing technologies has opened a new area of research into phytoplankton, both for individual species and populations, including new methods to characterize and define species. Full genome sequencing of eukaryotic algae species is still limited (~194 sequenced species; NCBI, 2021), as many algae genomes, particularly dinoflagellates (LaJeunesse et al., 2005) are large (e.g., up to 185 Gbp in *Lingulodinium polyedrum*) and complicated by high rates of gene duplication, lateral gene transfer, endosymbiontic and horizontal gene transfer events, repeated regions, and duplication (Casabianca et al., 2017; Wisecaver and Hackett, 2011). The power in fully characterized genomes can be demonstrated by the diatoms *Thalasiossira pseudonana* (Armbrust et al., 2004) and *Phaeodactylum tricornutum* (Bowler et al., 2008), which paved the way for protist genome studies. The current 100 Diatom Genomes project (JGI, 2021) is an exciting advance in this field.

Genome sequencing and annotation give us insights into the ecology, physiology, and evolutionary history of an organism or group of organisms. The presence or absence of specific genetic elements can dictate the niche opportunities for phytoplankton, for example, the ability to grow under low light (Kashtan et al., 2014) or high temperature (Alcorta et al., 2020), as well as shaping the nature of interactions between phytoplankton and associated microbial populations. Genomes can also reflect the history of an organism, such that interactions with viruses are imprinted on the genome in the form of CRISPR/Cas arrays or horizontal gene transfer (HGT) events, evident through the abundance and position of genomic islands, as examples. With multiple full genome sequences available, we can move beyond understanding genome content, to comparing genomes and describing genome aspects that lead to an organisms' unique physiologies, niche adaptation, rates of diversification, and speciation. This is comparative genomics.

Comparative genomics

Comparative genomics is broadly defined but typically involves the comparison of multiple genomic regions derived from different organisms, either different species or different individuals, that is, strains/cultures of the same species. Though a complete genome is not strictly necessary for comparative genomics, as a subset of core genes (discussed below) can be sufficient, and missing genes can be inferred from probabilistic models; a complete genome can greatly improve confidence and allow for greater insights. Ultimately, the aim of comparative genomics is to describe evolutionary histories between groups of organisms and to identify genomic regions and essential features that drive differences in physiology and environmental niche adaptation.

Core and pangenome

An essential part of comparative genomics is describing the stable core genome, variable shell genome, and pangenome (Shi and Falkowski, 2007). The stable core comprises genes that are highly conserved between the organisms studied. These genes generally encode the major metabolic functions and ribosomal apparatus. The variable shell genome contains the genes that vary in presence/absence or copy number; many of these genes may have been involved in HGT or homologous recombination (Avrani and Lindell, 2015; Pantůček et al., 2018). The pangenome encompasses all the genes in the core and variable shell genomes of each individual, thus it increases as the number of genomes in the comparison increases, and congruently, the core genome decreases, until the size of each becomes stable, and it can be reasonably assumed that the pangenome contains all possible genes for the species complex (Fig. 12.1).

Genetic drift between groups of organisms is best reflected via an analysis of those stable core genes. Due to their role in core functions, such as

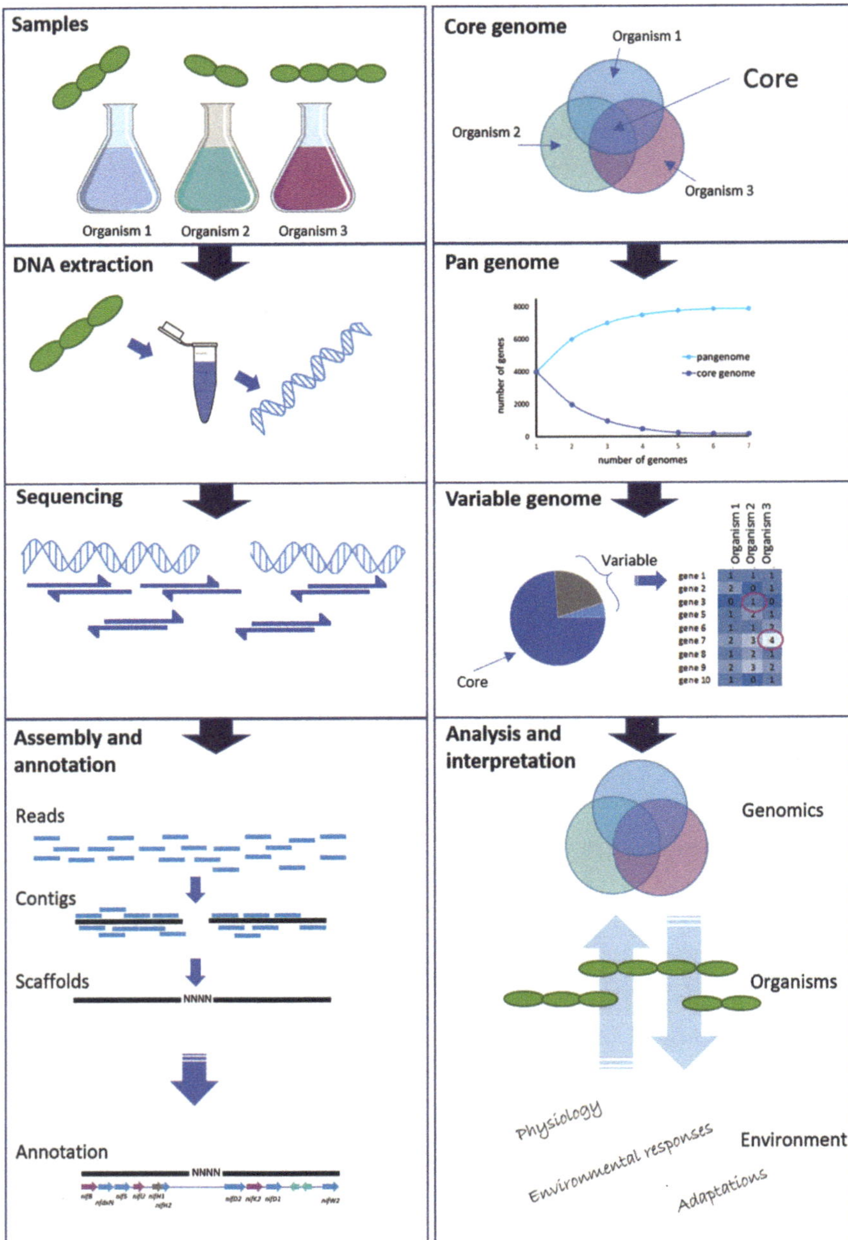

Figure 12.1 Workflow of comparative genomics, from monoclonal cultures of unique individuals, DNA extraction, sequencing, genome assembly and annotation, to typical analyses of comparative genomics: core genome, pangenome, variable, genome, and analysis and interpretation of results.

DNA replication and protein translation, mutations within these genes does not occur unrestricted: loss of cellular viability is not an option, but in non-critical regions, variation can be quite expansive. For this reason, mutations in these protein-encoding regions are analogous to those within ribosomal operons and can be used to infer evolutionary relationships. Average nucleotide identity (ANI) or average amino acid identity (AAI) between these core proteins can be a reliable predictor for identifying boundaries between species and genus complexes (Richter and Rosselló-Móra, 2009). At higher taxonomic levels, however, these indices become highly variable and fail to reflect the evolutionary histories that brought about new families, orders, and classes of organisms. Ultimately robust phylogenetic analyses of these core proteins are necessary for reconstructing evolutionary histories, but for large taxonomic groups can be restricted by a lack of computational resources.

Variable shell genome and unique genes

The variable shell genome is made up of regions of the genome shared among fewer than all the organisms in the comparative analysis or are unique to a single organism. These genes can confer distinctive morphological orphysiological features to an organism that may bestow unique advantages or shift the organisms' environmental niche (Shi and Falkowski, 2007). Even the acquisition of a single gene can yield adaptations to a specific strain (Rocap et al., 2003). Comparative genomics often focuses on these shell or variable genomes to describe exceptional qualities of an organism.

Horizontal gene transfer is often invoked as an explanation for the occurrence of genes present in the variable shell genome. For many genes, this has been clearly demonstrated, in particular operational and nonribosomal information genes (Shi and Falkowski, 2007). However, many genes within the variable shell genome can arise, due to gene duplication or recombination within the organism (Kashtan et al., 2014). Duplications are common amongst genes responsible for core metabolic functions, such as photosynthesis, with these paralogous genes often shown to be responsible for conferring some additional niche opportunity or resistance (Larsson et al., 2011) For complete or near-complete genomes, additional tools are available to identify exogenous chromosomal elements (plasmids), as well as genomic islands, that provide clear indications as to whether genes associated with these elements have been acquired due to HGT.

Genome evolution

As single-celled organisms live and divide, sexually or asexually, they are exposed to mutagens (such as oxygen radicals and UV), undergo homologous recombination, HGT events, and virus attacks, which can all alter their DNA, through single nucleotide polymorphisms (SNPs) and/or gene loss and gene gain. Over evolutionary timescales these processes lead to an accumulation of changes in the genome, leading to genome evolution (Lenski, 2017). Genome evolution leads to the acquisition of unique character traits and, eventually, species divergence (Tomitani et al., 2006; Cohan and Perry, 2007).

The process of genome evolution in prokaryotes is regulated by two strategies: the expanding genome "generalist" and the reducing genome "specialist" (Sriswasdi et al., 2017). Generalist genomes are typically large with high levels of homologous recombination. This leads to populations where strains have very similar genomes, and gene acquisition and sharing is common. Specialist genomes are small; gene loss is more common and high intraspecific diversity exists within populations. In general, it seems that generalist genome evolution gives rise to new species that have long-term persistence, whereas specialists arise more frequently but only have short-term ecotype success (Sriswasdi et al., 2017).

These two genome strategies have been described by two hypotheses: The Red Queen and The Black Queen hypotheses (Van Valen, 1973; Morris et al., 2012; Strotz et al., 2018). The Red Queen hypothesis suggests that under sustained competition, random mutation provides an advantage, arising primarily through homologous recombination, which thus becomes the driving evolutionary force (Van Valen, 1973; Strotz et al., 2018). Conversely, the Black Queen hypothesis prioritizes gene loss and gain, where subsets of the population specialize. This is particularly relevant in relation to common good biomolecules that may be produced in abundance, allowing subsets of the population to arise that do not produce it. These individuals benefit through a lower metabolic cost, but are dependent on the individuals that produce the common good (Morris et al., 2012).

Horizontal gene transfer

Horizontal gene transfer (HGT) is the transfer of genetic material between contemporary organisms. HGT is a common event in prokaryotes (Dagan and Martin, 2007) and, although less likely to occur in complex eukaryotes, was an important aspect of their early evolution and continues to occur in single-celled eukaryotes, such as microalgae. There are three main routes of HGT in prokaryotes: transformation (direct uptake of a foreign DNA fragment); transduction (insertion of foreign genes, mediated by a virus); and conjugation (transfer of genetic material via direct contact between two cells). Bacterial plasmids are required for conjugal HGT; this can occur between cells of the same species, different species, and between bacteria and yeast, and fungi and plants. Virus interactions, which lead to HGT via transduction, are usually host-specific and occur less extensively.

HGT events are typically detected using a phylogenetic analysis of unrelated genomes. Discontinuity in the phylogeny of a gene compared to other genes in the genome suggests an HGT event.

Intraspecific diversity

Intraspecific diversity in a population arises from differences between strains, either cooccurring within a population or between populations. This diversity arises through localized adaptations and genetic changes, but has not yet resulted in genetic isolation or speciation (Pietrasiak et al., 2019). Intraspecific diversity facilitates rapid adaptation to short-term changes in the environment and is an advantage for populations by increasing their robustness (Muraille, 2018). The collective response that arises from intraspecific diversity allows populations to respond to environmental changes through variable strain dominance and persist at high densities through disturbances (Godhe and Rynearson, 2017). The variable strain dominance increases the speed that populations can adapt to new environments (Sandrini et al., 2016).

Cyanobacteria

Cyanobacteria are an ancient monophyletic lineage of oxygenic photosynthetic prokaryotes. They are the most morphologically, phylogenetically, and ecologically diverse phylum of bacteria and exist in a broad range of habitats: from the poles to the equator, in extreme environments, from marine, freshwater and terrestrial types (Mishra et al., 2019; Whitton and Potts, 2000). Cyanobacteria also form a major part of phytoplankton assemblages and play major roles in biogeochemical carbon, nitrogen, and phosphorus cycles (Liu et al., 1997; Capone et al., 1997; Karl et al., 1997; Zehr, 2011).

Cyanobacteria have relatively small haploid circular genomes, thus making them easy to sequence and analyze; to date there are ~1864 sequenced cyanobacterial genomes (NCBI, 2021). The comparative genomics of cyanobacteria has greatly expanded our knowledge of functional pathways and adaptive physiologies in genomes in the photic zone.

Cyanobacterial species exist as a population of strains. Though very closely related and morphologically indistinguishable, these strains can be differentiated by their genetic makeup, physiological responses to environmental stimuli, or biochemical composition. This intraspecific diversity (strain diversity) is a distinct advantage for species, allowing rapid adaption to environmental changes via shifts in the proportion of strains within a population, and thus increasing the robustness of the species at the population and ecosystem levels (Muraille, 2018).

Comparative genomics studies regarding cyanobacteria are numerous. A large proportion of these studies define species complexes, whereas others explore the ecological implications of species diversity. Examples include definitions of the *Anabaena/Dolichospermum/Aphanizomenon* (ADA) cluster, along with ecological investigations of the *Microcystis, Prochlorococcus*, and *Raphidiopsis* genera.

Comparative genomics has been used to understand the relationship of the species within three Nostocales genera: *Anabaena, Dolichospermum*, and *Aphanizomenon*. These closely related species continue to be difficult to identify, as high morphological plasticity and cryptic diversity have all contributed to a history of taxonomic confusion. Recent comparative genomics has shown that many species are polyphyletic and described these genera as a single cluster, named "ADA," and suggested they form a single genus (Driscoll et al., 2018; Dreher et al., 2021). Members of this clade also produce diverse secondary metabolites, such as the toxins microcystin, anatoxin, and saxitoxin, which occur in monophyletic subgroups (Osterholm et al., 2020). These studies all suggest that a revision of the taxonomy of the ADA cluster is required.

Another highly studied organism is *Prochlorococcus*; it is one of the most numerous microorganisms in tropical and subtropical oceans. Comparative genomics has revealed a basis to multiple cooccurring ecotypes, with each subpopulation possessing a distinct "genomic backbone" or core genes and a set of flexible genes (Kashtan et al., 2014). These subpopulations coexist and may represent stable niche partitioning that have diverged millions of years ago. Two main ecotypes, based on low-light and high-light adaption, have also been described (Rocap et al., 2003; Delmont and Eren, 2018; Kashtan et al., 2014). The dominance of these two ecotypes change seasonally, demonstrating population dynamics through intraspecific diversity (Kashtan et al., 2014).

Microcystis is a globally occurring harmful freshwater cyanobacterium, with 51 taxonomically accepted species. Comparative genomics has been used to define multiple species that cooccur (Pérez-Carrascal et al., 2019; Willis and Woodhouse, 2020) and multiple strains within this species complex. As a freshwater species *Microcystis* occurs in geographically isolated lakes, reservoirs, and rivers; however, despite this, isolates based on genetic similarity, have been found in multiple locations globally (Willis and Woodhouse, 2020); multiple species can also cooccur. This strategy of cosmopolitan distribution and continued migration is an example of a species employing the Red Queen strategy, with an expanding genome, but speciation defined at the boundary of homologous recombination has resulted in cooccurring species. However, debate continues on the most relevant species boundary within *Microcystis* (Dick et al., 2021).

Case Study: Understanding intraspecific diversity in Raphidiopsis raciborskii through comparative genomics

Raphidiopsis raciborskii

Raphidiopsis raciborskii (basionym *Cylindrospermopsis raciborskii*) is a harmful cyanobacterium of the order Nostocales, occurring globally in tropical to subtropical freshwater ecosystems (for a review see Burford et al., 2016). It is filamentous and diazotrophic (nitrogen-fixing), with many strains producing the alkaloid cyanotoxin cylindrospermopsin (CYN), which is harmful to humans and animals (Falconer et al., 1999). *R. raciborskii* is considered an invasive species as its habitat range has been increasing in recent decades (Sinha et al., 2012). Summer blooms of *R. raciborskii* cause water quality problems, including a decrease in biodiversity, increase in water turbidity, and water toxicity (reducing recreational access to water bodies and drinking water resources). Due to the negative impacts of this species, it has been the subject of extensive research, and it is now considered a model species for cyanobacteria harmful algae blooms (cyanoHAB) research.

The major cause for concern of *R. raciborskii* blooms is toxin production, and how environmental drivers affect the toxin yield of a bloom. The existence of toxic and nontoxic strains, and their cooccurrence has long been recognized (Saker et al., 1999; Chonudomkul et al., 2004; Piccini et al., 2011). More recently, the production of CYN has been shown to be constitutive (Willis et al., 2015; Pierangelini et al., 2015; Orr et al., 2018) over the short term, with CYN cell quotas (qCYN) linked to cell division. However, CYN cell quotas (qCYN) have been observed to shift over longer time periods of years

(Willis et al., 2020; Xiao et al., 2020). Short-term constitutive production means that qCYN is not affected by environmental factors, with toxin production unresponsive to environmental biotic or abiotic factors. Additionally, qCYN varies between strains, with up to three-fold variation in cooccurring strains (Willis et al., 2016). qCYN can range between 0 to 55.8 fg $cell^{-1}$, leading to blooms with up to 180 mg L^{-1} CYN (e.g. Saker et al., 1999; Orr et al., 2010). The variation in qCYN between strains means that the toxin yield, and thus risk, of a bloom is not correlated to cell number. This makes bloom risk difficult to predict and can only be assessed via direct toxin quantification of the water (e.g. Bittencourt-Oliveira et al., 2012; Fathalli et al., 2011; Burford et al., 2014).

R. raciborskii has high intraspecific diversity; thus the ecophysiological characteristics, including toxin yield, of populations have temporal and spatial variability. This intraspecific diversity and strain variability facilitate rapid adaptation to new environments, allowing *R. raciborskii* to be both a persistent and invasive species. Understanding strain variation and population structure of strains throughout a bloom is needed to better understand the interaction between strains and their environment, leading to changes in strain dominance and changes in toxin yield. Comparative genomics is currently being used to determine how strains differ at the genomic level and linking this to their response to environmental factors.

Raphidiopsis raciborskii full genome sequence

The first full genome sequence of *R. raciborskii* was strain CS-505 (Stucken et al., 2010), describing a genome size of 3.89 Mbp (±0.2 mbp) and a plasmid of 30 Kbp, the cylindrospermopsin gene cluster, a reduced *nifHDK* nitrogen-fixing gene cluster, and pathways for the major metabolic functions. Understanding the genome content of *R. raciborskii* allowed elucidation of physiological functions and a greater understanding of how *R. raciborskii* differs from other cyanobacteria. The genome is the second smallest of the Nostocales, with the closely related *R. brookii* (the smallest at 3.2 Mbp) and is twice as small as its next closest relative, *Anabaena* (8.6 Mbp). The *R. raciborskii* genome is closer in size to the single-celled *Synechococcus* genomes (3.57 Mbp) than any other filamentous cyanobacterium. However, the GC content of ~40% is similar to other filamentous cyanobacteria.

The *R. raciborskii* genome contains a reduced nitrogen-fixing gene cluster, with only 38 genes, rather than the full 49 genes usually associated with diazotrophy, and some rearrangement of the *nifH* nitrogenase operons. This reduction of the nitrogen-fixation gene cluster is evidence of a reducing genome and suggests *R. raciborskii* may lose the ability to fix nitrogen, as is the case for *R. brookii.*

Subsequent genome sequencing of *R. raciborskii* strains have led to the current 27 publicly available genomes (NCBI, 2021), with 18 of these isolates from Australia and eight from South America. There have been four intraspecific comparative genomic studies using *R. raciborskii*; investigating toxic versus nontoxic strains (Sinha et al., 2014), intraspecific diversity of cooccurring strains (Willis et al., 2018), phylogenomics (Abreu et al., 2018), and phylogeography (Vico et al., 2020).

Populations and intraspecific diversity

In a species with high intraspecific diversity, understanding the extent of the diversity is required to understand the behavior of populations in response to environmental variables and to lead to methods to develop predictive models. An understanding of intraspecific diversity also guides strain selection for laboratory studies (Baxter et al., 2020).

The extent to which changes in the proportion of strains occur in a population has been illustrated by a mesocosm experiment. Five days after the addition of phosphorus, the proportion of toxic strains in the population had increased, the toxin yield also increased, but there was no increase in population size compared to the addition of nitrogen and phosphorus (Burford et al., 2014). These population shifts are a result of the maximum growth rate of different strains occurring under different environmental conditions, and thus environmental change led to improved conditions for a portion of the strains in the population. Studies show the maximum growth rate of different strains occur at slightly different points along environmental gradients, such as nutrient concentration (Willis et al., 2015) or light and temperature (Xiao et al., 2017a).

Intraspecific diversity: comparative genomics

Undertaking comparative genomics of intraspecific diversity requires multiple isolates from a population. Until recently, a bloom population was considered to be a dominant strain, and thus a single isolate was cultured from a sample. However, the isolation of 24 *Raphidiopsis raciborskii* isolates from a single 250 mL water sample found that each isolate varied from the other in qCYN, growth rate, and morphology (straight or coiled, trichome length, cell size) (Willis et al., 2016). Studies with more disparate strains have also revealed differences in growth response to nutrients (Willis et al., 2015, 2019a; Amaral et al., 2014; Aubriot and Bonilla, 2018), CO_2 (Willis et al., 2017, 2019b), and light and temperature (Xiao et al., 2017a).

A genome comparison of strains from disparate water bodies showed substantial variation, with three strains exhibiting a core genome of 82% and ~3% strain-specific genes (Sinha et al., 2014). The strain unique genes were largely involved in abiotic stress responses. The comparison of Australian, Brazilian, and Singapore strains (including *R. brookii*) was ~62% (2031 genes), whereas the pangenome was 4716 genes, which is 1.77-fold higher than the average genome. This large genome variability and level of strain-unique genes is significant when considering the small size and reducing strategy of the *R. raciborskii* genome (Stucken et al., 2010; Larsson et al., 2011).

In-depth comparative genomics of nine cooccurring strains illustrated the extent of intraspecific diversity within a population (Willis et al., 2018). In this study, nine of the 24 isolates from a single sample (Willis et al., 2016), were sequenced. These isolates could not be distinguished based on HIP_1 fingerprinting or 16S-ITS sequences, and hence are all considered identical species. Despite this, their phenotypes differed in growth rate, toxin content, cell size, and morphology (straight or coiled). The average nucleotide identity (ANI) between strains was 0.9976 (ranging from 0.9950–0.9987, between pairs of genomes). Genome sizes differed between the straight and coiled morphotypes, with the genomes of the three coiled strains larger with 3.9 Mbp and 3658 ± 11 genes compared to 3.8 Mbp and 3514 ± 20 genes in the straight genomes. The average genome size was 3.8–3.9 Mbp with 3562 ± 74 genes encoding for 3462 ± 34 proteins. See Fig. 12.2.

The core genome comprised ~83% of orthologous groups (OGs), which is similar to the previously reported 82–84% for three geographically distinct strains (Sinha et al., 2014). The core genome of the nine cooccurring strains was slightly larger at 2796 (OGs), compared to the 2767 OGs in the three more ecologically disparate strains. Their pangenome contained 4171 OGs with 874 variable and 433 unique

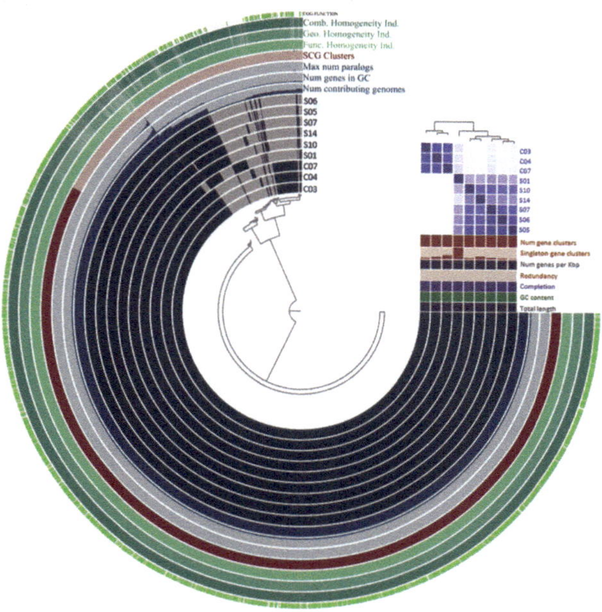

Figure 12.2 Genomic diversity of nine cooccurring *R. raciborskii* strains isolated from Lake Wivenhoe, Australia. Image built with Anv'io.

OGs; the variable genome was 16% of the genome. Approximately 1% of the genome was unique genes. The similarity in the size of the variable genome between the cooccurring strains and the geographically isolated strains indicates this is a feature of *R. raciborskii* genome strategy, abiding by the "Black Queen hypothesis" and the concept of a reducing specialist genome (Willis and Woodhouse, 2020). Genome reduction typically infers niche specialization; however, the large variable genome of *R. raciborskii* and high intraspecific diversity, suggests a broad range of environmental niches, facilitating flexible adaptation amongst closely related strains.

The study of intraspecific diversity also revealed an interesting set of variable genes, which provide insights into the phenotypic plasticity and adaptation of strains in this genus. In the case of the nine cooccurring genomes, these differences were mostly involved in amino sugar and carbohydrate biosynthesis, as well as DNA modification.

Coiled and straight strains formed two distinct groups, perhaps subpopulations. The coiled morphotype genomes contained 209 genes that were exclusive to these genomes, compared to only 105 that were unique to the straight genomes. Differences in carbohydrate metabolism were present in the coiled morphotypes, including an 8 Kbp gene cluster resembling an O-antigen B-band biosynthetic locus from *Pseudomonas aeruginosa* PAO1 (serotype 05) (Larkin and Imperiali, 2009; Willis et al., 2018). This biosynthesis locus is critical for the formation of diacetylated-aminuronic acids, which are localized within complex cell walls of several rod-shaped unicellular pathogenic bacteria. Based on the absence of this gene cluster from the genomes of the straight strains, it is possible that this additional gene cluster contributes to the coiled morphology of the trichomes. The presence or absence of the O-antigen B-band has been demonstrated to influence cell surface hydrophobicity, and hence the virulence of *P. aeruginosa* (Makin and Beveridge, 1996). Given this and the aquatic habitat of *R. raciborskii*, it is foreseeable that the inclusion of such a sugar within the outer cell membrane may alter the hydrophobicity of the filament in a manner that promotes coiling.

The variability in DNA modification suggests the potential for differences in gene expression, and thus transcriptional differences between strains, creating even greater phenotypic diversity of strains. Overall, the variable and unique genes of the *R. raciborskii* genomes are involved in mechanisms that sense and react to abiotic factors, or phenotypic traits potentially pertaining to the adaptability of *R. raciborskii* to different environments promoting habitat expansion.

Strain phenotypic variation can also arise from transcriptional regulation and post-translational regulation. This is particularly relevant to the regulation of core genes, such as phosphorus regulation, that has been shown to differ in gene expression between strains under the same conditions (Willis et al., 2019a). Genomic variation can describe the differences between strains, whereas differences in the transcription of certain pathways can define transient differences in response to environmental drivers. Understanding the extent of genome conservation between cooccurring and ecologically distinct strains, will help understand the range of diversity within and/or between populations.

The high level of genome variability in the *R. raciborskii* species complex suggests on-going genome adaptation. Comparative genomics has shown that this occurs principally via HGT and phage interactions in *R. raciborskii*. The main differences observed between strains was gene presence/absence, with SNPs not playing a major role.

Phage interactions also appear to differ in frequency between straight and coiled morphotypes, with straight trichome strains having more exposure to phage, and thus potentially greater strain variability. The three coiled genomes in the nine-genome study were less variable than the straight genomes (Willis et al., 2018).

Linking genotypes to ecological strategies

The comparative genomic study of nine cooccurring *R. raciborskii* strains highlighted the extent of genome heterogeneity. The large and variable genome of closely related cooccurring strains suggests a species with high plasticity, composed of many strains arising from frequent gene gain and loss events. This suggests that *R. raciborskii* has an adaptive genotype. Linking genotypes to ecological strategies is the next step in using comparative genomic studies for ecological applications, such as predictive modeling. Genome comparison has highlighted differences in carbon, light, and phosphorus utilization strategies as potentially forming strain defining phenotypes.

Differences in the growth response of *R. raciborskii* strains from within a waterbody to a range of temperature and light conditions has also been demonstrated (Xiao et al., 2017a). Additionally, individual strains vary in the scale of their response to changes in light and temperature, highlighting the ability to adapt to variable growth conditions. Using this information, strain dominance has been predicted using a Monte Carlo approach (Xiao et al., 2017b).

Two strains of *R. raciborskii* were shown to vary in their cellular carbon to phosphorus ratio, and capacity to withstand phosphorus starvation for longer (Willis et al., 2017). Building on this work, Xiao et al. (2020) showed that strains of *R. raciborskii* isolated from within a waterbody traded cellular phosphorus storage for optimizing growth rates, such that within a population faster growing strains had a reduced capacity to store phosphorus, whereas slower growing strains contained more phosphorus. Variation in gene expression for a range of phosphorus utilization strategies has also been identified at the strain level (Willis et al., 2019a). However, there has not yet been a comprehensive study that has combined comparative genomics with physiological characterization to conclusively link genotypes with ecotypes.

Conclusion

Comparative genomics has changed our understanding of species diversity, both interspecific and intraspecific. It has allowed links to be made between genome content, physiology, and environmental niches. Genetic characterization from full genome sequences provides extensive information, whereas individual and comparative genomics highlight important differences between them. Genome strategies and mechanisms of genetic change, such as SNPs or HGT, give insights into a microorganism's evolutionary pathway and environmental adaptation. Understanding intraspecific diversity can elucidate the physiological and ecological diversity within a species complex and between them, proposing pathways to strategies of prediction of population dynamics.

Continued advances in omic and metaomic techniques will allow ever greater analyses to take place, with increased numbers of samples at higher resolutions and smaller temporal and spatial scales. These omic approaches will allow us to continue to unravel the complex interactions between an organisms' genome and their environment.

References

Abreu, V.A.C., Popin, R.V., Alvarenga, D.O., Schaker, P.D.C., Hoff-Risseti, C., Varani, A.M., Fiore, M.F., 2018. Genomic and genotypic characterization of Cylindrospermopsis raciborskii: toward an intraspecific phylogenetic evaluation by comparative genomics. Front. Microbiol. 9, 306.

Alcorta, J., Alarcón-Schumacher, T., Salgado, O., Díez, B., 2020. Taxonomic novelty and distinctinve genomic features of hot spring cyanobacteria. Front. Genet. 11, 568223.

Amaral, V., Bonilla, S., Aubriot, L., 2014. Growth optimisation of the invasive cyanobacterium Cylindrospermopsis raciborskii in response to phosphate fluctuations. Eur. J. Phycol. 49, 134–141.

Armbrust, E.V., Berges, J.A., Bowler, C., Green, B.R., Martinez, D., Putnam, N.H., Zhou, S., Allen, A.E., Apt, K.E., Bechner, M., Brzezinski, M.A., Chaal, B.K., Chiovitti, A., Davis, A.K., Demarest, M.S., Detter, J.C., Glavina, T., Goodstein, D., Hadi, M.Z., Hellsten, U., Hildebrand, M., Jenkins, B.D., Jurka, J., Kapitonov, V.V., Kröger, N., Lau, W.W.Y., Lane, T.D., Larimer, F.W., Lippmeier, J.C., Lucas, S., Medina, M., Monstant, A., Obronik, M., Parker, M.S., Palenik, B., Pazour, G.J., Richardson, P.M., Rynearson, T.A., Saito, M.A., Sxhwartz, D.C., Thamatrakoln, K., Valentin, K., Vardi, A., Wilkerson, F.P., Rokhsar, D.S., 2004. The genome of the diatom *Thalassiosira Pseudonana*: ecology, evolution, and metabolism. Science 306, 79–86.

Aubriot, L., Bonilla, S., 2018. Regulation of phosphate uptake reveals cyanobacterial bloom resilience to shifting N:P ratios. Freshw. Biol. https://doi.org/10.1111/fwb.13066.

Avrani, S., Lindell, D., 2015. Convergent evolution toward an improved growth rate and a reduced resistance range in *Prochlorococcus* strains resistant to phage. Proc. Natl. Acad. Sci. USA, E2191–E2200.

Baxter, K., Jameson, I.D., Willis, A., 2020. Towards defining global ecotypes of the toxic cyanobacterium *Raphidiopsis raciborskii*. Appl. Phycol. https://doi.org/10.1080/26388081.2020.1740891.

Bittencourt-Oliveira, M.C., Buch, B., Hereman, T.C., Arruda-Neto, J.D.T., Moura, A.N., Zocchi, S.S., 2012. Effects of light intensity and temperature on Cylindrospermopsis raciborskii (Cyanobacteria) with straight and coiled trichomes: growth rate and morphology. Braz. J. Biol. 72 (2), 343–351.

Bowler, C., Allen, A.E., Badger, J.H., Grimwood, J., Jabbari, K., Kuo, A., Maheswari, U., Martens, C., Maumus, F., Otillar, R.P., Rayko, E., Salamov, A., Vandepoele, K., Beszteri, B., Gruber, A., Heijde, M., Katinka, M., Mock, T., Valentin, K., Verret, F., Berges, J.A., Brownlee, C., Cadoret, J.-P., Chiovitti, A., Choi, C.J., Coesel, S., De Martino, A., Detter, J.C., Durkin, C., Falciatore, A., Fournet, J., Haruta, M., Huysman, M.J.J., Jenkins, B.D., Jiroutova, K., Jorgensen, R.E., Joubert, Y., Kaplan, A., Kröger, N., Kroth, P.G., La Roche, J., Lindquist, E., Lommer, M., Martin-Jézéquel, V., Lopez, P.J., Lucas, S., Mangogna, M., McGinnis, K., Medlin, L.K., Montsant, A., Oudot-Le Secq, M.-P., Napoli, C., Obornik, M., Parker, M.S., Petit, J.-L., Porcel, B.M., Poulsen, N., Robison, M., Rychlewski, L., Rynearson, T.A., Schmutz, J., Shapiro, H., Siaut, M., Stanley, M., Sussman, M.R., Taylor, A.R., Vardi, A., von Dassow, P., Vyverman, W., Willis, A., Wyrwicz, L.S., Rokhsar, D.S., Weissenbach, J., Armbrust, E.V., Green, B.R, Van de Peer, Y., Grigoriev, I.V., 2008. The *Phaeodactylum* genome reveals the evolutionary history of diatom genomes. Nature 456, 239–244.

Burford, M., Davis, T., Orr, P., Rati, S., Willis, A., Neilan, B., 2014. Nutrient-related changes in ecotype dominance of the cyanobacterium *Cylindrospermopsis raciborskii*. FEMS Microbiol. Ecol. 89 (1), 135–148.

Burford, M., Beardall, J., Willis, A., Orr, P., Magalhaes, V., Rangel, L., Azevedo, S., Neilan, B., 2016. Understanding the winning strategies used by the toxic cyanobacterium *Cylindrospermopsis raciborskii*. Harmful Algae 54, 44–53.

Capone, D.G., Zehr, J.P., Paerl, H.W., Bergman, B., Carpenter, E.J., 1997. *Trichodesmium*, a globally significant marine cyanobacterium. Science 276, 1221–1229.

Casabianca, S., Cornetti, L., Capellacci, S., Vernais, C., Penna, A., 2017. Genome complexity of harmful algae. Harmful Algae 63, 7–12.

Chonudomkul, D., Yongmanitchai, W., Theeragool, G., Kawachi, M., Kasai, F., Kaya, K., Watanabe, M.M., 2004. Morphology, genetic diversity, temperature tolerance and toxicity of *Cylindrospermopsis raciborskii* (Nostocales, Cyanobacteria) strains from Thailand and Japan. FEMS Microbiol. Ecol. 48 (3), 345–355.

Cohan, F.M., Perry, E.B., 2007. A systematics for discovering the fundamental units of bacterial diversity. Curr. Biol. 17, R373–R386.

Dagan, T., Martin, W., 2007. Ancestral genome sizes specify the minimum rate of lateral gene transfer during prokaryote evolution. Proc. Natl. Acad. Sci. USA 104, 870–875.

Delmont, T.O., Eren, A.M., 2018. Linking pangenomes and metagenomes: the *Prochlorococcus* metapangenome. PeerJ 6, e4320. https://doi.org/10.7717/peerj.4320.

Dick, G.J., Duhaime, M.D., Evans, J.T., Errara, R.M., Godwin, C.M., Kharbush, J.J., Nitschky, H.S., Powers, M.A., Vanderploeg, H.A., Schmidt, K.C., Smith, D.J., Yancey, C.E., Zwiers, C.C., Denef, V.J., 2021. The genetic and ecophysiological diversity of *Microcystis*. Environ. Microbiol. https://doi.org/10.1111/1462-2920.15615.

Dreher, T.W., Davis, E.W., Mueller, R.S., 2021. Complete genomes derived by directly sequencing freshwater bloom populations emphasize the significance of the genus level ADA clade within the Nostocales. Harmful Algae 103, 102005.

Driscoll, C.B., Meyer, K.A., Šulčius, S., Brown, N.M., Dick, G.J., Cao, H., Gasiūnas, G., Timinskas, A., Yin, Y., Landry, Z.C., Otten, T.G., Davis, T.W., Watson, S.B., Derher, T.W., 2018. A closely-related clade of globally distributed bloom-forming cyanobacteria within the Nostocales. Harmful Algae 77, 93–107.

Falconer, I.R., Hardy, S.J., Humpage, A.R., Froscio, S.M., Tozer, G.J., Hawkins, P.R., 1999. Hepatic and renal toxicity of the blue-green alga (Cyanobacterium) *Cylindrospermopsis raciborskii* in male swiss albino mice. Environ. Toxicol. 14, 143–150.

Fathalli, A., Jenhani, A.B.R., Moreira, C., Welker, M., Romdhane, M., Antunes, A., Vasconcelos, V., 2011. Molecular and phylogenetic characterization of potentially toxic cyanobacteria in Tunisian freshwaters. Syst. Appl. Microbiol. 34, 303–310.

Godhe, A., Rynearson, T., 2017. The role of intraspecific variation in the ecological and evolutionary success of diatoms in changing environments. Philos. Trans. R. Soc. B 372, 20160399.

Karl, D., Letelier, R., Tupas, L., Dore, J., Christian, J., Hebel, D., 1997. The role of nitrogen fixation in biogeochemical cycling in the subtropical North Pacific Ocean. Nature 388, 533–538.

Kashtan, N., Roggensack, S.E., Rodrigue, S., Thompson, J.W., Biller, S.J., Coe, A., Ding, H., Marttinen, P., Malmstrom, R.R., Stocker, R., Follows, M.J., Stepanauskas, R., Chisholm, S.W., 2014. Single-cell genomics reveals hundreds of coexisting subpopulations in wild *Prochlorococcus*. Science 344, 416–420.

LaJeunesse, T.C., Lambert, G., Anderson, R.A., Coffroth, M.A., Galbraith, D.W., 2005. *Symbiodinium* (Pyrrhophyta) genome size (DNA content) are smallest among dinoflagellates. J. Phycol. 41, 880–886.

Larkin, A., Imperiali, B., 2009. Biosynthesis of UDP-GlcNAc.A by WbpB, WdpE, and WbpD: enzymes in the Wbp pathway responsible for O-antigen assembly in *Pseudomonas aeruginosa* PAO1. Biochemistry 48, 5446–5455.

Larsson, J., Nylander, J.A.A., Bergman, B., 2011. Genome fluctuations in cyanobacteria reflect evolutionary, developmental and adaptive traits. BMC Evol. Biol. 11, 187.

Lenski, R.E., 2017. Experimental evolution and the dynamics of adaptation and genome evolution in microbial populations. ISME J. 11, 2181–2194.

Liu, H., Nolla, H.A., Campbell, L., 1997. *Prochlorococcus* growth rate and contribution to primary production in the equatorial and subtropical North Pacific Ocean. Aquat. Microb. Ecol. 12, 39–47.

Makin, S.A., Beveridge, T.J., 1996. The influence of A-band and B-band lipopolysaccharide on the surface characteristics and adhesion of *Pseudomonas aeruginosa* to surfaces. Microbiology 142, 229–307.

Mishra, A.K., Tiwari, D.N., Rai, A.N., 2019. Cyanobacteria: From Basic Science to Applications. Academic Press, London, UK.

Morris, J.J., Lenski, R.E., Zinser, E.R., 2012. The Black Queen hypothesis: evolution of dependencies through adaptive gene loss. mBio 3, e00036.

Muraille, E., 2018. Diversity generator mechanisms are essential components of biological systems: the two queen hypothesis. Front. Microbiol. 9, 223.

NCBI, National Center for Biotechnology Information, 2021. https://ncbi.nlm.nih.gov.

Orr, P.T., Rasmussen, J.P., Burford, M.A., Eaglesham, G.K., Lennox, A.M., 2010. Evaluation of quantitative real-time PCR to characterise spatial and temporal variations in cyanobacteria, *Cylindrospermopsis raciborskii* (Woloszynska) Seenaya et Subba Raju and cylindrospermopsin concentrations in three subtropical Australian reservoirs. Harmful Algae 9, 243–254.

Orr, P.T., Willis, A., Burford, M.A., 2018. Application of first order rate kinetics to explain changes in bloom-toxicity – the importance of understanding cell toxin quotas. J. Ocean. Limnol. 36, 1063–1074.

Osterholm, J., Popin, R.V., Fewer, D.P., Sivonen, K., 2020. Phylogenomic analysis of secondary metabolism in the toxic cyanobacterial genera *Anabaena*, *Dolichospermum* and *Aphanizomenon*. Toxins 12, 248.

Pantůček, R., Sedláček, I., Indráková, A., Vrbovská, V., Mašlaňová, I., Kovařovic, V., Švec, P., Králová, S., Krištofová, L., Kekláková, J., Petráš, P., Doškař, J., 2018. *Staphylococcus edaphicus* sp. nov., Isolated in Antartica, harbors the mecC gene and genomic islands with a suspected role in adaptation to extreme environments. Appl. Environ. Microbiol. 84, e01746.

Pérez-Carrascal, O.M., Terrat, Y., Giani, A., Fortin, N., Greer, C.W., Tromas, N., Shapiro, B.J., 2019. Coherence of Microcystis species revealed through population genomics. ISME J. 13, 2887–2900.

Piccini, C., Aubriot, L., Fabre, A., Amaral, V., Gonzalez-Piana, M., Giani, A., Figueredo, C.C., Vidal, L., Kruk, C., Bonilla, S., 2011. Genetic and eco-physiological differences of South American *Cylindrospermopsis raciborskii* isolates support the hypothesis of multiple ecotypes. Harmful Algae 10, 644–653.

Pierangelini, M., Sinha, R., Willis, A., Burford, M.A., Orr, P.T., Beardall, J., Neilan, B., 2015. Constitutive cylindrospermopsin pool size in *Cylindrospermopsis raciborskii* under different light and CO2 partial pressure conditions. Appl. Environ. Microbiol. 81 (9), 3069–3076.

Pietrasiak, N., Osorio-Santos, K., Shalygin, S., Martin, M.P., Johansen, J.R., 2019. When is a lineage a species? A case study in *Myxacorys* gen. nov. (Synechococcales: Cyanobacteria) with the description of two new species from the Americas. J. Phycol. 55, 976–996.

Richter, M., Rosselló-Móra, R., 2009. Shifting the genomic gold standard for the prokaryotic species definition. Proc. Natl. Acad. Sci. 106, 19126–19131.

Rocap, G., Larimer, F.W., Lamerdin, J., Malfatti, S., Chain, P., Ahlgren, N.A., Arellano, A., Coleman, M., Hauser, L., Hess, W.R., Johnson, Z.I., Land, M., Lindell, D., Post, A.F., Regala, W., Shah, M., Shaw, S.L., Steglich, C., Sullivan, M.B., Ting, C.S., Tolonen, A., Webb, E.A., Zinser, E.R., Chisholm, S.W., 2003. Genome divergence in two *Prochlorococcus* ecotypes reflects oceanic niche differentiation. Nature 424, 1042–1047.

Saker, M.L., Neilan, B.A., Griffiths, D.J., 1999. Two morphological forms of *Cylindrospermopsis raciborskii* (Cyanobacteria) isolated from Soloman Dam, Palm Island, Queensland. J. Phycol. 35, 599–606.

Sandrini, G., Ji, X., Verspagen, J.M.H., Tann, R.P., Slot, P.C., Luimstra, V.M., Schuurmans, J.M., Matthijs, H.C.P., Huisman, J., 2016. Rapid adaptation of harmful cyanobacteria to rising CO2. Proc. Natl. Acad. Sci. 113, 9315–9320.

Shi, T., Falkowski, P.G., 2007. Genome evolution in cyanobacteria: the stable core and the variable shell. Proc. Natl. Acad. Sci. USA 105, 7.

Sinha, R., Pearson, L.A., Davis, T.W., Burford, M.A., Orr, P.T., Neilan, B.A., 2012. Increased incidence of *Cylindrospermopsis raciborskii* in temperate zones – is climate change responsible? Water Res. 46, 1408–1419.

Sinha, R., Pearson, L.A., Davis, T.W., Muenchhoff, J., Pratma, R., Jex, A., Burford, M.A., Neilan, B.A., 2014. Comparative genomics of *Cylindrospermopsis raciborskii* strains with differential toxicities. BMC Genomics 15, 83.

Sriswasdi, S., Yang, C-C., Iwasaki, W., 2017. Generalist species drive microbial dispersion and evolution. Nat. Commun. 8, 1162.

Strotz, L.C., Simões, M., Girard, M.G., Breitkreuz, L., Kimmig, J., Lieberman, B.S., 2018. Getting somewhere with the Red Queen: chasing a biologically modern definition of the hypothesis. Biol. Lett. 14, 20170734.

Stucken, K., John, U., Cembella, A., Murillo, A.A., Soto-Liebe, K., Fuentes-Vladés, J.J., Friedel, M., Plominsky, A.M., Vásquez, M., Glöckner, G., 2010. The smallest known genomes of multicelluarl and toxic cyanobacteria: comparison, minimal gene sets for linked traits and the evolutionary implications. PLoS ONE 5, e9235.

Tomitani, A., Knoll, A.H., Cavanaugh, C.M., Ohno, T., 2006. The evolutionary diversification of cyanobacteria: molecular-phylogenetic and paleontological perspectives. Proc. Natl. Acad. Sci. 103, 5442–5447.

Van Valen, L., 1973. A new evolutionary law. Evol. Theory 1, 1–30.

Vico, P., Bonilla, S., Cremella, B., Aubriot, L., Iriarte, A., Piccini, C., 2020. Biogeography of the cyanobacterium *Raphidiopsis* (*Cylindrospermopsis*) *raciborskii*: integrating genomics, phylogenetic and toxicity data. Mol. Phylogenet. Evol. 148, 106824.

Whitton, B.A., Potts, M., 2000. The Ecology of Cyanobacteria. Kluwer, Dordrecht, the Netherlands.

Willis, A., Bent, S.J., Jameson, I.D., 2020. Morphological changes and genome evoltuion in Raphidiopsis raciborskii CS-506 after 23 years in living culture. Appl. Phycol. https://doi.org/10.1080/26388081.2020.1753573.
Willis, A., Woodhouse, N.J., 2020. Defining cyanobacterial species: description and diversity through genomics. Crit. Rev. Plant Sci. 39, 101–124.
Willis, A., Adams, M.P., Chuang, A., Orr, P.T., O'Brien, K.R., Burford, M., 2015. Constitutive toxin production under various nitrogen and phosphorus regimes of three ecotypes of *Cylindrospermopsis raciborskii* ((Woloszynska) Seenayya et Subba Raju). Harmful Algae 47, 27–34.
Willis, A., Chuang, A.W., Woodhouse, J., Neilan, B., Burford, M.A., 2016. Intraspecific variation in growth, morphology and toxin quotas for the cyanobacterium *Cylindrospermopsis raciborskii*. Toxicon 119, 307–310.
Willis, A., Posselt, A.J., Burford, M.A., 2017. Variations in carbon-to-phosphate ratio of two Australian strains of *Cylindrospermopsis raciborskii*. Eur. J. Phycol. 52, 303–310.
Willis, A., Woodhouse, J.N., Ongley, S.E., Jex, A.R., Burford, M.A., Neilan, B.A., 2018. Genome variation in nine co-occurring toxic *Cylindrospermopsis raciborskii* strains. Harmful Algae 73, 157–166.
Willis, A., Chuang, A.W., Dyhrman, S., Burford, M.A., 2019a. Differential expression of phosphorus acquisition genes in response to phosphorus stress in two *Raphidiopsis raciborskii* strains. Harmful Algae 82, 19–25.
Willis, A., Chuang, A.W., Orr, P.T., Beardall, J., Burford, M.A., 2019b. Subtropical freshwater phytoplankton show a greater response to increased temperature than to increased pCO2. Harmful Algae 90, 101705.
Wisecaver, J.H., Hackett, J.D., 2011. Dinoflagellate genome evolution. Annu. Rev. Microbiol. 65, 369–387.
Xiao, M., Adams, M.P., Willis, A., Burford, M.A., O'Brien, K.R., 2017a. Variation within and between cyanobacterial species and strains affects competition: implications for phytoplankton modelling. Harmful Algae 69, 38–47.
Xiao, M., Willis, A., Burford, M.A., 2017b. Differences in cyanobacterial strain responses to light and temperature reflect species plasticity. Harmful Algae 62, 84–93.
Xiao, M., Hamilton, D.P., Chuang, A., Burford, M.A., 2020. Intra-population strain variation in phosphorus storage strategies of the freshwater cyanobacterium *Raphidiopsis raciborskii*. FEMS Microbiol. Ecol. 96, fiaa092.
Zehr, J.P., 2011. Nitrogen fixation by marine cyanobacteria. Trends Microbiol. 19, 162–173.

CHAPTER 13

Transcriptomic and metatranscriptomic approaches in phytoplankton: insights and advances

Bethany C. Kolody[a], Matthew J. Harke[b], Sharon E. Hook[c], and Andrew E. Allen[d,e]
[a]Innovative Genomics Institute, University of California Berkeley, Berkeley, CA, United States
[b]Gloucester Marine Genomics Institute, Gloucester, MA, United States
[c]CSIRO Oceans and Atmosphere, Hobart, TAS, Australia
[d]Scripps Institution of Oceanography, University of California, San Diego, CA, United States
[e]Microbial and Environmental Genomics Group, J. Craig Venter Institute, La Jolla, CA, United States

Introduction

As molecular approaches in biology have expanded in recent decades (Morozova et al., 2009), the genome entered the common vernacular as the molecular plan from which phenotype and physiology arise. Yet, cells with identical genetic code can have vastly different shapes and functions. For example, the DNA (Craig Venter et al., 2001) underlying the morphology of a human neuron also underlies the morphology of an adipose cell. Furthermore, single-celled organisms are capable of remodeling their morphology and physiology in biogeochemically and ecologically significant ways without altering their genome. Consider, for example, bacterial sporulation (Piggot and Hilbert, 2004) or a mixotrophic protist shifting from photosynthetic to heterotrophic metabolism (Stoecker et al., 2017).

To probe the molecular and physiological underpinnings of these phenomena, the genome, or parts list, is inadequate. Techniques for assessing gene expression (e.g., northern blotting and RT-qPCR) evolved as a means of assessing how the genomic molecular plan can ultimately be translated into a protein, providing an understanding of how genotype gives rise to phenotype (Crick, 1970). Most early explorations into gene expression focused on just a few genes, but in the same way genomics began to gain momentum in the late 1990s, so too did the field of gene expression. Just as genomics measured the comprehensive gene repertoire of an organism, "transcriptomics" was coined (Piétu et al., 1999) as the inventory of messenger RNA (mRNA), providing a "global" view of gene expression. In 1991, 600 expressed sequence tags (ESTs) were generated using mRNA from the human brain, adding to the repertoire of known genes (Adams et al., 1991). By 1995 the first sequencing-based transcriptomic method, serial analysis of gene expression (SAGE) was debuted in a landmark proof-of-concept, profiling gene expression patterns of the pancreas (Velculescu et al., 1995). Transcriptomics rapidly became less painstaking and more automated. In 2006 the first transcriptome studies using next-generation sequencing technology appeared (Morozova et al., 2009), and has since been applied to study diverse organisms (McGettigan, 2013), including microalgae (Lin, 2021).

Microalgae are responsible for more than 45% of global primary production (Field et al., 1998), but their microscopic stature, vast diversity, and difficulty to culture have inhibited their study. Many species that contribute importantly to ecosystem health and could offer useful or life-saving natural products have yet to be identified (Caron et al., 2017; Sibbald and Archibald, 2017). Genomics has expanded our understanding of algal relationships, as well as metabolic potential and physiology, but is limited by scarce cultured representatives and inability to distinguish active genes in a given condition (Lin, 2021). Transcriptomics is particularly well-suited to the study of microalgae, many of which have large, unwieldy genomes (Caron et al., 2017; Sibbald and Archibald, 2017) (Fig. 13.1.6). Although prokaryotic algae have manageable genome sizes (for example, *Prochlorococcus marinus* has a genome or 1.75 Mb; Dufresne et al., 2003), even diatoms with atypically small genomes (Connolly et al., 2008; Osuna-Cruz et al., 2020), such as *Thalassiosira pseudonana* and *Phaeodactylum tricornutum* are on the order of 30 Mb (Armbrust et al., 2004; Bowler et al., 2010, 2008), and dinoflagellate genomes can range in size from 1.5 Gb to over 100 Gb (Hennon and Dyhrman, 2020; Murray et al., 2016; Shoguchi et al., 2013) and are permanently condensed across the cell cycle (Akbar et al., 2018). This often poses a problem for studying the main culprits of marine harmful algal blooms (HABs) (McLean, 2013). Sequencing transcriptomes from such organisms across varied conditions is a complementary alternative that allows a glimpse at much of their metabolic potential, with the bonus of potentially inferring the conditions in which genes are active. Additionally, in eukaryotic algae, targeting sequences upstream of genes (Weis et al., 2013) allow us to infer the subcellular localization of organellar transcripts (Smith, 2013).

Transcriptomics has fostered insights into phytoplankton diversity, ecology, and physiology that could not be captured in any other way. An advantage of transcriptomics over more targeted approaches is it captures, in a relatively unbiased manner, the full transcriptional response of the cell to environmental stimuli, providing new information on gene responses, which may have otherwise been missed. Transcriptomic studies have illuminated many aspects of phytoplankton physiology, from response and acclimation to nutrient and vitamin fluxes (Alexander et al., 2020; Allen et al., 2008; Bertrand et al., 2015; Caron et al., 2017; Cohen et al., 2018, 2017; Dyhrman et al., 2012, 2006; Marchetti et al., 2012; Remmers et al., 2018; Smith et al., 2019, 2016; Whitney and Lomas, 2016), to circadian clocks (Okamoto and Hastings, 2003; Toepel et al., 2008) and light sensing (Caron et al., 2017; Tardu et al., 2016), life stages (Jang et al., 2019; Mars Brisbin and Mitarai, 2019) and trophic strategy (Liu et al., 2015a; McKie-Krisberg et al., 2018), the evolutionary history of serial endosymbiosis (Ebenezer et al., 2019; Geimer et al., 2009; Wong et al., 2018) and kleptoplastidy (Lasek-Nesselquist et al., 2015; Onuma et al., 2020), interactions with other algae (Astier et al., 2020) and protists (Bachvaroffi, 2019; Caron, 2016), the phytoplankton microbiome (Durham et al., 2019; Landa et al., 2017), harmful algal blooms (Haley et al., 2017; Harke and Gobler, 2013; Metegnier et al., 2020), natural production potential (Beedessee et al., 2020; O'Neill et al., 2015b), and real-time viral infection dynamics (Derelle et al., 2017; Lawrence et al., 2017).

International collaborative sequencing initiatives, including Marine Microbial Eukaryote Transcriptome Sequencing Project (MMETSP), have catalogued the metabolic potential of protists, for which there are cultured representatives, and, consequently, enabled molecular investigations of eukaryotic algal evolution and physiology. The MMETSP also greatly expanded the algal reference gene repertoire, enabling metatranscriptomics, the application of transcriptomics to whole communities, on eukaryotic algae (Alexander et al., 2015; Caron et al., 2017).

Metatranscriptomics further expanded the known gene repertoire of algae via direct sequencing of environmental samples (Keeling et al., 2014). It has allowed us to directly observe the physiology of uncultivated lineages that make up the majority of algal biomass and biodiversity. In contrast to metabarcoding methods, which rely on amplification of a single marker gene by PCR, metatranscriptomics reports community structure without the biases unavoidable in exponential amplification. Unlike transcriptomics, metatranscriptomics can be used to infer interactions between phytoplankton and with other community members (bacteria, archaea, fungi) (Seymour et al., 2017). In addition to ship-based and in-situ sampling, metatranscriptomics can also be applied to incubation and mesocosm experiments, where physiology can be manipulated experimentally.

In this chapter, we review the insights gained by using transcriptomics to study algae, including enhanced understanding of algal diversity, endosymbiosis, light and nutrient sensing and response, ecological interactions, viral/host dynamics, secondary metabolite production, harmful algal bloom formation, life stages, trophic strategy, and responses to contaminants. We then make recommendations to enable researchers to overcome common hurdles in adopting the techniques, and finally, discuss prospects for the field.

How has transcriptomics changed our understanding of phytoplankton ecology?

Algal diversity

Transcriptomic sequencing of cultured isolates alongside global aquatic sampling efforts have vastly expanded our knowledge of the diversity of algae. The Marine Microbial Eukaryote Transcriptome Sequencing Project (MMETSP) was the largest concerted effort to create a transcriptomic database to further the molecular research of microbial eukaryotes, including algae. It resulted in 650 assembled, functionally annotated transcriptomes available from cultured lineages of mostly marine algae (Keeling et al., 2014). It increased the sequencing data available for microbial eukaryotes by 50-fold, increasing coverage of both understudied lineages and dominant lineages, such as diatoms and dinoflagellates. Marine observational studies utilizing the MMETSP as a reference have benefited from significantly increased rates of gene annotation, e.g., 58–62% of reads identified (Alexander et al., 2015), as opposed to the 12–14% identified using genomic resources only (Marchetti et al., 2012). This resource also changed our perception of marine community structure by alleviating taxonomic biases. For example, metatranscriptomics on the oligotrophic ocean observed a community dominated by diatoms and haptophytes. But, after mapping to the MMETSP, the same community could be seen to be 36–40% dinoflagellate reads, with diatoms and haptophytes being next most represented (Alexander et al., 2015). The MMETSP has also provided insights into the conserved functional response of diverse algal phyla (Harke et al., 2017) and illuminated the presence of a set of core genes amongst prymnesiophyte algae (Koid et al., 2014). In addition to algae themselves, the MMETSP identified viruses infecting cultured organisms. For example, novel nucleocytoplasmic large DNA viruses (NCLDV) were catalogued and found to infect diverse eukaryotes not previously known to be hosts (Gallot-Lavallée and Blanc, 2017). The MMETSP also elucidated interesting aspects of algal physiology. For example, together with 1000 plant transcriptome project (Matasci et al., 2014), more than 60 homologues of chlorophyte cation channelrhodopsins (CCRs) were uncovered, which allow for phototaxis in cryptophytes (Govorunova et al., 2017). The MMETSP also enabled elucidation of historical gains and losses of metabolic function over evolutionary history. For example, algae, like primates, once lost the ability to produce vitamin C. However, after acquiring plastids, photosynthetic eukaryotes developed an alternative mechanism for vitamin C biosynthesis, in which the terminal enzyme, L-gulonolactone oxidase (GULO), was replaced with L-galactonolactone dehydrogenase (GLDH) (Wheeler et al., 2015). The MMETSP also furthered our understanding of the extent of core physiological traits shared among diverse lineages, such as silicon transport. Silicon transport was found to not be unique to classically silicifying algae, like diatoms. Rather, the SIT silicon transporter family was identified in a diversity of silicified and nonsilicified microeukaryotes, and may have initially evolved to avoid silicon toxicity in Precambrian oceans (Marron et al., 2016).

Insights into algal diversity have also been made independently of the MMETSP. Even lesser studied lineages, such as cryptophytes, now have representative transcriptomes available. For example, sequencing the transcriptome of *Rhodomonas* sp., which is used as invertebrate feed in aquaculture, identified putative genes involved in alloxanthin and polyunsaturated fatty acid synthesis (Li et al., 2016). Single-cell transcriptomics was also used to profile uncultured Rhizaria (Krabberød et al., 2017), which often host algal symbionts. This lineage is highly diverse and abundant, but difficult to culture (Krabberød et al., 2017).

Direct sampling efforts have also greatly expanded the phytoplankton gene repertoire (Fig. 13.1). Tara Oceans, described in Chapter 11, catalogued 47 million marine planktonic genes from 370 globally distributed metagenomes, and used 187 metatranscriptomes to investigate expression (Salazar et al., 2019). While metagenomic sampling efforts have documented an impressive gene repertoire of uncultivated taxa, metatranscriptomic samples have highlighted the more ecologically relevant subset that is being expressed in given conditions. For example, metatranscriptomics showed that the marine cyanobacteria, *Prochlorococcus* and *Synechococcus*,

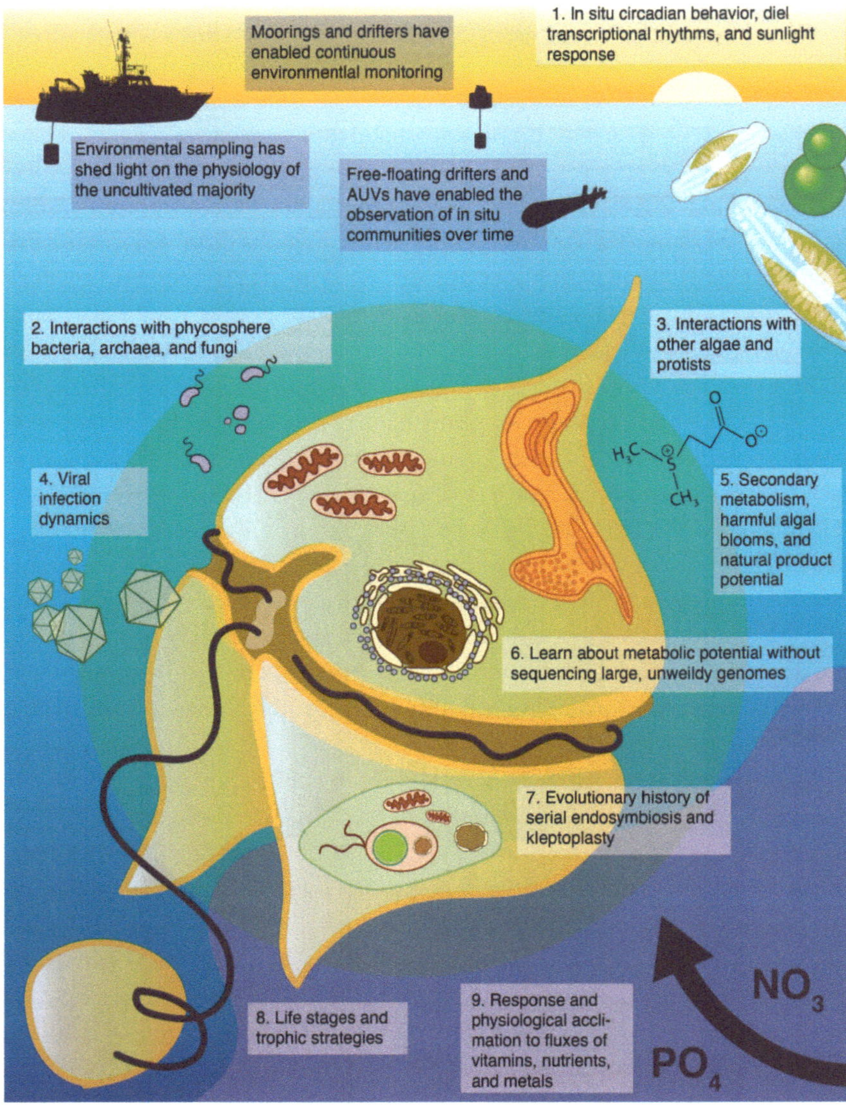

Figure 13.1 Transcriptomic insights into phytoplankton ecology and physiology.

contribute more to total community expression than would have been inferred from their genomic abundance (Dupont et al., 2015).

Eukaryotic algal phylogeny has been enhanced by environmental sampling efforts and sequencing of cultured isolates, as morphology has often not been sufficient to deduce phylogeny. The advent of molecular techniques, especially transcriptomics, has revised fundamental relationships between algal lineages in recent years (Burki et al., 2020; Caron et al., 2017). These revisions have culminated in the creation of nested SAR (Stramenopila, Alveolata, Rhizaria), and TSAR (SAR + Telonema) groups, rather than classification of Rhizaria (Simpson et al., 2004), Stramenoplia, and Alveolata (Baldauf, 2008) as independent supergroups. Conversely, Haptista and Cryptista, formerly components of Hacrobia, are now considered distinct supergroups, with Cryptista more closely related to Archaeplastida (Burki et al., 2020). Finally, a new supergroup, Hemimastigophora, was identified via cultivation-independent transcriptomics (Lax et al., 2018). Phylotranscriptomic analysis of dinoflagellates revealed that the theca, the overlapping cellulose plates found in armored dinoflagellates, had a single origin, that nonphotosynthetic dinoflagellates retain plastids that are important for metabolic function, and that histone-like proteins (HLPs) were acquired horizontally from bacteria, at least twice (Janouškovec et al., 2017). Finally, nematocysts, projectile organelles that excrete a harpoon-like object to lance prey, were also found to be evolved from a single independent origin in dinoflagellates (Gavelis et al., 2017).

Endosymbiosis

Transcriptomics has also enabled major insights into physiological adaptations resulting from serial endosymbioses that structure the subcellular organization of diverse algae (Fig. 13.1.7). Eukaryotic algae contain plastids that result from primary, secondary, and tertiary endosymbiosis of other algal lineages, in which the engulfed "prey" was partially retained as a photosynthetic energy source (McFadden, 2001). Over evolutionary time, genes move between the nucleus and organelles, resulting in a complex metabolic mosaic. For example, the major freshwater model organism, *Euglena gracilis*, descended from a protozoan that engulfed a red alga by secondary endosymbiosis (O'Neill et al., 2015a). The red algal endosymbiont was since lost, but not before donating genes to the *E. gracilis* nuclear genome. Modern *E. gracilis* plastids are the result of a subsequent endosymbiosis of a green alga (O'Neill et al., 2015a). Indeed, paired transcriptomics and proteomics found that multiple photosynthetic eukaryotes contributed material to the *E. gracilis* nuclear genome, possibly from transient endosymbioses (Ebenezer et al., 2019).

The transcriptional regulation of endocytosed plastids and nucleomorphs could be simpler than expected. Genes from the complex, triple-envelope plastid of *E. gracilis* are constitutively expressed (Geimer et al.,

2009). The Chlorarachniophyte, *Bigelowiella natans*, also acquired its plastid from endosymbiosis of green algae. Its nucleomorph (nucleus retained from the green alga) did not change expression across light treatments (Rangsrikitphoti and Durnford, 2019). However, the nucleomorphs of both *B. natans* and the cryptomonad flagellate, *Guillardia theta*, have increased antisense transcription relative to nuclei (Wong et al., 2018).

The collection of transcriptomic data from algae also altered evolutionary paradigms that were developed in land plants. For example, based on data from land plants, plastids were thought to accumulate mutations more rapidly than mitochondria. Transcriptomic data from algae has shown that for diverse microeukaryotes, the reverse is true (Smith, 2015). Additionally, it was thought that ancestral cell division machinery was widely replaced with host-derived dynamin-dependent division in eukaryotic mitochondria, because that is the case for mammals and yeast and plant models. Analysis of diverse protists from the MMETSP showed that frequently, including in Stramenopiles, the division methods of the bacterial endosymbiont are retained (Leger et al., 2015).

Transcriptomics has also offered novel insights into kleptoplasts (temporal plastids that are engulfed and retained temporarily (days to months)), within the lifetime of a single organism. Kleptoplasty has been observed in seaslugs (Trench, 1969), flatworms (Van Steenkiste et al., 2019), ciliates (Johnson et al., 2007), dinoflagellates (Minnhagen et al., 2008), and foraminifera (Bernhard and Bowser, 1999). Transcriptomic studies suggest that hosts modulate kleptoplast transcription. For example, upon ingestion by the ciliate *Mesodinium rubrum*, the cryptophyte *Geminigera cryophile*, downregulates genes involved in protein transport (Lasek-Nesselquist et al., 2015). The dinoflagellate, *Nusuttodinium aeruginosum*, ingests cryptomonad *Chroomonas sp.* and remodels the transcriptional profile of the cryptomonad nucleus to enlarge the plastid. After ingestion, the cryptomonad nucleus greatly diminishes light responsiveness. Cryptomonad genes involved in metabolism, translation, and DNA replication are upregulated, and those involved in sensory systems and motility are downregulated (Onuma et al., 2020). The Antarctic Ross Sea dinoflagellate (RSD), which harbors temporary kleptoplasts derived from haptophytes, also significantly modifies the physiology of the stolen plastid. Although the plastid isn't fully integrated, host-encoded genes are targeted to the kleptoplast and are necessary to maintain photosystem I (Hehenberger et al., 2019).

Light sensing and transcriptional response

Transcriptomics has provided insights into algal light-sensing (Fig. 13.1.1). Photosynthetically available radiation falls in the visible spectrum, but divergent wavelengths of light are also useful for signaling cues for plants and algae (Kottke et al., 2017; Li et al., 2011). Phytochromes are a class of photoreceptors that sense red and far-red light. In plants, they control aspects of development, including seed germination, growth, and flowering time (Li et al., 2011), and in cyanobacteria, they are involved in phototaxis and regulation of growth rate (Wiltbank and Kehoe, 2019). Recent genomic and transcriptomic studies suggest that phytochromes were lost from the cyanobacterial endosymbiont during chloroplast evolution and regained in ancient eukaryotic algae (Rockwell and Lagarias, 2017). Environmental transcriptomic data has also shown that phytochromes are more widespread than previously thought (Caron et al., 2017), and that canonical plant phytochromes are structurally and functionally related to those in prasinophytes (Duanmu et al., 2014) and originated in a eukaryotic ancestor of Archeaplastida (Duanmu et al., 2014; Li et al., 2015). Cryptochromes, on the other hand, are blue-light photoreceptors that are believed to have evolved from the DNA photolyase protein family. They have diversified to serve a variety of purposes beyond DNA repair, including mediating light-responsive plant growth and entraining circadian clocks (Heijde et al., 2010). They have been implicated in life-cycle control, DNA-repair, and circadian clock regulation in green algae (Kottke et al., 2017), and regulate the transcriptional light response in photosynthetic organisms, ranging from tomatoes (Facella et al., 2008) to diatoms (Coesel et al., 2009). In the green alga, *Dunaliella Salina*, both red and blue light affected the abundance of transcripts involved in carotenoid metabolism, leading to an eventual increase in total carotenoid content (Li et al., 2020b). In the red alga, *Cyanidioschyzon merolae*, blue light regulated the abundance of protein synthesis-related transcripts, whereas red light regulated the abundance of photosynthesis-related transcripts (Tardu et al., 2016).

Transcriptomics has also created insights into circadian physiology. Circadian clocks allow organisms to have a physiological sense for the time of day, and have been extensively modeled in eukaryotes, including *Arabidopsis thaliana* (McClung, 2006), *Mus musculus*, and *Drosophila melanogaster* (O'Neill et al., 2011), *Ostreococcus tauri* (Corellou et al., 2009) and *Synechococcus elongatus* (Mackey and Golden, 2007). However, the circadian mechanisms of many phytoplankton are unknown. In *Synechococcus* sp. strain

PCC 7942, virtually all 800 promoters examined were observed to have periodic transcription (Liu et al., 1995). A microarray experiment found that the circadian clock controls the transcript levels of about 3% of dinoflagellate, *Pyrocystis lunula*, genes. Phytochromes, cryptochromes, and homologs of known core clock genes were not identified, leading to speculation that dinoflagellates independently evolved a distinct timekeeping mechanism (Okamoto and Hastings, 2003). However, in the dinoflagellate, *Lingulodinium polyedrum*, proteins involved in bioluminescence change dramatically on a diel cycle, whereas associated mRNA levels remain constant, implicating post-translational regulation (Okamoto and Hastings, 2003).

Recently, metatranscriptomic profiling has been implemented to track the same community of algae over time, leading to new insights into diel metabolism. In a study employing a free-drifting robotic sampler in the California current, all active phytoplankton lineages modulated their transcription on a diel cycle. *Ostreococcus* spp. (Ottesen et al., 2013) cycled the largest portion of its transcriptome (nearly 25%), and dinoflagellates cycled the smallest (<1%). Cellular (or molecular) functions were similarly partitioned across organisms, with photosynthetic machinery peaking in a conserved cascade across morning hours, and cell-division related machinery peaking in the evening, perhaps to minimize DNA exposure to UV-stress (Kolody et al., 2019). In a similar study conducted in the North Pacific subtropical gyre (NPSG), functional partitioning over the day/night cycle was also observed in natural populations of the haptophyte, *Emiliania huxleyi* (Hernández Limón et al., 2020), and metabolite (Boysen et al., 2020) and pigment pools (Becker et al., 2020) also changed dramatically over day and night. Additionally, coordinated diel transcription was observed in a population of symbiotic diatom-diazotrophic association (DDA), which represent keystone species in the NPSG. Diel cycling coordinated diverse genes, including those involved in cell division, symbiont maintenance, and resource acquisition (Harke et al., 2019).

Culture-based transcriptomes recovered larger percentages of diel cycling genes than observed in nature. Sixty percent of the transcriptome oscillated in the Stramenopile, *Nannochloropsis oceanica*, during a day/night cycle (Poliner et al., 2015), and 61% of transcripts cycled across day/night in the haptophyte, *Emiliania huxleyi*, with those related to anabolism peaking around dawn, and those related to catabolism peaking from mid-day to dusk (Hernández Limón et al., 2020). Eighty percent of *Prochlorococcus* sp. (Zinser et al., 2009) transcripts and nearly all *Ostreococcus* sp. (Noordally et al., 2018) transcripts are rhythmically expressed with a 24-h period. However, protein abundances largely do not reflect diel changes in the mRNA pool (Guerreiro et al., 2014; Noordally et al., 2018; Seaton et al., 2018). For slow-turnover proteins that are decoupled from diel transcription, the daily timing of gene expression could allow algae to adapt to seasonal changes in photoperiod. Because protein synthesis rates are typically 3–5× higher at day than at night, proteins with late-day peaking transcription would be more abundant in long photoperiods, due to reduced translation rates after sunset (Kolody et al., 2019; Seaton et al., 2018).

Ecological interactions

As discussed in Chapter 15, the natural habitat of algae is not an axenic culture, and interactions with other microbes (Figs. 13.1.2, 13.1.3) play a major role in their ecology and physiology. Transcriptomics has yielded insights into parasitism, symbiosis, and algal interactions within their microbiomes. For example, the syndinian, *Amoebophyra*, which parasitizes dinoflagellates, is primarily known from environmental sequencing. Transcriptomics revealed that *Amoebophyra* sp., associated with the dinoflagellate, *Karlodinium veneficum*, can recode all 3 stop codons as sense-codons, whereas its host cannot (Bachvaroffi, 2019), and in a separate study, *Amoebophyra* differentially expressed two-thirds of its genes across infection stages (Farhat et al., 2018). Transcriptomics also yielded insights into an ecologically significant but poorly studied symbiosis: that between algae and Rhizaria (Caron, 2016), unicellular eukaryotes of which most are nonphotosynthetic. Symbiosis of photosynthetic organisms (such as dinoflagellates, chlorophytes, and haptophytes) with rhizaria is likely modulated by interactions between glycans and c-type lectins (Balzano et al., 2015). Additionally, transcriptomics has illuminated signaling between algal cells. For example, transcriptomic experiments have resulted in new insights into the capacity of several diatom lineages to produce the gaseous signaling molecule, nitric oxide (Astier et al., 2020). Nitric oxide synthase (NOS) was identified in the de novo transcriptome of *Pseudo-nitzschia multistriata*, and homologous sequences were subsequently found in MMETSP *Thalassiosira*, *Skeletonema*, and *Cylindrotheca* (Di Dato et al., 2015).

Transcriptomics has also highlighted the importance of interactions between algae and phycosphere microbes (Seymour et al., 2017). For example, *Sulfitobacter* sp. promotes diatom cell division (Amin et al., 2015) and the presence of bacteria downregulates S-adenosylmethionine synthetase (SAMS) production in the toxic dinoflagellate, *Alexandrium tamarense*

(Moustafa et al., 2010). Phycosphere colonizers respond differently to diverse algae. For example, the roseobacter, *Ruegeria pomeroyi*, remodeled over 25% of its transcriptome when cocultured with a diatom (*Thalassiosira pseudonana*), rather than dinoflagellate (*Alexandrium tamarense*) (Landa et al., 2017). Transcriptomics experiments have also begun to elucidate the role of phycosphere microbes in nutrient limitation. For example, both iron and cobalamin (vitamin B12), which is uniquely produced by bacteria and archaea, interactively colimit phytoplankton growth in the Southern Ocean. Phytoplankton growth rates are thus determined by complex feedback loops involving both phytoplankton and bacterial and archaeal growth strategies (Bertrand et al., 2015).

Virus/host dynamics

Several novel algal viruses have been discovered using transcriptomics (Fig. 13.1.4). The first chlorarachniophyte viruses were reported in a metatranscriptomic study identifying RNA viruses from cultured chlorophytes and chlorarachniophytes (Charon et al., 2020). Additionally, an RNA virome from Yangshan Deep-Water Harbour near Shanghai doubled the known diversity of RNA viruses (Wolf et al., 2020). Abiotic factors promoting infection have also been investigated via transcriptomics. For example, metatranscriptomic investigations in the California current revealed that silica limitation promotes viral infection of diatoms, leading to higher mortality (Kranzler et al., 2019). Biotic factors shaping viral infection dynamics over time can also be observed using transcriptomics. A cooccurrence analysis of metatranscriptomes from Quantuck Bay, NY and Narragansett Bay, RI revealed the infection dynamics of marine algal viruses, identifying both "steady-state" and "boom-and-bust" ecologies (Moniruzzaman et al., 2017). Because total RNA is sequenced in a transcriptome, DNA viruses observed in transcriptomes can be assumed to be actively infecting. Similarly, lytic RNA viruses are undetectable in genomes, but can be detected in transcriptomes. A metatranscriptomic time-series, following algae off the California coast, observed viruses infecting the majority of abundant algal lineages (Kolody et al., 2019). Surprisingly, double-stranded DNA (dsDNA) virus total mRNA rose in tandem with total RNA of putative hosts, rather than being anticorrelated, as would be expected from kill-the-winner theory (Thingstad, 2000). This was also the case for Pelagibacter virus 37-F6 (Martinez-Hernandez et al., 2020) and cyanophages (Aylward et al., 2017), which both also had diel cycling in total mRNA. This unexpected dynamic may arise, because dsDNA viruses are large and need to maintain host metabolism to reproduce (Kolody et al., 2019). For example, the dsDNA P-SSP7 cyanophage (Sabehi and Lindell, 2012) transcribes a photosystem II D1 protein alongside its own replication machinery, likely to sustain host metabolism (Lindell et al., 2007). Additionally, transcriptomics studies have demonstrated that infection strategies are sometimes consistent across different host species. For example, the T4-like cyanophage, Syn9 launched nearly identical transcriptional programs when infecting three distinct *Synechoccocus* strains (Doron et al., 2016).

Transcriptomic studies have also shown how viruses modulate eukaryotic host metabolism. For example, the transcriptome of the heterokont, *Aureococcus anophagefferens* CCMP1984, was extensively remodeled throughout infection by a giant virus (Moniruzzaman et al., 2018). During infection, host genes related to photosynthesis and the formation of cytoskeleton, fatty acids, and carbohydrates were suppressed, whereas those related to protein synthesis, polyamine biosynthesis, cellular respiration, transcription, and RNA processing were upregulated (Moniruzzaman et al., 2018). Cells of the coccolithophore, *Emiliania huxleyi*, infected by the large virus, *E. huxleyi* virus (EhV), remodel their transcriptomes towards fatty acid synthesis necessary for viral assembly (Rosenwasser et al., 2014).

Secondary metabolite production

Marine algae produce numerous secondary metabolites, although few have been identified. Their purpose is often unknown, but some are allelopathic (inhibitory towards predators or competitors) (Leflaive and Ten-Hage, 2007) and others may act as signaling molecules for members of the algal microbiome (Durham et al., 2019). For example, the algal organosulfur metabolite, dimethylsulfoniopropionate (DMSP; Fig. 13.1.5), causes changes in the composition and gene expression of marine bacterioplankton communities, enriching for Gammaproteobacteria and Bacteriodetes and genes involved in heterotrophy (Vila-Costa et al., 2010). DMSP also causes the heterotrophic marine bacterium, *Ruegeria pomeroyi* DSS-3, to synthesize a quorum-sensing molecule and shifts its metabolism towards a more cooperative lifestyle (Johnson et al., 2016).

Other algal secondary metabolites are of industrial interest, e.g., for UV-protection or cosmetic benefits to skin (Ariede et al., 2017), vitamins, and nutrients for food and feed additives (Priyadarshani and Rath, 2012), or energy production (Priyadarshani and Rath, 2012). Algal secondary metabolites also have promising anticancer (Lefranc et al., 2019) and antibiotic properties (Bhowmick et al., 2020). Transcriptomics is enabling identification of novel natural products, as well as investigations into their ecological roles and potential uses. For example, the MMETSP enabled large-scale screening for bioactive natural products. Enzymes involved in the synthesis of monogalactosyldiacylglycerols (MGDGs), which have anti-inflammatory and anticancer properties, and sulfoquinovosyldiacylglycerols (SQDG), which inhibit an enzyme relevant for Alzheimer's disease, were found in diverse lineages (Riccio et al., 2020). Potential sources of natural products have also turned up in unexpected places. *Euglena gracilis* is a well-utilized model organism, but key metabolic details were not known until transcriptional profiling experiments were conducted. Although easily cultured, the *E. gracilis* genome could not be sequenced, because it contains the modified base, J (glucosylated hydroxythymidine). However, transcriptomics revealed that *E. gracilis* has 3 separate vitamin C biosynthesis pathways, making it a promising commercial target (O'Neill et al., 2015b).

Harmful algal blooms (HABs)

Transcriptomics has illuminated many facets of harmful algal bloom (HAB) ecology as methods have evolved and sequencing costs have decreased. Much of this research has focused on the role environmental factors have on the occurrence and rates of toxin synthesis. Although targeted qRT-PCR-based screening of known genes of interest has been helpful for understanding and monitoring HAB physiology (e.g., Harke et al., 2012), more global approaches, such as microarray and RNA-Seq studies, provide deep insight into phytoplankton responses to environmental stimuli.

In freshwater systems, studies on *Microcystis* employed transcriptomics to explore the response of this cyanobacterium to nutrients, as eutrophication has long been known as a primary driver of HABs in freshwater ecosystems. Transcriptomics revealed how cells navigate variance in exogenous nutrient resources and the impact of nutrient availability on toxin synthesis (Harke and Gobler, 2013). The approach also demonstrated that urea additions trigger increased transcript abundance of transposases, with opposite trends when under phosphorus (P) reduction, suggesting the role nutrient chemistry may play in genome architecture (Steffen et al., 2014). Metatranscriptomics has also explored the ecology of *Microcystis* and other cooccurring bacteria, demonstrating the bottom up role nutrients have in community composition and how climate change factors may favor cyanobacterial blooms (Steffen et al., 2015). An investigation of *Microcystis* carbon uptake systems suggested this species will do well in future climate scenarios (Sandrini et al., 2015).

Metatranscriptomics has also illuminated community dynamics across spatial gradients, finding changing community composition with changing resources. These studies confirmed early observations of the important P scavenging and storage capabilities of *Microcystis*, which suggest that management strategies focusing on P alone may not reduce *Microcystis* abundance. Time-series investigations have explored the timing of metabolic responses within the cell to nutrient deprivation and restoration, discovering links to N supply and the choreography of metabolic responses in time (Harke and Gobler, 2015). A recent study using diel metatranscriptomics analysis revealed the timing of cellular processes during natural bloom conditions, such as cell division, carbon fixation, and nitrogen (N) assimilation, further illuminating the temporal patterns in response to the environment (Davenport et al., 2019). Tang et al. (2018) also found a seasonal shift in transcriptional responses of natural bloom populations of *Microcystis* from Lake Taihu (Tang et al., 2018). Transcriptomics has revealed the role of temperature in synthesis of the toxin microcystin, whereby reduced temperatures induced transcription of all *mcy* genes (Martin et al., 2020). In addition, this study confirmed prior observations of a 48-h time lag between transcription and end-product (microcystin) concentrations (Martin et al., 2020).

Other genera of HAB producing cyanobacteria have also received attention. For instance, studies have investigated the response of the N_2 fixing cyanobacterium *Dolichospermum* (previously *Anabaena*) to P stress (Teikari et al., 2015), N stress (Flaherty et al., 2011), the role small RNAs have in regulating N_2 fixation (He et al., 2020), and controls of heterocyst differentiation (Ehira, 2013; Mitschke et al., 2011). In addition, changing concentrations of Ca^{2+} modulated regulation of N- and C-related processes within the cell (Walter et al., 2016).

The transcriptome can also be used to understand physiological changes that occur during bloom formation. The algal transcriptome has been shown to be highly responsive to nutrients in particular (Bowler et al., 2010; Rogato et al., 2015). However, recent studies have found that rates of physiological processes predicted from culture-based studies in the laboratory were not predictive of interspecies dynamics in field-based populations (Xiao et al., 2020), highlighting the need to perform experiments in situ. Several recent studies have performed in-situ transcriptomic studies. For example, Hook and colleagues utilized metatranscriptomes to show increased activity of dinoflagellates near aquaculture leases in Southern Tasmania (Australia), that seemingly corresponded to the measured nutrient gradient. The transcripts with increased abundance near the leases were unique to that area, suggesting either that unique dinoflagellates are active near the leases, or that the dinoflagellates in the system have unique metabolic activity near the nutrient source (Hook et al., 2021).

A recent study examined changes in community composition and corresponding changes of the transcriptomic profiles in successive *Alexandrium minutum* blooms in Brest, France (Metegnier et al., 2020). They mapped reads from the metatranscriptome to different taxa included in the MMSTEP database (Keeling et al., 2014) to examine how each organism changed with bloom progression. The transcriptome of *Alexandrium* showed different physiological processes than the other algae present, in that photosynthesis-related transcripts, including gluconeogenesis and other carbohydrate metabolism related transcripts, were underrepresented; transcripts related to ion transfer, heterotrophy, and cell-cell signaling was overrepresented (Metegnier et al., 2020). By contrast, nutrient transport transcripts did not change, despite limiting nutrient concentrations in the river. However, the transcriptomes of the haptophytes and diatoms contained abundant transcripts encoding photosynthesis-related gene products, including light harvesting proteins, glucose metabolism, and gluconeogenesis, as well as those involved in translation (Metegnier et al., 2020).

The Raphidophyte, *Heterosigma akashiwo*, has been extensively studied as it forms ichthyotixic blooms. Transcriptomics has illuminated many facets of its nutritional ecology in response to N and P stress, conditions often occurring during bloom conditions (Haley et al., 2017). In response to predicted rise in atmospheric CO_2, *H. akashiwo* modulates its physiology by altering expression of carbon-concentrating mechanisms and beta-carbonic anhydrases. Elevated CO_2, appears to impact cell motility, suggesting future ocean conditions may modify the swimming behavior of this species (Hennon et al., 2019). Ji et al. (2018) profiled the changes in the metatranscriptome of *H. akashiwo* during an algal bloom. They found that though, initially, the algal community was diverse, this diversity constricts during bloom; the *H. akashiwo* and dinoflagellates increased in abundance, whereas nutrients (notably nitrate and P) quickly became limiting under bloom conditions (Ji et al., 2018). Transcriptomic changes indicate rapid change in P assimilation genes, suggesting that DOP might be driving bloom uptake (Ji et al., 2018).

Rather than an influx of a particular nutrient, some harmful algal blooms are associated with altered nutrient ratios. For example, *Prymnesium parvum* is a mixotrophic haptophyte that forms ichthyotoxic blooms that are often associated with altered N:phosphorous ratios (Beszteri et al., 2012). Genes involved in photosynthesis and polyketide synthase genes possibly involved in toxin production were upregulated in low P, but not at low N (Liu et al., 2015b).

Although HABs have historically been monitored using toxin identification, pigment identification, or morphologically (Anderson et al., 2012), quantification based on qPCR is emerging for use as a means of monitoring HABs (Main et al., 2018) (see Chapter 12). In the future, it is possible that environmental RNA will be used to not only monitor but predict HAB formation. The most obvious example of the potential use is monitoring for increased transcription of toxin encoding genes, which have been recently characterized in a variety of species (Harke et al., 2016a; Murray et al., 2016; Wang et al., 2020). This approach, especially if coupled with remotely deployable, in-situ PCR and other sensors (see Craw et al., this volume), could be instrumental in protecting water supplies, aquaculture leases, and ecologically sensitive areas. Other potential targets for bloom forecasting include metabolic changes. For example, many HAB species utilize mixotrophy or take up urea (Burkholder et al., 2008). Marker transcripts, such as increased abundance of urea transporter (Jing et al., 2017), or formididase, which in involved in utilization of organic N (Burkholder et al., 2008; Glibert et al., 2018; Jing et al., 2017), may also indicate imminent bloom formation. Changing metabolism between phototrophy and mixotrophy may also be indicated by decreases in the abundance of transcripts-encoding photosynthesis genes, photopigments, and nitrate assimilation (Graupner et al., 2018) and increases in transcripts related to ion transfer, mixotrophy, and cell-cell signaling (Metegnier et al., 2020).

Domoic acid (DA) is a potent neurotoxin produced by diatoms in the genus *Pseudo-nitzschia* (PN) (Bates et al., 1998). In recent years, toxic PN blooms have occurred with increased frequency and duration. For example, conditions in 2015 resulted in the largest recorded PN bloom to date in the North Pacific (McCabe et al., 2016; Ryan et al., 2017). There is an urgent need to improve region-wide DA toxin risk management to safeguard public health, promote viable and sustainable fisheries, and protect living marine resources. Despite decades of research into the molecular mechanisms that control DA production, the genes responsible for DA biosynthesis remained enigmatic until recently. Using transcriptomics, Brunson et al. (2018) elucidated the DA biosynthetic pathway in the genome of the HAB-forming diatom *Pseudo-nitzschia multiseries* (Brunson et al., 2018). RNA sequencing was conducted on cultures of *P. multiseries* grown under DA-inducing conditions, phosphate limitation and high pCO_2 (Sun et al., 2011), and resulted in identification of select transcripts that are upregulated under DA-inducing conditions. Four of the ten most highly upregulated transcripts are mapped to a compact ~8kb cluster in the *P. multiseries* genome, with conserved organization. These four transcripts (terpene synthase (*dabA*), a hypothetical protein (*dabB*), an α-ketoglutarate (αKG)-dependent iron (II) dioxygenase (*dabC*), and a cytochrome P450 (*dabD*)) could also be identified in the transcriptome of DA-producing *P. australis* (Keeling et al., 2014). The biochemical function of each of the encoded Dab enzymes was subsequently investigated *in vitro* by using *Escherichia coli* and *Saccharomyces cerevisiae*-based expression systems. Therefore hypotheses regarding environmental regulation of DA production can now be tested through molecular monitoring of *dab* transcripts. Profiling for *dab* expression in the environment will likely provide important information regarding environmental inputs to PN HAB toxicity in addition to informing predictive models of DA outbreaks.

Life stages and trophic strategy

Transcriptomics has shed light on the physiology of diverse algal life stages and trophic strategies (Fig. 13.1.8). Pairing transcriptomics with traditional laboratory techniques has furthered algal photosynthesis research. For example, screening for light-stress responsive genes helped identify the photoprotective mechanism used in diatom light-harvesting complexes (Zhu and Green, 2010). Transcriptomics was also used to elucidate the carbon-concentrating mechanism (CCM) in the model green alga, *Chlamydomonas reinhardtii* (Yamano and Fukuzawa, 2009).

However, many algae are not purely photosynthetic. Mixotrophy is a well-documented trophic strategy among phytoplankton and evidence is accumulating that canonically photosynthetic algae are capable of supplementing their metabolism by other means (Flynn et al., 2013; Kamjunke and Tittel, 2008). Dinoflagellates are known for their widespread heterotrophic and mixotrophic (both predatory and osmotrophic) lifestyles, but facultative osmotrophy is also common among diverse phytoplankton lineages, having been described in algal groups spanning cyanobacteria, chrysophytes, chlorophytes, diatoms, and xanthophytes (Kamjunke and Tittel, 2008). Transcriptomics has been used to better understand the impact of light/dark cycles on mixotrophs and has shown taxonomic differences in light responsiveness. For example, though the mixotroph,

Ochromonas sp., had largely unchanged gene expression across light/dark cycle as long as bacterial prey was present, *Prymnesium parvum* showed changes in gene expression, despite preying on ciliates (Liu et al., 2016). Transcriptomics is also beginning to elucidate metabolic shifts across trophic modes. For instance, *P. parvum* displays transcriptome remodeling in axenic cultures versus in the presence of bacterial or ciliate prey and didn't grow faster in the presence of prey, despite upregulating genes involved in organic carbon and fatty acid acquisition (Liu et al., 2015a).

Transcriptomics has also elucidated physiological differences across growth states. For example, in the mixotrophic dinoflagellate, *Yihiella yeosuensis*, antipredation genes (polyketide synthesis, cell wall biogenesis) are upregulated in cysts, whereas motility-related genes are upregulated in vegetative cells (Jang et al., 2019). Flagellated *Phaeocystis globosa* also have morphological shifts associated with blooms, in which they aggregate in a secreted mass of mucus. These colonies, which negatively impact fisheries, appear to be a defensive measure and are characterized by upregulation of dimethylsulfide and acrylate production, as well as pathogen interaction pathways (Mars Brisbin and Mitarai, 2019). Colonies may form as a defensive measure (Mars Brisbin and Mitarai, 2019), or in response to an influx of iron (Bender et al., 2018). Coccolithophores are also major bloom formers. *Emiliania huxleyi* blooms are often terminated by viral infection, which induces a life-cycle shift to noncalcifying haploid cells. Haploid cells have pronounced differences in their transcriptomes (Von Dassow et al., 2009), including increased expression of genes related to protein turnover, motility, and endocytosis. Analysis shows that life stages are functionally distinct and could potentially occupy different niches in the environment (Rokitta et al., 2011).

Nutrient response

Much of the algal transcriptomics literature has focused on the response of phytoplankton to nutrient pulses, and adaptation to nutrient stress (Fig. 13.1.9). Phytoplankton undergo extensive transcriptome remodeling in response to shifting availability of macronutrients, such as nitrogen (N) (Alexander et al., 2020; Remmers et al., 2018; Smith et al., 2019), phosphorous (P) (Caron et al., 2017; Dyhrman et al., 2012, 2006; Whitney and Lomas, 2016) and micronutrients, such as iron (Allen et al., 2008; Cohen et al., 2018, 2017; Marchetti et al., 2012; Smith et al., 2016), copper (Hippmann et al., 2017), sulfur (Bochenek et al., 2013), silica (Mock et al., 2008), calcium (Nam et al., 2019), and vitamin B12 (Bertrand et al.,

2012). N and P responses were species-specific in a study of bloom-forming diatoms, dinoflagellates, and haptophytes (Harke et al., 2017) and across diatom species (Bender et al., 2014). One response that is shared across algae is the upregulation specific nutrient transporters in limiting conditions. For example, the pelagophyte, *Aureococcus anophagefferense*, upregulates N and P transporters when each is depleted (Frischkorn et al., 2014). Mixotrophic organisms, such as dinoflagellates, have the common response of using osmotrophy and endocytosis to supplement missing nutrients. For instance, in addition to reducing cell size and chlorophyll content and upregulating N transporters, the dinoflagellate, *Prorocentrum shikohuense*, activates phagotrophy to overcome N limitation (Li et al., 2020a). Transcriptomic studies of the marine dinoflagellate, *Prorocentrum donghaiense*, suggest that endocytosis is also important for supplementing phosphorous when it is limiting. *Phaeocystis antarctica* upregulates motility and endocytosis-related genes in low iron, indicating that it may resort to mixotrophy when nutrients are limiting (Rizkallah et al., 2020).

The source of a nutrient may influence whether a bloom will form (Glibert, 2017; Glibert et al., 2018). For example, *P. donghaiense* responded differently to dissolved organic versus inorganic phosphorous, with dissolved organic phosphorous being more important for bloom formation (Zhang et al., 2019a). Agricultural runoff is thought to be a major cause of dinoflagellate blooms, due to the highly available forms of nutrients they contain (Van Dolah, 2000). For example, the dinoflagellate, *Prorocentrum minimum*, preferentially uptakes the anthropogenic N source, urea, over nitrate when it is available (Matantseva et al., 2016). Indeed, *P. minimum* transcription increased 12.5-fold between urea-starved and N-replete conditions (Pechkovskaya et al., 2020).

Iron is an important micronutrient for algae since it is an essential component of their photosynthetic and electron transport machinery (Morrissey and Bowler, 2012). Several lineages have adaptations to iron limitation. An analysis of 47 diatom transcriptomes found that high-affinity iron uptake and ferritin genes are present across all four diatom classes are likely vertically inherited (Groussman et al., 2015). *Phaeocystis antarctica* grows more slowly and makes less particulate organic carbon (POC) in low iron conditions, but can adjust its photophysiology to maintain similar absolute electron transport rates across photosystem II (Koch et al., 2019). *P. antarctica* also substitutes the iron-containing molecule, ferredoxin, for its flavin-containing homolog, flavodoxin and cytochrome c6 for plastocyanin when iron is limiting, a technique that is also common in diatoms

(La Roche et al., 1996). However, *P. antarctica* goes a step further by also substituting iron-intensive enzymes involved N assimilation and fatty acid desaturation for iron-economical homologues (Rizkallah et al., 2020). It has even been speculated that some lineages, such as diatoms, could utilize iron-independent proteorhodopsins to maintain photosynthesis in low iron conditions (Marchetti et al., 2015).

Contaminants

Transcriptomics has also been a useful tool in exploring how phytoplankton respond to contaminants. For instance, Lu et al. (2019) investigated the response of phytoplankton communities to the common antimicrobial compound ciprofloxacin, revealing not only shifts in community structure to favoring cyanobacteria, but through metatranscriptomic analysis, revealed that drivers of this community shift by observing enhanced P transport and photosynthesis by cyanobacteria, but suppressed transcription of genes related to replication, transcription, translation, and cell proliferation in eukaryotes (Lu et al., 2019). Copper sulfate, a commonly used algicide for controlling biomass, was investigated in *Microcystis* (Harke et al., 2016b), finding that genes related to photosynthesis and oxidative phosphorylation were impacted by Cu^{2+} toxicity (Wang et al., 2019). Finally, transcriptomics illuminated a bacterial formate/nitrite transporter homolog, which is upregulated in cadmium-resistant *Euglena gracilis* cells, and appears to confer stress tolerance (Deloménie et al., 2007).

Transcriptomics limitations and how to avoid them

Given the utility of transcriptomic approaches towards the study of algae, we anticipate many readers of this chapter will want to adopt such approaches. In this section, we outline some of the common hurdles that are encountered, as well as strategies to mitigate them. See Fig. 13.2.

Collection

Biomass for transcriptomics can either be collected from cultured organisms or the environment. In the first case, it is important to consider the experimental question (e.g., using environmentally relevant nutrient concentrations). Many questions, such as those regarding cell cycle (Gillard et al., 2008) or diel changes (Fujiwara et al., 2020; Zones et al., 2015) in particular, may also be better answered by transcriptomics and give a cleaner

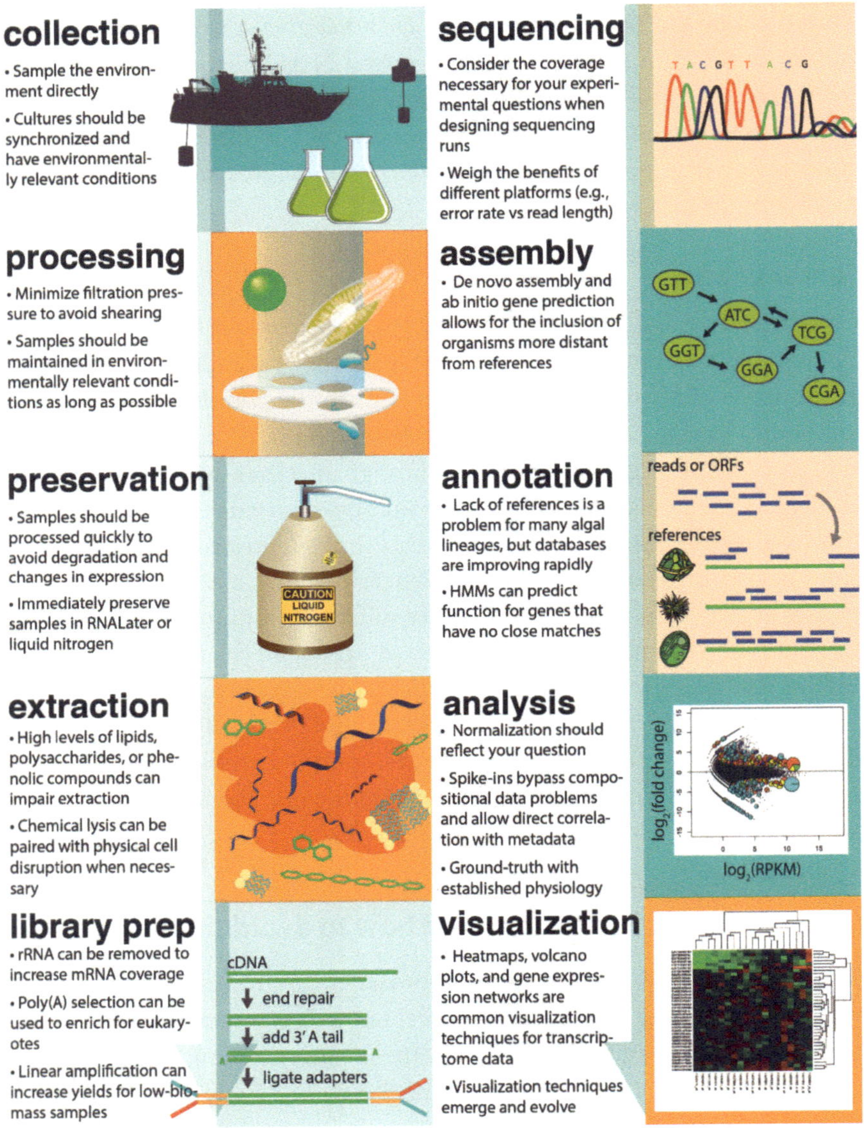

Figure 13.2 Transcriptomic pipeline and best practices.

and more interpretable signal if cultures are synchronized (Mubeen et al., 2019). This may be more feasible in some algae than others (Tamiya, 1966).

As algae are single-celled organisms, to obtain enough material to extract RNA, either large volumes must be filtered, or very dense cell densities must be utilized (Falciatore and Bowler, 2002; Ku and Sebé-Pedrós, 2019). Though such high cell densities may be representative of algal blooms or conditions found in eutrophic environments, they are not typical of oligotrophic or mesotrophic environments (Ward et al., 2012). Improved extraction methods, library preparation kits, and the advent of single-celled transcriptomics (scRNA-Seq) may help alleviate some of these limitations (Ku and Sebé-Pedrós, 2019).

Processing and preservation

RNA is highly responsive to the needs of the cell, so care must be taken to process and preserve RNA for sequencing as quickly as possible, and in as environmentally relevant conditions as is feasible to observe the relevant RNA pool. The median mRNA half-life in the cyanobacterium, *Prochlorococcus*, was found to be 2.4 minutes (Steglich et al., 2010) using a combination of microarrays, quantitative RT-PCR, and a fitting method for RNA decay rates. In eukaryotes, a majority of studies using in vivo labeling or transcriptional inhibition report the median mRNA half-life as between 10 to 25 minutes (Wada and Becskei, 2017), but it has been reported at as low as 2 minutes, with the use of controllable synthetic promoters (Baudrimont et al., 2017). Therefore water/culture samples should be filtered as quickly as is possible without creating damaging amounts of pressure, and RNA should be preserved with preservation buffers (e.g., RNA*later*™ buffer, ThermoFisher Scientific) or liquid nitrogen immediately (Akbar et al., 2018).

Extraction

To sequence total RNA, it must first be extracted from cells, a process which could impact organisms differentially (especially important when considering metatranscriptomes) and lead to downstream biases. For example, some algae have cell wall characteristics that make them more difficult to lyse (Falciatore and Bowler, 2002; Ku and Sebé-Pedrós, 2019), and it can be challenging to disrupt the cell wall without damaging RNA for some organisms (e.g., dinoflagellates; Akbar et al., 2018). As a consequence, specialized cell lysis techniques, relative to those used in animal cells, are often helpful (Ku and Sebé-Pedrós, 2019; Osborn and Hook, 2013). In some cases, commercial lysis buffers are often insufficient and can be paired with mechanical cell disruption (e.g., by micropestle or silica beads). High levels of lipids, polysaccharides, or phenolic compounds (accumulated as a stress response or as a UV-protectant) can impede RNA extraction from algae, but optimized protocols have been proposed to counteract these issues (Mundt et al., 2019).

Library preparation

Transcriptomics sequences total RNA, including messenger RNA (mRNA), small RNA (smRNA), microRNA (miRNA), ribosomal RNA (rRNA), and transfer RNA (tRNA) (Morozova et al., 2009). Typically, mRNA is the focus of transcriptome experiments, but smRNA and miRNA are being increasingly studied for their role in post-transcriptional regulation (Molnár et al., 2007). rRNA can also be used for taxonomic classification, but most approaches involve chemical depletion strategies, due to its great abundance, to increase coverage of mRNA transcripts (Huang et al., 2020). Algae can have multiple chloroplasts of distinct prokaryotic or eukaryotic origin, and speciated billions of years ago, meaning that there is great diversity within the various algal lineages and great distance from other taxa (Caron et al., 2017; Sibbald and Archibald, 2017). This can have practical implications on rRNA depletion methods. Eukaryotic mRNA can also be enriched for by poly(A) selection (Cooper et al., 2014; Wolf et al., 2018), which has the additional effect of depleting prokaryotic reads (Dreyfus and Régnier, 2002); however, this requires higher amounts of input material. Poly(A) selection of metatranscriptomes has been shown to greatly increase eukaryotic coverage, but comes at the expense of organellar transcripts, which are not polyadenylated (Dupont et al., 2015). During library preparation, linear amplification can also be employed to increase yields of low-biomass samples, while inducing minimal biases (Hoeijmakers et al., 2011). Choice of protocol for rRNA removal, cDNA synthesis, and library preparation has been shown to alter reported organism representation and gene expression (Alberti et al., 2014). Ultimately, methods should be chosen to reflect the quantity of input RNA available and whether study objectives are descriptive or quantitative (Alberti et al., 2014).

Sequencing

Various next generation sequencing platforms are currently employed for transcriptomics, including Illumina (Solexa), SOLiD, Ion Torrent, Single molecule real-time (SMRT) sequencing (PacBio), and Nanopore(Ansorge, 2009). Increased sophistication of methodologies, and deeper sequencing have directly improved our understanding of algal physiology. For example, initial microarray profiling of the ichthyotoxic haptophyte, *Prymnesium parvum*, revealed three phosphate transporters upregulated in response to P stress and did not observe a change in nitrogen acquisition genes under nitrogen starvation (Beszteri et al., 2012). A follow-up study using RNA-seq was sensitive enough to show that both N and P transporters had higher expression levels in deplete conditions (Liu et al., 2015b). Similarly, advances in methodologies are likely to unveil physiological nuances that we cannot currently observe with RNA-seq alone.

When choosing a sequencing platform and sequencing depth, consider the length of reads and coverage required. With current methods, 10–25 million nonribosomal reads per sample have achieved reliable results (Wolf et al., 2018), although increased depth may enable higher-resolution investigations of population or strain level differences, e.g., by distinguishing single nucleotide polymorphisms (SNPs) (Le Gac et al., 2016). Longer reads are better as they allow for more reliable taxonomic and functional annotation (Wolf et al., 2018), but can come at the cost of higher error rates (Ansorge, 2009). Increasingly, combining the benefits of short reads (low error rates) with long reads has grown in popularity.

Assembly

Quality-controlled transcriptomic reads can be analyzed in multiple ways. A transcriptome can be created via de novo assembly (Grabherr et al., 2013, 2011). Alternately, an approach that is commonly used to investigate well-described algae is to map reads directly to the genomes or transcriptomes of a known reference and retain only reads mapping closely (Alexander et al., 2015; Cohen et al., 2017; Marchetti et al., 2012). The drawback of this approach for environmental metatranscriptomics, currently, is that it fails to capture algal species for which close references do not exist. An alternative is to call open reading frames (ORFs) directly on metatranscriptomic reads and annotate these ORFs directly to infer taxonomy and function. Because many sequencing strategies rely on short reads (e.g., 300–500 bp; Ansorge, 2009; Huang et al., 2012), in the latter case, reads are typically more informative if assembled into contigs prior to annotation. Assembly can be difficult for transcriptomes, because of population-level heterogeneity and gene-level splice variants, but bioinformatic methods have progressed to rise to the challenge (Martin and Wang, 2011). To capture even more distant lineages, ORFs can be translated into amino acid sequences prior to annotation. In a California current phytoplankton microbiome, 7× more reads mapped to ab initio ORFs using this technique than mapped to MMETSP references in nucleotide-space (Kolody et al., 2019).

Annotation

All transcriptomic studies are plagued by transcripts with unknown function (Grossart et al., 2020), which confound their interpretation (Grossart et al., 2020; Murray et al., 2016). However, in protists these problems may be especially acute, as the organisms are comparatively understudied (Grossart et al., 2020; Sibbald and Archibald, 2017) and have diverse and unique metabolisms (Murray et al., 2016). The inability to annotate genes from uncultured eukaryotes does not appear to be due to coverage or assembly problems (Akbar et al., 2018; Gibson et al., 2013), and could represent novel pathways. For example, a quarter of genes found in the toxic dinoflagellate, *Azadinium spinosum*, had no known homologues in other dinoflagellate genera (Meyer et al., 2015). Transcripts encoding unique metabolic function in organisms indigenous to the Southern Hemisphere and other understudied regions of the world may remain unidentified for quite some time. The lack of available genomes is particularly detrimental for dinoflagellates (Stephens et al., 2018). The metabolic functions performed by dinoflagellates in particular, such as mixotrophy or some symbioses, are often unique, and may not be well represented by model organisms (Murray et al., 2016). For example, only 28% of *Alexandrium minutum* (Yang et al., 2010) and 24% of *Oxyrrhis marina* transcriptomes (Lowe et al., 2011) could be annotated. Some gene functions are more conserved than others. Most yeast genes involved in the regulation of meiosis are conserved in dinoflagellates, but few of those involved in forming the synaptonemal complex or spores are (Morse, 2019). Though canonical genes in model organisms can be used to annotate homologs in distant lineages, increased culturing efforts and careful genetic experimentation are necessary to elucidate the novel proteins that make diverse phytoplankton lineages unique.

Analysis

Transcriptomic techniques in algae are used to infer the metabolic and biochemical potential of an organism for which sequencing the genome is not feasible, and to infer the physiological responses to changing conditions. In the first case, only genes relevant to the given culturing or environmental conditions will be detectible in the resulting transcriptome, so combining multiple transcriptomes across varied conditions will yield a more complete picture of the organism's metabolic potential (Akbar et al., 2018; Maheswari et al., 2010). Additionally, if the organism is not grown axenically or relies on a prey species, these species must be removed bioinformatically. Consequently, we recommend feeding a prey species that is as taxonomically distant from the organisms of interest (and its plastids) as possible.

Inferring physiology from a transcriptome or metatranscriptome is trickier still. For example, a nutrient transporter upregulated in an environmental metatranscriptome could be interpreted either as the algae making use of exogenous supply or the very opposite: increasing transporter capability to cope with low nutrient availability. Though environmental metadata is critical in these situations, it is also important to ground-truth interpretation with experimental results where possible. For example, in the case of diatom nitrate transporters, the interpretation is indeed ambiguous. Model diatoms upregulate the nitrate transporter, NRT2, in response to both nitrate pulses and long-term nitrogen starvation (McCarthy et al., 2017; Poulsen and Kröger, 2005; Smith et al., 2019). This also exemplifies why transcriptomics experiments should extract RNA at multiple timepoints to distinguish primary and secondary responses and associated feedbacks to a stress or stimulus (Akbar et al., 2018; Cramer et al., 2011).

Experimental findings can also act as positive control that can be used as a jumping-off point for other results. For example, a set of California current metatranscriptomes investigated diel cycling transcripts across many lineages that had no cultured representatives and found conserved timing of the expression of photosynthetic machinery. Studies in the lineages that did have cultured isolates peaked at the same time of day as observed in nature, lending credibility to results found in more distant cases (Kolody et al., 2019).

Ground-truthing physiological claims is especially important, given the common decoupling of transcription, protein abundance, and protein activity. Gene expression inferred from a metatranscriptome is sometimes a linear and proportional indicator of microbial metabolism (Helbling et al., 2012). However, the relationship between mRNA and protein levels is complex and is not always proportional or uniform (Maier et al., 2009). Posttranscriptional regulation of mRNA is complex in eukaryotes. Every step of mRNA metabolism is specifically regulated, and this elaborate web of regulation exerts at least as much control as the processes driving mRNA synthesis (Hieronymus and Silver, 2004; Keene and Tenenbaum, 2002; Moore, 2005). The length of untranslated regions in mRNA transcripts illustrates this fact. Most mRNA binding factors target transcript-specific sequences in the untranslated regions (UTRs) on the 3' and 5' ends of a transcript. Though short in prokaryotes (3–10 nucleotides long for 5' UTR) (Lodish et al., 2008), in eukaryotes, these regions can be hundreds to thousands of nucleotides (Lin and Li, 2012), demonstrating the complexity of posttranscriptional regulation.

Whether transcription is reflective of physiology depends upon both the given organism and turnover rate of the given transcript and protein pools. Indeed, a transcriptome is not a perfectly accurate picture of the genes being expressed in the cell; it is an estimate of mRNA pools, which reflect both mRNA synthesis and degradation rates. For instance, 73% of protein abundance is explained by mRNA abundance in yeast (Lu et al., 2007), but for many organisms, this percentage is unknown. Half-lives of transcripts vary, depending on their function. For example, *Karenia brevis* mRNAs have half-lives ranging from 42 mins to 144 hours, and transcripts involved in posttranscriptional regulation, such as the pentatricopeptide repeat (PPR) proteins had dramatically shorter half-lives, suggesting that regulation in dinoflagellates is largely at the protein level (Morey and Van Dolah, 2013). Genes with mRNA pools that turn over more rapidly generally better reflect concurrent protein levels, but the specific physiology being probed may also determine which transcripts are informative. For example, transcripts with a half-life greater than 24 hours are likely not involved in regulating physiological changes that are related to photoperiod, but those known to have short half-lives, such as components of photosystem II that peak when photosynthetically available radiation (and UV damage) is greatest, likely represent true changes at the protein level (Kolody et al., 2019).

Normalization also poses a problem for the interpretation of transcriptional data. Due to technical variation in coverage across sequencing runs, transcriptomics data is typically normalized to the total number of reads in a given run (McGettigan, 2013). Normalization strategies should be tailored to the question being asked, and it is important to consider all underlying assumptions (Evans et al., 2018). For example, questions about the physiology of a specific organism in a metatranscriptome should normalize reads to total abundance of that organism, whereas questions about total community activity of a given enzyme function should utilize normalization to total reads in each sample. Library normalization assumes that total expression is the same for each library, and taxonomic normalization assumes that total expression is the same for a given taxon in each library. Because of this, both methods make it impossible to make inferences about absolute differences in transcript abundance across samples. Instead, investigators can comment on relative differences in expression, e.g., "phosphorous transporters are a larger proportion of total expression in sample A than sample B."

Unfortunately, without an indication of the total transcriptional pool, results can be counterintuitive. For example, suppose *dabA*, a diatom gene involved in biosynthesis of the harmful algal bloom secondary metabolite, domoic acid (Brunson et al., 2018), has higher relative abundance in an early-bloom sample than a late-bloom sample. It may be tempting to conclude that absolute *dabA* expression also decreased over the course of the bloom. However, because cell density increased as the bloom progressed, it is possible that the reverse is true, and there is actually more absolute *dabA* copies per liter of seawater, and more domoic acid, in the late-bloom condition. This compositional data problem prevents the meaningful correlation of relative transcript abundances with any form of absolute metadata. Compositional data problems can be skirted somewhat by bioinformatic workarounds, such as establishing a reference frame and using log ratio transforms (Morton et al., 2019). One way to address this problem is to attain an estimate of the total transcript pool or an analog, such as biomass or cell counts for each sample. For example, some studies have normalized metatranscriptomes to metagenomes (Dupont et al., 2015; Wemheuer et al., 2015). Another popular approach is to spike a known quantity of control RNA into sequencing libraries, which can be used to approximately account for differences in sequencing depth across samples. This method has been employed to estimate mRNA copies per volume of sample or per cell in several metatranscriptomic projects (Aylward et al., 2017; Bertrand et al., 2015; Gifford et al., 2011; Moran et al., 2013; Satinsky et al., 2014). An External RNA Controls Consortium (ERCC), consisting of 92 mRNAs with known concentrations has also been used as an internal standard in several scRNA-seq experiments (Ziegenhain et al., 2018).

Future prospects

Significant advances have been made with the advent of transcriptomics and next generation sequencing. However, progress will only continue with the advent of new technology, more ample algal references, and a larger pool of functionally identified genes. As of 2017, 90% of eukaryotic genomes on the Genomes Online Database are from animals, fungi, and plants, and 90% of the protist genomes that are sequenced belong to parasites (Sibbald and Archibald, 2017). However, as sequencing prices continue to drop, we will be able to sequence whole eukaryotic algal genomes, rather than relying on transcriptomes for references. New and diverse reference species are progressively being added to databases. For example, the United

States Department of Energy has recently pledged to sequence 100 new diatom genomes (https://jgi.doe.gov/csp-2021-100-diatom-genomes/). Additionally, structural prediction of protein folding from genomic data has recently been greatly advanced by the AI system AlphaFold (Wei, 2019), which may mitigate the hypothetical protein problem.

Interpretation of metatranscriptomic data will be empowered by the accumulation careful experiments with model organisms that elucidate novel pathways. New methods, such as the incorporation of spike-ins to estimate RNA copies per unit volume, can potentially allow for more relevant interpretation of transcriptomics datasets.

Additionally, emerging technologies, sych as metabolomics, lipidomics, proteomics, and methyl-proteomics are increasingly accessible and will help to triangulate the true physiology creating observed transcriptomic patterns. Multiple omics technologies are already commonly combined. For example, transcriptomic data used to validate gene models used in *Emiliania* pan genome (Read et al., 2013) and a combined lipidomic and metatranscriptomic study highlighted a number of photophysiological patterns in surface microbial communities of the NPSG (Becker et al., 2020). Proteomic data will be an especially important complement in taxa that regulate much of their physiology posttranscriptionally, such as dinoflagellates (Morse et al., 2018). Though some changes in dinoflagellate transcription have been observed, such as those to growth state (Yang et al., 2010), nutrient status (Morey et al., 2011; Shi et al., 2017), redox state (Okamoto and Hastings, 2003), exposure to metals (Okamoto et al., 2001) and the presence of bacteria (Moustafa et al., 2010), dinoflagellates appear to have much more stable transcriptomes than other phytoplankton species (Kolody et al., 2019), and some core aspects of physiology, such as control of protein synthesis by the circadian clock, appear to operate entirely posttranscriptionally (Morse et al., 2018; Roy et al., 2014). MicroRNAomic profiling has also proven useful for dinoflagellates, and identified gene targets predicted to decrease sulfolipid metabolism as a mechanism for coping with phosphorous stress (Shi et al., 2017). Nanoscale secondary ion mass spectrometry (nanoSIMS) allows for spatial metabolic information on the cellular level (Matantseva et al., 2016), and could be paired with single-cell transcriptomics to give a more nuanced picture of microbial interactions.

Single-cell transcriptomics

Single-cell transcriptomics (scRNA-seq) is an emerging technology that examines the total RNA of individually sorted cells (Ku et al., 2020).

Though single-cell genomics typically recovers only partial and incomplete genomes, especially for eukaryotes (Seeleuthner et al., 2018), scRNA-seq seems to fare better, possibly because of a greater cellular quantity of RNA than DNA. Coverage varies based on methodology (Ku and Sebé-Pedrós, 2019), but early scRNA-seq studies on ciliates were able to recover 80–102% of genes detected in culture-based transcriptomes (Kolisko et al., 2014).

scRNA-seq has been offered as a means to probe the metabolism of diverse uncultured taxa (Kolisko et al., 2014). Single-cell transcriptomics on an isolate from deep water offshore Monterey Bay, CA, identified *Abedinium*, a new, early-branching dinoflagellate lineage (Cooney et al., 2020). scRNA-seq could also be beneficial in traditional laboratory experiments, especially for taxa that are difficult to synchronously culture or phenomena that affect cells in different ways (Ku and Sebé-Pedrós, 2019). One of the primary benefits of scRNA-seq would be in exploring how the transcriptome changes with different cellular states/ life stages. Currently, in several batch cultures, cell "age" is difficult to discern (Ku and Sebé-Pedrós, 2019). Cell division would seldom be synchronous, and therefore, cells collected are likely to be of different ages and to have experienced different environmental stimuli. scRNA-seq may also be particularly helpful for understanding viral infection of algae. Traditional RNA-seq sums expression patterns across individuals, which obscures infection dynamics of different cells at different times. Already, scRNA-seq of *Emiliania huxleyi* cells found heterogeneous viral transcription across different host cells (Ku et al., 2020).

The emergence of scRNASeq may also allow for a better understanding of the role of individual organisms within complex community interactions (Ku and Sebé-Pedrós, 2019). Many types of algae form associations with other species. The best known example of this may be the *Symbiodinium* that form species complexes with reef-building corals (Yellowlees et al., 2008), but the nitrogen-fixing cyanobacteria that form complexes with diatoms are another example. These species complexes perform ecological and biogeochemical functions as part of a community (Harke et al., 2019; Zielinski et al., 2016). The complete pathways would be difficult to unravel in the field, even with a metatranscriptomic approach, given that bacterial transcripts are unlikely to be captured by methods relying on poly(A) pull down that are commonly utilized in eukaryotic studies (Harke et al., 2019). It would also be unlikely to replicate these complex "community-based" physiological pathways in culture studies (Murray et al., 2016). Furthermore, the complexity of dinoflagellate genomes, inability to culture them axenically, and lateral gene transfer mean that differentiating their genomes from bacterial genomes is a challenge: there is frequently no genome reference to which to compare reference transcriptomes (Murray et al., 2016), although reference transcriptomes exist for a diverse array of species (Keeling et al., 2014).

Environmental monitoring

Algal transcriptomics could be used for environmental monitoring (Fig. 13.1). Automated environmental samplers are becoming increasingly sophisticated. For example, the environmental sample processor (ESP), which autonomously collects and preserves cellular material, can be deployed on a free-drifting float (Kolody et al., 2019; Ottesen et al., 2014, 2013) or continuously monitor a single location via attachment to a mooring (Preston et al., 2011). Newer ESP models are even capable of being deployed as part of a coordinated system of sampling robots (Zhang et al., 2021) and on long-range (Zhang et al., 2019b) and deep-diving (Das and Trembath-Reichert, 2020) autonomous underwater vehicles. This diversity of sampling strategies allows for unprecedented monitoring of remote or dangerous locations (Birch, 2018). This monitoring can allow for high throughput screening methods, a better understanding of the physiological changes that occur during harmful algal blooms, as well as bloom forecasting.

High throughput screening tools are needed to assess the impacts of environmental stressors, especially in combination (Villeneuve and Garcia-Reyero, 2011). These tools have advantages over more standardized toxicity tests, in that they provide information about the mode of action, allowing for comparisons of cumulative impacts with multiple modes of action, and provide more sensitivity than conventional tests (Tice et al., 2013; Xia et al., 2008). The algal transcriptome could be used as one of these high throughput screening tools. Previous studies have demonstrated that, as in other taxonomic groups, the transcriptomic profile in algae exposed to different contaminant stressors are unique (Osborn and Hook, 2013). There are also unique changes in the transcriptomes of a variety algae exposed to nutrient stress or light limitation (Frischkorn et al., 2014; Harke et al., 2017; Harke and Gobler, 2013). Diatoms have also been shown to modify their transcriptomes in response to ocean acidification and increased temperature (Thangaraj and Sun, 2021). Changes in the algal transcriptome could be evaluated after short exposures to compare species differences in response to nutrient addition, for example, or to predict responses to herbicides. The relationships between environmental stimuli, cell growth, and toxin production could be used to build a systems biology "network of the cell," to help resource managers predict the onset of harmful algal blooms (Hennon and Dyhrman, 2020). This will help resource managers predict the formation of blooms or predict the increased toxicity of bloom species.

Conclusions

Microalgae are key primary producers, and their ecology and physiology have downstream impacts for the entirety of the aquatic food web. Traditionally limited by their small size and great diversity, research on these organisms has accelerated, due to the advent of transcriptomics and metatranscriptomics. Together, these techniques have yielded insights into algal evolutionary history and plastid acquisition, core physiological features (e.g., sunlight sensing and harvesting, responses to nutrient fluxes), secondary metabolism, viral infection, and ecological interactions. Transcriptomic methods do not come without caveats, but these can be overcome at every step of the pipeline. Finally, new technology, such as scRNA-seq and progress in other realms of algal research herald a bright future for the study of these important organisms.

Acknowledgments

We acknowledge support for research and synthesis related to transcriptomics in microalgae from National Science Foundation Grants NSF-OCE-0727997, NSF-OCE-1756884, and NSF-ANT-1043671; United States Department of Energy Genomics Science program Grants DE-SC00006719, DE-SC0008593 and DE-SC0018344; and Gordon and Betty Moore Foundation Grants GBMF3828 and GBMF5006 (to A.E.A.).

References

Adams, M.D., Kelley, J.M., Gocayne, J.D., Dubnick, M., Polymeropoulos, M.H., Xiao, H., Merril, C.R., Wu, A., Olde, B., Moreno, R.F., 1991. Complementary DNA sequencing: expressed sequence tags and human genome project. Science 252, 1651–1656.

Akbar, M.A., Ahmad, A., Usup, G., Bunawan, H., 2018. RNA-seq as an emerging tool for marine dinoflagellate transcriptome analysis: process and challenges. Processes 6, 1–12.

Alberti, A., Belser, C., Engelen, S., Bertrand, L., Orvain, C., Brinas, L., Cruaud, C., Giraut, L., Da Silva, C., Firmo, C., Aury, J.M., Wincker, P., 2014. Comparison of library preparation methods reveals their impact on interpretation of metatranscriptomic data. BMC Genomics 15, 1–13.

Alexander, H., Rouco, M., Haley, S.T., Dyhrman, S.T., 2020. Transcriptional response of Emiliania huxleyi under changing nutrient environments in the North Pacific Subtropical Gyre. Environ. Microbiol. 22, 1847–1860.

Alexander, H., Rouco, M., Haley, S.T., Wilson, S.T., Karl, D.M., Dyhrman, S.T., 2015. Functional group-specific traits drive phytoplankton dynamics in the oligotrophic ocean. Proc. Natl. Acad. Sci. 112, E5972–E5979.

Allen, A.E., LaRoche, J., Maheswari, U., Lommer, M., Schauer, N., Lopez, P.J., Finazzi, G., Fernie, A.R., Bowler, C., 2008. Whole-cell response of the pennate diatom Phaeodactylum tricornutum to iron starvation. Proc. Natl. Acad. Sci. 105, 10438–10443.

Amin, S.A., Hmelo, L.R., Van Tol, H.M., Durham, B.P., Carlson, L.T., Heal, K.R., Morales, R.L., Berthiaume, C.T., Parker, M.S., Djunaedi, B., Ingalls, A.E., Parsek, M.R., Moran, M.A., Armbrust, E.V., 2015. Interaction and signalling between a cosmopolitan phytoplankton and associated bacteria. Nature 522, 98–101.

Anderson, D.M., Cembella, A.D., Hallegraeff, G.M., 2012. Progress in understanding harmful algal blooms: paradigm shifts and new technologies for research, monitoring, and management. In: Giovannoni, S.J. (Ed.), Annual Review of Marine Science, Vol. 4, pp. 143–176.

Ansorge, W.J., 2009. Next-generation DNA sequencing techniques. Nat. Biotechnol. 25, 195–203.

Ariede, M.B., Candido, T.M., Jacome, A.L.M., Velasco, M.V.R., de Carvalho, J.C.M., Baby, A.R., 2017. Cosmetic attributes of algae-a review. Algal Res. 25, 483–487.

Armbrust, E.V., Berges, J.A., Bowler, C., Green, B.R., Martinez, D., Putnam, N.H., Zhou, S.G., Allen, A.E., Apt, K.E., Bechner, M., Brzezinski, M.A., Chaal, B.K., Chiovitti, A., Davis, A.K., Demarest, M.S., Detter, J.C., Glavina, T., Goodstein, D., Hadi, M.Z., Hellsten, U., Hildebrand, M., Jenkins, B.D., Jurka, J., Kapitonov, V.V., Kroger, N., Lau, W.W.Y., Lane, T.W., Larimer, F.W., Lippmeier, J.C., Lucas, S., Medina, M., Montsant, A., Obornik, M., Parker, M.S., Palenik, B., Pazour, G.J., Richardson, P.M., Rynearson, T.A., Saito, M.A., Schwartz, D.C., Thamatrakoln, K., Valentin, K., Vardi, A., Wilkerson, F.P., Rokhsar, D.S., 2004. The genome of the diatom Thalassiosira pseudonana: ecology, evolution, and metabolism. Science 306, 79–86.

Astier, J., Rossi, J., Chatelain, P., Klinguer, A., Besson-Bard, A., Rosnoblet, C., Jeandroz, S., Nicolas-Francès, V., Wendehenne, D., 2020. Nitric oxide production and signalling in algae. J. Exp. Bot.

Aylward, F.O., Boeuf, D., Mende, D.R., Wood-Charlson, E.M., Vislova, A., Eppley, J.M., Romano, A.E., DeLong, E.F., 2017. Diel cycling and long-term persistence of viruses in the ocean's euphotic zone. Proc. Natl. Acad. Sci. 201714821.

Bachvaroffi, T.R., 2019. A precedented nuclear genetic code with all three termination codons reassigned as sense codons in the syndinean Amoebophrya sp. Ex Karlodinium veneficum. PLoS ONE 14, 1–21.

Baldauf, S.L., 2008. An overview of the phylogeny and diversity of eukaryotes. J. Syst. Evol. 46, 263–273.

Balzano, S., Corre, E., Decelle, J., Sierra, R., Wincker, P., Da Silva, C.D., Poulain, J., Pawlowski, J., Not, F., 2015. Transcriptome analyses to investigate symbiotic relationships between marine protists. Front. Microbiol. 6, 1–14.

Bates, S., Garrison, D., Horner, R., 1998. Bloom dynamics and physiology producing Pseudo-nitzschia species. NATO ASI Ser. G Ecol., 267–292.

Baudrimont, A., Voegeli, S., Viloria, E.C., Stritt, F., Lenon, M., Wada, T., Jaquet, V., Becskei, A., 2017. Multiplexed gene control reveals rapid mRNA turnover. Sci. Adv. 3.

Becker, K.W., Harke, M.J., Mende, D.R., Muratore, D., Weitz, J.S., DeLong, E.F., Dyhrman, S.T., Van Mooy, B.A.S., 2020. Combined pigment and metatranscriptomic analysis reveals highly synchronized diel patterns of phenotypic light response across domains in the open oligotrophic ocean. ISME J., 1–32.

Beedessee, G., Kubota, T., Arimoto, A., Nishitsuji, K., Waller, R.F., Hisata, K., Yamasaki, S., Satoh, N., Kobayashi, J., Shoguchi, E., 2020. Integrated omics unveil the secondary metabolic landscape of a basal dinoflagellate. BMC Biol. 18, 1–16.

Bender, S.J., Durkin, C.A., Berthiaume, C.T., Morales, R.L., Armbrust, E.V., 2014. Transcriptional responses of three model diatoms to nitrate limitation of growth. Front. Mar. Sci. 1, 1–15.
Bender, S.J., Moran, D.M., McIlvin, M.R., Zheng, H., McCrow, J.P., Badger, J., DiTullio, G.R., Allen, A.E., Saito, M.A., 2018. Colony formation in *Phaeocystis antarctica*: connecting molecular mechanisms with iron biogeochemistry. Biogeosciences 15, 4923–4942.
Bernhard, J.M., Bowser, S.S., 1999. Benthic foraminifera of dysoxic sediments: chloroplast sequestration and functional morphology. Earth-Sci. Rev. 46, 149–165.
Bertrand, E.M., Allen, A.E., Dupont, C.L., Norden-Krichmar, T.M., Bai, J., Valas, R.E., Saito, M.A., 2012. Influence of cobalamin scarcity on diatom molecular physiology and identification of a cobalamin acquisition protein. Proc. Natl. Acad. Sci. USA 109, 1762–1771.
Bertrand, E.M., McCrow, J.P., Moustafa, A., Zheng, H., McQuaid, J.B., Delmont, T.O., Post, A.F., Sipler, R.E., Spackeen, J.L., Xu, K., Bronk, D.A., Hutchins, D.A., Allen, A.E., 2015. Phytoplankton-bacterial interactions mediate micronutrient colimitation at the coastal Antarctic sea ice edge. Proc. Natl. Acad. Sci. USA 112, 9938–9943.
Beszteri, S., Yang, I., Jaeckisch, N., Tillmann, U., Frickenhaus, S., Glöckner, G., Cembella, A., John, U., 2012. Transcriptomic response of the toxic prymnesiophyte Prymnesium parvum (N. Carter) to phosphorus and nitrogen starvation. Harmful Algae 18, 1–15.
Bhowmick, S., Mazumdar, A., Moulick, A., Adam, V., 2020. Algal metabolites: an inevitable substitute for antibiotics. Biotechnol. Adv. 107571.
Birch, J., 2018. Collecting and processing samples in remote and dangerous places: the Environmental Sample Processor as a case study. Pure Appl. Chem. 90, 1625–1630.
Bochenek, M., Etherington, G.J., Koprivova, A., Mugford, S.T., Bell, T.G., Malin, G., Kopriva, S., 2013. Transcriptome analysis of the sulfate deficiency response in the marine microalga Emiliania huxleyi. New Phytol. 199, 650–662.
Bowler, C., Allen, A.E., Badger, J.H., Grimwood, J., Jabbari, K., Kuo, A., Maheswari, U., Martens, C., Maumus, F., Otillar, R.P., Rayko, E., Salamov, A., Vandepoele, K., Beszteri, B., Gruber, A., Heijde, M., Katinka, M., Mock, T., Valentin, K., Verret, F., Berges, J.A., Brownlee, C., Cadoret, J.P., Chiovitti, A., Choi, C.J., Coesel, S., De Martino, A., Detter, J.C., Durkin, C., Falciatore, A., Fournet, J., Haruta, M., Huysman, M.J.J., Jenkins, B.D., Jiroutova, K., Jorgensen, R.E., Joubert, Y., Kaplan, A., Kroger, N., Kroth, P.G., La Roche, J., Lindquist, E., Lommer, M., Martin-Jézéquel, V., Lopez, P.J., Lucas, S., Mangogna, M., McGinnis, K., Medlin, L.K., Montsant, A., Oudot-Le Secq, M.-P., Napoli, C., Obornik, M., Schnitzler Parker, M., Petit, J.-L., Porcel, B.M., Poulsen, N., Robison, M., Rychlewski, L., Rynearson, T.A., Schmutz, J., Shapiro, H., Siaut, M., Stanley, M., Sussman, M.R., Taylor, A.R., Vardi, A., von Dassow, P., Vyverman, W., Willis, A., Wyrwicz, L.S., Rokhsar, D.S., Weissenbach, J., Armbrust, E.V., Green, B.R., Van de Peer, Y., Grigoriev, I.V., 2008. The Phaeodactylum genome reveals the evolutionary history of diatom genomes. Nature 456, 239–244.
Bowler, C., Vardi, A., Allen, A.E., 2010. Oceanographic and biogeochemical insights from diatom genomes. Annu. Rev. Mar. Sci. 2, 333–365.
Boysen, A., Carlson, L., Durham, B., Groussman, R., Aylward, F., Ribalet, F., Heal, K., DeLong, E., Armbrust, E.V., Ingalls, A., 2020. Diel oscillations of particulate metabolites reflect synchronized microbial activity in the North Pacific Subtropical Gyre.
Brunson, J.K., McKinnie, S.M.K., Chekan, J.R., McCrow, J.P., Miles, Z.D., Bertrand, E.M., Bielinski, V.A., Luhavaya, H., Oborník, M., Smith, G.J., Hutchins, D.A., Allen, A.E., Moore, B.S., 2018. Biosynthesis of the neurotoxin domoic acid in a bloom-forming diatom. Science 361, 1356–1358.
Burkholder, J.A.M., Glibert, P.M., Skelton, H.M., 2008. Mixotrophy, a major mode of nutrition for harmful algal species in eutrophic waters. Harmful Algae 8, 77–93.
Burki, F., Roger, A.J., Brown, M.W., Simpson, A.G.B., 2020. The new tree of eukaryotes. Trends Ecol. Evol. 35, 43–55.

Caron, D.A., 2016. The rise of Rhizaria. Nature 532, 444–445.
Caron, D.A., Alexander, H., Allen, A.E., Archibald, J.M., Armbrust, E.V., Bachy, C., Bell, C.J., Bharti, A., Dyhrman, S.T., Guida, S.M., Heidelberg, K.B., Kaye, J.Z., Metzner, J., Smith, S.R., Worden, A.Z., 2017. Probing the evolution, ecology and physiology of marine protists using transcriptomics. Nat. Rev. Microbiol. 15, 6–20.
Charon, J., Marcelino, V.R., Wetherbee, R., Verbruggen, H., Holmes, E.C., 2020. Metatranscriptomic identification of diverse and divergent RNA viruses in green and chlorarachniophyte algae cultures. Viruses 12, 1–27.
Coesel, S., Mangogna, M., Ishikawa, T., Heijde, M., Rogato, A., Finazzi, G., Todo, T., Bowler, C., Falciatore, A., 2009. Diatom PtCPF1 is a new cryptochrome/photolyase family member with DNA repair and transcription regulation activity. EMBO Rep. 10, 655–661.
Cohen, N.R., Ellis, K.A., Lampe, R.H., McNair, H., Twining, B.S., Maldonado, M.T., Brzezinski, M.A., Kuzminov, F.I., Thamatrakoln, K., Till, C.P., Bruland, K.W., Sunda, W.G., Bargu, S., Marchetti, A., 2017. Diatom transcriptional and physiological responses to changes in iron bioavailability across ocean provinces. Front. Mar. Sci. 4.
Cohen, N.R., Gong, W., Moran, D.M., McIlvin, M.R., Saito, M.A., Marchetti, A., 2018. Transcriptomic and proteomic responses of the oceanic diatom Pseudo-nitzschia granii to iron limitation. Environ. Microbiol. 20, 3109–3126.
Connolly, J.A., Oliver, M.J., Beaulieu, J.M., Knight, C.A., Tomanek, L., Moline, M.A., 2008. Correlated evolution of genome size and cell volume in diatoms (Bacillariophyceae). J. Phycol. 44, 124–131.
Cooney, E.C., Okamoto, N., Cho, A., Hehenberger, E., Richards, T.A., Santoro, A.E., Worden, A.Z., Leander, B.S., Keeling, P.J., 2020. Single cell transcriptomics of Abedinium reveals a new early-branching dinoflagellate lineage. Genome Biol. Evol. 12, 2417–2428.
Cooper, E.D., Bentlage, B., Gibbons, T.R., Bachvaroff, T.R., Delwiche, C.F., 2014. Metatranscriptome profiling of a harmful algal bloom. Harmful Algae 37, 75–83.
Corellou, F., Schwartz, C., Motta, J.-P., Djouani-Tahri, E.B., Sanchez, F., Bouget, F.-Y., 2009. Clocks in the green lineage: comparative functional analysis of the circadian architecture of the picoeukaryote Ostreococcus. Plant Cell 21, 3436–3449.
Craig Venter, J., Adams, M.D., Myers, E.W., Li, P.W., Mural, R.J., Sutton, G.G., Smith, H.O., Yandell, M., Evans, C.A., Holt, R.A., Gocayne, J.D., Amanatides, P., Ballew, R.M., Huson, D.H., Wortman, J.R., Zhang, Q., Kodira, C.D., Zheng, X.H., Chen, L., Skupski, M., Subramanian, G., Thomas, P.D., Zhang, J., Gabor Miklos, G.L., Nelson, C., Broder, S., Clark, A.G., Nadeau, J., McKusick, V.A., Zinder, N., Levine, A.J., Roberts, R.J., Simon, M., Slayman, C., Hunkapiller, M., Bolanos, R., Delcher, A., Dew, I., Fasulo, D., Flanigan, M., Florea, L., Halpern, A., Hannenhalli, S., Kravitz, S., Levy, S., Mobarry, C., Reinert, K., Remington, K., Abu-Threideh, J., Beasley, E., Biddick, K., Bonazzi, V., Brandon, R., Cargill, M., Chandramouliswaran, I., Charlab, R., Chaturvedi, K., Deng, Z., di Francesco, V., Dunn, P., Eilbeck, K., Evangelista, C., Gabrielian, A.E., Gan, W., Ge, W., Gong, F., Gu, Z., Guan, P., Heiman, T.J., Higgins, M.E., Ji, R.R., Ke, Z., Ketchum, K.A., Lai, Z., Lei, Y., Li, Z., Li, J., Liang, Y., Lin, X., Lu, F., Merkulov, G.V., Milshina, N., Moore, H.M., Naik, A.K., Narayan, V.A., Neelam, B., Nusskern, D., Rusch, D.B., Salzberg, S., Shao, W., Shue, B., Sun, J., Yuan Wang, Z., Wang, A., Wang, X., Wang, J., Wei, M.H., Wides, R., Xiao, C., Yan, C., Yao, A., Ye, J., Zhan, M., Zhang, W., Zhang, H., Zhao, Q., Zheng, L., Zhong, F., Zhong, W., Zhu, S.C., Zhao, S., Gilbert, D., Baumhueter, S., Spier, G., Carter, C., Cravchik, A., Woodage, T., Ali, F., An, H., Awe, A., Baldwin, D., Baden, H., Barnstead, M., Barrow, I., Beeson, K., Busam, D., Carver, A., Center, A., Lai Cheng, M., Curry, L., Danaher, S., Davenport, L., Desilets, R., Dietz, S., Dodson, K., Doup, L., Ferriera, S., Garg, N., Gluecksmann, A., Hart, B., Haynes, J., Haynes, C., Heiner, C., Hladun, S., Hostin, D.,

Houck, J., Howland, T., Ibegwam, C., Johnson, J., Kalush, F., Kline, L., Koduru, S., Love, A., Mann, F., May, D., McCawley, S., McIntosh, T., McMullen, I., Moy, M., Moy, L., Murphy, B., Nelson, K., Pfannkoch, C., Pratts, E., Puri, V., Qureshi, H., Reardon, M., Rodriguez, R., Rogers, Y.H., Romblad, D., Ruhfel, B., Scott, R., Sitter, C., Smallwood, M., Stewart, E., Strong, R., Suh, E., Thomas, R., Ni Tint, N., Tse, S., Vech, C., Wang, G., Wetter, J., Williams, S., Williams, M., Windsor, S., Winn-Deen, E., Wolfe, K., Zaveri, J., Zaveri, K., Abril, J.F., Guigo, R., Campbell, M.J., Sjolander, K.V., Karlak, B., Kejariwal, A., Mi, H., Lazareva, B., Hatton, T., Narechania, A., Diemer, K., Muruganujan, A., Guo, N., Sato, S., Bafna, V., Istrail, S., Lippert, R., Schwartz, R., Walenz, B., Yooseph, S., Allen, D., Basu, A., Baxendale, J., Blick, L., Caminha, M., Carnes-Stine, J., Caulk, P., Chiang, Y.H., Coyne, M., Dahlke, C., Deslattes Mays, A., Dombroski, M., Donnelly, M., Ely, D., Esparham, S., Fosler, C., Gire, H., Glanowski, S., Glasser, K., Glodek, A., Gorokhov, M., Graham, K., Gropman, B., Harris, M., Heil, J., Henderson, S., Hoover, J., Jennings, D., Jordan, C., Jordan, J., Kasha, J., Kagan, L., Kraft, C., Levitsky, A., Lewis, M., Liu, X., Lopez, J., Ma, D., Majoros, W., McDaniel, J., Murphy, S., Newman, M., Nguyen, T., Nguyen, N., Nodell, M., Pan, S., Peck, J., Peterson, M., Rowe, W., Sanders, R., Scott, J., Simpson, M., Smith, T., Sprague, A., Stockwell, T., Turner, R., Venter, E., Wang, M., Wen, M., Wu, D., Wu, M., Xia, A., Zandieh, A., Zhu, X., 2001. The sequence of the human genome. Science 291, 1304–1351.

Cramer, G.R., Urano, K., Delrot, S., Pezzotti, M., Shinozaki, K., 2011. Effects of abiotic stress on plants: a systems biology perspective. BMC Plant Biol. 11, 1–14.

Crick, F., 1970. Central dogma of molecular biology. Nature 227, 561–563.

Das, J., Trembath-Reichert, E., 2020. Deep diving with Clio. Sci. Robot. 5, 1–3.

Davenport, E.J., Neudeck, M.J., Matson, P.G., Bullerjahn, G.S., Davis, T.W., Wilhelm, S.W., Denney, M.K., Krausfeldt, L.E., Stough, J.M.A., Meyer, K.A., Dick, G.J., Johengen, T.H., Lindquist, E., Tringe, S.G., McKay, R.M.L., 2019. Metatranscriptomic analyses of diel metabolic functions during a microcystis bloom in Western Lake Erie (United States). Front. Microbiol. 10, 2081.

Deloménie, C., Foti, E., Floch, E., Diderot, V., Porquet, D., Dupuy, C., Bonaly, J., 2007. A new homolog of FocA transporters identified in cadmium-resistant Euglena gracilis. Biochem. Biophys. Res. Commun. 358, 455–461.

Derelle, E., Yau, S., Moreau, H., Grimsley, N.H., 2017. Prasinovirus attack of Ostreococcus is furtive by day but savage by night. J. Virol. 92. https://doi.org/10.1128/JVI.01703-17.

Di Dato, V., Musacchia, F., Petrosino, G., Patil, S., Montresor, M., Sanges, R., Ferrante, M.I., 2015. Transcriptome sequencing of three Pseudo-nitzschia species reveals comparable gene sets and the presence of Nitric Oxide Synthase genes in diatoms. Sci. Rep. 5, 1–14.

Doron, S., Fedida, A., Hernndez-Prieto, M.A., Sabehi, G., Karunker, I., Stazic, D., Feingersch, R., Steglich, C., Futschik, M., Lindell, D., Sorek, R., 2016. Transcriptome dynamics of a broad host-range cyanophage and its hosts. ISME J. 10, 1437–1455.

Dreyfus, M., Régnier, P., 2002. The poly(A) tail of mRNAs: bodyguard in eukaryotes, scavenger in bacteria. Cell 111, 611–613.

Duanmu, D., Bachy, C., Sudek, S., Wong, C.H., Jiménez, V., Rockwell, N.C., Martin, S.S., Ngan, C.Y., Reistetter, E.N., Van Baren, M.J., Price, D.C., Wei, C.L., Reyes-Prieto, A., Lagarias, J.C., Worden, A.Z., 2014. Marine algae and land plants share conserved phytochrome signaling systems. Proc. Natl. Acad. Sci. USA 111, 15827–15832.

Dufresne, A., Salanoubat, M., Partensky, F., Artiguenave, F., Axmann, I.M., Barbe, V., Duprat, S., Galperin, M.Y., Koonin, E.V., Le Gall, F., Makarova, K.S., Ostrowski, M., Oztas, S., Robert, C., Rogozin, I.B., Scanlan, D.J., Tandeau de Marsac, N., Weissenbach, J., Wincker, P., Wolf, Y.I., Hess, W.R., 2003. Genome sequence of the cyanobacterium Prochlorococcus marinus SS120, a nearly minimal oxyphototrophic genome. Proc. Natl. Acad. Sci. USA 120, 10020–10025.

Dupont, C.L., McCrow, J.P., Valas, R., Moustafa, A., Walworth, N., Goodenough, U., Roth, R., Hogle, S.L., Bai, J., Johnson, Z.I., Mann, E., Palenik, B., Barbeau, K.A., Craig Venter, J., Allen, A.E., 2015. Genomes and gene expression across light and productivity gradients in eastern subtropical Pacific microbial communities. ISME J. 9, 1076–1092.
Durham, B.P., Boysen, A.K., Carlson, L.T., Groussman, R.D., Heal, K.R., Cain, K.R., Morales, R.L., Coesel, S.N., Morris, R.M., Ingalls, A.E., Armbrust, E.V., 2019. Sulfonate-based networks between eukaryotic phytoplankton and heterotrophic bacteria in the surface ocean. Nat. Microbiol. 4, 1706–1715.
Dyhrman, S.T., Haley, S.T., Birkeland, S.R., Wurch, L.L., Cipriano, M.J., McArthur, A.G., 2006. Long serial analysis of gene expression for gene discovery and transcriptome profiling in the widespread marine coccolithophore Emiliania huxleyi. Appl. Environ. Microbiol. 72, 252–260.
Dyhrman, S.T., Jenkins, B.D., Rynearson, T.A., Saito, M.A., Mercier, M.L., Alexander, H., Whitney, L.A.P., Drzewianowski, A., Bulygin, V.V., Bertrand, E.M., Wu, Z., Benitez-Nelson, C., Heithoff, A., 2012. The transcriptome and proteome of the diatom thalassiosira pseudonana reveal a diverse phosphorus stress response. PLoS ONE 7.
Ebenezer, T.E., Zoltner, M., Burrell, A., Nenarokova, A., Novák Vanclová, A.M.G., Prasad, B., Soukal, P., Santana-Molina, C., O'Neill, E., Nankissoor, N.N., Vadakedath, N., Daiker, V., Obado, S., Silva-Pereira, S., Jackson, A.P., Devos, D.P., Lukeš, J., Lebert, M., Vaughan, S., Hampl, V., Carrington, M., Ginger, M.L., Dacks, J.B., Kelly, S., Field, M.C., 2019. Transcriptome, proteome and draft genome of Euglena gracilis. BMC Biol. 17, 1–23.
Ehira, S., 2013. Transcriptional regulation of heterocyst differentiation in Anabaena sp. strain PCC 7120. Russ. J. Plant Physiol. 60, 443–452.
Evans, C., Hardin, J., Stoebel, D.M., 2018. Selecting between-sample RNA-Seq normalization methods from the perspective of their assumptions. Brief. Bioinform. 19, 776–792.
Facella, P., Lopez, L., Carbone, F., Galbraith, D.W., Giuliano, G., Perrotta, G., 2008. Diurnal and circadian rhythms in the tomato transcriptome and their modulation by cryptochrome photoreceptors. PLoS ONE 3.
Falciatore, A., Bowler, C., 2002. Revealing the molecular secrets of marine diatoms. Annu. Rev. Plant Biol. 53, 109–130.
Farhat, S., Florent, I., Noel, B., Kayal, E., Da Silva, C., Bigeard, E., Alberti, A., Labadie, K., Corre, E., Aury, J.M., Rombauts, S., Wincker, P., Guillou, L., Porcel, B.M., 2018. Comparative time-scale gene expression analysis highlights the infection processes of two amoebophrya strains. Front. Microbiol. 9, 1–19.
Field, C.B., Behrenfeld, M.J., Randerson, J.T., Falkowski, P., 1998. Primary production of the biosphere: integrating terrestrial and oceanic components. Science 281.
Flaherty, B.L., Nieuwerburgh, F.V., Head, S.R., Golden, J.W., 2011. Directional RNA deep sequencing sheds new light on the transcriptional response of Anabaena sp. strain PCC 7120 to combinded-nitrogen deprivation. BMC Genomics 12, 332.
Flynn, K.J., Stoecker, D.K., Mitra, A., Raven, J.A., Glibert, P.M., Hansen, P.J., Granéli, E., Burkholder, J.M., 2013. Misuse of the phytoplankton-zooplankton dichotomy: the need to assign organisms as mixotrophs within plankton functional types. J. Plankton Res. 35, 3–11.
Frischkorn, K.R., Harke, M.J., Gobler, C.J., Dyhrman, S.T., 2014. De novo assembly of Aureococcus anophagefferens transcriptomes reveals diverse responses to the low nutrient and low light conditions present during blooms. Front. Microbiol. 5, 1–16.
Fujiwara, T., Hirooka, S., Ohbayashi, R., Onuma, R., Miyagishima, S.Y., 2020. Relationship between cell cycle and diel transcriptomic changes in metabolism in a unicellular red alga1[OPEN]. Plant Physiol. 183, 1484–1501.
Gallot-Lavallée, L., Blanc, G., 2017. A glimpse of nucleo-cytoplasmic large DNA virus biodiversity through the eukaryotic genomics window. Viruses 9.

Gavelis, G.S., Wakeman, K.C., Tillmann, U., Ripken, C., Mitarai, S., Herranz, M., Özbek, S., Holstein, T., Keeling, P.J., Leander, B.S., 2017. Microbial arms race: ballistic "nematocysts" in dinoflagellates represent a new extreme in organelle complexity. Sci. Adv. 3.

Geimer, S., Belicová, A., Legen, J., Sláviková, S., Herrmann, R.G., Krajčovič, J., 2009. Transcriptome analysis of the Euglena gracilis plastid chromosome. Curr. Genet. 55, 425–438.

Gibson, A.K., Smith, Z., Fuqua, C., Clay, K., Colbourne, J.K., 2013. Why so many unknown genes? Partitioning orphans from a representative transcriptome of the lone star tick Amblyomma americanum. BMC Genomics 14, 1–15.

Gifford, S.M., Sharma, S., Rinta-Kanto, J.M., Moran, M.A., 2011. Quantitative analysis of a deeply sequenced marine microbial metatranscriptome. ISME J. 5, 461–472.

Gillard, J., Devos, V., Huysman, M.J.J., De Veylder, L., D'Hondt, S., Martens, C., Vanormelingen, P., Vannerum, K., Sabbe, K., Chepurnov, V.A., Inzé, D., Vuylsteke, M., Vyverman, W., 2008. Physiological and transcriptomic evidence for a close coupling between chloroplast ontogeny and cell cycle progression in the pennate diatom Seminavis robusta. Plant Physiol. 148, 1394–1411.

Glibert, P.M., 2017. Eutrophication, harmful algae and biodiversity — challenging paradigms in a world of complex nutrient changes. Mar. Pollut. Bull. 124, 591–606.

Glibert, P.M., Heil, C.A., Wilkerson, F.P., Dugdale, R.C., 2018. Nutrients and harmful algal blooms: dynamic kinetics and flexible nutrition. In: Berdalet, E., Burford, M.A., Pitcher, G.C., Zhou, M. (Eds.), Global Ecology and Oceanography of Harmful Algal Blooms. Global Ecology and Oceanography of Harmful Algal Blooms, pp. 92–111.

Govorunova, E.G., Sineshchekov, O.A., Li, H., Spudich, J.L., 2017. Microbial rhodopsins: diversity, mechanisms, and optogenetic applications. Annu. Rev. Biochem. 86, 845–872.

Grabherr, M.G., Haas, B.J., Yassour, M., Levin, J.Z., Thompson, D.A., Amit, I., Adiconis, X., Fan, L., Raychowdhury, R., Zeng, Q., 2011. Full-length transcriptome assembly from RNA-Seq data without a reference genome. Nat. Biotechnol. 29, 644–652.

Grabherr, M.G., Haas, Brian J., Yassour, Moran, Levin, Joshua Z., Thompson, Dawn A., Amit, Ido, Adiconis, Xian, Fan, Lin, Raychowdhury, Raktima, Zeng, Qiandong, Chen, Zehua, Mauceli, Evan, Hacohen, Nir, Gnirke, Andreas, Rhind, Nicholas, Birren, Bruce W., Nusbaum, Chad, Lindblad-Toh, Kerstin, Friedman, Nir, Regev, Aviv, 2013. Trinity: reconstructing a full-length transcriptome without a genome from RNA-Seq data. Nat. Biotechnol. 29, 644–652.

Graupner, N., Jensen, M., Bock, C., Marks, S., Rahmann, S., Beisser, D., Boenigk, J., 2018. Evolution of heterotrophy in chrysophytes as reflected by comparative transcriptomics. FEMS Microbiol. Ecol. 94.

Grossart, H.-P., Massana, R., McMahon, K.D., Walsh, D.A., 2020. Linking metagenomics to aquatic microbial ecology and biogeochemical cycles. Limnol. Oceanogr. 65, S2–S20.

Groussman, R.D., Parker, M.S., Armbrust, E.V., 2015. Diversity and evolutionary history of iron metabolism genes in diatoms. PLoS ONE 10, 1–25.

Guerreiro, A.C.L., Benevento, M., Lehmann, R., van Breukelen, B., Post, H., Giansanti, P., Maarten Altelaar, A.F., Axmann, I.M., Heck, A.J.R., 2014. Daily rhythms in the cyanobacterium Synechococcus elongatus probed by high-resolution mass spectrometry–based proteomics reveals a small defined set of cyclic proteins. Mol. Cell. Proteomics 13, 2042–2055.

Haley, S.T., Alexander, H., Juhl, A.R., Dyhrman, S.T., 2017. Transcriptional response of the harmful raphidophyte Heterosigma akashiwo to nitrate and phosphate stress. Harmful Algae 68, 258–270.

Harke, M.J., Berry, D.L., Ammerman, J.W., Gobler, C.J., 2012. Molecular response of the bloom-forming cyanobacterium, Microcystis aeruginosa, to phosphorus limitation. Microb. Ecol. 63, 188–198.

Harke, M.J., Frischkorn, K.R., Haley, S.T., Aylward, F.O., Zehr, J.P., Dyhrman, S.T., 2019. Periodic and coordinated gene expression between a diazotroph and its diatom host. ISME J. 13, 118–131.

Harke, M.J., Gobler, C.J., 2013. Global transcriptional responses of the toxic Cyanobacterium, Microcystis aeruginosa, to nitrogen stress, phosphorus stress, and growth on organic matter. PLoS ONE 8.

Harke, M.J., Gobler, C.J., 2015. Daily transcriptome changes reveal the role of nitrogen in controlling microcystin synthesis and nutrient transport in the toxic cyanobacterium, Microcystis aeruginosa. BMC Genomics 16, 1068.

Harke, M.J., Jankowiak, J.G., Morrell, B.K., Gobler, C.J., 2016a. Transcriptomic responses in the bloom-forming cyanobacterium Microcystis induced during exposure to zooplankton. Appl. Environ. Microbiol. 83.

Harke, M.J., Juhl, A.R., Haley, S.T., Alexander, H., Dyhrman, S.T., 2017. Conserved transcriptional responses to nutrient stress in bloom-forming algae. Front. Microbiol. 8.

Harke, M.J., Steffen, M.M., Gobler, C.J., Otten, T.G., Wilhelm, S.W., Wood, S.A., Paerl, H.W., 2016b. A review of the global ecology, genomics, and biogeography of the toxic cyanobacterium, Microcystis spp. Harmful Algae 54, 4–20.

He, P., Cai, X., Chen, K., Fu, X., 2020. Identification of small RNAs involved in nitrogen fixation in Anabaena sp. PCC 7120.

Hehenberger, E., Gast, R.J., Keeling, P.J., 2019. A kleptoplastidic dinoflagellate and the tipping point between transient and fully integrated plastid endosymbiosis. Proc. Natl. Acad. Sci. USA 116, 17934–17942.

Heijde, M., Zabulon, G., Corellou, F., Ishikawa, T., Brazard, J., Usman, A., Sanchez, F., Plaza, P., Martin, M., Falciatore, A., Todo, T., Bouget, F.Y., Bowler, C., 2010. Characterization of two members of the cryptochrome/photolyase family from Ostreococcus tauri provides insights into the origin and evolution of cryptochromes. Plant Cell Environ.

Helbling, D.E., Ackermann, M., Fenner, K., Kohler, H.P.E., Johnson, D.R., 2012. The activity level of a microbial community function can be predicted from its metatranscriptome. ISME J. 6, 902–904.

Hennon, G.M.M., Dyhrman, S.T., 2020. Progress and promise of omics for predicting the impacts of climate change on harmful algal blooms. Harmful Algae 91, 7.

Hennon, G.M.M., Williamson, O.M., Hernández Limón, M.D., Haley, S.T., Dyhrman, S.T., 2019. Non-linear physiology and gene expression responses of harmful alga Heterosigma akashiwo to rising CO2. Protist 170, 38–51.

Hernández Limón, M.D., Hennon, G.M.M., Harke, M.J., Frischkorn, K.R., Haley, S.T., Dyhrman, S.T., 2020. Transcriptional patterns of Emiliania huxleyi in the North Pacific Subtropical Gyre reveal the daily rhythms of its metabolic potential. Environ. Microbiol. 22, 381–396.

Hieronymus, H., Silver, P.A., 2004. A systems view of mRNP biology. Genes Dev. 18, 2845–2860.

Hippmann, A.A., Schuback, N., Moon, K.M., McCrow, J.P., Allen, A.E., Foster, L.J., Green, B.R., Maldonado, M.T., 2017. Contrasting effects of copper limitation on the photosynthetic apparatus in two strains of the open ocean diatom Thalassiosira oceanica. PLoS ONE.

Hoeijmakers, W.A.M., Bártfai, R., Françoijs, K.J., Stunnenberg, H.G., 2011. Linear amplification for deep sequencing. Nat. Protoc. 6, 1026–1036.

Hook, S.E., White, C., Ross, D.J., 2021. A metatranscriptomic analysis of changing dynamics in the plankton communities adjacent to aquaculture leases in southern Tasmania, Australia. Mar. Genomics, 100858.

Huang, W., Li, L., Myers, J.R., Marth, G.T., 2012. ART: a next-generation sequencing read simulator. Bioinformatics 28, 593–594.

Huang, Y., Sheth, R.U., Kaufman, A., Wang, H.H., 2020. Scalable and cost-effective ribonuclease-based rRNA depletion for transcriptomics. Nucleic Acids Res. 48, E20.

Jang, S.H., Jeong, H.J., Chon, J.K., 2019. De novo transcriptome of the newly described phototrophic dinoflagellate Yihiella yeosuensis: comparison between vegetative cells and cysts. Mar. Biol. 166, 1–21.

Janouškovec, J., Gavelis, G.S., Burki, F., Dinh, D., Bachvaroff, T.R., Gornik, S.G., Bright, K.J., Imanian, B., Strom, S.L., Delwiche, C.F., Waller, R.F., Fensome, R.A., Leander, B.S., Rohwer, F.L., Saldarriaga, J.F., 2017. Major transitions in dinoflagellate evolution unveiled by phylotranscriptomics. Proc. Natl. Acad. Sci. 114, E171–E180.

Ji, N.J., Lin, L.X., Li, L., Yu, L.Y., Zhang, Y.Q., Luo, H., Li, M.Z., Shi, X.G., Wang, D.Z., Lin, S.J., 2018. Metatranscriptome analysis reveals environmental and diel regulation of a Heterosigma akashiwo (raphidophyceae) bloom. Environ. Microbiol. 20, 1078–1094.

Jing, X.L., Lin, S.J., Zhang, H., Koerting, C., Yu, Z.G., 2017. Utilization of urea and expression profiles of related genes in the dinoflagellate Prorocentrum donghaiense. PLoS ONE 12.

Johnson, M.D., Oldach, D., Delwiche, C.F., Stoecker, D.K., 2007. Retention of transcriptionally active cryptophyte nuclei by the ciliate Myrionecta rubra. Nature 445, 426–428.

Johnson, W.M., Kido Soule, M.C., Kujawinski, E.B., 2016. Evidence for quorum sensing and differential metabolite production by a marine bacterium in response to DMSP. ISME J. 10, 2304–2316.

Kamjunke, N., Tittel, J., 2008. Utilisation of leucine by several phytoplankton species. Limnologica.

Keeling, P.J., Burki, F., Wilcox, H.M., Allam, B., Allen, E.E., Amaral-Zettler, L.A., Armbrust, E.V., Archibald, J.M., Bharti, A.K., Bell, C.J., Beszteri, B., Bidle, K.D., Cameron, C.T., Campbell, L., Caron, D.A., Cattolico, R.A., Collier, J.L., Coyne, K., Davy, S.K., Deschamps, P., Dyhrman, S.T., Edvardsen, B., Gates, R.D., Gobler, C.J., Greenwood, S.J., Guida, S.M., Jacobi, J.L., Jakobsen, K.S., James, E.R., Jenkins, B., John, U., Johnson, M.D., Juhl, A.R., Kamp, A., Katz, L.A., Kiene, R., Kudryavtsev, A., Leander, B.S., Lin, S., Lovejoy, C., Lynn, D., Marchetti, A., McManus, G., Nedelcu, A.M., Menden-Deuer, S., Miceli, C., Mock, T., Montresor, M., Moran, M.A., Murray, S., Nadathur, G., Nagai, S., Ngam, P.B., Palenik, B., Pawlowski, J., Petroni, G., Piganeau, G., Posewitz, M.C., Rengefors, K., Romano, G., Rumpho, M.E., Rynearson, T., Schilling, K.B., Schroeder, D.C., Simpson, A.G.B., Slamovits, C.H., Smith, D.R., Smith, G.J., Smith, S.R., Sosik, H.M., Stief, P., Theriot, E., Twary, S.N., Umale, P.E., Vaulot, D., Wawrik, B., Wheeler, G.L., Wilson, W.H., Xu, Y., Zingone, A., Worden, A.Z., 2014. The marine microbial eukaryote transcriptome sequencing project (MMETSP): illuminating the functional diversity of eukaryotic life in the oceans through transcriptome sequencing. PLoS Biol. 12, e1001889.

Keene, J.D., Tenenbaum, S.A., 2002. Eukaryotic mRNPs may represent posttranscriptional operons. Mol. Cell 9, 1161–1167.

Koch, F., Beszteri, S., Harms, L., Trimborn, S., 2019. The impacts of iron limitation and ocean acidification on the cellular stoichiometry, photophysiology, and transcriptome of Phaeocystis antarctica. Limnol. Oceanogr. 64, 357–375.

Koid, A.E., Liu, Z., Terrado, R., Jones, A.C., Caron, D.A., Heidelberg, K.B., 2014. Comparative transcriptome analysis of four prymnesiophyte algae. PLoS ONE 9.

Kolisko, M., Boscaro, V., Burki, F., Lynn, D.H., Keeling, P.J., 2014. Single-cell transcriptomics for microbial eukaryotes. Curr. Biol. 24, R1081–R1082.

Kolody, B.C., McCrow, J.P., Allen, L.Z., Aylward, F.O., Fontanez, K.M., Moustafa, A., Moniruzzaman, M., Chavez, F.P., Scholin, C.A., Allen, E.E., Worden, A.Z., Delong, E.F., Allen, A.E., 2019. Diel transcriptional response of a California Current plankton microbiome to light, low iron, and enduring viral infection. ISME J., 2817–2833.

Kottke, T., Oldemeyer, S., Wenzel, S., Zou, Y., Mittag, M., 2017. Cryptochrome photoreceptors in green algae: unexpected versatility of mechanisms and functions. J. Plant Physiol. 217, 4–14.
Krabberød, A.K., Orr, R.J.S., Bråte, J., Kristensen, T., Bjørklund, K.R., Shalchian-Tabrizi, K., 2017. Single cell transcriptomics, mega-phylogeny, and the genetic basis of morphological innovations in Rhizaria. Mol. Biol. Evol. 34, 1557–1573.
Kranzler, C.F., Krause, J.W., Brzezinski, M.A., Edwards, B.R., Biggs, W.P., Maniscalco, M., McCrow, J.P., Van Mooy, B.A.S., Bidle, K.D., Allen, A.E., Thamatrakoln, K., 2019. Silicon limitation facilitates virus infection and mortality of marine diatoms. Nat. Microbiol. 4, 1790–1797.
Ku, C., Ku, C., Sheyn, U., Sheyn, U., Sebé-Pedrós, A., Sebé-Pedrós, A., Ben-Dor, S., Schatz, D., Tanay, A., Tanay, A., Rosenwasser, S., Vardi, A., 2020. A single-cell view on alga-virus interactions reveals sequential transcriptional programs and infection states. Sci. Adv. 6.
Ku, C., Sebé-Pedrós, A., 2019. Using single-cell transcriptomics to understand functional states and interactions in microbial eukaryotes. Philos. Trans. R. Soc. B, Biol. Sci. 374.
La Roche, J., Boyd, P.W., McKay, R.M.L., Geider, R.J., 1996. Flavodoxin as an in situ marker for iron stress in phytoplankton. Nature 382, 802–805.
Landa, M., Burns, A.S., Roth, S.J., Moran, M.A., 2017. Bacterial transcriptome remodeling during sequential co-culture with a marine dinoflagellate and diatom. ISME J. 11, 2677–2690.
Lasek-Nesselquist, E., Wisecaver, J.H., Hackett, J.D., Johnson, M.D., 2015. Insights into transcriptional changes that accompany organelle sequestration from the stolen nucleus of Mesodinium rubrum. BMC Genomics 16, 1–14.
Lawrence, S.A., Floge, S.A., Davy, J.E., Davy, S.K., Wilson, W.H., 2017. Exploratory analysis of Symbiodinium transcriptomes reveals potential latent infection by large dsDNA viruses. Environ. Microbiol. 19, 3909–3919.
Lax, G., Eglit, Y., Eme, L., Bertrand, E.M., Roger, A.J., Simpson, A.G.B., 2018. Hemimastigophora is a novel supra-kingdom-level lineage of eukaryotes. Nature 564, 410–414.
Le Gac, M., Metegnier, G., Chomérat, N., Malestroit, P., Quéré, J., Bouchez, O., Siano, R., Destombe, C., Guillou, L., Chapelle, A., 2016. Evolutionary processes and cellular functions underlying divergence in Alexandrium minutum. Mol. Ecol. 25, 5129–5143.
Leflaive, J., Ten-Hage, L., 2007. Algal and cyanobacterial secondary metabolites in freshwaters: a comparison of allelopathic compounds and toxins. Freshw. Biol. 52, 199–214.
Lefranc, F., Koutsaviti, A., Ioannou, E., Kornienko, A., Roussis, V., Kiss, R., Newman, D., 2019. Algae metabolites: from: in vitro growth inhibitory effects to promising anticancer activity. Natl. Prod. Rep. 36, 810–841.
Leger, M.M., Petru, M., Žárský, V., Eme, L., Vlček, Č., Harding, T., Lang, B.F., Eliáš, M., Doležal, P., Roger, A.J., 2015. An ancestral bacterial division system is widespread in eukaryotic mitochondria. Proc. Natl. Acad. Sci. USA 112, 10239–10246.
Li, F.W., Melkonian, M., Rothfels, C.J., Villarreal, J.C., Stevenson, D.W., Graham, S.W., Wong, G.K.S., Pryer, K.M., Mathews, S., 2015. Phytochrome diversity in green plants and the origin of canonical plant phytochromes. Nat. Commun. 6.
Li, H., Li, L., Yu, L., Yang, X., Shi, X., Wang, J., Li, J., Lin, S., 2020a. Transcriptome profiling reveals versatile dissolved organic nitrogen utilization, mixotrophy, and N conservation in the dinoflagellate Prorocentrum shikokuense under N deficiency. Sci. Total Environ., 143013.
Li, H., Wang, W., Wang, Z., Lin, X., Zhang, F., Yang, L., 2016. De novo transcriptome analysis of carotenoid and polyunsaturated fatty acid metabolism in Rhodomonas sp. J. Appl. Phycol. 28, 1649–1656.
Li, J., Li, G., Wang, H., Wang Deng, X., 2011. Phytochrome signaling mechanisms. Arab. B. 9, e0148.

Li, Y., Cai, X., Gu, W., Wang, G., 2020b. Transcriptome analysis of carotenoid biosynthesis in Dunaliella salina under red and blue light. J. Oceanol. Limnol. 38, 177–185.
Lin, X., 2021. The application of transcriptomics, metagenomics, and metatranscriptomics in algal research. In: Gao, K., Hutchins, D.A., Beardall, J. (Eds.). Springer Singapore, Singapore, pp. 285–291.
Lin, Z., Li, W.H., 2012. Evolution of 5' untranslated region length and gene expression reprogramming in yeasts. Mol. Biol. Evol. 29, 81–89.
Lindell, D., Jaffe, J.D., Coleman, M.L., Futschik, M.E., Axmann, I.M., Rector, T., Kettler, G., Sullivan, M.B., Steen, R., Hess, W.R., Church, G.M., Chisholm, S.W., 2007. Genome-wide expression dynamics of a marine virus and host reveal features of co-evolution. Nature 449, 83–86.
Liu, Y., Tsinoremas, N.F., Johnson, C.H., Lebedeva, N.V., Golden, S.S., Ishiura, M., Kondo, T., 1995. Circadian orchestration of gene expression in cyanobacteria. Genes Dev. 9, 1469–1478.
Liu, Z., Campbell, V., Heidelberg, K.B., Caron, D.A., 2016. Gene expression characterizes different nutritional strategies among three mixotrophic protists. FEMS Microbiol. Ecol. 92, 1–11.
Liu, Z., Jones, A.C., Campbell, V., Hambright, K.D., Heidelberg, K.B., Caron, D.A., 2015a. Gene expression in the mixotrophic prymnesiophyte, Prymnesium parvum, responds to prey availability. Front. Microbiol. 6, 1–12.
Liu, Z., Koid, A.E., Terrado, R., Campbell, V., Caron, D.A., Heidelberg, K.B., 2015b. Changes in gene expression of Prymnesium parvum induced by nitrogen and phosphorus limitation. Front. Microbiol. 6, 1–13.
Lodish, H., Berk, A., Kaiser, C.A., Krieger, M., Scott, M.P., Bretscher, A., Ploegh, H., Matsudaira, P., 2008. Molecular Cell Biology. Macmillan.
Lowe, C.D., Mello, L.V., Samatar, N., Martin, L.E., Montagnes, D.J.S., Watts, P.C., 2011. The transcriptome of the novel dinoflagellate Oxyrrhis marina (Alveolata: Dinophyceae): response to salinity examined by 454 sequencing. BMC Genomics 12.
Lu, P., Vogel, C., Wang, R., Yao, X., Marcotte, E.M., 2007. Absolute protein expression profiling estimates the relative contributions of transcriptional and translational regulation. Nat. Biotechnol. 25, 117–124.
Lu, T., Zhu, Y., Ke, M., Peijnenburg, W., Zhang, M., Wang, T., Chen, J., Qian, H., 2019. Evaluation of the taxonomic and functional variation of freshwater plankton communities induced by trace amounts of the antibiotic ciprofloxacin. Environ. Int. 126, 268–278.
Mackey, S.R., Golden, S.S., 2007. Winding up the cyanobacterial circadian clock. Trends Microbiol. 15, 381–388.
Maheswari, U., Jabbari, K., Petit, J.L., Porcel, B.M., Allen, A.E., Cadoret, J.P., De Martino, A., Heijde, M., Kaas, R., La Roche, J., Lopez, P.J., Martin-Jézéquel, V., Meichenin, A., Mock, T., Schnitzler Parker, M., Vardi, A., Armbrust, E.V., Weissenbach, J., Katinka, M., Bowler, C., 2010. Digital expression profiling of novel diatom transcripts provides insight into their biological functions. Genome Biol. 11.
Maier, T., Güell, M., Serrano, L., 2009. Correlation of mRNA and protein in complex biological samples. FEBS Lett. 583, 3966–3973.
Main, C.R., Greenfield, D.I., Doll, C., Wang, Y.F., Whereat, E.B., Mortensen, R., Pettay, D.T., Coyne, K.J., 2018. Critical comparison of molecular methods for detection and enumeration of the harmful algal species, Heterosigma akashiwo. Environ. Water Samples 30, 2425–2434.
Marchetti, A., Catlett, D., Hopkinson, B.M., Ellis, K., Cassar, N., 2015. Marine diatom proteorhodopsins and their potential role in coping with low iron availability. ISME J. 9, 2745–2748.
Marchetti, A., Schruth, D.M., Durkin, C.A., Parker, M.S., Kodner, R.B., Berthiaume, C.T., Morales, R., Allen, A.E., Armbrust, E.V., 2012. Comparative metatranscriptomics identifies molecular bases for the physiological responses of phytoplankton to varying iron availability. Proc. Natl. Acad. Sci. 109, E317–E325.

Marron, A.O., Ratcliffe, S., Wheeler, G.L., Goldstein, R.E., King, N., Not, F., De Vargas, C., Richter, D.J., 2016. The evolution of silicon transport in eukaryotes. Mol. Biol. Evol. 33, 3226–3248.
Mars Brisbin, M., Mitarai, S., 2019. Differential gene expression supports a resource-intensive, defensive role for colony production in the bloom-forming haptophyte, Phaeocystis globosa. J. Eukaryot. Microbiol. 66, 788–801.
Martin, J.A., Wang, Z., 2011. Next-generation transcriptome assembly. Nat. Rev. Genet. 12, 671–682.
Martin, R.M., Moniruzzaman, M., Stark, G.F., Gann, E.R., Derminio, D.S., Wei, B., Hellweger, F.L., Pinto, A., Boyer, G.L., Wilhelm, S.W., 2020. Episodic decrease in temperature Increases mcy gene transcription and cellular microcystin in continuous cultures of Microcystis aeruginosa PCC 7806. Front. Microbiol. 11.
Martinez-Hernandez, F., Luo, E., Tominaga, K., Ogata, H., Yoshida, T., DeLong, E.F., Martinez-Garcia, M., 2020. Diel cycling of the cosmopolitan abundant Pelagibacter virus 37-F6: one of the most abundant viruses on earth. Environ. Microbiol. Rep. 12, 214–219.
Matantseva, O., Skarlato, S., Vogts, A., Pozdnyakov, I., Liskow, I., Schubert, H., Voss, M., 2016. Superposition of individual activities: urea-mediated suppression of nitrate uptake in the dinoflagellate prorocentrum minimum revealed at the population and single-cell levels. Front. Microbiol. 7, 1–14.
Matasci, N., Hung, L.H., Yan, Z., Carpenter, E.J., Wickett, N.J., Mirarab, S., Nguyen, N., Warnow, T., Ayyampalayam, S., Barker, M., Burleigh, J.G., Gitzendanner, M.A., Wafula, E., Der, J.P., dePamphilis, C.W., Roure, B., Philippe, H., Ruhfel, B.R., Miles, N.W., Graham, S.W., Mathews, S., Surek, B., Melkonian, M., Soltis, D.E., Soltis, P.S., Rothfels, C., Pokorny, L., Shaw, J.A., DeGironimo, L., Stevenson, D.W., Villarreal, J.C., Chen, T., Kutchan, T.M., Rolf, M., Baucom, R.S., Deyholos, M.K., Samudrala, R., Tian, Z., Wu, X., Sun, X., Zhang, Y., Wang, J., Leebens-Mack, J., Wong, G.K.S., 2014. Data access for the 1,000 Plants (1KP) project. GigaScience 3, 1–10.
McCabe, R.M., Hickey, B.M., Kudela, R.M., Lefebvre, K.A., Adams, N.G., Bill, B.D., Gulland, F.M.D., Thomson, R.E., Cochlan, W.P., Trainer, V.L., 2016. An unprecedented coastwide toxic algal bloom linked to anomalous ocean conditions. Geophys. Res. Lett. 43, 10,366–10,376.
McCarthy, J.K., Smith, S.R., McCrow, J.P., Tan, M., Zheng, H., Beeri, K., Roth, R., Lichtle, C., Goodenough, U., Bowler, C.P., Dupont, C.L., Allen, A.E., 2017. Nitrate reductase knockout uncouples nitrate transport from nitrate assimilation and drives repartitioning of carbon flux in a model pennate diatom. Plant Cell 29, 2047–2070.
McClung, C.R., 2006. Plant circadian rhythms. Plant Cell Online 18, 792–803.
McFadden, G.I., 2001. Primary and secondary endosymbiosis and the origin of plastids. J. Phycol. 37, 951–959.
McGettigan, P.A., 2013. Transcriptomics in the RNA-seq era. Curr. Opin. Chem. Biol. 17, 4–11.
McKie-Krisberg, Z.M., Sanders, R.W., Gast, R.J., 2018. Evaluation of mixotrophy-associated gene expression in two species of polar marine algae. Front. Mar. Sci. 5, 1–12.
McLean, T.I., 2013. "Eco-omics": a review of the application of genomics, transcriptomics, and proteomics for the study of the ecology of Harmful Algae. Microb. Ecol. 65, 901–915.
Metegnier, G., Paulino, S., Ramond, P., Siano, R., Sourisseau, M., Destombe, C., Le Gac, M., 2020. Species specific gene expression dynamics during harmful algal blooms. Sci. Rep. 10.
Meyer, J.M., Rödelsperger, C., Eichholz, K., Tillmann, U., Cembella, A., McGaughran, A., John, U., 2015. Transcriptomic characterisation and genomic glimps into the toxigenic dinoflagellate Azadinium spinosum, with emphasis on polykeitde synthase genes. BMC Genomics 16, 1–14.

Minnhagen, S., Carvalho, W.F., Salomon, P.S., Janson, S., 2008. Chloroplast DNA content in Dinophysis (Dinophyceae) from different cell cycle stages is consistent with kleptoplasty. Environ. Microbiol. 10, 2411–2417.
Mitschke, J., Vioque, A., Haas, F., Hess, W.R., Muro-Pastor, A.M., 2011. Dynamics of transcriptional start site selection during nitrogen stress-induced cell differentiation in Anabaena sp. PCC 7120, pp. 20130–20135.
Mock, T., Samanta, M.P., Iverson, V., Berthiaume, C., Robison, M., Holtermann, K., Durkin, C., BonDurant, S.S., Richmond, K., Rodesch, M., Kallas, T., Huttlin, E.L., Cerrina, F., Sussman, M.R., Armbrust, E.V., 2008. Whole-genome expression profiling of the marine diatom Thalassiosira pseudonana identifies genes involved in silicon bioprocesses. Proc. Natl. Acad. Sci. USA 105, 1579–1584.
Molnár, A., Schwach, F., Studholme, D.J., Thuenemann, E.C., Baulcombe, D.C., 2007. miRNAs control gene expression in the single-cell alga Chlamydomonas reinhardtii. Nature 447, 1126–1129.
Moniruzzaman, M., Gann, E.R., Wilhelm, S.W., 2018. Infection by a giant virus (AaV) induces widespread physiological reprogramming in Aureococcus anophagefferens CCMP1984-a harmful bloom algae. Front. Microbiol. 9, 1–16.
Moniruzzaman, M., Wurch, L.L., Alexander, H., Dyhrman, S.T., Gobler, C.J., Wilhelm, S.W., 2017. Virus-host relationships of marine single-celled eukaryotes resolved from metatranscriptomics. Nat. Commun. 8, 1–10.
Moore, M.J., 2005. From birth to death: the complex lives of eukaryotic mRNAs. Science 309, 1514–1518.
Moran, M.A., Satinsky, B., Gifford, S.M., Luo, H., Rivers, A., Chan, L.K., Meng, J., Durham, B.P., Shen, C., Varaljay, V.A., Smith, C.B., Yager, P.L., Hopkinson, B.M., 2013. Sizing up metatranscriptomics. ISME J. 7, 237–243.
Morey, J.S., Monroe, E.A., Kinney, A.L., Beal, M., Johnson, J.G., Hitchcock, G.L., Van Dolah, F.M., 2011. Transcriptomic response of the red tide dinoflagellate, Karenia brevis, to nitrogen and phosphorus depletion and addition. BMC Genomics 12.
Morey, J.S., Van Dolah, F.M., 2013. Global analysis of mRNA half-lives and de novo transcription in a Dinoflagellate, Karenia brevis. PLoS ONE 8.
Morozova, O., Hirst, M., Marra, M.A., 2009. Applications of new sequencing technologies for transcriptome analysis. Annu. Rev. Genomics Hum. Genet. 10, 135–151.
Morrissey, J., Bowler, C., 2012. Iron utilization in marine cyanobacteria and eukaryotic algae. Front. Microbiol. 3, 1–13.
Morse, D., 2019. A transcriptome-based perspective of Meiosis in Dinoflagellates. Protist 170, 397–403.
Morse, D., Tse, S.P.K., Lo, S.C.L., 2018. Exploring dinoflagellate biology with high-throughput proteomics. Harmful Algae 75, 16–26.
Morton, J.T., Marotz, C., Washburne, A., Silverman, J., Zaramela, L.S., Edlund, A., Zengler, K., Knight, R., 2019. Establishing microbial composition measurement standards with reference frames. Nat. Commun. 10.
Moustafa, A., Evans, A.N., Kulis, D.M., Hackett, J.D., Erdner, D.L., Anderson, D.M., Bhattacharya, D., 2010. Transcriptome profiling of a toxic dinoflagellate reveals a gene-rich protist and a potential impact on gene expression due to bacterial presence. PLoS ONE 5.
Mubeen, U., Giraldi, L.A., Jüppner, J., Giavalisco, P., 2019. A multi-omics extraction method for the in-depth analysis of synchronized cultures of the green alga chlamydomonas reinhardtii. J. Vis. Exp. 2019, 1–9.
Mundt, F., Heinrich, S., Hanelt, D., 2019. RNA isolation from taxonomically diverse photosynthetic protists. Limnol. Oceanogr., Methods 17, 190–199.
Murray, S.A., Suggett, D.J., Doblin, M.A., Kohli, G.S., Seymour, J.R., Fabris, M., Ralph, P.J., 2016. Unravelling the functional genetics of dinoflagellates: a review of approaches and opportunities. Perspect. Phycol. 3, 37–52.

Nam, O., Park, J.M., Lee, H., Jin, E.S., 2019. De novo transcriptome profile of coccolithophorid alga Emiliania huxleyi CCMP371 at different calcium concentrations with proteome analysis. PLoS ONE 14, 1–24.
Noordally, Z.B., Hindle, M., Martin, S.F., Seaton, D.D., Simpson, T.I., Le Bihan, T., Millar, A.j., 2018. Circadian protein regulation in the green lineage I. A phospho-dawn anticipates light onset before proteins peak in daytime. Running. BioRxiv.
Okamoto, O.K., Hastings, J.W., 2003. Novel dinoflagellate clock-related genes identified through microarray analysis. J. Phycol. 39, 519–526.
Okamoto, O.K., Robertson, D.L., Fagan, T.F., Hastings, J.W., Colepicolo, P., 2001. Different regulatory mechanisms modulate the expression of a dinoflagellate iron-superoxide dismutase. J. Biol. Chem. 276, 19989–19993.
O'Neill, E.C., Trick, M., Henrissat, B., Field, R.A., 2015a. Euglena in time: evolution, control of central metabolic processes and multi-domain proteins in carbohydrate and natural product biochemistry. Perspect. Sci. 6, 84–93.
O'Neill, E.C., Trick, M., Hill, L., Rejzek, M., Dusi, R.G., Hamilton, C.J., Zimba, P.V., Henrissat, B., Field, R.A., 2015b. The transcriptome of Euglena gracilis reveals unexpected metabolic capabilities for carbohydrate and natural product biochemistry. Mol. BioSyst. 11, 2808–2820.
O'Neill, J.S., van Ooijen, G., Dixon, L.E., Troein, C., Corellou, F., Bouget, F.-Y., Reddy, A.B., Millar, A.J., 2011. Circadian rhythms persist without transcription in a eukaryote. Nature 469, 554–558.
Onuma, R., Hirooka, S., Kanesaki, Y., Fujiwara, T., Yoshikawa, H., Miyagishima, S.ya., 2020. Changes in the transcriptome, ploidy, and optimal light intensity of a cryptomonad upon integration into a kleptoplastic dinoflagellate. ISME J. 14, 2407–2423.
Osborn, H.L., Hook, S.E., 2013. Using transcriptomic profiles in the diatom Phaeodactylum tricornutum to identify and prioritize stressors. Aquat. Toxicol. 138, 12–25.
Osuna-Cruz, C.M., Bilcke, G., Vancaester, E., De Decker, S., Bones, A.M., Winge, P., Poulsen, N., Bulankova, P., Verhelst, B., Audoor, S., Belisova, D., Pargana, A., Russo, M., Stock, F., Cirri, E., Brembu, T., Pohnert, G., Piganeau, G., Ferrante, M.I., Mock, T., Sterck, L., Sabbe, K., De Veylder, L., Vyverman, W., Vandepoele, K., 2020. The Seminavis robusta genome provides insights into the evolutionary adaptations of benthic diatoms. Nat. Commun. 11.
Ottesen, E.A., Young, C.R., Eppley, J.M., Ryan, J.P., Chavez, F.P., Scholin, C.A., DeLong, E.F., 2013. Pattern and synchrony of gene expression among sympatric marine microbial populations. Proc. Natl. Acad. Sci. 110, E488–E497.
Ottesen, E.A., Young, C.R., Gifford, S.M., Eppley, J.M., Marin, R., Schuster, S.C., Scholin, C.A., DeLong, E.F., 2014. Multispecies diel transcriptional oscillations in open ocean heterotrophic bacterial assemblages. Science 345.
Pechkovskaya, S.A., Knyazev, N.A., Matantseva, O.V., Emelyanov, A.K., Telesh, I.V., Skarlato, S.O., Filatova, N.A., 2020. Dur3 and nrt2 genes in the bloom-forming dinoflagellate Prorocentrum minimum: transcriptional responses to available nitrogen sources. Chemosphere 241, 125083.
Piétu, G., Mariage-Samson, R., Fayein, N.A., Matingou, C., Eveno, E., Houlgatte, R., Decraene, C., Vandenbrouck, Y., Tahi, F., Devignes, M.D., Wirkner, U., Ansorge, W., Cox, D., Nagase, T., Nomura, N., Auffray, C., 1999. The genexpress IMAGE knowledge base of the human brain transcriptome: a prototype integrated resource for functional and computational genomics. Genome Res. 9, 195–209.
Piggot, P.J., Hilbert, D.W., 2004. Sporulation of Bacillus subtilis. Curr. Opin. Microbiol. 7, 579–586.
Poliner, E., Panchy, N., Newton, L., Wu, G., Lapinsky, A., Bullard, B., Zienkiewicz, A., 2015. Transcriptional coordination of physiological responses in Nannochloropsis oceanica CCMP1779 under light / dark cycles, pp. 1097–1113.

Poulsen, N., Kröger, N., 2005. A new molecular tool for transgenic diatoms: control of mRNA and protein biosynthesis by an inducible promoter-terminator cassette. FEBS J. 272, 3413–3423.

Preston, C.M., Harris, A., Ryan, J.P., Roman, B., Marin, R., Jensen, S., Everlove, C., Birch, J., Dzenitis, J.M., Pargett, D., Adachi, M., Turk, K., Zehr, J.P., Scholin, C.A., 2011. Underwater application of quantitative PCR on an ocean mooring. PLoS ONE 6.

Priyadarshani, I., Rath, B., 2012. Commercial and industrial applications of micro algae – a review 3, pp. 89–100.

Rangsrikitphoti, P., Durnford, D.G., 2019. Transcriptome profiling of Bigelowiella natans in response to light stress. J. Eukaryot. Microbiol. 66, 316–333.

Read, B.A., Kegel, J., Klute, M.J., Kuo, A., Lefebvre, S.C., Maumus, F., Mayer, C., Miller, J., Monier, A., Salamov, A., Young, J., Aguilar, M., Claverie, J.M., Frickenhaus, S., Gonzalez, K., Herman, E.K., Lin, Y.C., Napier, J., Ogata, H., Sarno, A.F., Shmutz, J., Schroeder, D., De Vargas, C., Verret, F., Von Dassow, P., Valentin, K., Van De Peer, Y., Wheeler, G., Dacks, J.B., Delwiche, C.F., Dyhrman, S.T., Glöckner, G., John, U., Richards, T., Worden, A.Z., Zhang, X., Grigoriev, I.V., Allen, A.E., Bidle, K., Borodovsky, M., Bowler, C., Brownlee, C., Mark Cock, J., Elias, M., Gladyshev, V.N., Groth, M., Guda, C., Hadaegh, A., Iglesias-Rodriguez, M.D., Jenkins, J., Jones, B.M., Lawson, T., Leese, F., Lindquist, E., Lobanov, A., Lomsadze, A., Malik, S.B., Marsh, M.E., MacKinder, L., Mock, T., Mueller-Roeber, B., Pagarete, A., Parker, M., Probert, I., Quesneville, H., Raines, C., Rensing, S.A., Riaño-Pachón, D.M., Richier, S., Rokitta, S., Shiraiwa, Y., Soanes, D.M., Van Der Giezen, M., Wahlund, T.M., Williams, B., Wilson, W., Wolfe, G., Wurch, L.L., 2013. Pan genome of the phytoplankton Emiliania underpins its global distribution. Nature 499, 209–213.

Remmers, I.M., D'Adamo, S., Martens, D.E., de Vos, R.C.H., Mumm, R., America, A.H.P., Cordewener, J.H.G., Bakker, L.V., Peters, S.A., Wijffels, R.H., Lamers, P.P., 2018. Orchestration of transcriptome, proteome and metabolome in the diatom Phaeodactylum tricornutum during nitrogen limitation. Algal Res. 35, 33–49.

Riccio, G., De Luca, D., Lauritano, C., 2020. Monogalactosyldiacylglycerol and sulfolipid synthesis in microalgae. Mar. Drugs 18, 1–18.

Rizkallah, M.R., Frickenhaus, S., Trimborn, S., Harms, L., Moustafa, A., Benes, V., Gäbler-Schwarz, S., Beszteri, S., 2020. Deciphering patterns of adaptation and acclimation in the transcriptome of Phaeocystis antarctica to changing iron conditions. J. Phycol. 56, 747–760.

Rockwell, N.C., Lagarias, J.C., 2017. Phytochrome diversification in cyanobacteria and eukaryotic algae. Curr. Opin. Plant Biol. 37, 87–93.

Rogato, A., Amato, A., Iudicone, D., Chiurazzi, M., Ferrante, M.I., d'Alcala, M.R., 2015. The diatom molecular toolkit to handle nitrogen uptake. Mar. Genomics 24, 95–108.

Rokitta, S.D., de Nooijer, L.J., Trimborn, S., de Vargas, C., Rost, B., John, U., 2011. Transcriptome analyses reveal differential gene expression patterns between the life-cycle stages of emiliania huxleyi (haptophyta) and reflect specialization to different ecological niches. J. Phycol. 47, 829–838.

Rosenwasser, S., Mausz, M.A., Schatz, D., Sheyn, U., Malitsky, S., Aharoni, A., Weinstock, E., Tzfadia, O., Ben-Dor, S., Feldmesser, E., Pohnert, G., Vardi, A., 2014. Rewiring host lipid metabolism by large viruses determines the fate of Emiliania huxleyi, a bloom-forming alga in the ocean. Plant Cell 26, 2689–2707.

Roy, S., Beauchemin, M., Dagenais-Bellefeuille, S., Letourneau, L., Cappadocia, M., Morse, D., 2014. The Lingulodinium circadian system lacks rhythmic changes in transcript abundance. BMC Biol. 12, 1–8.

Ryan, J.P., Kudela, R.M., Birch, J.M., Blum, M., Bowers, H.A., Chavez, F.P., Doucette, G.J., Hayashi, K., Marin, R., Mikulski, C.M., Pennington, J.T., Scholin, C.A., Smith,

G.J., Woods, A., Zhang, Y., 2017. Causality of an extreme harmful algal bloom in Monterey Bay, California, during the 2014–2016 northeast Pacific warm anomaly. Geophys. Res. Lett. 44, 5571–5579.

Sabehi, G., Lindell, D., 2012. The P-SSP7 cyanophage has a linear genome with direct terminal repeats. PLoS ONE 7.

Salazar, G., Paoli, L., Alberti, A., Huerta-Cepas, J., Ruscheweyh, H.J., Cuenca, M., Field, C.M., Coelho, L.P., Cruaud, C., Engelen, S., Gregory, A.C., Labadie, K., Marec, C., Pelletier, E., Royo-Llonch, M., Roux, S., Sánchez, P., Uehara, H., Zayed, A.A., Zeller, G., Carmichael, M., Dimier, C., Ferland, J., Kandels, S., Picheral, M., Pisarev, S., Poulain, J., Acinas, S.G., Babin, M., Bork, P., Boss, E., Bowler, C., Cochrane, G., de Vargas, C., Follows, M., Gorsky, G., Grimsley, N., Guidi, L., Hingamp, P., Iudicone, D., Jaillon, O., Kandels-Lewis, S., Karp-Boss, L., Karsenti, E., Not, F., Ogata, H., Pesant, S., Poulton, N., Raes, J., Sardet, C., Speich, S., Stemmann, L., Sullivan, M.B., Sunagawa, S., Wincker, P., 2019. Gene expression changes and community turnover differentially shape the global ocean metatranscriptome. Cell 179, 1068–1083.

Sandrini, G., Cunsolo, S., Schuurmans, J.M., Matthijs, H.C., Huisman, J., 2015. Changes in gene expression, cell physiology and toxicity of the harmful cyanobacterium Microcystis aeruginosa at elevated CO2. Front. Microbiol. 6, 401.

Satinsky, B.M., Crump, B.C., Smith, C.B., Sharma, S., Zielinski, B.L., Doherty, M., Meng, J., Sun, S., Medeiros, P.M., Paul, J.H., Coles, V.J., Yager, P.L., Moran, M.A., 2014. Microspatial gene expression patterns in the Amazon River Plume. Proc. Natl. Acad. Sci. 111, 11085–11090.

Seaton, D.D., Graf, A., Baerenfaller, K., Stitt, M., Millar, A.J., Gruissem, W., 2018. Photoperiodic control of the Arabidopsis proteome reveals a translational coincidence mechanism. Mol. Syst. Biol. 14, e7962.

Seeleuthner, Y., Mondy, S., Lombard, V., Carradec, Q., Pelletier, E., Wessner, M., Leconte, J., Mangot, J.F., Poulain, J., Labadie, K., Logares, R., Sunagawa, S., De Berardinis, V., Salanoubat, M., Dimier, C., Kandels-Lewis, S., Picheral, M., Searson, S., Acinas, S.G., Boss, E., Follows, M., Gorsky, G., Grimsley, N., Karp-Boss, L., Krzic, U., Not, F., Ogata, H., Raes, J., Reynaud, E.G., Sardet, C., Speich, S., Stemmann, L., Velayoudon, D., Weissenbach, J., Pesant, S., Poulton, N., Stepanauskas, R., Bork, P., Bowler, C., Hingamp, P., Sullivan, M.B., Iudicone, D., Massana, R., Aury, J.M., Henrissat, B., Karsenti, E., Jaillon, O., Sieracki, M., De Vargas, C., Wincker, P., 2018. Single-cell genomics of multiple uncultured stramenopiles reveals underestimated functional diversity across oceans. Nat. Commun. 9, 1–10.

Seymour, J.R., Amin, S.A., Raina, J.B., Stocker, R., 2017. Zooming in on the phycosphere: the ecological interface for phytoplankton-bacteria relationships. Nat. Microbiol. 2.

Shi, X., Lin, X., Li, L., Li, M., Palenik, B., Lin, S., 2017. Transcriptomic and microRNAomic profiling reveals multi-faceted mechanisms to cope with phosphate stress in a dinoflagellate. ISME J. 11, 2209–2218.

Shoguchi, E., Shinzato, C., Kawashima, T., Gyoja, F., Mungpakdee, S., Koyanagi, R., Takeuchi, T., Hisata, K., Tanaka, M., Fujiwara, M., Hamada, M., Seidi, A., Fujie, M., Usami, T., Goto, H., Yamasaki, S., Arakaki, N., Suzuki, Y., Sugano, S., Toyoda, A., Kuroki, Y., Fujiyama, A., Medina, M., Coffroth, M.A., Bhattacharya, D., Satoh, N., 2013. Draft assembly of the Symbiodinium minutum nuclear genome reveals Dinoflagellate gene structure. Curr. Biol. 23, 1399–1408.

Sibbald, S.J., Archibald, J.M., 2017. More protist genomes needed. Nat. Ecol. Evol. 1, 1–3.

Simpson, A.G.B., Roger, A.J., Metazoa, A., 2004. The real 'kingdoms' of eukaryotes. Curr. Biol. 14, R693–R696.

Smith, D.R., 2013. RNA-Seq data: a goldmine for organelle research. Brief. Funct. Genomics 12, 454–456.

Smith, D.R., 2015. Mutation rates in plastid genomes: they are lower than you might think. Genome Biol. Evol. 7, 1227–1234.

Smith, S.R., Dupont, C.L., McCarthy, J.K., Broddrick, J.T., Oborník, M., Horák, A., Füssy, Z., Cihlář, J., Kleessen, S., Zheng, H., McCrow, J.P., Hixson, K.K., Araújo, W.L., Nunes-Nesi, A., Fernie, A., Nikoloski, Z., Palsson, B.O., Allen, A.E., 2019. Evolution and regulation of nitrogen flux through compartmentalized metabolic networks in a marine diatom. Nat. Commun. 10, 4552.

Smith, S.R., Gillard, J.T.F., Kustka, A.B., McCrow, J.P., Badger, J.H., Zheng, H., New, A.M., Dupont, C.L., Obata, T., Fernie, A.R., Allen, A.E., 2016. Transcriptional orchestration of the global cellular response of a model pennate diatom to diel light cycling under iron limitation. PLoS Genet. 12.

Steffen, M.M., Belisle, B.S., Watson, S.B., Boyer, G.L., Bourbonniere, R.A., Wilhelm, S.W., 2015. Metatranscriptomic evidence for co-occurring top-down and bottom-up controls on toxic cyanobacterial communities. Appl. Environ. Microbiol. 81, 3268–3276.

Steffen, M.M., Dearth, S.P., Dill, B.D., Li, Z., Larsen, K.M., Campagna, S.R., Wilhelm, S.W., 2014. Nutrients drive transcriptional changes that maintain metabolic homeostasis but alter genome architecture in Microcystis. ISME J. 8, 2080–2092.

Steglich, C., Lindell, D., Futschik, M., Rector, T., Steen, R., Chisholm, S.W., 2010. Short RNA half-lives in the slow-growing marine cyanobacterium Prochlorococcus. Genome Biol. 11.

Stephens, T.G., Ragan, M.A., Bhattacharya, D., Chan, C.X., 2018. Core genes in diverse dinoflagellate lineages include a wealth of conserved dark genes with unknown functions. Sci. Rep. 8, 1–11.

Stoecker, D.K., Hansen, P.J., Caron, D.A., Mitra, A., 2017. Mixotrophy in the marine plankton. Annu. Rev. Mar. Sci. 9, 311–335.

Sun, J., Hutchins, D.A., Feng, Y., Seubert, E.L., Caron, D.A., Fu, F.X., 2011. Effects of changing pCO2 and phosphate availability on domoic acid production and physiology of the marine harmful bloom diatom Pseudo-nitzschia multiseries. Limnol. Oceanogr. 56, 829–840.

Tamiya, H., 1966. Synchronous cultures of algae. Annu. Rev. Plant Physiol. 17, 1–27.

Tang, X., Krausfeldt, L.E., Shao, K., LeCleir, G.R., Stough, J.M.A., Gao, G., Boyer, G.L., Zhang, Y., Paerl, H.W., Qin, B., Wilhelm, S.W., 2018. Seasonal gene expression and the ecophysiological implications of toxic microcystis aeruginosa blooms in Lake Taihu. Environ. Sci. Technol. 52, 11049–11059.

Tardu, M., Dikbas, U.M., Baris, I., Kavakli, I.H., 2016. RNA-seq analysis of the transcriptional response to blue and red light in the extremophilic red alga, Cyanidioschyzon merolae. Funct. Integr. Genomics 16, 657–669.

Teikari, J., Osterholm, J., Kopf, M., Battchikova, N., Wahlsten, M., Aro, E.M., Hess, W.R., Sivonen, K., 2015. Transcriptomic and proteomic profiling of Anabaena sp. Strain 90, 5212–5222.

Thangaraj, S., Sun, J., 2021. Transcriptomic reprogramming of the oceanic diatom Skeletonema dohrnii under warming ocean and acidification. Environ. Microbiol. 23, 980–995.

Thingstad, T.F., 2000. Elements of a theory for the mechanisms controlling abundance, diversity, and biogeochemical role of lytic bacterial viruses in aquatic systems. Limnol. Oceanogr. 45, 1320–1328.

Tice, R.R., Austin, C.P., Kavlock, R.J., Bucher, J.R., 2013. Improving the human hazard characterization of chemicals: a Tox21 update. Environ. Health Perspect. 121, 756–765.

Toepel, J., Welsh, E., Summerfield, T.C., Pakrasi, H.B., Sherman, L.A., 2008. Differential transcriptional analysis of the cyanobacterium Cyanothece sp. strain ATCC 51142 during light-dark and continuous-light growth. J. Bacteriol. 190, 3904–3913.

Trench, R.K., 1969. Chloroplasts as functional endosymbionts in the mollusc tridachia crispata (Bërgh), (Opisthobranchia, Sacoglossa). Nature 222, 1071–1072.

Van Dolah, F.M., 2000. Marine algal toxins: origins, health effects, and their increased occurrence. Environ. Health Perspect. 108, 133–141.

Van Steenkiste, N.W.L., Stephenson, I., Herranz, M., Husnik, F., Keeling, P.J., Leander, B.S., 2019. A new case of kleptoplasty in animals: marine flatworms steal functional plastids from diatoms. Sci. Adv. 5, 1–9.

Velculescu, V.E., Zhang, L., Vogelstein, B., Kinzler, K.W., 1995. Serial analysis of gene expression. Science 270, 484–487.

Vila-Costa, M., Rinta-Kanto, J.M., Sun, S., Sharma, S., Poretsky, R., Moran, M.A., 2010. Transcriptomic analysis of a marine bacterial community enriched with dimethylsulfoniopropionate. ISME J. 4, 1410–1420.

Villeneuve, D.L., Garcia-Reyero, N., 2011. Vision & strategy predictive ecotoxicology in the 21st century. Environ. Toxicol. Chem. 30, 1–8.

Von Dassow, P., Ogata, H., Probert, I., Wincker, P., Da Silva, C., Audic, S., Claverie, J.M., de Vargas, C., 2009. Transcriptome analysis of functional differentiation between haploid and diploid cells of Emiliania huxleyi, a globally significant photosynthetic calcifying cell. Genome Biol. 10.

Wada, T., Becskei, A., 2017. Impact of methods on the measurement of mRNA turnover. Int. J. Mol. Sci. 18.

Walter, J., Lynch, F., Battchikova, N., Aro, E.M., Gollan, P.J., 2016. Calcium impacts carbon and nitrogen balance in the filamentous cyanobacterium Anabaena sp. PCC 7120, pp. 3997–4008.

Wang, H., Guo, R.Y., Lim, W.A., Allen, A.E., Ki, J.S., 2020. Comparative transcriptomics of toxin synthesis genes between the non-toxin producing dinoflagellate Cochlodinium polykrikoides and toxigenic Alexandrium pacificum. Harmful Algae 93.

Wang, T., Hu, Y., Zhu, M., Yin, S., 2019. Integrated transcriptome and physiology analysis of Microcystis aeruginosa after exposure to copper sulfate. J. Oceanol. Limnol. 38, 102–113.

Ward, B.A., Dutkiewicz, S., Jahn, O., Follows, M.J., 2012. A size-structured food-web model for the global ocean. Limnol. Oceanogr. 57, 1877–1891.

Wei, G.-W., 2019. Protein structure prediction beyond AlphaFold. Nat. Mach. Intell.

Weis, B.L., Schleiff, E., Zerges, W., 2013. Protein targeting to subcellular organelles via mRNA localization. Biochim. Biophys. Acta, Mol. Cell Res. 1833, 260–273.

Wemheuer, B., Wemheuer, F., Hollensteiner, J., Meyer, F.D., Voget, S., Daniel, R., 2015. The green impact: bacterioplankton response towards a phytoplankton spring bloom in the southern North Sea assessed by comparative metagenomic and metatranscriptomic approaches. Front. Microbiol. 6, 1–13.

Wheeler, G., Ishikawa, T., Pornsaksit, V., Smirnoff, N., 2015. Evolution of alternative biosynthetic pathways for vitamin C following plastid acquisition in photosynthetic eukaryotes. eLife 2015, 1–25.

Whitney, L.P., Lomas, M.W., 2016. Growth on ATP elicits a P-stress response in the picoeukaryote Micromonas pusilla. PLoS ONE 11, 1–18.

Wiltbank, L.B., Kehoe, D.M., 2019. Diverse light responses of cyanobacteria mediated by phytochrome superfamily photoreceptors. Nat. Rev. Microbiol. 17, 37–50.

Wolf, T., Kämmer, P., Brunke, S., Linde, J., 2018. Two's company: studying interspecies relationships with dual RNA-seq. Curr. Opin. Microbiol. 42, 7–12.

Wolf, Y.I., Silas, S., Wang, Y., Wu, S., Bocek, M., Kazlauskas, D., Krupovic, M., Fire, A., Dolja, V.V., Koonin, E.V., 2020. Doubling of the known set of RNA viruses by metagenomic analysis of an aquatic virome. Nat. Microbiol. 5, 1262–1270.

Wong, D.K., Grisdale, C.J., Fast, N.M., 2018. Evolution and diversity of pre-mRNA splicing in highly reduced nucleomorph genomes. Genome Biol. Evol. 10, 1573–1583.

Xia, M., Huang, R., Witt, K.L., Southall, N., Fostel, J., Cho, M.H., Jadhav, A., Smith, C.S., Inglese, J., Portier, C.J., Tice, R.R., Austin, C.P., 2008. Compound cytotoxicity profiling using quantitative high-throughput screening. Environ. Health Perspect. 116, 284–291.

Xiao, M., Hamilton, D.P., O'Brien, K.R., Adams, M.P., Willis, A., Burford, M.A., 2020. Are laboratory growth rate experiments relevant to explaining bloom-forming cyanobacteria distributions at global scale? Harmful Algae 92, 2.

Yamano, T., Fukuzawa, H., 2009. Carbon-concentrating mechanism in a green alga, Chlamydomonas reinhardtii, revealed by transcriptome analyses. J. Basic Microbiol. 49, 42–51.

Yang, I., John, U., Beszteri, S., Glöckner, G., Krock, B., Goesmann, A., Cembella, A.D., 2010. Comparative gene expression in toxic versus non-toxic strains of the marine dinoflagellate Alexandrium minutum. BMC Genomics 11.

Yellowlees, D., Rees, T.A.V., Leggat, W., 2008. Metabolic interactions between algal symbionts and invertebrate hosts. Plant Cell Environ. 31, 679–694.

Zhang, S.F., Yuan, C.J., Chen, Y., Lin, L., Wang, D.Z., 2019a. Transcriptomic response to changing ambient phosphorus in the marine dinoflagellate Prorocentrum donghaiense. Sci. Total Environ. 692, 1037–1047.

Zhang, Y., Ryan, J.P., Hobson, B.W., Kieft, B., Romano, A., Barone, B., Preston, C.M., Roman, B., Raanan, B., Pargett, D., Dugenne, M., White, A.E., Freitas, F.H., Poulos, S., Wilson, S.T., Delong, E.F., Karl, D.M., Birch, J.M., Bellingham, J.G., Scholin, C.A., 2021. A system of coordinated autonomous robots for Lagrangian studies of microbes in the oceanic deep chlorophyll maximum. Sci. Robot. 6, 9138.

Zhang, Y., Ryan, J.P., Kieft, B., Hobson, B.W., McEwen, R.S., Godin, M.A., Harvey, J.B., Bellingham, J.G., Birch, J.M., Scholin, C.A., Chavez, F.P., 2019b. Targeted sampling by autonomous underwater vehicles. Front. Mar. Sci. 6, 1–12.

Zhu, S.H., Green, B.R., 2010. Photoprotection in the diatom Thalassiosira pseudonana: role of LI818-like proteins in response to high light stress. Biochim. Biophys. Acta, Bioenerg. 1797, 1449–1457.

Ziegenhain, C., Vieth, B., Parekh, S., Hellmann, I., Enard, W., 2018. Quantitative single-cell transcriptomics. Brief. Funct. Genomics 17, 220–232.

Zielinski, B.L., Allen, A.E., Carpenter, E.J., Coles, V.J., Crump, B.C., Doherty, M., Foster, R.A., Goes, J.I., Gomes, H.R., Hood, R.R., McCrow, J.P., Montoya, J.P., Moustafa, A., Satinsky, B.M., Sharma, S., Smith, C.B., Yager, P.L., Paul, J.H., 2016. Patterns of transcript abundance of eukaryotic biogeochemically-relevant genes in the Amazon River Plume. PLoS ONE 11, 20.

Zinser, E.R., Lindell, D., Johnson, Z.I., Futschik, M.E., Steglich, C., Coleman, M.L., Wright, M.A., Rector, T., Steen, R., McNulty, N., Thompson, L.R., Chisholm, S.W., 2009. Choreography of the transcriptome, photophysiology, and cell cycle of a minimal photoautotroph, Prochlorococcus. PLoS ONE 4.

Zones, J.M., Blaby, I.K., Merchant, S.S., Umen, J.G., 2015. High-resolution profiling of a synchronized diurnal transcriptome from Chlamydomonas reinhardtii reveals continuous cell and metabolic differentiation. Plant Cell 27, 2743–2769.

CHAPTER 14

From genes to ecosystems: using molecular information from diatoms to understand ecological processes

John A. Berges[a], Erica B. Young[a], Kimberlee Thamatrakoln[b], and Alison R. Taylor[c]

[a]Department of Biological Sciences and School of Freshwater Sciences, University of Wisconsin-Milwaukee, Milwaukee, WI, United States
[b]Department of Marine and Coastal Sciences, Rutgers University, New Brunswick, NJ, United States
[c]Department of Biology and Marine Biology, The University of North Carolina Wilmington, Wilmington, NC, United States

Introduction

> *We cannot cheat on DNA. We cannot get round photosynthesis. We cannot say I am not going to give a damn about phytoplankton. All these tiny mechanisms provide the preconditions of our planetary life. To say we do not care is to say in the most literal sense that "we choose death."*
>
> — *Barbara Ward*

Economist, author, and advocate of sustainable development, Barbara Ward, may have been the first person to incorporate nucleic acids, phytoplankton, and global ecology into a single speech (Ward, 1973), but these concepts are now central to oceanographic research (e.g., Bowler et al., 2009; Karlusich et al., 2020). Understanding the ecological function of the oceans relies on being able to observe, describe, and ultimately predict physiological states of organisms (especially microbial primary producers) and their roles in biogeochemical processes (Rothstein et al., 2006). Contextualizing oceanic temporal and spatial scales is daunting, so work has traditionally focused on extrapolating from small-scale measurements, coupled with modeling (Hellweger, 2020; Vandenkoornhuyse et al., 2010). Recent initiatives, such as the BioGeoSCAPES program (www.biogeoscapes.org), and others have aimed to integrate traditional datasets (e.g., nutrient distributions) with genomic, transcriptomic, and proteomic (i.e., "omics") data to understand ocean metabolism (Anderson, 2020; Karsenti et al., 2011).

We argue that a deeper understanding of ecosystem processes can be gained when molecular-based data is contextualized with physiological processes. Physiological measurements of phytoplankton have contributed substantially towards understanding ecological concepts, such as community succession and competition (Margalef, 1978; Smayda, 1980). Moreover, specific physiological rate measurements have been translated to ecological or cellular functions, e.g., ^{14}C uptake (Williams et al., 2002) for primary production, photochemical yield from chlorophyll *a* fluorescence emissions for photosynthesis (Kolber and Falkowski, 1993), flow cytometric measurements for growth (Li, 2009), and enzyme assays for cellular activity (e.g., Berges et al., 1995; Sala et al., 2019). Connecting omics and physiology will inevitably help bridge the gap to ecology; as Brock (1966) observed, "*Ecology is physiology under the worst possible conditions.*"

In this chapter, we use case studies that illustrate attempts to use molecular data to infer or predict processes of ecological significance and highlight the importance of physiology in making these connections meaningful. We focus on diatoms (critical primary producers in the world oceans that dominate phytoplankton communities in most coastal regions) (Field et al., 1998; Smayda, 1980; Smetacek, 2018), and for which we also have considerable phylogenetic, biochemical, physiological information from genomic resources (e.g., Armbrust et al., 2004; Bowler et al., 2008; Mock et al., 2017).

Methodological approaches and relevant scales are important when trying to connect omics measurements to physiology and ecology. Genomics (DNA sequencing) is useful for taxonomy (based on gene identities, de Vargas et al., 2015) and identifying genes involved in specific functions (Rastogi et al., 2020; Ward and van Oostende, 2016). Transcriptomics (RNA sequencing) provides information about which genes are actively transcribed in response to different conditions (e.g., Keeling et al., 2014; Salazar et al., 2019). The pool of proteins (proteomics) allows examination of metabolic processes, including translational and posttranslational mechanisms (e.g., Nunn et al., 2013). Metabolomics integrates all of the above and characterizes the cellular chemical environment. In field settings, omic measurements necessarily involve bulk community sampling (thus adding the prefix 'meta'), and require partitioning the data among taxa (Johnson et al., 2018). However, our ability to do this is limited by the information contained within databases, which are largely derived from culture-based studies (see below).

Processes involving nucleic acids have long been explained using metaphors of "language," or "codes," and while such analogies risk blurring the concepts they aim to clarify, or awaken unintended associations (see Weigmann, 2004), we suggest that we can put knowledge about phytoplankton into a metaphor of language (cf. Rynearson and Palenik, 2011). In such a metaphor, we are challenged with understanding conversation, i.e., the functioning of an ecosystem. Omics data represents a vocabulary and basic word function, and so physiology serves as an advanced grammar and style manual, and the bridge needed to interpret omics data in ecological terms. In this chapter, we illustrate both the power and limitations of using molecular information to extrapolate to ecological processes using three case studies, which incorporate physiological principles: (1) examining inorganic nutrient metabolism focusing on silicon, (2) determining cell function and metabolism, focusing specifically on cell death processes, and (3) exploring higher-order environmental and community-level interactions through cell communication and allelopathy.

Case study: silicon metabolism

Physiological basis of silicon cycling in the ocean

For decades, oceanographers have used rates of inorganic nutrient acquisition as indicators of the physiological state of the resident phytoplankton community to draw conclusions about biogeochemical processes within an ecosystem. This bulk, community-level approach has formed the foundation of our understanding of biogeochemical cycling in the ocean. However, with the introduction of high-throughput, next generation sequencing, and advances in computing and technology, application of molecular-based, taxon-resolved data is revealing a level of heterogeneity and complexity in community interactions not fully realized until now. Laboratory, culture-based studies highlight the diversity within diatoms, with different species having different cellular mechanisms for, for example, iron acquisition (Kustka et al., 2007), iron storage (Marchetti et al., 2009), and photosynthesis (Strzepek and Harrison, 2004). This molecular-level diversity leads to interspecies competition in natural communities that can manifest as differential responses to nutrient limitation (Cohen et al., 2017; Lampe et al., 2018) or resource partitioning (Alexander et al., 2015).

As a case study for the importance of physiology to in contextualizing molecular data related to nutrient metabolism, we focus here on silicon (Si). Diatoms take up dissolved Si, predominantly in the form of silicic acid [$Si(OH_4)$], from the surrounding water and process it intracellularly into biogenic silica (bSi). Biogenic Si is complexed with organic materials into beautifully structured, intricate frustules or cell wall "houses of glass" (Fig. 14.1), with species-specific nanoscale patterning. Like all living organisms, diatoms require nitrogen (N) and phosphorus (P), but also have an obligate requirement for Si for cell wall formation. As the dominant group of silicifying organisms in the ocean, diatoms are responsible for >90% of the total bSi production, with a collectively minor contribution by other siliceous organisms, such as radiolarians, silicoflagellates, sponges, and some picocyanobacteria (Tréguer et al., 2021).

With the silica component of a diatom cell wall accounting for up to 50% of the dry weight of the cell (Lewin and Guillard, 1963), diatoms have a high silicon demand for growth. However, physiological plasticity allows diatoms to alter cellular Si:C ratios in response to environmental conditions. To maintain maximum growth in low Si environments, diatoms produce thinner frustules, with up to 5 times lower Si:C and bSi per frustule surface area (Martin-Jézéquel et al., 2000; McNair et al., 2018; Paasche, 1973). Prolonged or chronic limitation eventually transitions the community from kinetic Si limitation to growth limitation (Brzezinski et al., 1990), but this physiological plasticity allows diatoms to subsist temporarily under low Si supply, and recover frustule Si content after Si resupply. In contrast, iron (Fe) limitation is reported to increase silicification, leading to frustules with higher Si:C (Hutchins and Bruland, 1998; Marchetti and Harrison, 2007; Takeda, 1998). This may help explain high Si:C burial in sediments underlying predominantly Fe-limited (e.g., HNLC) regions of the ocean (Pichevin et al., 2014).

Diatoms contribute ~20% of global primary productivity, and thus form *the most critical* link between the carbon and silicon cycles. However, contributions of diatoms to organic matter cycling relies on a readily available source of Si in the euphotic zone, and limitation of diatom growth and productivity by Si has been widely observed (Brzezinski et al., 2008; Krause et al., 2019; Nelson and Dortch, 1996; Nelson et al., 2001). A major goal of upper ocean Si cycling studies has been to evaluate Si limitation, relying on bioassay approaches (Beardall et al., 2001), and physiological measurements of Si uptake and incorporation rates using the radiotracer ^{32}Si (Brzezinski et al., 2008, 2011). Finer resolution of the response to Si is obtained by measuring rates of Si uptake across a range of Si concentrations, and fitting data to a Michaelis-Menten hyperbola (Brzezinski and Nelson, 1996), deriving estimates of the half-saturation constant for uptake, K_s, and the maximum uptake rate, V_m. These two kinetic parameters have provided a mechanistic, physiological understanding of nutrient competition in natural systems (Tilman, 1977; Tilman et al., 1982), and have been used to parameterize Si uptake as an external control on diatom growth in selection-based biogeochemical models (Chai et al., 2007; Follows et al., 2007; Moore et al., 2002).

Though determining the concentration dependence of silica production in field assemblages of diatoms is straightforward, interpretation of the resulting data is not. If Si addition increases silica production (ρ, μmol Si L^{-1} d^{-1}), that rate is a clear and quantitative biogeochemical index of the extent to which ambient Si restricts biogenic silica production (Brzezinski et al., 2003), although the relationship to the underlying physiology of the resident diatoms is less clear. Ideally, the relationship between the specific rate of silica production ($V_b = \rho/$ μmol Si L^{-1} bSi) and $Si(OH)_4$ conforms to Michaelis—Menten-type saturation kinetics, allowing estimates of K_s and V_m to be used to evaluate the efficiency of $Si(OH)_4$ use by resident diatoms and the presence and severity of Si limitation (Krause et al., 2012; Martin-Jézéquel et al., 2000; Nelson and Dortch, 1996). However, when biphasic or nonsaturable Si uptake kinetics are observed, the interpretation is less clear (Leynaert et al., 2009; Thamatrakoln and Hildebrand, 2008). Recent work employing a new bSi labeling method for resolving the contribution to bulk silica production by individual species (McNair et al., 2015) confirm that diatoms in natural communities can sacrifice silica production to maintain maximum growth (McNair et al., 2018), highlighting that bulk community metrics only provide a partial picture of the physiological state of individual cells within a community.

Silicon metabolism in diatoms and the molecular basis of silicification

Prior to the evolution of silicifying organisms, the world's oceans were nearly saturated with silicic acid (1.7–2.5 mM; Holland, 1984; Iler, 1979). Today, because of utilization largely by diatoms, global oceanic averages have declined to ~70 μM, but is generally <10 μM, and sometimes <1 μM in productive surface waters (Martin-Jézéquel et al., 2000; Tréguer et al., 1995). Thus diatoms are faced with the challenge of obtaining enough silicic acid from a limiting environment to support the large demand for frustule synthesis.

Following uptake, diatoms must overcome the unique chemistry of silicic acid, that is, that it autopolymerizes into solid silica at concentrations

>2 mM (Iler, 1979). With intracellular concentrations ranging from 19 to 340 mM (Martin-Jézéquel et al., 2000), it is hypothesized that maintenance of supersaturated levels of intracellular dissolved Si is achieved through binding to organic compounds (Azam et al., 1974; Hildebrand, 2000; Hildebrand and Wetherbee, 2003; Sullivan, 1979) or sequestration within "silicon transport vesicles," or STVs (Annenkov et al., 2013, 2020). However, electron spectroscopic imaging (Rogerson et al., 1987) and more recently, cryo-electron microscopy (Kumar et al., 2020), show silicic acid is distributed throughout the cytoplasm. Moreover, the energetic costs of generating the requisite number of STVs to support cell wall synthesis call into question a primary or dominant role of vesicles in silicon uptake and storage (Thamatrakoln and Kustka, 2009), but this is still being debated (Annenkov et al., 2020; Grachev et al., 2017). Once inside the cell, Si is transported through an unknown mechanism to a specialized vesicle called the silica deposition vesicle (SDV), where Si polymerization and complexation with organic compounds proceed to form the complete frustule prior to exocytosis to the extracellular space (Fig. 14.1; Li et al., 1984; Pickett-Heaps et al., 1990; Vrieling et al., 1999).

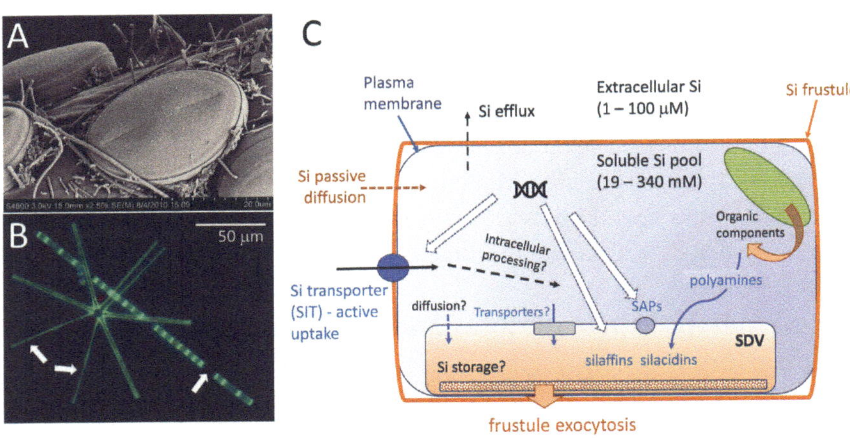

Figure 14.1 Silicon metabolism in diatoms. **A.** Scanning electron microscopy image of epiphytic *Cocconeis* cells showing silica frustules. **B.** Incorporation of fluorescent dye PDMPO in phytoplankton *Asterionella* and *Aulacoseira*. Newly formed frustules fluoresce green where older frustules (arrows) are unlabeled. **C.** Schematic of Si uptake and components of frustule synthesis in the silica deposition vesicle (SDV).

This tightly coordinated sequence of events is facilitated by a sophisticated, genetically encoded network of biochemical pathways, the full identification of which are still unknown. Analysis of genes involved in

Si metabolism have contributed some understanding of silicification on a cellular molecular level. One of the more well-characterized aspects of silicon metabolism is uptake of Si at the cell surface. At low external Si concentrations (<20 μM), silicic acid uptake is mediated largely by specialized membrane-bound proteins called silicon transporters (SITs; Fig. 14.1), which are encoded by a novel, multigene family of membrane-bound proteins (Hildebrand et al., 1997). At high external Si, silicic acid can freely diffuse across the cell membrane (Thamatrakoln and Hildebrand, 2008), in which case, SITs may serve more of a regulatory function (Shrestha and Hildebrand, 2015). In the model diatom, *Thalassiosira pseudonana*, expression of two distinct SITs (SIT1 and SIT2) is induced by Si limitation, supporting a role in uptake and transport at low Si; expression of SIT3, is comparatively low, with no correlation between expression and silicon concentration, suggesting SIT3 may be more involved in silicon sensing (Brembu et al., 2017; Chen et al., 2018; Du et al., 2014; Shrestha and Hildebrand, 2015; Thamatrakoln and Hildebrand, 2007). However, SIT gene expression does not appear to be solely controlled by Si concentration. Phylogenetic analysis of SITs revealed the presence of several major clades (Marron et al., 2016), with one clade specifically induced by Si limitation and another responsive to Fe limitation (Durkin et al., 2012). This latter finding may contribute to physiological observations of increased silica production in Fe-limited diatoms (Marchetti and Harrison, 2007).

Little is known about the intracellular processing of silicon from the point of uptake at the cell membrane to the point it crosses the silicalemma into the SDV, creating a significant gap in our understanding of biosilicification. In addition, after nearly four decades of research, the SDV and its associated components have yet to be isolated and purified. Despite these gaps, extensive microscopic examinations (Pickett-Heaps et al., 1990) and characterization of proteins isolated directly from diatom cell walls have provided insight into the molecular basis for silica biomineralization, frustule structure formation, and patterning (Frigeri et al., 2006; Heintze et al., 2020; Kröger et al., 1994, 2001; Swift and Wheeler, 1992; Tesson and Hildebrand, 2010). Silica polymerization in the SDV is facilitated by long-chain polyamines (Kröger et al., 2002; Poulsen and Kröger, 2004) and catalyzed by silaffins and silacidins (Kröger et al., 2001, 2002). Silicalemma-associated proteins, SAPs (Tesson et al., 2017), and silicanin-1 (Kotzsch et al., 2016) help guide the micromorphogenesis of frustule pattern formation (Hildebrand et al., 2018). Scanning electron microscopy (SEM) and transmission electron microscopy (TEM) have provided detailed imaging of Si (Baines et al., 2012) and frustule surface ultrastructure (Heintze et al., 2020; Poulsen and Kröger, 2004). Lablling with the fluorescent dye 2-(4-pyridyl)-5{[4-dimethylaminoethyl-aminocarbamoyl)-methoxy] phenyl} oxazole (PDMPO), which is occluded in newly formed silica (McNair et al., 2015) allows direct visualization of biogenic Si deposition in relation to the cell division (McNair et al., 2018), nanoscale patterning (Heintze et al., 2020), and in response to nutrient limitation (Durkin et al., 2012). For the few taxa with detailed genetic information, and a genetic transformation system, using reporter-genes, including GFP in combination with elegant microscopy, has allowed visualization and localization of Si-related gene-products in diatoms, including SITs and silicanin (Heintze et al., 2020; Sapriel et al., 2009).

The promise and problems of using omics to understand silicon biogeochemistry

Gene by gene analysis has provided insights into discrete aspects of silicon uptake and silica production, but with the current rate of progress, it will take decades to fully resolve the molecular mechanisms underlying diatom biosilicification. However, when the first diatom whole genome sequences became available over 15 years ago (Armbrust et al., 2004) and a second followed (Bowler et al., 2008), we were finally given a partial dictionary and an opportunity to unlock mysteries of diatom biology. Among the most significant and unexpected findings was the presence of a complete urea cycle which, until then, was thought to be restricted to metazoans (Allen et al., 2011; Armbrust et al., 2004). This discovery was only possible because the urea cycle had already been well-characterized in heterotrophic organisms enabling the identification of homologous genes in diatoms. Searching for homologous genes based on other organisms is a common approach to deciphering genomic data. However, because diatom silicification is unique, we cannot refer to homologous genes in other groups, so the genetic underpinnings remain largely obscured, possibly within the >40% of diatom genes annotated as "hypothetical" or "putative proteins" (i.e., words lacking definitions). Contributing to the challenges is the intimate connection between silicification and the cell cycle (Brzezinski et al., 1990; Heintze et al., 2020), which in turn is related to numerous cellular metabolic processes, including nitrogen assimilation, photosynthesis, and carbon and energy metabolism. Nevertheless, in combination with physiological experiments, genomic data has been instrumental in identifying genes and proteins that are either responsive to Si (i.e., during Si starvation or Si replenishment experiments), coordinately expressed with known silicification-related genes (e.g., SITs or silaffins), or associated with periods of active silicon uptake and biogenic silica production (Brembu et al., 2017; Chen et al., 2018; Du et al., 2014; Frigeri et al., 2006; Mock et al., 2008; Scheffel et al., 2011; Smith et al., 2016). However, among these Si responsive genes in *T. pseudonana*, >60% have no known homologs (Shrestha et al., 2012). Though exploration of silicification-related genes, such as SITs has benefited from details in model diatom species, poor conservation of other major Si-related gene groups across diatom taxa limits our understanding (Sapriel et al., 2009; Thamatrakoln and Hildebrand, 2007). For example, 44% of Si-regulated genes in *Phaeodactylum tricornutum* have not been found in any other diatom and have unknown functions (Sapriel et al., 2009), whereas functionally homologous silaffins compared in two model diatoms, *T. pseudonana* and *Cylindrotheca fusiformis*, have limited amino acid sequence similarity (Poulsen and Kröger, 2004). Another study identified >500 proteins differentially expressed in response to silicon availability with no homology to any known proteins and lacking any conserved domains (Du et al., 2014). Thus although these genome-enabled efforts provide useful insight into many "words" in the dictionary, we still lack a comprehensive molecular and physiological framework of silicon metabolism in diatoms.

Though it may be possible to tease apart connections between Si metabolism and other cellular regulation with detailed physiological measurements, the current stumbling block is the lack of information on many of the genes and proteins identified in omics-based studies. Ways to apply such information are also being demonstrated. The first metatranscriptomic analysis of a eukaryotic phytoplankton community revealed the complex gene expression network underscoring the ability of diatoms to rapidly respond to iron enrichment (Marchetti et al., 2012). A recent integrated metagenomics and physiological analysis of *Prochlorococcus* responses to N, P, and Fe limitation provide an excellent example of how omics approaches can provide global-scale biogeochemical context for ecosystem processes (Ustick et al., 2021). Similar approaches with diatoms, but including Si, could start to bridge the gap between genes related to Si metabolism and Si biogeochemistry.

Case study: cell death

Cell death: an enigmatic process with profound ecological implications

There has been an evolution in our considerations of phytoplankton population dynamics, from a traditional view, where loss terms were limited to grazing, sedimentation, and a vague intrinsic loss term (Reynolds, 1984; Walsh, 1983), to one where we now accept the involvement of a range of pathogens, including viruses (Bigalke and Pohnert, 2019; Scholz et al., 2016; Suttle et al., 1990), and the idea that programmed cell death (PCD) occurs in phytoplankton (Bidle, 2015, 2016; Franklin et al., 2006), and we speculate about its origins and meanings (Durand et al., 2016; Nedelcu, 2009). The ecological implications of cell death for diatoms are significant: it's a process which rapidly converts particulate to dissolved organic matter, a mechanism of aggregation and rapid sedimentation, a driver of bloom demise in nature, and a mechanism for survival (group selection) of clonal populations (Armbrecht et al., 2014; Berges and Falkowski, 1998; Durand et al., 2016; Franklin et al., 2006; Sarthou et al., 2005; Timmermans et al., 2007).

There are conceptual difficulties defining PCD in a unicellular organism; Franklin et al. (2006) went so far as to title one section of their paper "Using the term 'PCD' – a health warning," and recommended instead the use of "cell death." But fundamentally, PCD could be defined as "*cell death resulting from specific gene expression within a moribund cell*" (Franklin et al., 2006). In nature, cells that sediment or are grazed are usually considered to simply disappear from the water column (Fig. 14.2A). From the presence of empty frustules and chlorotic cells in the water column, we can infer processes external to the cell (extrinsic) that could lead to cell fragments (particulate organic matter), and it is clear that virus infection can lead to lysis and generation of dissolved organic matter (Fig. 14.2A). But PCD implies a process where we could detect physiologically competent cells with photosynthetically active chloroplasts undergoing PCD that might be distinguished by some markers of cell death (Fig. 14.2A). There are various morphological and biochemical markers available (see Galluzzi et al., 2009, 2015 for detailed analyses in animal systems), but a definition based in gene expression demands evidence at the molecular level: transcription, translation, and perhaps posttranslational activation of specific genes/proteins. Thus this is a case where connections between genes, cell physiology, and an ecosystem-level process are not merely useful, but necessary.

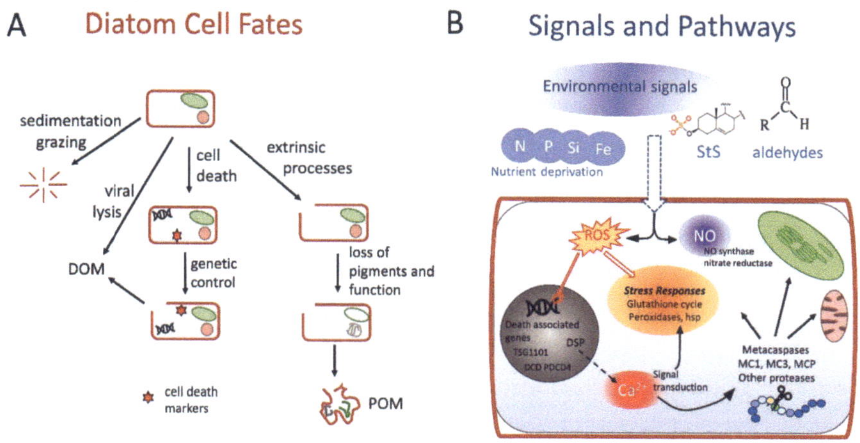

Figure 14.2 Summary of processes associated with cell death in marine diatoms. **A**. Alternative cell fates of diatoms in nature. Conventional losses of cells, due to grazing or sedimentation, likely result in complete loss of cells from the water column. Other extrinsic processes result in cells that have lost membrane or organelle integrity and pigmentation and ultimately result in pools of particulate organic matter (POM). In contrast, cell death processes have expression of specific genes, show specific characteristics (markers, red star) and, along with viral infections, result in lysis that leads to pools of dissolved organic matter (DOM). **B**. Signals and processes involved in cell death in a diatom cell. External signals (nutrient deprivation, or compounds, such as aldehydes or sterol sulphates, StS, released by processes such as grazing) lead to internal signaling pathways, mediated by compounds, such as reactive oxygen species (ROS), nitric oxide (NO), and intracellular calcium, which results in gene expression in pathways, including stress responses, and those associated with cell death (see Table 14.1).

The cell death process in phytoplankton and diatoms: specific molecular markers

Based on early molecular work on cell death, there was considerable optimism that many elements were conserved and generalizations were possible (Aravind et al., 1999). However, this has been challenged by the immense taxonomic diversity represented within phytoplankton (e.g., Berges and Choi, 2014; Bidle, 2015, 2016). In examining cell death in phytoplankton, green algae have been important models, and have been used extensively in culture (Sathe et al., 2019; Segovia et al., 2003). Critical information has also been gained from exploring the interactions between the coccolithophore, *Emiliania huxleyi*, and its associated viruses in both laboratory and field settings. This is a fascinating "arms race" involving coopting of cell death pathways and glycosphingolipids (Bidle and Vardi, 2011; Vardi et al., 2009, 2012), processes, which also occur in metazoan systems, where cell death proteases have become subverted to process viral proteins (Connolly and Fearnhead, 2017). Diatoms have figured prominently in the study of phytoplankton cell death, largely due to the availability of genome sequences (Armbrust et al., 2004; Bowler et al., 2008), which allowed searching for homologous cell death genes. One element that was identified and associated with cell death were the metacaspases.

Metacaspases illustrate some key issues in interpretation of molecular data concerning cell death. In the first place, although these proteases share homology with capases that have long been associated with mammalian PCD, all forms of metacaspases characterized so far differ in two important respects: they are calcium-dependent (caspases are calcium-independent) and are kinetically favored to cleave after arginine residues (caspases cleave after aspartate residues, as indicated in the name, "Cysteine-dependent ASPartate-directed proteASES"). This latter feature is particularly problematic because "caspase," "metacaspase," or "caspase-like" activity has been universally measured in phytoplankton using aspartate-containing substrates (e.g., Segovia et al., 2003). This leads to the conclusion that either phytoplankton metacaspases have low but measurable activity against aspartate-based substrates, or that use of these substrates measures activity of enzymes other than metacaspases; in either case, use of such substrates is clearly inappropriate (Minina et al., 2020). Secondly, metacaspases have diverse evolutionary origins that complicate the interpretation of their functions among different taxa. In higher plants, metacaspases are divided, based on the arrangement of P10 and P20 subunits, into Type I effector and Type II executor forms. Type II metacaspases appear to have arisen from a primary endosymbiotic event in green plants (Klemenčič and Funk, 2019). Diatoms, which are derived from a secondary endosymbiosis event, do not have Type II metacaspases, but instead possess Type III forms, which have a different subunit arrangement, and a metacaspase-like protease that only has a P20 subunit and is likely of bacterial origin (Choi and Berges, 2013). In the model diatom *T. pseudonana*, of the six metacaspase genes, TpMC2 is a Type I, TpMC1 and TpMC3 are Type III, and TpMC4, 5, and 6 are metacaspase-like proteases (Choi and Berges, 2013). Thus it is unclear whether metacaspases in a given organism are related to cell death, or whether only specific metacaspases might be.

Cell death in diatoms: experimental studies

Along with metacaspases, a number of other marker genes have been identified that vary with specific stress and species. To explore this, we consider the factors that have been used to induce cell death experimentally.

Aging/senescence. Despite the fact that almost everyone who has grown an algal culture has observed it entering "plateau/senescence" and "decline/death" phases, we know remarkably little about the mechanisms that cause these phenomena. Often the cause (e.g., nutrient exhaustion or build-up of secondary products of metabolism) is undocumented or unknown. Genes associated with stress have been implicated in studies of senescence, for example, in *Skeletonema marinoi*, upregulation of glutathione pathway genes (glutathione synthase, transferase, reductase, and peroxidase) and heat-shock proteins (Lauritano et al., 2015) (Fig. 14.2B). The mechanism in *S. marinoi* may involve sterols sulfates (StS), which can induce an oxidative burst and production of nitric oxide (NO; Gallo et al., 2017). Aging cultures with increases in StS also showed DNA damage (TUNEL staining) and phosphatidyl serine inversion in membranes (annexin staining), both markers of cell death in mammalian cells (Galluzzi et al., 2009).

In a different species, *S. costatum*, Chung et al. (2005) used a cDNA subtraction library to identify death-specific protein (ScDSP), a gene specifically induced during cell death. Because ScDSP contains peptide domains associated with calcium-binding, the authors concluded ScDSP was involved in a Ca^{2+}-dependent signal transduction cascade (Chung et al., 2005). However, a subsequent study demonstrated a role for NO and photosynthesis in ScDSP upregulation (Chung et al., 2008), and a homolog of ScDSP in *T. pseudonana* was found to play a role in photosynthetic electron transport independent of cell death (Thamatrakoln et al., 2013).

Macronutrient limitation (N, P, Si). As with all living organisms, diatom growth can be limited by N and P, but also, uniquely, by Si (see Silicon case study). Each has significant ecological implications, for example, Si limitation may lead to increased grazing (Zhang et al., 2017) or enhanced susceptibility to viral infection (Kranzler et al., 2019). In cultures of *T. weissflogii*, Si deprivation led to several-fold increases in general protease activity, as well as caspase-like activity, coinciding with losses of cells (Jiang et al., 2014). Si-deprived *S. marinoi* showed increased expression of antioxidant genes, as well as a subset of putative metacaspase genes and ScDSP genes, coincident with hallmarks of cell death: changes in morphological features, positive annexin staining and increased caspase-like activities (Wang et al., 2017). However, expression of three other genes, known to be in involved in cell death in other organisms (TSG1101, a tumor susceptibility gene related to ubiquitin conjugation; a DCD domain-containing protein found in plant proteins; and PDCD4, a tumor suppressor in animals and a stress response protein in plants) did not respond in *S. marinoi* (Orefice et al., 2015).

The commonly limiting macronutrients, N and P, have also been explored in limitation experiments. N-deprivation was used in early work on cell death (Berges and Falkowski, 1998), but *T. weissflogii* demonstrated only a slow decline in cell abundance and a slight increase in general protease activities without cell death. In *T. pseudonana*, N-deprivation showed a rapid increase in dead cells after 10 days of deprivation, yet no obvious changes in caspase-like activity, raising questions about the role of N limitation in cell death (Franklin et al., 2012). However, based on proteomic responses of *T. pseudonana*, Lin et al. (2017) concluded that N-limitation induced cell death and was associated with 122 differentially expressed proteins related to reactive oxygen species (ROS) accumulation, oxidative stress responses and cell death pathways. Potential markers included an upregulated protein with a NB-ARC domain (associated with plant resistance and regulation of cell death in animals), but a homolog of the previously mentioned DSP gene was actually down-regulated during starvation. In contrast, Wang et al. (2020) observed different responses to N and P limitation in *S. marinoi*, with stronger effects of N limitation, notably an increase in DSP expression, a number of antioxidant genes (including ALDH, GSHS and GOX), a heat shock protein (HSP90) and PDCD4, a homolog of a protein associated with abiotic stress in plants and tumor suppression in animals. Six metacaspase genes showed varying patterns, with only one consistently rising as nutrient stress increased. One response that was greater under P limitation that N limitation was in TSG101, a eukaryotic cell growth regulator. Contrasting results make generalization difficult, but clear candidate for marker genes do not emerge from these studies.

Fe limitation. Fe limitation has been observed in large regions of open ocean and has significant effects on phytoplankton physiology, namely through impacts on photosynthetic metabolism. Bidle and Bender (2008) examined the response of *T. pseudonana* to Fe starvation, demonstrating that both aging and Fe limitation led to morphological and physiological hallmarks of cell death and followed expression of six metacaspase genes (TpMC1-6). Overall, four metacaspases were upregulated during cell death, and one (TpMC2, the only Type I, effector-like metacaspase) was expressed in cultures deprived of Fe, and not in aging cultures. In both cases, a broad-spectrum caspase inhibitor was able to rescue dying cultures. A genome-wide transcriptome analysis of Fe-limited versus Fe-replete cultures of *T. pseudonana*, identified 1312 differentially expressed genes; among the 632 upregulated genes were several associated with oxidative stress, such as peroxioredoxin, two homologs of the previously mentioned ScDSP, and a subset of metacaspases (Thamatrakoln et al., 2012). Direct challenge of cultures with the oxidative stress-inducer hydrogen peroxide at sublethal doses produced similar patterns in metacaspase expression, leading to the conclusion that Fe limitation induced metacaspase activity and cell death through an oxidative stress mechanism. A study of proteins in *T. pseudonana* suggested that responses to Fe deprivation are strongly affected by the state of the culture (aging/senescence; Luo et al., 2014), suggesting that it may be difficult to associate markers of cell death with specific causes.

Aldehydes. Aldehyde compounds are produced by many diatoms and have been implicated in grazing interactions and the regulation of microbial activity (Edwards et al., 2015; see Sensing and interactions case study). Aldehydes have also been implicated in promoting the growth and survival of diatoms (e.g., Ribalet et al., 2007), which may explain observations that copepod grazing on *S. marinoi* affects transcription of stress and cell-death related genes, including heat-shock proteins and NO synthase, and results in changes in metabolites and physiology (Amato et al., 2018). Vardi et al. (2008) used the model aldehyde decadienal (DD) released from diatom cells by grazing and thought to be a chemical signal, to examine cell death in *P. tricornutum* by overexpressing the nitric oxide associated protein, PtNOA1, that DD activates. Overexpression resulted in increased NO, cell death, altered expression of superoxide dismutases and upregulation of metacaspases. van Creveld et al. (2015) subsequently used redox-sensitive dyes to show that the mitochondrion is a key mediator of this process. However, it is worth noting that a more recent study was unable to replicate the oxylipin-NO-PCD pathway in a different strain of *P. tricornutum*, highlighting the need to test oxylipins at more environmentally relevant concentrations and to consider their potential as signaling infochemicals, as opposed to reactive disruptors of cell physiology (Dolch et al., 2017).

From cell death genes to cell death ecology: prospects and problems

Overall, considering the range of experiments performed, few generalizations emerge: during cell death, we have tended to identify genes in cell stress pathways, especially related to oxidative stress, and a few homologous genes from recognized cell death pathways (e.g., metacaspases) have been identified, but there is little consistency in expressions or clarity in the overall pathways to which they belong (Fig. 14.2, Table 14.1). One problem could be that we are not always classifying death correctly; cells may be dying in processes other than PCD. In mammalian studies, emphasis is placed on running appropriate positive and negative controls for PCD (see Galluzzi et al., 2009), but we have little idea what these might be in diatoms. It might prove useful to have negative controls, in which cells are killed in ways that are clearly not programmed; Garrison and Tang (2014), for example, used a propeller to generate short-lived, episodic turbulence that resulted in increased number of intact dead cells (Evans blue staining) in *T. pseudonana* and *S. costatum*. Cell death pathways remain incompletely defined and largely unexplored. Nedelcu (2009) provided an excellent overview of cell death genes in unicellular eukaryotes, including diatoms (e.g., adaptor proteins, pro- and anti-cell death regulators, nuclear factors, and endoG-like proteases), but most have not been evaluated in the context of their roles in a cell death network. Finally, cell death is a process that operates at an individual level, but because the measurements we make are typically at the population-level, this may require measurements of subpopulations or even single cells. Novel single-cell molecular (Ku and Sebé-Pedrós, 2019) and mathematical (Hellweger et al., 2016; Hellweger, 2020; Kreft et al., 2017) approaches exist to do this, and the diatom life cycle may actually provide some unique possibilities to pursue such ideas with cell heterogeneity (e.g., Laney et al., 2012).

Case study: environmental sensing and community-level interactions

Sensory biology of diatoms: the importance of timescales

Signal perception and transduction in diatoms coordinates an array of physiological processes, such as nutrient acquisition and acclimation, photosynthetic metabolism and photoprotection, buoyancy regulation, stress surveillance and defense, inter- and intra-specific cell-cell interactions, and cell death. The signaling repertoire of diatoms determines what they perceive and how they respond to the environment, and ultimately shapes abundance and productivity. Our understanding of the molecular processes and signaling pathways underlying sensory biology in diatoms remains rudimentary (Falciatore et al., 2020). The deep evolutionary origins of eukaryote cell

Table 14.1 Summary of molecular markers of cell death identified in diatoms species under specific stress conditions.

Condition	Species	Experimental approach	Molecular markers identified	Reference
Aging/senescence	*Skeletonema costatum*	cDNA subtraction	Death Specific Protein (ScDSP)	Chung et al., 2005
Aging/senescence	*Skeletonema marinoi*	mRNA transcription	glutathione synthase, transferase, reductase and peroxidase, heat shock proteins	Lauritano et al., 2015
Si limitation	*Skeletonema marinoi*	mRNA transcription	TSG1101, DCD, PDCD4, DSP	Orefice et al., 2015
Si limitation	*Skeletonema marinoi*	mRNA transcription	Various antioxidant genes; metacaspases SmMC1-6	Wang et al., 2017
Fe limitation	*Thalassiosira pseudonana*	mRNA transcription	Metacaspases, TpMC1-6	Bidle and Bender, 2008
Fe limitation	*Thalassiosira pseudonana*	Genome-wide transcription	peroxioredoxin, DSP, metacaspases TPMC1-6, protein kinase with leucine-rich repeat domains	Thamatrakoln et al., 2012
Fe limitation	*Thalassiosira pseudonana*	overexpression	DSP	Thamatrakoln et al., 2013
Fe limitation	*Thalassiosira pseudonana*	proteomics	heat-shock proteins, serine protease inhibitor, DSP	Luo et al., 2014
N or P limitation	*Thalassiosira pseudonana*	proteomics	N: protein with a NB-ARC domain, NTPase in the NACHT family, DSP, serine protease inhibitor. P: alkaline phosphatase	Lin et al., 2017
N or P limitation	*Skeletonema marinoi*	mRNA transcription	DSP, ALDH, GSHS, GOX, HSP90, PDCD4, metacaspases SmMC1-6, TSG101	Wang et al., 2020
Aldehydes	*Phaeodactylum tricornutum*	overexpression	nitric oxide associated protein, PtNOA1	Vardi et al., 2008

signaling pathways (Marchadier et al., 2016; Cai et al., 2014), in combination with the long history of animal and plant signaling research, has provided a road map for studying diatom sensory biology. However, it is important not to overlook novel diatom-specific signaling processes. For example, the discovery of novel diatom-specific cyclins that are expressed at the G1-to-S phase of the cell cycle, and upregulated by resupply of P, highlights the link between nutrient availability and cell cycle progression (Huysman et al., 2010). Comparative genomics and transcriptomics have been applied to characterize different signal transduction pathways in diatoms (Mojib and Kubanek, 2019; Montsant et al., 2007; Port et al., 2014). Nevertheless, the ability to sense environmental cues or biotic stimuli over short timescales requires mechanisms that are not initially dependent on transcriptional responses. For example, large diatoms can regulate sinking rates, rapidly alternating between sinking and near neutral buoyancy within seconds (Gemmell et al., 2016). This behavior is modulated by light (Du Clos et al., 2019) and can be induced by microscale turbulence (Arrieta et al., 2020). The physiological mechanisms that allow for such large and rapid variations in sinking rate likely involve mechanosensory-induced changes in cytoplasmic Ca^{2+} (Falciatore et al., 2000) that initiate changes in cell volume (Lavoie and Raven, 2020). As we outline below, some of the most promising progress to understand rapid environmental sensing by diatoms has been made by combining transformation and reverse genetics with physiological and single-cell experiments.

The ability of diatoms to sense and respond to nutrients provides an excellent case study. The resource landscape for unicellular planktonic organisms is predominantly defined by point sources and turbulence that result in ephemeral microscale patches (Stocker, 2012). How diatoms have adapted to rapidly sense and respond to these fleeting opportunities for nutrient acquisition in heterogeneous microscale environments is largely unknown. Bioelectrical mechanisms (i.e., action potentials that give rise to rapid changes in membrane voltage) are among the fastest processes that link signal perception with downstream responses in animals and plants. Diatoms are also able to generate fast action potentials, the depolarizing phase of which is based on Na^+ and Ca^{2+} influx through voltage-gated ion channels in the plasma membrane (Fig. 14.3; Taylor, 2009). This fundamental signaling capability has deep eukaryote evolutionary origins (Brunet and Arendt, 2016; Moran et al., 2015; Verret et al., 2010). Consistent with this, diatoms encode sensory information in response to mechanostimuli, light, nutrients, and infochemicals through rapid and dynamic changes in cytosolic concentrations of Ca^{2+}, $[Ca^{2+}]_{cyt}$ (Falciatore et al., 2000; McLachlan et al., 2012; Vardi et al., 2006).

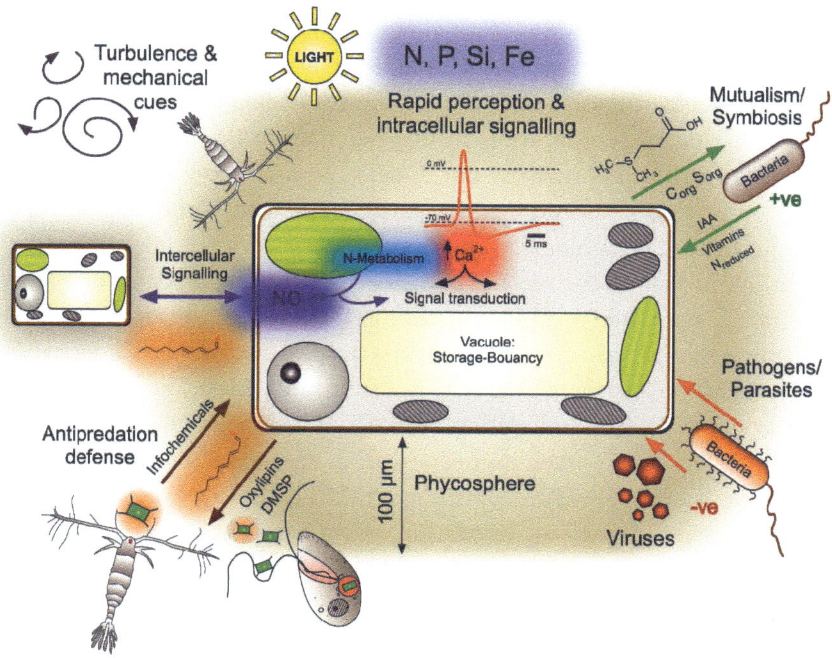

Figure 14.3 Conceptual diagram summarizing aspects of signaling and microbial interactions in the diatom phycosphere. The relevant spatial scale of many of these interactions is ~100 μm. Diatoms are able to perceive transient abiotic and biotic cues over short temporal scales (ms-s), which can include rapid membrane depolarization (i.e., action potentials) and influx of Ca^{2+} from the surrounding medium. Many signal transduction cascades are initiated by transient elevations or oscillations in cytoplasmic Ca^{2+}, which in turn regulate a variety of target proteins and metabolic pathways, ultimately influencing gene expression. The phycosphere also plays a critical role in determining diatom-microbe associations. Exchange of metabolites at the diatom cell surface mediates these interactions and several are known to act as infochemicals that can facilitate beneficial associations, mediate cell-cell communication, or act in chemical defense.

Diatoms possess fast sensory systems that retain conserved eukaryote tools, but have also evolved alternative pathways relevant to their unique evolutionary history and environmental physiology. Helliwell et al. (2019) recently demonstrated a novel family of single domain voltage dependent ion channels (EukCats) required to generate the action potential in diatoms. CRISPR–Cas9 gene editing was used to generate EukCatA knockout mutants in cell lines of *P. tricornutum* modified to express a genetically encoded cytoplasm localized Ca^{2+} sensor (R-GECO1). These mutants were unable to generate action potentials, failed to initiate elevations in $[Ca^{2+}]_{cyt}$, and exhibited impaired motility, a phenotype that could be rescued by transformation with the wild-type *P. tricornutum* EukCatA gene (Helliwell et al., 2019). The wide distribution of EukCatAs among diatoms suggests these channels play an important role in diatom environmental sensing. This example of reverse genetic tools in diatoms demonstrates that we are now poised to significantly increase our understanding of how diatoms encode a range of environmental cues with sufficient temporal resolution and specificity to initiate physiologically appropriate responses.

A case in point is the recent discovery of a novel Ca^{2+}-signaling pathway specifically associated with P limitation in diatoms. As a macronutrient that controls ocean primary productivity, research has focused on understanding diatom adaptive strategies for P acquisition, utilization, and conservation, such as phospholipid remodeling (Van Mooy et al., 2009) and P turnover (Martin et al., 2011). Although the diatom responses to P availability are detectable within 24–48 h (Martin et al., 2011), the relevant timescale for sensing extracellular P, or indeed any other nutrient or chemical cue, is much shorter (milliseconds to seconds). Using R-GECO1 Ca^{2+} sensor lines of *P. tricornutum* and *T. pseudonana*, Helliwell et al. (2021) observed rapid (on the order of seconds) $[Ca^{2+}]_{cyt}$ elevations in P-limited cells in response to resupply of a range of environmentally relevant inorganic and organic P sources and concentrations. This $[Ca^{2+}]_{cyt}$ response was completely absent in P-replete cultures and N- or trace metal limited cultures. Interestingly, the P-responsive $[Ca^{2+}]_{cyt}$ elevations were dependent on the presence of extracellular Ca^{2+} in the medium, but did not involve the aforementioned PtEUKCATA1 channel (Helliwell et al., 2019). The Ca^{2+} channel blocker ruthenium red selectively blocked the P-Ca^{2+} signaling, suggesting an alternative and yet to be determined Ca^{2+} influx pathway. Further proteomic analysis following P resupply revealed significant upregulation of proteins associated with P-acquisition, including alkaline phosphatases, consistent with previous studies. However, quite unexpectedly, there was also upregulation of core proteins involved in N uptake and assimilation. The authors concluded that the early Ca^{2+}-dependent, P-induced stimulation of N-uptake and corresponding assimilatory pathway is necessary to prime the cell in readiness to resume biosynthesis and growth. Importantly, the unforeseen cross-talk between P-resupply response and N-metabolism in P-limited diatoms was only revealed through the combination of physiological and proteomic experiments at timescales relevant to signal perception and metabolic responses. Moreover, the study represents the first description of a Ca^{2+}-dependent P-sensing mechanism in eukaryotes, and raises important considerations for the selection and interpretation of marker genes used to interrogate the physiological state of diatoms *in situ*. Clearly, studies that integrate cell physiology, reverse genetics, and omics have greater power to unlock the secrets of diatom sensory capability, and in doing so yield important discoveries in eukaryote biology.

Diatom stress and defense signaling: the challenge of deciphering redox cross-talk

Signaling cross-talk in diatoms is difficult to detect and characterize, even when models of signal transduction (e.g., Ca^{2+}-based signaling) and regulation of metabolic networks are well understood. This becomes even more challenging when considering cell signals, whose targets are not easily predicted through conserved protein motifs. Here we consider the interplay between oxylipins, nitric oxide (NO), and cell stress.

Many diatoms are known to produce oxylipins (reviewed in Bates et al., 2019), polyunsaturated aldehydes that can serve an allelopathic role in chemical defense against predators and competitors (Pohnert, 2005; Johnson et al., 2020; Ruocco et al., 2020a). Nutrient limitation and oxidative stress enhance diatom oxylipin production (Lupette et al., 2018; Ribalet et al., 2007), and they also act as a stress-responsive infochemical in intraspecific cell-cell communication (Leflaive and Ten-Hage, 2009). High resolution analytical chemistry has identified diatom polyunsaturated aldehyde synthesis pathways, via enzyme (e.g., lipoxygenase) (Ruocco et al., 2020b) and nonenzymatic (i.e., lipid peroxidation) pathways (Lupette et al., 2018), although the signals that cause their production and downstream cellular consequences are not fully understood. Grazing is one of the most striking signals to elicit diatom oxylipin-mediated defense. Mechanical disruption of diatom cells can trigger rapid production and release of defense oxylipins, aided by phospholipase activity (Pohnert, 2002). Though zooplankton (copepods) predation damage has been the focus of diatom-grazer interactions involving chemical defense, microzooplankton grazers of diatoms can also elicit increased oxylipin production that suppresses predator ingestion (Johnson et al., 2020). When incubated with copepod predators, the chain-forming diatom, *S. marinoi*, responded with a decrease in chain length after 48 h that significantly decreased grazing losses (Bergkvist et al., 2012). A protist grazer did not induce the same switch in chain morphology, supporting the dependence of the diatom defense response on copepod-specific diffusible cues, but how these cues are perceived by the diatom are unknown. From transcriptomic and metabolomic analysis of *S. marinoi*, Amato et al. (2018) proposed involvement of G-protein coupled receptor-mediated processes, NO signaling, and oxylipins. However, the analysis was conducted on cultures 3–5 d after introduction of the copepods, by which time significant chain remodeling was accomplished. The pattern of early sensory responses of diatoms to copepod predation remain to be elucidated.

Short-term (within seconds) cellular responses to oxylipins have been described in the model diatom, *P. tricornutum*. Micromolar concentrations of the diatom-derived aldehyde 2E,4E/Z-decadienal (DD) activates a Ca^{2+}-dependent cell death pathway involving NO production (Vardi et al., 2008) that can serve as an intraspecific stress surveillance cue in diatoms. Predation on *P. triconutum* by the protist grazer, *Oxyrrhis marina*, stimulates NO and oxylipin production within 2 h (Johnson et al., 2020). Grazing on cells overexpressing PtNOA (that exhibit elevated levels of NO) was greatly reduced, consistent with the observed suppression of grazing by treating cultures with exogenous NO (Johnson et al., 2020). Thus reactive NO and oxylipins appear to play important roles in early stress perception and response in diatoms. Although the signaling potential of oxylipins may be unclear, there is increasing evidence that NO plays a central role in diatom environmental sensing and acclimation responses.

Nitric oxide (NO) is a potent signal involved in a diverse array of cellular functions. This small free radical gas molecule can readily diffuse across membranes, where it interacts with a multitude of targets through S-nitrosylation. NO has emerged as a widespread signal in plants, algae, and phytoplankton, including diatoms and coccolithophores (Fig. 14.3; Bidle, 2015). However, there is no universal understanding of how NO is produced among eukaryotes. Regardless, endogenous enzymatic production of NO in photosynthetic organisms is mainly accomplished through plastid-associated reduction of nitrite catalyzed by nitrate reductase (Foresi et al., 2010; Hancock and Neill, 2019; Santolini et al., 2017). NO is thus a signal intimately linked with N-metabolism and abiotic stress in photosynthetic organisms. In *Chlamydomonas reinhardtii*, NO (derived from intracellular nitrite) can act as a signal to initiate remodeling of photosynthesis during N starvation (Wei et al., 2014). Dolch et al. (2017) reported that NO treatment in *P. tricornutum* reduced cell growth and increased triacylglycerol (TAG) accumulation without cell death. Using PtNOA overexpressing lines, they also confirmed that PtNOA regulates endogenous NO production via conversion of nitrite by nitrate reductase. In the sea-ice diatom, *Fragilariopsus cylindrica*, salinity and temperature stress induce NO production also via a nitrite-NR pathway (Kennedy et al., 2020). Overall, these studies point to a central role for NO in diatom responses to exogenous infochemicals and environmental conditions.

Is NO a "master regulator" at the intersection of multiple metabolic and signaling pathways in diatoms? NO interacts with a dynamic cellular redox system and can rapidly react with other ROS species, antioxidants, lipids, and an array of target proteins through S-nitrosylation of -SH groups. Understanding the outcome of NO production in environmental sensing and metabolic reprogramming of diatoms is clearly a challenge. Novel methods in mass spectrometry-based proteomic profiling of S-nitrosylated proteins (Lamotte et al., 2014) could provide important leads to understand critical targets of NO-mediated posttranslational modification. Moreover, the fate of diatoms in response to NO production (i.e., successful acclimation or a cell destined for death) likely involves nuanced spatiotemporal dynamics of the signal *and* its targets. This is also true when considering the role of NO in microbial interactions with diatoms. Careful experimental design and new physiological tools are needed to characterize complex and interconnected ROS and NO-associated signaling and acclimation responses.

Communication and microbial dynamics in the diatom phycosphere

Metagenomics has spurred advances in understanding diatom-microbe associations. Remarkable conservation of diatom-associated bacteria has been demonstrated (Behringer et al., 2018), although perhaps unsurprisingly, time and location of sampling are dominant factors that structure these communities (Ajani et al., 2018). The boundary layer or phycosphere of the cell (Seymour et al., 2017) is where cell-cell signaling and microscale ecology of diatom-microbe interactions are determined (Fig. 14.3). These interactions at the cell surface can be competitive, mutualistic, or pathogenic (Amin et al., 2012). The dynamics of nutrients, carbon exchange, and infochemicals in this proximal zone govern bacterial diversity in the phycosphere, which is markedly different than in the surrounding environment. What are the signaling mechanisms through which diatoms influence surface colonization and stabilize mutually beneficial associations? How do these intersect with defense and suppression of microbial pathogens? How can omics help describe these relationships?

Diatom-bacteria associations are consolidated through metabolic exchanges of vitamins, organic acids, carbon, and sulphur compounds. Transcriptomic and metabolomic analyses are needed to characterize microbial interactions in the phycosphere (trophic linkages, mutualism, pathogenesis, parasitism) that have the potential to significantly influence dynamics of ocean biogeochemical cycles and the fate of carbon. Diatom-bacteria mutualisms based on auxotrophy are widespread (Amin et al., 2012; Helliwell, 2017), and likely determine a significant portion of resource turnover in the phycosphere. Understanding the mechanisms that establish and maintain these microbial autotroph-heterotroph associations is needed to link them with biogeochemical models. For example, the diatom *T. pseudonana* requires vitamin B12, which is provided by associated *Rosebacteria*, such as *Ruegeria pomeroyi*, in exchange for a C_3-sulphonate produced by the diatom (Durham et al., 2015). In coculture experiments, the presence of *R. pomeroyi* alters the expression of a suite of *T. pseudonana* genes, many of which are homologous to conserved bacterial recognition and second messenger pathways in higher plants (Durham et al., 2017). Significantly, the availability of high-quality vascular plant databases enable the identification of these candidate diatom genes and proteins that mediate the host response. Analysis of the bacterial transcripts indicated upregulation of transport and metabolism of diatom derived organic compounds, supporting the trophic benefits of the association (Durham et al., 2015, 2017). The subsequent identification of *Roseobacter* transport orthologues (indicative of active trophic mutualism) in metatranscriptome data from a surface water transect in the eastern N. Pacific illustrates the potential to leverage laboratory-based studies to define targets that identify microbial associations and metabolite exchange in the environment. In spite of these detailed findings, we note that over half the diatom genes differently expressed in response to *R. pomeroyi* are hypothetical (cf. Supplementary Table 2; Durham et al., 2017), illustrating again the dependence on databases for functional annotation that limits a broader understanding of diatom ecological physiology.

The complexity and diversity of diatom-microbial associations is further illustrated by studies on the coastal diatom *Pseudo-nitzschia multiseries* and its bacterial affiliates (Amin et al., 2015). Removal of the bacterial consortium through antibiotic treatment significantly decreased *P. multiseries* growth. The resulting axenic *P. multiseries* was then cocultured with a variety of strains isolated from its bacterial consortium, with variable effects, ranging from enhanced growth to cell death, depending on the bacterial strain.

The *Sulfitobacter* strain (SA11) had the most significant positive effect on diatom growth, although this was both species-specific (no enhancements were observed with *T. pseudonana*) and strain-specific (a nonresponsive *P. multiseries* strain was identified). Transcriptomics and culture metabolomics identified potential mechanisms promoting the symbiotic diatom-bacterial association, including bacterial production of the growth-enhancing hormone, indole acetic acid (IAA), and release of reduced nitrogen, supporting increased diatom growth, and diatom release of DMSP and tryptophan that facilitate bacterial metabolism (Amin et al., 2015).

In addition to stabilization of favorable microbial associations in the phycosphere, diatoms must also selectively elicit defense mechanisms to deter or suppress unfavorable colonizers and pathogens (Shibl et al., 2020). For example, some diatoms can release eicosanoid oxylipins that suppress lytic algicidal bacteria (Meyer et al., 2018). However, characterization of diatom-pathogen interactions remains sparse, and the impact of parasite or virus infection on diatom bloom dynamics and export into the mesopelagic is poorly understood (Scholz et al., 2016). Nevertheless, there are promising advances in understanding diatom-virus dynamics. Infection of *Chaetoceros socialis* by a single-stranded RNA virus induces heavily silicified spore formation, in which production of viable virions is dramatically suppressed (Pelusi et al., 2020). This represents a biotic trigger of a diatom defense strategy that employs a dramatic shift in life history phase and links viral infection to a mechanism that would facilitate sinking. Nutrient availability also appears to influence host-virus interactions. Using metatranscriptomics and targeted PCR of extracellular diatom viruses in the California current ecosystem (CCE), Kranzler et al. (2019) were able to diagnose early, active, and lytic phases of infection among diatom populations and identified Si limitation as the strongest predictor of viral infection. This was recapitulated in Si-limited cultures of planktonic marine diatom *Chaetoceros tenuissimus*, which were significantly more susceptible to virus infection and mortality (Kranzler et al., 2019). In contrast, Fe limitation was found to delay infection and reduce mortality (Kranzler et al., 2021), in both natural communities in the CCE and laboratory culture. Taken together with increased silica production by Fe-limited diatoms, this also highlights a potential mechanism for virally-infected cells to escape the surface layer and be transported to depth. These studies demonstrate the utility of metatranscriptomics to uncover potential biogeochemical drivers and consequences of cell-pathogen interactions that are subsequently confirmed in laboratory model systems.

Conclusion: challenges and prospects

In this chapter, we have used three case studies to illustrate some of the important information we have gained about molecular level processes in diatoms. Genomic-, transcriptomic-, and proteomic-based approaches have provided an enormous amount of detailed information relevant to ecological processes in diatom communities, with real advances and critical problems in each case. There are also a number of common issues.

For all omics methods, there is a critical dependence on reference databases for taxonomic and functional interpretations. Sequence-based ecological studies nearly always use a comparative approach, but the critical reference sequences are often lacking for diatom and other unicellular eukaryotes (Keeling et al., 2014). Even for more well-studied human-associated bacteria, inconsistent conclusions from taxonomic (e.g., 16S rRNA) and functional (e.g., shotgun metagenomics) data have been reported (Xu et al., 2014), and interpretation of the same proteomics datasets can also vary, depending on the database used (Timmins-Schiffman et al., 2017). Ongoing efforts to address database limitations include the Marine Microbial Eukaryotic Transcription Sequencing Project to assemble, annotate, and publish transcriptomes (Keeling et al., 2014) and the Ocean Protein Portal (Saito et al., 2019).

Another major current limitation is the snap-shots in time, represented by most omics data that does not capture dynamics of genetic potential, gene expression, and turnover on temporal scales relevant to microbial populations. For example, the transcript levels of two important genes may be similar in quantitative transcriptomics, but the stability and turnover of the mRNA and functional proteins may be very different, resulting in distinct impacts on cell and population functions. Considerable analytical effort has focused on gene transcription changes, but cases of translational and posttranslational regulation of proteins in diatoms are becoming more apparent (Veluchamy et al., 2015). Furthermore, even in comparatively simple cyanobacteria, transcription can be difficult to quantitatively relate to protein abundance or activity (Waldbauer et al., 2012). Coupling higher temporal frequency omics sampling with high throughput measurements of specific processes may offer some more nuanced interpretation needed to extrapolate to ecosystem processes. This may work best for processes mediated by a distinct functional group e.g. N_2 fixation where omics of *nif* genes can be coupled with high-throughput nitrogenase activity measurements (Tang et al., 2020) and could be emulated for Si in diatoms if the molecular regulation of silicification could be resolved.

As noted above, understanding processes such as cell death may require measurements below the population level. This could be achieved through single-cell approaches, utilizing flow cytometric imaging combined with genetic and biochemical markers that can probe diatom physiological states (Colin et al., 2017; Merola et al., 2017), combined with novel noninvasive physiological analyses, such as for photosynthesis (PAM imaging). Omics approaches, including genetically encoded probes, e.g., redox sensors (Kudla et al., 2018), combined with single-cell metabolic interrogation, using advanced chemical imaging, such as Raman microscopy and NanoSIMS (reviewed by Taylor, 2019), may facilitate greater spatial resolution of diatom responses and will allow us to probe cell-to-cell heterogeneity of diatom processes, including signaling, metabolism, and productivity. Emerging technologies may also provide profitable applications for diatom ecology. Computational structural biology aided by artificial intelligence may be a new frontier. The recent community-wide assessment of computational techniques for protein structure and prediction (predictioncenter.org/casp14/) demonstrates advances in computational biology that have, until now, been largely driven by drug discovery and protein design, but may be applied to oceanography. The prospect of faster and more accurate structural prediction of proteins coded by genes of unknown function (Senior et al., 2020) has the potential to transform our understanding and interpretation of the functional genetic responses of diatoms to their biotic and abiotic environment.

There is also the issue of experimental approach and integration of approaches. The trend of using omics approaches in descriptive observational research limits interpretations for ecosystem functions (Bell, 2019), and commonly used correlative approaches are unsatisfying. In visioning the future of microbial oceanography, Huber and Preston (2019) identified in-situ experimentation and analysis, as a key goal. A more promising combination of careful experimental design, physiological and biochemical measurements to support omics data in field- or laboratory-based manipulations can generate data for modeling that reflect the biochemical pathways and the modes of regulation (e.g., Su et al., 2006; Nunn et al., 2013). Whatever the applications and approaches for data collection, modeling will play a central role in bridging gaps between genes and the ecosystem, though lack of functional information continues to hamper ability to integrate molecular data into conceptual models of biogeochemical processes, and generate working hypotheses about diatoms roles in ecosystem functions. Kreft et al. (2017) explored three approaches to modeling, comparing metabolic flux, gene centric, and agent-based (or individual-based) models to capture single-cell activities to population-level processes. Metabolic flux models offer the most direct use of molecular data, but require steady-state and fixed species-composition assumptions. Gene-centric models, are easy to integrate with existing biogeochemical models, but cope poorly with multiple environmental variables. Agent-based models are the only model types that deal effectively with cell-to-cell heterogeneity, but are limited by the availability of rate formulations and parameters. Model approaches integrating genetic, metabolic, and physiological parameters (e.g., Follows et al., 2007; van Tol and Armbrust, 2021) can help describe the impacts of diatoms on ecosystem processes.

Returning to our metaphor of molecular information as language, with genomic data representing the vocabulary and basic grammar (e.g., word functions), the aim is to understand and predict the ecological functioning and biogeochemistry of the oceans, trying to understand conversations, including advanced ideas, such as idiom and inuendo. Brussaard et al. (2016) emphasized the importance of integrated approaches and argued that interdisciplinary teams must "learn... each other's language." Physiology probably represents the advanced grammar that are critical to understanding. For example, the analogy of cultural and local context, which constitute the "physiological" variables for understanding the influences of climate, food, and cultural neighbors may mean that the common French language of Haiti versus the French Alps might be virtually mutually uninterpretable. In another view, we could consider our attempts to study ocean ecosystems like a guest in a busy cocktail party. Metaomic data are like recordings from a crowded room, where everyone is engaged in different conversations and it is virtually impossible to hear anyone. Focusing on a smaller group, where we can listen and follow the topic, may be more profitable; for example, choosing a smaller taxonomic subset such as diatoms, and focusing on smaller-scale connections, such as the connection of a gene to a physiological outcome. Learning the language of this group or process is more feasible, with a narrower context of high- or low-frequency words. We may learn to identify essential versus redundant words, but the order of words and their context is critical for understanding. In this metaphor, we are just now learning our first short sentences and a few key "marker" genes using omics, and it will be the physiological context that will reveal the complex sentences that will help us interpret and predict the broader ecosystem function of our oceans.

References

Ajani, P.A., Kahlke, T., Siboni, N., Carney, R., Murray, S.A., Seymour, J.R., 2018. The microbiome of the cosmopolitan diatom *Leptocylindrus* reveals significant spatial and temporal variability. Front. Microbiol. 9, 2758. https://doi.org/10.3389/fmicb.2018.02758.

Alexander, H., Jenkins, B.D., Rynearson, T.A., Dyhrman, S.T., 2015. Metatranscriptome analyses indicate resource partitioning between diatoms in the field. Proc. Natl. Acad. Sci. USA 112, E2182–E2190.

Allen, A.E., Dupont, C.L., Obornik, M., Horak, A., Nunes-Nesi, A., Mccrow, J.P., Zheng, H., Johnson, D.A., Hu, H., Fernie, A.R., Bowler, C., 2011. Evolution and metabolic significance of the urea cycle in photosynthetic diatoms. Nature 473, 203–207.

Amato, A., Sabatino, V., Nylund, G.M., Bergkvist, J., Basu, S., Andersson, M.X., Sanges, R., Godhe, A., Kiorboe, T., Selander, E., Ferrante, M.I., 2018. Grazer-induced transcriptomic and metabolomic response of the chain-forming diatom *Skeletonema marinoi*. ISME J. 12, 1594–1604. https://doi.org/10.1038/s41396-018-0094-0.

Amin, S.A., Hmelo, L.R., van Tol, H.M., Durham, B.P., Carlson, L.T., Heal, K.R., Morales, R.L., Berthiaume, C.T., Parker, M.S., Djunaedi, B., Ingalls, A.E., Parsek, M.R., Moran, M.A., Armbrust, E.V., 2015. Interaction and signalling between a cosmopolitan phytoplankton and associated bacteria. Nature 522, 98–101. https://doi.org/10.1038/nature14488.

Amin, S.A., Parker, M.S., Armbrust, E.V., 2012. Interactions between diatoms and bacteria. Microbiol. Mol. Biol. Rev. 76, 667–684. https://doi.org/10.1128/MMBR.00007-12.

Anderson, R., 2020. GEOTRACES: accelerating research on the marine biogeochemical cycles of trace elements and their isotopes. Annu. Rev. Mar. Sci. 12, 49–85.

Annenkov, V.V., Basharina, T.N., Danilovtseva, E.N., Grachev, M.A., 2013. Putative silicon transport vesicles in the cytoplasm of the diatom *Synedra acus* during surge uptake of silicon. Protoplasma 250, 1147–1155.

Annenkov, V.V., Gordon, R., Zelinskiy, S.N., Danilovtseva, E.N., 2020. The probable mechanism for silicon capture by diatom algae: assimilation of polycarbonic acids with diatoms-is endocytosis a key stage in building of siliceous frustules? J. Phycol. 56, 1729–1737.

Aravind, L., Dixit, V., Koonin, E., 1999. The domains of death: evolution of the apoptosis machinery. Trends Biochem. Sci. 24, 47–53.

Armbrecht, L.H., Smetacek, V., Assmy, P., Klaas, C., 2014. Cell death and aggregate formation in the giant diatom *Coscinodiscus wailesii* (Gran & Angst, 1931). J. Exp. Mar. Biol. Ecol. 452, 31–39.

Armbrust, E.V., Berges, J.A., Bowler, C., Green, B.R., Martinez, D., Putnam, N.H., Zhou, S., Allen, A.E., Apt, K.E., Bechner, M., Brzezinski, M.A., Chaal, B.K., Chiovitti, A., Davis, A.K., Demarest, M.S., Detter, J.C., Glavina, T., Goodstein, D., Hadi, M.Z., Hellesten, U., Hildebrand, M., Jenkins, B.D., Jurka, J., Kapitonov, V.V., Kroger, N., Lau, W.W.Y., Lane, T.W., Larimer, F.W., Lippmeier, J.C., Lucas, S., Medina, M., Monstant, A., Obornik, M., Parker, M.S., Palenik, B., Pazour, G.J., Richardson, P.M., Rynearson, T.A., Saito, M.A., Schwartz, D.C., Thamatrakoln, K., Valentin, K., Vardi, A., Wilkerson, F.P., Rokhsar, D.S., 2004. The genome of the diatom *Thalassiosira pseudonana*: ecology, evolution and metabolism. Science 306, 79–86.

Arrieta, J., Jeanneret, R., Roig, P., Tuval, I., 2020. On the fate of sinking diatoms: the transport of active buoyancy-regulating cells in the ocean. Philos. Trans. A Math. Phys. Eng. Sci. 378, 20190529. https://doi.org/10.1098/rsta.2019.0529.

Azam, F., Hemmsingsen, B.B., Volcani, B.E., 1974. Role of silicon in diatom metabolism V. Silicic acid transport and metabolism in the heterotrophic diatom *Nitzschia alba*. Arch. Microbiol. 97, 103–114.

Baines, S.B., Twining, B.S., Brzezinski, M.A., Krause, J., Vogt, S., Assael, D., McDaniel, H., 2012. Significant silicon accumulation by marine picocyanobacteria. Nat. Geosci. 5, 886–891.

Bates, S.S., Lundholm, N., Hubbard, K.A., Montresor, M., Leaw, C.P., 2019. Toxic and harmful marine diatoms. In: Seckbach, J., Gordon, R. (Eds.), Diatoms: Fundamentals and Applications. Scrivener, pp. 389–434.

Beardall, J., Young, E.B., Roberts, S., 2001. Approaches for determining phytoplankton nutrient limitation. Aquat. Sci. 63, 44–69.

Behringer, G., Ochsenkuhn, M.A., Fei, C., Fanning, J., Koester, J.A., Amin, S.A., 2018. Bacterial communities of diatoms display strong conservation across strains and time. Front. Microbiol. 9, 659. https://doi.org/10.3389/fmicb.2018.00659.

Bell, T., 2019. Next-generation experiments linking community structure and ecosystem functioning. Environ. Microbiol. Rep. 11, 20–22.

Berges, J.A., Choi, C.J., 2014. Cell death in algae: physiological processes and relationships with stress. Persp. Phycol. 1, 103–112.

Berges, J.A., Cochlan, W.P., Harrison, P.J., 1995. Laboratory and field responses of algal nitrate reductase to diel periodicity in irradiance, nitrate exhaustion, and the presence of ammonium. Mar. Ecol. Prog. Ser. 124, 259–269.

Berges, J.A., Falkowski, P.G., 1998. Physiological stress and cell death in marine phytoplankton: induction of proteases in response to nitrogen or light limitation. Limnol. Oceanogr. 43, 129–135.

Bergkvist, J., Thor, P., Jakobsen, H.H., Wängberg, S.-Å., Selander, E., 2012. Grazer-induced chain length plasticity reduces grazing risk in a marine diatom. Limnol. Oceanogr. 57, 318–324. https://doi.org/10.4319/lo.2012.57.1.0318.

Bidle, K.D., 2015. The molecular ecophysiology of programmed cell death in marine phytoplankton. Annu. Rev. Mar. Sci. 7. https://doi.org/10.1146/annurev-marine-010213-135014.

Bidle, K.D., 2016. Programmed cell death in unicellular phytoplankton. Curr. Biol. 26, R594–R607.

Bidle, K.D., Bender, S.J., 2008. Iron starvation and culture age activate metacaspases and programmed cell death in the marine diatom *Thalassiosira pseudonana*. Eukaryotic Cell 7, 223–236.

Bidle, K.D., Vardi, A., 2011. A chemical arms race at sea mediates algal host-virus interactions. Curr. Opin. Microbiol. 14, 449–457.

Bigalke, A., Pohnert, G., 2019. Algicidal bacteria trigger contrasting responses in model diatom communities of different composition. MicrobiologyOpen 8, e818. https://doi.org/10.1002/mbo3.818.

Bowler, C., Allen, A.E., Badger, J.H., Grimwood, J., Jabbari, K., Kuo, A., Maheswari, U., Martens, C., Maumus, F., Otillar, R.P., Rayko, E., Salamov, A., Vandepoele, K., Beszteri, B., Gruber, A., Heijde, M., Katinka, M., Mock, T., Valentin, K., Verret, F., Berges, J.A., Brownlee, C., Cadoret, J.-P., Chiovitti, A., Choi, C.J., Coesel, S., De Martino, A., Detter, J.C., Durkin, C., Falciatore, A., Fournet, J., Haruta, M., Huysman, M.J.J., Jenkins, B.D., Jiroutova, K., Jorgensen, R.E., Joubert, Y., Kaplan, A., Kroger, N., Kroth, P.G., La Roche, J., Lindquist, E., Lommer, M., Martin-Jezequel, V., Lopez, P.J., Lucas, S., Mangogna, M., McGinnis, K., Medlin, L.K., Montsant, A., Oudot-Le Secq, M.-P., Napoli, C., Obornik, M., Parker, M.S., Petit, J.-L., Porcel, B.M., Poulsen, N., Robison, M., Rychlewski, L., Rynearson, T.A., Schmutz, J., Shapiro, H., Siaut, M., Stanley, M., Sussman, M.R., Taylor, A.R., Vardi, A., Von Dassow, P., Vyverman, W., Willis, A., Wyrwicz, L.S., Rokhsar, D.S., Weissenbach, J., Armbrust, E.V., Green, B.R., Van De Peer, Y., Grigoriev, I.V., 2008. The *Phaeodactylum* genome reveals the evolutionary history of diatom genomes. Nature 456, 239–244.

Bowler, C., Karl, D.M., Colwell, R.R., 2009. Microbial oceanography in a sea of opportunity. Nature 459, 180–184.

Brembu, T., Chauton, M.S., Winge, P., Bones, A.M., Vadstein, O., 2017. Dynamic responses to silicon in *Thalasiossira pseudonana* - identification, characterisation and classification of signature genes and their corresponding protein motifs. Sci. Rep. 7, 4865. https://doi.org/10.1038/s41598-017-04921-0.

Brock, T.D., 1966. Principles of Microbial Ecology. Prentice Hall, NJ.

Brunet, T., Arendt, D., 2016. From damage response to action potentials: early evolution of neural and contractile modules in stem eukaryotes. Philos. Trans. R. Soc. Lond. B, Biol. Sci. 371, 1685. https://doi.org/10.1098/rstb.2015.0043.

Brussaard, C.P.D., Bidle, K.D., Pedros-Alio, C., Legrand, C., 2016. The interactive microbial ocean. Nat. Microbiol. 2, 16255. https://doi.org/10.1038/nmicrobiol.2016.255.

Brzezinski, M.A., Baines, S.B., Balch, W.M., Beucher, C.P., Chai, F., Dugdale, R.C., Krause, J.W., Landry, M.R., Marchi, A., Measures, C.I., Nelson, D.M., Parker, A.E., Poulton, A.J., Selph, K.E., Strutton, P.G., Taylor, A.G., Twining, B.S., 2011. Co-limitation of diatoms by iron and silicic acid in the equatorial Pacific. Deep-Sea Res. II 58, 493–511.

Brzezinski, M.A., Dummousseaud, C., Krause, J.W., Mesasures, C.I., Nelson, D.M., 2008. Iron and silicic acid concentrations together regulate Si uptake in the equatorial Pacific Ocean. Limnol. Oceanogr. 53, 875–889.

Brzezinski, M.A., Jones, J.L., Bidle, K., Azam, F., 2003. The balance between silica production and silica dissolution in the sea: insights from Monterey Bay, California applied to the global data set. Limnol. Oceanogr. 48, 1846–1854.

Brzezinski, M.A., Nelson, D.M., 1996. Chronic substrate limitation of silicic acid uptake rates in the western Sargasso Sea. Deep-Sea Res. II 43, 437–453.

Brzezinski, M.A., Olson, R.J., Chisholm, S.W., 1990. Silicon availability and cell-cycle progression in marine diatoms. Mar. Ecol. Prog. Ser. 67, 83–96.

Cai, X., Wang, X., Clapham, D.E., 2014. Early evolution of the eukaryotic Ca^{2+} signaling machinery: conservation of the CatSper channel complex. Mol. Biol. Evol. 31, 2735–2740. https://doi.org/10.1093/molbev/msu218.

Chai, F., Jiang, M.-S., Chao, Y., Dugdale, R.C., Chavez, F., Barber, R.T., 2007. Modeling responses of diatom productivity and biogenic silica export to iron enrichment in the equatorial Pacific Ocean. Glob. Biogeochem. Cycles 21, GB3S90. https://doi.org/10.1029/2006GB002804.

Chen, X.-H., Li, Y.-Y., Zhang, H., Liu, J.-L., Xie, Z.-X., Lin, L., Wang, D.-Z., 2018. Quantitative proteomics reveals common and specific responses of a marine diatom *Thalassiosira pseudonana* to different macronutrient deficiencies. Front. Microbiol. 9, 2761. https://doi.org/10.3389/fmicb.2018.02761.

Choi, C.J., Berges, J.A., 2013. New types of metacaspases in phytoplankton reveal diverse origins of cell death proteases. Cell Death Disease 4, e490. https://doi.org/10.1038/cddis.2013.21.

Chung, C.C., Hwang, S.P.L., Chang, J., 2005. Cooccurrence of ScDSP gene expression, cell death, and DNA fragmentation in a marine diatom, *Skeletonema costatum*. Appl. Environ. Microbiol. 71, 8744–8751.

Chung, C.C., Hwang, S.P., Chang, J., 2008. Nitric oxide as a signaling factor to upregulate the death-specific protein in a marine diatom, *Skeletonema costatum*, during blockage of electron flow in photosynthesis. Appl. Environ. Microbiol. 74, 6521–6527. https://doi.org/10.1128/aem.01481-08.

Cohen, N.R., Ellis, K.A., Lampe, R.H., McNair, H., Twining, B.S., Maldonado, M.T., Brzezinski, M.A., Kuzminov, F.I., Thamatrakoln, K., Till, C.P., Bruland, K.W., Sunda, W.G., Bargu, S., Marchetti, A., 2017. Diatom transcriptional and physiological responses to changes in iron bioavailability across ocean provinces. Front. Mar. Sci. 4, 360. https://doi.org/10.3389/fmars.2017.00360.

Colin, S., Coelho, L.P., Sunagawa, S., Bowler, C., Karsenti, E., Bork, P., Pepperkok, R., de Vargas, C., 2017. Quantitative 3D-imaging for cell biology and ecology of environmental microbial eukaryotes. eLife 6, e26066. https://doi.org/10.7554/eLife.26066.

Connolly, P.F., Fearnhead, H.O., 2017. Viral hijacking of host caspases: an emerging category of pathogen-host interactions. Cell Death Differ. 24, 1401–1410.

de Vargas, C., Audic, S., Henry, N., Decelle, J., Mahe, F., Logares, R., Lara, E., Berney, C., Le Bescot, N., Probert, I., Carmichael, M., Poulain, J., Romac, S., Colin, S., Aury, J.M., Bittner, L., Chaffron, S., Dunthorn, M., Engelen, S., Flegontova, O., Guidi, L., Horak, A., Jaillon, O., Lima-Mendez, G., Lukes, J., Malviya, S., Morard, R., Mulot, M., Scalco, E., Siano, R., Vincent, F., Zingone, A., Dimier, C., Picheral, M., Searson, S., Kandels-Lewis, S., Acinas, S.G., Bork, P., Bowler, C., Gorsky, G., Grimsley, N., Hingamp, P., Iudicone, D., Not, F., Ogata, H., Pesant, S., Raes, J., Sieracki, M.E., Speich, S., Stemmann, L., Sunagawa, S., Weissenbach, J., Wincker, P., Karsenti, E., Tara Oceans Consortium, 2015. Eukaryotic plankton diversity in the sunlit ocean. Science 348, 1261605. https://doi.org/10.1126/science.1261605.

Dolch, L.J., Lupette, J., Tourcier, G., Bedhomme, M., Collin, S., Magneschi, L., Conte, M., Seddiki, K., Richard, C., Corre, E., Fourage, L., Laeuffer, F., Richards, R., Reith, M., Rebeille, F., Jouhet, J., McGinn, P., Marechala, E., 2017. Nitric oxide mediates nitrite-sensing and acclimation and triggers a remodeling of lipids. Plant Physiol. 175, 1407–1423. https://doi.org/10.1104/pp.17.01042.

Du, C., Liang, J.-R., Chen, D.-D., Xu, B., Zhuo, W.-H., Gao, Y.-H., Chen, C.-P., Bowler, C., Zhang, W., 2014. iTRAQ-based proteomic analysis of the metabolism mechanism associated with silicon response in the marine diatom *Thalassiosira pseudonana*. J. Proteome Res. 13, 720–734.

Du Clos, K.T., Karp-Boss, L., Villareal, T.A., Gemmell, B.J., 2019. *Coscinodiscus wailesii* mutes unsteady sinking in dark conditions. Biol. Lett. 15, 20180816. https://doi.org/10.1098/rsbl.2018.0816.

Durand, P.M., Sym, S., Michod, R.E., 2016. Programmed cell death and complexity in microbial systems. Curr. Biol. 26, R587–R593.

Durham, B.P., Dearth, S.P., Sharma, S., Amin, S.A., Smith, C.B., Campagna, S.R., Armbrust, E.V., Moran, M.A., 2017. Recognition cascade and metabolite transfer in a marine bacteria-phytoplankton model system. Environ. Microbiol. 19, 3500–3513. https://doi.org/10.1111/1462-2920.13834.

Durham, B.P., Sharma, S., Luo, H., Smith, C.B., Amin, S.A., Bender, S.J., Dearth, S.P., Van Mooy, B.A., Campagna, S.R., Kujawinski, E.B., Armbrust, E.V., Moran, M.A., 2015. Cryptic carbon and sulfur cycling between surface ocean plankton. Proc. Natl. Acad. Sci. USA 112, 453–457.

Durkin, C.A., Marchetti, A., Bender, S.J., Truong, T., Morales, R., Mock, T., Armbrust, E.V., 2012. Frustule-related gene transcription and the influence of diatom community composition on silica precipitation in an iron-limited environment. Limnol. Oceanogr. 57, 1619–1633.

Edwards, B.R., Bidle, K.D., Van Mooy, B.A.S., 2015. Dose-dependent regulation of microbial activity on sinking particles by polyunsaturated aldehydes: implications for the carbon cycle. Proc. Natl. Acad. Sci. 112, 5909–5914.

Falciatore, A., d'Alcala, M.R., Croot, P., Bowler, C., 2000. Perception of environmental signals by a marine diatom. Science 288, 2363–2366. https://doi.org/10.1126/science.288.5475.2363.

Falciatore, A., Jaubert, M., Bouly, J.P., Bailleul, B., Mock, T., 2020. Diatom molecular research comes of age: model species for studying phytoplankton biology and diversity. Plant Cell 32, 547–572. https://doi.org/10.1105/tpc.19.00158.

Field, C., Behrenfeld, M., Randerson, J., Falkowski, P., 1998. Primary production of the biosphere: integrating terrestrial and oceanic components. Science 281, 237–240.

Follows, M.J., Dutkiewicz, S., Grant, S., Chisholm, S.W., 2007. Emergent biogeography of microbial communities in a model ocean. Science 315, 1843–1846.

Foresi, N., Correa-Aragunde, N., Parisi, G., Calo, G., Salerno, G., Lamattina, L., 2010. Characterization of a nitric oxide synthase from the plant kingdom: NO generation from the green alga *Ostreococcus tauri* is light irradiance and growth phase dependent. Plant Cell 22, 3816–3830. https://doi.org/10.1105/tpc.109.073510.

Franklin, D.J., Airs, R.L., Fernandes, M., Bell, T.G., Bongaerts, R.J., Berges, J.A., Malin, G., 2012. Identification of senescence and death in *Emiliania huxleyi* and *Thalassiosira pseudonana*: cell staining, chlorophyll alterations, and dimethylsulfoniopropionate (DMSP) metabolism. Limnol. Oceanogr. 57, 305–317.

Franklin, D.J., Brussaard, C.D.P., Berges, J.A., 2006. What is the role and nature of programmed cell death in phytoplankton ecology? Eur. J. Phycol. 41, 1–14.

Frigeri, L.G., Radabaugh, T.R., Haynes, P.A., Hildebrand, M., 2006. Identification of proteins from a cell wall fraction of the diatom *Thalassiosira pseudonana*: insights into silica structure formation. Mol. Cell. Proteomics 5, 182–193.

Gallo, C., d'Ippolito, G., Nuzzo, G., Sardo, A., Fontana, A., 2017. Autoinhibitory sterol sulfates mediate programmed cell death in a bloom-forming marine diatom. Nat. Commun. 8, 1292. https://doi.org/10.1038/s41467-017-01300-1:1292.

Galluzzi, L., Aaronson, S.A., Abrams, J., Alnemri, E.S., Andrews, D.W., Baehrecke, E.H., Bazan, N.G., Blagosklonny, M.V., Blomgren, K., Borner, C., Bredesen, D.E., Brenner, C., Castedo, M., Cidlowski, J.A., Ciechanover, A., Cohen, G.M., De Laurenzi, V., De Maria, R., Deshmukh, M., Dynlacht, B.D., El-Deiry, W.S., Flavell, R.A., Fulda, S., Garrido, C., Golstein, P., Gougeon, M.L., Green, D.R., Gronemeyer, H., Hajnoczky, G., Hardwick, J.M., Hengartner, M.O., Ichijo, H., Jaattela, M., Kepp, O., Kimchi, A., Klionsky, D.J., Knight, R.A., Kornbluth, S., Kumar, S., Levine, B., Lipton, S.A., Lugli, E., Madeo, F., Malomi, W., Marine, J.C., Martin, S.J., Medema, J.P., Mehlen, P., Melino, G., Moll, U.M., Morselli, E., Nagata, S., Nicholson, D.W., Nicotera, P., Nunez, G., Oren, M., Penninger, J., Pervaiz, S., Peter, M.E., Piacentini, M., Prehn, J.H., Puthalakath, H., Rabinovich, G.A., Rizzuto, R., Rodrigues, C.M., Rubinsztein, D.C., Rudel, T., Scorrano, L., Simon, H.U., Steller, H., Tschopp, J., Tsujimoto, Y., Vandenabeele, P., Vitale, I., Vousden, K.H., Youle, R.J., Yuan, J., Zhivotovsky, B., Kroemer, G., 2009. Guidelines for the use and interpretation of assays for monitoring cell death in higher eukaryotes. Cell Death Differ. 16, 1093–1107.

Galluzzi, L., Bravo-San Pedro, J.M., Vitale, I., Aaronson, S.A., Abrams, J.M., Adam, D., Alnemri, E.S., Altucci, L., Andrews, D., Annicchiarico-Petruzzelli, M., Baehrecke, E.H., Bazan, N.G., Bertrand, M.J., Bianchi, K., Blagosklonny, M.V., Blomgren, K., Borner, C., Bredesen, D.E., Brenner, C., Campanella, M., Candi, E., Cecconi, F., Chan, F.K., Chandel, N.S., Cheng, E.H., Chipuk, J.E., Cidlowski, J.A., Ciechanover, A., Dawson, T.M., Dawson, V.L., De Laurenzi, V., De Maria, R., Debatin, K.M., Di Daniele, N., Dixit, V.M., Dynlacht, B.D., El-Deiry, W.S., Fimia, G.M., Flavell, R.A., Fulda, S., Garrido, C., Gougeon, M.L., Green, D.R., Gronemeyer, H., Hajnoczky, G., Hardwick, J.M., Hengartner, M.O., Ichijo, H., Joseph, B., Jost, P.J., Kaufmann, T., Kepp, O., Klionsky, D.J., Knight, R.A., Kumar, S., Lemasters, J.J., Levine, B., Linkermann, A., Lipton, S.A., Lockshin, R.A., Lopez-Otin, C., Lugli, E., Madeo, F., Malorni, W., Marine, J.C., Martin, S.J., Martinou, J.C., Medema, J.P., Meier, P., Melino, S., Mizushima, N., Moll, U., Munoz-Pinedo, C., Nunez, G., Oberst, A., Panaretakis, T., Penninger, J.M., Peter, M.E., Piacentini, M., Pinton, P., Prehn, J.H., Puthalakath, H., Rabinovich, G.A., Ravichandran, K.S., Rizzuto, R., Rodrigues, C.M., Rubinsztein, D.C., Rudel, T., Shi, Y., Simon, H.U., Stockwell, B.R., Szabadkai, G., Tait, S.W., Tang, H.L., Tavernarakis, N., Tsujimoto, Y., Vanden Berghe, T., Vandenabeele, P., Villunger, A., Wagner, E.F., et al., 2015. Essential versus accessory aspects of cell death: recommendations of the NCCD 2015. Cell Death Differ. 22, 58–73.

Garrison, H.S., Tang, K.W., 2014. Effects of episodic turbulence on diatom mortality and physiology, with a protocol for the use of Evans Blue stain for live-dead determinations. Hydrobiologia 738, 155–170.

Gemmell, B.J., Oh, G., Buskey, E.J., Villareal, T.A., 2016. Dynamic sinking behaviour in marine phytoplankton: rapid changes in buoyancy may aid in nutrient uptake. Proc. R. Soc. B 283, 20161126. https://doi.org/10.1098/rspb.2016.1126.

Grachev, M.A., Bedoshvili, Y.D., Gerasimov, E.Y., Zaikovskii, V.I., Gneusheva, K.V., Likhoshway, Y.V., 2017. Silica-containing inclusions in the cytoplasm of diatom *Synedra acus*. Dokl. Biochem. Biophys. 472, 44–48.

Hancock, J.T., Neill, S.J., 2019. Nitric oxide: its generation and interactions with other reactive signaling compounds. Plants 8, 41. https://doi.org/10.3390/plants8020041.

Heintze, C., Formanek, P., Pohl, D., Hauptstein, J., Rellinghaus, B., Kröger, N., 2020. An intimate view into the silica deposition vesicles of diatoms. BMC Mat. 2, 11. https://doi.org/10.1186/s42833-020-00017-8.

Helliwell, K.E., 2017. The roles of B vitamins in phytoplankton nutrition: new perspectives and prospects. New Phytol. 216, 62–68.

Helliwell, K.E., Chrachri, A., Koester, J.A., Wharam, S., Verret, F., Taylor, A.R., Wheeler, G.L., Brownlee, C., 2019. Alternative mechanisms for fast Na^{+}/Ca^{2+} signaling in eukaryotes via a novel class of single-domain voltage-gated channels. Curr. Biol. 29, 503–1511.e1506. https://doi.org/10.1016/j.cub.2019.03.041.

Helliwell, K.E., Harrison, E.L., Christie-Oleza, J.A., Rees, A.P., Kleiner, F.H., Gaikwad, T., Downe, J., Aguilo-Ferretjans, M.M., Al-Moosawi, L., Brownlee, C., Wheeler, G.L., 2021. A novel Ca^{2+} signaling pathway coordinates environmental phosphorus sensing and nitrogen metabolism in marine diatoms. Curr. Biol. 31, 978–989. https://doi.org/10.1016/j.cub.2020.11.073.

Hellweger, F.L., 2020. Combining molecular observations and microbial ecosystem modeling: a practical guide. Annu. Rev. Mar. Sci. 12, 267–289.

Hellweger, F.L., Clegg, R.J., Clark, J.R., Plugge, C.M., Kreft, J.U., 2016. Advancing microbial sciences by individual-based modelling. Nat. Rev. Microbiol. 14, 461–471.

Hildebrand, M., 2000. Silicic acid transport and its control during cell wall silicification in diatoms. In: Bäuerlein, E. (Ed.), Biomineralization: from Biology to Biotechnology and Medical Applications. Wiley-VCH, Weinheim.

Hildebrand, M., Lerch, S.J.L., Shrestha, R.P., 2018. Understanding diatom cell wall silicification - moving forward. Front. Mar. Sci. 5, 125. https://doi.org/10.3389/fmars.2018.00125.

Hildebrand, M., Volcani, B.E., Gassmann, W., Schroeder, J.I., 1997. A gene family of silicon transporters. Nature 385, 688–689.

Hildebrand, M., Wetherbee, R., 2003. Components and control of silicification in diatoms. In: Muller, W.E.G. (Ed.), Progress in Molecular and Subcellular Biology. Springer-Verlag, Berlin Heidelberg.

Holland, H.D., 1984. The Chemical Evolution of the Atmosphere and Oceans. Princeton University Press, Princeton, NJ.

Huber, J.A., Preston, C., 2019. Take to the high seas: microbiology labs below the ocean surface. Environ. Microbiol. Rep. 11, 23–25.

Hutchins, D.A., Bruland, K.W., 1998. Iron-limited diatom growth and Si:N uptake ratios in a coastal upwelling regime. Nature 393, 561–564.

Huysman, M.J.J., Martens, C., Vandepoele, K., Gillard, J., Rayko, E., Heijde, M., Bowler, C., Inze, D., Van de Peer, Y., De Veylder, L., Vyverman, W., 2010. Genome-wide analysis of the diatom cell cycle unveils a novel type of cyclins involved in environmental signaling. Genome Biol. 11, R17. https://doi.org/10.1186/gb-2010-11-2-r17.

Iler, R.K., 1979. The Chemistry of Silica: Solubility, Polymerization, Colloid and Surface Properties, and Biochemistry. John Wiley and Sons, New York.

Jiang, Y., Yin, K., Berges, J.A., Harrison, P.J., 2014. Effects of silicate resupply to silicate-deprived *Thalassiosira weissflogii* (Bacillariophyceae) in stationary or senescent phase: short-term patterns of growth and cell death. J. Phycol. 50, 602–606.

Johnson, L.K., Alexander, H., Brown, C.T., 2018. Re-assembly, quality evaluation, and annotation of 678 microbial eukaryotic reference transcriptomes. GigaScience 8, 1–12.

Johnson, M.D., Edwards, B.R., Beaudoin, D.J., Van Mooy, B.A.S., Vardi, A., 2020. Nitric oxide mediates oxylipin production and grazing defense in diatoms. Environ. Microbiol. 22, 629–645. https://doi.org/10.1111/1462-2920.14879.

Karlusich, J.J.P., Ibarbalz, F.M., Bowler, C., 2020. Phytoplankton in the Tara Ocean. Annu. Rev. Mar. Sci. 12, 233–265.

Karsenti, E., Acinas, S.G., Bork, P., Bowler, C., De Vargas, C., Raes, J., Sullivan, M., Arendt, D., Benzoni, F., Claverie, J.M., Follows, M., Gorsky, G., Hingamp, P., Iudicone, D., Jaillon, O., Kandels-Lewis, S., Krzic, U., Not, F., Ogata, H., Pesant, S., Reynaud, E.G., Sardet, C., Sieracki, M.E., Speich, S., Velayoudon, D., Weissenbach, J., Wincker, P., Tara Oceans Consortium, 2011. A holistic approach to marine eco-systems biology. PLoS Biol. 9, e1001177. https://doi.org/10.1371/journal.pbio.1001177.

Keeling, P.J., Burki, F., Wilcox, H.M., Allam, B., Allen, E.E., Amaral-Zettler, L.A., Armbrust, E.V., Archibald, J.M., Bharti, A.K., Bell, C.J., Beszteri, B., Bidle, K.D., Cameron, C.T., Campbell, L., Caron, D.A., Cattolico, R.A., Collier, J.L., Coyne, K., Davy, S.K., Deschamps, P., Dyhrman, S.T., Edvardsen, B., Gates, R.D., Gobler, C.J., Greenwood, S.J., Guida, S.M., Jacobi, J.L., Jakobsen, K.S., James, E.R., Jenkins, B., John, U., Johnson, M.D., Juhl, A.R., Kamp, A., Katz, L.A., Kiene, R., Kudryavtsev, A., Leander, B.S., Lin, S., Lovejoy, C., Lynn, D., Marchetti, A., Mcmanus, G., Nedelcu, A.M., Menden-Deuer, S., Miceli, C., Mock, T., Montresor, M., Moran, M.A., Murray, S., Nadathur, G., Nagai, S., Ngam, P.B., Palenik, B., Pawlowski, J., Petroni, G., Piganeau, G., Posewitz, M.C., Rengefors, K., Romano, G., Rumpho, M.E., Rynearson, T., Schilling, K.B., Schroeder, D.C., Simpson, A.G., Slamovits, C.H., Smith, D.R., Smith, G.J., Smith, S.R., Sosik, H.M., Stief, P., Theriot, E., Twary, S.N., Umale, P.E., Vaulot, D., Wawrik, B., Wheeler, G.L., Wilson, W.H., Xu, Y., Zingone, A., Worden, A.Z., 2014. The Marine Microbial Eukaryote Transcriptome Sequencing Project (MMETSP): illuminating the functional diversity of eukaryotic life in the oceans through transcriptome sequencing. PLoS Biol. 12, e1001889. https://doi.org/10.1371/journal.pbio.1001889.

Kennedy, F., Martin, A., McMinn, A., 2020. Insights into the production and role of nitric oxide in the antarctic sea-ice diatom *Fragilariopsis cylindrus*. J. Phycol. 56, 1196–1207. https://doi.org/10.1111/jpy.13027.

Klemenčič, M., Funk, C., 2019. Evolution and structural diversity of metacaspases. J. Exp. Bot. 70, 2039–2047. https://doi.org/10.1093/jxb/erz082.

Kolber, Z., Falkowski, P.G., 1993. Use of active fluorescence to estimate phytoplankton photosynthesis in situ. Limnol. Oceanogr. 38, 1646–1665.

Kotzsch, A., Pawolski, D., Milentyev, A., Shevchenko, A., Scheffel, A., Poulsen, N., Shevchenko, A., Kröger, N., 2016. Biochemical composition and assembly of biosilica-associated insoluble organic matrices from the diatom *Thalassiosira pseudonana*. J. Biol. Chem. 291, 4982–4997.

Kranzler, C.F., Brzezinski, M.A., Cohen, N.R., Lampe, R.H., Maniscalco, M., Till, C.P., Mack, J., Latham, J.R., Bruland, K.W., Twining, B.S., Marchetti, A., Thamatrakoln, K., 2021. Impaired viral infection and reduced mortality of diatoms in iron-limited oceanic regions. Nat. Geosci. 14, 231–237. https://doi.org/10.1038/s41561-021-00711-6.

Kranzler, C.F., Krause, J.W., Brzezinski, M.A., Edwards, B.R., Biggs, W.P., Maniscalco, M., McCrow, J.P., Van Mooy, B.A.S., Bidle, K.D., Allen, A.E., Thamatrakoln, K., 2019. Silicon limitation facilitates virus infection and mortality of marine diatoms. Nat. Microbiol. 4, 1790–1797. https://doi.org/10.1038/s41564-019-0502-x.

Krause, J.W., Brzezinski, M.A., Villareal, T.A., Wilson, C., 2012. Increased kinetic efficiency for silicic acid uptake as a driver of summer diatom blooms in the North Pacific Subtropical Gyre. Limnol. Oceanogr. 57, 1084–1098.

Krause, J.W., Schultz, I.K., Rowe, K.A., Dobbins, W., Winding, M.H.S., Sejr, M.K., Duarte, C.M., Agustí, S., 2019. Silicon limitation drives bloom termination and potential carbon sequestration in an Arctic bloom. Sci. Rep. 9, 8149. https://doi.org/10.1038/s41598-019-44587-4.

Kreft, J.U., Plugge, C.M., Prats, C., Leveau, J.H.J., Zhang, W., Hellweger, F.L., 2017. From genes to ecosystems in microbiology: modeling approaches and the importance of individuality. Front. Microbiol. 8, 2299. https://doi.org/10.3389/fmicb.2017.02299.

Kröger, N., Bergsdorf, C., Sumper, M., 1994. A new calcium binding glycoprotein family constitutes a major diatom cell wall component. EMBO J. 13, 4676–4683.

Kröger, N., Deutzmann, R., Sumper, M., 2001. Silica-precipitating peptides from diatoms: the chemical structures of silaffin-1a from *Cylindrotheca fusiformis*. J. Biol. Chem. 276, 26066–26070.

Kröger, N., Lorenz, S., Brunner, E., Sumper, M., 2002. Self assembly of highly phosphorylated silaffins and their function in biosilica morphogenesis. Science 298, 584–586.

Ku, C., Sebé-Pedrós, A., 2019. Using single-cell transcriptomics to understand functional states and interactions in microbial eukaryotes. Philos. Trans. - R. Soc. B, Biol. Sci. 374, 20190098. https://doi.org/10.1098/rstb.2019.0098.

Kudla, J., Becker, D., Grill, E., Hedrich, R., Hippler, M., Kummer, U., Parniske, M., Romeis, T., Schumacher, K., 2018. Advances and current challenges in calcium signaling. New Phytol. 218, 414–431. https://doi.org/10.1111/nph.14966.

Kumar, S., Rechav, K., Kaplan-Ashiri, I., Gal, A., 2020. Imaging and quantifying homeostatic levels of intracellular silicon in diatoms. Sci. Adv. 6, eaaz7554. https://doi.org/10.1126/sciadv.aaz7554.

Kustka, A.B., Allen, A.E., Morel, F.M.M., 2007. Sequence analysis and transcriptional regulation of iron acquisition genes in two marine diatoms. J. Phycol. 43, 715–729.

Lamotte, O., Bertoldo, J.B., Besson-Bard, A., Rosnoblet, C., Aime, S., Hichami, S., Terenzi, H., Wendehenne, D., 2014. Protein S-nitrosylation: specificity and identification strategies in plants. Front. Chem. 2, 114. https://doi.org/10.3389/fchem.2014.00114.

Lampe, R.H., Cohen, N.R., Ellis, K.A., Bruland, K.W., Maldonado, M.T., Peterson, T.D., Till, C.P., Brzezinski, M.A., Bargu, S., Thamatrakoln, K., Kuzminov, F.I., Twining, B.S., Marchetti, A., 2018. Divergent gene expression among phytoplankton taxa in response to upwelling. Environ. Microbiol. 20, 3069–3082.

Laney, S.R., Olson, R.J., Sosik, H.M., 2012. Diatoms favor their younger daughters. Limnol. Oceanogr. 57, 1572–1578.

Lauritano, C., Orefice, I., Procaccini, G., Romano, G., Ianora, A., 2015. Key genes as stress indicators in the ubiquitous diatom *Skeletonema marinoi*. BMC Genomics 16, 411. https://doi.org/10.1186/s12864-015-1574-5.

Lavoie, M., Raven, J.A., 2020. How can large-celled diatoms rapidly modulate sinking rates episodically? J. Exp. Bot. 71, 3386–3389. https://doi.org/10.1093/jxb/eraa129.

Leflaive, J., Ten-Hage, L., 2009. Chemical interactions in diatoms: role of polyunsaturated aldehydes and precursors. New Phytol. 184, 794–805. https://doi.org/10.1111/j.1469-8137.2009.03033.x.

Lewin, J.C., Guillard, R.R.L., 1963. Diatoms. Annu. Rev. Microbiol. 17, 373–414.

Leynaert, A., Longphuirt, S.N., Claquin, P., Chauvaud, L., Ragueneau, O., 2009. No limit? The multiphasic uptake of silicic acid by benthic diatoms. Limnol. Oceanogr. 54, 571–576.

Li, C.-W., Volcani, B.E., Miller, A., Phillips, D.C., Williams, R.J.P., 1984. Aspects of silicification in wall morphogenesis of diatoms. Philos. Trans. R. Soc. Lond. B 304, 519–528.

Li, W.K.W., 2009. From cytometry to macroecology: a quarter century quest in microbial oceanography. Aquat. Microb. Ecol. 57, 239–251.

Lin, Q., Liang, J.R., Huang, Q.Q., Luo, C.S., Anderson, D.M., Bowler, C., Chen, C.P., Li, X.S., Gao, Y.H., 2017. Differential cellular responses associated with oxidative stress and cell fate decision under nitrate and phosphate limitations in *Thalassiosira pseudonana*: comparative proteomics. PLoS ONE 12, e0184849. https://doi.org/10.1371/journal.pone.0184849.

Luo, C.S., Liang, J.R., Lin, Q., Li, C., Bowler, C., Anderson, D.M., Wang, P., Wang, X.W., Gao, Y.H., 2014. Cellular responses associated with ROS production and cell fate decision in early stress response to iron limitation in the diatom *Thalassiosira pseudonana*. J. Proteome Res. 13, 5510–5523.

Lupette, J., Jaussaud, A., Vigor, C., Oger, C., Galano, J.M., Reversat, G., Vercauteren, J., Jouhet, J., Durand, T., Marechal, E., 2018. Non-enzymatic synthesis of bioactive isoprostanoids in the diatom *Phaeodactylum* following oxidative stress. Plant Physiol. 178, 1344–1357. https://doi.org/10.1104/pp.18.00925.

Marchadier, E., Oates, M.E., Fang, H., Donoghue, P.C., Hetherington, A.M., Gough, J., 2016. Evolution of the calcium-based intracellular signaling system. Genome Biol. Evol. 8, 2118–2132. https://doi.org/10.1093/gbe/evw139.

Marchetti, A., Harrison, P.J., 2007. Coupled changes in the cell morphology and the elemental (C, N, and Si) composition of the pennate diatom *Pseudo-nitzschia* due to iron deficiency. Limnol. Oceanogr. 52, 2270–2284.

Marchetti, A., Parker, M.S., Moccia, L.P., Lin, E.O., Arrieta, A.L., Ribalet, F., Murphy, M.E.P., Maldonado, M.T., Armbrust, E.V., 2009. Ferritin is used for iron storage in bloom-forming marine pennate diatoms. Nature 457, 467–470.

Marchetti, A., Schruth, D.M., Durkin, C.A., Parker, M.S., Kodner, R.B., Berthiaume, C.T., Morales, R., Allen, A.E., Armbrust, E.V., 2012. Comparative metatranscriptomics identifies molecular bases for the physiological responses of phytoplankton to varying iron availability. Proc. Natl. Acad. Sci. USA 109, E317–E325.

Margalef, R., 1978. Life forms of phytoplankton as survival alternatives in an unstable environment. Oceanol. Acta 1, 493–509.

Marron, A.O., Ratcliffe, S., Wheeler, G.L., Goldstein, R.E., King, N., Not, F., De Vargas, C., Richter, D.J., 2016. The evolution of silicon transport in eukaryotes. Mol. Biol. Evol. 33, 3226–3248.

Martin, P., Van Mooy, B.A., Heithoff, A., Dyhrman, S.T., 2011. Phosphorus supply drives rapid turnover of membrane phospholipids in the diatom *Thalassiosira pseudonana*. ISME J. 5, 1057–1060. https://doi.org/10.1038/ismej.2010.192.

Martin-Jézéquel, V., Hildebrand, M., Brzezinski, M.A., 2000. Silicon metabolism in diatoms: implications for growth. J. Phycol. 36, 821–840.

McLachlan, D.H., Underwood, G.J.C., Taylor, A.R., Brownlee, C., 2012. Calcium release from intracellular stores is necessary for the photophobic response in the benthic diatom *Navicula perminuta* (Bacillariophyceae). J. Phycol. 48, 675–681. https://doi.org/10.1111/j.1529-8817.2012.01158.x.

McNair, H.M., Brzezinski, M.A., Krause, J.W., 2015. Quantifying diatom silicification with the fluorescent dye, PDMPO. Limnol. Oceanogr., Methods 13, 587–599.

McNair, H.M., Brzezinski, M.A., Krause, J.W., 2018. Diatom populations in an upwelling environment decrease silica content to avoid growth limitation. Environ. Microbiol. 20, 4184–4193.

Merola, F., Memmolo, P., Miccio, L., Savoia, R., Mugnano, M., Fontana, A., D'Ippolito, G., Sardo, A., Iolascon, A., Gambale, A., Ferraro, P., 2017. Tomographic flow cytometry by digital holography. Light Sci. Appl. 6, e16241. https://doi.org/10.1038/lsa.2016.241.

Meyer, N., Rettner, J., Werner, M., Werz, O., Pohnert, G., 2018. Algal oxylipins mediate the resistance of diatoms against algicidal bacteria. Mar. Drugs 16, 486. https://doi.org/10.3390/md16120486.

Minina, E.A., Staal, J., Alvarez, V.E., Berges, J.A., Berman-Frank, I., Beyaert, R., Bidle, K.D., Bornancin, F., Casanova, M., Cazzulo, J.J., Choi, C.J., Coll, N.S., Dixit, V.M., Dolinar, M., Fasel, N., Funk, C., Gallois, P., Gevaert, K., Gutierrez-Beltran, E., Hailfinger, S., Klemenčič, M., Koonin, E.V., Krappmann, D., Linusson, A., Machado, M.F.M.,

Madeo, F., Megeney, L.A., Moschou, P.N., Mottram, J.C., Nyström, T., Osiewacz, H.D., Overall, C.M., Pandey, K.C., Ruland, J., Salvesen, G.S., Shi, Y., Smertenko, A., Stael, S., Ståhlberg, J., Suárez, M.F., Thome, M., Tuominen, H., Van Breusegem, F., Van Der Hoorn, R.a.L., Vardi, A., Zhivotovsky, B., Lam, E., Bozhkov, P.V., 2020. Classification and nomenclature of metacaspases and paracaspases: no more confusion with caspases. Mol. Cell 77, 927–929.

Mock, T., Otillar, R.P., Strauss, J., McMullan, M., Paajanen, P., Schmutz, J., Salamov, A., Sanges, R., Toseland, A., Ward, B.J., Allen, A.E., Dupont, C.L., Frickenhaus, S., Maumus, F., Veluchamy, A., Wu, T., Barry, K.W., Falciatore, A., Ferrante, M.I., Fortunato, A.E., Glockner, G., Gruber, A., Hipkin, R., Janech, M.G., Kroth, P.G., Leese, F., Lindquist, E.A., Lyon, B.R., Martin, J., Mayer, C., Parker, M., Quesneville, H., Raymond, J.A., Uhlig, C., Valas, R.E., Valentin, K.U., Worden, A.Z., Armbrust, E.V., Clark, M.D., Bowler, C., Green, B.R., Moulton, V., Van Oosterhout, C., Grigoriev, I.V., 2017. Evolutionary genomics of the cold-adapted diatom *Fragilariopsis cylindrus*. Nature 541, 536–540.

Mock, T., Samanta, M.P., Iverson, V., Berthiaume, C., Robison, M., Holtermann, K., Durkin, C., Splinter Bondurant, S., Richmond, K., Rodesch, M., Kallas, T., Huttlin, E.L., Cerrina, F., Sussman, M.R., Armbrust, E.V., 2008. Whole-genome expression profiling of the marine diatom *Thalassiosira pseudonana* identifies genes involved in silicon bioprocesses. Proc. Natl. Acad. Sci. USA 105, 1579–1584.

Mojib, N., Kubanek, J., 2019. Comparative transcriptomics supports the presence of G protein-coupled receptor-based signaling in unicellular marine eukaryotes. Limnol. Oceanogr. 65, 762–774. https://doi.org/10.1002/lno.11345.

Montsant, A., Allen, A.E., Coesel, S., Martino, A.D., Falciatore, A., Mangogna, M., Siaut, M., Heijde, M., Jabbari, K., Maheswari, U., Rayko, E., Vardi, A., Apt, K.E., Berges, J.A., Chiovitti, A., Davis, A.K., Thamatrakoln, K., Hadi, M.Z., Lane, T.W., Lippmeier, J.C., Martinez, D., Parker, M.S., Pazour, G.J., Saito, M.A., Rokhsar, D.S., Armbrust, E.V., Bowler, C., 2007. identification and comparative genomic analysis of signaling and regulatory components in the diatom *Thalassiosira pseudonana*. J. Phycol. 43, 585–604. https://doi.org/10.1111/j.1529-8817.2007.00342.x.

Moore, J.K., Doney, S.C., Glover, D.M., Fung, I.Y., 2002. Iron cycling and nutrient-limitation patterns in surface waters of the World Ocean. Deep-Sea Res. II 49, 463–507.

Moran, Y., Barzilai, M.G., Liebeskind, B.J., Zakon, H.H., 2015. Evolution of voltage-gated ion channels at the emergence of Metazoa. J. Exp. Biol. 218, 515–525. https://doi.org/10.1242/jeb.110270.

Nedelcu, A.M., 2009. Comparative genomics of phylogenetically diverse unicellular eukaryotes provide new insights into the genetic basis for the evolution of the programmed cell death machinery. J. Mol. Evol. 68, 256–268.

Nelson, D.M., Brzezinski, M.A., Sigmon, D.E., Franck, V.M., 2001. A seasonal progression of Si limitation in the Pacific sector of the Southern Ocean. Deep-Sea Res. II 48, 3973–3995.

Nelson, D.M., Dortch, Q., 1996. Silicic acid depletion and silicon limitation in the plume of the Mississippi River: evidence from kinetic studies in spring and summer. Mar. Ecol. Prog. Ser. 136, 163–178.

Nunn, B.L., Faux, J.F., Hippmann, A.A., Maldonado, M.T., Harvey, H.R., Goodlett, D.R., Boyd, P.W., Strzepek, R.F., 2013. Diatom proteomics reveals unique acclimation strategies to mitigate Fe limitation. PLoS ONE 8, e75653. https://doi.org/10.1371/journal.pone.0075653.

Orefice, I., Lauritano, C., Procaccini, G., Ianora, A., Romano, G., 2015. Insights into possible cell-death markers in the diatom *Skeletonema marinoi* in response to senescence and silica starvation. Mar. Genomics 24, 81–88. https://doi.org/10.1016/j.margen.2015.06.008.

Paasche, E., 1973. Silicon and the ecology of marine plankton diatoms. II. Silicate-uptake kinetics in five diatom species. Mar. Biol. 19, 262–269.

Pelusi, A., De Luca, P., Manfellotto, F., Thamatrakoln, K., Bidle, K.D., Montresor, M., 2020. Virus-induced spore formation as a defense mechanism in marine diatoms. New Phytol. 229, 2251–2259. https://doi.org/10.1111/nph.16951.

Pichevin, L.E., Ganeshram, R.S., Geibert, W., Thunell, R., Hinton, R., 2014. Silica burial enhanced by iron limitation in oceanic upwelling margins. Nat. Geosci. 7, 541–546.

Pickett-Heaps, J., Schmid, A.M.M., Edgar, L.A., 1990. The cell biology of diatom valve formation. In: Round, F.E., Chapman, D.J. (Eds.), Progress in Phycological Research. Biopress Ltd., Bristol.

Pohnert, G., 2002. Phospholipase A2 activity triggers the wound-activated chemical defense in the diatom *Thalassiosira rotula*. Plant Physiol. 129, 103–111. https://doi.org/10.1104/pp.010974.

Pohnert, G., 2005. Diatom/copepod interactions in plankton: the indirect chemical defense of unicellular algae. ChemBioChem 6, 946–959. https://doi.org/10.1002/cbic.200400348.

Port, J., Parker, M., Kodner, R.B., Wallace, J.C., Armbrust, E.V., Faustman, E.M., 2014. Identification of G protein-coupled receptor signaling pathway proteins in marine diatoms using comparative genomics. BMC Genomics 14, 503. https://doi.org/10.1186/1471-2164-14-503.

Poulsen, N., Kröger, N., 2004. Silica morphogenesis by alternative processing of silaffins in the diatom *Thalassiosira pseudonana*. J. Biol. Chem. 279, 42993–42999.

Rastogi, A., Vieira, F.R.J., Deton-Cabanillas, A.F., Veluchamy, A., Cantrel, C., Wang, G.H., Vanormelingen, P., Bowler, C., Piganeau, G., Hu, H.H., Tirichine, L., 2020. A genomics approach reveals the global genetic polymorphism, structure, and functional diversity of ten accessions of the marine model diatom *Phaeodactylum tricornutum*. ISME J. 14, 347–363.

Reynolds, C.S., 1984. The Ecology of Freshwater Phytoplankton. Cambridge University Press, Cambridge.

Ribalet, F., Berges, J.A., Ianora, A., Casotti, R., 2007. Growth inhibition of cultured marine. phytoplankton by toxic algal-derived polyunsaturated aldehydes. Aquat. Toxicol. 85, 219–227. https://doi.org/10.1016/j.aquatox.2007.09.006.

Rogerson, A., Defreitas, A.S.W., Mcinnes, A.G., 1987. Cytoplasmic silicon in the centric diatom *Thalassiosira pseudonana* by electron spectroscopic imaging. Can. J. Microbiol. 33, 128–131.

Rothstein, L.M., Cullen, J.J., Abbott, M., Chassignet, E.P., Denman, K., Doney, S.C., Ducklow, H., Fennel, K., Follows, M., Haidvogel, D., Hoffman, E., Karl, D.M., Kindle, J., Lima, I., Maltrud, M., McClain, C., McGillicuddy, D., Olascoaga, M.J., Spitz, Y., Wiggert, J., Yoder, J., 2006. Modeling ocean ecosystems: the PARADIGM program. Oceanography 19, 22–51. https://doi.org/10.5670/oceanog.2006.89.

Ruocco, N., Albarano, L., Esposito, R., Zupo, V., Costantini, M., Ianora, A., 2020a. Multiple roles of diatom-derived oxylipins within marine environments and their potential biotechnological applications. Mar. Drugs 18, 342. https://doi.org/10.3390/md18070342.

Ruocco, N., Nuzzo, G., d'Ippolito, G., Manzo, E., Sardo, A., Ianora, A., Romano, G., Iuliano, A., Zupo, V., Costantini, M., Fontana, A., 2020b. Lipoxygenase pathways in diatoms: occurrence and correlation with grazer toxicity in four benthic species. Mar. Drugs 18, 66. https://doi.org/10.3390/md18010066.

Rynearson, T.A., Palenik, B., 2011. Learning to read the oceans: genomics of marine phytoplankton. Adv. Mar. Biol. 60, 1–39.

Saito, M.A., Bertrand, E.M., Duffy, M.E., Gaylord, D.A., Held, N.A., Hervey, W.J., Hettich, R.L., Jagtap, P.D., Janech, M.G., Kinkade, D.B., Leary, D.H., McIlvin, M.R.,

Moore, E.K., Morris, R.M., Neely, B.A., Nunn, B.L., Saunders, J.K., Shepherd, A.I., Symmonds, N.I., Walsh, D.A., 2019. Progress and challenges in ocean metaproteomics and proposed best practices for data sharing. J. Proteome Res. 18, 1461–1476. https://doi.org/10.1021/acs.jproteome.8b00761.

Sala, M.M., Piontek, J., Endres, S., Romani, A.M., Dyhrman, S.T., Steen, A.D., 2019. Extracellular enzymes in aquatic environments: exploring the link between genomic potential and biogeochemical consequences. Front. Microbiol. 10, 1463. https://doi.org/10.3389/fmicb.2019.01463.

Salazar, G., Paoli, L., Alberti, A., Huerta-Cepas, J., Ruscheweyh, H.J., Cuenca, M., Field, C.M., Coelho, L.P., Cruaud, C., Engelen, S., Gregory, A.C., Labadie, K., Marec, C., Pelletier, E., Royo-Llonch, M., Roux, S., Sanchez, P., Uehara, H., Zayed, A.A., Zeller, G., Carmichael, M., Dimier, C., Ferland, J., Kandels, S., Picheral, M., Pisarev, S., Poulain, J., Tara Oceans, C., Acinas, S.G., Babin, M., Bork, P., Bowler, C., de Vargas, C., Guidi, L., Hingamp, P., Iudicone, D., Karp-Boss, L., Karsenti, E., Ogata, H., Pesant, S., Speich, S., Sullivan, M.B., Wincker, P., Sunagawa, S., 2019. Gene expression changes and community turnover differentially shape the global ocean metatranscriptome. Cell 179, 1068–1083.e21. https://doi.org/10.1016/j.cell.2019.10.014.

Santolini, J., Andre, F., Jeandroz, S., Wendehenne, D., 2017. Nitric oxide synthase in plants: where do we stand? Nitric Oxide 63, 30–38. https://doi.org/10.1016/j.niox.2016.09.005.

Sapriel, G., Quinet, M., Heijde, M., Jourdren, L., Tanty, V., Luo, G., Le Crom, S., Lopez, P.J., 2009. Genome-wide transcriptome analyses of silicon metabolism in *Phaeodactylum tricornutum* reveal the multilevel regulation of silicic acid transporters. PLoS ONE 4, e7458. https://doi.org/10.1371/journal.pone.0007458.

Sarthou, G., Timmermans, K.R., Blain, S., Treguer, P., 2005. Growth physiology and fate of diatoms in the ocean: a review. J. Sea Res. 53, 25–42.

Sathe, S., Orellana, M.V., Baliga, N.S., Durand, P.M., 2019. Temporal and metabolic overlap between lipid accumulation and programmed cell death due to nitrogen starvation in the unicellular chlorophyte *Chlamydomonas reinhardtii*. Phycol. Res. 67, 173–183.

Scheffel, A., Poulsen, N., Shian, S., Kröger, N., 2011. Nanopatterned protein microrings from a diatom that direct silica morphogenesis. Proc. Natl. Acad. Sci. USA 108, 3175–3180.

Scholz, B., Guillou, L., Marano, A.V., Neuhauser, S., Sullivan, B.K., Karsten, U., Kupper, F.C., Gleason, F.H., 2016. Zoosporic parasites infecting marine diatoms - a black box that needs to be opened. Fungal Ecol. 19, 59–76. https://doi.org/10.1016/j.funeco.2015.09.002.

Segovia, M., Haramaty, L., Berges, J.A., Falkowski, P.G., 2003. Cell death in the unicellular chlorophyte *Dunaliella tertiolecta*: a hypothesis on the evolution of apoptosis in higher plants and metazoans. Plant Physiol. 132, 99–105.

Senior, A.W., Evans, R., Jumper, J., Kirkpatrick, J., Sifre, L., Green, T., Qin, C., Žídek, A., Nelson, A.W.R., Bridgland, A., Penedones, H., Petersen, S., Simonyan, K., Crossan, S., Kohli, P., Jones, D.T., Silver, D., Kavukcuoglu, K., Hassabis, D., 2020. Improved protein structure prediction using potentials from deep learning. Nature 577, 706–710. https://doi.org/10.1038/s41586-019-1923-7.

Seymour, J.R., Amin, S.A., Raina, J.B., Stocker, R., 2017. Zooming in on the phycosphere: the ecological interface for phytoplankton-bacteria relationships. Nat. Microbiol. 2, 17065. https://doi.org/10.1038/nmicrobiol.2017.65.

Shibl, A.A., Isaac, A., Ochsenkuhn, M.A., Cardenas, A., Fei, C., Behringer, G., Arnoux, M., Drou, N., Santos, M.P., Gunsalus, K.C., Voolstra, C.R., Amin, S.A., 2020. Diatom modulation of select bacteria through use of two unique secondary metabolites. Proc. Natl. Acad. Sci. USA 117, 27445–27455. https://doi.org/10.1073/pnas.2012088117.

Shrestha, R., Tesson, B., Norden-Krichmar, T., Federowicz, S., Hildebrand, M., Allen, A., 2012. Whole transcriptome analysis of the silicon response of the diatom *Thalassiosira pseudonana*. BMC Genomics 13, 499. https://doi.org/10.1186/1471-2164-13-499.

Shrestha, R.P., Hildebrand, M., 2015. Evidence for a regulatory role of diatom silicon transporters in cellular silicon responses. Eukaryotic Cell 14, 29–40. https://doi.org/10.1128/EC.00209-14.

Smayda, T.J., 1980. Phytoplankton species succession. In: Morris, I. (Ed.), The Physiological Ecology of Phytoplankton. Blackwell, Oxford, pp. 493–570.

Smetacek, V., 2018. Seeing is believing: diatoms and the ocean carbon cycle revisited. Protist 169, 791–802.

Smith, S.R., Glé, C., Abbriano, R.M., Traller, J.C., Davis, A., Trentacoste, E., Vernet, M., Allen, A.E., Hildebrand, M., 2016. Transcript level coordination of carbon pathways during silicon starvation-induced lipid accumulation in the diatom *Thalassiosira pseudonana*. New Phytol. 210, 890–904.

Stocker, R., 2012. Marine microbes see a sea of gradients. Science 338, 628–633. https://doi.org/10.1126/science.1208929.

Strzepek, R.F., Harrison, P.J., 2004. Photosynthetic architecture differs in coastal and oceanic diatoms. Nature 431, 689–692.

Su, Z.C., Mao, F.L., Dam, P., Wu, H.W., Olman, V., Paulsen, I.T., Palenik, B., Xu, Y., 2006. Computational inference and experimental validation of the nitrogen assimilation regulatory network in cyanobacterium *Synechococcus* sp. WH 8102. Nucleic Acids Res. 34, 1050–1065.

Sullivan, C.W., 1979. Diatom mineralization of silicic acid IV. Kinetics of soluble Si pool formation in exponentially growing synchronized cultures of *Navicula pelliculosa*. J. Phycol. 15, 210–216.

Suttle, C.A., Chan, A.M., Cottrell, M.T., 1990. Infection of phytoplankton by viruses and reduction of primary productivity. Nature 347, 467–469.

Swift, D.M., Wheeler, A.P., 1992. Evidence of an organic matrix from diatom biosilica. J. Phycol. 28, 202–209.

Takeda, S., 1998. Influence of iron availability on nutrient consumption ratio of diatoms in oceanic waters. Nature 393, 774–777.

Tang, W., Cerdán-García, E., Berthelot, H., Polyviou, D., Wang, S., Baylay, A., Whitby, H., Planquette, H., Mowlem, M., Robidart, J., Cassar, N., 2020. New insights into the distributions of nitrogen fixation and diazotrophs revealed by high-resolution sensing and sampling methods. ISME J. 14, 2514–2526. https://doi.org/10.1038/s41396-020-0703-6.

Taylor, A.R., 2009. A fast Na^+/Ca^{2+} based action potential in a marine diatom. PLoS ONE 4, e4966. https://doi.org/10.1371/journal.pone.0004966.

Taylor, G.T., 2019. Windows into microbial seascapes: advances in nanoscale imaging and application to marine sciences. Annu. Rev. Mar. Sci. 11, 465–490. https://doi.org/10.1146/annurev-marine-121916-063612.

Tesson, B., Hildebrand, M., 2010. Extensive and intimate association of the cytoskeleton with forming silica in diatoms: control over patterning on the meso- and micro-scale. PLoS ONE 5, e14300. https://doi.org/10.1371/journal.pone.0014300.

Tesson, B., Lerch, S.J.L., Hildebrand, M., 2017. Characterization of a new protein family associated with the silica deposition vesicle membrane enables genetic manipulation of diatom silica. Sci. Rep. 7, 13457. https://doi.org/10.1038/s41598-017-13613-8.

Thamatrakoln, K., Bailleul, B., Brown, C.M., Gorbunov, M.Y., Kustka, A.B., Frada, M., Joliot, P.A., Falkowski, P.G., Bidle, K.D., 2013. Death-specific protein in a marine diatom regulates photosynthetic responses to iron and light availability. Proc. Natl. Acad. Sci. USA 110, 20123–20128.

Thamatrakoln, K., Hildebrand, M., 2007. Analysis of *Thalassiosira pseudonana* silicon transporters indicates distinct regulatory levels and transport activity through the cell cycle. Eukaryotic Cell 6, 271–279.

Thamatrakoln, K., Hildebrand, M., 2008. Silicon uptake in diatoms revisited: a model for saturable and nonsaturable uptake kinetics and the role of silicon transporters. Plant Physiol. 146, 1397–1407.

Thamatrakoln, K., Korenovska, O., Niheu, A.K., Bidle, K.D., 2012. Whole-genome expression analysis reveals a role for death-related genes in stress acclimation of the diatom *Thalassiosira pseudonana*. Environ. Microbiol. 14, 67–81.

Thamatrakoln, K., Kustka, A.B., 2009. When to say when: can excessive drinking explain silicon uptake in diatoms? BioEssays 31, 322–327.

Tilman, D., 1977. Resource competition between planktonic algae: an experimental and theoretical approach. Ecology 58, 338–348.

Tilman, D., Kilham, S.S., Kilham, P., 1982. Phytoplankton community ecology: the role of limiting nutrients. Annu. Rev. Ecol. Syst. 13, 349–372.

Timmermans, K.R., Veldhuis, M.J.W., Brussaard, C.P.D., 2007. Cell death in three marine diatom species in response to different irradiance levels, silicate, or iron concentrations. Aquat. Microb. Ecol. 46, 253–261.

Timmins-Schiffman, E., May, D.H., Mikan, M., Riffle, M., Frazar, C., Harvey, H.R., Noble, W.S., Nunn, B.L., 2017. Critical decisions in metaproteomics: achieving high confidence protein annotations in a sea of unknowns. ISME J. 11, 309–314.

Tréguer, P., Nelson, D.M., Van Bennekom, A.J., Demaster, D.J., Leynaert, A., Queguiner, B., 1995. The silica balance in the world ocean: a re-estimate. Science 268, 375–379.

Tréguer, P.J., Sutton, J.N., Brzezinski, M., Charette, M.A., Devries, T., Dutkiewicz, S., Ehlert, C., Hawkings, J., Leynaert, A., Liu, S.M., Llopis Monferrer, N., López-Acosta, M., Maldonado, M., Rahman, S., Ran, L., Rouxel, O., 2021. Reviews and syntheses: the biogeochemical cycle of silicon in the modern ocean. Biogeosciences 18, 1269–1289. https://doi.org/10.5194/bg-18-1269-2021.

Ustick, L.J., Larkin, A.A., Garcia, C.A., Garcia, N.S., Brock, M.L., Lee, J.A., Wiseman, N.A., Moore, J.K., Martiny, A.C., 2021. Metagenomic analysis reveals global-scale patterns of ocean nutrient limitation. Science 372, 287–291. https://doi.org/10.1126/science.abe6301.

van Creveld, S.G., Rosenwasser, S., Schatz, D., Koren, I., Vardi, A., 2015. Early perturbation in mitochondria redox homeostasis in response to environmental stress predicts cell fate in diatoms. ISME J. 9, 385–395.

Van Mooy, B.A.S., Fredricks, H.F., Pedler, B.E., Dyhrman, S.T., Karl, D.M., Koblizek, M., Lomas, M.W., Mincer, T.J., Moore, L.R., Moutin, T., Rappe, M.S., Webb, E.A., 2009. Phytoplankton in the ocean use non-phosphorus lipids in response to phosphorus scarcity. Nature 458, 69–72. https://doi.org/10.1038/nature07659.

van Tol, H.M., Armbrust, E.V., 2021. Genome-scale metabolic model of the diatom *Thalassiosira pseudonana* highlights the importance of nitrogen and sulfur metabolism in redox balance. PLoS ONE 16, e0241960. https://doi.org/10.1371/journal.pone.0241960.

Vandenkoornhuyse, P., Dufresne, A., Quaiser, A., Gouesbet, G., Binet, F., Francez, A.J., Mahe, S., Bormans, M., Lagadeuc, Y., Couee, I., 2010. Integration of molecular functions at the ecosystemic level: breakthroughs and future goals of environmental genomics and post-genomics. Ecol. Lett. 13, 776–791.

Vardi, A., Bidle, K.D., Kwityn, C., Hirsh, D.J., Thompson, S.M., Callow, J.A., Falkowski, P.G., Bowler, C., 2008. A diatom gene regulating nitric-oxide signaling and susceptibility to diatom-derived aldehydes. Curr. Biol. 18, 895–899.

Vardi, A., Formiggini, F., Casotti, R., De Martino, A., Ribalet, F., Miralto, A., Bowler, C., 2006. A stress surveillance system based on calcium and nitric oxide in marine diatoms. PLoS Biol. 4, e60. https://doi.org/10.1371/journal.pbio.0040060.

Vardi, A., Haramaty, L., Van Mooy, B.A., Fredricks, H.F., Kimmance, S.A., Larsen, A., Bidle, K.D., 2012. Host-virus dynamics and subcellular controls of cell fate in a natural coccolithophore population. Proc. Natl. Acad. Sci. USA 109, 19327–19332.

Vardi, A., van Mooy, B.A., Fredricks, H.F., Popendorf, K.J., Ossolinski, J.E., Haramaty, L., Bidle, K.D., 2009. Viral glycosphingolipids induce lytic infection and cell death in marine phytoplankton. Science 326, 861–865.

Veluchamy, A., Rastogi, A., Lin, X., Lombard, B., Murik, O., Thomas, Y., Dingli, F., Rivarola, M., Ott, S., Liu, X., Sun, Y., Rabinowicz, P.D., Mccarthy, J., Allen, A.E., Loew, D., Bowler, C., Tirichine, L., 2015. An integrative analysis of post-translational histone modifications in the marine diatom *Phaeodactylum tricornutum*. Genome Biol. 16, 102. https://doi.org/10.1186/s13059-015-0671-8.

Verret, F., Wheeler, G., Taylor, A.R., Farnham, G., Brownlee, C., 2010. Calcium channels in photosynthetic eukaryotes: implications for evolution of calcium-based signalling. New Phytol. 187, 23–43. https://doi.org/10.1111/j.1469-8137.2010.03271.x.

Vrieling, E.G., Gieskes, W.W.C., Beelen, T.P.M., 1999. Silicon deposition in diatoms: control by the pH inside the silicon deposition vesicle. J. Phycol. 35, 548–559.

Waldbauer, J.R., Rodrigue, S., Coleman, M.L., Chisholm, S.W., 2012. Transcriptome and proteome dynamics of a light-dark synchronized bacterial cell cycle. PLoS ONE 7, e43432. https://doi.org/10.1371/journal.pone.0043432.

Walsh, J.J., 1983. Death in the sea: enigmatic loss processes. Cont. Shelf Res. 12, 1–86.

Wang, H., Chen, F., Mi, T., Liu, Q., Yu, Z., Zhen, Y., 2020. Responses of marine diatom *Skeletonema marinoi* to nutrient deficiency: programmed cell death. Appl. Environ. Microbiol. 86, e02460. https://doi.org/10.1128/AEM.02460-19.

Wang, H., Mi, T., Zhen, Y., Jing, X., Liu, Q., Yu, Z., 2017. Metacaspases and programmed cell death in *Skeletonema marinoi* in response to silicate limitation. J. Plankton Res. 39, 729–743.

Ward, B., 1973. Speech for Stockholm. In: Strong, M.F. (Ed.), Who Speaks for Earth?. W.W. Norton & Company, London, pp. 19–32.

Ward, B.B., van Oostende, N., 2016. Phytoplankton assemblage during the North Atlantic spring bloom assessed from functional gene analysis. J. Plankton Res. 38, 1135–1150.

Wei, L., Derrien, B., Gautier, A., Houille-Vernes, L., Boulouis, A., Saint-Marcoux, D., Malnoe, A., Rappaport, F., de Vitry, C., Vallon, O., Choquet, Y., Wollman, F.A., 2014. Nitric oxide-triggered remodeling of chloroplast bioenergetics and thylakoid proteins upon nitrogen starvation in *Chlamydomonas reinhardtii*. Plant Cell 26, 353–372. https://doi.org/10.1105/tpc.113.120121.

Weigmann, K., 2004. The code, the text and the language of God. EMBO Rep. 5, 116–118. https://doi.org/10.1038/sj.embor.7400069.

Williams, P.J.L.B., Thomas, D.N., Reynolds, C.S. (Eds.), 2002. Phytoplankton Productivity: Carbon Assimilation in Marine and Freshwater Ecosystems. Blackwell Science, Oxford.

Xu, Z., Malmer, D., Langille, M.G., Way, S.F., Knight, R., 2014. Which is more important for classifying microbial communities: who's there or what they can do? ISME J. 8, 2357–2359. https://doi.org/10.1038/ismej.2014.157.

Zhang, S., Liu, H., Ke, Y., Li, B., 2017. Effect of the silica content of diatoms on protozoan grazing. Front. Mar. Sci. 4. https://doi.org/10.3389/fmars.2017.00202.

CHAPTER 15

Global marine phytoplankton revealed by the *Tara* Oceans expedition

Flora Vincent[a], Federico M. Ibarbalz[b], and Chris Bowler[c,d]

[a]Department of Plant and Environmental Sciences, Weizmann Institute of Science, Rehovot, Israel
[b]Universidad de Buenos Aires, CONICET, Centro de Investigaciones del Mar y la Atmósfera (CIMA), Buenos Aires, Argentina
[c]Fédération de Recherche Global Ocean Systems Ecology and Evolution, FR2022/Tara Oceans GOSEE, Paris, France
[d]Institut de Biologie de l'ENS (IBENS), Département de biologie, École normale supérieure, CNRS, INSERM, Université PSL, Paris, France

Introduction

The ocean represents the largest continuous ecosystem on Earth, and at least two thirds of the biomass within is made up of plankton (Bar-On et al., 2018). Marine plankton communities are composed of prokaryotes (bacteria and archaea), single-celled eukaryotes, viruses, and microscopic animals (including larval stages of larger organisms). The photosynthetic plankton are termed phytoplankton, and consist of bacteria (e.g., *Prochlorococcus* and *Synechococcus*) and single-celled eukaryotes (e.g., diatoms, dinoflagellates, and haptophytes), as well as a range of photosymbionts (most notably, radiolarians in symbiosis with photosynthetic dinoflagellates). By converting carbon dioxide into organic biomass, phytoplankton collectively sustain the vast majority of marine food webs and are fundamental for the conversion of around 50 gigatons of inorganic carbon into organic material (Field et al., 1998; Yamasaki, 2003). They also play major roles in the biogeochemical cycles of oxygen, nitrogen, phosphorus, silica, and sulfur.

Phytoplankton are evolutionarily diverse (Adl et al., 2012) and have varying ecologies and life histories (Eilertsen and Wyatt, 2000). The larger sized groups are defined as microphytoplankton (20–200 microns in size range) and consist essentially of diatoms and dinoflagellates. Because of their comparatively large size, they have been studied for several centuries (Pierella Karlusich et al., 2020a). The advent of more sophisticated sampling and microscopy technologies in the 20th century subsequently revealed the abundance and diversity of other groups, now classified as nanophytoplankton (2–20 microns) and picophytoplankton (0.2–2 microns). Their most prominent members include haptophytes (including coccolithophores), pelagophytes, and prasinophytes, as well as the minuscule picocyanobacteria, *Prochlorococcus* and *Synechococcus*. The picocyanobacteria were only discovered in the 1970s–1980s, yet are now considered to be some of the most abundant organisms on the planet. *Synechococcus*, 0.8 to 1.5 μm in diameter, was discovered in 1979 in the Arabian Sea (Johnson and Sieburth, 1979; Waterbury et al., 1979), whereas *Prochlorococcus* is even smaller, at just 0.5 to 0.7 μm in diameter, and was discovered in 1986 following the introduction of flow cytometry into oceanographic research (Chisholm et al., 1988).

Exploration of the global distribution and diversity of marine phytoplankton is being further accelerated by genomics-based approaches. The application of DNA sequencing in a metagenomics framework in marine biology and oceanography at a large spatial scale began with J. Craig Venter's Global Ocean Sampling expeditions from 2003 to 2008. This technique involves the sequencing *en masse* of DNA extracted from a community of organisms, and their subsequent disentanglement to yield information relating to the taxonomy of the organisms present in the community and the potential functions of the genes they contain (Venter et al., 2004). The beginning of the 21st century additionally witnessed the sequencing of complete genomes of several marine phytoplanktonic species, including *Prochlorococcus* (Dufresne et al., 2003; Rocap et al., 2003), *Synechococcus* (Palenik et al., 2003), the diatoms *Thalassiosira pseudonana* (Armbrust et al., 2004) and *Phaeodactylum tricornutum* (Bowler et al., 2008), and the prasinophytes *Ostreococcus tauri* (Derelle et al., 2006), *O. lucimarinus* (Palenik et al., 2007), *Micromonas commoda* and *M. pusilla* (Worden et al., 2009). More recently, the Marine Microbial Eukaryote Transcriptome Sequencing Project (MMETSP) sequenced expressed genes from cultured protists. It resulted in over 650 assembled and functionally annotated transcriptomes from 396 distinct strains within 210 different genera from undersampled branches of the eukaryotic tree of life, most of them corresponding to phytoplankton species (Keeling et al., 2014). These resources provide further possibilities to improve our capacity to extract useful information from metagenomes.

The current chapter is focused on the *Tara* Oceans expeditions (2009–2013) and relies on the knowledge we have acquired from their corresponding studies since then. Covering a 140,000 kilometers voyage across all the oceans of the planet, *Tara* Oceans performed a study of ocean plankton of unprecedented scope. It was aimed at exploring the wide variety of planktonic organisms (from viruses to small animals) in the epipelagic (between 0 and 200 m) and mesopelagic (200 to 1000 m) oceans on a global scale. Around 35,000 samples were collected from 210 stations in 20 biogeographic provinces. The purpose of this chapter is to present several key aspects of this expedition as they relate to phytoplankton: the main sampling principles used during the expedition; the characterization of their diversity, spatial distribution patterns, and biological interactions by omics and microscopic imagery; and the major discoveries to date from the expedition.

The *Tara* Oceans expedition and the holistic characterization of phytoplankton

The standard sampling program of *Tara* Oceans was designed to study a wide variety of marine ecosystems: upwellings, biodiversity hotspots, low pH or oxygen-poor zones. A total of 210 stations were defined (Fig. 15.1A), on which a more precise environmental characterization was conducted, to contextualize the morphological and genetic sampling of plankton. Moreover, continuous in-line measurements and remote sensing data from satellites were used to further expand context by characterizing oceanographic boundaries and properties around a discrete sampling event. A variety of methods served to specifically explore phytoplankton in the ocean and to relate their distributions to photosynthesis, as summarized below.

Sampling context

The expedition targeted well-defined mesoscale features, for example, the study of the plankton community in Agulhas rings by Villar et al. (2015), or in the Marquesas archipelago by Caputi et al. (2019). To provide a richer environmental context for the morphological and genomic study of phytoplankton (see below), data were collected about altimetry (sea surface height), sea surface temperature, nutrients, and satellite ocean-color (Pesant et al., 2015). In addition, temperature-salinity profiles available around sampling stations were compiled from autonomous Argo floats.

Physical, optical, and imaging in-line sensors

Continuous measurements of the physical, chemical, and biological properties of surface waters were taken throughout the expeditions. The in-line surface sampling system installed on SV *Tara* (Fig. 15.1B) comprised a

SeaBird temperature and conductivity sensor (TSG), a WETLabs (AC-S) spectrophotometer (described in Boss et al., 2013), a WETLabs chlorophyll fluorometer, and a fast repetition rate fluorometer (FRRF) to assess photosynthetic efficiency. During the *Tara* Oceans Polar Circle expedition in 2013, additional sensors for pH, pCO2, optical backscattering (3 wavelengths), fluorescence emission (ALFA), and surface photosynthetically active radiation (PAR) were added to the in-line system. An Imaging Flow Cytobot (IFCB) also recorded images of microplankton every 25 minutes.

CTD-rosette

Environmental conditions were determined from the surface down to a depth of 1000 m using a suite of sensors mounted on a rosette vertical sampling system (RVSS). The sensors included conductivity and temperature sensors, chlorophyll and CDOM fluorometers, a transmissometer, and a one-wavelength backscatter meter for particles within a 0.5–20 μm size range, and an underwater vision profiler (UVP) for particles and organisms of >100 μm size. An oxygen sensor and an in situ ultraviolet spectrophotometer (ISUS) nitrate sensor were also mounted on the RVSS. A Secchi disk was also deployed at each sampling station. In addition to sensors mounted on the RVSS, seawater was collected using Niskin bottles for discrete measurements, including pigment concentrations from HPLC analysis, the carbonate system, nutrients (NO2-, PO43-, N02-/NO3-, Si), dissolved organic carbon (DOC), fluorescence colored dissolved organic matter (fCDOM), and dissolved oxygen isotopes (Fig. 15.1B). DOC, CDOM, and dissolved oxygen isotopes are available only for the Polar Circle expedition. Nutrient predictions were also derived from the World Ocean Atlas, and estimates of iron were collected from ECCO2-DARWIN and PISCES2 biogeochemical models (see Caputi et al., 2019, for an example). These data have been quality controlled and assembled at https://doi.org/10.1594/PANGAEA.875582.

Parameters relevant for photosynthesis

Parameters of particular relevance for evaluating primary production included an in-line AC-S system, which provided estimates based on spectral decomposition of particulate absorption spectra measured *in situ* for chlorophyll (Chl) a, Chl b, Chl c, photoprotective carotenoids, and photosynthetic carotenoids, together with a collection of discrete samples for pigment analysis by HPLC back on land. A comparison of AC-S data with HPLC-derived pigments has been performed using data from the *Tara* Oceans

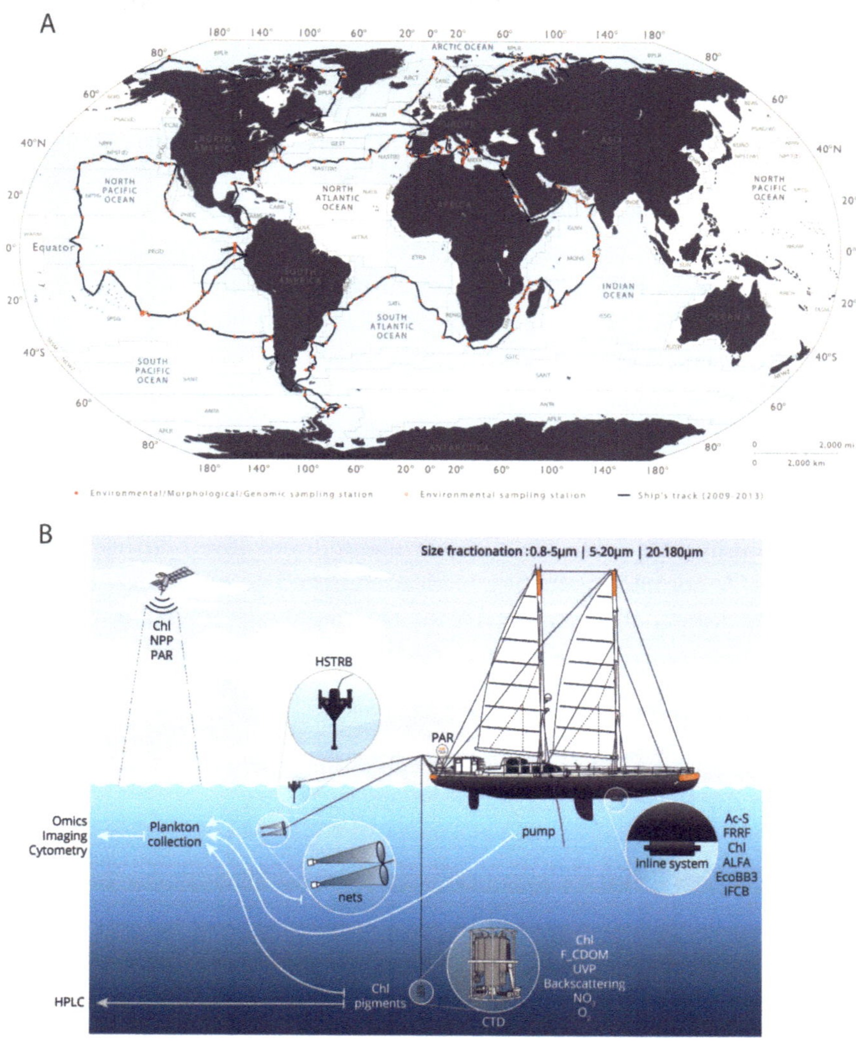

Figure 15.1 Overview of the *Tara* trajectory and sampling scheme during *Tara* Oceans. (A) Trajectory of the *Tara* schooner during *Tara* Oceans expedition. (B) Basic sampling scheme for each sampling station.

Polar Circle expedition, which revealed advantages and limitations of both methods (Chase et al., 2013). Throughout the expedition, an in-line system measured surface chlorophyll fluorescence, and a fluorometer mounted on the RVSS measured chlorophyll fluorescence with each CTD cast. At each sampling station, remote sensing of chlorophyll satellite data was collected, together with derived estimates of net primary production (NPP). PAR was derived from in-situ measurements using a PAR sensor during the Polar Circle expedition, whereas during the previous *Tara* Oceans expedition, PAR was predicted based on the advanced moderate resolution imaging spectroradiometer (AMODIS). Sun elevation above the horizon was noted at each sampling site, and season index and photoperiod were derived from sampling dates. An additional dataset of interest for assessing photosynthesis and its impacts concerns carbon export from the photic zone. These estimates were based on particle concentration and size distributions in the water column down to 1000 m obtained from the UVP. From these datasets, carbon export was derived using Stokes' law, as described in detail in Guidi et al. (2016).

Integral sampling of phytoplankton in Tara Oceans

Although not specifically targeted, phytoplankton were sampled comprehensively using the different collection methods employed during the expeditions. A complete sampling station consisted of collecting plankton from three distinct environmental features, typically the surface water layer, the deep chlorophyll maximum (DCM) layer, and mesopelagic zone. A sampling station lasted typically 48 hours, and the sequence of sampling deployments generally followed the same order (Pesant et al., 2015). Sampling devices consisted essentially of a high-volume peristaltic pump, Niskin bottles on the RVSS, and instrumented plankton nets.

Plankton collection

The plankton sampled during the *Tara* Oceans expedition cover six orders of magnitude in size (10 nm–10 mm) and correspond to viruses, giant viruses (giruses), prokaryotes (bacteria and archaea), unicellular eukaryotes (protists), and multicellular eukaryotes (metazoans) (Fig. 15.1B). These five groups form the bulk of biomass throughout the oceans (Bar-On and Milo, 2019). Various sampling methods were used to capture organism diversity, and were preserved using a range of fixation methods, as described extensively in Pesant et al. (2015).

Plankton omics

Genomics samples were stored in liquid nitrogen on SV *Tara* and were transferred in dry ice to Genoscope in Evry (France) for DNA extraction and sequencing. Different nucleic acid extraction methods were applied to obtain DNA and RNA from the different size fractions collected during the expedition, and the reader is advised to consult Alberti et al. (2017) for a comprehensive description. In addition to sequencing samples enriched in viruses, bacteria, micro-eukaryotes, and small animals, hundreds of single-amplified genomes (SAGs) from single cells of taxonomically and/or biogeographically relevant organisms were also generated (Alberti et al., 2017; Seeleuthner et al., 2018), and besides metagenomics and metatranscriptomics, metabarcoding from amplicons of 16S and 18S rDNA was also performed (for primer sequences and general methods, see Alberti et al., 2017). All sequencing was performed using Illumina machines (Illumina, USA).

16S rDNA fragments derived from metagenomes were identified and assembled for generating 16S rDNA miTags (i.e., metagenomic Illumina tags for 16S rDNA) (Sunagawa et al., 2015; Salazar et al., 2019). 18S rRNA diversity was assessed by metabarcoding of the V9 region. The sequencing effort for each sample was considerable to overcome some of the limitations of metabarcoding-based approaches (discussed in Alberti et al., 2017; Carradec et al., 2018).

Plankton imaging

The abundance, composition, and morphological characteristics (e.g., size, shape, and optical density) of plankton and nonliving suspended particles were determined using various methods, collectively covering the entire spectrum of plankton sizes and taxonomic groups. The optical methods used to analyze samples from the *Tara* Oceans expedition are summarized below:

1. Underwater vision profiler (UVP5, Hydroptics, France); operated in situ and designed to detect and count objects of >100 μm in size and to identify those of >600 μm in size.
2. Zooscan (Hydroptics, France): a benchtop instrument designed for the analysis of objects in the size-range 300–5000 μm, and best suited to study preserved samples of organisms, such as large hard-shelled protists (e.g., Rhizaria) and metazoa. The instrument was deployed on land using >200 μm size fractions fixed in 4% formaldehyde.
3. FlowCAM (Fluid Imaging Technologies, USA): designed for the analysis of objects in the size range 50–300 μm, and best suited to study live or preserved organisms, such as protists, small metazoa, and larval stages of larger metazoa. The instrument was used on board with live samples.

4. Imaging Flow Cytobot (IFCB; Woods Hole Oceanographic Institute, USA): designed for the analysis of objects in the size range 10–150 μm, and best suited to study live small-size protists, such as flagellates, ciliates, and diatoms.
5. Environmental high-content fluorescence microscopy (e-HCFM): based on an automated Leica SP8 TCS confocal laser scanning microscope that enables 3D multicolor imaging of cells (Colin et al., 2017). Designed for analysis of objects in the size range 2–500 μm, and best suited to study preserved samples of nano- and microplankton cells. The instrument was developed at EMBL and analyzed 5–20 and 20–180 μm size fractions fixed in 1% paraformaldehyde and 0.25% glutaraldehyde that had been stored at 4 °C until analysis.
6. Inverted light microscopy (ILM): used for the analysis of objects in the size range 3–180 μm, and best suited for the study of live or preserved protists, such as flagellates, ciliates, and diatoms. The instrument was deployed on land using unfiltered samples to determine total cell counts, as well as on 20–180 μm size fractions. All samples had been fixed as for e-HCFM. Samples fixed in 10% neutralized formaldehyde solution or in 2% Lugol and maintained in the dark were also used for light microscopy.
7. Flow cytometry: used to determine counts of picocyanobacteria, heterotrophic bacteria, and eukaryotic picophytoplankton (see Marie et al., 2014, for details).
8. Transmission electron microscopy (TEM) and scanning electron microscopy (SEM): designed for the analysis of objects in the size range 10 nm–200 μm, and best suited to study preserved samples of viruses, prokaryotes, and small eukaryotes. The instruments were deployed on land using unfixed samples (for SEM) or samples that had been fixed in neutralized formaldehyde or glutaraldehyde (for TEM).

The majority of images from *Tara* Oceans are now discoverable, as well as tools for collaborative annotation, on the EcoTaxa website (http://ecotaxa.obs-vlfr.fr/ and http://ecotaxa.sb-roscoff.fr/).

Global diversity and biogeography patterns of phytoplankton

Global detection of phytoplankton

Tara Oceans has so far studied the composition of the plankton community principally using ribosomal marker genes. The 18S rRNA gene of eukaryotes was analyzed through metabarcoding, in which a gene fragment (here the V9 variable region) is universally amplified from an environmental sample through PCR, and then massively sequenced. The fraction of reads assigned to a taxon can be used as a proxy for its relative abundance, and the number of different reads or clusters of reads as a measure of richness, although several factors affect these estimates. The former is influenced by the number of marker gene copies per genome (de Vargas et al., 2015), the latter by artifacts from PCR and sequencing methods, or the definition of an operational taxonomic unit (OTU, a proxy for species in DNA-based analyses) for example.

The 18S rRNA gene metabarcoding data consisted of approximately 1.3 billion reads obtained from ~890 samples from 144 different geographical sites. Reads were clustered into approximately 220,000 distinct OTUs of unicellular eukaryotes, of which fewer than 30% could be assigned to phytoplankton and known hosts of photosymbionts (de Vargas et al., 2015; Ibarbalz et al., 2019). Diatoms and photosynthetic dinoflagellates, followed by haptophytes and green algae (more precisely those from the group Mamiellophyceae) accounted for most photosynthetic OTUs (Fig. 15.2A). Compared to morphology-based estimates, the actual numbers of taxa based on OTUs are approximately 10-fold higher. Cryptophytes, dictyochophytes, and algae belonging to the ochrophytes other than diatoms were other well represented phytoplankton groups.

Repartitioning by size fractions

Size fractionation of the plankton community allowed us to observe that eukaryotic reads of phototrophs in the 0.8–5-μm size fraction were heterogeneously distributed among different groups, such as haptophytes, dinoflagellates, chlorophytes (especially clade VII prasinophytes and Mamiellophyceae), pelagophytes, dictyochophytes, cryptophytes, and even diatoms (see Leblanc et al., 2018, for a case study on small diatoms). Then, at 5–20 μm and 20–180 μm, diatoms and dinoflagellates were the most frequent among eukaryotic phototrophs. Finally, the presence of groups with the capacity of forming chains (e.g., diatoms from the genera *Hyalosira* and *Fragilaria*), colonies (e.g., *Phaeocystiys* from the haptophytes) or with an epizoic lifestyle (e.g., diatom genus *Pseudohimantidium*) explains their detection in the 180–2000 μm size fraction (Fig. 15.2B).

Prokaryotic abundance and diversity was captured from the 0.22–3-μm size fraction through 16S rDNA fragments derived from metagenomic DNA shotgun sequencing. These fragments were identified, assem-

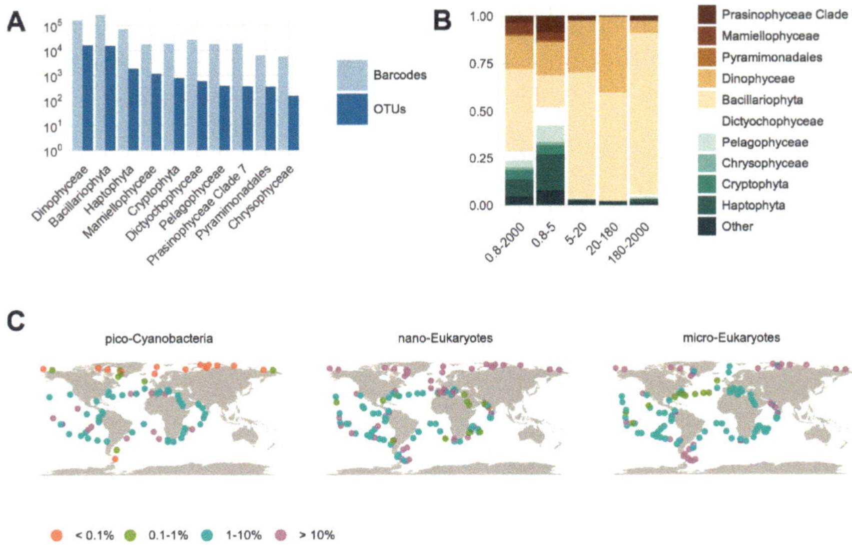

Figure 15.2 Diversity and size distribution of major phytoplankton groups. (A) Observed total number of barcodes and OTUs for the major eukaryotic phytoplankton groups in the complete *Tara* Oceans 18S rRNA (V9) dataset. (B) Relative abundance of major eukaryotic phytoplankton groups in differentiated size fractions. Data corresponds to sea surface samples. For size fractions 0.8–5 and 5–20 microns, Arctic samples are not available. (C) Phytoplankton relative contribution to conventional size classes in plankton. Circles represent sampling stations across the global ocean with nonzero detection. Their color refers to the relative abundance among prokaryotic 16S rRNA gene reads in the picoplankton (usually 0.2–2 um, here 0.2–1.6 or 0.2–3 um); eukaryotic 18S rRNA gene reads in the nanoplankton (usually 2–20 um, here 3–20 or 5–20 um); and eukaryotic 18S rRNA gene reads in the microplankton (usually 20–200 um, here 20–180 um). Colors vary according to a reduced log10-scale.

bled, and then contrasted against a reference database, generating so-called miTags (metagenomic Illumina tags, i.e., 16S rDNA fragments derived from metagenomes sequenced with an Illumina platform; see Sunagawa et al., 2015). Marine free-living, planktonic cyanobacteria are the smallest and most abundant phytoplankton group in today's ocean, but their taxonomic diversity (as measured by the traditional yard-sticks of ribosomal RNA or morphology) is surprisingly low. The genera *Prochlorococcus* and *Synechococcus* are the major representatives, for which the typical sequence dissimilarity threshold of 3% on the rRNA does not resolve well for many distinct "species." However, it has been possible to define clades through molecular phylogenetic analyses, namely 13 clades for *Prochlorococcus* and approximately 20 for *Synechococcus*. More recently, an alternative marker gene has been used (*petB*, a gene-encoding cytochrome b6) and mined from the *Tara* Oceans metagenomes, from which ecologically significant taxonomic units have emerged (Farrant et al., 2016). This was further advanced for *Prochlorococcus* in a comprehensive study that analyzed the metagenomes in the context of the *Prochlorococcus* pan-genome (Delmont and Eren, 2018), reminding that pangenomics, which distinguishes between core and accessory genes, can play a crucial role in understanding the evolutionary dynamics of this lineage. Nevertheless, marine photosynthetic bacteria still appear to be far less diversified than their eukaryotic counterparts.

Photosynthesis and heterotrophy are not sharply separated in planktonic organisms and blend across the unicellular life forms in what has been called mixotrophy (Flynn et al., 2013; Caron, 2016; Mitra et al., 2016; Selosse et al., 2017; Stoecker et al., 2017). Actually, this represents an ongoing paradigm shift that proposes a new functional classification for marine protists (Mitra et al., 2016). Faure et al. (2019) carried out a search for mixotrophs in the *Tara* Oceans metabarcoding dataset and obtained more than 300,000 mixotrophic OTUs belonging to diverse lineages of the eukaryotic tree of life. With the largest number of OTUs corresponding to radiolarians that host intracellular photosymbionts, this study supports the idea of mixotrophy being ubiquitous. However, the lack of knowledge on the extension of (facultative) photosymbiosis within plankton lineages prevents a full description of the contributions of these organisms to primary production.

Additionally to the community analysis via marker genes, *Tara* Oceans dedicated efforts to chemically fix samples on board and analyze them by confocal microscopy for automated high-content imaging (Colin et al., 2017). This resulted in a classification (at a high taxonomic level) of objects from the 5–20-μm size fraction (estimated accuracy: 93.8% at the phylum or class level), with the following average cell abundances among the eukaryotic phytoplankton: 35% diatoms, 34% dinoflagellates, 16% haptophytes, and 14% other eukaryotic phytoplankton. These proportions differed from the metabarcoding analysis, particularly when focusing on the low read abundances for haptophytes in the same size fraction (5–20 μm). Further comparative studies of imaging- and sequence-based data and a better assessment of their limitations will be required for addressing these discrepancies.

Biogeography and community variation across environmental gradients

Today there is an overall understanding of the global distribution of the different phytoplankton groups in relation to their size. As part of the picophytoplankton (0.2–2 μm), cyanobacteria from the genus *Prochlorococcus* are found in warm, nutrient-poor waters in the tropical and subtropical ocean, whereas those from the genus *Synechococcus* have a much wider distribution (Partensky et al., 1999). Multiple groups among eukaryotic nanophytoplankton (2–20 μm) can be detected potentially year-round in temperate regions. Microphytoplankton (20–200 μm), by contrast, grow mainly at higher latitudes at the beginning of spring and sometimes also in the late summer, as well as in upwelling regions, typically in the form of large blooms. These patterns have been characterized over a few centuries of observations (Pierella Karlusich et al., 2020a) and were also captured by the *Tara* Oceans global sampling (Fig. 15.2C). What is more, current knowledge is advanced enough to allow numerical models to agree with observations on this aspect (Follows et al., 2007).

From the *Tara* Oceans global sampling, *Prochlorococcus* dominated between 40°N and 40°S, in warm, oligotrophic areas of the open ocean (Fig. 15.3A). According to a finer analysis based on the *petB* gene, different ecologically significant taxonomic units appear to repartition across regions characterized by particular environmental conditions (Fig. 15.3A; Farrant et al., 2016), particularly as a function of temperature and iron concentrations (Farrant et al., 2016). Beyond those latitudes, the cyanobacteria are represented by *Synechococcus*, which is also abundant in waters more proximal to land and in areas enriched by local upwellings (Fig. 15.3A). Flombaum et al. (2013) found that temperature is the main control on regional distributions of both *Prochlorococcus* and *Synechococcus*, with a lower temperature boundary for *Synechococcus*, which exhibits populations in Arctic waters, but is absent in subzero waters around Antarctica (Letelier and Karl, 1989; Robineau et al., 1999).

Among eukaryotic phytoplankton, Pelagophytes are found preferentially in warm areas, similarly to picocyanobacteria (Fig. 15.3B and 15.4A), whereas diatoms (Bacillariophyta) are particularly prevalent at high latitudes and in upwelling environments (Fig. 15.3B; Pierella Karlusich et al., 2020b). Diatoms have a known ability to bloom and outgrow other marine phytoplankton when nitrate and silicate are abundant (see examples in Fig. 15.4A; Malviya et al., 2016). The variation of read abundances of chlorophytes, dinoflagellates, and haptophytes as entire groups is less

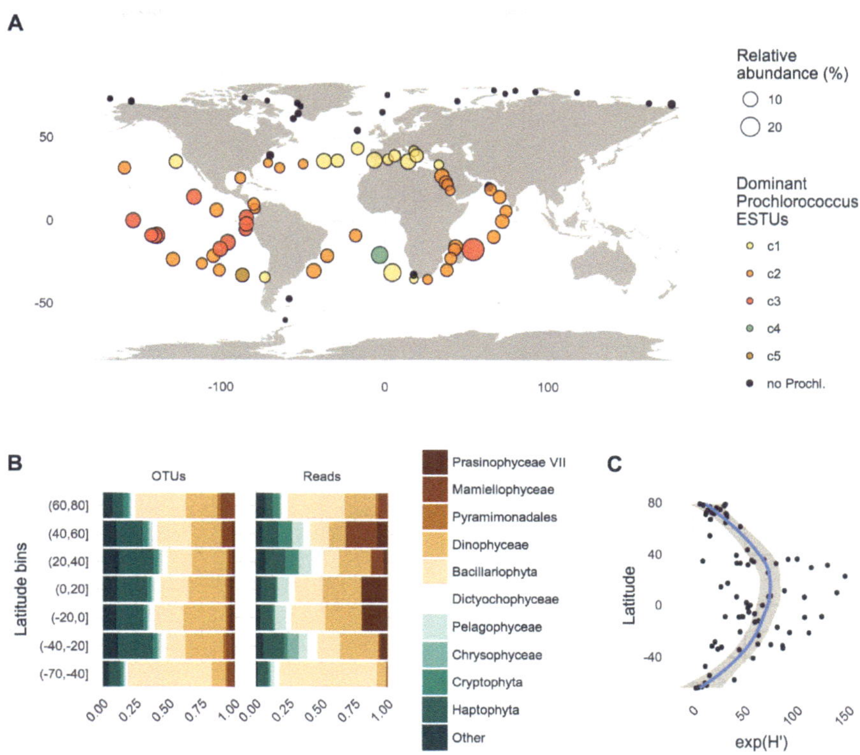

Figure 15.3 Global phytoplankton patterns in the surface ocean. (A) Distribution and abundance of picocyanobacteria (*Prochlorococcus* and *Synechococcus*) in the surface ocean. Circle size refers to relative abundances (%) for total picocyanobacteria retrieved from 16S rRNA gene fragments (miTags) for size fraction 0.22–1.6/3 um. Colors refer to different dominant ecologically significant taxonomic units (ESTUs) of *Prochlorococcus* defined based on a *petB* gene analysis from metagenomes by Farrant et al. (2016). In black, samples with no *Prochlorococcus* detection. (B) Latitudinal averages in the proportion of OTUs (proxy for relative richness) and reads (proxy for relative abundance) for major eukaryotic phytoplankton groups. (C) Latitudinal trend in eukaryotic phytoplankton diversity as measured by the exponential Shannon index that provides an estimate for the effective number of species. For (B) and (C), data correspond to the 18S rRNA gene (V9) metabarcoding analysis.

predictable. For them, a closer analysis of the taxonomic composition can reveal contrasting trends among subgroups, as observed for *Phaeocystis* (Haptophyta), whose colonies become abundant toward colder latitudes (Sow et al., 2020). In the case of chlorophytes, results from *Tara* Oceans and the Ocean Sampling Day initiative show that *Mamiellophyceae* are prominent in coastal waters (Monier et al., 2016; Vannier et al., 2016), whereas

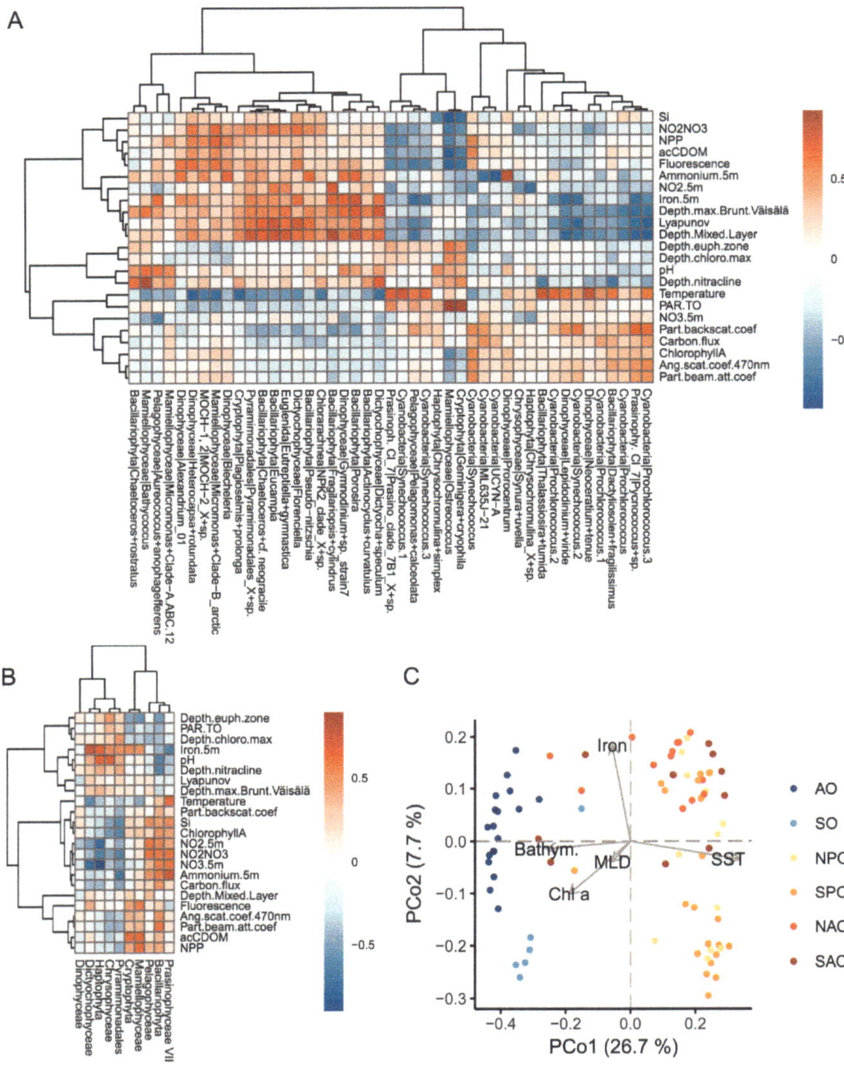

Figure 15.4 Phytoplankton distribution and environmental parameters. (A) Relation of dominant OTUs with parameters relevant for primary production. A diverse selection among the most abundant OTUs, classified at varying taxonomic levels, are shown. Color gradient represents Spearman's rho correlation coefficient. (B) Principal coordinates analysis for eukaryotic phytoplankton communities from the surface ocean. Phytoplankton relative abundances were retrieved from 18S rRNA gene metabarcoding data (V9 region) for size fraction 0.8–2000 um. Colors refer to oceanic regions: Arctic Ocean (AO), Southern Ocean (SO), North and South Pacific Ocean (NPO, SPO) and North and South Atlantic Ocean (NAO, SAO).

clade VII prasinophytes dominate oceanic waters (Dos Santos et al., 2017). Coccolithophores (also Haptophyta) will require further analyses to obtain thorough biogeographical patterns. The *Tara* Oceans survey did not succeed in reflecting their relevance in the oceanic ecosystem, particularly at low to middle latitudes (O'Brien et al., 2013). This may relate to the fact that this group occurs in high densities only during specific conditions and for a relatively brief time (Brown, 1994). The survey did, however, catch a broad switch from the Isochrysidales (which includes *Emiliania huxleyi* and many noncalcifying species) to the calcifying order Coccolithales toward the cold waters of the Southern Ocean (data not shown). With respect to dinoflagellates, their large contribution to the photosynthetic community in terms of read abundance together with a regular global distribution (Le Bescot et al., 2016) (Fig. 15.4A) suggest broad ecological strategies employed by different genera, although biases due to their very large rRNA gene copy numbers cannot be ruled out.

When analyzed at the OTU level, an ordination analysis of the eukaryotic phytoplankton community results broadly in the separation between polar and lower latitude (i.e., tropical/subtropical) samples along the main axis, with a few samples from temperate regions completing what could be a temperature gradient (Fig. 15.4B). Along the second axis, polar assemblages split between Arctic and Southern Oceans, while among warmer stations, certain samples from the Pacific Ocean predicted to be low in iron concentration differentiate from a mixed cluster of Atlantic and Pacific locations (Caputi et al., 2019). Although the distribution and success of phytoplankton species across the ocean is determined by a combination of processes at both ecological and evolutionary scales, this analysis indicates that temperature and the concentration of nutrients seem to play central roles in the assembly of phytoplankton communities.

Large-scale gradients of alpha-diversity and climate change

Primary producers as well as animals and heterotrophic microbes are distributed on land and in the ocean according to macroecological patterns, among which the latitudinal diversity gradient (LDG) is one of the most extensively described (Kreft and Jetz, 2007; Tittensor et al., 2010; Ibarbalz et al., 2019). This gradient is thought to be generated by a combination of processes that operate at both ecological and evolutionary scales (Pontarp et al., 2019). Several studies point to solar radiation and temperature as major drivers (Clarke and Gaston, 2006; Brown, 2014). Eukaryotic marine phytoplankton retrieved from the *Tara* Oceans expedition exhibit a bell-shaped

LDG (Fig. 15.3C) with a decrease in the number of species toward the poles (Ibarbalz et al., 2019).

Today, the ocean is absorbing heat and anthropogenic CO_2 at fast rates, which is affecting the physical, chemical, and biological properties of the environment of marine plankton (Hoegh-Guldberg et al., 2014). Oceanic conditions towards the end of the century have been predicted in a compilation of Earth system models that assume different CO_2 emission scenarios established by the IPCC (Bopp et al., 2013). Recently they were combined with statistical models of plankton alpha-diversity that were built based on the *Tara* Oceans metabarcoding survey (Ibarbalz et al., 2019). These projections suggest general increases of phytoplankton diversity, particularly towards the high latitudes in the Northern Hemisphere, in line with a process of "tropicalization" of temperate and subpolar oceans and potential disruptions in key ecosystem services and functions, such as the fishing industry and carbon export.

Phytoplankton in interactions

The role of marine microbes within planktonic ecosystems has typically been studied under the prism of bottom-up research, namely, understanding how resources and abiotic factors affect their abundance, diversity, and functions. However, how species interact with each other is critical to form the ecosystems that sustain life on Earth. Top-down direct interactions (such as symbiosis, viral infection, epibiosis) drive coevolution, influence species distribution, contribute to ecosystem stability, and affect global biogeochemical cycles. Phytoplankton are an extremely good case study for exploring biotic interactions. They are pivotal in marine microbial communities and are known to interact with numerous other organisms in the ocean. These interactions can provide insights about biogeographical patterns, impact plankton evolution, and global biogeochemical processes.

Interactions are important predictors of phytoplankton distribution

Cooccurrence networks based on metaomics data are traditionally used to study microbial communities and interactions (Faust and Raes, 2012; Li et al., 2016), e.g., in human and soil microbiomes (Barberán et al., 2012; Faust et al., 2012), as well as in marine and lake bacterioplankton (Fuhrman and Steele, 2008; Eiler et al., 2011; Milici et al., 2016). Such networks provide an opportunity to expand community analysis towards an understanding of the relational roles played by different organisms, many of which are uncultured and uncharacterized (Proulx et al., 2005; Chaffron et al., 2010). Over large spatial scales, nonrandom patterns, according to which organisms frequently or never occur in the same samples, are the result of diverse processes, such as biotic interactions, habitat filtering, historical effects, and neutral processes (Fuhrman, 2009). Quantifying the relative importance of each process is still in its infancy. However, these networks can be useful to reveal niche spaces, to identify potential biotic interactions, and to guide more focused studies.

Using the *Tara* Oceans dataset and cooccurrence approaches, the first planktonic "interactome" was constructed based on 9292 taxa across two depths, seven organismal size ranges, and four organismal domains (Lima-Mendez et al., 2015). Random forest-based models revealed that 95% of planktonic (OTU) abundances were better explained by biotic, rather than abiotic, variables. Environmental factors are therefore incomplete predictors of community structure. By converting species taxonomy into functional types, such as autotrophs or heterotrophs, the *Tara* Oceans interactome showed strong connections between phytoplankton and grazers, as expected. More surprising, was the high abundance of associations between Syndiniales (potential parasites) and microphytoplankton (excluding diatoms), highlighting the role of alveolated parasitoids as top-down effectors of microphytoplankton population structure and functioning (Fig. 15.5).

Potentially lethal biotic interactions constrain diatom biogeography

Diatoms are one of the most abundant primary producers in the ocean and are amongst the most diverse groups, as shown by the *Tara* Oceans metabarcoding dataset (Fig. 15.2A) (de Vargas et al., 2015; Malviya et al., 2016). Studying how biotic interactions shape their distribution can therefore help us better understand this highly successful phytoplankton group, independently from environmental conditions. Species cooccurrence can happen when two taxa respond similarly to a common environmental condition ("niche effect"), rather than establish a real interaction. Based on the *Tara* Oceans interactome, only 13% of diatom potential associations could be explained by a niche effect (Fig. 15.6).

Unlike other phytoplankton groups, diatoms occupy niches that are less populated with potential parasites, pathogens, and predators (Hendry et al., 2018; Vincent and Bowler, 2020), reflecting their unique ability to exclude other organisms and thrive, thus supporting Smetacek's 3 "P"s' hypothesis

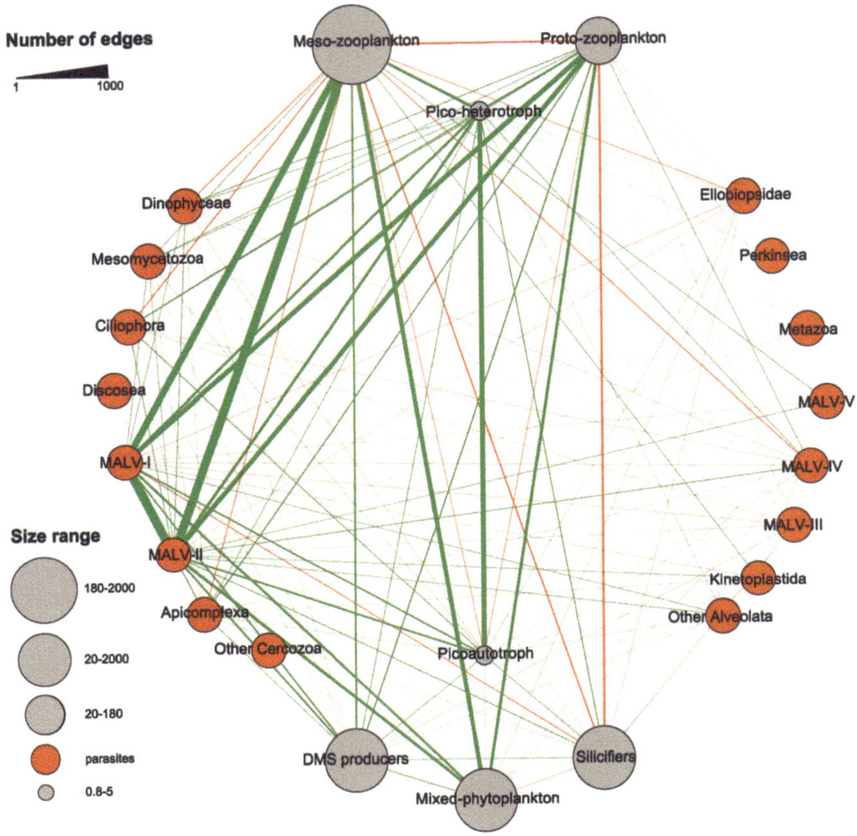

Figure 15.5 Interactions between pathogens and different plankton functional types (PFTs) in the *Tara* Oceans interactome. Subnetwork of metanodes that encapsulate barcodes affiliated to parasites or PFTs. The PFTs mapped onto the network are phytoplankton DMS producers, mixed phytoplankton, phytoplankton silicifiers, picoeukaryotic heterotrophs, protozooplankton and mesozooplankton. Edge width reflects the number of edges in the taxon graph between the corresponding metanodes. Overrepresented links (multiple-test corrected $P < 0.05$) are colored in green if they represent copresences, and in red if they represent exclusions; gray means nonoverrepresented combinations. When both copresences and exclusions were significant, the edge is shown as copresence. (Adapted from Lima-Mendez et al., 2015.)

(Smetacek, 2012), according to which mortality agents are major drivers of phytoplankton evolution (Fig. 15.7). Diatoms have developed an efficient arsenal composed of silicified cell walls, spines, toxic oxylipins, and chain formation to increase size. The observed exclusion pattern was proposed to reflect the worldwide impact of the diatom arms race against lethal bi-

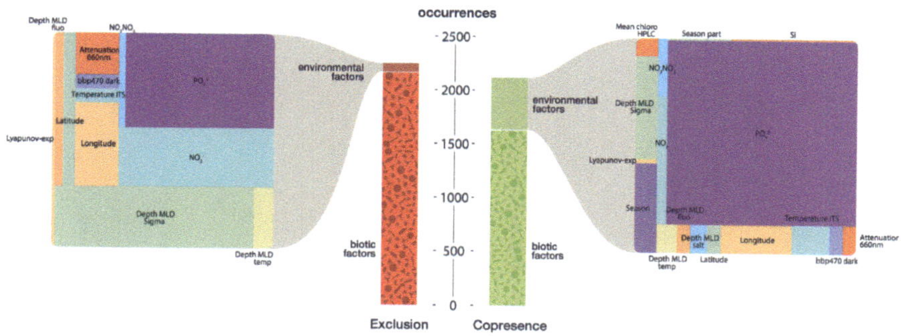

Figure 15.6 Biotic versus abiotic drivers of diatom interactions. The central scale shows the number of diatom copresences and exclusions that are best explained by biotic (bottom) and abiotic (top) factors. The left and right panels are close-up views of the abiotic drivers, where the size of the rectangles is relative to the number of correlations that are best explained by each parameter. Abbreviations: PO4, phosphate; MLD, mixed-layer depth (layer in which active turbulence homogenizes water, estimated by density [sigma] and temperature); NO2, nitrite; attenuation 660 nm, light scattering by suspended particles; bbp470, backscattering coefficient of particles; HPLC chloro, chlorophyll pigment measurement (high-performance liquid chromatography adjusted); Lyapunov-exp, ocean perturbation (Lyapunov exponent); Si, silicate; Season, categorical variable for season. A full description of the environmental parameters is available on the PANGAEA website (https://doi.pangaea.de/10.1594/PANGAEA.840718). From Vincent and Bowler (2020).

otic interactions. Nonetheless, only 6.5% edges of the diatom cooccurrence network have been confirmed using literature references. To a certain extent, this high fraction of unmatched interactions should be seen as the "unknown" proportion of microbial diversity emerging from metabarcoding surveys. Part of it is truly unknown, new, and very exciting; part of it is due to biases in data gathering and processing, and part of it is due to the lack of an extensive reference database.

Single consortia study and biogeography of photosynthetic interactions

The combination of imaging and molecular datasets from *Tara* Oceans has revealed new interactions and the potential benefits underlying such associations. Several approaches based on omics data can thus enable the discovery of biotic interactions. The first source is to identify potential interactions in the *Tara* Oceans interactome that are "biologically relevant." Guided by the predictions from the cooccurrence network and abundance patterns, acoel flatworms (*Symsagittifera* sp.) together with their photosynthetic green

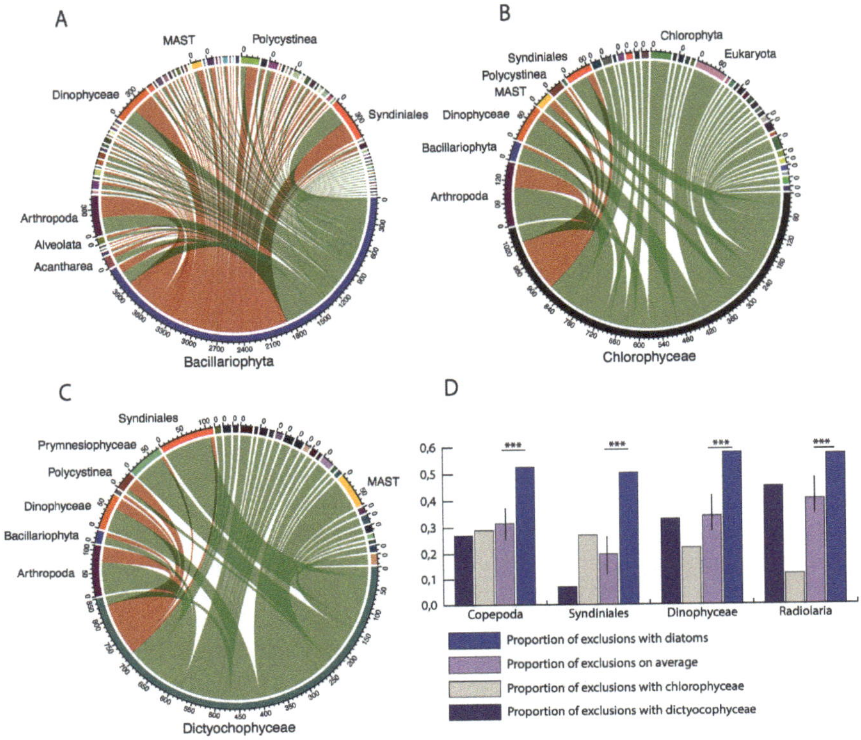

Figure 15.7 Major patterns of interactions for a few phytoplankton groups. (A to D) Circular representation of copresences (green bands) and exclusions (red bands) within subnetworks extracted from the *Tara* Oceans interactome (46) for diatoms (A), chlorophyceae (B), and dictyochophyceae (silicifying biflagellate mixotrophs) (C), with other taxa. The thickness of the band corresponds to the number of interactions, and major partners are labeled around the circles if they represent more than 100 associations. Data from all size fraction networks are represented here. (D) Comparison of proportions of exclusions showing that diatoms significantly exclude potential predators, parasites, and competitors such as copepods, Syndiniales, Dinophyceae, and Radiolarians, compared to control groups. (Adapted from Vincent and Bowler, 2020.)

microalgal endosymbionts (*Tetraselmis* sp.) were collected in microplankton samples from *Tara* Oceans Station 22 in the Mediterranean Sea (Fig. 15.8A), imaged and taxonomically identified (Lima-Mendez et al., 2015).

The second source is to start from images taken as part of the *Tara* Oceans analysis pipeline that reveal intriguing associations. One such example is the epibiotic association between the diatom *Fragilariopsis doliolus* and genera of tintinnid ciliates (Fig. 15.8B), first observed in stations from an Agulhas ring in the South Atlantic Ocean (Vincent et al., 2018). After

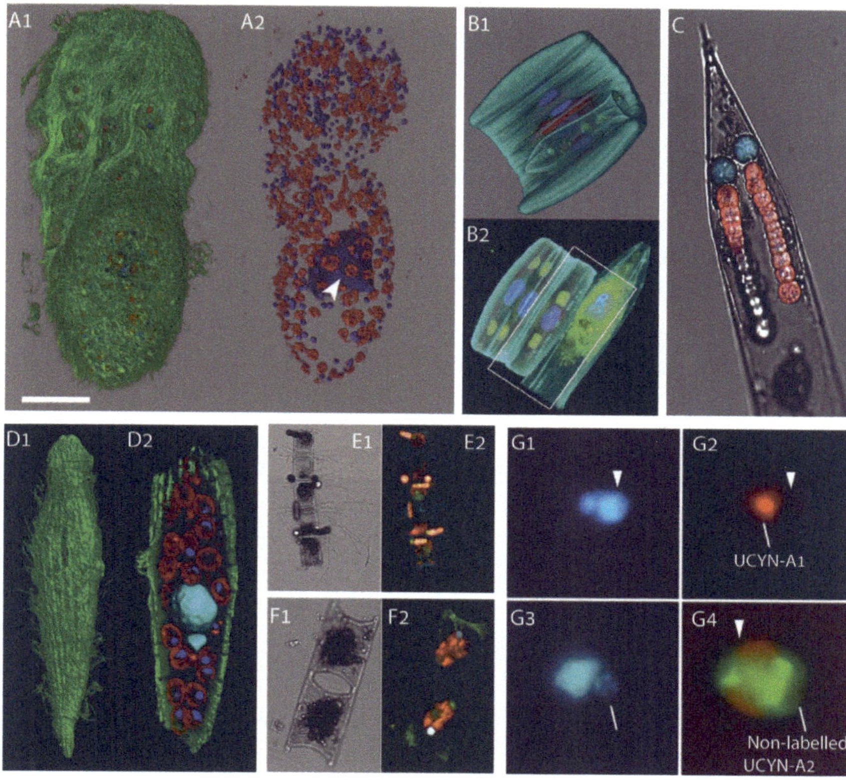

Figure 15.8 Examples of phytoplankton interactions observed in *Tara* Oceans samples. (A1) Dorsal view of an entire acoel flatworm specimen (~300 mm). The epidermis (green) is completely covered with cilia and displays some pore holes. (A2) The removal of the green channel reveals the widespread distribution of small unicellular algae (red areas) inside the acoel body. The worm's nuclei display a clear signal (compact round blue shapes), whereas the algal nuclei are dimmer. A dinoflagellate theca (arrowhead) is located in the central syncytium, likely indicating predation (Lima-Mendez et al., 2015). (B1–B2) Diatoms (*Fragilariopsis* sp. cells organized in a chain) and a heterotrophic tintinnid ciliate (*Salpingella* sp., shell with trumpet-shaped oral opening) form a symbiotic relationship (scale bars 10 μm) (Vincent et al., 2018). (C) Intracellular cyanobacterial symbionts (*Richelia* sp.) are seen within the pennate diatom, *Rhizosolenia*. Two cyanobacterial trichomes are visible with their heterocysts (Decelle et al., 2015). (D1) Three-dimensional (3D) reconstructions of symbiotic specimens imaged with confocal laser scanning microscopy (CLSM). The ciliate exhibits numerous cilia, highlighted in green (DiOC6). (D2) The nuclei of the ciliate (cyan) and the symbiotic microalgae (blue) were reconstructed from the Hoechst fluorescence signal, and the chloroplasts of the microalgae are highlighted by the red chlorophyll autofluorescence. Imaging observations of diazotrophs in *Tara* Oceans samples (Mordret et al., 2016). (E1–E2) Images were obtained by environmental high-content fluorescence microscopy (eHCFM; Colin et al., 2017). *Chaetoceros* with *Calothrix*. (F1–F2) *Hemiaulus* with *Richelia* (Pierella Karlusich et al., 2021). (G1–G4) Left panels correspond to 4′,6-diamidino-2-phenylindole signal (blue-labeled DNA); right panels correspond to the combined signal of the prymnesiophyte-specific probes (green-labeled host under blue light excitation) and the UCYN-A probe (red-labeled symbiont under green-light excitation). Two UCYN-A symbiotic pairs, indicating the specific labeling of UCYN-A1 (upper) and *Braarudosphaera bigelowii* (lower) with their specific partners, the small prymnesiophyte closely related to *B. bigelowii* and UCYN-A2 respectively; Prymnesiophyte partners are indicated by arrowheads (Cornejo-Castillo et al., 2016).

initial isolation and molecular identification, the distribution of both partners was recovered in the OTU dataset and guided the choice of stations in which to perform further imaging and in depth ribotyping. The study revealed that *F. doliolus*, one of the most abundant diatoms in the global ocean, forms unique barrel-shaped chains that enable interactions with tintinnids. The consortia were particularly prevalent in nutrient-replete conditions, which are rich in potential predators, supporting the hypothesis of a mutualistic symbiosis, wherein diatoms acquire motility and tintinnids benefit from the armor-like protection of the diatom's silicified barrel. Promising research is ahead to explore the physical and chemical underpinnings of such interactions and their biological mechanisms.

Another study combining single consortia isolation, imaging, and sequencing is the photosynthetic dinoflagellate genus *Symbiodinium*, which sustains coral reefs by establishing mutualistic endosymbioses with a wide diversity of benthic hosts (Fig. 15.8D). Mordret et al. (2016) reported one of the few examples of symbiosis within oceanic plankton; the pelagic host is a new calcifying ciliate species closely related to *Tiarina fusus* (Colepidae).

Further studies have provided an updated global snapshot of biogeographical diversity of potential marine interactions. Decelle et al. (2018) reported the first global picture of the diversity and activity of *Symbiodinium* in the open ocean. *Symbiodinium* is the largest and most prevalent group of endosymbiotic dinoflagellates known. Clades A and C were significantly more abundant and widely distributed (encompassing 0.1% of phytoplankton reads). Both lineages were transcriptionally active, showing expression of core metabolic pathways, such as photosynthesis, carbon fixation, glycolysis, and ammonium uptake. Both small and large plankton size fractions contained *Symbiodinium*, suggesting the putative existence of a free-living population (small fraction) and a host-associated lifestyle (larger fraction) (Decelle et al., 2018).

Diazotrophs are important marine nitrogen fixers that can form symbioses with diatoms or haptophytes (Foster and Zehr, 2006), providing their host with bioavailable nitrogen in nitrogen-limited areas, yet their global distribution remains understudied. Using over 2,000,000 images (Fig. 15.8E, F) and molecular markers from 1300 metagenomes, several new high-density regions of diazotrophs were discovered, highlighting new sources and sinks of diazotroph-fueled production (Pierella Karlusich et al., 2020b). Distributional and abundance analysis reinforce the prior view that larger-sized diazotrophs (>10 µm) dominate the tropical belts, whereas unicellular diazotrophs were located in surface and mesopelagic samples.

Multiple cooccurring diazotrophic lineages were observed, suggesting that diazotroph niches commonly overlap.

Finally, knowledge was gained regarding the uncultivated unicellular cyanobacterium, "*Candidatus Atelocyanobacterium thalassa*," commonly known as UCYN-A. It was first detected through the amplification of transcripts of the *nifH* gene (encoding the dinitrogenase reductase subunit of nitrogenase) (Zehr et al., 1998). Unlike many other cyanobacteria, UCYN-A cannot perform photosynthesis. Instead, it lives in a mutualistic partnership with an uncultivated unicellular alga, a calcifying prymnesiophyte closely related *to Braarudosphaera bigelowii* (Thompson et al., 2012). Sequences from UCYN-A and the identified hosts were found in the metagenomic and metatranscriptomic datasets from the *Tara* Oceans and Malaspina expeditions (Cabello et al., 2016; Cornejo-Castillo et al., 2016, 2019), revealing new information about the diversity, abundance, and evolution of this diazotrophic group (Fig. 15.8G).

Viruses of phytoplankton

The *Tara* Oceans data provides an unprecedented opportunity to search for novel viruses infecting photosynthetic hosts, both bacteria and eukaryotes. Marine viruses are the most abundant biological entity in the ocean and are considered as major evolutionary drivers of microbial life (Suttle, 2007). They have gained important attention in the last few decades, because they seem to play a large role in marine ecosystems, through control of host metabolism and abundance. Viruses are thought to lyse 20–40% of bacteria per day, thus releasing carbon and other nutrients that affect the food web (reviewed in Suttle, 2007), and recent subcellular estimates show that up to 25% of massive *E. huxleyi* blooms can be terminated by viruses (Vincent et al., 2021).

Viruses can potentially complex micronutrients, such as iron (Bonnain et al., 2016). A survey of the *Tara* Oceans metagenomes for genes-encoding viral structural proteins with putative iron-binding sites showed that they are widespread and abundant (Caputi et al., 2019). Viruses can also alter biogeochemical cycling by reprogramming the host metabolism, creating the virocell (Forterre, 2013). During infection, virocells express viral auxiliary metabolic genes (AMGs) that hijack host metabolism toward pathways that optimize production of new viral particles. This reprogramming is well characterized in cyanobacteria–cyanophage systems, in which these genes supplement photosynthetic electron transport, while redirecting energy from carbon fixation to the pentose phosphate pathway (Puxty et al.,

2016). Comprehensive studies of the AMG content of marine viral communities have been published based on samples from *Tara* Oceans (Roux et al., 2016; Fridman et al., 2017; Warwick-Dugdale et al., 2019), where photosynthesis, carbon, sulfur and nitrogen metabolism appear among the most abundant functions, likely to directly modulate biogeochemical cycles. Beyond these large-scale patterns, ocean viruses have now also been linked *in silico* to ecologically important marine microbial hosts, providing foundational hypotheses that can be tested by experimental virus–host linkage methods. These data have provided global virus–host infection maps (Roux et al., 2016), and for cyanobacteria, they have advanced our understanding of how viral infections associate with diel dynamics of host communities. Specifically, a multiomics analysis that leverages *Tara* Oceans viral genome data revealed a peak of cyanophage gene expression in the afternoon or dusk, followed by an increase of genomes from the virions at night, confirming that cyanophages drive the diel release of cyanobacteria-derived organic matter into the environment (Yoshida et al., 2018).

Nucleocytoplasmic large DNA viruses are a group of eukaryotic viruses with a large double-stranded DNA genome, ranging from 100 kb to 1.26 Mb (Mihara et al., 2018), infecting eukaryotes. They are also capable of affecting algal host metabolism (Monier et al., 2017). Some of them are known to have important roles in marine ecosystems, affecting the population dynamics of bloom-forming algae (Pagarete et al., 2011). Their absolute abundances have been estimated across a section of the *Tara* Oceans transect, showing that the ratio of nucleocytoplasmic large DNA viruses to eukaryotes is within the range of the ratio of phages to bacteria in seawater (Hingamp et al., 2013). Their large global richness indicates that they might affect a broad range of hosts (Mihara et al., 2018). Using metagenomic data, the biogeography of NCLDVs is revealed in Endo et al. (2020), showing a heterogeneous distribution of NCLDVs across oceans, and a higher proportion of unique NCLDVs in the polar biomes. Most interestingly, the community structures of NCLDV families correlate with specific eukaryotic lineages, including many photosynthetic groups. Most diverse NCLDVs families were Mimiviridae and Phycodnaviridae, infecting phytoplankton such as *Phaeocystis*, *Prymnesium*, *Chrysochromulina* or *Bathycoccus*, *Micromonas* and *Ostreococcus*, respectively, and reveals potential chrysophyte viruses as a novel clade of Mimiviridae. Beyond cataloguing genomic viral diversity, their role in the ecosystem can be explored by leveraging the large-scale biogeochemical data collected during *Tara* Oceans. In a further study (Kaneko et al., 2021) associations were tested between viral community composition and carbon export efficiency (CEE) measured at a global scale. A regression model based on relative abundance of viral marker genes explained 67% of the variation in CEE. Viruses with high importance in the model were predicted to infect ecologically important hosts, including many photosynthetic eukaryotic hosts, such as *Mamiellales* and *Prymnesiales*. RNA viruses have not yet been thoroughly investigated in the *Tara* Oceans data, mostly due to the lack of proper methodologies to assess the abundance and activity of those viruses in the ocean.

Conclusion

Thanks to the consistent sampling and analysis protocols deployed at global scale by *Tara* Oceans; a view has emerged of the relative abundances and diversity ranges of each of the phytoplankton groups in the ocean. The patterns confirm previous observations, e.g., that *Prochlorococcus* dominates the open ocean at low latitudes, whereas diatoms predominate in nutrient-rich regions and at higher latitudes. More detailed analyses have furthermore extended our knowledge, e.g., concerning the size spectra occupied by each group. Dinoflagellates emerge as a broadly distributed highly diverse group spanning a wide size spectrum, although many are not photosynthetic and others perform photosynthesis only facultatively. More broadly, mixotrophs appear to be an important component of phytoplankton communities beyond the obligately autotrophic groups, such as diatoms. Photosymbionts additionally emerge as abundant and previously underappreciated contributors to photosynthetic primary production in the global ocean; several new photosymbioses have been identified, and the global extent of known interactions are better appreciated. Finally, exploration of organism cooccurrence patterns using *Tara* Oceans data reveals the different behavior of diatoms with respect to other phytoplankton groups, which provides a basis for understanding the mechanisms that underpin the different ecological niches they occupy compared with other groups. It furthermore supports Victor Smetacek's observation that diatoms have particularly strong defenses against pathogens, parasites, and predators.

Future studies should better integrate DNA sequence-derived information with more traditional data based on photosynthetic pigment analysis, microscopy, and remote sensing. This is essential to go beyond estimates of relative abundance to measures of contributions to primary production, biomass, and biovolumes. We also need a universal marker gene that is specific for photosynthetic organisms, and which is less prone to copy number variations than is 18S rRNA. This is particularly important to better constrain dinoflagellate communities that have multiple nonphotosynthetic members, and whose genomes have huge numbers of rRNA gene repeats. Finally, the availability of gene catalogues from ocean phytoplankton, together with reference genomes, will allow a deeper exploration of the biology of these evolutionarily disparate groups of ecologically successful organisms.

Acknowledgments

We would like to thank all of our colleagues from the *Tara* Oceans consortium and the Tara Ocean Foundation for their inspirational vision. Work in the Bowler laboratory has been funded by European Research Council ERC Advanced Awards Diatomite and Diatomic under the European Union's Horizon 2020 research and innovation programme (grant agreement numbers 294823 and 835067), the Fonds Français pour l'Environnement Mondial, and the French Government "Investissements d'Avenir" Programmes MEMO LIFE (grant ANR-10-LABX-54), Université de Recherche Paris Sciences et Lettres (Université PSL) (grant ANR-1253 11-IDEX-0001-02), and OCEANOMICS (grant ANR-11-BTBR-0008).

References

Adl, S.M., Simpson, A.G.B., Lane, C.E., Lukes, J., Bass, D., Bowser, S.S., et al., 2012. The revised classification of eukaryotes. J. Eukaryot. Microbiol. 59, 429–493.

Alberti, A., Poulain, J., Engelen, S., Labadie, K., Romac, S., Ferrera, I., Albini, G., et al., 2017. Viral to metazoan marine plankton nucleotide sequences from the Tara Oceans expedition. Sci. Data, 170093.

Armbrust, E.V., Berges, J.A., Bowler, C., Green, B.R., Martinez, D., Putnam, N.H., et al., 2004. The genome of the diatom Thalassiosira Pseudonana: ecology, evolution, and metabolism. Science 306, 79–86.

Bar-On, Y.M., Milo, R., 2019. The biomass composition of the oceans: a blueprint of our blue planet. Cell 179, 1451–1454.

Bar-On, Y.M., Phillips, R., Milo, R., 2018. The biomass distribution on Earth. Proc. Natl. Acad. Sci. USA 115, 6506–6511.

Barberán, A., Bates, S.T., Casamayor, E.O., Fierer, N., 2012. Using network analysis to explore co-occurrence patterns in soil microbial communities. ISME J. 6, 343–351.

Bonnain, C., Breitbart, M., Buck, K.N., 2016. The Ferrojan horse hypothesis: iron-virus interactions in the ocean. Front. Mar. Sci. 3, 82.

Bopp, L., Resplandy, L., Orr, J.C., Doney, S.C., Dunne, J.P., Gehlen, M., et al., 2013. Multiple stressors of ocean ecosystems in the 21st century: projections with CMIP5 models. Biogeosciences 10, 6225–6245.

Boss, E., Picheral, M., Leeuw, T., Chase, A., Karsenti, E., Gorsky, G., et al., 2013. The characteristics of particulate absorption, scattering and attenuation coefficients in the surface ocean; contribution of the Tara Oceans expedition. Methods Oceanogr. 7, 52–62.

Bowler, C., Allen, A.E., Badger, J.H., Grimwood, J., Jabbari, K., Kuo, A., et al., 2008. The Phaeodactylum genome reveals the evolutionary history of diatom genomes. Nature 456, 239–244.

Brown, C.W., 1994. Global distribution of coccolithophore blooms. Oceanography 8, 59–60.

Brown, J.H., 2014. Why are there so many species in the tropics? J. Biogeogr. 41, 8–22.
Cabello, A.M., Cornejo-Castillo, F.M., Raho, N., Blasco, D., Vidal, M., Audic, S., et al., 2016. Global distribution and vertical patterns of a prymnesiophyte-cyanobacteria obligate symbiosis. ISME J. 10, 693–706.
Caputi, L., Carradec, Q., Eveillard, D., Kirilovsky, A., Pelletier, E., Pierella Karlusich, J.J., et al., 2019. Community-level responses to iron availability in open ocean plankton ecosystems. Glob. Biogeochem. Cycles 33, 391–418.
Caron, D.A., 2016. Mixotrophy stirs up our understanding of marine food webs. Proc. Natl. Acad. Sci. USA 113, 2806–2808.
Carradec, Q., Pelletier, E., Da Silva, C., Alberti, A., Seeleuthner, Y., Blanc-Mathieu, R., et al., 2018. A global ocean atlas of eukaryotic genes. Nat. Commun. 9, 1038.
Chaffron, S., Rehrauer, H., Pernthaler, J., von Mering, C., 2010. A global network of coexisting microbes from environmental and whole-genome sequence data. Genome Res. 20, 947–959.
Chase, A., Boss, E., Zaneveld, R., Bricaud, A., Claustre, H., Ras, J., et al., 2013. Decomposition of in situ particulate absorption spectra. Methods Oceanogr. 7, 110–124.
Chisholm, S.W., Olson, R.J., Zettler, E.R., Goericke, R., Waterbury, J.B., Welschmeyer, N.A., 1988. A novel free-living prochlorophyte abundant in the oceanic euphotic zone. Nature 334, 340–343.
Clarke, A., Gaston, K.J., 2006. Climate, energy and diversity. Proc. R. Soc. B, Biol. Sci. 273, 2257–2266.
Colin, S., Coelho, L.P., Sunagawa, S., Bowler, C., Karsenti, E., Bork, P., et al., 2017. Quantitative 3D-imaging for cell-biology and ecology of environmental microbial eukaryotes. eLife 6, 1–15.
Cornejo-Castillo, F.M., Cabello, A.M., Salazar, G., Sánchez-Baracaldo, P., Lima-Mendez, G., Hingamp, P., et al., 2016. Cyanobacterial symbionts diverged in the late Cretaceous towards lineage-specific nitrogen fixation factories in single-celled phytoplankton. Nat. Commun. 7, 11071.
Cornejo-Castillo, F.M., Muñoz-Marín, M. del C., Turk-Kubo, K.A., Royo-Llonch, M., Farnelid, H., Acinas, S.G., Zehr, J.P., 2019. UCYN-A3, a newly characterized open ocean sublineage of the symbiotic N 2-fixing cyanobacterium Candidatus Atelocyanobacterium thalassa. Environ. Microbiol. 21, 111–124.
de Vargas, C., Audic, S., Henry, N., Decelle, J., Mahe, F., Logares, R., et al., 2015. Eukaryotic plankton diversity in the sunlit ocean. Science 348, 1261605.
Decelle, J., Carradec, Q., Pochon, X., Henry, N., Romac, S., Mahé, F., et al., 2018. Worldwide occurrence and activity of the reef-building coral symbiont symbiodinium in the open ocean. Curr. Biol. 28, 3625–3633.
Decelle, J., Colin, S., Foster, R.A., 2015. Photosymbiosis in marine planktonic protists. In: Marine Protists. Springer/Springer Japan, pp. 465–500.
Delmont, T.O., Eren, E.M., 2018. Linking pangenomes and metagenomes: the Prochlorococcus metapangenome. PeerJ 2018, e4320.
Derelle, E., Ferraz, C., Rombauts, S., Rouzé, P., Worden, A.Z., Robbens, S., et al., 2006. Genome analysis of the smallest free-living eukaryote Ostreococcus tauri unveils many unique features. Proc. Natl. Acad. Sci. USA 103, 11647–11652.
Dos Santos, A.L., Gourvil, P., Tragin, M., Noël, M.H., Decelle, J., Romac, S., Vaulot, D., 2017. Diversity and oceanic distribution of prasinophytes clade VII, the dominant group of green algae in oceanic waters. ISME J. 11, 512–528.
Dufresne, A., Salanoubat, M., Partensky, F., Artiguenave, F., Axmann, I.M., Barbe, V., et al., 2003. Genome sequence of the cyanobacterium Prochlorococcus marinus SS120, a nearly minimal oxyphototrophic genome. Proc. Natl. Acad. Sci. USA 100, 10020–10025.
Eiler, A., Heinrich, F., Bertilsson, S., 2011. Coherent dynamics and association networks among lake bacterioplankton taxa. ISME J. 6, 330–342.

Eilertsen, H.C., Wyatt, T., 2000. Phytoplankton models and life history strategies. South African J. Mar. Sci. 22, 323–338.
Endo, H., Blanc-Mathieu, R., Li, Y., Salazar, G., Henry, N., Labadie, K., de Vargas, C., Sullivan, M.B., Bowler, C., Wincker, P., Karp-Boss, L., Sunagawa, S., Ogata, H., 2020. Biogeography of marine giant viruses reveals their interplay with eukaryotes and ecological functions. Nat. Ecol. Evol. 4, 1639–1649.
Farrant, G.K., Doré, H., Cornejo-Castillo, F.M., Partensky, F., Ratin, M., Ostrowski, M., et al., 2016. Delineating ecologically significant taxonomic units from global patterns of marine picocyanobacteria. Proc. Natl. Acad. Sci. USA 113, E3365–E3374.
Faure, E., Not, F., Benoiston, A.S., Labadie, K., Bittner, L., Ayata, S.D., 2019. Mixotrophic protists display contrasted biogeographies in the global ocean. ISME J. 13, 1072–1083.
Faust, K., Raes, J., 2012. Microbial interactions: from networks to models. Nat. Rev. Microbiol. 10, 538–550.
Faust, K., Sathirapongsasuti, J.F., Izard, J., Segata, N., Gevers, D., Raes, J., Huttenhower, C., 2012. Microbial co-occurrence relationships in the human microbiome. PLoS Comput. Biol. 8, e1002606.
Field, C.B., Behrenfeld, M.J., Randerson, J.T., Falkowski, P., 1998. Primary production of the biosphere: integrating terrestrial and oceanic components. Science 281, 237–240.
Flombaum, P., Gallegos, J.L., Gordillo, R.A., Rincón, J., Zabala, L.L., Jiao, N., et al., 2013. Present and future global distributions of the marine Cyanobacteria Prochlorococcus and Synechococcus. Proc. Natl. Acad. Sci. USA 110, 9824–9829.
Flynn, K.J., Stoecker, D.K., Mitra, A., Raven, J.A., Glibert, P.M., Hansen, P.J., et al., 2013. Misuse of the phytoplankton-zooplankton dichotomy: the need to assign organisms as mixotrophs within plankton functional types. J. Plankton Res. 35, 3–11.
Follows, M.J., Dutkiewicz, S., Grant, S., Chisholm, S.W., 2007. Emergent biogeography of microbial communities in a model ocean. Science 315, 1843–1846.
Forterre, P., 2013. The virocell concept and environmental microbiology. ISME J. 7, 233–236.
Foster, R.A., Zehr, J.P., 2006. Characterization of diatom – cyanobacteria symbioses on the basis of nifH, hetR and 16S rRNA sequences. Environ. Microbiol. 8, 1913–1925.
Fridman, S., Flores-Uribe, J., Larom, S., Alalouf, O., Liran, O., Yacoby, I., et al., 2017. A myovirus encoding both photosystem i and II proteins enhances cyclic electron flow in infected Prochlorococcus cells. Nat. Microbiol. 2, 1350–1357.
Fuhrman, J., Steele, J., 2008. Community structure of marine bacterioplankton: patterns, networks, and relationships to function. Aquat. Microb. Ecol. 53, 69–81.
Fuhrman, J.A., 2009. Microbial community structure and its functional implications. Nature 459, 193–199.
Guidi, L., Chaffron, S., Bittner, L., Eveillard, D., Larhlimi, A., Roux, S., Coelho, L.P., 2016. Plankton networks driving carbon export in the oligotrophic ocean. Nature 532, 465.
Hendry, K.R., Marron, A.O., Vincent, F.J., Conley, D.J., Gehlen, M., Ibarbalz, F.M., et al., 2018. Competition between silicifiers and non-silicifiers in the past and present ocean and its evolutionary impacts. Front. Mar. Sci. 5, 1–21.
Hingamp, P., Grimsley, N., Acinas, S.G., Clerissi, C., Subirana, L., Poulain, J., et al., 2013. Exploring nucleo-cytoplasmic large DNA viruses in Tara Oceans microbial metagenomes. ISME J. 7, 1678–1695.
Hoegh-Guldberg, O., Cai, Rongshuo, Poloczanska, E.S., Brewer, P.G., Sundby, Svein, Hilmi, Karim, et al., 2014. Climate Change 2014: Impacts, Adaptation, and Vulnerability. Part B: Regional Aspects. Contribution of Working Group II to the Fifth Assessment Report of the Intergovernmental Panel on Climate Change.
Ibarbalz, F.M., Henry, N., Brandão, M.C., Martini, S., Busseni, G., Byrne, H., et al., 2019. Global trends in marine plankton diversity across kingdoms of life. Cell 179, 1084–1097.
Johnson, P.W., Sieburth, J.M.N., 1979. Chroococcoid cyanobacteria in the sea: a ubiquitous and diverse phototrophic biomass. Limnol. Oceanogr. 24, 928–935.

Kaneko, H., Blanc-Mathieu, R., Endo, H., Chaffron, S., Delmont, T.O., Gaia, M., et al., 2021. Eukaryotic virus composition can predict the efficiency of carbon export in the global ocean. iScience 24, 102002.

Keeling, P.J., Burki, F., Wilcox, H.M., Allam, B., Allen, E.E., Amaral-Zettler, L.A., et al., 2014. The marine microbial eukaryote transcriptome sequencing project (MMETSP.: illuminating the functional diversity of eukaryotic life in the oceans through transcriptome sequencing. PLoS Biol. 12, e1001889.

Kreft, H., Jetz, W., 2007. Global patterns and determinants of vascular plant diversity. Proc. Natl. Acad. Sci. USA 104, 5925–5930.

Le Bescot, N., Mahé, F., Audic, S., Dimier, C., Garet, M.J., Poulain, J., et al., 2016. Global patterns of pelagic dinoflagellate diversity across protist size classes unveiled by metabarcoding. Environ. Microbiol. 18, 609–626.

Leblanc, K., Quéguiner, B., Diaz, F., Cornet, V., Michel-Rodriguez, M., Durrieu de Madron, X., Bowler, C., Malviya, S., Thyssen, M., Grégori, G., Rembauville, M., Grosso, O., Poulain, J., de Vargas, C., Pujo-Pay, M., Conan, P., 2018. Nanoplanktonic diatoms are globally overlooked but play a role in spring blooms and carbon export. Nat. Commun. 9, 953.

Letelier, R., Karl, D., 1989. Phycoerythrin-containing cyanobacteria in surface waters of the Drake Passage during February 1987. Antarct. J. US 24, 185–188.

Li, C., Lim, K.M.K., Chng, K.R., Nagarajan, N., 2016. Predicting microbial interactions through computational approaches. Methods 102, 12–19.

Lima-Mendez, G., Faust, K., Henry, N., Decelle, J., Colin, S., Carcillo, F., et al., 2015. Determinants of community structure in the global plankton interactome. Science 348, 1262073.

Malviya, S., Scalco, E., Audic, S., Vincent, F., Veluchamy, A., Poulain, J., 2016. Insights into global diatom distribution and diversity in the world' s ocean. Proc. Natl. Acad. Sci. USA 113, 1516–1525.

Marie, D., Rigaut-Jalabert, F., Vaulot, D., 2014. An improved protocol for flow cytometry analysis of phytoplankton cultures and natural samples. Cytometry, Part A 85, 962–968.

Mihara, T., Koyano, H., Hingamp, P., Grimsley, N., Goto, S., Ogata, H., 2018. Taxon richness of "Megaviridae" exceeds those of bacteria and archaea in the ocean. Microbes Environ. 33, 162–171.

Milici, M., Deng, Z.L., Tomasch, J., Decelle, J., Wos-Oxley, M.L., Wang, H., et al., 2016. Co-occurrence analysis of microbial taxa in the Atlantic ocean reveals high connectivity in the free-living bacterioplankton. Front. Microbiol. 7, 649.

Mitra, A., Flynn, K.J., Tillmann, U., Raven, J.A., Caron, D., Stoecker, D.K., et al., 2016. Defining planktonic protist functional groups on mechanisms for energy and nutrient acquisition: incorporation of diverse mixotrophic strategies. Protist 167, 106–120.

Monier, A., Chambouvet, A., Milner, D.S., Attah, V., Terrado, R., Lovejoy, C., et al., 2017. Host-derived viral transporter protein for nitrogen uptake in infected marine phytoplankton. Proc. Natl. Acad. Sci. USA 114, E7489–E7498.

Monier, A., Worden, A.Z., Richards, T.A., 2016. Phylogenetic diversity and biogeography of the Mamiellophyceae lineage of eukaryotic phytoplankton across the oceans. Environ. Microbiol. Rep. 8, 461–469.

Mordret, S., Henry, N., Richter, D.J., Romac, S., Henry, N., Colin, S., 2016. The symbiotic life of Symbiodinium in the open ocean within a new species of calcifying ciliate (Tiarina sp.). ISME J. 10, 1424–1436.

O'Brien, C.J., Peloquin, J.A., Vogt, M., Heinle, M., Gruber, N., Ajani, P., et al., 2013. Global marine plankton functional type biomass distributions: Coccolithophores. Earth Syst. Sci. Data 5, 259–276.

Pagarete, A., Le Corguillé, G., Tiwari, B., Ogata, H., de Vargas, C., Wilson, W.H., Allen, M.J., 2011. Unveiling the transcriptional features associated with coccolithovirus infection of natural Emiliania huxleyi blooms. FEMS Microbiol. Ecol. 78, 555–564.

Palenik, B., Brahamsha, B., Larimer, F.W., Land, M., Hauser, L., Chain, P., et al., 2003. The genome of a motile marine Synechococcus. Nature 424, 1037–1042.
Palenik, B., Grimwood, J., Aerts, A., Rouzé, P., Salamov, A., Putnam, N., et al., 2007. The tiny eukaryote Ostreococcus provides genomic insights into the paradox of plankton speciation. Proc. Natl. Acad. Sci. USA 104, 7705–7710.
Partensky, F., Blanchot, J., Vaulot, D., 1999. Differential distribution and ecology of Prochlorococcus and Synechococcus in oceanic waters: a review. Bull. Inst. Ocean Monaco 19, 457–476.
Pesant, S., Not, F., Picheral, M., Kandels-Lewis, S., Le Bescot, N., Gorsky, G., et al., 2015. Open science resources for the discovery and analysis of Tara Oceans data. Sci. Data 2, 150023.
Pierella Karlusich, Juan José, Ibarbalz, F.M., Bowler, C., 2020a. Exploration of marine phytoplankton: from their historical appreciation to the omics era. J. Plankton Res. 42, 595–612.
Pierella Karlusich, Juan José, Ibarbalz, F.M., Bowler, C., 2020b. Phytoplankton in the Tara Ocean. Annu. Rev. Mar. Sci. 12, 233–265.
Pierella Karlusich, Juan José, Pelletier, E., Lombard, F., Carsique, M., Dvorak, E., Colin, S., Picheral, M., Cornejo-Castillo, F.M., Acinas, S.G., Pepperkok, R., Karsenti, E., de Vargas, C., Wincker, P., Bowler, C., Foster, R.A., 2021. Global distribution patterns of marine nitrogen-fixers by imaging and molecular methods. Nat. Commun. 12, 4160.
Pontarp, M., Bunnefeld, L., Cabral, J.S., Etienne, R.S., Fritz, S.A., Gillespie, R., et al., 2019. The latitudinal diversity gradient: novel understanding through mechanistic eco-evolutionary models. Trends Ecol. Evol. 34, 211–223.
Proulx, S.R., Promislow, D.E.L., Phillips, P.C., 2005. Network thinking in ecology and evolution. Trends Ecol. Evol. 20, 345–353.
Puxty, R.J., Millard, A.D., Evans, D.J., Scanlan, D.J., 2016. Viruses inhibit CO2 fixation in the most abundant phototrophs on Earth. Curr. Biol. 26, 1585–1589.
Robineau, B., Legendre, L., Michel, C., Budeus, G., Kattner, G., Schneider, W., Pesant, S., 1999. Ultraphytoplankton abundances and chlorophyll a concentrations in ice-covered waters of northern seas. J. Plankton Res. 21, 735–755.
Rocap, G., Larimer, F.W., Lamerdin, J., Malfatti, S., Chain, P., Ahlgren, N.A., et al., 2003. Genome divergence in two Prochlorococcus ecotypes reflects oceanic niche differentiation. Nature 424, 1042–1047.
Roux, S., Brum, J.R., Dutilh, B.E., Sunagawa, S., Duhaime, M.B., Loy, A., et al., 2016. Ecogenomics and potential biogeochemical impacts of globally abundant ocean viruses. Nature 537, 689–693.
Salazar, G., Paoli, L., Alberti, A., Huerta-Cepas, J., Ruscheweyh, H.J., Cuenca, M., et al., 2019. Gene expression changes and community turnover differentially shape the global ocean metatranscriptome. Cell 179, 1068–1083.
Seeleuthner, Y., Mondy, S., Lombard, V., Carradec, Q., Pelletier, E., Wessner, M., et al., 2018. Single-cell genomics of multiple uncultured stramenopiles reveals underestimated functional diversity across oceans. Nat. Commun. 9, 1–10.
Selosse, M.A., Charpin, M., Not, F., 2017. Mixotrophy everywhere on land and in water: the grand écart hypothesis. Ecol. Lett. 20, 246–263.
Smetacek, V., 2012. Making sense of ocean biota: how evolution and biodiversity of land organisms differ from that of the plankton. J. Biosci. 37, 589–607.
Sow, S.L.S., Trull, T.W., Bodrossy, L., 2020. Oceanographic fronts shape phaeocystis assemblages: a high-resolution 18S rRNA gene survey from the ice-edge to the equator of the South Pacific. Front. Microbiol. 11, 1847.
Stoecker, D.K., Hansen, P.J., Caron, D.A., Mitra, A., 2017. Mixotrophy in the marine plankton. Annu. Rev. Mar. Sci. 9, 311–335.
Sunagawa, S., Coelho, L.P., Chaffron, S., Kultima, J.R., Labadie, K., Salazar, G., et al., 2015. Structure and function of the global ocean microbiome. Science 348, 1–10.

Suttle, C.A., 2007. Marine viruses — major players in the global ecosystem. Nat. Rev. Microbiol. 5, 801–812.

Thompson, A.W., Foster, R.A., Krupke, A., Carter, B.J., Musat, N., Vaulot, D., et al., 2012. Unicellular cyanobacterium symbiotic with a single-celled eukaryotic alga. Science 337, 1546–1550.

Tittensor, D.P., Mora, C., Jetz, W., Lotze, H.K., Ricard, D., Berghe, E., Vanden, Worm B., 2010. Global patterns and predictors of marine biodiversity across taxa. Nature 466, 1098–1101.

Vannier, T., Leconte, J., Seeleuthner, Y., Mondy, S., Pelletier, E., Aury, J.M., et al., 2016. Survey of the green picoalga Bathycoccus genomes in the global ocean. Sci. Rep. 6, 1–11.

Venter, J.C., Remington, K., Heidelberg, J.F., Halpern, A.L., Rusch, D., Eisen, J.A., et al., 2004. Environmental genome shotgun sequencing of the Sargasso Sea. Science 304, 66–74.

Villar, E., Farrant, G.K., Follows, M., Garczarek, L., Speich, S., Audic, S., et al., 2015. Environmental characteristics of Agulhas rings affect interocean plankton transport. Science 348, 1261447.

Vincent, F., Bowler, C., 2020. Diatoms are selective segregators in global ocean planktonic communities. mSystems 5, e00444.

Vincent, F., Sheyn, U., Porat, Z., Schatz, D., Vardi, A., 2021. Visualizing active viral infection reveals diverse cell fates in synchronized algal bloom demise. Proc. Natl. Acad. Sci. USA 118, e2021586118.

Vincent, F.J., Colin, S., Romac, S., Scalco, E., Bittner, L., Garcia, Y., et al., 2018. The epibiotic life of the cosmopolitan diatom Fragilariopsis doliolus on heterotrophic ciliates in the open ocean. ISME J. 12, 1094–1108.

Warwick-Dugdale, J., Solonenko, N., Moore, K., Chittick, L., Gregory, A.C., Allen, M.J., et al., 2019. Long-read viral metagenomics captures abundant and microdiverse viral populations and their niche-defining genomic islands. PeerJ 2019, e6800.

Waterbury, J.B., Watson, S.W., Guillard, R.R.L., Brand, L.E., 1979. Widespread occurrence of a unicellular, marine, planktonic, cyanobacterium [10]. Nature 277, 293–294.

Worden, A.Z., Lee, J.H., Mock, T., Rouzé, P., Simmons, M.P., Aerts, A.L., et al., 2009. Green evolution and dynamic adaptations revealed by genomes of the marine picoeukaryotes micromonas. Science 324, 268–272.

Yamasaki, A., 2003. An overview of CO2 mitigation options for global warming - emphasizing CO2 sequestration options. J. Chem. Eng. Jpn. 36, 361–375.

Yoshida, T., Nishimura, Y., Watai, H., Haruki, N., Morimoto, D., Kaneko, H., et al., 2018. Locality and diel cycling of viral production revealed by a 24 h time course cross-omics analysis in a coastal region of Japan. ISME J. 12, 1287–1295.

Zehr, J.P., Mellon, M.T., Zani, S., 1998. New nitrogen-fixing microorganisms detected in oligotrophic oceans by amplification of nitrogenase (nifH) genes. Appl. Environ. Microbiol. 64, 3444–3450.

Index

E

H

I

N

O

P

Q

R

S

Z

FSC
www.fsc.org
MIX
Papier | Fördert gute Waldnutzung
FSC® C083411

Kolibri 360 GmbH
Ferdinand-Jühlke-Straße 7
99095 Erfurt, Deutschland
produktsicherheit@kolibri360.de